About WEF

Formed in 1928, the Water Environment Federation (WEF) is a not-for-profit technical and educational organization with 35,000 individual members and 75 affiliated Member Associations representing water quality professionals around the world. WEF and its Member Associations proudly work to achieve our mission of preserving and enhancing the global water environment.

For information on membership, publications, and conferences, contact:

Water Environment Federation
601 Wythe Street
Alexandria, VA 22314-1994 USA
(703) 684-2400
http://www.wef.org

About ASCE/EWRI

Founded in 1852, the American Society of Civil Engineers (ASCE) represents more than 146,000 members of the civil engineering profession worldwide, and is America's oldest national engineering society. Created in 1999, the Environmental and Water Resources Institute (EWRI) is an Institute of the American Society of Civil Engineers. EWRI services are designed to complement ASCE's traditional civil engineering base and to attract new categories of members (non-civil engineer allied professionals) who seek to enhance their professional and technical development.

For more information on membership, publications, and conferences, contact:

ASCE/EWRI
1801 Alexander Bell Drive
Reston, VA 20191-4400 USA
(703) 295-6000
http://www.asce.org

DESIGN OF MUNICIPAL WASTEWATER TREATMENT PLANTS

DESIGN OF MUNICIPAL WASTEWATER TREATMENT PLANTS

WEF Manual of Practice No. 8
ASCE Manuals and Reports on Engineering Practice No. 76
Fifth Edition

Prepared by the Design of Municipal Wastewater Treatment Plants Task Force of the Water Environment Federation and the American Society of Civil Engineers/Environmental and Water Resources Institute

WEF Press

Water Environment Federation   Alexandria, Virginia

American Society of Civil Engineers/Environmental and Water Resources Institute
Reston, Virginia

New York   Chicago   San Francisco   Lisbon   London   Madrid
Mexico City   Milan   New Delhi   San Juan   Seoul
Singapore   Sydney   Toronto

The **McGraw·Hill** Companies

Design of Municipal Wastewater Treatment Plants

1 2 3 4 5 6 7 8 9 0 FGR/FGR 0 1 5 4 3 2 1 0 9

ISBN:	P/N 978-0-07-166359-5 of set 978-0-07-166358-8	P/N 978-0-07-166360-1 of set 978-0-07-166358-8	P/N 978-0-07-166361-8 of set 978-0-07-166358-8
MHID:	P/N 0-07-166359-2 of set 0-07-166358-4 Volume 1	P/N 0-07-166360-6 of set 0-07-166358-4 Volume 2	P/N 0-07-166361-4 of set 0-07-166358-4 Volume 3

This book is printed on acid-free paper.

Prepared by **Design of Municipal Wastewater Treatment Plants Task Force** of the **Water Environment Federation** and the **American Society of Civil Engineers/Environmental and Water Resources Institute**

Terry L. Krause, P.E., BCEE, *Chair*

Roderick D. Reardon, Jr., P.E., BCEE, *Volume 1 Leader*
Albert B. Pincince, Ph.D., P.E., BCEE, *Volume 2 Leader*
Thomas W. Sigmund, P.E., *Volume 3 Leader*

Solomon Abel, P.E.
Kenneth N. Abraham, P.E., P. Eng.
Mohammad M. Abu-Orf
Orris E. Albertson
Charles M. Alix, P.E.
George P. Anipsitakis, Ph.D., P.E.
Richard G. Atoulikian, PMP, P.E.
David M. Bagley, Ph.D., P.E.
Katherine Bangs
Michael W. Barnett, Ph.D.
Britt D. Bassett, P.E., BCEE
Somnath Basu, Ph.D., P.E., BCEE
Laura B. Baumberger, P.E.
Robert Beggs, Ph.D., P.E.
Mario Benisch
Jeff Berk, P.E.
Vanessa Bertollini
George Bevington
Katya Bilyk, P.E.
Paul A. Bizier, P.E., BCEE
Linda Blankenship, P.E., BCEE
David Bloxom, P.E.
Joshua Philip Boltz, Ph.D., P.E.
Brian L. Book, P.E.
Robert C. Borneman, P.E., BCEE
Lucas Botero
Edward Boyajian
Ken Brischke
Jeanette Brown, P.E., DEE, D. WRE
Scott L. Buecker, P.E.
Marie Sedran Burbano, Ph.D., P.E.
Ron Burdick
Misti Burkman, P.E.
Peter Burrowes
Onder Caliskaner, Ph.D., P.E.
Alan James Callier
Anne M. Carayon, P.E.
Scott Carr
Leonard W. Casson, Ph.D., P.E., BCEE
Peter V. Cavagnaro, P.E., BCEE
Richard H. Cisterna, P.E.
James H. Clark, P.E.
Patrick E. Clifford
Patrick F. Coleman, Ph.D., P. Eng.
Anne Conklin
Timothy A. Constantine
Kevin D. Conway, P.E.
Rhodes R. Copithorn
John B. Copp, Ph.D.
George V. Crawford, P. Eng.
Ronald W. Crites, P.E.
Brent E. Crowther, P.E.
Ky Dangtran, Ph.D.
Michael E. Davis, Ph.D.
Chris DeBarbadillo, P.E.
Carlos De Leon

Michael J. Dempsey
Steven K. Dentel, Ph.D., P.E., BCEE
Laxman Mani Devkota, Ph.D., P.E., M. ASCE
Petros Dimitriou-Christidis, Ph.D., P.E.
Paul A. Dombrowski
Alexandra Doody, LEED AP
Brian Dooley
Kimberly R. Drake, RLA
Ronald Droste
Derya Dursun, Ph.D.
Brian Dyson, Ph.D.
Chris Easter
Robert W. Emerick, Ph.D., P.E.
Murali Erat
Angela S. Essner, P.E.
Adam Evans, P.E.
Kristin Evans, Ph.D., P.E.
Richard Finger
Alvin C. Firmin, P.E., BCEE
Kari Beth Fitzmorris, Sc.D.
James D. Fitzpatrick
Amanda L. Fox
Val S. Frenkel, Ph.D., P.E., D. WRE
Morgan R. Gagliano
James Gallovich, P.E.
M. Truett Garrett, Jr., Sc.D., P.E.
Trevor Ghylin
Boris Ginzburg
Mikel E. Goldblatt
Albert W. Goodman, P.E.
David C. Hagan, P.E.
John Harrison, P.E.
Brian Hemphill, P.E.
Gene Heyer, P.E., PMP
Webster Hoener
Michael Hribljan, M.Eng., P. Eng
Sarah Hubbell
Gary L. Hunter, P.E.
Sidney Innerebner, Ph.D., P.E.
Samuel S. Jeyanayagam, Ph.D., P.E., BCEE
Bruce R. Johnson, P.E., BCEE
Gary R. Johnson, P.E., BCEE
Terry L. Johnson, Ph.D., P.E.
John C. Kabouris, Ph.D., P.E.
Amit Kaldate, Ph.D.
Brian M. Karmasin
Dimitri Katehis
Ishin Kaya, P. Eng.
Raymond J. Kearney, P.E., BCEE
Justyna Kempa-Teper, Ph.D., P. Eng.
Philip C. Kennedy, AICP
Wayne L. Kerns
Carl M. Koch, Ph.D., P.E., BCEE
John E. Koch, P.E., BCEE
Tom A. Kraemer, P.E.
Thomas E. Kunetz, P.E.
May Kyi
Peter LaMontagne
Cory Lancaster
Damon Lau
Nathan Lester
Scott D. Levesque, P.E.
Jian Li, Ph.D., P. Eng., P.E.
Helen X. Littleton
Terry J. Logan, Ph.D.
Frank Loge, Ph.D.
Carlos Lopez
Becky J. Luna, P.E.
Venkatram Mahendraker, Ph.D., P. Eng.
Arthur P. Malm, P.E.
Chris Marlowe, CIH, CSP
F. Jason Martin, P.E.
Russell Mau, Ph.D., P.E.
William C. McConnell
John H. McGettigan, P.E., LEED AP

Charles M. McGinley, P.E.
James P. McQuarrie, P.E.
Jon H. Meyer
Indra N. Mitra, Ph.D., P.E.
Greg Moen, P.E.
Eberhard Morgenroth, Ph.D.
Audra N. Morse
Erin Mosley, P.E.
Lynne H. Moss
Christopher Muller, Ph.D.
Naoko Munakata
Sudhir N. Murthy
J. B. Neethling, Ph.D., P.E., BCEE
Robert Nerenberg, Ph.D., P.E.
James J. Newton, P.E., BCEE
John W. Norton, Jr., Ph.D., LEED AP
David W. Oerke, P.E.
Carroll J. Oliva
Rebecca Overacre
Lokesh Padhye
Tim Page-Bottorff
Sanath Bandara Palipana, B.E., G. Dip., M. Env. Eng. Sc., C.P. Eng.
Sanjay Patel
Vikram M. Pattarkine, Ph.D.
Jeff Peeters, M. Eng., P. Eng.
Marie-Laure Pellegrin, Ph.D.
Ana J. Pena-Tijerina, Ph.D., P.E.
Chris J. Peot
Robert E. Pepperman
Matt Peyton
Heather M. Phillips, P.E.
Scott D. Phipps
Richard J. Pope, P.E., BCEE
Benjamin T. Porter, P.E.
Raymond C. Porter
Russell Porter, P.E.
Douglas Prentiss
Chris Quigley, Ph.D., P.E.
Douglas L. Ralston
Tanja Rauch-Williams, Ph.D., P.E.
Joseph C. Reichenberger, P.E., BCEE
Joel C. Rife, P.E.
Ignasi Rodriguez-Roda, Ph.D.
Frank Rogalla
James M. Rowan, P.E.
A. Robert Rubin, Ph.D.
Andrew Salveson, P.E.
Julian Sandino, P.E., Ph.D.
Hari Santha
Patricia A. Scanlan
Perry L. Schafer, P.E., BCEE
James W. Schettler, P.E.
Harold E. Schmidt, Jr., P.E., BCEE
Kenneth Schnaars
Ralph B. "Rusty" Schroedel, Jr., P.E., BCEE
Paul J. Schuler
Robert J. Scott
Dipankar Sen, Ph.D.
Rick Shanley
Andrew R. Shaw
Gary Shimp
Ronald R. Skabo, P.E.
Marsha Slaughter, P.E.
Mark M. Smith, P.E.
Vic Smith, P.E., LEED AP
Henri Spanjers
Julia Spicher
Tom Spooren
George Sprouse, Ph.D., P.E., BCEE
Robert B. Stallings
Roger V. Stephenson, Ph.D., P.E., BCEE
Tracy Stigers
Kendra D. Sveum
Steven Swanback
Jay L. Swift, P.E.

Imre Takacs
Stephen Tarallo
Rudy J. Tekippe, Ph.D., P.E., BCEE
David Terrill, P.E.
Daniel L. Thomas, Ph.D., P.E.
Peter J. H. Thomson, P.E.
Andrea Turriciano, P.E.
Dave Ubert
Chip Ullstad, P.E., BCEE
Art K. Umble, Ph.D., P.E., BCEE
K. C. Upendrakumar, P.E.
Don Vandertulip, P.E.
Ifetayo Venner, P.E.
Miguel Vera
Cindy Wallis-Lage
Matthew Ward, P.E.
Thomas E. Weiland, P.E.
James E. Welp
Michael J. Whalley, M. Eng., P. Eng.
Jane W. Wheeler
G. Elliott Whitby, Ph.D.
Drury Denver Whitlock
Todd O. Williams, P.E., BCEE
Hannah T. Wilner
Michael J. Wilson, P.E.
Philip C. Y. Wong
David W. York, Ph.D., P.E.
Thor A. Young, P.E., BCEE

Under the Direction of the **Municipal Subcommittee** of the **Technical Practice Committee**

2009

Water Environment Federation
601 Wythe Street
Alexandria, VA 22314-1994 USA
http://www.wef.org

American Society of Civil Engineers/Environmental and Water Resources Institute
1801 Alexander Bell Drive
Reston, VA 20191-4400 USA
http://www.asce.org

Manuals of Practice of the Water Environment Federation

The WEF Technical Practice Committee (formerly the Committee on Sewage and Industrial Wastes Practice of the Federation of Sewage and Industrial Wastes Associations) was created by the Federation Board of Control on October 11, 1941. The primary function of the Committee is to originate and produce, through appropriate subcommittees, special publications dealing with technical aspects of the broad interests of the Federation. These publications are intended to provide background information through a review of technical practices and detailed procedures that research and experience have shown to be functional and practical.

Water Environment Federation
Technical Practice Committee
Control Group

R. Fernandez, *Chair*
J. A. Brown, *Vice-Chair*
B. G. Jones, *Past Chair*

A. Babatola
L. W. Casson
K. D. Conway
V. D'Amato
A. Ekster
S. Innerebner
R. C. Johnson
S. Moisio
T. Page-Bottorff
S. Passaro
R. C. Porter
E. P. Rothstein
A. T. Sandy
A. Tyagi
A. K. Umble
T. O. Williams

Manuals and Reports on Engineering Practice

(As developed by the ASCE Technical Procedures Committee, July 1930, and revised March 1935, February 1962, and April 1982)

A manual or report in this series consists of an orderly presentation of facts on a particular subject, supplemented by an analysis of limitations and applications of these facts. It contains information useful to the average engineer in his or her everyday work, rather than findings that may be useful only occasionally or rarely. It is not in any sense a "standard," however; nor is it so elementary or so conclusive as to provide a "rule of thumb" for nonengineers.

Furthermore, material in this series, in distinction from a paper (which expresses only one person's observations or opinions), is the work of a committee or group selected to assemble and express information on a specific topic. As often as practicable, the committee is under the direction of one or more of the Technical Divisions and Councils, and the product evolved has been subjected to review by the Executive Committee of the Division or Council. As a step in the process of this review, proposed manuscripts are often brought before the members of the Technical Divisions and Councils for comment, which may serve as the basis for improvement. When published, each work shows the names of the committees by which it was compiled and indicates clearly the several processes through which it has passed in review, in order that its merit may be definitely understood.

February 1962 (and revised in April 1982) the Board of Direction voted to establish a series entitled "Manuals and Reports on Engineering Practice," to include the Manuals published and authorized to date, future Manuals of Professional Practice, and Reports on Engineering Practice.

All such Manual or Report material of the Society would have been refereed in a manner approved by the Board Committee on Publications and would be bound, with applicable discussion, in books similar to past Manuals. Numbering would be consecutive and would be a continuation of present Manual numbers. In some cases of reports of joint committees, bypassing of Journal publications may be authorized.

Contents

Preface

This manual, updated from the 4th edition, continues its goal to be one of the principal references of contemporary practice for the design of municipal wastewater treatment plants (WWTPs). The manual was written for design professionals familiar with wastewater treatment concepts, the design process, and the regulatory basis of water pollution control. It is not intended to be a primer for the inexperienced or the generalist. The manual is intended to reflect current plant design practices of wastewater engineering professionals, augmented by performance information from operating facilities. The design approaches and practices presented in the manual reflect the experiences of more than 300 authors and reviewers from around the world.

This three-volume manual consists of 27 chapters, with each chapter focusing on a particular subject or treatment objective. The successful design of a municipal WWTP is based on consideration of each unit process and the upstream and downstream effects of that unit's place and performance in the overall scheme of the treatment works. The chapters that compose Volume 1 generally cover design concepts and principles that apply to the overall WWTP. Volume 2 contains those chapters that discuss liquid-train-treatment operations or processes. Volume 3 contains the chapters that deal with the management of solids generated during wastewater treatment.

In the 11 years since the publication of the 4th edition of this manual, key technical advances in wastewater treatment have included the following:

- Membrane bioreactors replaced conventional secondary treatment processes in a smaller footprint;
- Advancements within integrated fixed-film/activated sludge (IFAS) systems and moving-bed biological-reactors systems;
- Disinfection alternatives to chlorine;
- Biotrickling filtration for odor control;
- Increased use of ballasted flocculation;
- Sidestream nutrient removal to reduce the loading on the main nutrient-removal process; and
- Use and application of modeling wastewater treatment processes for the basis of design and evaluations of alternatives.

In response to these advancements, this edition includes some significant changes from the 4th edition. As with prior editions, technologies that are no longer considered current industry practice have been deleted, such as vacuum filters for sludge dewatering. While not intended to be all-inclusive, the following list describes some of the other pertinent processes and newer processes or concepts:

- Concept of sustainability,
- Energy management,
- Odor control and air emissions,
- Chemically assisted/ballast flocculation clarification,
- Membrane bioreactors,
- IFAS processes,
- Enhanced nutrient-control systems,
- Sidestream treatment, and
- Approaches to minimizing biosolids production.

Additionally, the focus of the manual has been sharpened. Like earlier editions, this manual presents current design guidelines and practices of municipal wastewater engineering professionals. Design examples also are provided, in some instances, to show how the guidelines and practice can be applied. However, information on process fundamentals, case histories, operations, and other related topics is covered to a lesser extent than in the previous edition. Readers are referred to other publications for information on those topics.

This 5th edition of this manual was produced under the direction of Terry L. Krause, P.E., BCEE, *Chair;* Roderick D. Reardon, Jr., P.E., BCEE, *Volume 1 Leader;* Albert B. Pincince, Ph.D., P.E., BCEE, *Volume 2 Leader;* and Thomas W. Sigmund, P.E., *Volume 3 Leader*.

Principal authors of the publication are:

Chapter 1 Terry L. Krause, P.E., BCEE
Hannah T. Wilner

Chapter 2 Julian Sandino, P.E., Ph.D.
Hannah T. Wilner

Rachel Carlson
Albert W. Goodman, P.E.
Indra N. Mitra, Ph.D., P.E.
Ignasi Rodriguez-Roda, Ph.D.
Chip Ullstad, P.E., BCEE
Don Vandertulip, P.E.

Drury Denver Whitlock
Michael J. Wilson, P.E.

Chapter 3
Alvin C. Firmin, P.E., BCEE
William C. McConnell

Orris E. Albertson
Kimberly R. Drake, RLA
Brian M. Karmasin
Cory Lancaster

Chapter 4
Jane W. Wheeler
Philip C. Kennedy, AICP

Kimberly R. Drake, RLA
Sanath Bandara Palipana, B.E., G. Dip., M. Env. Eng. Sc., C.P. Eng.

Chapter 5
Ralph B. "Rusty" Schroedel, Jr., P.E., BCEE
George V. Crawford, P. Eng.

Peter V. Cavagnaro, P.E., BCEE
Patrick E. Clifford
Michael E. Davis, Ph.D.
Arthur P. Malm, P.E.
John H. McGettigan, P.E., LEED AP
Erin Mosley, P.E.
John W. Norton, Jr., Ph.D., LEED AP
Don Vandertulip, P.E.

Chapter 6
Joseph C. Reichenberger, P.E., BCEE

Katherine Bangs
James Gallovich, P.E.
David Terrill, P.E.

Chapter 7
Raymond C. Porter
Charles M. Alix, P.E.

Petros Dimitriou-Christidis, Ph.D., P.E.
Chris Easter

Charles M. McGinley, P.E.
Richard J. Pope, P.E., BCEE
Chris Quigley, Ph.D., P.E.
Mark M. Smith, P.E.
Tom Spooren
Matthew Ward, P.E.

Chapter 8 Tim Page-Bottorff

Chris Marlowe, CIH, CSP
Douglas Prentiss

Chapter 9 Dave Ubert
David Bloxom, P.E.

Vic Smith, P.E., LEED AP
Hannah T. Wilner

Chapter 10 Ronald R. Skabo, P.E.
Wayne L. Kerns

Misti Burkman, P.E.
Damon Lau
Robert J. Scott

Chapter 11 Joel C. Rife, P.E.
Lucas Botero

Chapter 12 Thomas E. Weiland, P.E.
Anne M. Carayon, P.E.

Chapter 13 Joshua Philip Boltz, Ph.D., P.E.
Eberhard Morgenroth, Ph.D.

Chris DeBarbadillo, P.E.
Michael J. Dempsey
Trevor Ghylin
John Harrison, P.E.
James P. McQuarrie, P.E.
Robert Nerenberg, Ph.D., P.E.

Chapter 14	Roger V. Stephenson, Ph.D., P.E., BCEE Rudy J. Tekippe, Ph.D., P.E., BCEE
	Patrick F. Coleman, Ph.D., P. Eng. Anne Conklin George V. Crawford, P. Eng. Samuel S. Jeyanayagam, Ph.D., P.E., BCEE Bruce R. Johnson, P.E., BCEE Roderick D. Reardon, Jr., P.E., BCEE George Sprouse, Ph.D., P.E., BCEE
Chapter 15	Art K. Umble, Ph.D., P.E., BCEE Amanda L. Fox
	Kenneth N. Abraham, P.E., P. Eng. Dipankar Sen, Ph.D.
Chapter 16	Val S. Frenkel, Ph.D., P.E., D. WRE Onder Caliskaner, Ph.D., P.E.
Chapter 17	Dimitri Katehis Cindy Wallis-Lage
	Timothy A. Constantine Heather M. Phillips, P.E.
Chapter 18	Ronald W. Crites Robert Beggs, Ph.D., P.E.
	Brian L. Book, P.E. Kristin Evans, Ph.D., P.E.
Chapter 19	Jay L. Swift, P.E. Russell Porter, P.E.
	Somnath Basu, Ph.D., P.E., BCEE Leonard W. Casson, Ph.D., P.E., BCEE Robert W. Emerick, Ph.D., P.E. Gary L. Hunter, P.E. Frank Loge, Ph.D.

Lokesh Padhye
Andrew Salveson, P.E.
Justyna Kempa-Teper, Ph.D., P. Eng.
Andrea Turriciano, P.E.
G. Elliott Whitby, Ph.D.

Chapter 20 Jeanette Brown, P.E., DEE, D. WRE

Chapter 21 Paul A. Bizier, P.E., BCEE
George P. Anipsitakis, Ph.D., P.E.

Chapter 22 Harold E. Schmidt, Jr., P.E., BCEE
Derya Dursun, Ph.D.
Mikel E. Goldblatt

Chapter 23 Jeff Berk, P.E.
Benjamin T. Porter, P.E.

May Kyi
Brian Hemphill, P.E.
Adam Evans, P.E.
Greg Moen, P.E.

Chapter 24 Carl M. Koch, Ph.D., P.E., BCEE
Angela S. Essner, P.E.

Laura B. Baumberger, P.E.
Morgan R. Gagliano
David C. Hagan, P.E.
John C. Kabouris, Ph.D., P.E.
Peter LaMontagne
Nathan Lester
Rebecca Overacre
Rick Shanley
Julia Spicher
Tracy Stigers
Steven Swanback

Chapter 25 Sudhir N. Murthy
Perry L. Schafer, P.E., BCEE

Charles M. Alix, P.E.
Anne Conklin
Kari Beth Fitzmorris, Sc.D.
Terry J. Logan, Ph.D.
Christopher Muller, Ph.D.
Chris J. Peot
James W. Schettler, P.E.
Miguel Vera
Todd O. Williams, P.E., BCEE

Chapter 26 Peter Burrowes
Ky Dangtran, Ph.D.

Scott Carr
Webster Hoener
Raymond J. Kearney, P.E., BCEE
James M. Rowan, P.E.
Hari Santha
James E. Welp

Chapter 27 Lynne H. Moss

Alexandra Doody, LEED AP
Tom A. Kraemer, P.E.
Terry J. Logan, Ph.D.
Robert E. Pepperman

Glossary Kendra D. Sveum
Matt Peyton

The following also contributed to the development of this manual: Murali Erat (Chapter 15), Sarah Hubble (Chapter 15), Vikram Pattarkine (Chapter 15), Frank Rogalla (Chapter 13), and Stephen Tarallo (Chapter 13).

Authors' and reviewers' efforts were supported by the following organizations:

AECOM, Philadelphia, Pennsylvania; Alexandria, Virginia; and Sheboygan, Wisconsin
Advanced Bioprocess Development, Ltd., Manchester, England
Aqualia, Madrid, Spain
Associated Engineering, Calgary, Alberta, Canada

Bassett Engineering, Inc., Montoursville, Pennsylvania
Beaumont Cherry Valley Water District, Beaumont, California
Binkley and Barfield, Inc., Consulting Engineers, Houston, Texas
Black & Veatch, Los Angeles, California; Sacramento, California; Atlanta, Georgia; Gaithersburg, Maryland; Kansas City, Missouri; and Cincinnati, Ohio
Brown and Caldwell, Davis, California; Rancho Cordova, California; Walnut Creek, California; Washington, D.C.; and Seattle, Washington
Caboolture Shire Council, Caboolture, QLD
Carollo Engineers, Phoenix, Arizona; Walnut Creek, California; Broomfield, Colorado; Littleton, Colorado; Sarasota, Florida; Winter Park, Florida; Portland, Oregon; Dallas, Texas; and Seattle, Washington
CDM, Phoenix, Arizona; Los Angeles, California; Maitland, Florida; Chicago, Illinois; Louisville, Kentucky; Baton Rouge, Louisiana; Cambridge, Massachusetts; Kansas City, Missouri; Manchester, New Hampshire; Edison, New Jersey; Albuquerque, New Mexico; Providence, Rhode Island; Austin, Texas; Dallas, Texas; and San Antonio, Texas
CH2M HILL, Englewood, Colorado; Tampa, Florida; Chicago, Illinois; Boston, Massachusetts; Kansas City, Missouri; Henderson, Nevada; Parsippany, New Jersey; Knoxville, Tennessee; Austin, Texas; Salt Lake City, Utah; Chantilly, Virginia; Richmond, Virginia; Bellevue, Washington; and Milwaukee, Wisconsin
CH2M Hill Canada, Ltd., Kitchener, Ontario, Canada; and Toronto, Ontario, Canada
Chastain-Skillman, Inc., Lakeland, Florida; and Orlando, Florida
City of Missoula, Missoula, Montana
City of Phoenix, Phoenix, Arizona
City of Stamford, Stamford, Connecticut
Consoer Townsend Envirodyne Engineers, Nashville, Tennessee
County Sanitation Districts of Los Angeles County, Whittier, California
Degremont Technologies – Infilco (Suez Environnement), Richmond, Virginia
District of Columbia Water and Sewer Authority, Washington, D.C.
DLT&V Systems Engineering, Oceanside, California
Donohue and Associates, Chesterfield, Missouri
Eco-logic Eng., Rocklin, California
Eimco Water Technologies, Salt Lake City, Utah
EnerTech Environmental, Inc., Los Angeles, California
Entex Technologies, Chapel Hill, North Carolina
Enviro Enterprises, Inc., La Barge, Wyoming
Environmental Group Services, Baltimore, Maryland
Environmental Operating Solutions, Inc., Bourne, Massachusetts

Forsgren Associates, Inc., Rexburg, Idaho
Freese and Nichols, Inc., Fort Worth, Texas
GE Water and Process Technologies, Oakville, Ontario, Canada; and Portland, Oregon
Georgia Institute of Technology, Atlanta, Georgia
Gloversville Johnstown Joint Wastewater Treatment Plant, Johnstown, New York
Gray and Osborne, Seattle, Washington
Greeley and Hansen, L.L.C., Phoenix, Arizona; Wilmington, Delaware; Tampa, Florida; Sarasota, Florida; Chicago, Illinois; Gary, Indiana; Indianapolis, Indiana; Landover, Maryland; Las Vegas, Nevada; New York City, New York; and Philadelphia, Pennsylvania; Richmond, Virginia; Roanoke, Virginia; and Springfield, Virginia
Green Bay Metropolitan Sewerage District, Green Bay, Wisconsin
Hazen and Sawyer, P.C., New York, New York; and Raleigh, North Carolina
HDR Engineering, Inc., Folsom, California; Irvine, California; Riverside, California; Portland, Oregon; Dallas, Texas; Bellevue, Washington; and Burlington, Washington
Herbert Rowland and Grubic, Inc., State College, Pennsylvania
Jiann-Ping Hsu College of Public Health, Georgia Southern University, Statesboro, Georgia
Johnson Controls, Inc., Milwaukee, Wisconsin
Kennedy/Jenks Consultants, Palo Alto, California; Sacramento, California; and San Francisco, California
Knowledge Automation Partners, Inc., Wellesley, Massachusetts
Lake County Public Works, Libertyville, Illinois
Lettinga Associates Foundation, Wageningen, The Netherlands
Logan Environmental, Inc., Beaufort, South Carolina
Louisiana State University, Baton Rouge, Louisiana
Loyola Marymount University, Los Angeles, California
Malcolm Pirnie, Inc., Columbus, Ohio; Wakefield, Massachusetts; and White Plains, New York
Metcalf and Eddy, Inc., Philadelphia, Pennsylvania
MWH Americas, Inc., Arcadia, California; Denver, Colorado; Tampa, Florida; Chicago, Illinois; Boston, Massachusetts; and Cleveland, Ohio
MWH EMEA, Inc., Mechelen, Belgium
Metropolitan Water Reclamation District of Greater Chicago, Chicago, Illinois; and Schaumburg, Illinois
Nanyang Technological University, Singapore
North Carolina State University Department of Biological and Agricultural Engineering, Raleigh, North Carolina

Regional Municipality of Waterloo, Ontario, Canada
Short Elliott Hendrickson, Inc., Sheboygan, Wisconsin
St. Croix Sensory, Inc., Lake Elmo, Minnesota
Stantec Consulting Ltd., Windsor, Ontario, Canada
Stearns and Wheler, Bowie, Maryland
Tetra Tech, Inc., Pasadena, California
Total Safety Compliance, Mesa, Arizona
Trinity River Authority of Texas, Arlington, Texas
University of California, Davis, California
University of Delaware, Newark, Delaware
University of Girona, Girona, Spain
University of Illinois, Urbana, Illinois
University of Notre Dame, Notre Dame, Indiana
University of Wyoming, Laramie, Wyoming
Westin Engineering, Elk Grove, California
Woodard & Curran, Inc., Cheshire, Connecticut

Chapter 20

Introduction to Solids Management

1.0 INTRODUCTION

Solids management is an important aspect of wastewater treatment design because of the interrelationships between the liquid and solids processes. Volume 3 covers the solids generated during sedimentation and/or biological and chemical treatment of raw wastewater. [For information on minor residuals streams (e.g., scum, grit, and screenings) and their removal from wastewater, see the chapter on preliminary treatment in Volume 2.]

It is impossible to completely separate liquid- and solids-handling processes, and engineers must consider their relationships when designing wastewater treatment plants. The liquid treatment processes chosen will affect both the amount of solids generated and their characteristics, which in turn will affect the choice of settling, conditioning, and dewatering processes. Furthermore, the recycle streams from solids treatment processes can affect liquid ones. For example, dewatering anaerobically digested biosolids produces a sidestream with high ammonia and phosphorus concentrations, which will increase these loadings to the plant's liquid treatment processes. This chapter includes a mass balance example that illustrates these relationships.

This volume describes accepted methods and procedures for planning, designing, and constructing solids-handling processes and equipment. It discusses current U.S. regulations; methods for determining solids quantities, and descriptions of typical characteristics associated with the various residuals generated in wastewater treatment plants. Finally, although there is a thorough discussion of degritting and screening as it relates to raw wastewater in Volume 2, there is a brief discussion in Volume 3 of how these processes relate to solids management.

Chapter 21 discusses methods for transporting and storing residuals and biosolids. Chapter 22 discusses solids conditioning, including the types of chemicals involved, factors that affect conditioning, chemical-feed systems, and dose optimization. Chapters 23 and 24 describe thickening and dewatering processes; they include information on process design conditions and criteria, key process variables, and ancillary equipment. Chapter 25 covers biological- and chemical-stabilization processes (e.g., aerobic and anaerobic digestion, composting, and alkaline stabilization). Chapter 26 discusses thermal processes (e.g., incineration) for stabilizing, drying, or destroying solids. Chapter 27 describes land application and other biosolids use and disposal practices.

2.0 DEFINITIONS

The Water Environment Federation has adopted the following terminology for wastewater residuals.

Sludge is any residual produced during primary, secondary or advanced wastewater treatment that has not undergone any process to reduce pathogens or vector attraction. Another common term for this is *raw sludge*. The term *sludge* should be used with a specific process descriptor (e.g., primary sludge, waste activated sludge, or secondary sludge).

Biosolids is any sludge that has been stabilized to meet the criteria in the U.S. Environmental Protection Agency's (U.S. EPA's) 40 *CFR* 503 regulations and, therefore, can be beneficially used. Stabilization processes include anaerobic digestion, aerobic digestion, alkaline stabilization, and composting. Additionally, heat drying produces biosolids that can also be used beneficially.

Solids and *residuals* are terms used when it is uncertain whether the material meets Part 503 criteria (e.g., during thickening, because stabilization may occur either before or after this process). In this volume, the terms *solids* and *residuals* will be used if general references to XXX and for general descriptors (e.g., solids handling).

Land application is the process of adding bulk or bagged biosolids to soil at *agronomic rates*—the amount needed to provide enough nutrients (e.g., nitrogen, phosphorus, and potash) for optimal plant growth while minimizing the likelihood that they pass below the root zone and leach to groundwater. Land application can involve agricultural land (e.g., fields used to produce food, feed, and fiber crops); pasture and rangeland; nonagricultural land (e.g., forests); public-contact sites (e.g., parks and golf courses); disturbed lands (e.g., mine spoils, construction sites, and gravel pits); and home lawns and gardens.

3.0 REGULATIONS

When designing any solids or biosolids project, engineers should take into account the prevailing local, state, and federal regulations. Most, if not all U.S. states, have adopted the federal regulations for managing wastewater residuals (40 *CFR* 503); some have promulgated regulations that impose even stricter requirements. Designers should begin by evaluating the regulatory consequences of any proposed action, because the choice of treatment process(es) often is governed as much by regulatory constraints as by process performance and cost.

3.1 40 *CFR* 503

3.1.1 Background

The U.S. Environmental Protection Agency's biosolids regulations (40 *CFR* 503) address the use and disposal of solids generated during the treatment of domestic wastewater and septage. They are organized into five subparts: general provisions, land application, surface disposal, pathogen and vector-attraction reduction, and incineration (*Fed. Reg.*, Feb. 19, 1993).

Part 503 biosolids standards typically are incorporated into a wastewater treatment plant's National Pollutant Discharge Elimination System (NPDES) permit. Such permits are issued by U.S. EPA or by states with agency-approved solids management programs. Any plant that treats domestic wastewater (e.g., facilities that generate, treat, or provide disposal for solids, including nondischarging and "sludge-only" facilities) must have a permit. That said, the Part 503 rule was written to be self-implementing, which means that treatment plants are expected to follow it even before their permits are issued. It can be enforced either by U.S. EPA or via citizen lawsuits.

The agency continues to review Part 503—especially the land-application provisions to ensure that current regulations protect public health and the environment.

If the treatment plant's solids will be disposed in municipal solid waste landfills or used as landfill cover material, they must comply with the requirements of 40 *CFR* 258 (municipal solid waste landfill regulations) rather than with Part 503.

3.1.2 *General Requirements*

Biosolids generators are responsible for complying with Part 503. The regulation establishes two sets of criteria for heavy metals—Pollutant Concentrations and Pollutant Ceiling Concentrations—and two sets of criteria for pathogen densities—Class A and Class B. It also allows for two approaches to reducing vector attraction: treating the solids or using physical barriers.

Biosolids (or material derived from solids) that meet the higher-quality criteria have fewer restrictions. The minimum requirements for a biosolids to qualify for land application are Pollutant Ceiling Concentrations, Class B requirements, and vector-attraction reduction requirements. A biosolids that meets Pollutant Concentration limits, Class A requirements, and vector-attraction reduction requirements can be land-applied without the additional precautions. However, all land-appliers must meet the minimum monitoring, recordkeeping, and reporting requirements (no matter which type of biosolids is used).

3.1.3 *Pollutant Limits*

Before a biosolids can be land-applied, its levels of heavy metals cannot exceed either Pollutant Concentration limits or Pollutant Ceiling Concentrations and Cumulative Pollutant Loading Rates (see Table 20.1). Bulk biosolids that will be applied to lawns and home gardens must meet Pollutant Concentration limits. Biosolids sold or given away in bags or other containers must meet either Pollutant Concentration limits or Pollutant Ceiling Concentrations. Users should be directed to apply this material at rates based on Annual Pollutant Loading Rates.

TABLE 20.1 Pollutant limits for land-applied biosolids (all limits are on a dry weight basis).

Pollutant	Ceiling concentration limits,[a] mg/kg	Cumulative pollutant loading rates, kg/ha	"High-quality" pollutant concentration limits,[b] mg/kg	Annual pollutant loading rates, kg/ha·a
Arsenic	75	41	41	2.0
Cadmium	85	39	39	1.9
Copper	4 300	1 500	1 500	75
Lead	840	300	300	15
Mercury	57	17	17	0.85
Molybdenum	75	—	—	—
Nickel	420	420	420	21
Selenium	100	100	36	5.0
Zinc	7 500	2 800	2 800	140

[a] Absolute values.
[b] Monthly averages.

3.1.4 Pathogen Limits

Part 503 labels biosolids either Class A or Class B based on their pathogen levels. Both types have been treated to reduce pathogens and minimize their ability to attract vectors (e.g., rats). However, Class B biosolids still contain detectible levels of pathogens, while Class A biosolids are essentially pathogen-free. [If its metals levels are also low, then a Class A material is labeled an "exceptional quality" (EQ) biosolids.]

Both Class A and Class B biosolids can be land-applied, but land-applying a Class B material involves buffer requirements, public-access limits, and crop-harvesting restrictions (see Table 20.2). These rules are intended to protect public health and enable microorganisms in the soil to degrade remaining pathogens.

Any biosolids being applied to lawns and home gardens—or sold or given away in bags or other containers—must meet Class A criteria. For information on selling or distributing biosolids (e.g., composted or heat-dried products), see Chapter 27. For information on product characterization and marketing approaches, also see Chapter 27.

3.1.4.1 Class A Requirements

To be considered "Class A", biosolids must meet specific fecal coliform and salmonella limits at the time of use or disposal. Also, the requirements of one of the following alternatives must be met:

1. Time/temperature (see Table 20.3),
2. Alkaline treatment,

TABLE 20.2 Allowable uses of Class A and Class B biosolids.

Class	Allowable uses	Restrictions
A and A EQ	• Home lawns and gardens • Public contact sites • Urban landscaping • Agriculture • Forestry • Soil and site rehabilitation • Landfill disposal • Surface land disposal	Unrestricted
B	• Agriculture • Forestry • Soil and site rehabilitation • Landfill disposal • Surface land disposal	• Food crops: no harvesting after application for 14 to 38 months. • Feed crops: no harvesting for 30 days after application. • Public access: restricted access for 30 to one year. • Turf: no harvest for one year after application.

3. Prior testing for enteric virus/viable helminth ova,
4. No prior testing for enteric virus/viable helminth ova, and
5. Biosolids have been treated by a process to further reduce pathogens (PFRP) or equivalent processes (see Table 20.4).

TABLE 20.3 Time and temperature guidelines for producing Class A biosolids.

Total solids	Temperature (*t*)	Time (*D*)	Equation	Notes
≥7%	≥50°C	≥20 min	$D = \frac{131\,700\,000}{10^{0.14t}}$	No heating of small particles by warmed gases or immersible liquid
≥7%	≥50°C	>15 sec	$D = \frac{131\,700\,000}{10^{0.14t}}$	Small particles heated by warmed gases or immersible liquid
<7%	>50°C	≥15 sec to <30 min	$D = \frac{131\,700\,000}{10^{0.14t}}$	
<7%	≥50°C	≥30 min	$D = \frac{131\,700\,000}{10^{0.14t}}$	

TABLE 20.4 Processes to further reduce pathogens.

Aerobic Digestion	The process is conducted by agitating sludge with air or oxygen to maintain aerobic conditions at residence times ranging from 60 days at 15°C to 40 days at 20°C, with a volatile solids reduction of at least 38%.
Air Drying	Liquid sludge is allowed to drain and/or dry on underdrained sand beds, or on paved or unpaved basins in which the sludge depth is a maximum of 9 inches. A minimum of 3 months is needed, for 2 months of which temperatures average on a daily basis above 0°C.
Anaerobic Digestion	The process is conducted in the absence of air at residence times ranging from 60 days at 20°C to 15 days at 35°C to 55°C, with a volatile solid reduction of at least 38%.
Composting	Using the within-vessel, static aerated pile, or window composting methods, the solid waste is maintained at minimum operating conditions of 40°C for 5 days. For 4 hours during this period the temperature exceeds 55°C.
Lime Stabilization	Sufficient lime is needed to produce a pH of 12 after 2 hours of contact.
Other Methods	Other methods or operating conditions maybe acceptable if pathogens and vector attraction of the waste (volatile solids) are reduced to an extent equivalent to the reduction achieved by any of the above methods.
Composting	Using the within-vessel composting method, the solid waste is maintained at operating conditions of 55°C or greater for 3 days. Using the static aerated pile composting method, the solid waste is maintained and operating conditions of 55°C or greater for 3 days. Using the windrow composting method, the solid waste attains a temperature of 55°C or greater for at least 15 days during the composting period. Also, during the high temperature period, there will be a minimum of five turnings of the windrow.
Heat Drying	Dewatered sludge cake is dried by direct or indirect contact with hot gasses, and moisture content is reduced to 10% or lower. Sludge particles reach temperatures well in excess of 80°C, or the wet bulb temperature of the gas stream in contact with the sludge at the point where it leaves the dryer is an excess of 80°C.
Heat Treatment	Liquid sludge is heated to temperatures of 180°C for 30 minutes.
Thermophilic Aerobic Digestion	Liquid sludge is agitated with air or oxygen to maintain aerobic conditions at residence times of 10 days at 55°C to 60°C, with a volatile solids reduction of at least 38%.
Other Methods	Other methods or operating conditions may be acceptable if pathogens and vector attraction of the waste (volatile solids) are needed to an extent equivalent to the reduction achieved by any of the above methods. Any of the processes listed below, if added to a PSRP, further reduce pathogens.
Beta Ray Irradiation	Sludge is irradiated with beta rays from an accelerator at dosages of at least 1.0 megarad at room temperature (ca. 20°C).
Gamma Ray Irradiation	Sludge is irradiated with gamma rays from certain isotopes, such as ^{60}Cobalt and ^{137}Cesium, at dosages of at least 1.0 megarad at room temperature (ca. 20°C).
Pasteurization	Sludge is maintained for at least 30 minutes at a minimum temperature of 70°C.
Other Methods	Other methods or operating conditions may be acceptable if pathogens are reduced to an extent equivalent to the reduction achieved by any of the above add-on methods.

Source: 40 CFR 257, Appendix II.

3.1.4.2 Class B Requirements

Solids must meet at least Class B pathogen requirements before being used or disposed. Solids that do not meet Class B criteria cannot be land-applied, but they may be placed in a surface-disposal unit that is covered daily.

If Class B biosolids or domestic septage are land-applied, site restrictions also must be met (see Section 3.1.6).

3.1.4.3 Pathogen Treatment Processes

Processes that significantly reduce pathogen levels in biosolids include aerobic and anaerobic digestion, air drying, alkaline stabilization, and composting. Processes that further reduce pathogens include beta ray irradiation, composting, gamma ray irradiation, heat drying, heat treatment, pasteurization, and thermophilic anaerobic digestion.

Municipal wastewater treatment plants typically use one of the following four processes to produce Class A or Class B biosolids.

3.1.4.3.1 Heat Drying

Heat drying and pelletizing processes typically produce Class A biosolids. Dryers typically have temperatures higher than 70°C, and retain biosolids for at least 30 minutes, thereby meeting the requirements for Class A pathogen reduction. If recycling shrinks retention times to less than 30 minutes, the PFRP time and temperature criteria for heat dryers may apply. Drying processes also must meet the requirements of the thermal equation in Part 503.

Heat dryers that produce a marketable biosolids can easily meet vector-attraction reduction requirements. Basically, if the material does not contain unstabilized primary sludge, then it must be at least 75% solids. If it does, then it must be at least 90% solids.

3.1.4.3.2 Digestion

Aerobic and anaerobic digestion systems typically can produce Class B biosolids if operated as designed. Modified anaerobic digestion systems (e.g., thermophilic anaerobic digestion) may produce Class A biosolids; such systems are considered PFRPs.

3.1.4.3.3 Composting

In-vessel composting or static aerated-pile systems can meet Class A pathogen-reduction requirements if the biosolids' temperature is maintained at 55°C or higher for 3 days. Windrow composting systems can meet Class A requirements if the biosolids' temperature is maintained at 55°C or higher for at least 15 days and the windrow is turned at least five times in that period. Other composting systems may produce Class A biosolids if they meet the time and temperature or pathogen-testing requirements.

To meet vector-attraction reduction requirements, composting systems must heat biosolids to more than 40°C for 14 days; the average temperature during that period must be higher than 45°C.

Also, after treatment, composted biosolids must be monitored periodically for pathogen regrowth. If the pathogen level increases, the material may not meet land application requirements and may require other means of disposal.

3.1.4.3.4 Alkaline Stabilization

One patented alkaline-stabilization process meets Class A pathogen-reduction requirements by elevating pH above 12 for 72 hours while elevating temperature above 52°C for 12 hours or longer, followed by air drying to produce a material with more than 50% solids. Other alkaline-stabilization approaches meet Class A standards by pasteurizing biosolids via the time and temperature criteria. Still others meet Class A standards via PFRP equivalency requirements.

3.1.5 Vector-Attraction Reduction Requirements

Vectors (e.g., flies, rodents, and birds) are attracted to volatile solids. Materials with lower volatile solids concentrations are less likely to attract vectors, which spread infectious disease agents. There are 10 options for reducing vector attraction. All biosolids must meet at least one of them before they can be beneficially used.

Part 503's pathogen and vector-attraction reduction requirements are complex. For more information, see U.S. EPA's related guidance documents [especially *A Plain English Guide to EPA Part 503 Biosolids Rule* (1994)], the regulation itself, and the preamble that accompanied the rule when it was originally published in the *Federal Register* (1993).

3.1.6 Management Practices

When using bulk biosolids that have not met Pollutant Concentration Limits, Class A pathogen requirements, and vector-attraction reduction (Section 3.1.5), land-appliers must not apply them

- to flooded, frozen, or snow-covered ground, where the material can enter wetlands or other U.S. waters (unless authorized to do so by the permitting authority);
- at rates above agronomic rates (except in reclamation projects when authorized to do so by the permitting authority);
- where they could adversely affect a threatened or endangered species;
- within 10 m of U.S. waters (unless authorized to do so by the permitting authority).

Also, biosolids sold or given away in a container must come with a label or an information sheet that provides the name and address of the person who prepared the analysis information on proper use, including the annual application rate (which ensures that annual pollutant loading rates are within regulatory limits).

When using Class B biosolids, land-appliers must ensure that

- public access to the site is restricted for 30 days after application if the land has little public exposure (e.g., agricultural lands, reclamation sites, and forests) and for 1 year after application if the land has great public exposure (e.g., public parks, golf courses, cemeteries, and ball fields);
- animals are not grazed on the site for 30 days after application;
- no food, feed, or fiber crops are harvested from the site for 30 days after application;
- no food crops whose harvested parts are aboveground but touch the soil (e.g., melons, cucumbers, and squash) are harvested for 14 months after application;
- no food crops whose harvested parts are underground (e.g., potatoes, carrots, and radishes) are harvested for 20 months after application if biosolids remain on the surface for at least 4 months before being incorporated into soil, or for 38 months after application if biosolids are incorporated into soil in less than 4 months;
- turf is not harvested for 1 year after application if the turf will be put on land with high public exposure (e.g., a lawn) unless the permitting authority specifies otherwise;

3.1.7 *Monitoring Requirements*

The minimum frequency of pollutant, pathogen, and vector-attraction reduction monitoring depends on the amount of solids used or disposed annually. Permitters may impose more frequent monitoring requirements, but after 2 years of monitoring, they may reduce the monitoring frequencies for pollutants (and sometimes for pathogens). However, monitoring frequencies may not drop below once per year.

3.1.8 *Recordkeeping Requirements*

Recordkeeping requirements depend on which pathogen-reduction option, vector-attraction reduction method, and pollutant limits are met, as well as the ultimate use of the biosolids or product derived from biosolids. In general, the biosolids or product preparer is responsible for certifications and records related to pollutant concentrations, pathogen reduction option, and vector-attraction reduction method. Meanwhile, the biosolids or product applier is responsible for certifications and records concerning field operations, application rates, management practices, and site restrictions. Unless otherwise noted, records should be kept for 5 years.

3.1.9 Reporting Requirements

Once a year, all Class I solids management facilities and publicly owned treatment works (POTWs) with a design flowrate of at least 4 000 m^3/d (1 mgd) or a service population of at least 10 000 people should submit the data in required records to the permitting authority.

3.1.10 Incineration

Part 503's requirements for solids incinerators address feed solids, the furnace itself, furnace operations, and exhaust gases. The rule does not apply to the incineration of hazardous solids (as defined in 40 *CFR* 261) or solids containing more than 50 ppm of polychlorinated biphenyls. It also does not apply to incinerators that cofire solids with other wastes, although an incinerator can burn a mix of solids and municipal solid waste (up to 30% as an "auxiliary fuel") and still be regulated under Part 503.

Furthermore, this rule does not apply to the ash produced by a solids incinerator. Design engineers should be aware that ash disposal can be a significant problem. Some states regulate the ash as a hazardous waste (although federal regulations do not).

As part of the NPDES permit application, treatment plants must conduct performance tests of their existing incineration facilities to determine pollution-control efficiencies for heavy metals and conduct air-dispersion modeling for site-specific conditions. Continuous emissions monitoring equipment also should be installed.

3.1.11 Prohibited Disposal Method (Ocean Disposal)

The Part 503 regulations do not address disposing solids in the ocean. Ocean disposal once was acceptable in the United States and practiced widely by communities on the Atlantic coast. However in 1988, the U.S. Congress passed the Ocean Dumping Ban Act, which made this practice unlawful after 1991. Although some still argue about the scientific basis of this decision and the environmental effects of ocean disposal, this method has ceased to be an option in the United States and so is not discussed in this manual. (Ocean disposal may be an option in other countries if studies indicate that the environmental effects are either negligible or beneficial.)

3.2 State Regulations

Many states have promulgated solids regulations that are as strict or stricter than Part 503. For example, Connecticut does not allow any biosolids to be land-applied. Some states do not allow raw sludge to be landfilled, limit the land application of liquid biosolids, or require liquid biosolids to be thoroughly mixed into the soil. That is why design engineers must review state (and local) regulations before making decisions about the solids-management train.

For details on state biosolids regulations, inquire with the state environmental management agency or check the National Biosolids Partnership's Web site (www.biosolids.org).

4.0 ENVIRONMENTAL MANAGEMENT SYSTEMS

The National Biosolids Partnership (NBP) has developed an environmental management system (EMS) for facilities that produce biosolids intended for beneficial use. The program is designed to help organizations establish good biosolids management practices and become certified for following them consistently. The NBP Code of Good Practice requires that certified organizations agree to

- commit to compliance with all applicable federal, state, and local requirements for biosolids production at the wastewater treatment facility, and management, transportation, storage, and use or disposal of biosolids away from the facility;
- providc biosolids that meet the applicable standards for their intended use or disposal;
- develop an EMS for biosolids that includes a method for independent third parties to verify that ongoing biosolids operations are effective;
- better monitor biosolids production and management practices;
- maintain good housekeeping practices for biosolids production, processing, transport, and storage, as well as during final use or disposal operations;
- develop response plans for unanticipated events (e.g., inclement weather, spills, and equipment malfunctions);
- enhance the environment by committing to sustainable, environmentally acceptable biosolids management practices and operations via an EMS;
- prepare and implement a preventive-maintenance plan for equipment used to manage solids and biosolids;
- seek continual improvement in all aspects of biosolids management; and
- provide effective communication methods with gatekeepers, stakeholders, and interested citizens about the key elements of each EMS, including information about system performance.

5.0 SOLIDS QUANTITIES

The amount of solids generated during wastewater treatment is an important design parameter because it affects the sizing of solids treatment processes and all related equipment. Solids-generation rates also affect the size of liquid treatment processes. For example, the size of a secondary treatment process to maintain a desired solids retention time (SRT) depends on how much solids will be produced.

5.1 Estimating Solids Quantities

While engineers generally recognize the importance of solids production in wastewater treatment plant design, many still do not understand the wide variation of quality and quantity of solids produced at treatment plants and how difficult it is to estimate solids quantities accurately. The best source of information for estimating solids production is plant-specific data that reflects the nature of the wastewater being treated and the treatment processes being used. If such data do not exist, default approaches or sophisticated mathematical models can be used; however, designers should understand that these estimates may differ significantly from actual results and, therefore, apply conservative safety factors to them.

In general, domestic wastewater typically produces about 0.23 kg/m^3 (1 dry ton/mil gal) of solids. Treatment plants using processes that destroy solids (e.g., digestion or heat treatment)n will generate less, and those using chemical addition will produce more. That said, 0.25 kg/m^3 is a convenient benchmark for cursory comparisons.

A good approach to estimating solids production is to provide a mass balance for the entire treatment plant that relates solids production to design parameters for each treatment process (see Figure 20.1). The mass balance should show key constituents [e.g., flow, total suspended solids (TSS), and biochemical oxygen demand (BOD)] and the process assumptions used in the calculations. It also should include solids generated during nitrogen- and phosphorus-removal processes.

Recycle streams can be included in one of two ways. In the first approach, engineers assume that a fixed percentage of solids or BOD is recycled from downstream processes to the head of the plant. They then iterate the solids balance until the recycled quantities assumed at the head of the plant equal the sum of recycled quantities computed for each process.

The second approach is to estimate the treatment plant's net solids production based on historical data, anticipated influent strength, or experience at similar facilities. Engineers then use this information to determine the amount of solids leaving the treatment plant, and typically apply it to the output end of the dewatering process. They then back-calculate solids loading to a specific process via the mass balance.

In either approach, engineers typically must estimate the quantities of primary, secondary, and chemical solids separately. They also must take into account expected fluctuations in wastewater characteristics that result from changes in industrial contribution, stormwater flows, seasonal weather conditions, and an expanded collection

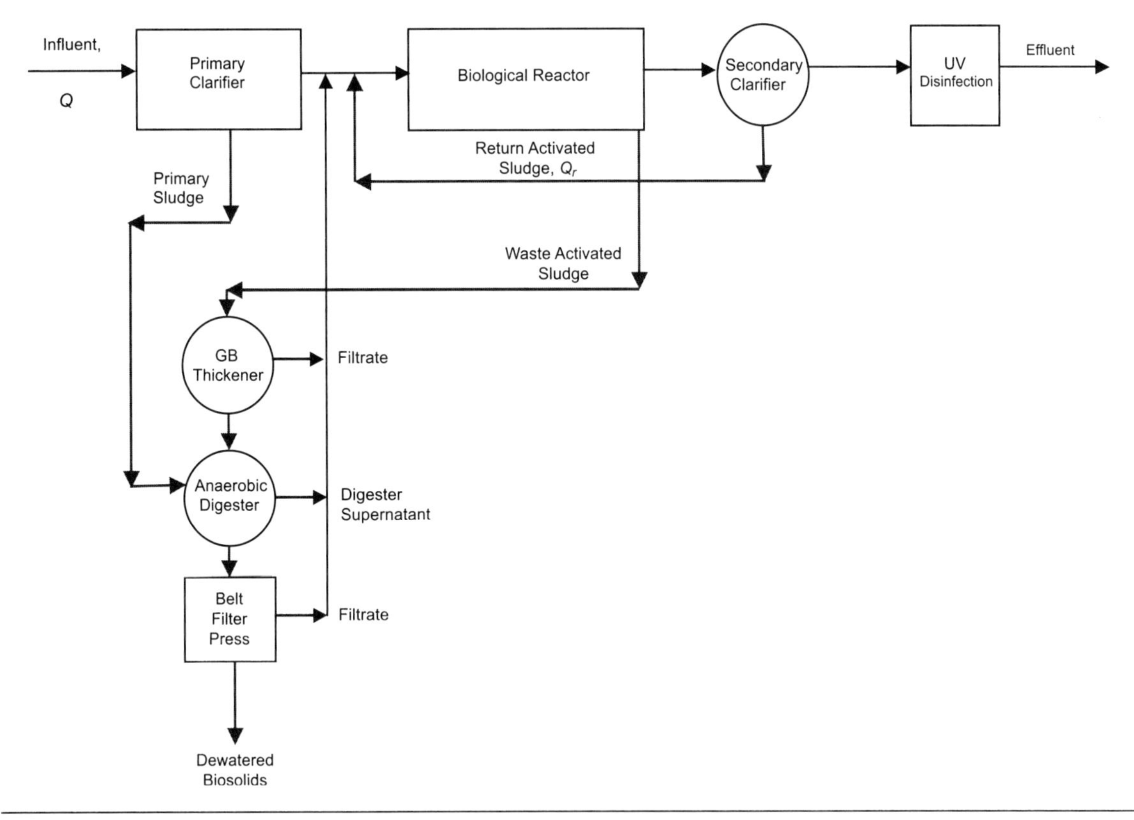

FIGURE 20.1 Process flow schematic for mass-balance example.

area. Engineers need to understand peak solids production and diurnal variations to size solids-handling processes properly.

5.2 Primary Solids Production

Most wastewater treatment plants use primary sedimentation tanks to remove settleable solids from wastewater. Primary sedimentation is a relatively efficient method for reducing BOD and TSS loading to secondary treatment processes. The amount of solids removed via primary sedimentation typically is related to either the surface overflow rate or hydraulic retention time (HRT). The relationship between primary solids production and HRT is as follows (Koch et al., 1990):

$$\text{Primary production} = \text{Plant flow} \times \text{Influent TSS} \times \text{Removal rate} \quad (20.1)$$

$$\text{Removal rate} = T/(a + bT) \quad (20.2)$$

Where

Removal rate = removal (%);

T = detention time (minutes);

a = constant (0.406 minutes); and

b = constant (0.0152).

This expression was developed by fitting a curve to data from 18 large wastewater treatment plants. While a suitable model for many treatment plants, its predicted value can vary greatly (see Figure 20.2). In fact, the equation provides a reasonable approximation when data from many plants are used, but the correlation coefficient often is rather poor when plotted for only one plant.

Figure 20.3 presents some typical curves relating primary solids production to the surface overflow rate (Great Lakes, 2004). This is the most common method for estimating primary solids production. However, other factors (e.g., hydraulic short-circuiting, poor flow distribution, density currents, and other mechanical factors) can affect performance significantly. Without proper clarifier design, actual plant data may indicate only a weak correlation between performance and either surface overflow rate or detention time. In fact, it is not unusual to briefly find occasional negative removal efficiencies for highly loaded primary sedimentation tanks. (For primary sedimentation design guidance, see Chapter 12.)

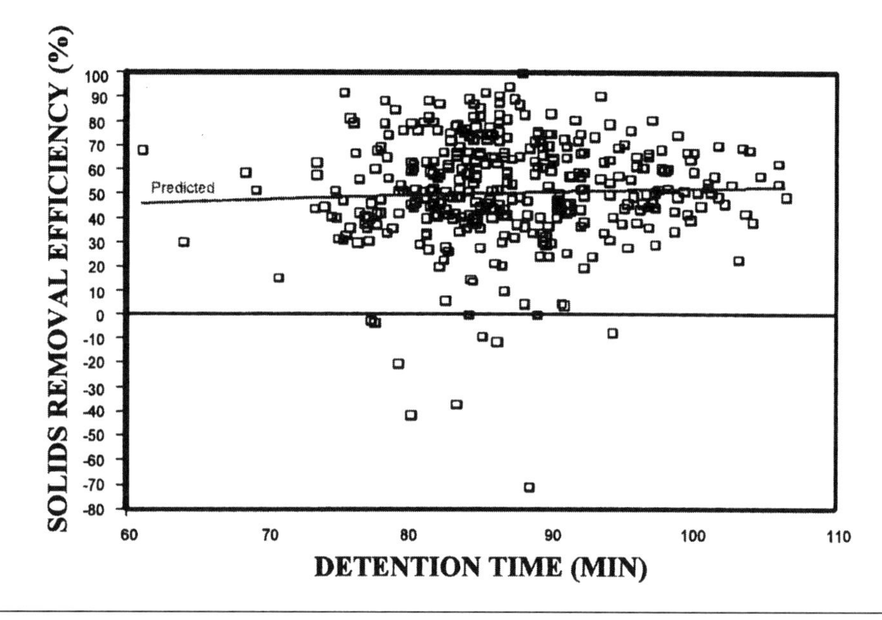

FIGURE 20.2 Primary tank performance, Cedar Creek—daily data.

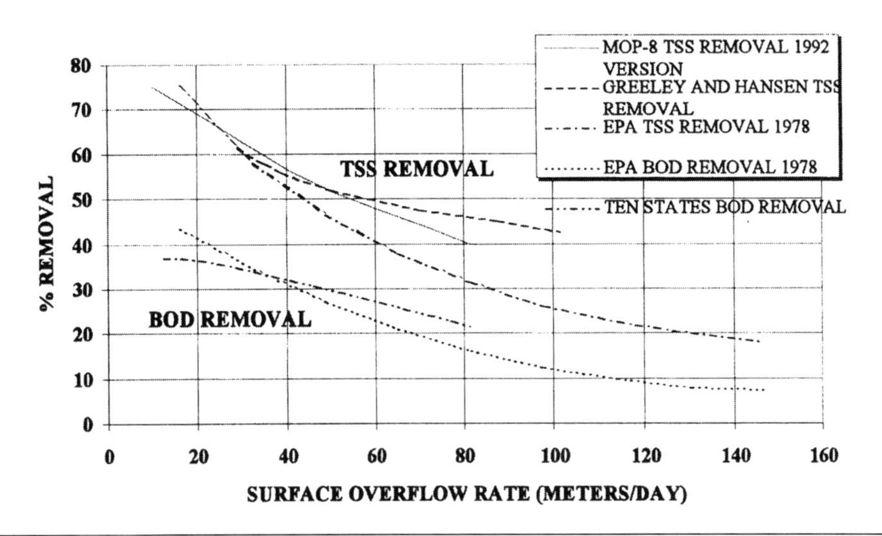

FIGURE 20.3 Primary treatment performance—TSS and BOD removal.

The degree of BOD or chemical oxygen demand (COD) removal across a primary sedimentation tank affects organic loading to the secondary treatment process and, hence, secondary solids production. Typically, BOD removal is about 50% of TSS removal, although wastewater characteristics can alter this ratio in either direction (Koch et al., 1990).

The retention time and surface overflow rate approaches are essentially equivalent for sedimentation tanks with similar depths. For tanks less than 4 or 5 m (12 to 15 ft) deep, the retention time approach may be better. Plants that add chemicals to enhance primary treatment performance or remove phosphorus will produce more solids.

In chemically enhanced primary treatment (CEPT), chemicals (e.g., ferric chloride) are used to remove more suspended solids and BOD from wastewater. Adding about 20 mg/L of ferric chloride and 0.2 mg/L of polymers to the headworks before primary sedimentation has been shown to increase primary solids production by about 45% (Chaudhary et al., 1989). About 30% of this increase was the result of better suspended solids removal; 65% stems from chemical precipitation and removal of colloidal material (Chaudhary et al., 1989). (For a more detailed discussion of this subject, see Chapter 12.)

5.3 Secondary Solids Production

Secondary solids are produced by biological treatment processes [e.g., activated sludge, biological nutrient removal, trickling filters, rotating biological contactors (RBCs), and other attached-growth systems] that convert soluble wastes or substrates (measured as BOD or COD) into microorganisms or biomass. Secondary solids also include some of the particulate that remains after primary sedimentation and becomes incorporated into the biomass. The quantity of secondary solids produced is a function of many factors [e.g., the efficiency of the primary treatment process, the ratio of TSS to 5-day BOD (BOD_5), the amount of soluble BOD or COD in the wastewater, and the design parameters of the secondary treatment process].

In an activated-sludge process, the length of time that secondary solids remain in the process (i.e., SRT) significantly affects the amount of secondary solids produced because the longer solids are retained, the more endogenous decay (self-destruction of biomass) occurs. Temperature also affects secondary solids production. At higher temperatures, solids production should be decreased by a higher growth rate and more endogenous respiration.

The kinetic relationship between secondary solids production and SRT theoretically can be expressed as follows:

$$Y_{\mathrm{obs}} = \frac{Y}{1 + k_d(\theta_c)} \tag{20.3}$$

Where

Y_{obs} = observed yield (g biomass/g substrate);
Y = yield (g biomass/g substrate);
k_d = endogenous decay rate (g biomass/g biomass-d^{-1}); and
θ_c = SRT or MCRT (d).

The substrate concentration is represented by either BOD_5 or COD; it typically is expressed as grams BOD_5 or COD consumed by the process, although some designers prefer to express it as the concentration applied rather than removed. The biomass concentration is expressed in either TSS or volatile suspended solids (VSS). (Typically, the ratio of VSS to TSS is in the range of 0.7 to 0.8.) The yield also can include effluent biomass and biomass wasted from the secondary treatment process. Some design engineers use the phrases *total yield* (including effluent solids) and *net yield* (excluding effluent solids) to distinguish between types of yields. Unfortunately, these phrases have been used interchangeably in the literature, along with *observed yield* and *apparent yield*, which can make plant comparisons confusing. The ranges listed in Table 20.5 have been reported for the yield and endogenous decay coefficients and are expressed as net production (i.e., 1 g TSS versus 1 g substrate removed).

Engineers also can obtain values for yield and endogenous decay coefficients by plotting solids production versus SRT (see Figure 20.4).

TABLE 20.5 Typical values for solids yield coefficients.

Coefficient	Basis	Range	Typical
Y	g VSS/g BOD_5	0.4–0.8	0.6
Y	g VSS/g COD	0.25–0.4	0.4
k_d	d^{-1}	0.04–0.075	0.06

Figure 20.5 includes the Monod curve, which is named for the scientist who pioneered the application of Michaelis-Menten enzyme kinetics to microbial growth (Monod, 1949). This curve fits the above equation to the data using linear-regression techniques. The Monod curve's values for yield and endogenous decay rate coefficients are 0.731 g VSS/g BOD_5 removed and 0.055 d^{-1}, respectively. These values are within the range of reported typical values.

The data in Figure 20.5 emphasize the data variations typical of operating plants. In many plants, it is difficult to see a clear relationship between SRT and solids production. In fact, some researchers have reported that solids production does not appear to be affected by SRT (Wilson et al., 1984, and Zabinski et al., 1984). A relationship between yield and SRT typically becomes apparent, however, when multiple plants are plotted.

In theory, the coefficients' values also should vary with temperature. For a given SRT, the growth rate and amount of endogenous respiration should increase with

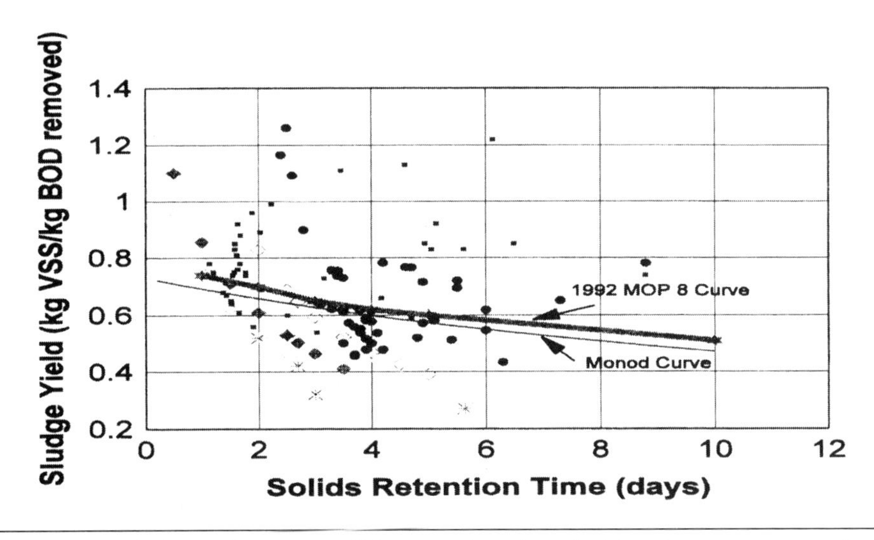

FIGURE 20.4 Solids yield versus solids retention time (with primary treatment).

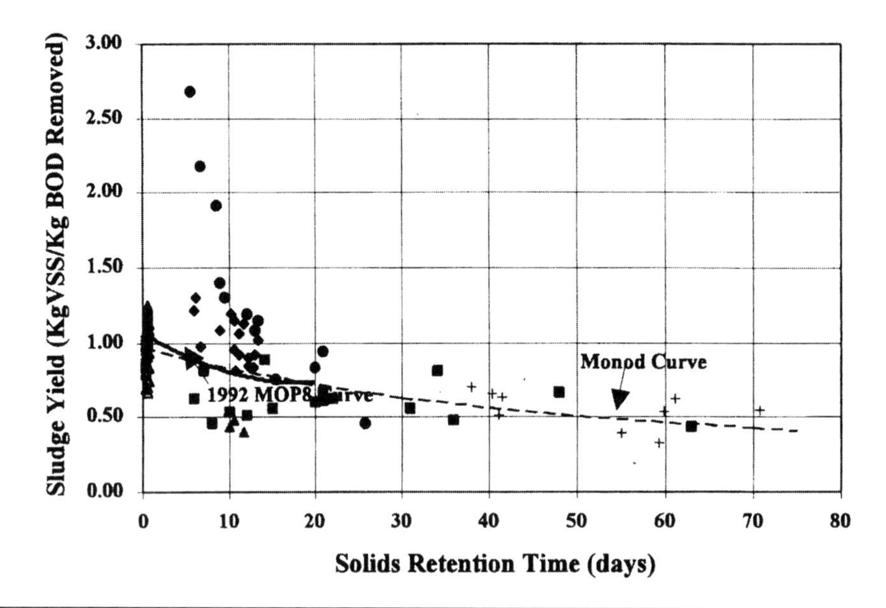

FIGURE 20.5 Solids yield versus solids retention time (without primary treatment).

temperature, thereby lowering solids production. The 1992 edition of this manual showed a series of curves relating solids production to SRT at three temperatures. The curve for 20°C in the 1992 edition also was plotted in Figure 20.5; it closely follows the curve fitted to the operating data. Although wastewater temperature varies over the year, the effect on solids production may be masked because many plants adjust SRT seasonally. Many plants see little difference in solids production throughout the year, and some plants have reported higher solids production during the warmer summer months (Koch et al., 1990).

One method for deriving yield coefficients is to plot the solids production per kilogram of aerated volatile solids versus the food-to-microorganism (F:M) ratio (U.S. EPA, 1979). A plot of these two expressions will show the value of yield as the slope and the value of endogenous decay rate as the intercept. A plot of typical data presented by U.S. EPA (1979) in *Process Design Manual Sludge Treatment and Disposal* yields values within the range shown in Table 20.5. This approach is often used to obtain kinetic parameters from pilot-plant data.

Another approach for estimating secondary solids production is to separate production into three terms (Koch et al., 1990):

$$\text{Net production} = \text{Inerts} + a\ \text{VSS} + b\ \text{SBOD} \tag{20.4}$$

Where

a = volatile solids coefficient,
b = soluble BOD coefficient,
VSS = volatile suspended solids, and
SBOD = soluble BOD.

The volatile solids coefficient varies between 0.6 and 0.8, and the soluble BOD coefficient varies between 0.3 and 0.5. Both coefficients decrease with increasing SRT and are dimensionless.

This expression separates biomass production into the organic fraction (volatile solids and soluble BOD) and inorganic fraction (inerts) of influent. Each term on the right side of the equation represents a different portion of the wastewater influent strength (nonvolatile solids and volatile solids that include particulate BOD_5 and soluble BOD_5). The volatile solids and soluble BOD coefficients also can be related to temperature and SRT.

A more sophisticated form of Equation 20.4 is used in Activated Sludge Models Nos. 1 and 2 developed by International Association of Water Pollution Research and Control (IAWPRC). In these models, solids production is modeled by separately calculating nondegradable TSS, degradable TSS, soluble substrates, and active and inactive biomass. Different growth rates are applied to the heterotrophic and autotrophic microorganisms in the anaerobic, anoxic, and aerobic sections of the process. Daigger et al. (1992) used Activated Sludge Model No. 1 to develop equations for estimating solids for both short SRTs, where they assumed partial degradation or organic particulate, and long SRTs, where they assumed complete destruction of degradable TSS as follows:

For SRTs longer than 3 days,

$$Y_{obs} = TSS/BOD_5 + Y_H SBOD_5/BOD_5 \tag{20.5}$$

If SRT is shorter than 3 days,

$$Y_{obs} = \frac{TSS[1 - f_v + f_v f_{nv} + (f_v - f_v f_{nv})(1 - f_D)]}{BOD_5} + \frac{Y_H(SBOD_5 + (BOD_5 + SBOD_5)f_D}{BOD_5} \tag{20.6}$$

Where

Y_{obs} = g TSS generated/g BOD_5 removed;
TSS = influent suspended solids concentration (mg/L);
BOD_5 = influent BOD_5 concentration removed (mg/L);
$SBOD_5$ = influent soluble BOD_5 concentration (mg/L);
f_v = volatile fraction of TSS (%);
f_i = inactive biomass fraction (%);
f_{nv} = nonbiodegradable VSS fraction (%);
f_D = fraction biodegradable VSS that is degraded at a given SRT (%); and
Y_H = heterotrophic cell yield (g suspended solids generated/g soluble BOD_5 removed).

Typical values for the parameters in Equations 20.5 and 20.6 and can be found elsewhere (Daigger et al., 1992).

For a TSS : BOD_5 ratio of 0.6, which indicates a plant with primary treatment, Equations 20.5 and 20.6 yield values that typically follow the curves in Figure 20.4 for an SRT longer than 2 days. For an SRT shorter than 2 days, Equations 20.5 and 20.6 yield values higher than the Figure 20.4 curves, although close to the data points in Figure 20.4. Using two models to predict solids production at different SRTs may provide a better fit to the data in Figure 20.4. However, it is necessary to assume values for many more parameters to apply to Equations 20.5 and 20.6.

Design engineers also should consider the effect of COD : BOD_5 ratio on solids yield. Plants with high COD : BOD_5 ratios tend to yield higher quantities of solids (U.S. EPA, 1987). Another factor that can affect secondary solids production is the need for phosphorus removal. If the secondary treatment process includes chemical or biological phosphorus removal, yields typically will be higher than those in Figure 20.4. Methods for estimating biological phosphorus yield can be found in other references (U.S. EPA, 1987) and are discussed in the "Combined Solids Production" section below.

All of these approaches can be used to estimate secondary solids production from suspended-growth activated-sludge systems. A similar approach can be applied to attached-growth secondary treatment systems (e.g., RBCs and trickling filters). In general, solids production for attached-growth systems can be expressed as follows:

$$\begin{aligned} \text{Net production} &= Y - k_d \\ &= Y\,\text{BOD}_\text{r} + k_d\,\text{AB} \end{aligned} \tag{20.7}$$

Where

Y = observed yield (g biomass/g substrate);
k_d = endogenous decay rate (g biomass/g biomass·d);
BOD_r = BOD removed; and
AB = attached biomass.

The yield coefficient has units similar to the BOD-removed coefficient for activated-sludge systems (i.e., grams VSS per gram BOD_5 removed). The attached-biomass coefficient has the same units as the attached-biomass coefficient for activated-sludge systems. The amount of attached biomass typically is directly related to the amount of surface area available to support attached growth. In addition, plots of solids production per unit of surface area versus BOD_5 removal per unit of surface area can be used to derive plant-specific kinetic constants from plant data. The values of the yield coefficient for attached-growth systems are similar to those for suspended-growth systems; however, the values of the attached-biomass coefficient for attached growth tend to be higher to reflect the longer effective SRT in attached systems. The attached-biomass coefficient for an attached-growth system typically ranges from 0.03 to 0.40 d^{-1} (U.S. EPA, 1979).

In hybrid systems that use both suspended and attached growth, solids production varies depending on how much biomass is attached and how much is suspended. Some researchers have reported that the character and amount of solids produced is dominated by the attached growth (Newberg et al., 1988).

5.4 Combined Solids Production

Treatment plants without primary treatment processes generate combined solids. This mix of primary and secondary solids will be significantly greater in quantity than the secondary solids produced at plants with both treatment processes (see Figure 20.5) (Koch et al., 1997, and Schultz et al., 1982). Design engineers can estimate how much combined sludge will be produced by adjusting the yield coefficients used to estimate secondary solids production to account for the additional solids. The values of the yield and endogenous decay rate coefficients for the Monod curve (see Equation 20.3), as shown in Figure 20.5, are 0.975 g VSS/g BOD_5 removed and 0.017 7 d^{-1}, respectively.

The high SRTs represent extended aeration plants and oxidation ditch plants. Although Figure 20.5 shows that high SRTs tend to lower solids production, several plants have solids-production levels significantly above this curve, so engineers should be cautious when using this information to design. A comparison of Figures 20.4 and 20.5 shows that the absence of primary treatment increases the solids yield coefficients. In IAWPRC's model-derived equations, the absence of primary sedimentation is accounted for by the TSS/BOD_5 term in the equation, which is higher for plants without primary sedimentation.

Biological nitrogen removal typically will increase solids production as a result of the nitrification and denitrification processes (see Table 20.6) (U.S. EPA, 1987). However, the extra solids produced often are offset by the additional endogenous respiration that occurs at higher SRTs. If a second substrate (e.g., methanol) is added for denitrification, even more solids will be produced. Engineers can estimate this extra mass based on the amount of substrate added and expected biomass yield using the methods previously described in section 5.3.

Biological phosphorus removal may generate more solids as a result of the inorganic salts that accumulate with phosphorus in the biomass (U.S. EPA, 1987). To estimate how much more solids will be produced, engineers can multiply the mass of additional phosphorus removed by 4.5, which is based on a molecular weight of 140 for phosphorus crystals in biomass (U.S. EPA, 1987).

5.5 Chemical Solids Production

Design engineers can estimate the quantity of chemical solids produced based on anticipated chemical reactions. Solids production typically increases in direct proportion to the amount of chemical added; however, competing reactions must be considered. For example, adding ferric chloride will generate more solids than estimated by considering the reaction of ferric chloride to form ferric hydroxide, which preferentially reacts with phosphate, yielding more precipitates than the hydroxide reaction alone. As a general rule, engineers can assume 1 g more solids per 1 g of ferric chloride added. Similarly, adding lime can significantly increase solids production.

Both chemicals also are used in the chemically enhanced primary treatment (CEPT) process. Adding inorganic chemicals for CEPT or phosphorous removal increases the mass and characteristics of primary solids.

5.6 Mass Balance Example

Mass balance calculations yield the data that engineers need to design solids thickening, dewatering, and stabilization processes. The following is a simplified example of

TABLE 20.6 Nitrification and denitrification yield factors.

Process	Bases	Typical values
Nitrification	g VSS/g NH_4-N removed	0.17
Dentrification	g VSS/g NO_3-N removed	0.8

a mass balance for an activated sludge plant with headworks degritting, primary clarification, gravity belt thickening, anaerobic digestion, belt filter press dewatering, and ultraviolet disinfection (see Figure 20.1 and Table 20.7). Mixed-liquor suspended solids are wasted from the biological reactors.

Mass balance is an iterative process, and this example shows two iterations. The first establishes the recycle flow and concentration. If the second iteration's results are not within ±5% of those of the first iteration, engineers should do a third iteration.

It is easy to set up a spreadsheet that incorporates the various formulas needed for numerous iterations. (NOTE: mg/L is identical to g/m^3.)

5.6.1 Step 1: Determine the Mass of BOD and TSS in Influent

a. Mass (kg/d) = Concentration (g/m^3) × Q (m^3/d)/1 000g/kg
b. Mass (lb/d) = Concentration (mg/L) × 8.34 × Q (mgd)

Table 20.7 Solids characteristics for mass balance example.*

	In SI units	In U.S. customary units
Q [m^3/d (mgd)]	90 850	24.0
QP	227 125	60.0
Influent		
BOD [g/m^3(mg/L)]	300	300
TSS [g/m^3(mg/L)]	335	335
TSS after grit [g/m^3(mg/L)]	286	286
Solids characteristics	%	%
Primary	4.8	4.8
Thickened WAS	5.5	5.5
TSS digested	5.3	5.3
Specific Gravity	1	1
Biodegradable fraction of WAS	65	65
Effluent characteristics		
BOD [g/m^3(mg/L)]	10	10
TSS [g/m^3(mg/L)]	14	14
UBOD	1.42 g/g	1.42 g/g

* BOD = biochemical oxygen demand; Q = influent flow; QP = peak flow; TSS = total suspended solids; UBOD = ultimate BOD; and WAS = waste activated sludge.

Influent Mass	kg/d	lb/d
BOD	27 255	60 048
TSS	30 435	67 054
TSS after grit	25 983	57 246

5.6.2 *Step 2: Estimate Soluble BOD in Effluent*

a. Biodegradable portion = Effluent TSS × 65%
b. UBOD = Biodegradable portion × 1.42
c. BOD of effluent TSS = 0.68 (obtained using $k = 0.23\ d^{-1}$) × UBOD
d. Effluent soluble BOD escaping treatment = Effluent BOD–BOD of Effluent TSS

	g/m^3	mg/L
Determine BOD of Effluent TSS (Biodegradable portion is 65%)	9.1	9.1
UBOD	12.9	12.9
BOD of effluent TSS	8.8	8.8
Effluent soluble BOD escaping treatment	1.2	1.2

5.6.3 *Step 3: Conduct First Iteration*

5.6.3.1 *Step 3.1: Primary Settling*

a. Assume 33% removal of BOD and 70% removal of TSS
b. Calculate mass of BOD and TSS removed and mass of BOD and TSS that will go to bioreactors.
c. Mass (kg/d) = Concentration (g/m^3) × Q (m^3/d)/1 000g/kg
d. Mass (lb/d) = Concentration (mg/L) × 8.34 × Q (mgd)
e. Calculate concentration of BOD in primary effluent
f. Calculate volatile fraction of primary solids

Primary settling	In SI units	In U.S. customary units
BOD removed [kg/d (lb/d)]	8 994	19 816
BOD to secondary [kg/d (lb/d)]	18 261	40 232
TSS removed [kg/d (lb/d)]	18 188	40 072
TSS to secondary [kg/d (lb/d)]	7 795	17 174
Primary effluent BOD [g/m^3 (mg/L)]	200	200

Volatile fraction of primary solids		
Volatile fraction of influent TSS	0.67	0.67
Volatile fraction of grit	0.10	0.10
Volatile fraction of incoming TSS discharge to secondary process	0.85	0.85
VSS in influent before grit [kg/d (lb/d)]	20 391	44 926
VSS removed in grit [kg/d (lb/d)]	445	981
VSS in secondary influent [kg/d (lb/d)]	6 626	14 598
VSS in primary solids [kg/d (lb/d)]	13 320	29 348
Volatile fraction in primary solids	0.73	0.73

5.6.3.2 *Step 3.2: Secondary Process*

a. Set operating parameters: mixed-liquor suspended solids (MLSS), Y_{obs}
b. Calculate mass quantities of BOD and TSS in effluent

Operating parameters	In SI units	In U.S. customary units
MLSS [g/m^3 (mg/L)]	3 500	3 500
Volatile fraction	0.8	0.8
Y_{obs}	0.3125	0.3125
Mixed-liquor volatile suspended solids (MLVSS) [g/m^3 (mg/L)]	2 800	

Effluent mass quantities		
BOD [kg/d (lb/d)]	909	2 002
TSS [kg/d (lb/d)]	1 272	2 802

c. Estimate the amount of TSS produced in the biological process (assume primary solids flow is small relative to plant flow)
d. TSS produced = $[Y_{obs} \times Q \times (S_o - S)]/10\,00$ g/kg where S_o = concentration of BOD in primary effluent and S = concentration of soluble BOD in the final effluent.
e. TSS produced = $Y_{obs} \times Q \times (S_o - S) \times 8.34$
f. Estimate total amount to be wasted assuming a volatile solids concentration of 80%
g. Estimate mass of waste solids
h. Estimate flowrate of waste solids

	In SI units	In U.S. customary units
TSS produced in the biological process [kg/d (lb/d)]	5 400	11 907
VSS wasted at 80% volatile [kg/d (lb/d)]	6 750	14 883
Fixed solids (by difference)	1 350	2 977
Mass of waste activated sludge (WAS) [kg/d (lb/d)]	5 478	12 079
Flowrate [m^3/d (gal/d)]	1 565	413 465

5.6.3.3 Step 3.3: Gravity Belt Thickening

a. Operating parameters

Gravity belt thickeners	In SI units	In U.S. customary units
Thickened solids (%)	4.8	4.8
Solids recovery (%)	92	92
Specific gravity	1	1

b. Determine flowrate of thickened solids [Flowrate = (mass of WAS × 0.92)/(1 000 × 0.048)]

	In SI units	In U.S. customary units
Flowrate [m^3/d (gal/d)]	105	27 737

c. Determine recycle flowrate
   i. Flowrate of WAS − Flowrate of thickened sludge
   ii. Calculate mass of TSS to digester mass = (Mass of WAS × 0.92)
   iii. Calculate mass of TSS to headworks mass = (Mass of WAS − Mass to digester)
   iv. Calculate concentration of TSS in recycle TSS = (Mass TSS × 1 000 g/kg)/Recycle flowrate
   v. Determine BOD concentration of TSS (BOD = TSS × 0.65 × 1.42 × 0.68)
   vi. Calculate mass of BOD in recycle [BOD = (Concentration × Flowrate)/1 000 g/kg]

	In SI units	In U.S. customary units
Recycle flowrate [m^3/d (gal/d)]	1 460	385 728
TSS to digester [kg/d (lb/d)]	5 040	10 871
TSS to headworks [kg/d (lb/d)]	438	1 208
TSS concentration in recycle [g/m^3 (mg/L)]	300	485
Determine BOD concentration of TSS [g/m^3 (mg/L)]	188	305
BOD [kg/d (lb/d)]	275	606

5.6.3.4 *Step 3.4: Anaerobic Digestion*

a. Set operating parameters

	In SI units	In U.S. customary units
VSS destruction (%)	47	47
Gas production [m^3/kg (ft^3/lb) VSS destroyed]	0.9	15
BOD in digester supernatant (mg/L)	1 000	1 000
TSS in digester supernatant (mg/L)	5 000	5 000
TSS concentration in digested solids (%)	5	5

b. Determine total solids fed to the digester and flowrate
c. TSS mass = Mass primary solids + mass thickened WAS (TWAS)
d. Calculate VSS mass fed to digester (assume 80% volatile)
e. Calculate VSS in mixture fed to digester and calculate VSS destroyed (assuming 50% destruction)
f. Calculate mass flow of primary solids to digester (4.8% solids)
g. Calculate mass flow of TWAS to digester
h. Calculate total mass flow
i. Calculate fixed solids by difference
j. Calculate mass of TSS in digested solids
k. Calculate gas production

	In SI units	In U.S. customary units
TSS mass, from primary solids and TWAS	23 228	51 218
Total flowrate [m^3/d (gal/d)]	471	124 307
VSS mass fed to the digester [kg/d (lb/d)]	17 352	38 262
Percent VSS mass fed to digester	74.7%	74.7%
VSS destroyed [kg/d (lb/d)]	8 156	17 983
Mass flow to digester-primary solids [kg/d (lb/d)]	378 920	835 519
TWAS mass flow [kg/d (lb/d)]	91 632	202 047
Total mass flow [kg/d (lb/d)]	470 552	1 037 567

Fixed solids [kg/d (lb/d)]	5 876	12 956
TSS mass in digested solids [kg/d (lb/d)]	14 552	32 087
Gas [kg/d (lb/d)]	7 340	11 248

l. Calculate mass balance around digester

	In SI units	In U.S. customary units
Mass input	470 552	1 037 567
Less gas	7 340	11 248
Mass output	463 212	1 026 318

5.6.3.5 *Step 3.5: Flowrate Distribution of Supernatant and Digested Solids*

a. (S/concentration supernatant) + (Total mass in digested sludge − S)/solids in sludge = mass output
b. Calculate mass of digested solids (mass = TSS mass in digested sludge − S)
c. Calculate supernatant flow {flow = S/(concentration of solids in supernatant (%)] × 1 000 kg/m^3}
d. Calculate sludge flow [flow = mass digested solids/(% solids × 1000 kg/m^3)]

	In SI units	In U.S. customary units
Supernatant (%)	0.5%	0.5%
Solids (%)	5.0%	5.0%
S	957	2 137
Digested solids	13 595	29 950
Supernatant flow	191	50 538
Digested solids flow	272	71 830

e. Determine BOD and TSS of supernatant mass flow
f. BOD = (Supernatant flow × 1000 g/m^3)/1 000 g/kg
g. TSS = (Supernatant flow × 1000 g/m^3)/1 000 g/kg

	In SI units	In U.S. customary units
BOD [kg/d (lb/d)]	191	422
TSS [kg/d (lb/d)]	957	2 109

5.6.3.6 *Step 3.6: Solids Dewatering*

a. Establish characteristics

	In SI units	In U.S. customary units
Solids cake (%)	22%	22%
sp gr	1.06	1.06
Solids capture (%)	96%	96%
Filtrate BOD concentration	2 000	2 000

b. Determine solids cake and filtrate characteristics
c. Recycle solids = digested solids × capture rate
d. Volume = recycle solids/(sp gr × cake solids × 1 000)

	SI	U.S.
Solids [kg/d (lb/d)]	13 051	28 778
Volume (m^3/d)	56.0	14 785

e. Determine filtrate characteristics
f. Flow = (Digested sludge flow – Volume of sludge cake)
g. BOD mass = (Filtrate BOD concentration × flow)/1 000
h. TSS mass = Digested solids × Percent not captured

	In SI units	In U.S. customary units
Flow	216	57 045
BOD mass	432	952
TSS mass	544	1 198

5.6.3.7 *Step 3.7: Summary of Recycle Flows*

	In SI units	In U.S. customary units
Recycle flow	1 867	493 311
Recycle TSS	1 939	4 275
Recycle BOD	898	1 981

5.6.4 *Step 4: Conduct Second Iteration*

5.6.4.1 *Step 4.1: New Influent Concentration and Mass of BOD and TSS to Primary Sedimentation*

a. Calculate new mass of TSS entering primary sedimentation (mass = Influent TSS + Recycle TSS)
b. Calculate new mass of BOD entering primary sedimentation (mass = Influent BOD + Recycle BOD)
c. Calculate BOD removal (assuming 33%)
d. Calculate TSS removal (assuming 70%)

	In SI units	In U.S. customary units
Influent TSS to primary tanks = Influent + Recycle (kg/d)	27 919	61 562
Influent BOD to primary tanks = Influent + Recycle (kg/d)	28 153	62 077
BOD removed [kg/d (lb/d)]	9 290	20 485
BOD to bioreactors [kg/d (lb/d)]	18 862	41 592
TSS removed [kg/d (lb/d)]	19 543	43 093
TSS to bioreactors [kg/d (lb/d)]	8 376	18 469

5.6.4.2 *Step 4.2: Secondary Process*

a. Using the target F:M ratio and original MLVSS concentration, calculate bioreactor volume
b. Set target SRT
c. Calculate new flowrate (influent flow + recycle flow)
d. Calculate new bioreactor influent BOD concentration {BOD = [BOD mass to bioreactors (kg/d) × 1 000 g/kg]/Flowrate (m^3/d)}
e. Calculate new concentration of MLVSS {MLVSS = $[(\text{SRT} \times Q)/V] \times [Y \times (S_o - S)]/[1 + (\text{kd} \times \text{SRT})]$}
f. Calculate MLSS (assuming 80% volatile solids)
g. Calculate new cell growth {New cells = $[Q \times Y_{\text{obs}} \times (S_o - S)]/1000$}
h. Calculate mass of TSS MLSS + new cells
i. Calculate WAS to thickening WAS = Mass of TSS—Mass of effluent TSS
j. Calculate flowrate [Flowrate = (WAS × 1 000)/MLSS]

	In SI units	In U.S. customary units
Target F:M ratio	0.35	0.35
Bioreactor volume [m^3 (mil. gal)]	18 559	4.90
SRT (days)	10	10
Y	0.5	0.5
k_d	0.06	0.06
Flowrate [m^3/d (mgd)]	92 717	24.5
BOD concentration	203	203
New concentration of MLVSS	3 044	3 044
MLSS	3 805	3 805
Mass of new cells	5 649	5 649
TSS mass [kg/d (lb/d)]	7 062	15 571
WAS to thickening	5 790	12 767
Flowrate	1 522	3 355

5.6.4.3 *Step 4.3: Gravity Belt Thickening*

a. Determine flowrate of thickened sludge;

$$\text{Flowrate} = (\text{mass of WAS} \times 0.92)/(1\,000 \times 0.048) \quad (20.8)$$

b. Determine recycle flowrate
   i. Flowrate of WAS – Flowrate of thickened sludge
   ii. Calculate mass of TSS to digester Mass = Mass of WAS × 0.92
   iii. Calculate mass of TSS to influent Mass = Mass of WAS—Mass to digester)
   iv. Calculate concentration of TSS in recycle TSS = Mass TSS × 1000 g/kg)/ Recycle flowrate
   v. Determine BOD concentration of TSS; BOD = TSS × 0.65 × 1.42 × 0.68
   vi. Calculate the mass of BOD in recycle; BOD = (Concentration × Flowrate)/ 1000 g/kg

	In SI units	In U.S. customary units
Flowrate	111	29 316
Recycle flowrate	1 411	372 664
TSS to digester [kg/d (lb/d)]	5 327	11 745
TSS recycle to headworks [kg/d (lb/d)]	463	1 021
TSS [g/m^3 (mg/L)]	328	328
BOD concentration in TSS [g/m^3 (mg/L)]	206	206
BOD mass [kg/d (lb/d)]	291	641

5.6.4.4 Step 4.4: Anaerobic Digestion

a. Set operating parameters

Operating parameters	In SI units	In U.S. customary units
SRT (days)	10	10
VSS destruction (%)	47	47
Gas production [m^3/kg (cu ft/lb)] VSS destroyed	0.9	15
BOD in digester supernatant (mg/L)	1 000	1 000
TSS in digester supernatant (mg/L)	5 000	5 000
TSS concentration in digested solids (%)	5	5

b. Determine total solids fed to the digester and flowrate
c. TSS Mass = Mass primary solids + Mass TWAS
d. Calculate VSS mass fed to digester, assume 80% volatile
e. Calculate VSS in mixture fed to digester and calculate VSS destroyed, assuming 50% destruction
f. Calculate mass flow of primary solids to digester (4.8% solids)
g. Calculate mass flow of TWAS to digester
h. Calculate total mass flow
i. Calculate fixed solids by difference
j. Calculate mass of TSS in digested solids
k. Calculate gas production

	In SI units	In U.S. customary units
TSS mass, from primary solids and TWAS	24 872	54 842
Total flowrate [m^3/d (gal/d)]	459	121 233
VSS mass fed to the digester [kg/d (lb/d)]	17 552	38 702
Percent VSS mass fed to digester	71%	71%
VSS destroyed [kg/d (lb/d)]	8 249	18 190
Mass flow to digester-PS [kg/d (lb/d)]	407 191	897 857
TWAS mass flow [kg/d (lb/d)]	96 848	213 549
Total mass flow [kg/d (lb/d)]	504 039	1 111 406
Fixed solids [kg/d (lb/d)]	7 320	16 140
TSS mass in digested solids [kg/d (lb/d)]	16 096	35 491
Gas [kg/d (lb/d)]	10 177	11 378

1. Calculate mass balance around digester

	In SI units	In U.S. customary units
Mass input	504 039	1 111 406
Less gas	10 177	11 378
Mass output	493 861	1 100 028

5.6.4.5 Step 4.5: Flowrate Distribution of Supernatant and Digested Solids

a. (S/concentration supernatant) + (Total mass in digested sludge − *S*)/solids in sludge = mass output
b. Calculate mass of digested solids (mass) = TSS mass in digested solids − *S*
c. Calculate supernatant flow, flow = S/concentration of solids in supernatant (%) × 1 000 kg/m^3
d. Calculate solids flow (flow) = mass digested solids/(% solids × 1 000 kg/m^3)

	In SI units	In U.S. customary units
Supernatant (%)	0.5%	0.5%
Solids (%)	5.0%	5.0%
S	955	2 168
Digested solids	15 141	33 323
Supernatant flow	191	50 470
Digested solids flow	303	79 994

e. Determine BOD and TSS of supernatant mass flow
f. BOD = (Supernatant flow × 1 000 g/m^3)/1 000 g/kg
g. TSS = (Supernatant flow × 1 000 g/m^3)/1 000 g/kg

	In SI units	In U.S. customary units
BOD [kg/d (lb/d)]	191	191
TSS [kg/d (lb/d)]	955	955

5.6.4.6 Step 4.6: Solids Dewatering

a. Establish characteristics

	In SI units	In U.S. customary units
Solids cake (%)	22%	22%
sp gr	1.06	1.06
Solids capture (%)	96%	96%
Filtrate BOD concentration	2 000	2 000

b. Determine solids cake and filtrate characteristics
c. Recycle solids = digested solids × Capture rate
d. Volume = Recycle solids/(sp gr × cake solids × 1 000)

	In SI units	In U.S. customary units
Solids [kg/d (lb/d)]	14 535	32 049
Volume (m^3/d)	62.3	16 465

e. Determine filtrate characteristics
f. Flow = Digested sludge flow – Volume of solids cake
g. BOD mass = (Filtrate BOD concentration × flow)/1 000
h. TSS mass = Digested solids × Percent not captured

	In SI units	In U.S. customary units
Flow	240	63 529
BOD (mass)	481	1 061
TSS (mass)	606	1 335

5.6.5 *Step 5: Create Summary of Recycle Flows*

	First Iteration		Second Iteration		
	In SI units	In U.S. customary units	In SI units	In U.S. customary units	Percent difference
Recycle flow	1 867	493 311	1 842	486 663	1%
Recycle TSS	1 939	4 275	2 024	4 463	−4%
Recycle BOD	898	1 981	963	2 123	−7%

6.0 SOLIDS CHARACTERISTICS

When designing solids-handling facilities (e.g., conveyance, conditioning, and thickening or dewatering systems), engineers must know the characteristics and volumes of the solids involved. There are several types of solids (e.g., primary, secondary, mixed primary, chemical, and biosolids), and their characteristics depend on many factors (e.g., the percentage of industrial wastes, ground garbage, and sidestreams in wastewater; the use of chemical precipitants and coagulants; process control; peak loads and weather conditions; and the treatment process chosen).

There also are numerous references available that can help designers obtain detailed information on solids sources, characteristics, and quantities. This manual, however, focuses on wastewater treatment plant design, so it only addresses topics that significantly affect the design process.

6.1 Primary Solids

Most wastewater treatment plants use primary sedimentation to remove settleable solids, which thicken via gravity. Called *primary solids*, this material consists of organic solids, grit, and inorganic fines. Primary solids typically are pumped downstream for more thickening, stabilizing, and dewatering. Use or disposal then follows.

The composition of primary solids varies widely—from day to day, hour to hour, within a plant, and between plants. Table 20.8 notes the typical composition (ASCE, 1998; U.S. EPA, 1979).

The total solids concentration depends on the rate at which solids are removed from the primary sedimentation tank. If they are removed rapidly, a lower solids concentration can be expected. Some plants remove solids more slowly, essentially using the primary sedimentation tanks as gravity thickeners.

Plants that receive both sanitary and stormwater or have a high contribution of infiltration and inflow will produce primary solids that vary greatly in both volume and volatile solids concentration. Those with inadequate grit removal may produce more primary solids, but they only contain 60% VSS because of all the inorganics and grit.

Plants that add inorganic chemicals to primary sedimentation tanks will produce solids with lower VSS and higher phosphorus concentrations.

The heavy metal content of primary solids depends on the types of industries that discharge to the plant. It is higher for plants that add inorganic chemicals (e.g., ferric chloride or lime) to primary tanks.

TABLE 20.8 Primary solids characteristics.

Parameter	Concentration (dry-weight basis)
Total solids	2.0–8.0
Total volatile solids, % of TS	60–80
Grease, % of TS	5.0–8.0
Phosphorus, % of TS	0.8–2.8
Protein, % of TS	20–30
Cellulose, % of TS	8–15
Nitrogen, % of TS	1.5–4.0
pH	pH 5.0–8.0

6.2 Secondary Solids

Secondary solids are those generated when soluble wastes and other particles in primary effluent are converted to biomass via aerobic biological-treatment processes (e.g., activated sludge, trickling filters, and RBCs). Typically, biological sludges are more difficult to thicken or dewater than primary solids and most chemical solids.

Table 20.9 indicates the typical composition of secondary solids (ASCE, 1998, and U.S. EPA, 1979). Treatment plants with high F:M ratios tend to produce secondary solids with higher nitrogen levels than conventional activated-sludge plants do because less endogenous respiration occurs. Plants with biological phosphorus-removal processes produce solids with higher phosphorus levels. Plants that treat significant amounts of industrial wastewater can produce solids with higher heavy metal concentrations.

6.3 Combined Solids

Typically, the flow produced by combining primary and secondary solids has properties more like secondary solids. However, it depends on the proportions of each type and their compositions.

6.4 Chemical Solids

Chemical solids are the result of adding metal salts or lime to wastewater to improve suspended solids removal or precipitate phosphorus. Typically, lime improves thickening and dewatering performance, while iron and aluminum salts can improve dewatering. The characteristics of chemical solids are affected by wastewater chemistry, pH, mixing, reaction time, and opportunities for flocculation. That said, they typically contain more heavy metals than other solids because of the heavy metals in coagulant and those that co-precipitate with iron and aluminum.

Table 20.9 Secondary solids characteristics.

Parameter	Concentration (dry-weight basis)
Total solids %	0.4–1.2
Total volatile solids, % of TS	60–85
Grease, % of TS	5–12
Phosphorus, % of TS	1.5–3.0
Protein, % of TS	32–41
Nitrogen, % of TS	2.4–7.0
pH	pH 6.5–8.0

7.0 PRETREATMENT OPTIONS

Solids are removed from primary and sedimentation tanks after being pretreated to facilitate pumping and subsequent handling. The most common pretreatment processes include degritting, grinding, and screening. Grinding and screening are becoming more popular as plants strive to increase the quality of their biosolids.

7.1 Degritting

Grit typically can be removed more easily and efficiently from raw wastewater rather than from the primary sludge stream. So, current design practice favors installing grit-removal and -processing facilities at the headworks, where raw wastewater first enters a treatment plant. This practice reduces wear on influent pumping systems (if grit removal is upstream of the wet well) and primary solids pumping, piping, thickening, and digestion systems. Efficiently removing grit from raw wastewater also reduces grit accumulation in thickening and digestion tanks.

However, some wastewater treatment plant operators have found it more convenient to capture grit and primary solids together, rather than separately. These plants typically use hydrocyclones (or other induced tangential-flow devices) to remove grit from primary solids. To remove 100- to 150-mesh grit particles effectively via hydrocyclones, the solids concentration should be less than 1% (preferably closer to 0.5%). Some of the grit-removal devices discussed in Chapter 11 also can be used to remove grit from primary sludge.

7.2 Grinding

Grinding processes cut or shear large solids into smaller particles to prevent operating problems in downstream processes (see Table 20.10). Chopper pumps typically are used in this application; they both chop and move solids.

TABLE 20.10 Processes following grinding.

Process	Objectives of grinding
Heat treatment	To prevent clogging of high-pressure pumps and heat exchangers
Nozzle-disk and solid-bowl centrifuge	To prevent clogging in nozzles and between disks; nozzle-disk units may also require fine screens
Chlorine oxidation	To enhance chlorine contact with sludge particles
Pumping with progressing cavity pumps	To prevent clogging of small pumps of 3 L/s (50 gpm) or less capacity
Belt filter presses	To prevent damage to filter belts from large particles

Today, more treatment plants are using in-line grinders to try to reduce equipment cleaning or maintenance downtime. They can shear solids into 6- to 13-mm (0.25- to 0.5-in.) particles, depending on design requirements, and can handle either dilute or thickened solids.

A solids macerator–grinder works like a meat grinder (see Figure 20.6). Its multiple-blade cutter rotates rapidly over a perforated grid plate through which solids are forced. The size of the holes and the speed of the blade determine how small the particles will become. Holes range from 11 mm (0.44 in.) in diameter to slots 26 to 38 mm (0.6 to 1.5 in.) wide.

Grinders can produce nominal pressure increases, although a design engineer should assume that the net head gain through the unit is zero. They typically are installed on the suction side of solids pumps to prevent pump clogging. If the grinder must be located on the discharge side of the pump, the discharge pressure must be low

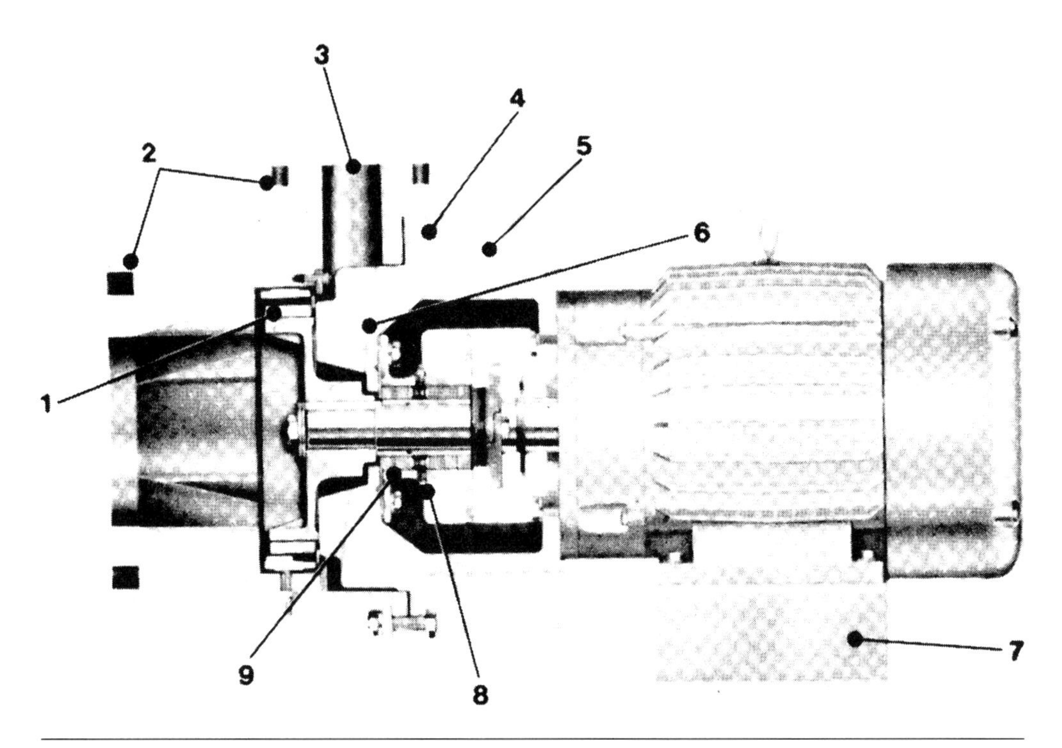

FIGURE 20.6 Macerator–grinder: (1) carbide impeller tips, (2) discharge and suction flanges, (3) discharge port, (4) canopy construction, (5) lifting holes, (6) deflection surface, (7) mounting pedestal (horizontal or vertical), (8) seal-flushing connection, and (9) seal.

(consult the manufacturer for installation guidelines). A high-pressure pump discharge will shorten the lives of grinder seals and shafts.

Although grinders will handle large organic particles readily, rocks and metal objects can cause extensive damage. So, they must be protected from tools or rocks dropped into the sedimentation tanks. One grinder design uses a sump, formed by a standard cross in the solids line ahead of the grinder. The bottom of the sump has a basket that periodically can be lifted out through the top section of the cross. Heavy objects flowing along the bottom of the pipe drop into the basket, thereby protecting the grinder.

More recently, vendors have developed slow-speed hydraulic or electric grinders that sense blockages. The macerator/in-line solids grinder has two sets of counter-rotating, intermeshing cutters that trap and shear solids; producing a consistent particle size (see Figure 20.7). The cutters are stacked on two steel or stainless-steel drive shafts with intermediate spacers between successive cutters. The spacers are made of the same material as the cutters. The drive shafts counter-rotate at different speeds, producing a self-cleaning action in the cutters.

FIGURE 20.7 In-line grinder.

7.3 Screening

If large openings are used to screen the raw wastewater, a lot of material can get through. This, of course, is for a large flow. If a smaller screen opening is used to screen sludge (which is a much lower flowrate), more material is removed. However, raw wastewater screening technology has improved significantly over the past couple of decades, which allows smaller and smaller openings to be used to screen raw wastewater. Many older plants have screens with relatively large openings; therefore, it is more cost-effective to simply screen the sludge. Furthermore, smaller debris can agglomerate into larger pieces. This is observed in aeration basins where the turbulence causes "roping" of fibrous materials into much larger solids in the sludge that are more easily removed than the finer material in the influent. Front-rake screen models are preferred in this application to keep moving parts from being permanently submerged in solids.

Another option is to use in-line screens, which are available with capacities up to about 32 L/s (68 cu ft/min). Screen openings are typically 5 mm, although openings up to 10 mm can be used. Screenings can contain up to 30% to 40% dry solids; solids throughput decreases as solids concentration increases. So, for example, if a treatment plant was screening primary sludge with a dry solids concentration of about 5%, the compacted screenings would be produced at a rate of about 0.3 to 1.0 kg of dry solids per cubic meter of solids flow. Maximum inlet pressure is about 100 kPa (14.5 lb/in.^{2}). Although the pressure drop across the screen depends on the feed rate, solids concentration, and screen openings, it typically ranges from about 15 to 50 kPa (2 to 7 lb/in.^{2}). If the pressure drops too much, causing solids to blind the inlet side of the screen, design engineers can add an in-line booster pump downstream of the screen.

8.0 REFERENCES

American Society of Civil Engineers (1998) *Manual of Practice on Transport of Water and Wastewater Residuals.*

Chaudhary, R.; et al. (1989) Evaluation of Chemical Addition in the Primary Plant at Los Angeles 11 Hyperion Treatment Plant. *Proceedings of the 62nd Annual Water Pollution Control Federation Technical Exposition and Conference;* San Francisco, California, Oct 15–19; Water Pollution Control Federation: Washington, D.C.

Daigger, G. T.; Butzz, J. A. (1998) *Upgrading Wastewater Treatment Plants,* 2nd ed.; Water Quality Management Library, Vol. 2; CRC Press: Boca Raton, Florida.

Great Lakes–Upper Mississippi River Board of State and Provincial Public Health and Environmental Managers (2004) *Recommended Standards for Wastewater Facilities;* Health Research Inc.; Health Education Services Division: Albany, New York.

Koch, C.; et al. (1990) Spreadsheets for Estimating Sludge Production. *Water Environ. Technol.*, **2** (11), 65.

Koch, C.; et al. (1997) A Critical Evaluation of Procedures for Estimating Biosolids Production. *Proceedings of the Joint Water Environment Federation and American Water Works Association Specialty Conference; Residuals and Biosolids Management: Approaching 2000*; Philadelphia, Pennsylvania; Water Environment Federation: Alexandria, Virginia; American Water Works Association: Denver, Colorado.

Monod, J. (1949) The Growth of Bacterial Cultures. *Annu. Rev. Microbiol.*, **3**, 371.

Newberg, J. W.; et al. (1988) Unit Process Trade-Offs for Combined Trickling Filters and Activated Sludge Processes. *J. Water Pollut. Control Fed.*, **60**, 1863.

Schultz, J.; et al. (1982) Realistic Sludge Production for Activated Sludge Plants without Primary Clarifiers. *J. Water Pollut. Control Fed.*, **54**, 1355.

Standards for Use and Disposal of Sewage Sludge (1993). *Fed. Regist.*, **58** (32), 9248–9415; Feb 19.

U.S. Environmental Protection Agency (1979) *Process Design Manual: Sludge Treatment and Disposal;* EPA-625/1-79-011; U.S. Environmental Protection Agency: Washington, D.C.

U.S. Environmental Protection Agency (1987) *Design Manual: Dewatering Municipal Wastewater Sludges;* EPA-625/1-87-014; U.S. Environmental Protection Agency: Washington, D.C.

Wilson, T.; et al. (1984) Operating Experiences at Low Solids Retention Time. *Water Sci. Technol.* (G.B.), **16**, 661.

Zabinski, A.; et al. (1984) Low SRT: An Operator's Tool for Better Operation and Cost Savings. *Proceedings of the 57th Annual Water Pollution Control Federation Technical Exposition and Conference;* New Orleans, Louisiana, Sep 30–Oct 5; Water Pollution Control Federation: Washington, D.C.

9.0 SUGGESTED READINGS

Olstein, M.; et al. (1996) *Benchmarking Wastewater Treatment Plant Operations;* Interim Report; Water Environment Research Foundation: Alexandria, Virginia.

Patrick, R.; et al. (1997) *Benchmarking Wastewater Operations, Collection, Treatment and Biosolids Management;* Project 96-CTS-5; Water Environment Research Foundation: Alexandria, Virginia.

Tabak, H. H.; et al. (1981) Biodegradability Studies with Organic Priority Pollutant Compounds. *J. Water Pollut. Control Fed.*, **53**, 1503.

U.S. Environmental Protection Agency (1982) *Fate of Priority Pollutants in Publicly Owned Treatment Plants;* EPA-440/1-82-303; U.S. Environmental Protection Agency, Technology Transfer: Cincinnati, Ohio.

U.S. Environmental Protection Agency (1986) *Superfund Public Health Evaluation Manual.* EPA-540/1-86-060; U.S. Environmental Protection Agency, Technology Transfer: Cincinnati, Ohio.

U.S. Environmental Protection Agency (1992) *Control of Pathogens and Vector Attraction in Sewage Sludge;* EPA-625/R-92-013; U.S. Environmental Protection Agency: Washington, D.C.

U.S. Environmental Protection Agency (1994) *A Plain English Guide to EPA's Part 503 Biosolids Rule;* EPA-83Z/R-93/003; U.S. Environmental Protection Agency: Washington, D.C.

U.S. Environmental Protection Agency (1995) *Process Design Manual for Suspended Solids Removal;* EPA-625/1-75-003a; U.S. Environmental Protection Agency: Washington, D.C.

Weiss, G., Ed. (1986) *Hazardous Chemical Data Book;* Noyes Data Corp.: Park Ridge, New Jersey.

Chapter 21

Solids Storage and Transport

1.0 INTRODUCTION

This chapter discusses methods for transporting residuals and biosolids. Grit dewatering and screenings are covered in Chapter 11. Chapters 22, 23, 24, and 25 discuss methods for thickening, dewatering, and treating solids.

The solids transport methods discussed here are currently prevalent in the United States. When designing solids transportation systems, engineers should consider how long they are expected to be in service and favor equipment that is flexible enough to remain useful despite changing technology, regulations, economics, and solids characteristics. They also should investigate full-scale working systems whenever possible to determine actual operating conditions and costs, and then make allowances for uncertainties.

1.1 Flow Characteristics

The flow characteristics (rheology) of wastewater residuals cannot be defined simply. They vary widely from process to process and from plant to plant (Wagner, 1990; Levine, 1986 and 1987; Borrowman, 1985; Carthew et al., 1983; Mulbarger et al., 1981; and U.S. EPA, 1979). And because a residual's rheological properties directly influence pipeline friction losses, headloss characteristics also vary extensively.

Solids content is an important rheological parameter. Generally, the higher a fluid's solids concentration, the higher its shear stress, density, and viscosity. Viscosity increases exponentially as solids concentration increases (Brar et al., 2005, and references therein).

A residual's rheological characteristics also are strongly affected by the kind of treatment the material has undergone (Guibaud et al., 2004, Brar et al., 2005). For example, a 3% raw, fresh, nonhydrolyzed solids has a higher apparent viscosity than a 4% thermally alkaline hydrolyzed solids (Brar et al., 2005). Similarly, digested solids can be pumped more easily than raw or undigested solids with the same moisture content (see Figure 21.1) (U.S. EPA, 1979). To derive approximate headlosses for solids, design engineers should calculate the headloss for water and the multiply the result by the factor in Figure 21.1 corresponding to the residual's solids concentration. This will provide a rough estimate when velocities are between 0.8 and 2.4 m/s (2.5 and 8 ft/s) and no thixotropic behavior or serious obstructions (e.g., from grease) are anticipated.

For design purposes, residuals can be divided into several distinct categories based on solids concentration (see Table 21.1). [For more detailed information on solids concentrations and residuals behavior, see *Conveyance of Wastewater Treatment Plant Residuals* (ASCE, 2000)]. *Dilute residuals* contain less than 5% solids. Such residuals typically

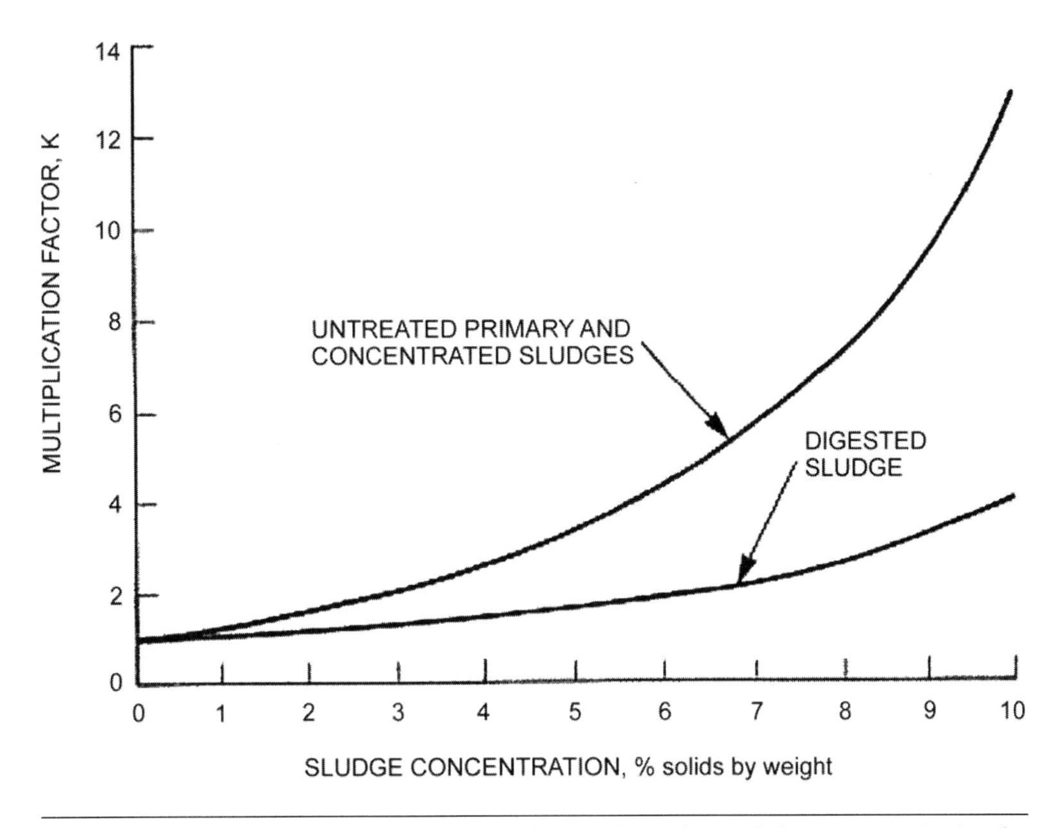

FIGURE 21.1 Approximate multiplication factors (based on solids concentration) to be applied to headlosses calculated for water in laminar flow (taken from U.S. EPA, 1979).

include waste activated sludge (WAS), which contains less than 2% solids, and primary sludge, which contains less than 5% solids.

Thickened residuals, which typically are produced via a mechanical thickening process, contain more solids. Such residuals range from WAS with a 3% solids content to primary sludge with a 10% solids content. They have a much higher viscosity and cannot be handled reliably by centrifugal pumps.

Dewatering tends to make residuals thixotropic, and their rheology is dependent on both time and applied stress. While engineers can design pumping systems that handle thixotropic materials, the limited availability of related data makes site-specific studies important.

Further dewatering makes residuals granular. Such residuals cannot be pumped; instead, they must be transported via conveyors or similar devices.

TABLE 21.1 Classification of water and wastewater thixotropic residuals by type and solids content.

			Thixotropic solids		Granular-compactable	
Solids type[a]	Temperature, °C	Thickened solids, % TS	Low-medium viscosity, % TS	Medium-high viscosity, % TS	Wet solids, % TS	Dry solids
RPS	10–20	5–10	20–26	24–40	35–65	65+
RWAS	10–25	3–7	10–16	14–25	22–65	65+
R (PS + WAS)[b]	10–25	4–8	13–20	18–31	28–65	65+
DPS	20–30	5–10	20–28	24–40	35–65	65+
SWAS	20–30	3–7	10–16	14–25	22–65	65+
D (PS + WAS)[b]	20–30	4–8	13–20	18–31	28–65	65+
Alum $[AI(OH)^3]^{(2)}$	5–25	3–7	8–15	13–30	25–60	65
Iron $[Fe(OH_3)]^{(2)}$	5–25	3–7	8–15	15–35	30–60	65
Lime ($CaCO_3$)	5–25	15–30	8–15	25–50	70–80	80+

[a] R = raw, D = digested, PS = Primary sludge, and WAS = waste activated sludge.
[b] 50 : 50 mixture of PS and WAS.
[c] $AIPO_4$ and $FePO_4$ will behave similar to the hydroxide fraction.
[d] Fluid bed combustion ash from gravity separation-dewatering.

Fluids may be Newtonian or non-Newtonian. Newtonian fluids (e.g., water). Residuals containing more than 3% solids do not follow Newtonian behavior.

The Herschel-Bulkley model is an equation that models the rheological behavior of both Newtonian and non-Newtonian fluids:

$$S = S_y + K\,(dv/dy)^n \tag{21.1}$$

Where

S = shear stress (in Pa, N/m^2, or $kg/m\text{-}s^2$),
S_y = yield stress (in Pa),
dv/dy = shear rate (velocity gradient in s^{-1}), and
K and n = fluid constants.

When $S_y = 0$ and n = 1, the model describes a Newtonian fluid and Equation 21.1 becomes:

$$S = \mu\,(dv/dy) \tag{21.2}$$

where μ = absolute viscosity (in Pa-s).

When $S_y \neq 0$ and $n = 1$, the model describes a Bingham plastic fluid:

$$S = S_y + Rc\,(dv/dy) \tag{21.3}$$

Where
Rc = coefficient of rigidity (Pa-s)

When $S_y = 0$ and $n \neq 0$, the model describes a pseudoplastic, Ostwald de Vaele, or shear-thinning fluid, and the equation becomes:

$$S = K\,(dv/dy)^n \tag{21.4}$$

Where
K = fluid consistency index, and
n = flow behavior index.

Researchers are almost equally divided on whether the Bingham plastic model or the pseudoplastic model describes the rheological behavior of thickened and dewatered residuals more appropriately.

When designing solids transport systems with kinetic pumps (e.g., centrifugal), engineers need to be accurate rather than conservative. Such systems are most efficient when the system curve matches the pump curve. However, engineers should be conservative when designing transport systems for thicker residuals (either liquid or dewatered) that rely on positive-displacement pumps.

1.2 Flow and Solids Monitoring

If the residuals contain less than 3% or 4% solids, engineers can use electromagnetic meters to obtain volumetric flowrates, and optical sensors to measure mass flowrates.

Flowrates for thickened or dewatered residuals flowing in a pipe cannot be directly measured by current instrumentation. Instead, engineers can calculate solids quantities indirectly by performing a mass balance around a particular process. In a dewatering process, for example, they can calculate the quantity of cake produced based on influent flow and solids concentrations and effluent filtrate, centrate, or pressate levels. Suspended solids meters register air bubbles as solids (Radney, 2008), so to avoid interference, designers should include degassing tanks to release entrapped air.

Design engineers also can indirectly measure the flowrate of thicker residuals based on positive-displacement pump operations. For example, if designers know the rotational speed and volumetric discharge per rotation of a progressing cavity pump, they can accurately calculate the fluid flowrate (assuming that the pump cavity fill rate is 100%). Manufacturers of hydraulically driven reciprocating piston pumps offer an internal flow measuring system that is accurate within 5%.

2.0 LIQUID RESIDUALS AND BIOSOLIDS STORAGE

2.1 Storage Requirements

The storage needs for liquid residuals and biosolids depend on where and why they are being stored. Liquid residuals may be need to be stored to equalize flows or provide more operational flexibility. For example, if dewatering equipment is operated periodically, liquid residuals would have to be stored between operating periods. The volume required would be process-specific. If biosolids are stored between stabilization and land application, then the storage volume needed would depend on agricultural needs and climatic issues.

If biosolids will be land-applied or surface disposed, then the treatment plant must have enough storage available to allow for times when the material cannot be used (e.g., frozen or snow-covered application sites, crop-fertilization limits, field-rotation requirements, and regulatory restrictions). Design engineers need to take these situations into account when designing the storage systems. For example, Figure 21.2 indicates the approximate number of days per year when climatic conditions do not allow effluent applications (U.S. EPA, 1995). Assuming that effluent and biosolids applications would be affected by the same climatic conditions, designers can use the information in this figure as a basis for estimating storage requirements. Likewise, Table 21.2 shows the months in which residuals can be applied in the north central United States (data that can be extrapolated to other areas). Designers also can obtain climatic data for U.S. sites from the National Oceanic and Atmospheric Administration's National Climatic Data Center in Asheville, North Carolina (http://lwf.ncdc.noaa.gov/oa/ncdc.html).

Chapter 25 identifies storage requirements specific to certain stabilization processes. For example, many autothermal thermophilic aerobic digester systems have a pre-digestion holding tank, so the digester(s) can be batch-fed. In addition, they often have a post-digestion tank in which biosolids can cool and further stabilize. Sizing requirements for such tanks are process-specific, so this information is included in the design guidelines for stabilization processes.

Regulations also may affect process storage systems. Some regulations require storage volumes ranging from 3 to 60 days' worth of solids (FDEP, 2008). Such storage may

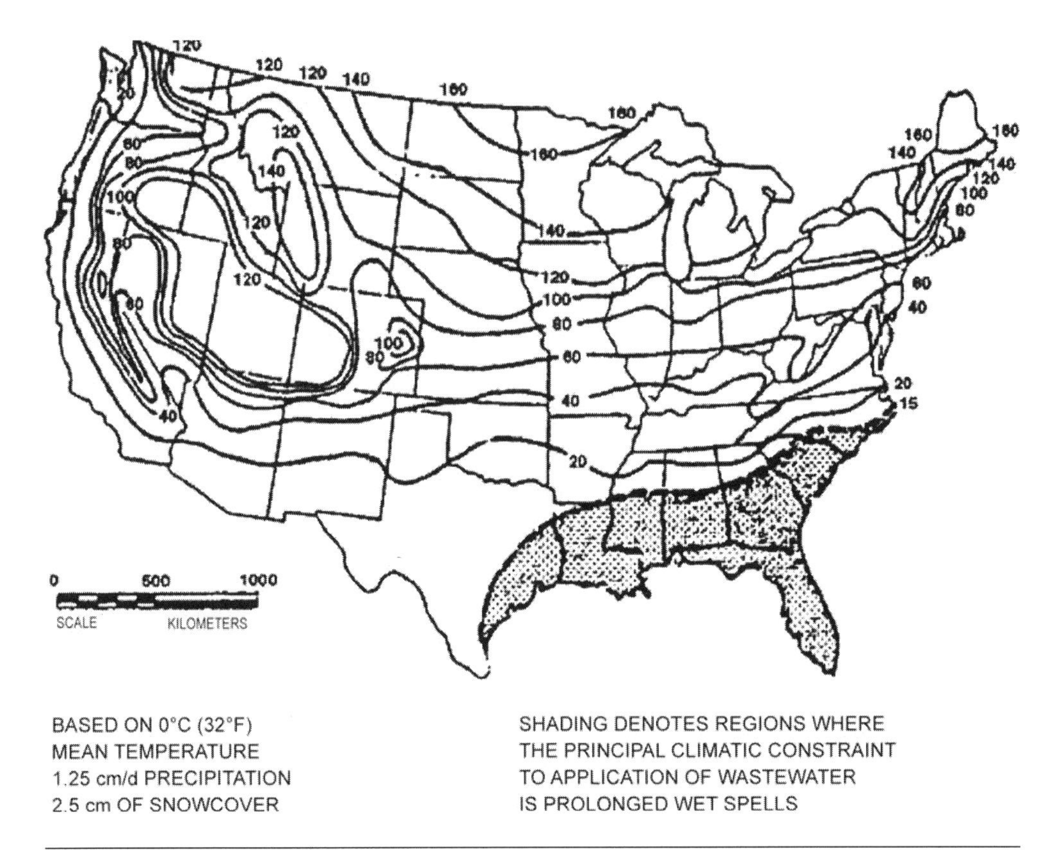

FIGURE **21.2** Storage days required as estimated from the use of the EPA-1 computer program for wastewater-to-land programs. Estimated storage based only on climatic factors.

simply be excess capacity in the digesters or other process tanks, or it may be separate storage tanks.

2.2 Storage Tanks

2.2.1 Typical Design Criteria

Although more costly per unit volume than earthen basins, storage tanks can be a good choice when solids volumes are small, land costs are high, or other restrictions make earthen basins infeasible. These tanks typically are cylindrical with either a flat or sloped bottom (see Figure 21.3) (U.S. EPA, 1979). The U.S. Army Corps of Engineers recommends a 4 : 1 floor slope and a minimum depth of 4.5 m (15 ft) (U.S. Army Corps, 1984). Cylindrical tanks are preferred because they do not have corners, which may

TABLE 21.2 General guide to months available for applying biosolids to various crops in the north central United States (U.S. EPA, 1979).[a]

Month	Corn	Soybeans	Cottons[c]	Forages[d]	Small grains[b]	
					Winter	Spring
January	S[e]	S	S/I	S	C	S
February	S	S	S/I	S	C	S
March	S/I	S/I	S/I	S	C	S/I
April	S/I	S/I	P, S/I	C	C	P, S/I
May	P, S/I	P, S/I	C	C	C	C
June	C	P, S/I	C	H, S	C	C
July	C	C	C	H, S	H, S/I	H, S/I
August	C	C	C	H, S	S/I	S/I
September	C	H, S/I	C	S	S/I	S/I
October	H, S/I	S/I	S/I	H, S	P, S/I	S/I
November	S/I	S/I	S/I	S	C	S/I
December	S	S	S/I	S	C	S

[a] Application may not be allowed due to frozen, flooded, and snow-covered soils.
[b] Wheat, barley, oats, or rye.
[c] Cotton, only grown south of southern Missouri.
[d] Established legumes (alfalfa, clover, trefoil, etc.), grass (orchard grass, timothy, brome, reed canary grass, etc.), or legume-grass mixture.
[e] S = surface application
S/I = surface/incorporation application
C = growing crop present; application would damage crop
P = crop planted; land not available until after harvest
H = after crop harvested, land is available again: for forages (e.g., legumes and grass), availability is limited and application must be light so regrowth is not suffocated.

become "dead spots". The tanks can be constructed of either concrete or steel. Steel tanks are susceptible to corrosion, however, which design engineers should consider when designing a solids storage system.

Storage tanks should be mixed to ensure that the discharged residuals are homogeneous (Spinosa and Vesilind, 2001). Mixer manufacturers recommend mixing energies ranging from 10 to 12 kW/1 million L (40 to 50 hp/1 million gal) (Lottman, 2008). The key is to keep solids suspended without inducing excessive air into the residuals.

The storage tank also may need aeration, especially if the residuals are unstabilized (U.S. EPA, 1979). If so, the tank's oxygen requirements should be similar to those for aerobic digesters. (For information on sizing aerobic-digester mixing equipment, see Chapter 25.) If the material was stabilized before storage, however, the tank's oxygen requirements (to maintain aerobic conditions) will be significantly less. Maintaining a minimum dissolved oxygen level of about 0.5 mg/L should prevent anaerobic activity

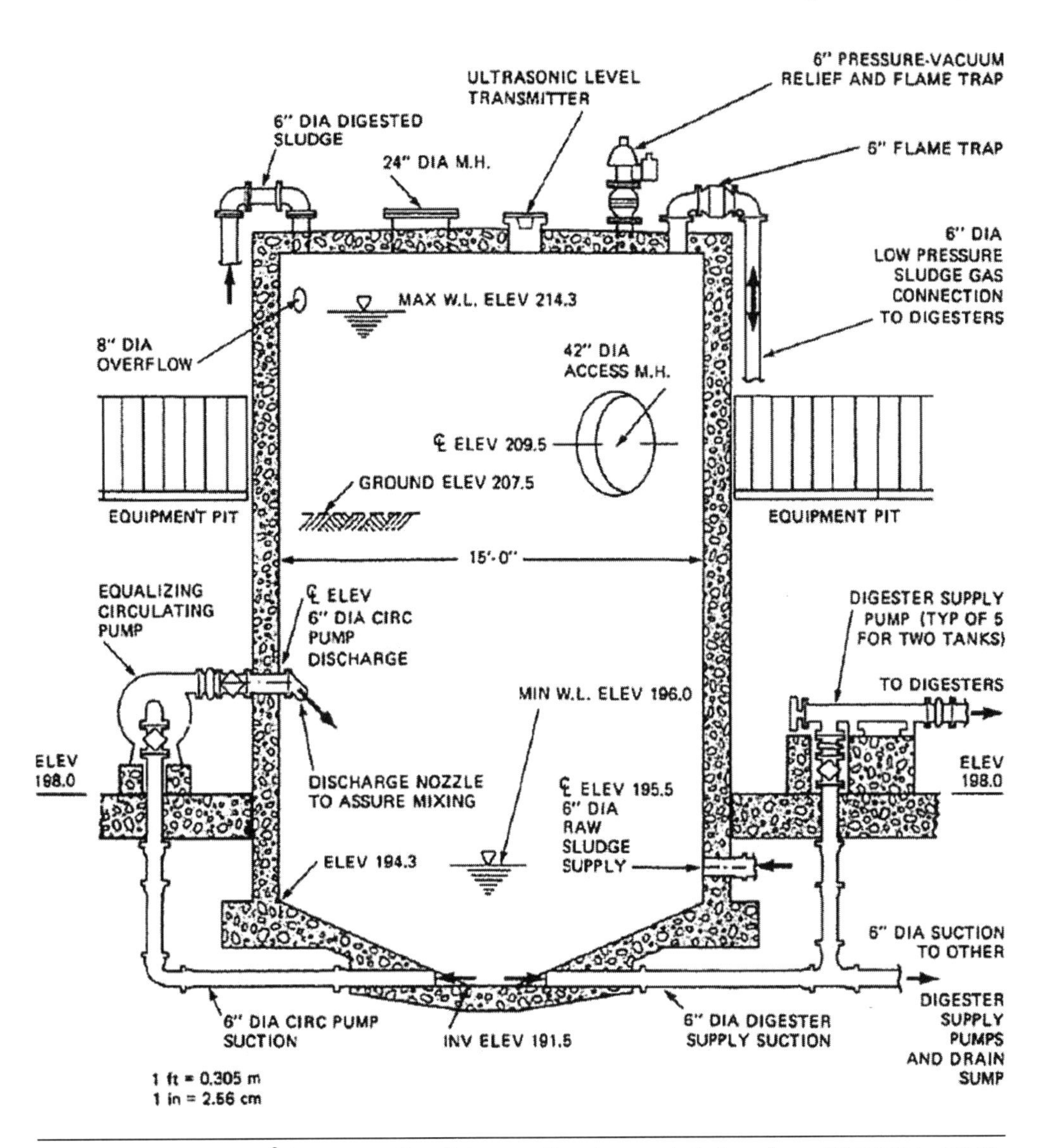

FIGURE 21.3 A 98-m^3 (26 000-gal) solids equalization tank (taken from U.S. EPA, 1979).

as long as the basin has adequate mixing. Otherwise, nuisance odors may be generated and more odor control may be required.

2.2.2 *Spill Prevention*

Aboveground tanks could spill or accidentally release solids via gravity. Engineers can prevent such accidents by designing the tank so all inflows and outflows occur above the tank's maximum high-water level. This minimizes the potential for damaged piping or valves to cause a catastrophic release of solids as pumps add or remove material. However, this method has other issues (e.g., freeze protection) that engineers must consider.

Alternatively, design engineers could provide appropriate emergency cut-offs on all piping and valves that are below a gravity-discharge tank's maximum high-liquid level.

If storage facilities are near surface waters or other sensitive areas, a containment wall or berm may be advisable. The berm should be designed to retain or retard the movement of spilled solids. A structural wall may not be necessary; an earthen berm may be sufficient. The containment berm should detain a spill long enough for it to be cleaned up but also include some means of removing excess water from rainfall or other sources.

2.2.3 *Odor Control*

Odors may be an issue, depending on the type of residuals and their storage time. Well-stabilized biosolids (those with minimal volatile solids) might be stored for several days without odor control. However, residuals characteristics may change over time, so it is preferable to cover storage tanks to minimize odors.

The air in the tank can be vented to an odor-control system, but it may be adversely affected by such compounds as dimethyl sulfide, dimethyl disulfide, and longer-chain mercaptans. Odorous volatile fatty acid compounds also may increase, yielding a sour odor (WEF, 2004).

One facility found that storing WAS and primary sludge separately reduced odors significantly, while chemical addition had little effect (Hentz et al., 2000). Holding-tank operations also can affect the character and intensity of odor emissions from downstream processes. Design engineer should take all of this into consideration when designing odor-control systems for solids storage tanks.

2.3 Storage Lagoons

Many smaller facilities use lagoons to treat solids. These treatment lagoons typically will store 2 years' worth of solids or more, and will be designed as an ultimate treatment system. (For more information on designing solids treatment lagoons, see Chapter 18.)

This chapter focuses on earthen basins designed to store solids for shorter periods (e.g., winter). These systems are not treatment lagoons and should not be used to treat raw sludge. Even digested solids stored in them could generate odors, unless the basins are aerated and well-mixed.

There are three types of solids storage basins: aerobic, facultative, and anaerobic (Lue-Hing, et al., 1998).

2.3.1 *Aerobic Basins*

Aerobic basins are designed to provide aeration and maintain a minimum dissolved oxygen concentration throughout the basin. So, their aeration requirements should be similar to those for aerobic digesters, except that they take into account prior solids stabilization. (For information on calculating aeration requirements for aerobic digesters, see Chapter 25.)

2.3.2 *Facultative Basins*

Facultative basins are unmixed and typically consist of three layers: a 0.3- to 1.0-m-deep aerobic surface layer, a deeper anaerobic zone, and a solids storage zone at the bottom. Both the aerobic and anaerobic zones are biologically active; anaerobic stabilization substantially reduces solids volume. The aerobic zone receives oxygen via surface transfer from the atmosphere, algal photosynthesis, and (if provided) surface-mix aerators. The oxygenation rate is low, however, so the U.S. Environmental Protection Agency (U.S. EPA, 1979) recommends that these basins only be used for anaerobically digested solids.

These basins typically are designed based on a volatile solids loading rate of 0.097 5 $kg/m^2 \cdot d$ (Lue-Hing, et al., 1998). They are typically 5 m (15 ft) deep to provide enough space for sufficiently thick aerobic and anaerobic layers. Aeration is typically provided by surface mixers, which are operated periodically to break up scum on the pond surface and optimize oxygen transfer. Most of the satisfactory installations use brush-type surface mixers, according to U.S. EPA.

Because facultative solids lagoons can "sour", odor control can be a major issue. When designing such storage basins, engineers should consider prevailing wind patterns, and minimize odor potential via proper loading and surface mixing.

2.3.3 *Anaerobic Basins*

Anaerobic digesters are similar to earthen basins but differ in that oxygen transfer from the surface is not considered in design. Therefore, earthen basins can be deeper than facultative ponds. Also, because maintaining an aerobic zone is not a key parameter, solids loading to earthen basins can be higher than with facultative ponds. Anaerobic ponds have essentially the same advantages and disadvantages as facultative ponds (Lue-Hing, et al., 1998).

3.0 LIQUID RESIDUALS TRANSPORT

There are two basic methods for transporting liquid residuals at wastewater treatment plants: pumping and trucking. Residuals typically are pumped onsite and trucked offsite.

3.1 Trucking

Although not typically part of the design process, trucking is often used to transport solids and biosolids—especially to land-application sites. Liquid residuals also may be trucked to another site for further treatment. Trucking liquid residuals may be a bigger challenge than transporting dewatered or dried biosolids, and there is some basic information designers should know.

Liquid residuals typically are transported via tanker trailer trucks. Such trucks typically have nominal capacities ranging from 22 680 to 34 020 L (6 000 to 9 000 gal). However, depending on weight restrictions, a tanker trailer may not be filled completely. In the United States, the maximum overall weight of a tractor-trailer is limited to 36 288 kg (80 000 lb). So, if a tractor weighs between 5443 and 6804 kg (12 000 and 15 000 lb), and an empty trailer weighs between 4 990 and 7 711 kg (11 000 and 17 000 lb), the contents cannot weigh more than about 21 773 to 25 855 kg (48 000 to 57 000 lb). This is equivalent to 21 000 to 25 500 L (5 700 to 6 800 gal) of dilute residuals.

Also, a 5 850-kg (12 900-lb), 26 460-L (7 000-gal) trailer typically is 13.1 m (40 feet) long and 2.4 m (8 ft) wide. Some are equipped with baffles to control backward and forward liquid surges.

3.2 Pumping

Pumping systems are an intrinsic part of solids management at wastewater treatment plants. They typically transport solids from

- Primary and secondary clarifiers to thickening, conditioning, or digestion systems;
- Thickening and digestion systems to dewatering operations;
- Biological processes to further treatment units; and
- Degritting facilities to temporary storage areas.

While specifying only one type of pump for all of a plant's solids-transport systems might seem advantageous, the wide range of conditions involved typically exceeds the capabilities of any given pump. Fortunately, many types of pumps are available (see Table 21.3).

3.3 Design Approach

When designing pumping systems, design engineers should begin by asking: What sort of residuals will be pumped? Kinetic pumps—especially recessed-impeller pumps—can handle some types of residuals, but other types may require positive-displacement pumps.

TABLE 21.3 Sludge pump applications by principle.

Principle	Common types	Typical applications
Kinetic (rotodynamic) pumps	Nonclog mixed-flow pump[a] Recessed-impeller pump (vortex pump, torgue flow pump) Screw centrifugal pump Grinder pump	Grit slurry,[b] incinerator ash slurry[c] Unthickened primary sludge[b,c] Return activated sludge[c,d] Waste activated sludges from attached-growth biological processes[c] Circulation of anaeronic digester[b] Drainage, filtrate, and centrate Dredges on sludge lagoons
Positive-displacement pumps	Plunger pumps Progressing cavity pump Air-operated diaphragm pump Rotary lobe pump Pneumatic ejector Peristaltic pump Reciprocating piston	Waste activated sludge Thickened sludges (all types) Unthickened primary sludge Feed to dewatering machines Unthickened secondary sludges Dewatered cakes[e]
Other	Air lift pump Archimedes screw pump	Return activated sludge

[a] Limited solids capability; useful in larger sizes for return activated sludges. In most other applications, recessed-impeller pumps are more common.
[b] Abrasion is moderate to severe. Abrasion-resisting alloy cast iron is usually specified.
[c] May contain precipitates from aluminum or iron salts added for phosphorus removal.
[d] Particular need for reliable flow meters, for process control, in this application.
[e] Reciprocating piston pumps and progressing cavity pumps only.

Kinetic pumps have lower capital costs (especially in large sizes), lower maintenance costs, and smaller footprints. They also are available in submersible form (although conventional dry-well pumps are preferred for most applications).

Positive-displacement pumps have better process control because the pumping rate is less affected by fluid viscosity. They function better over the entire head range from zero to maximum without damaging the pump or motor, or changing drive speed. They work better under high pressure and at low flows. They also are less sensitive to non-ideal suction conditions (e.g., entrained air and gas) and less likely to disrupt fragile floc particles in return activated sludge (RAS) and flocculated sludge.

The traditional approach to designing residuals transport systems is to minimize the pumping distance and apply a conservative multiplier to headlosses calculated for equivalent flows of water. However, this approach can be inaccurate. Such inaccuracies may not matter for short pumping distances, but they can be problematic for longer distances or critical applications.

The need to pump residuals long distances has increased in the last 20 years, so researchers have been developing methods to predict site-specific friction losses in pumping systems more accurately (Mulbarger et al., 1981; Carthew et al., 1983; Wagner, 1990; Honey and Pretorius 2000; Murakami et al., 2001). Study results have shown that, once rheological properties have been determined, the Bingham plastic model (Carthew et al., 1983; Mulbarger et al., 1981) or the pseudoplastic model (Honey and Pretorius, 2000; Murakami et al., 2001) for non-Newtonian fluids may describe how wastewater residuals flow. They also can predict the critical velocity at which laminar flow changes to turbulent flow. [In the turbulent flow range, dilute residuals obey conventional flow relationships for Newtonian fluids (Mulbarger et al., 1981).]

When designing solids pumping systems for smaller plants, engineers should be careful to ensure that velocities will be adequate without undersizing piping, which increases the risk of line blockage.

3.3.1 *Dilute Residuals*

Clarifiers often produce a relatively dilute settled sludge (maximum concentrations of 1.2 to 1.5% are typical for activated sludge). At velocities greater than 0.3 to 0.6 m/s (1 to 2 ft/sec), such solids are in the turbulent flow regime and have a headloss essentially equal to that of water (Mulbarger, 1997). At lower velocities, the flow becomes laminar, and headlosses increase sharply. So, engineers should design dilute-residual pumping systems to maintain a minimum velocity of 0.6 to 0.75 m/s (2 to 2.5 ft/sec) whenever possible to ensure turbulent flow.

3.3.2 *Thickened Residuals*

The concentration at which residuals can be defined as "thickened" depends on the type of solids and the preceding treatment processes (see Table 21.1)(ASCE, 2000).

When designing pumping systems for thickened solids, design engineers can use the Darcy-Weisbach and Manning equations for water to determine headloss—regardless of the solids' flow regime (laminar, transition, or turbulent)—and apply a solids correction factor to the final calculation. The correction factor for residuals with up to 12% solids may be found in Figures 21.4 and 21.5 [taken from Sanks et al. (1998) and Metcalf and Eddy (2003), respectively]. As a simplified alternative, designers can use Figures 21.6 and 21.7, which indicate the headloss multiplier for worst-case design conditions in 150- and 200-mm (6- and 8-in.) forcemains, respectively.

If design engineers use the curves in Figures 21.6 and 21.7, they should choose pumps and motors that will operate satisfactorily over the entire headloss range from

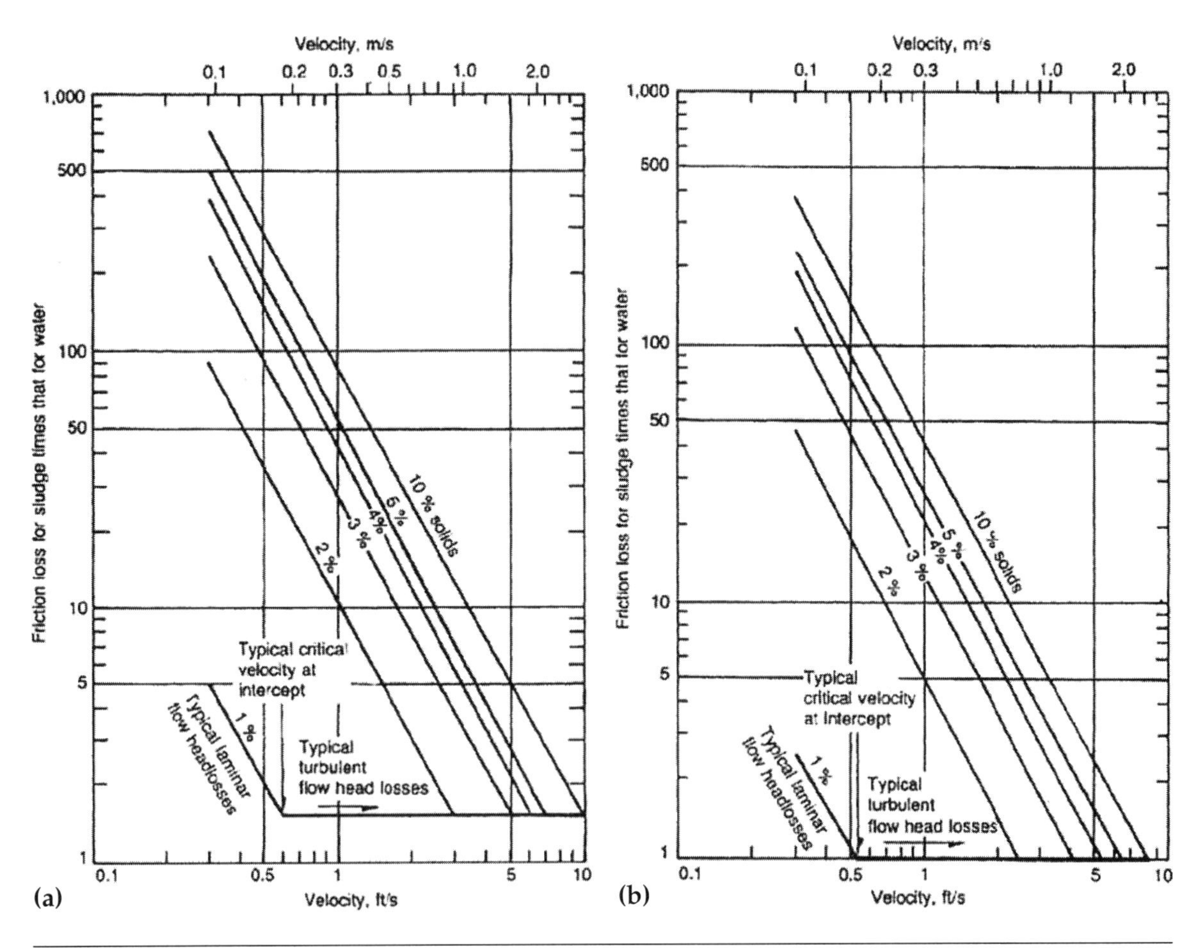

FIGURE 21.4 Multiplication factor for residuals headloss: (a) routine design and (b) worst-case design (Sanks et al., [Eds.], 1998).

"water" to "worst-case". Head changes affect centrifugal pumps much more than positive-displacement pumps, so if centrifugal pumps (e.g., recessed-impeller) are used, engineers also should check the motors to avoid overloading if operating head drops significantly below design head. Motors may be overloaded if a pump becomes "runaway" (operates beyond the right terminus of its characteristic curve). Also, residuals occasionally can exceed the worst-case headloss curve. So in some instances, oversized motors and variable-frequency drives should be specified to provide the operational flexibility needed.

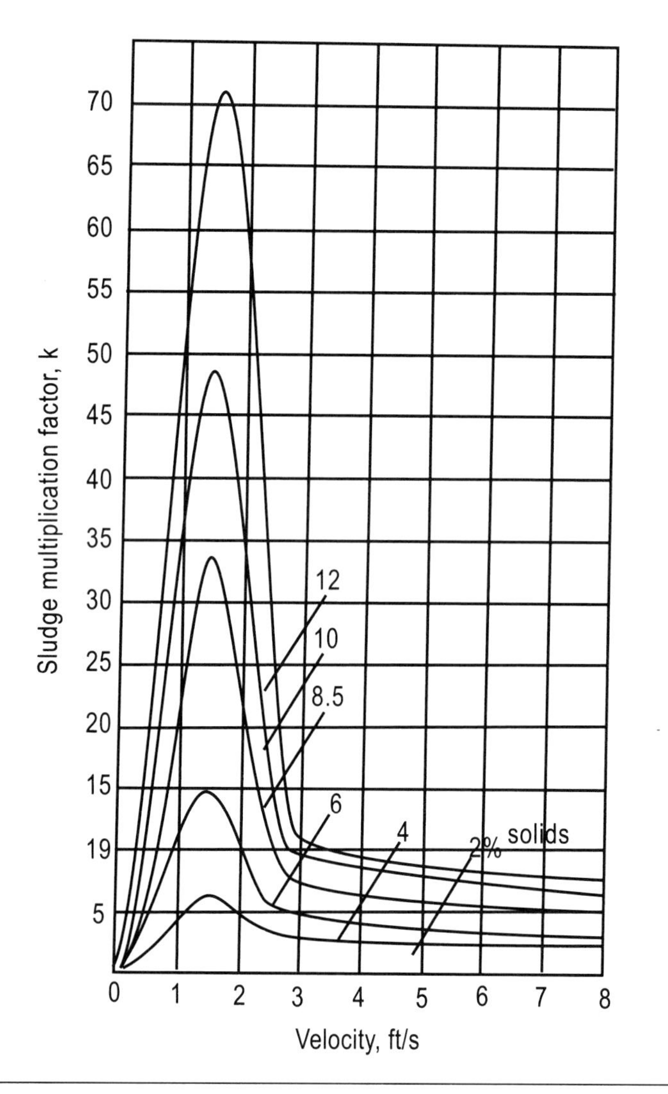

FIGURE 21.5 Multiplication factor for residuals headloss (from Metcalf & Eddy, *Wastewater Engineering: Treatment and Reuse*, 4th ed. Copyright © 2003, The McGraw-Hill Companies, New York, N.Y., with permission).

In addition to headloss, design engineers should consider the nature of the process that will receive the solids. Many positive-displacement pumps suitable for thickened residuals produce a pulsating flow, which may not be acceptable if the downstream process depends on steady flow or flow-proportioned chemical addition to operate properly.

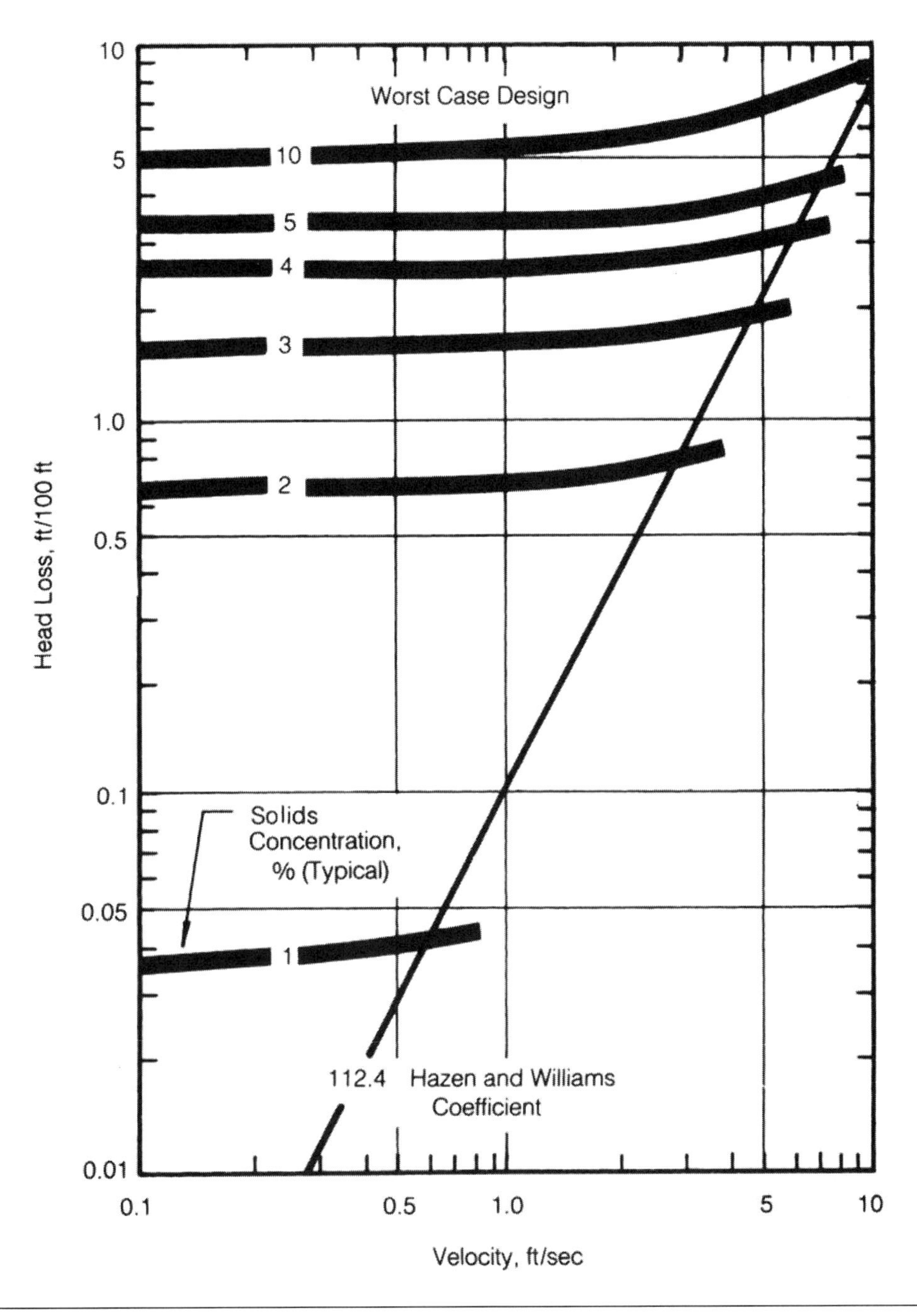

FIGURE 21.6 Predicted frictional headlosses for worst-case design of a 150-mm-diameter (6-in.-diameter) solids forcemain (in. × 25.4 = mm; ft × 0.304 8 = m) (Mulbarger et al., 1981).

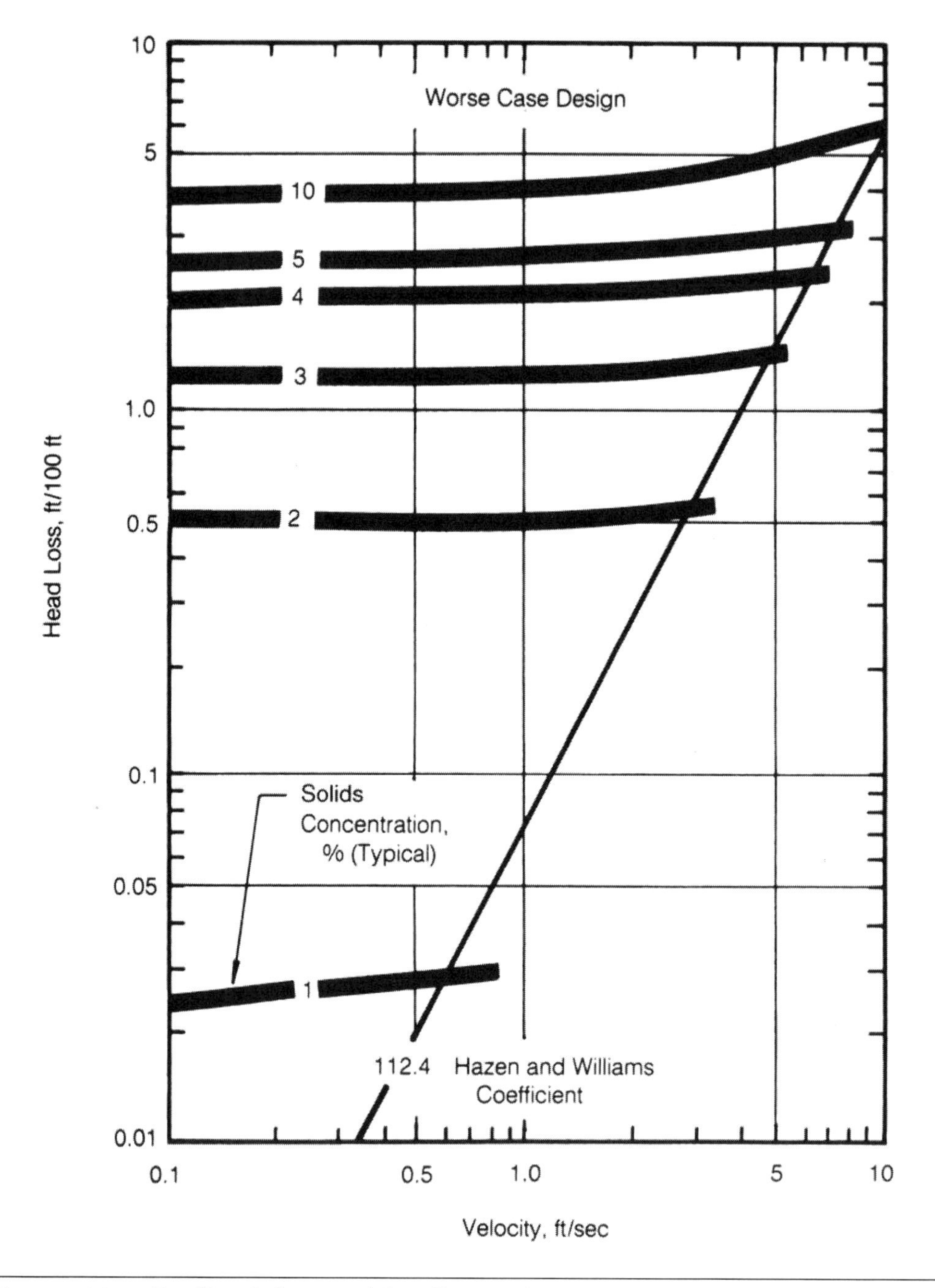

FIGURE 21.7 Predicted frictional headlosses for worst-case design of a 203-mm-diameter (8-in.-diameter) solids forcemain (in. × 25.4 = mm; ft × 0.3048 = m) (Mulbarger et al., 1981).

Researchers have derived practical equations for long-distance pumping of WAS, thickened residuals, and digested biosolids in the 2 to 5% solids range (Murakami et al., 2001). Such residuals behave like pseudoplastic fluids. Based on the assumptions that fluid viscosity depends solely on percent solids, residuals density is 1 000 kg/m^3 and temperature is 15°C, researchers proposed the following equations (Equations 21.5, 21.6, and 21.7):

Laminar flow

$$H_f\,(\mathrm{m}) = 1.90 \times 10^{-3} k^{0.88} \frac{L}{D^{1.20}} V^{0.20}$$

Where
$k = 0.059\ C^{2.74}$ for digested biosolids,
$k = 0.052\ C^{2.91}$ for thickened sludge, and
$k = 0.050\ C^{3.06}$ for waste activated sludge.

Turbulent flow

$$H_f\,(\mathrm{m}) = 9.06 \left(\frac{1}{C_H}\right)^{1.93\times(1-C/100)} \frac{L}{D^{1.18}} V^{1.82}$$

Where
$C_H = 110$ for mortar-lining, cast iron pipe, and
$C_H = 95$ for carbon steep pipe.

Solids concentration is approximately 5% or less.
Critical velocity *c*

$$V_C\,(\mathrm{m/s}) = 1.20 \frac{C_H}{100} k^{0.52}$$

Where
C = percent biosolids concentration,
L = pipe length (m), and
D = pipe diameter (m).

Another design approach is based on the assumption that the flow of thickened residuals follows the Bingham plastic model. To use this model, design engineers need to know a solids' yield stress (S_y) and coefficient of rigidity (R_c), which may be determined experimentally. If solids-specific data are not available, then designers can use

Figures 21.8 and 21.9 (ASCE, 2000) to estimate these values. [Similar graphs have been created by Battistoni (1997); Guibaud et al. (2004); Laera et al. (2007); and Mori et al. (2007).]

Once the rigidity coefficient and yield stress are known, designers can use the following two equations to calculate the upper and lower critical velocities:

$$V_{uc} = 1\,500\ R_c/\rho D + 1\,500/\rho D\ (R_c^2 + S_y\ \rho D^2/4\,500)^{1/2} \tag{21.8}$$

$$V_{lc} = 1\,000\ R_c/\rho D + 1\,000/\rho D\ (R_c^2 + S_y\ \rho D^2/3\,000)^{1/2} \tag{21.9}$$

Where

V_{uc} = upper critical velocity (m/s),
V_{lc} = lower critical velocity (m/s),
ρ = fluid density (kg/m^3),
D = pipe diameter (m), and
R_c = rigidity coefficient (N-s/m^2).

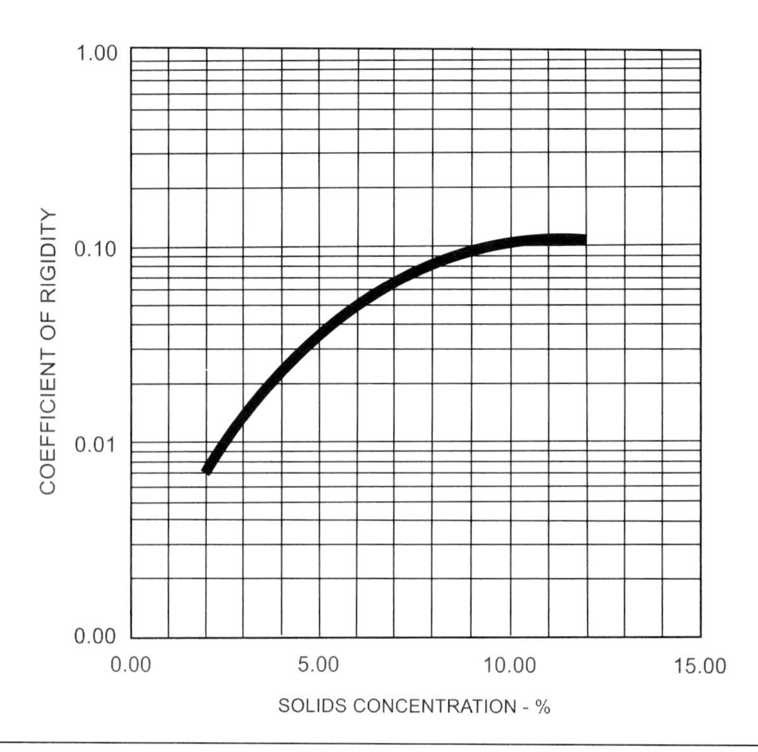

FIGURE 21.8 Coefficient of rigidity versus solids concentration (ASCE, 2000).

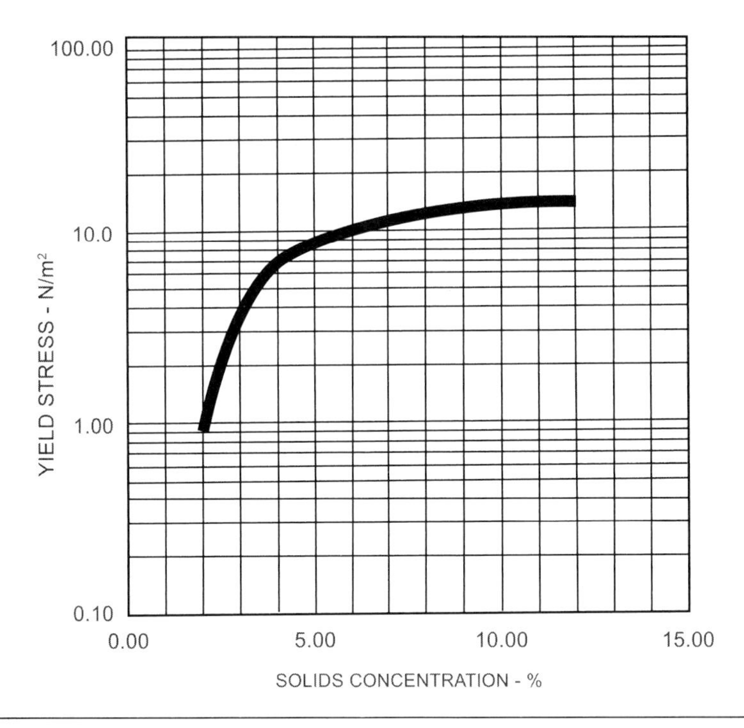

FIGURE 21.9 Yield stress versus solids concentration (ASCE, 2000).

Alternatively, designers can calculate the Reynolds number as follows:

$$\mathrm{Re} = \rho \mathrm{V} \mathrm{D}/\mathrm{Rc} \tag{21.10}$$

If $Re < 2\,000$, the flow is laminar. If $Re > 3\,000$, the flow is turbulent.

3.3.2.1 Laminar Flow

At velocities less than the lower critical velocity, or when $Re < 2\,000$, the residuals' flow will be in the laminar range and designers can calculate headloss using the Buckingham equation:

$$H/L = 32\,(S_y/6\,\rho\,g\,D + R_c\,V/\rho\,g\,D^2) \tag{21.11}$$

Where

H/L = headloss per unit length (m/m),

S_y = yield stress (Pa or N/m^2),

ρ = fluid density (kg/m^3),
g = gravitational acceleration (m/s^2),
D = pipe diameter (m),
R_c = rigidity coefficient (Pa-s or N-s/m^2), and
V = velocity (m/s).

Honey and Pretorius (2000) experimentally determined the rheological parameters of settled activated sludge. They measured solids concentration, particle density, and torque, and then derived shear stress and shear rate from the torque data. They then determined the fluid consistency coefficient (K) and the pseudoplastic model's flow-behavior index (n). (They suggest that the pseudoplastic model more accurately indicates the behavior of 5% settled activated sludge in laminar flow.) From there, they compared the following generalized Reynolds number with a critical Reynolds number for pseudoplastic fluids to determine the flow regime.

$$Re(g) = \rho\, V\, D/(8V/D)^{n-1}\, K\, [(3n+1)/4n]^n \tag{21.12}$$

$$Re(\text{critical}) = 6\,464\, n/(1+3n)^2\, [(1/(2+n)]^{(2+n)/(1+n)} \tag{21.13}$$

They then used the Darcy-Weisbach equation:

$$H_f = 4\, f\, L\, V^2/2gD \tag{21.14}$$

Where

f = the dimensionless Fanning friction factor, which is $16/Re(g)$ for pseudoplastic fluids in laminar flow.

In their study, Honey and Pretorius assumed that solids behaved as a thixotropic fluid. The fluid exerted a maximum headloss when the pump was turned on, and dropped to a lower, constant headloss after a certain time or travel distance in the pipeline. Then the thixotropic behavior disappeared.

3.3.2.2 Transition and Turbulent Flow

At velocities greater than the upper critical velocity, or when $Re > 3\,000$, solids' flow will be turbulent.

When designing pumping systems for turbulent solids, engineers can solve the Hazen-Williams equation for turbulent water and apply the solids correction factor to the result (Sanks 1998; Metcalf & Eddy 2003). When using this equation, design engineers should assume that C equals 140 under normal conditions and 112.4 under worst-case design conditions.

Designers also can use Reynolds and Hedstrom numbers to calculate headlosses. To find the Reynolds number, see Equation 21.10. The Hedstrom number is calculated as follows:

$$He = D^2\, \rho\, Sy/R_c^{\,2} \tag{21.15}$$

After calculating the Hedstrom and Reynolds numbers, the friction factor (also called the Fanning friction factor) is then calculated using Figure 21.10. Design engineers then should use the Darcy-Weisbach equation to calculate headloss:

$$\Delta P = 2 f \rho L V^2/D \tag{21.16}$$

where

ΔP = pressure headloss (Pa).

Design engineers should make sure they use the correct friction factor for residuals, because the friction factor for water taken from a Moody diagram is often quoted as four times that of solids (Figure 21.10).

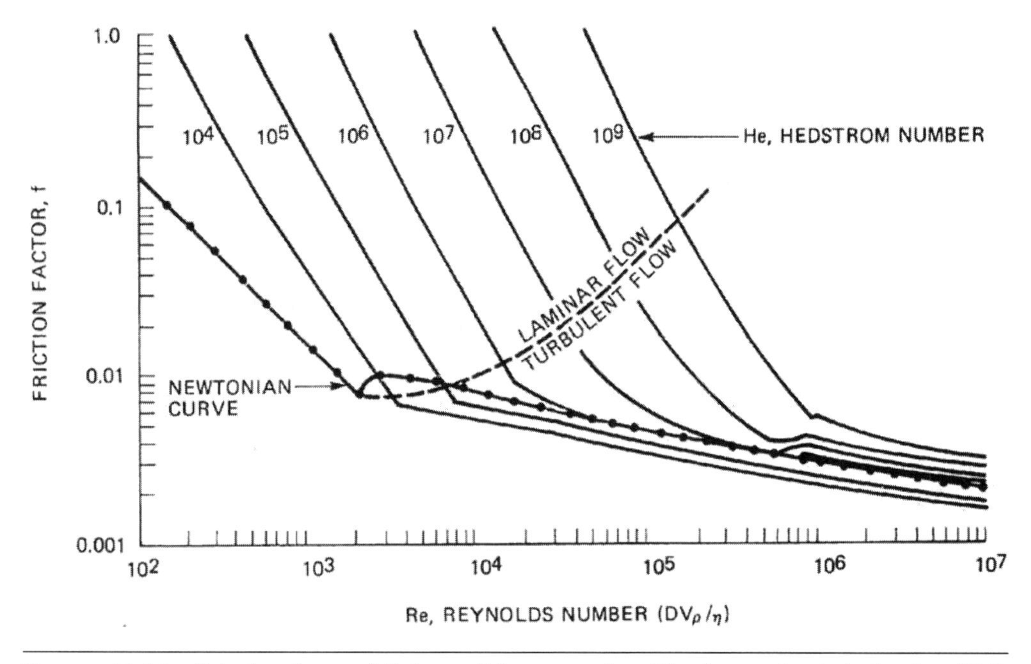

FIGURE 21.10 Friction factor (f) for solids, assuming Bingham plastic behavior (U.S. EPA, 1979).

Chilton and Stainsby (1998) used both analytical methods and numerical techniques to determine headlosses of four residuals flowing through a 150-mm (6-in.) pipe. They used the rheological parameters noted in a 1980 paper by Ackers and Allen. The materials were characterized only by their density, not by type or solids concentration.

Recently, Bechtel (2003 and 2005) used computational fluid dynamic (CFD) methods to analyze pipeline flow and then compared his results with

- An analytical solution and an early work of Mulbarger to determine pipeline headlosses in the laminar-flow range and
- The same analytical solution, the early work of Mulbarger, a graphical approach from Metcalf and Eddy (1991), the equations proposed by Chilton and Stainsby (1998), and Steffe's 1996 work to determine pipeline headlosses in the turbulent-flow range.

The analytical solution involved calculating the Reynolds and Hedstrom numbers and then determining the Fanning friction factor as follows:

$$f^4 + f^3(-16/Re - 8He/3Re^2) + 16He^4/3Re^8 = 0 \tag{21.17}$$

Bechtel found that that the Mulbarger curves overpredicted headlosses for solids at laminar flows. When solids were at turbulent flows, all of the models—except Mulbarger curves—predicted similar results.

3.4 Example 21.1

Calculate the friction-related headloss associated with pumping 5% thickened WAS 150 m (L) at a laminar flowrate. The pump's design flowrate is 400 L/min. The pipe's inside diameter should be at least 150 mm (6 in.), and the inside diameter of a mortar-lined ductile iron pipe is 155 mm (0.155 m) (D). So, the fluid velocity (V) is

$$V = 24/60/60/\pi \times 0.155^2 \times 4$$

$$= 0.353 \text{ m/s}$$

3.4.1 *Option 1: Using the Darcy-Weisbach Equation*

Design engineers could use the Darcy-Weisbach equation and the Moody Diagram for water and then multiply the result by the appropriate solids multiplication factor.

The Darcy-Weisbach equation is

$$H_f = f\,L\,V^2/D\,2g$$

Where

f = the friction factor for water as derived from a Moody Diagram.

Unlike the $C = 140$ used in the Hazen-Williams equation for ductile iron, the Moody Diagram for ductile iron pipe lacks an explicit ε coefficient. This coefficient may range from 0.13 to 0.33 mm.

In this case, the median of the range is selected, so:

$\varepsilon = 0.23$ mm

$D = 155$ mm

$\varepsilon/D = 0.001484$, relative roughness

The Reynolds number for water is:

$$Re = V\,D/v$$

Where

$v = 1.14 \times 10^{-6}$ m^2/s, kinematic viscosity of water at 15°C

So, $Re = 47\,996$ and from the Moody Diagram

$f = 0.026$

Substituting all the values from above, we get:

$$H_f = 0.026 \times 150 \text{ m} \times (0.353 \text{ m/s})^2/(0.155 \text{ m} \times 2 \times 9.81 \text{ m/s}^2)$$

$$H_f = 0.1598 \text{ m (for water flows)}$$

From Figure 21.4b (Sanks, 1998, Figure 19-4), for worst-case design, a solids multiplication factor equal to 35 is derived, so the final headloss for residuals containing 5% solids is:

$$H_f = 0.1598 \times 35 = 5.59 \text{ m of water column}$$

3.4.2 *Option 2: Using the Buckingham Equation*

The Reynolds number (Re) for sludge is:

$$Re = \rho\,V\,D/R_c$$

Where

ρ = 1 020 kg/m^3, sludge density

R_c = 0.035 kg/m.s, coefficient of rigidity determined using Figure 21.8

Re = 1595, which is less than 2000, so the flow regime is laminar and Equation 21.11 may be used. Substituting all the values from above gives

$$H/L = 32\ (9.5\ \text{Pa}/6 \times 1020 \times 9.81 \times 0.155 + 0.035 \times 0.353/1020 \times 9.81 \times 0.155^2)$$

$$H/L = 0.034\ \text{m of head per meter of pipe}$$

Because the pipe is 150 m long, total headloss is:

$$H_f = 5.10\ \text{m of water column}$$

3.4.3 *Option 3: Using Figure 21.6 (6-in. pipe, worst case)*

V = 0.353 m/s (1.158 ft/s), so Figure 21.6 indicates that headloss is 3.5 m/100 m (3.5 ft/100 ft). Because the total length is 150 m, the total headloss is:

$$H_f = 3.5 \times 1.5 = 5.25\ \text{m of water column}$$

In addition to friction-related headloss, design engineers should calculate static head, "minor" headlosses from valves and fittings, and velocity head. The sum of all these headlosses is the total dynamic head that the pump must provide. Design engineers also need to ensure that the available net positive suction head is sufficiently more than is needed. They also should consider changes in thickened solids characteristics and evaluate multiple duty points, if needed.

3.5 Kinetic Pumps

Kinetic (dynamic) pumps continuously add energy to the pumped fluid to make the velocity in the pump higher than the velocity at the discharge point and, therefore, increase pressure. Following are several types of these pumps and their common applications.

3.5.1 *Solids-Handling Centrifugal Pumps*

A wide variety of centrifugal pumps is available. Except for special designs (e.g., recessed impeller), however, these pumps only should be used with relatively dilute (less than 1% solids), trash-free residuals. They typically are used to transport RAS because of the pump's high volumetric flowrate and excellent efficiency. The minimal debris in RAS typically does not clog the pumps.

Centrifugal pumps are not recommended for primary sludge, primary scum, or thickened sludge for two reasons. First, there is no means to ensure that the pump's suction will positively draw thickened solids to the pump impeller. Second, the system head curve depends on solids concentration that is often inconsistent, leading to significant variations in liquid flowrate and pump power requirements.

3.5.2 *Recessed-Impeller Pumps*

The recessed-impeller pump (also called a torque-flow, vortex, or shear-lift pump) has a standard concentric casing with an axial suction opening and a tangential discharge opening. The impeller, which is recessed into the pump casing, can be open or semi-open with either straight radial blades or ones tapered to the shaft.

In residuals pumping applications, design engineers typically choose pumps with fully recessed, open impellers. When it rotates, the impeller creates a spiraling vortex field in the fluid within the casing. This vortex moves residuals through the pump, allowing large solids to pass easily. Most of the solids do not pass through the impeller vanes, thereby minimizing abrasion.

Recessed-impeller pumps work well on untreated residuals containing no more than 2.5% solids or on digested solids (biosolids) with about 4% solids. Although they can pump thicker residuals, varying friction losses cause erratic flowrates and heavy radial thrusts on the pump shaft. Positive-displacement pumps perform better in such applications. If design engineers use recessed-impeller pumps to transport thickened solids, they should provide flow meters and variable-speed drives to maintain a relatively constant flow. They also should specify the heaviest possible shafts and bearings. In addition, the pumps should be horizontally mounted to simplify maintenance, and include adequate clean-outs and flushing connections.

Although contact between solids and impeller vanes is minimal, design engineers should consider specifying abrasion-resistant, cast-iron (ASTM A532) volutes and impellers, especially if the residuals' grit and abrasives content is high or unknown. However, such impellers cannot be trimmed, so if using them, designers must size the pump(s) accurately.

Recessed-impeller pumps are available in both vertical (close-coupled or extended-shaft) and horizontal configurations that are suitable for either wet or dry wells. Wet-well pumps are available with hydraulic drives or submersible electric motors. They typically are available in sizes from 50 to 200 mm (2 to 8 in.), with capacities from 180 to 1800 L/min (50 to 500 gpm) at up to 64 m (210 ft) total dynamic head.

The primary drawback of recessed-impeller pumps (compared to other nonclog centrifugal units) is their significantly lower efficiency. A recessed-impeller pump's efficiency typically is between 5% and 20% lower than that of a comparable standard pump.

3.5.3 Screw/Combination Centrifugal Pumps

Screw/combination centrifugal pumps combine a screw-type impeller with a normal centrifugal impeller. They typically have a relatively high efficiency and relatively low net positive suction head (NPSH) requirements. In addition, the corkscrew action of screw impellers may provide more positive feed to the suction, so the pump handles thicker solids better.

3.5.4 Disc Pumps

Disc pumps operate on the principles of boundary layer and viscous drag. Their impellers are basically parallel discs installed at a certain distance apart. Fluid flows through the gap between the rotating discs, which transfer energy to the fluid and generate velocity and pressure gradients that force the fluid to flow through the pump. Pump wear is minimized because the fluid moves parallel to the discs and does not touch other pump parts.

Disc pumps traditionally are specified for residuals with up to 6% solids, slurries, viscous materials, and residuals with high entrained-air content (e.g., from DAF units). They can run dry and handle abrasive materials, which makes them excellent candidates for pumping grit.

3.5.5 Grinder Pumps

Special combination centrifugal pump grinders are also available (see Figure 21.11). These pumps combine a hardened steel cutting bar with a relatively typical centrifugal vortex-type pump. They can be used as digester recirculation pumps and prevent rag balls. However, operating experience indicates that such pumps require as much maintenance as grinders.

3.6 Positive-Displacement Pumps

There are several types of positive-displacement pumps that can be used to transport residuals.

3.6.1 Plunger Pumps

Plunger pumps have pistons driven by either an exposed eccentric crank shaft or a walking beam. They are available in simplex, duplex, triplex, and quadplex configurations. Plunger pumps have an output of 150 to 225 L/min (40 to 60 gpm) per plunger and can develop up to 70 m (230 ft) of discharge head. These pumps typically are designed for an efficiency of 40% to 50%, which leaves a power reserve to overcome changes in pumping head (Sanks et al., 1998).

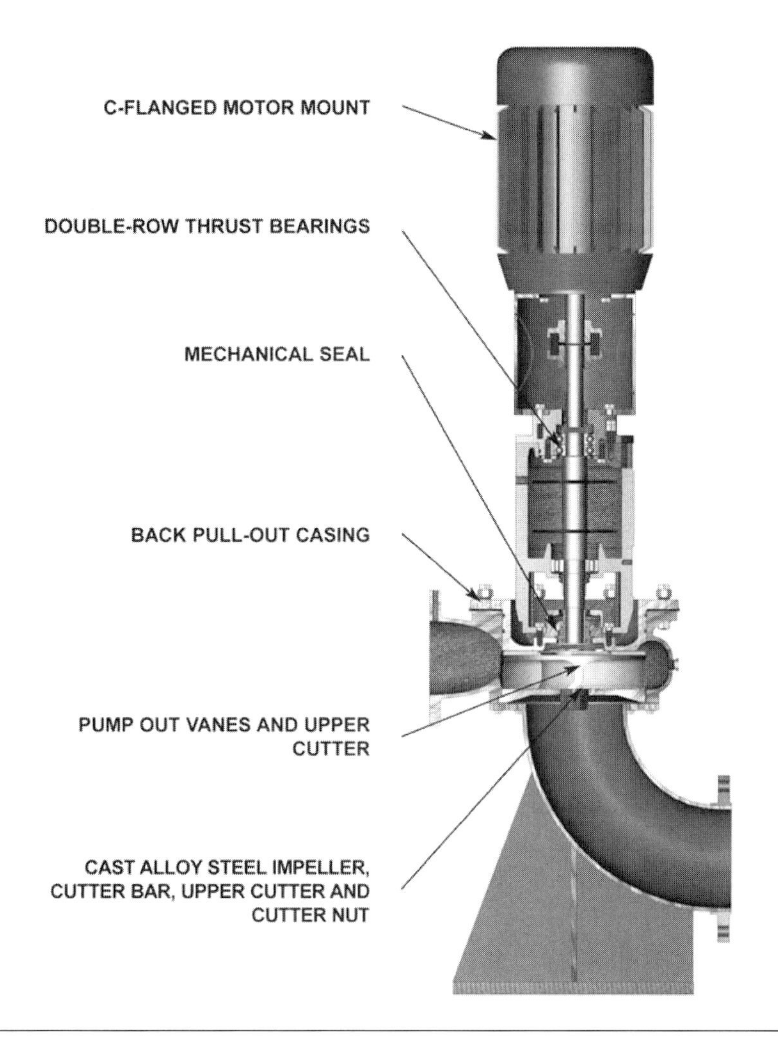

FIGURE 21.11 Chopper–grinder pump (courtesy of Vaughan Company, Inc.).

Plunger pumps have several advantages:

- They can transport residuals containing up to 15% solids if the equipment is designed for load conditions;
- They are available in cost-effective options up to 30 L/s (500 gpm) and 60 m (200 ft) of discharge head;
- Units with large port openings can operate at low pumping rates;
- They provide positive delivery unless some object prevents the ball check valves from seating;

- They provide constant-but-adjustable capacity in spite of large variations in pumping head;
- They can operate for a little while under "no-flow" conditions (e.g., a plugged suction line) without damage;
- The pulsating action of low-velocity simplex and duplex pumps sometimes helps concentrate residuals in feed hoppers and resuspend solids in pipelines; and
- They have relatively low operations and maintenance (O&M) costs.

Changing the stroke length changes the pump output. However, the pumps typically operate best at or near full stroke, so designers typically provide a variable-pitch V-belt drive or a variable-speed drive to control pumping capacity.

Plunger pumps have paired ball or flap check valves on the suction and discharge sides. A connecting rod joins the throw of the crankshaft to the piston. The piston is housed in an oil-filled crankcase (for lubrication) and sealed in a stuffing box gland and packing, which is kept moist by an annular pool of water directly above it. Unless the pool receives a constant supply of water, the packing will dry and fail rapidly, which can cause solids to spray throughout the immediate area.

Plunger pumps can operate with up to 3 m (10 ft) of suction lift, but this can reduce the solids concentration they can handle. Using a pump whose suction pressure is higher than its discharge pressure is impractical because flow would be forced past the check valves. Using special intake and discharge air chambers reduces noise and vibration and dampens pulsations of intermittent flow.

If designers use pulsation-dampening air chambers, they should be glass-lined to avoid destruction via hydrogen sulfide corrosion. If the pump operates while the discharge pipeline is obstructed, the pump, motor, or pipeline can be damaged; a simple shear pin arrangement can prevent this problem.

The number of pistons directly influences the variation in downstream flowrates. If a smooth discharge is required, design engineers should consider using triplex or quadplex pumps.

3.6.2 Progressing Cavity Pumps

Compared to plunger pumps, progressing cavity pumps operate more cleanly and discharge a smoother flow. They can provide a consistent flow despite changes in discharge head. However, an improperly selected and designed pump can lead to excessive maintenance problems and costs. Design engineers especially should guard against pump operation in no-flow conditions because it can quickly damage the stator.

A progressing cavity pump uses a worm-shaped metal rotor that turns eccentrically inside a pliable elastomeric stator. The stator's axial pitch is about 50% that of the rotor. The rotor seals against the stator, forming a sealing line or lines that move down the pump as the rotor turns. Cavities progress axially between these lines, moving solids from the suction end of the pump to the discharge end. As the stator wears, some "slippage" flow occurs at the sealing lines; this slippage causes further wear. To minimize slippage, design engineers should use enough cavities (multistage construction) to limit the pressure difference across the sealing lines.

The elastomeric stator is relatively soft and subject to abrasion, so progressing cavity pumps should be used in facilities with good grit-removal facilities. They should not be used to transport grit. Also, design engineers should minimize the rotor's rotational speed. In some applications (particularly with variable-speed drives), designers should select a pump larger than design flow requires to ensure that the pump still can meet design flow requirements after the stator has begun to wear.

One advantage of a progressing cavity pump is that the stator acts as a check valve, preventing backflow under most conditions. So, an actual check valve or antireverse ratchet is only required if the pump's static backpressure is more than 50 m or stator wear is expected due to significant grit concentrations. However, design engineers always should include isolation valves on both suction and discharge sides so the pump can be removed from service for routine maintenance.

Most progressing cavity pumps are tested with water. When used to transport solids, the pumps may need more motor horsepower. Design engineers should consult with pump manufacturers on each application to ensure that adequately sized motors are specified.

Solids capacity depends on pump size. Pumps sized for at least 3 L/s (50 gpm) at suitably low rotating speeds typically pass solids of about 20 mm (0.8 in.), so grinders are unnecessary. Smaller pumps, however, typically need grinders. If grinders are not included, then design engineers should specify protective covers on any required universal joints.

To minimize pump maintenance costs,

- make sure prior processes remove grit effectively;
- limit rotating speeds to about 250 rpm;
- make suction lines as short as possible and use open-throat, hopper-type suction ports (Jones, 1993);
- limit the pressure per stage to about 170 kPa (25 psi) (if higher pressures are needed, most manufacturers offer multistage pumps);
- carefully specify rotor material, stator material, and design of universal joints (where applicable);
- provide room to dismantle the pump efficiently;

- consider including reversing starters, which allow the pump to reverse flow direction and possibly clear minor blockages in suction piping; and
- reverse the pump's flow direction in high suction lift applications.

The pump discharge must have pressure safety switches to prevent blocked discharge lines from rupturing. Also, design engineers should use flow indicator switches or proprietary devices to prevent the pumps from running dry. Designers also should consider using pressure-relief assemblies or rupture disks to protect downstream piping.

3.6.3 Diaphragm Pumps

Diaphragm pumps typically transport solids from primary sedimentation tanks and gravity thickeners. These pumps are a relatively simple means of pumping thickened residuals and can handle grit with minimum wear. Manufacturers claim that the pump's pulsating action increases solids concentrations when transporting gravity-thickened residuals. However, pulsating flow may not be acceptable for some downstream treatment processes.

An air-operated diaphragm pump typically consists of a single-chambered, spring-return diaphragm; an air-pressure regulator; a solenoid valve; a gauge; a muffler; and a timer (see Figure 21.12). Compressed air flexes a membrane that is pushed or pulled to contract or enlarge an enclosed cavity. However, unless the wastewater treatment plant already uses compressed air, providing this service can significantly increase pumping costs. Also, the air exhausted from the pump valves is noisy.

Hydraulically or electric-motor driven diaphragm pumps also are available, and should be considered.

3.6.4 Rotary Lobe Pumps

A rotary lobe pump uses multilobed, intermeshed rotating impellers to transfer residuals containing up to 10% solids. Like progressing cavity pumps, rotary lobe pumps offer a relatively smooth flow and do not require check valves in many applications with low-to-moderate discharge static heads. However, both suction and discharge isolation valves are needed so the pump can be removed from service for maintenance.

Pumping efficiency depends on maintaining relatively close tolerances between rotating lobes, so all large or abrasive material should be removed from residuals before they enter the pumps. Rotary lobe pumps are more suitable for applications with efficient grit removal and should not be used to transport grit. Engineers also should design the pumps to rotate at the lowest possible speed to minimize abrasion. Both rotary lobe and progressing cavity pumps react similarly to abrasion, so engineers can

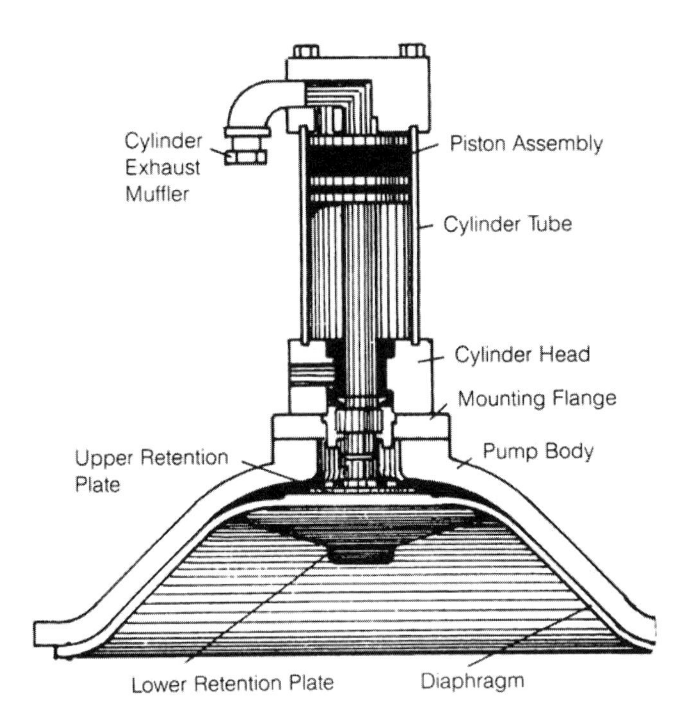

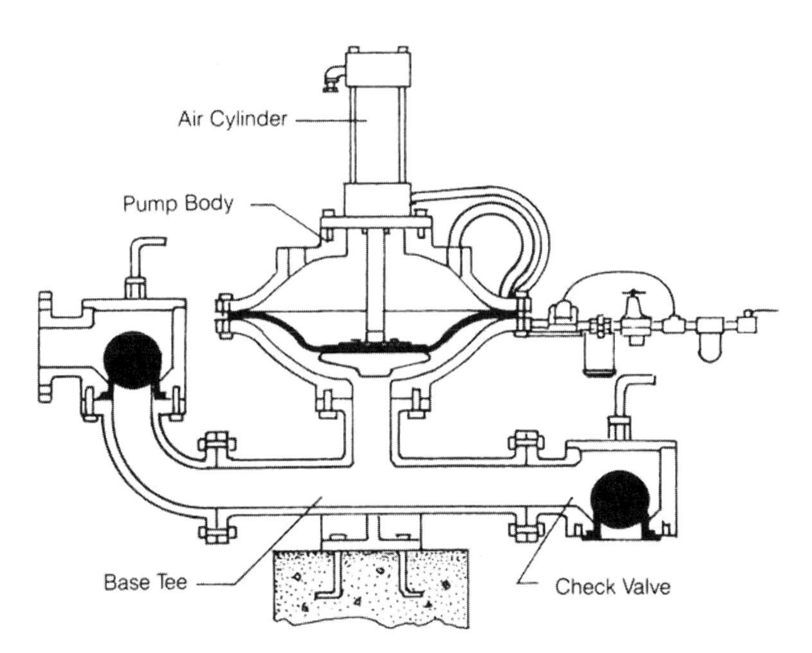

FIGURE 21.12 Air-operated diaphragm pump.

apply many of a progressing cavity's design considerations to a rotary lobe pump. Designers also should select appropriate lobe material for the residuals to be transported; otherwise, the lobes may fail prematurely. An advantage of these pumps over progressing cavity pumps is their ability to handle short periods of no flow without significant damage.

3.6.5 *Pneumatic Ejectors*

Rather than rotating elements and electric motors, a pneumatic ejector has a receiving container, inlet and outlet check valve, air supply, and liquid level detector. When liquid reaches a preset level, air is forced into the container and the stored volume is ejected. Then, the air supply cuts off and liquid flows through the inlet into the receiver.

Pneumatic ejectors can be used to convey residuals and scum. They also have been used in some facilities to transport grit and screenings. They are available in capacities from 110 to 570 L/min (30 to 150 gpm) and heads up to 30 m (100 ft).

3.6.6 *Peristaltic Hose Pumps*

Although more widely applied in the industrial sector, peristaltic hose pumps have been used to transport municipal solids. These self-priming pumps are available in capacities of 36 to 1 250 L/min (1 to 330 gpm) and heads up to 152 m (500 ft). They can be used to meter flow because their output is directly proportional to speed at either high or low discharge pressures. Peristaltic hose pumps are suitable for suction lift applications [up to 44.7 kPa (15 ft of water)] and can pump abrasive fluids. These devices are relatively simple, requiring only common tools and basic mechanical skills for assembly, servicing, and repair.

A peristaltic hose pump has no seals, valves, or bearings; it moves residuals by alternately compressing and relaxing a specially designed resilient hose. The hose is compressed between the inner wall of the pump housing and the compression shoes on the rotor. A liquid lubricant may be used to minimize sliding friction. The residuals only touch the hose's thick inner wall, which cushions entrained abrasives during compression; abrasives are released after compression. Replacement hoses can be expensive, however, so to maximize hose life, the pump's maximum rotational speed should be limited to 25 rpm.

The primary disadvantage of this pump is its pulsed flow (because the rotor typically only has two compression shoes). Depending on the rotational speed required to obtain the design pumping rate, the pulsing flow may not be suitable for downstream processes. This can be offset, however, by using pulsation dampeners on the discharge.

3.6.7 *Reciprocating Piston Pumps*

Reciprocating piston pumps are useful and cost-effective when dewatered cake must be transported to cake storage or loading facilities. They typically are not used ahead of the dewatering process. However, because these pumps can achieve discharge pressures up to 1.5×10^4 kPa (2 200 psi), they are the primary choice for pumping thickened sludge long distances. Given these pumps' high potential discharge pressures, however, engineers must design downstream piping systems properly.

3.7 Other Pumps

Following are other types of pumps used to transport residuals.

3.7.1 *Air-Lift Pumps*

An air-lift pump has an open riser pipe, the lower end of which is submerged in the liquid to be pumped. When an air-supply tube introduces compressed air at the bottom of the pipe, air bubbles form and mix with the residuals in the pipe. As the density of the air–residuals mixture decreases, denser material outside the pipe pushes the mixture up and out of the riser pipe.

Air-lift pumps often are used to transport WAS, RAS and similar residuals in smaller treatment plants, where high efficiency and a precisely controlled flowrate are not required. Air-lift pumps typically are used in high-volume, low-head applications, those with lifts less than 1.5 m (5 ft). Their capacity can be varied by optimizing the air-supply rate. Increasing the air supply beyond its optimum level, however, only decreases the volume of liquid discharged. The main advantages of air-lift pumps are the absence of moving parts and their simple construction and use.

The air-supply arrangement governs the solids-handling capability. Air-lift pumps with an external air supply and circumferential diffuser can pass solid particles as large as the riser pipe's internal diameter without clogging. Those with air supplied by a separately inserted pipe lack this non-clog feature.

3.7.2 *Archimedes Screw Pumps*

Archimedes screw pumps occasionally are used to transport RAS (see Figure 21.13). This pump has an open design for lifts up to 9 m (30 ft) and an enclosed design for lifts up to 12 m (40 ft) or more. It automatically adjusts its discharge rates in proportion to the depth of liquid in the inlet chamber until the water gets to the "fill point", and then becomes constant. In other words, the pump has an inherent variable-flow capacity and does not need motor-speed controllers.

An Archimedes screw pump has a fairly constant efficiency (70 to 75%) within 30% to 100% of its rated design capacity. The screw spirals' peripheral tip speeds are typically are less than 229 m/min (750 ft/min); those for centrifugal or recessed-impeller pumps are 1 070 to 1 220 m/min (3 500 to 4 000 ft/min). Also, the screw pumps are not pressurized. These characteristics are advantageous in RAS applications because they make the screw pump less likely to shear the activated sludge floc.

FIGURE 21.13 Archimedes screw pump (courtesy of Siemens Water Technologies).

The pump's principal disadvantage is its space requirements. If exposed to the sun and left idle for extended periods or unequipped with cooling water sprays, the pump can warp due to thermal expansion. Off-line units also may freeze in cold weather. Another potential disadvantage is that RAS often aerates in these systems. In some RAS applications, Archimedean screw pumps were no longer used because the RAS' high dissolved oxygen content was interfering with biological nutrient removal.

3.8 Long-Distance Pipelines

Many sites successfully pump residuals long distances. The work of Carthew et al. (1983) was driven by the design of a 29-km pipeline. Honey and Pretorius (2000) solved their example using a 2-km pipeline, and Murakami et al. (2001) quote a distance of 1 km in their manuscript. However, engineers must develop special design criteria to minimize potential operating problems. They should carefully determine residuals characteristics (e.g., viscosity, solids percentage, and type) and study the effects of flow velocity on fluid viscosity and pipe-friction losses (Mulbarger et al., 1981).

When designing long-distance systems, actual field data are critical. Several studies provide detailed information on the analysis and design of long-distance solids transport systems (Carthew et al., 1983; Mulbarger et al., 1981; Setterwall, 1972; Spaar, 1972; U.S. EPA, 1979). Figure 21.14 shows the test systems used in the field by Carthew et al. (1983) and Murakami et al. (2001).

Long-distance pumping typically creates high-pressure losses, so design engineers should choose pumps that can generate the high pressures needed. In the United Kingdom, for example, a vertical, positive-displacement, hydraulically driven ram pump transfers primary and activated sludge over a 2.2-km-long pipeline, working against pressures of up to 26 bar (377 psi) (Ram Pumps, 1999).

3.9 Common Design Deficiencies in Pumps and Piping

Several design errors in pumping and piping systems are particularly noteworthy:

- Incorrectly calculating friction head and not providing enough allowance for variations that occur during operation.
- Not providing adequate flushing and cleaning lines. Many residuals form grease deposits or scale in pipe, and flushing water connections and cleanout ports become more critical as residuals become thicker.
- Not providing enough suction to handle thickened sludge. Thixotropy and plasticity can greatly affect friction, so a good design includes a straight, short

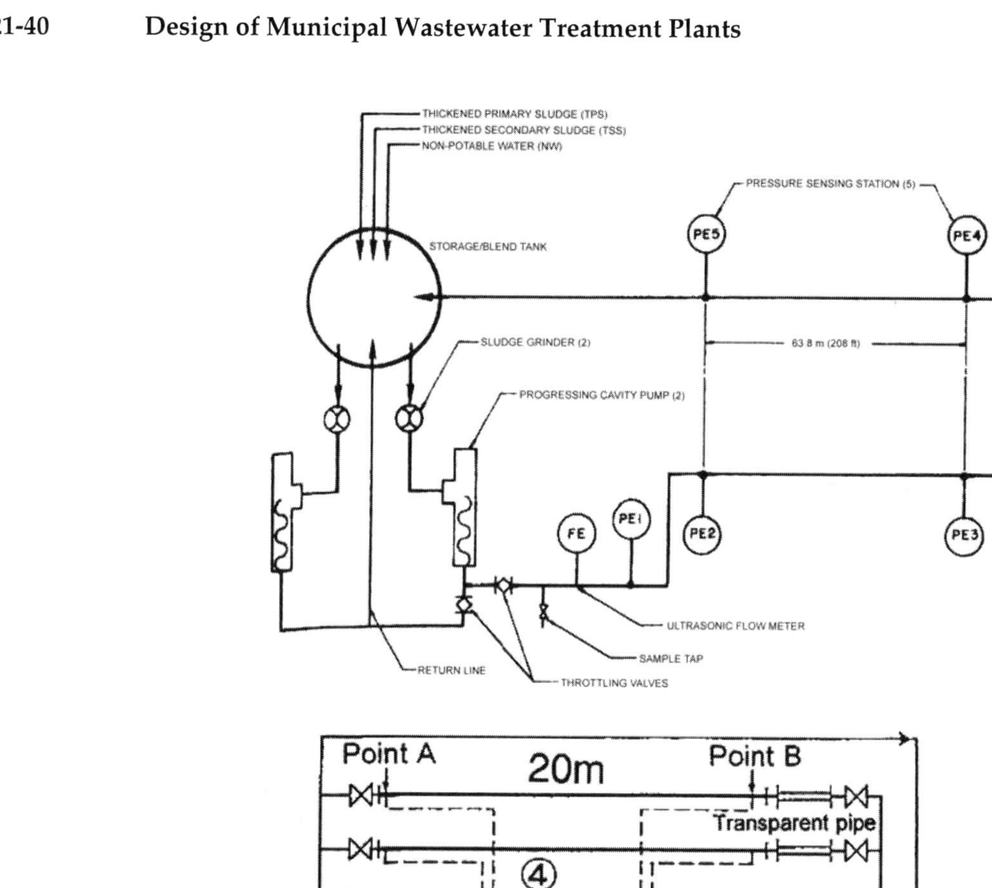

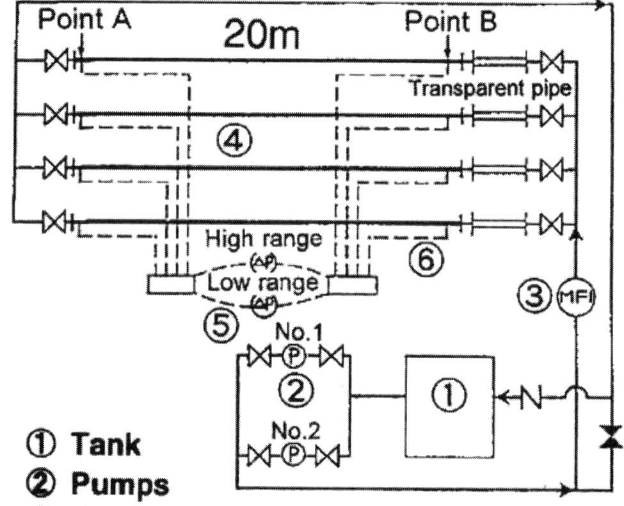

FIGURE 21.14 Experimental setups used by (top) Carthew et al., 1983, and (bottom) Murakami et al., 2001.

suction pipe to a pump set low enough to allow for substantial positive suction pressure.

- Operating progressing cavity pumps at excessive speed or pressure per stage will increase maintenance costs.
- Burying or encasing pipe elbows. Grit slurries and some (supposedly degritted) residuals can wear out elbows.
- Using one pump to withdraw solids from two or more tanks simultaneously. Ideally, each tank should have a dedicated pump, with interconnections that allow another pump to be used when the dedicated pump is out of service. Otherwise, the system should be valved so one pump can draw from multiple tanks sequentially.
- Creating a pipeline route with high spots, which trap air or gas. Designers should avoid high spots because air-relief valves are too troublesome in these applications.
- Using the wrong valves. In this application, design engineers should use plug valves, with at least 80% clear waterway area. In most cases, "full port" valves are preferable. Pinch valves may also be applicable, but the designer should carefully weigh their advantages against their disadvantages.
- Lack of rupture disks or other pressure-relief devices between isolation valves. [Other errors are cited in publications by Sanks et al. (1998) and U.S. EPA (1982).]

There are general design guidelines for any residuals pumping system. The minimum size of piping is dependent on designers. The minimum desirable size for residuals piping is 150-mm (6-in.), although some designers prefer 200-mm (8-in.) piping. In some cases, 100-mm (4-in.) piping may be acceptable, if it is lined to reduce friction. For smaller plants, the designer may need to consider intermittent pumping to ensure that velocities are maintained. In any pipe size, using a smooth lining minimizes the formation of struvite crystals in pumping anaerobically digested solids. Typically, ductile iron piping can be lined with cement, glass, or polyethylene. While other materials, such as polyvinyl chloride (PVC), may not require lining, operating pressures should be carefully considered.

When designing a transport system for thickened sludge, engineers also should consider both the process it is coming from and the one that will receive it. For example, dissolved air flotation units produce solids that contain a lot of entrained air, which can be problematic for many pumps. Or if a mechanical thickening process will be discharging solids directly into an open-throat progressing cavity pump, then design engineers must choose a pumping rate that exceeds the thickening unit's maximum discharge rate.

3.10 Standby Capacity

When determining whether standby transport capacity is needed, design engineers should consider the plant size, system's function, arrangement of units, anticipated service period, and time required for repair. For example, standby capacity for RAS pumping is important because a service interruption could quickly impair effluent quality. Primary and secondary sludge pumping also are critical functions, so designers typically either provide dual units or use units that can perform dual duty. For example, primary sludge pumps also typically serve as standbys for scum pumps.

If single units are used, they should be heavy duty, have readily available spare parts, and be easy to repair quickly (preferably in place). Design engineers should ensure that the pump comes with adequate spare parts.

4.0 DEWATERED CAKE STORAGE

4.1 Storage Requirements

Dewatered cake typically is stored somewhere before receiving more treatment (e.g., heat drying) or being hauled offsite for use or disposal. Because the cake contains so little moisture, experts assume that there is little hazard associated with storing it (NFPA Report on Comments A2007—NFPA 820). Most flammable liquids would have been removed during dewatering, and methane-generating microorganisms do not thrive in dry aerobic environments, so special safety precautions are not required. That said, the dewatered cake's viscous, sticky nature can complicate storage designs.

The amount of storage needed depends on what will happen to the cake afterward. Often, biosolids will only be held for a few days or weeks before being treated further or hauled offsite. In which case, they typically are stored in large roll-off containers, 18-wheel dump trailers, concrete bunkers with push walls, or bins with augurs.

However, if the biosolids will be land-applied or surface disposed, long-term storage may be required. In these cases, they often are stockpiled on concrete slabs or other impervious pads. When designing long-term storage facilities, engineers need to consider buffering, odor control, and accessibility. They also need to determine whether the storage facility should be open or covered. [For more information on calculating storage requirements for land application, see Section 2.1. For further guidance, see U.S. EPA's *Guide to Field Storage of Biosolids* (U.S. EPA, 2000).]

4.2 Odor-Control Issues

Odor control can be an issue with dewatered solids—especially when larger quantities are stored or the solids storage area is relatively close to neighbors. The odors that dewatered, anaerobically digested solids produce are primarily organo-sulfur compounds.

Storing the cake for 20 to 30 days at 25°C significantly cuts odor generation (Novak et al., 2004; WERF, 2008). Adding alum after digestion also reduced storage-related odors (Novak et al., 2004). However, the best way to minimize dewatered cake odors was to optimize the solids treatment processes before dewatering, according to the Water Environment Research Foundation.

5.0 DEWATERED CAKE TRANSPORT

Modern dewatering operations can produce cake containing 15% to 40% solids or more, depending on the conditioning chemicals and dewatering equipment used. The consistency of such cakes ranges from pudding to damp cardboard, so they will not exit the dewatering equipment by flowing via gravity into a pipe or channel. Instead, they must be transported via

- Positive-displacement pumps;
- Mechanical conveyors (e.g., flat or troughed belt, corrugated belt, or screw augers); or
- Gravity from the bottom of the dewatering equipment into a storage hopper or truck directly below.

Before choosing a cake transportation method, design engineers should analyze various options based on solids-management requirements, site or building constraints, reliability, O&M, and life-cycle costs.

5.1 Pumping

Both progressing cavity pumps and hydraulically driven reciprocating piston pumps can handle dewatered cake. Compared to belt or screw conveyors, these pumps better control odors (because the cake travels in an enclosed pipe), eliminate spills, and have fewer maintenance requirements. The pumps also have much smaller footprints and, therefore, are suitable in buildings with space constraints. They can even reduce noise levels in some cases. However, pumps often need more electricity than conveyors to move a given volume of cake.

Which pump to use depends on the application. Progressing cavity pumps provide a steady flow, while hydraulically driven reciprocating piston pumps pulsate (List et al., 1998). Progressing cavity pumps typically are preferred in applications where the cake is thinner and transport distances are short. Hydraulically driven reciprocating piston pumps are more expensive, but may handle greater pressures and thicker cake. (A solids process' discharge piping can be a high-pressure environment.)

5.2 Hydraulics

The hydraulic characteristics of dewatered cake (with more than 15% solids) have not been extensively studied or widely reported. Likewise, headloss-calculation methods for pumping such solids are limited. However, researchers have shown that dewatered cake may exhibit both plastic and pseudoplastic (thixotropic) behavior (List et al., 1998; Barbachem and Payne, 1995; Bassett et al., 1991). For a Bingham plastic, a minimum shearing stress is required to initiate flow. For thixotropic materials, the apparent viscosity and headloss gradient (dH/dL) decrease as the rate of shear increases or as the fluid travels a certain distance inside a pipe until time-independent behavior is reached (Honey and Pretorius, 2000). These two behaviors complicate hydraulic design.

Nonetheless, dewatered cake can be pumped—even though experts only recommend it for relatively short distances. There are many successful pumping installations in North America and Europe, and many others are being designed or constructed.

5.3 Flow and Headloss Characteristics

In most dewatered cake applications, headlosses are high—often in the range of 1380 to 6900 kPa (200 to 1 000 psi). It depends on the length, diameter, and configuration of the discharge piping. Headloss also depends on cake type and solids concentration, as well as the conditioning and dewatering methods that produced the cake. Current experience indicates that headlosses typically are sensitive to velocity and piping constrictions, particularly if the dewatered cake's solids concentration exceeds 30%.

Typical piping headlosses in cake-pumping applications range from 11.3 to 79.1 kPa/m (0.5 to 3.5 psi/ft); design engineers often use these values as a general guideline during preliminary design. Ideally, the final design would keep headlosses below 45.2 kPa/m (2 psi/ft).

5.4 Design Approach

Engineers should avoid constrictions (e.g., smaller-than-line-size valves and short radius bends) when designing discharge piping. The piping should be large enough that theoretical cake flow velocities never exceed 0.15 m/s (0.5 ft/sec)—although maximum velocities of 0.08 m/s (0.25 ft/sec) are preferred, especially if dewatered cake solids concentrations exceed 30%. The pipe should be designed to allow flushing and pigging.

Unlike liquid residuals, where pumping design equations are available for solids concentrations up to 12%, design approaches for pumping dewatered cake are more site-specific. Field testing is highly recommended, especially in relatively long-distance applications. A pipe pumping cake containing more than 30% solids should not be more than 152 m (500 ft) long, and line lubrication is highly recommended.

Four peer-reviewed case studies with actual field data were published in the 1990s. These studies explored pipeline headlosses for the following types of cake:

- Anaerobically digested, centrifugally dewatered cake containing 20% solids (Bassett et al., 1991);
- Anaerobically digested, centrifugally dewatered cake containing 22% solids (Barbachem and Pyne, 1995);
- Unspecified dewatered cake containing 28% solids with and without polymer injection for line lubrication (List et al., 1998); and
- Undigested, plate and frame pressed cake containing 34% solids (Barbachem and Pyne, 1995).

Comparing actual field data, researchers developed a simple headloss equation for 20% cake based on the pseudoplastic model, warning that it might only apply to the specific case (Bassett et al., 1991). Equation 21.18 may be valid for all pipe sizes from 100 to 300 mm (4 to 12 in.) with velocities ranging from 0.015 to 0.43 m/s (0.05 to 1.4 ft/sec):

$$\Delta P = 15.68/D^{1.28} + 0.245\ Q/D^2 \tag{21.18}$$

Where

ΔP = pressure drop (psi/ft),
D = inside diameter of the pipe (in.), and
Q = cake flow (gpm).

Table 21.4 illustrates the effect of pipe size on pipeline headloss (based on Equation 21.18). It shows that headloss increases almost 50% for each decrease in nominal pipe size. It also shows that pumping thicker cake (from 20% to 28%) causes a minimal increase in headloss (approximately 10%) when a large enough pipe (e.g., 300-mm) is used. (NOTE: The previous statement is based on a comparison of the results from two investigations.)

Barbachem and Pyne (1995) used the Re, He, and f number methodology described in Equations 21.10, 21.15, 21.17 and 21.14 to model their actual field data. Figure 21.15 combines headloss data reported from Bassett et al. (1991) and Barbachem and Pyne

TABLE 21.4 Headloss data for pumping cake at near-maximum recommended fluid velocities.

Pipe size	Percent solids concentration	Velocity (m/s)	H_f (m of water/10 m)	H_f (psi/ft)	Change in H_f from one size smaller
100 mm (4 in.)	20 (Bassett, 1991)	0.08	64.0	2.77	57%
150 mm (6 in.)	20 (Bassett, 1991)	0.08	41.0	1.77	46%
200 mm (8 in.)	20 (Bassett, 1991)	0.08	28.0	1.21	55%
300 mm (12 in.)	20 (Bassett, 1991)	0.08	18.0	0.78	Baseline
300 mm (12 in.)	28 (List et al., 1998)	0.10	19.5	0.86	10%

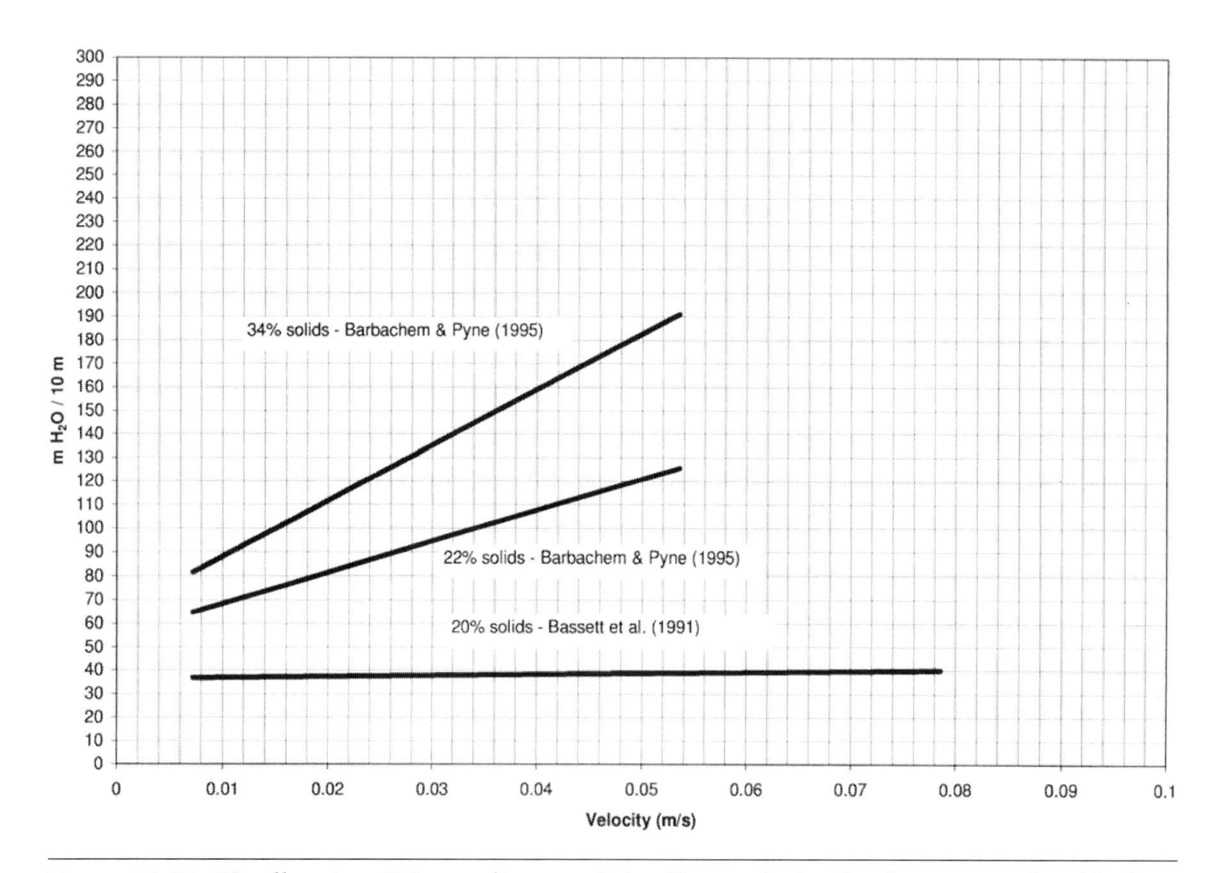

FIGURE 21.15 Headloss in a 150-mm-diameter (6-in.-diameter) pipe for three types of residuals at various velocities below the recommended maximum (created based on Bassett et al., 1991, and Barbachem and Pyne, 1995).

(1995) for flow in a 150-mm (6-in.) pipe. Although the data come from two investigations and so differences in data may be due to different experimental procedures, certain trends and results may be derived. Table 21.5, which one data point from each curve in Figure 21.15, leads to two significant conclusions. First (as expected), friction headlosses through a 150-mm pipe increase dramatically as solids concentrations increase. Second, the 150-mm pipe is too small and inappropriate for pumping dewatered cake containing more than 20% solids; excessive headlosses are created.

In addition to actual field data and considering the high compressibility of cake, List et al. (1998) offered a specific method for determining headlosses created in cake pumping applications. Assuming steady, non-accelerating pumping applications and a Bingham plastic behavior, the method is as follows:

1. Collect field data to determine a pumping pressure for a given flow.
2. Construct a graph depicting $(\Delta P/L)(D/4)$ on the y-axis and $8V/D$ on the x-axis [where ΔP = pressure loss (Pa), L = pipe length (m), D = pipe diameter (m), and V = fluid velocity (m/s)]
3. The intercept ($V = 0$) determines the critical shear stress {stress at the pipe wall [τ_w (Pa or N/m^2)]}.
4. The slope of the graph is the fluid's dynamic viscosity [μ (Pa-s or kg/m-s)]
5. Calculate a dimensionless factor Z as follows:

$$Z = 8\,\mu\,Q_m/\rho_\alpha\,A\,D\,\tau_w \tag{21.19}$$

Where

Q_m = mass flowrate (kg/s),
ρ_α = cake density at atmospheric pressure (kg/m^3), and
A = pipe cross-sectional area (m^2).

TABLE 21.5 Headloss data for pumping cake through a 150-mm (6-in.) pipe at near maximum recommended velocity.

Type of residuals	Velocity (m/s)	H_f (m of water/ 10 m)	H_f (psi/ft)	Change in H_f
20% anaerobically digested, centrifuged	0.08	41	1.77	Baseline
22% anaerobically digested, centrifuged	0.05	120	5.19	293%
34% undigested, plate and frame pressed	0.05	180	7.79	50.0%

Engineers then solve the following equation by trial and error until the left side equals the right side:

$$\rho^* - 1 + Z \ln[(Z + 1)/(Z + \rho^*)] = 4\, \tau_w L/KD \tag{21.20}$$

Where

ρ^* = is the density ratio equal to ρ/ρ_α, and
K = bulk modulus of cake (N/m^2) determined from a graph of bulk density versus consolidating pressure and assumed constant

They then calculate the pressure drop:

$$\Delta P = K\,(\rho^* - 1) \tag{21.21}$$

Assuming a pseudoplastic fluid behavior under the same conditions involved solving a differential equation, but the results were less than 4% different than those derived from the Bingham plastic method above.

As reported in List's article, the results obtained for pumping 28% cake through a 305-mm-diameter (12 in.-diameter), 150-m long pipe using Equation 21.21, using a differential equation assuming Bingham plastic fluid (not shown) and one for a pseudoplastic fluid (not shown) were all very similar.

If the compressibility effect is disregarded, the following two equations may be solved for a Bingham plastic and a pseudoplastic material, respectively:

$$\Delta P/L = 4\, \tau_w/D - 32\, \mu V/D^2 \tag{21.22}$$

$$\Delta P/L = -\, 4\, \tau_w/D - 4k/D(8V/D)^m \tag{21.23}$$

Where k and m are empirical parameters describing the material's properties.

The actual application involved accelerating flow generated by a reciprocating piston pump. The actual field tests of pumping 28% cake through a 250-mm (10-in.) pipe showed that maximum pressure near the pump was 3100 kPa, which was not very far from the calculated values for steady flow.

Table 21.6 is a compilation of data from a progressing cavity pump manufacturer on transporting cakes with different solids percentages (Bourke, 1997). Headlosses ranged from 0.25 to 3.0 psi per foot of straight 150-mm (6-in.) pipe (Bourke, 1997). The table shows that centrifugally dewatered cake is pumped more easily than rotary-drum dewatered cake. Centrifugally dewatered cake also is pumped more easily than belt-

TABLE 21.6 Headloss data [in m of water/10 m (psi/ft)] as reported in Bourke (1999) for fluid velocities less than 0.06 m/s.

Method	Percent solids	100 mm (4 in.)	150 mm (6 in.)	200 mm (8 in.)
Rotary drum	20–30	Slightly more than double of 150 mm	7.62–11.5 (0.33–0.50)	About half of 150 mm
Centrifuge	20–30	Slightly more than double of 150 mm	50% less than rotary drum	About half of 150 mm
Rotary drum and then heat treated	45	Slightly more than double of 150 mm	23.0–35.0 (1.00–1.50)	About half of 150 mm
Filter press	65	Slightly more than double of 150 mm	35.0–46.0 (1.50–2.00)	About half of 150 mm

pressed dewatered cake. That said, the headlosses reported in Table 21.6 seem somewhat underrated; designers probably should use more conservative values or require performance warranties from the pumping-system supplier.

A progressing cavity pump may put significant shear stress on cake, resulting in a thixotropic, shear-thinning behavior that decreases the fluid's apparent viscosity, when compared to cake transported by a hydraulically driven reciprocating piston pump. Progressing cavity pumps can handle cakes with low solids contents, while hydraulically driven reciprocating piston pumps handle thicker cakes more reliably. However, a progressing cavity pump may achieve near 100% cavity fill, so it pumps more efficiently than a hydraulically driven reciprocating piston pump. Also, a progressing cavity pump manufacturer typically quotes headlosses in the neighborhood of 1 psi/ft, while hydraulically driven reciprocating piston pump manufacturers typically use 2 psi/ft in their designs.

Density is an important parameter in all solids calculations. Residuals density at a certain temperature may be correlated with solids concentration; this is often helpful in design. Table 21.7 lists data correlating density with solids concentration at unspecified temperatures (presumably near room temperature).

5.4.1 *Example 21.2: Pumping Cake with 28% Solids*

Determine the friction headloss when pumping cake with 28% solids through a 305-mm-diameter (12-in.-diameter), 150-m-long pipeline. This example was adapted from List et al. (1998). Other experimentally determined input parameters include:

TABLE 21.7 Residuals density at various solids concentrations.

Type of residuals	Percent solids	Density (kg/m^3)	Reference
Diluted dewatered solids	3.51	1 015	Spinosa and Lotito, 2003
Settled activated sludge	5.00	1 015	Honey and Pretorius, 2000
Diluted dewatered solids	5.15	1 020	Spinosa and Lotito, 2003
Diluted dewatered solids	6.81	1 026	Spinosa and Lotito, 2003
Diluted dewatered solids	8.40	1 031	Spinosa and Lotito, 2003
Manure 1	9.10	1 037	El-Mashad et al., 2005
Diluted dewatered solids	9.49	1 034	Spinosa and Lotito, 2003
Diluted dewatered residuals	10.5	1 038	Spinosa and Lotito, 2003
Manure 2	10.7	1 044	El-Mashad et al., 2005
Dewatered solids	28.3	1 062	List et al., 1998

- Bulk modulus of cake (K) = 2 550 kN/m^2,
- Cake density at atmospheric pressure (ρ_α) = 1 060 kg/m^3,
- Cake mass flowrate (Qm) = 8.012 kg/s,
- Wall shear stress (τ_w) = 1 468 N/m^2,
- Cake's dynamic viscosity (μ) = 12 kg/m·s, and
- Fluid velocity (V) = 0.1036 m/s.

5.4.1.1 Solution Option 1

Following the methodology described previously in Section 5.4, engineers should use Equation 21.19 to calculate the following dimensionless factor:

$$Z = 0.222$$

They then solve Equation 21.20 by trial and error:

$$\rho^* = 2.15$$

Engineers then use Equation 21.21 to calculate the pressure drop:

$$\Delta P = 2\,933 \text{ kPa (299 m of water column).}$$

5.4.1.2 Solution Option 2

Assuming that the cake is behaving like a Bingham plastic fluid and disregarding the compressibility effect, engineers could use Equation 21.22 to get the following result:

$$\Delta P = 2\,826 \text{ kPa (288 m of water column)}$$

5.4.1.3 Solution Option 3

If the cake's rheological parameters are difficult to determine, an estimated conservative headloss value may be sufficient for design. This practice is valid because cake is transported by positive-displacement pumps that can deliver the same flowrate over a wide range of pressures; it is even more valid when pumping cake short distances through sufficiently large pipelines.

Typical headlosses in cake pumping systems may range between 0.5 and 3.5 psi/ft. Their headlosses are larger (in the high end of the range) when pipe diameter is small (e.g., 6 in.) and fluid velocity is near the maximum recommended (0.08 m/s). In this example, the pipe diameter is sufficiently large (12 in.) and fluid velocity is not much higher than the recommended maximum, so system headlosses should be in the low end of the range. On the other hand, the cake's solids content is 28%, and engineers should be careful not to select a design whose values would be too small. A headloss value of 2.00 psi/ft may be safely assumed.

Also, engineers could take Table 21.4's value for pumping cake with 20% solids through the 12-in. line and double it (to approximate the value for a cake with 28% solids). The estimated headloss for design then becomes 1.56 psi/ft.

Another approach is to extend the 34% and 22% curves of Figure 21.15 to a velocity of 0.1 m/s and record the headloss for each case at that particular velocity (300 m/10 m and 190 m/10 m, respectively). Pumping cake with 28% solids would fall somewhere between the two values. If the median value is selected, the headloss is 245 m/10 m of 6-in.-diameter pipe. For every pipe size increase, headloss is halved. Increasing from a 6-in. to a 12-in. pipe will halve the headloss three times (to 30.6 m/10 m or 1.33 psi/ft).

The headlosses calculated from Options 1 and 2 are 0.86 and 0.83 psi/ft, respectively. These are well below the estimated values in Option 3.

5.5 Line Lubrication for Long-Distance Pumping

A method for reducing pipe-friction losses (mainly for long-distance pumping applications), called *boundary layer injection*, involves injecting a liquid lubricant into the discharge pipe via an annular ring, which distributes the liquid equally around the pipe's perimeter to create a "boundary layer". The lubricant could be water-, polymer-, or oil-based. Field tests indicate that such lubrication can cut discharge pressure up to 80% (List et al., 1998). [For a detailed description of this process, see *Conveyance of Wastewater Treatment Plant Residuals* (ASCE, 2000).]

Consider the experience of a water reclamation facility in Georgia, which was pumping a cake with 25% solids through a glass-lined pipe. The original pressure drop was 67.8 kPa/m (3 psi/ft). After adding a small amount of water (0.5% of total solids flow) via boundary layer injection, the pressure drop lowered to 22.5 kPa/m, according to a study by a manufacturer of hydraulically driven reciprocating piston pumps (Crow and Cortopassi, 1994).

There are two types of boundary layer injection rings: the original configuration, which has individual injection points, and a newer design, in which lubricating media is injected through an annular groove around the pipe perimeter. Typically, the annular groove design is preferred because it requires less lubricant and has minimal effect on the cake's percent solids content (Wanstrom, 2008).

5.6 Controls

Pump controls typically are used to match the pumping and cake-production rates. For example, a progressing cavity pump's speed may be controlled by a variable-frequency drive, while a piston pump's speed is controlled by fluid flow in the hydraulic power unit. Meanwhile, ultrasonic or radar level sensors monitor the level in the cake-collection hopper and send high- and low-level signals to the pump controls, which automatically adjust pump speed to maintain a preset hopper level.

Most plant operators report stable cake production from belt presses or centrifuges; spot checks are used to prevent bridging or clogging. Little operator time is required to adjust the system to match cake-production rates.

Automatic pump controls involving capacitance probes, pressure switches, and no-flow sensors typically have proved unreliable. If the control fails to shut off the pump when the feed hopper holds little cake, the pump can run dry. If it fails to start the pump, cake can spill over the hopper or push back and pack centrifuges or other dewatering units. Either condition can result in expensive repairs and production loss.

5.7 Progressing Cavity Pumps

Progressing cavity pumps have low capital costs and can consistently transport thinner dewatered cakes over short distances. To ensure effective operations, pump manufacturers and treatment plant personnel recommend that design engineers

- Only use progressing cavity pumps to transport cakes containing about 20% solids or less
- Use a large-diameter pipe to reduce friction loss to more suitable levels;
- Limit the pumping distance to 50 m (164 ft) or less;
- Minimize suction piping length or eliminate this pipe;
- Restrict pump rotational speeds to 200 rpm or less (Bourke, 1997);
- Limit the pressure per stage to 52 kPa (75 psi);

- After determining final design criteria, ask pump manufacturers for specific recommendations on pump materials, number of stages, and models; and
- Use either adjustable-frequency or adjustable-hydraulic motor drives.

The expected service life of the pump's stator and rotor assembly is plant-specific, and drops markedly as the cake's solids concentration increases. Other factors that affect equipment wear include pump speed, grit content, running time, and operating pressure. Parts are expensive and labor-intensive to replace (compared to servicing kinetic pumps).

When progressing cavity pumps transport cake containing more than 20% solids, the cake can bridge over the screw auger in the feed hopper. It also can clog the throat section (interface of auger feed screw and rotor and stator assembly of pump). If uncorrected, the problems can cause the pump to run dry quickly, ruining the stator and rotor. To avoid this, designers can put temperature sensors in the stator to monitor for the high temperatures that indicate the pump is running dry.

Depending on the pumping system configuration, bridging could pack a centrifuge bowl or overflow a supply hopper. Some manufacturers use paddle-type bridge breakers to combat this problem (with moderate success). However, they may require intensive maintenance. A bridge breaker can be powered by the pump motor via gears or chains, or by a dedicated motor. Using a dedicated motor lets operators adjust paddle speed independent of the pump. A newer design uses a ribbon auger attached to a plate fixed either to the pump drive shaft or to a separate variable speed drive; it allows the pump to transport cake with higher concentrations of solids (Dillon, 2007). Another design uses a twin screw feeder also powered by a dedicated variable-speed drive (Doty, 2005). Both designs address cake bridging problems and maximize the pump cavity's fill rate.

5.8 Hydraulically Driven Reciprocating Piston Pumps

In the United States and Canada, hydraulically driven reciprocating piston pumps are the standard for transporting high solids and dewatered cake long distances. Municipal treatment plants have used them for this purpose for more than 20 years, and many of the units are still operating. Several manufacturers sell them in the United States.

Developed from concrete pumping technology, a hydraulically driven reciprocating piston pump consists of a twin screw auger feeder, a pumping assembly, and a hydraulic power unit (see Figure 21.16). It can handle both screenings and biosolids containing 5% to more than 40% dry solids that were dewatered via belt presses, centrifuges, plate and frame presses, screw presses, or rotary presses.

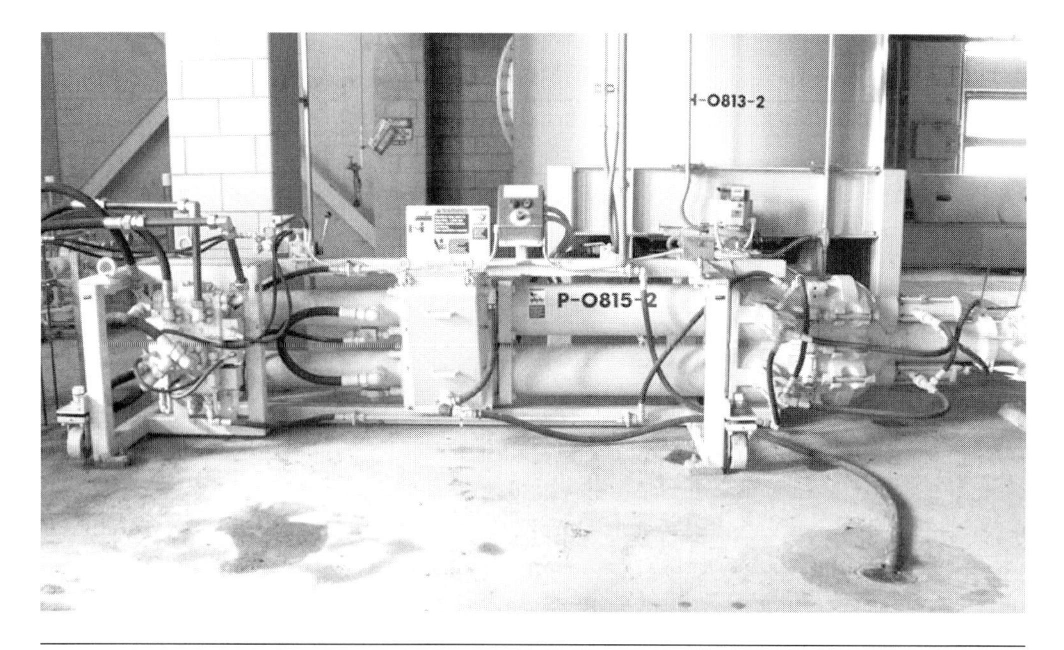

FIGURE 21.16 A hydraulically driven reciprocating piston pump moving a centrifugally dewatered cake (containing 31% solids) at the Norman Cole Jr. facility in Lorton, Virginia (courtesy of Schwing Bioset, Inc.).

The pump's principal advantage is that it can move high solids materials. Hydraulically driven reciprocating piston pumps have higher capital costs than other pumps, but can move very thick dewatered cakes that other types of pumps cannot.

The pump has two product-delivery cylinders with pistons powered by two isolated hydraulic-drive cylinders (rams). The delivery cylinders are synchronized so while one is being filled with dewatered cake, the other is delivering cake to the discharge line. This reduces the pulsing effect of one cylinder and piston, maintaining essentially uninterrupted flow.

Nearly all the hydraulically driven reciprocating piston pumps used at U.S. sites have twin screw auger feeders. These feeders develop 34 to 206 kPa (5 to 30 psi) of pressure in the pump assembly's charging unit. This pressure helps push cake into the emptying cylinder while the piston returns to its starting position. The feeders typically are driven by a hydraulic motor but also can be used with electric motor drives.

The pumping assembly's charging unit typically has either a poppet valve or a transfer tube. Each delivery cylinder has a suction intake and discharge exhaust poppet valve. The valves or tubes are hydraulically driven and synchronized to the piston strokes; they permit one cylinder to fill while the other discharges. Poppet valves provide a positive shutoff to prevent backflow, cost less, and need less maintenance than transfer tubes. The poppet valve also can be equipped with an internal flow monitor that will measure the volume of pumped biosolids to within ±5% (Wanstrom, 2008).

Hydraulically driven reciprocating piston pumps can run dry for indefinite intervals at slightly faster wear rates but without catastrophic damage. So if bridging or clogging occurs, plant operators have time to react before severe problems develop. A water-filled isolation box (e.g., water box) between the hydraulic and delivery halves of each cylinder allows this capability. The water cools the connecting rods, flows into the delivery cylinder, and lubricates it as the delivery piston moves forward on its discharge stroke.

5.8.1 Operating Experience and Design Considerations

Based on more than 20 years of operating experience with hydraulically driven reciprocating piston pumps at U.S. and Canadian wastewater treatment plants, design engineers should

- Minimize line pressure in the piping system when pumping material containing more than 25% solids,
- Use a boundary layer injection system in long-distance or high-solids applications,
- Maximize volumetric efficiency (see Section 6.8.3), and
- Minimize stroke rate (see Section 6.8.2).

Hydraulically driven reciprocating piston pumps are available with capacities up to 1 500 L/min (400 gpm) and discharge-pressure capabilities up to 13 800 kPa (2 000 psi). There also are a wide range of suction feed hopper-to-pump configurations. In addition, hydraulic power units are available in broad output ranges, depending on the discharge-line pressure required.

When designing such pump systems, engineers should discuss suitable equipment sizing, features, and options with manufacturers and staff at installations with similar requirements. They should ensure that the piping design and components (valves, etc.) are suitable for high-pressure service. When sizing the pump(s), they should

- Determine the cake-production rate;
- Estimate the pump's volumetric efficiency given the cake characteristics;
- Reduce stroke speed in accordance with the type of pump service (e.g., intermittent or continuous) expected;
- Examine the pump curves and check the stroke speed turndown ratio; and
- Choose a pump based on the volumetric-efficiency and service factors needed.

5.8.2 *Cake Production Rate*

Cake-production rate depends on installation-specific requirements, including the

- Number of hydraulically driven reciprocating piston pumps and dewatering machines;
- Operating schedule of both dewatering equipment and pumps (e.g., 24-hour service or single shift);
- Constraints downstream of piping (e.g., furnace capacity, storage capacity, or trucking schedules);
- Variable capacity requirements imposed on the pumps by dewatering processes; and
- Standby considerations during maintenance or emergencies.

5.8.3 *Volumetric Efficiency*

Volumetric efficiency is the ratio of solids volume pumped per piston stroke to the total volume displaced per piston stroke. If a hydraulically driven reciprocating piston pump were pumping water or residuals containing 1 to 4% solids, its volumetric efficiency would be essentially 100% because wastewater is nearly incompressible. This behavior is typical of a true Newtonian fluid.

Dewatered cake, however, neither physically resembles nor behaves like a true Newtonian fluid. It typically contains air, other entrained or dissolved gases, and concentrated organic material, all of which are compressible. So, when the piston begins applying pressure, the cake tends to compress. Until squeezed against the downstream resistance, the cake does not move forward with the pumping cylinder. When the cake finally moves forward, the piston already has displaced a certain volume of the cylinder. This displaced volume is part of the "lost" volumetric efficiency; the rest is due to the inability to completely fill the cylinder as the piston returns to its starting position. Even with the slight pressure provided by a twin screw auger or conical plow feeder [34 to 206 kPa (5 to 30 psi)] and the partial vacuum in the cylinder, dewatered cake resists moving into the cylinder bore. Such resistance typically increases as cake dryness increases, further lowering volumetric efficiency.

Using a pressure sensor in the transition between the twin screw auger and poppet housing can ensure cylinder-filling efficiency. This sensor will monitor pressure in the transition and, via a programmable logic controller, automatically increase or decrease auger speed to maintain a preset pressure. So, regardless of pump speed or fluctuations in solids concentration, optimum pressure is maintained on the suction poppet to promote the highest filling efficiency possible for a given material (Wanstrom, 2008).

Meanwhile, designers must account for volumetric-efficiency loss when sizing a hydraulically driven reciprocating piston pump. There is no theoretical model for predicting volumetric efficiency, but it typically ranges from 60 to 90%. Once design engineers know or can estimate cake characteristics, they should ask manufacturers to recommend an appropriate volumetric efficiency. As a rough, conservative estimate, a volumetric efficiency of 70% can be used for pumping dewatered cake containing 20 to 30% solids.

5.8.4 *Service Factor*

As with most wastewater treatment equipment in continuous service, operating hydraulically driven reciprocating piston pumps at lower speeds makes them more reliable and extends equipment life. So, design engineers should limit the pump-stroking speed (strokes per minute) to 50% of the maximum recommended for intermittent operation or 75% of the maximum recommended for continuous service, whichever is less.

5.8.5 *Hydraulic Power Unit Sizing*

Power unit sizing is principally a function of the pump's discharge pressure and the hydraulic oil flow needed to achieve the desired solids pumping rate. If pumping tests cannot be made and data are not available from other sources, design engineers should ask manufacturers for specific recommendations and rate the power unit conservatively.

5.9 Conveyors

Conveyors typically move wet or dry solids (e.g., primary grit, screenings, and dewatered cake) that are not easily pumped. Municipal treatment plants typically use either belt or screw conveyors.

5.9.1 *Belt Conveyors*

Belt conveyors move material on top of a moving, flexible belt (see Figure 21.17). Such belts typically are supported by rollers spaced 0.9 to 1.5 m (3 to 5 ft) apart on the carrying side and about 3 m (10 ft) apart on the return side of the conveyor. The rollers on the carrying side are called *load-side rollers*; those on the return side are called *idlers*. To increase capacity, load-side rollers may be angled so the belt will form a concave carrying surface.

The belt is driven by one or more drive drums or pulleys connected to a motor via a belt or chain drive. In simple conveyor systems, the *drive pulley* is located at the discharge (head) end of the belt and the *tail pulley* is at the loading end.

FIGURE 21.17 Belt conveyor (troughed with idlers) transporting screw press dewatered, anaerobically digested solids at the City of Tallahassee's T. P. Smith facility in Florida.

The belt must maintain a minimum tension to reduce sag between carrying idlers, provide contact force, and prevent slippage at the drive pulley. This tension can be maintained by several *take-up devices*, including a weighted pulley (called a *gravity take-up*), a spring-loaded pulley, or a screw adjustment for pulley position. The least costly option is a screw take-up on the tail pulley; it typically is used in conveyors less than 90 m (300 ft) long.

5.9.1.1 Belt Conveyor Applications

Conventional belt conveyors move solids via a continuous loop of reinforced rubber belt. They typically transport relatively dry material (15% or more solids). For this method to be economical, solids must be dry enough not to flow freely or seek a constant level (like a liquid does). Solids with a high angle of repose (i.e., the slope of the solids pile when measured from the horizontal) are suitable for transport via belt conveyors. Digested and belt-pressed primary and secondary solids can have an angle of repose of 40° or more, while wet concrete has an angle of repose of less than 25°. Because belt movement vibrates the material, design engineers must consider the solids' flow tendencies when deciding whether a belt conveyor is suitable. The angle that a pile of material retains while moving is called the *surcharge angle*. Belt-pressed solids may have a surcharge angle of more than 30°, while concrete's angle typically is less than 5°. That's why engineers should determine a material's characteristics before deciding which conveyor to use.

The transport distance and elevation change also influence the choice of conveyor. Belt conveyors have been used to move mining ores and construction spoil solids more than 14 km (8 mi), but in typical treatment plants, the distance could be less than 200 m (660 ft). If the distance is less than 6 m (20 ft), other conveyors may be more suitable.

Conventional belt conveyors are limited by both the rate of elevation change and horizontal direction changes that require multiple belts. The conveyor's maximum incline depends on the material involved and belt speed required. Faster speeds allow for higher angles so long as the speed exceeds the rate at which material flows or rolls back down the incline. However, faster speeds also increase O&M costs because they increase friction and shorten belt life. When moving dewatered cake, a belt conveyor's maximum incline angle is limited to about 15 to 20° above horizontal. The maximum incline is much less for solids that are watery or tend to flow easily.

Elevation gains also can be limited by curvature radius as a horizontal belt becomes an inclined one. The radius must be long enough so the belt will not lift from the idlers under any operating condition. Depending on the specific design, this radius could be 15 to 76 m (50 to 250 ft) or more. So, engineers need to consider an existing plant's physical dimensions when deciding what type of conveyor to use.

Belt conveyors have a low cost per linear meter (foot) of transport distance, but they may require significant space and be maintenance intensive. They also can be a source of odors. In addition, if the conveyor will be installed outside, weather conditions can affect operations.

5.9.1.2 Belt Conveyor Design and Operation Considerations

When considering a belt conveyor for a new or existing plant, design engineers should begin by establishing the following criteria:

- Material characteristics (e.g., angle of repose and surcharge, degree of matting or stickiness, average density, and range of variation of these characteristics);
- Material volumes and transport rates (i.e., daily or weekly variations in solids rate and hours of operation) so they can determine conveyor capacity;
- Belt construction material (acid, oil, and abrasion resistance); and
- Conveyor layout and power requirements so they can determine belt width and speed (conveyor activity); loading arrangement (chutes or conveyor skirt-boards); curve radius, incline angles, and total elevation gain (multiple conveyors may be warranted if these factors are limiting); idler type, spacing, and pulley and take-up arrangements (which influence belt friction and power requirements); and motor horsepower and belt tension.

The Conveyor Equipment Manufacturers Association (CEMA) publishes a handbook that includes procedures for establishing such criteria and sizing belts (CEMA, 1979). First, however, engineers must know the characteristics of the material to be conveyed. Some treatment plant solids are sticky, for example, so the belt should be cleaned to prevent spillage on return runs and the consequent loss of drive-pulley friction. Other treatment plants may have site-specific problems, such as water breakout, odors, or spillage.

The U.S. Environmental Protection Agency offers the following guidelines for problems unique to treatment plant solids (U.S. EPA, 1979):

- Belt transfer points should have both minimum drop heights and skirtboards with wipers to minimize splashing and spillage.
- Belt cleaning is potentially troublesome. Counterweighted rubber scrapers below the head pulley have been ineffective and required intensive maintenance. Scrapers with multiple "fingers" and adjustable tensions are suggested. Another option is a water spray followed by a rubber scraper (if the water can be collected and disposed easily).
- Design engineers should avoid accessories (e.g., snubber or counterweight pulleys) that touch the dirty side of the belt. Snubbers are pulleys positioned to increase the angle of contact between the belt and the drive pulley, thereby increasing friction and reducing drive slippage. Instead of snubbers or gravity take-ups, designers should use manual screw take-ups and, if necessary, multiple shorter belts.

- When designing conveyors, design engineers should include housekeeping facilities (e.g., frequent hose stations); oversized floor or paving drains with exaggerated grades below the conveyor; and nonskid tread plates rather than grates).

Because a wastewater treatment plant is a humid or wet environment and solids typically are corrosive and abrasive, engineers need to design belt conveyors carefully. Conveyor framing should be made from corrosion-resistant materials (e.g., 6061-T6 aluminum alloy). Idlers can be made of neoprene or PVC, and cable-supported neoprene rollers can be used for troughed sections. Roller bearings should be sealed with external grease fittings. Drive chains and motors require removable splash guards to protect them from spillage. Belt materials should include abrasion- and oil-resistant covers.

To prolong belt life, designers should set the belt's actual running tension conservatively below its rated tension and check the loaded conveyor's initial tension to avoid overstressing the belt. A vulcanized belt splice provides longer life than mechanical joints. Slower belt speeds also typically lengthen belt life, so about 30 m/min (100 ft/min)—approximately 50% of CEMA's maximum speed guideline—is suggested.

When designing conveyor sections that are outside buildings, engineers should provide for weather and wind protection. At a minimum, they should provide a half-diameter rain cover; however, a three-quarter cover with open access on the downwind side prevents wind-induced spillage and belt-training problems while allowing access to parts for maintenance. If odors must be controlled, designers can completely enclose the conveyor and provide ventilation. The enclosure must include hinged or easily removable partial cover plates to allow access for maintenance, cleaning, and periodic observation of the conveyor.

5.9.1.3 Special Belt Conveyors

Manufacturers have developed belt conveyors that overcome some of the limitations previously mentioned. For example, conveyors with cleats, buckets, or sidewalls attached to the belt can move material up steeper inclines. One patented conveyor allows both horizontal and vertical curves. Another uses two flat, converging belts to completely enclose the material and to permit steady inclined or vertical lifts. Design engineers considering one of these conveyors should discuss the application with various manufacturers and check similar installations to compare operating costs and avoid potential design problems.

For example, cleated belt conveyors have components and design considerations similar to those of flat ones, but their belts are actually a series of flexible overlapping pockets (cleats) connected individually to a drive chain underneath (see Figure 21.18). The cleats allow material to be transported up steeper inclines without

FIGURE 21.18 Cleated belt conveyor.

backflow and permit systems to have both vertical and horizontal curves. Design engineers could create a helix to maximize elevation gain in a relatively small space. During offloading, the cleats flatten out over the head pulley and assume the shape of a conventional belt.

However, cleated belts are substantially more expensive and more difficult to clean after offloading. Rotating brush and spray cleaners typically are required to prevent excessive spillage on the return run of the belt. Also, because the drive is more complex, it and the cleats wear and corrode more rapidly than conventional belt conveyors (U.S. EPA, 1979). In addition, changing a cleated conveyor belt is more costly and time consuming because each pocket must be removed and replaced individually.

5.9.2 *Screw Conveyors*

Screw conveyors push material via a helical blade (flight) mounted in a U-shaped trough or enclosed in a tube. The flights may be attached to a center shaft, or the screw may be shaftless. In shafted conveyors, a drive mechanism turns the center shaft, which is supported by the end bearings and intermediate hanger bearings needed to reduce shaft deflection. Both shafts and flights can be tapered.

Flights are manufactured in a wide variety of designs and can have full or partial cross-sections (or both). Two flights that are cut, folded, or otherwise shaped can mix or fold the material during transport. The *pitch*—the horizontal distance between flight blades—can vary along the shaft length.

5.9.2.1 *Screw Conveyor Applications*

Screw conveyors (augers) can be used to move solids horizontally, vertically, or along an incline. When properly designed and used, they are an economical and reliable transport method. Before selecting a screw conveyor, design engineers must evaluate the material to be moved. Its water content and flowability are particularly important for inclined and vertical conveyors.

Standard screw conveyors work best when moving material horizontally over a relatively short distance. Although some operating screw conveyors are more than 150 m (500 ft) long, most of the ones in treatment plants are 9 to 12 m (30 to 40 ft) long. The conveyors are available in sections that are about 3 to 4 m (10 to 12 ft) long, depending on shaft and flight size. Longer sections are either custom made or formed by joining standard lengths. They typically need intermediate hanger bearings to reduce shaft deflection.

Inclined screw conveyors are less efficient than horizontal ones and have different design criteria. For every degree of elevation beyond 10 degrees, a screw conveyor's capacity declines about 2% and its speed must increase significantly to compensate. Inclined and vertical conveyor speeds typically are 200 rpm or more, while a horizontal conveyor's velocity is 20 to 40 rpm in an abrasive application. Inclined conveyors also use different flight designs than horizontal ones.

Vertical screw conveyors are designed for uniform flight loading to avoid packing or binding the material. Like inclined conveyors, vertical units have faster shaft speeds and the screw's centrifugal action helps provide lift. These systems typically include special horizontal feeders. Engineers should consult with manufacturers when designing lifts taller than 6 m (20 ft). Although at least one manufacturer allows vertical lifts up to 21 m (70 ft) high, another recommends a practical limit of 7.6 m (25 ft).

Screw conveyors typically move grit or solids horizontally. Inclined conveyors are sometimes used for dewatered cake. Screws also are used as truck-loading hopper dischargers to spread a load across the entire truck trailer. Manual or automatic knife gates on the conveyor bottom function as multiple discharge points. Screws can control solids feeding from hoppers to either belt conveyors or the suction side of cake pumps.

5.9.2.2 Screw Conveyor Design and Operation Considerations

When designing screw conveyors, engineers first must define material properties, volume, and variability. Conveyor capacity is a direct function of screw speed, flight size or diameter (assuming shaft size remains fixed), and amount of trough loading. Conveyor flight pitch and any folding, cutting, or other special flight designs also affect capacity. Some manufacturer catalogs include tables and charts that help design engineers determine preliminary conveyor size based on a range of these variables. Designers should consult with manufacturers when developing a design for a specific application.

Some design criteria specific to treatment plants deserve consideration. For example, designers should avoid screw conveyors if the residuals contain sticks, large objects, or rope-type materials. Enclosed screw conveyors can reduce or eliminate spillage and housekeeping problems but are slightly more susceptible to jamming and are difficult to access for maintenance. For sticky solids like dewatered cake, a designer should avoid intermediate support or hanger bearings because they could cause plugging as material packs against them. Larger shafts, heavier shaft-wall thicknesses, or both allow greater screw lengths between support bearings; typically, enlarging the shaft is more effective than increasing the shaft wall thickness. If intermediate bearings are unavoidable, the flight design near the hanger can be modified to minimize the packing problem.

Other design criteria include construction materials and drive configuration. For example, a conveyor that will transport abrasive and corrosive materials (e.g., dewatered cake and grit) should have a flight facing made of hardened materials. Steel flights with hot-dipped galvanized troughs have been used successfully for dewatered cake. Where exposed to residuals, outlet knife gates should have stainless steel parts because any free water released from the solids will collect on the parts.

Ideally, the conveyor drive will be mounted at the unloading end so the shaft is in tension during operation and will not buckle during a jam. The motor can be connected directly or via a belt or drive chain. If the conveyor's daily capacity varies significantly, designers can use variable-speed drives to match capacity to transport requirements. End bearings should be heavy-duty roller bearings located outside the conveyor. Design engineers also typically specify shaft seals with a compression-type packing gland to prevent abrasive material or corrosive liquids from migrating to the outside of the conveyor or to the shaft bearing.

If the material packs, sticks, and has a high angle of repose (e.g., dewatered cake), the feed portion of the conveyor deserves special consideration. If the screw is fed from a hopper above, the hopper's sides must be steep enough and its opening large enough to prevent material from bridging across it. Exact figures depend on the cake, but as a general guideline, the hopper wall should be no more than 30 to 35 deg from the vertical and the bottom opening area be approximately 1 m^2 (10 ft^2), with the smallest dimension about 0.6 m (2 ft) long. To help remove material evenly across the hopper bottom, the flight diameter should gradually shrink as it travels across the screw feed area or the flight pitch should increase.

When used properly, screw conveyors have fewer O&M requirements than a belt conveyor. Because they can be completely enclosed, screw conveyors have substantially fewer housekeeping and odor-control requirements. Any intermediate hangers should have a hinged or other easily removed cover for bearing inspection or replacement. If odor control is a significant problem, the screw conveyor can be connected to a vent system at a point just past the discharge end.

Access to the screw conveyor for inspection, maintenance, or replacement can be from above or below, depending on installation requirements. Typically, the screw conveyor cover is attached in bolted sections that can be removed as required. For more frequent access to certain areas, such as intermediate hangers, the cover can be hinged on one side. A variety of easily removable cover arrangements are available. Access from below can be provided by removing trough sections or by installing special hinged trough sections available from most manufacturers.

Designers should provide hose stations and oversized drains so staff can clean up the conveyor when part must be replaced. Inclined or vertical conveyors should have a low-point drain so any backflow liquid can be drained manually.

5.10 Standby Capacity

Design engineers should consider several factors when determining the need for standby transport capacity. These factors include the function involved, plant size, anticipated service period, repair time, and arrangement of units. At larger treatment plants, dewatering operations are often critical and cannot be out of service for long periods, so cake pumping or conveying systems should be designed with standby or quick-replacement capability.

If single units are used, they should be heavy duty and quickly repairable, preferably in place. Spare parts should be readily available. Design engineers should ensure that the original pump supply includes adequate spare parts.

6.0 DRIED SOLIDS STORAGE

6.1 Design Considerations

Dried solids typically are stored either onsite or at a land-application site before disposal or beneficial use. They may be stored in stockpiles or silos.

Because dried solids contain a significant amount of combustible organic material that can be released as dust, temperature control is important. If silos are used, engineers should design them to promote cooling and maximize heat dissipation. Therefore, tall, narrow silos are better than wide ones. Narrow silos also make fires easier to control. However, if the silo is too narrow, it will make relief venting problematic. If multiple silos are used, there should be procedures to ensure that they are emptied cyclically to avoid exceeding safe residence times. Also, designers need to consider the stored product's thermal stability in case a prolonged plant shutdown or silo blockage occurs.

6.2 Safety Issues

It is critical that dried solids be stored safely. Dried solids have self-ignited on several occasions, and in at least one instance, resulted in the discharge of solids to a surface waterbody. In other cases, the dried particles have caused explosions.

If the storage area's temperature rises above a critical point, dried biosolids can begin to self heat. Design engineers can calculate the critical temperature using isothermal basket tests and considering the effects of storage volume and residence time. In England, wastewater treatment professionals performed a series of tests on 1 m^3 of biosolids and found that the self-heating temperature was typically above 60°C (HSE, 2005). When they extrapolated the data to 27 m^3 of biosolids, however, the critical temperature range dropped to between 50°C and 60°C. Results depend on the actual product and contaminants (e.g., oil), which can lower the critical temperature. Therefore, Health and Safety Executive recommends that biosolids should be cooled to no more than 40°C before being sent to a silo. (If the silo is particularly large, the critical temperature may be even lower.) Controlling biosolids temperature not only prevents silo fires but also avoids self-heating further along in the biosolids use or disposal process. Biosolids that are hotter than the critical temperature must not be removed from the treatment plant until they are so cool that there is no risk of a fire either in transit or at the use or disposal site. There are coolers are manufactured specifically for biosolids, and design engineers may want to consider using them in larger installations.

Dust generation is another issue, because biosolids dust can be an explosion hazard. Biosolids may generate little dust if the material is within specifications and the transport method minimizes attrition. However, if the storage silos could contain significant levels of dust, design engineers should try to minimize any potential explosion. The Occupational Safety and Health Administration requires that silos be designed to either contain the maximum explosion pressure or include passive explosion-relief vents (1995 Hazard Information Bulletin). If explosion-relief vents are used, it is important that they discharge to a safe area—preferably to a outdoor location away from normal working areas.

If biosolids are not stabilized before drying, water condensation may lead to bacterial decomposition inside the silo. Bacterial activity produces heat, which could cause a silo fire. Also, wet pellets may become sticky and difficult to handle. To minimize condensation, designers should ventilate the silo via small volumes of dehumidified air or larger volumes of atmospheric air.

Because fires can be a concern in storage silos, design engineers should include systems to identify and contain them. Multi-point temperature probes can monitor stored material, but they provide only localized measurements and so may miss a hot area. A biosolids fire will produce carbon monoxide and consume atmospheric oxygen, so engineers should include carbon monoxide monitoring in their silo designs. At first, the burning material will only produce small quantities of carbon monoxide, which may be further diluted by aspirated air, so engineers must set the detector to identify low carbon monoxide levels. The slow initial exothermic reaction will be followed by a rapid exothermic reaction producing large quantities of carbon monoxide, so engineers should determine the alarm set point for carbon monoxide detection (this typically is on the order of 100 ppm).

Spraying water into a burning silo may only produce a surface cake that bars further water penetration. Designers also should consider using an inert gas to contain a fire. The Occupational Safety and Health Administration specifically requires that all dried, hot biosolids be transported and stored in a nitrogen inert atmosphere that contains less than 5% oxygen. While this will help prevent fires, injecting an inert gas into a burning silo will not necessarily extinguish one. Such injections may have limited effect; it will prevent further propagation but thermal currents may divert the gas away from the hottest parts of the stored material. The temperature will drop right after the cold gas enters the silo, but this does not mean the fire has been extinguished. So, design engineers should include provisions for monitoring temperatures over time to determine whether a fire has been brought under control. Such "worst case" scenarios need to be considered in both design and operations.

7.0 DRIED SOLIDS TRANSPORT

7.1 Belt Conveyors

Belt conveyors are one of the most widely used and efficient means of transporting bulk materials. A chute deposits material onto the top of the belt at one end, and the belt transports it to the other end, where the material is discharged into another chute.

Belt conveyors range from 356 to 1 524 mm (14 to 60 in.) wide and can accommodate a broad range of capacities, speeds, and distances. The smallest belt conveyors can handle up to 76 kg/s (30 ton/hr) and operate at speeds up to 100 m/min (325 ft/min). The largest ones can handle up to 1 134 kg/s (4 500 ton/hr) and operate at speeds up to 200 m/min (650 ft/min). They work well transporting material 6 to 200 m (20 to 660 ft); if the material needs to move less than 6 m (20 ft), other conveying devices are preferred.

7.1.1 Belt Conveyor Applications

Belt conveyors are appropriate for transporting dried solids because they are gentle, efficient, and durable. They also prevent material degradation. When installing a belt conveyor, the belt should be angled to form a trough to prevent material from rolling off the belt. Belts used to transport dried solids typically are no more than 1 016 mm (40 in.) wide.

While belt conveyors are preferred when transporting dried solids horizontally, other conveyors should be considered when moving them vertically. Typically, a belt conveyor should not be inclined more than 20 deg when transporting dried solids—especially dried biosolids pellets, which tend to be free flowing. If dried materials must be moved at steeper angles, consider using sidewalls, cleats, and/or cover belts.

Although belt conveyors may have higher capital costs, they can outlast most other conveyors. In addition, they typically need less energy to move material than augering or dragging it.

7.1.2 Belt Conveyor Design and Operation Considerations

Several important material characteristics and design criteria should be considered when designing belt conveyors. Engineers can find procedures for establishing design and other criteria and belt sizing in the Conveyor Equipment Manufacturers Association's *Belt Conveyors for Bulk Materials* (1979). But first, they should determine the following:

- The material's bulk density, lump size, temperature, and moisture content;
- The conveyor's peak capacity (metric tons per hour);
- Conveyor layout, length, and elevation gain; and
- Loading and unloading requirements.

This information will enable design engineers to calculate the conveyor's size (width) and belt speed. They also can determine the energy required to move material on the conveyor—taking into account the transport distance, any elevation gain, and any friction loads induced by required belt accessories and cleaners.

There are several types of conveyor belts:

- Solid-woven cotton (layers of woven threads that may be used with or without treatment);
- Solid-woven PVC (a single-ply made from nylon and polyester and impregnated or coated with PVC);
- Stitched canvas [separate plies of fabric (usually cotton) stitched together and treated];
- Multiple ply (made from three or more plies of fabric bonded together by elastomeric material);
- Reduced ply (one or two plies of nylon or polyester); and
- Metal (wire or steel band).

Multiple-ply belts convey dried solids effectively. Three to five plies typically are sufficient for narrow belts, while wide belts require anywhere from 6 to 16 plies. The belts may be hot-vulcanized or mechanically spliced in the field. Engineers should ask equipment manufacturers about standard design features.

Carrying idlers typically consist of three sealed cylinders made of painted steel. The center cylinder (roller) is horizontal, while the two side rollers are inclined to force the belt into a trough shape. Return idlers typically consist of six to eight equally spaced neoprene or PVC disks mounted on painted steel rods beneath the conveyor's deck pans. Idlers are spaced to minimize belt sag but are closer together at loading points to minimize the related impact on the belt. Typically, they are about 0.9 to 1.5 m (3 to 5 ft) apart on the carrying side of the conveyor and about 3.1 m (10 ft) apart on the return run.

A *take-up device* is an adjustable pulley or roller arrangement that compensates for belt-length changes caused by wear, stretching, and varying loads. Adjustments may be manual or automatic via counterweights, springs, pneumatics, or hydraulics. Tail pulleys are often used; these devices are on the end of the conveyor opposite the drive pulley. Designers should avoid counterweight pulley take-ups because they require roller contact on the carrying side of the belt, thereby increasing housekeeping requirements.

There are many drive arrangement options. In most belt conveyor systems, the drive is at the discharge (head) end of the belt to limit the tension required and enhance belt service life. The most common drive is a gear motor, a system in which one or more drive pulleys or drums is connected to the motor via a chain, belt, or direct drive.

When transporting dried solids, belt speed typically should be no more than 40 m/min (125 ft/min) to keep material stationary, minimize belt wear, and increase belt life. Moving dried materials more rapidly can generate excessive fugitive dust and associated housekeeping demands.

Belt conveyors have options that allow for design flexibility. If the material must be weighed, for example, a scale can be mounted on the conveyor so operators can monitor material flow or control chemical addition. A single conveyor can have multiple loading chutes. Plow stations may be installed to divert material off the belt at intermediate points. Also, the conveyors may be designed with reversible motors.

Odors are a significant disadvantage of using a belt conveyor to transport dried solids. Enclosing the conveyor can keep dust and odors in and weather out. Also, proper and regular housekeeping is essential in reducing the buildup of dust and other materials.

A clean discharge is vital to the O&M of a belt conveyor. If the belt does not discharge cleanly, then when the carrying side of the belt contacts the return idlers, it may deposit material on them, causing excessive wear and extensive cleanup requirements. So, design engineers should provide a belt-cleaning system. Several cleaning systems are commonly available, including urethane cleaning blades (also available in tungsten carbide, ceramic, stainless steel, or rubber); brush cleaners; scrapers; and spray-wash systems. If using scrapers, designers should avoid the counterweighted rubber type mounted below the head pulley; they are ineffective. Scrapers with multiple "fingers" and adjustable tensioners are preferred.

All conveyors must include emergency stop switches and pull cables. They also should have speed switches—especially when system control is highly automated. Misalignment switches also may be desirable to indicate potential problems promptly.

Design engineers should make provisions for oiling and greasing all pulley and sheave axles or shafts. Facilities using belt conveyors also should include hose stations and drains to permit frequent washdowns of the area.

7.2 Screw Conveyors

Screw conveyors are one of the most economical options for moving dried biosolids. Ranging from 150 to 600 mm (6 to 24 in.) in diameter, these conveyors use a shaft-mounted spiral helix or a self-supporting helix to move material in a covered trough. They typically are installed horizontally or at inclines less than 45 degrees. Inlet and discharge openings may be located where needed; the system typically is supported at the ends, loading points, and intermediate points by either feet or saddles.

Shafted screw conveyors typically are available in 3- to 4-m-long sections that are bolted together. Internal intermediate hangers provide support, maintain alignment, and serve as bearing surfaces.

Shaftless screw spirals typically are furnished in one piece (either fabricated or welded) that can be up to 50 m long. They rely on polyethylene liners or steel wear bars for intermediate support.

Conveyor length is limited by the center shaft's torsional capacity, as well as the couplings (shafted screws) or spiral (shaftless screws). So, screw conveyors typically are no more than 14 m long, unless exceptional design considerations are addressed.

7.2.1 *Screw Conveyor Applications*

Screw conveyors can be used to feed, distribute, collect, or mix dried materials. They also can heat or cool the material while transferring it.

If material degradation is a concern (e.g., dried biosolids slated for beneficial reuse), shaftless screw conveyors are better than shafted screw conveyors because their slow turning spiral and continuous support along the trough generates less dust.

7.2.2 *Screw Conveyor Design*

When designing screw conveyors, engineers should begin by identifying capacity requirements, origination, and terminus points, and conclude with determining components and final layouts. Following are brief summaries of each design step.

7.2.2.1 Known Factors

Design engineers should identify the material's characteristics, such as type, incidence, and lump size, as well as the conveyor's capacity requirements (cubic meters per hour), transport distance, and elevation change. They also should note whether any other operation, such as mixing or cooling, is required during transfer and choose an appropriate conveyor (e.g., one with a ribbon flight or jacketed trough).

7.2.2.2 Materials Classification

Conveyed material should be classified according to CEMA standards (see Tables 21.8 and 21.9). The dewatered grit, dried biosolids, and other dried granular solids typically encountered at wastewater plants would be classified as $C_{1/2}35$ [$C_{1/2}$ = 13 mm (0.5 in.) and under granules; 3 = average flowability; 5 = mildly abrasive]. Designers also should identify other miscellaneous properties that may affect conveyor design, such as whether the sludge is especially corrosive.

7.2.2.3 Determine Conveyor Diameter and Speed

Conveyor manufacturers publish charts and tables listing the capacities of various screw conveyors based on trough loadings and conveyor speeds [revolutions per minute (rpm)]. Shafted screw conveyors transporting dried solids are typically lightly loaded (trough loads less than 30%), while shaftless ones can handle loads up to 70% or 80%. Also, the area of a shafted screw conveyor's trough is equal to the area of the screw spiral minus the shaft area. The area of a shaftless screw conveyor's trough is equal to that of the screw spiral. In other words, a shaftless screw conveyor has more dried

TABLE 21.8 Material classification code chart, example 1.

	Material characteristics included	
Major class	**Density**	**Bulk density, loose**
Size	Very fine	No. 200 sieve (0.074 mm) and under
		No. 100 sieve (0.15 mm) and under
		No. 40 sieve (0.41 mm) and under
	Fine	No. 6 sieve (3.35 mm) and under
	Granular	No. 6 sieve to 13 mm
		13 to 80 mm
		80 to 180 mm
	Lumpy	0 to 410 mm
		Over 410 mm to be specified
		X = actual maximum size
	Irregular	Stringy, fibrous, cylindrical, slabs, etc.
Flowability	Very free flowing	
	Free flowing	
	Average flowability	
	Sluggish	
Abrasiveness	Mildly abrasive	
	Moderate abrasive	
	Extremely abrasive	
Miscellaneous properties or hazards	Builds up and hardens	
	Generates static electricity	
	Decomposes-deteriorates in storage	
	Becomes plastic or tends to soften	
	Very dusty	
	Aerates and becomes fluid	
	Explosiveness	
	Stickiness-Adhesion	
	Contaminable, affecting use	
	Degradable, affecting use	
	Gives off harmful or toxic gas or fumes	
	Highly corrosive	
	Mildly corrosive	
	Hygroscopic	
	Interlocks, mats or agglomerates	
	Oil present	
	Packs under pressure	
	Very light and fluffy—may be windswept	
	Elevated temperature	

TABLE 21.9 Material classification codes for typical wastewater residuals.

Material	Material characteristics		
	Weight, kg/m^3	Material code	Trough loading, %
Dewatered cake (compactable)	800–960	E-450V	15
Grit	1 600–2 180	B6-47	15
Solids, dried	640–800	E-47TW	15
Solids, dry ground	720–880	B-46S	30

Material: dried sludge solids

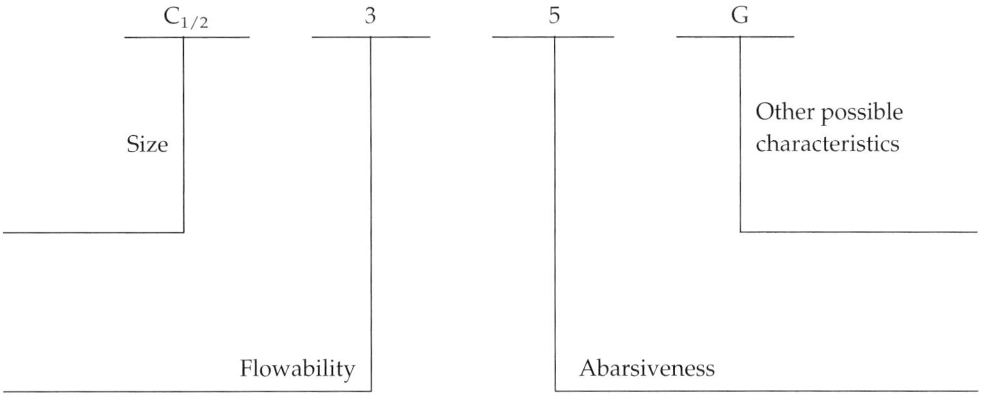

$C_{1/2}$ 3 5 G

Size

Flowability

Abarsiveness

Other possible characteristics

solids capacity at slower rotational speeds than an identically sized one with a shaft (see Table 21.10).

If the application involves inclination, design engineers should consult with the manufacturer.

Typically, a conveyor's screw diameter is equal to the pitch of its spiral (helix), so a 300-mm-diameter screw would have a 300-mm pitch. In theory, one rotation of the screw will transport the material one pitch (minus some inefficiency), meaning a conveyor operating at 30 rpm would move material 30 pitch lengths in 1 minute.

TABLE 21.10 Capacity differences between shafted and shaftless screw conveyors.

Type	Size	Speed, r/min	Trough area, cm^2	Loading	Capacity, m^3/h
Shafted	300 mm	30	650	30%	35
Shaftless	300 mm	30	700	70%	88

7.2.2.4 Compare Conveyor Diameter to Lump Size

Conveyor diameters typically are at least four to six times the diameter of 75% of the lumps encountered at the wastewater treatment plant. Design engineers should confirm that the conveyor diameter can move the maximum lump size safely.

7.2.2.5 Determine Conveyor Horsepower

Drive horsepower is calculated as a function of capacity (cubic meters per hour), density (kilograms per cubic meter), transport length (meters), diameter (meters), flighting design, elevation change, and friction losses.

7.2.2.6 Select Components for Torsional and Horsepower Requirements

Once design engineers determine the horsepower needed, they should select components (e.g., pipe shafts, drive shafts, and bearings) that resist or transmit loads induced by the conveyors. The torsional limits of conveyor shafts and flighting may require designers to divide long transport distances among two or more short conveyors.

7.2.3 Other Considerations

Screw conveyors are used to transport dried solids because they can be sealed to completely contain odors and dust. Carbon steel components typically are suitable, but galvanized steel also may be considered. Although the dried solids eventually will wear away most of the galvanizing or paint in the interior, the areas not continuously in contact with the conveyed material should be protected against corrosion.

The conveyors should not be operated with exposed shafting or flighting; all covers and lids should be kept closed. The covers may provide significant structural integrity.

Screw conveyors should include speed switches at the tail or non-drive ends to verify auger movement. Running a conveyor into non-operating equipment can cause severe damage.

These conveyors require relatively little maintenance or housekeeping, compared to other solids transport options. Routine maintenance typically consists of checking and adjusting the drive unit, and greasing hanger bearings (shafted screw conveyors) or replacing polyethylene liners (shaftless screw conveyors).

7.3 Drag Conveyors

Drag (*en masse*) conveyors have a wide range of uses in numerous industries. They have a long history of conveying such materials as biosolids, coal, grit, logs, rock salt, sawdust, and wood chips. These conveyors are highly adaptable and can be customized to transport most—if not all—treatment plant solids.

Drag conveyors have a slow moving chain-and-flight assembly that typically pushes material along a steel pan or trough, which may be rectangular or U-shaped. The troughs typically are constructed from structural "C" channels or bent plate. The chain typically is made of cast forged or fabricated steel. The flights, which typically are flat plates, "C" channels, or tubes, are shaped to match the trough and are bolted directly to chain links. Wider chains [up to 600 mm (24 in.) wide] can act as both chain and flight.

7.3.1 Drag Conveyor Applications

Most drag conveyors are designed to transport material horizontally. Special designs, however, can move material vertically or in "Z" lifts. A *"Z" lift* conveyor system is designed to transport material horizontally, then vertically, and then horizontally again.

Drag conveyors are used in many treatment plant processes, such as primary clarification, grit removal, and solids treatment. They also may be used as live bottom feeders in hoppers.

Drag conveyors are exceptionally strong and robust. Their lengths typically are only limited by the chain's strength and the weight of the material being transported. In practical terms, drag conveyors typically are not designed to move material more than about 40 m horizontally or 10 to 15 m vertically.

7.3.2 Drag Conveyor Design

Drag conveyor design primarily depends on the chain and the pulling loads encountered when pushing material the length of the conveyor. Following are design issues to consider for a drag conveyor's major components.

7.3.2.1 Chain Type

Chain designs and material characteristics vary greatly. Rollerless chains typically are suitable for most treatment plant applications. However, "Z" lift applications sometimes require roller chains to reduce overall chain load and kilowatt requirements. Also, drag chains—whose links can be up to 600 mm (24 in.) wide and can serve as both chain and flight—are suitable for horizontal and slightly inclined installations.

7.3.2.2 Chain Material

Malleable cast-iron chains are suitable for most drag conveyors used in wastewater treatment plants. However, steel chains with hardened components are more wear resistant than most cast chains.

7.3.2.3 Chain Pitch

Chain pitch is the distance between two successive rollers on a conveying chain. This distance typically depends on the desired size or spacing of cross rods or flights. Short conveyors can have 100-mm (4-in.) pitches, while long or heavily loaded conveyors often have 150- to 300-mm (6- to 12-in.) pitches.

7.3.2.4 Sprocket Size

Head and tail sprockets should be designed with as many teeth as practical because the number of teeth greatly influences chain and sprocket wear, as well as how smoothly the conveyor operates. As a general rule for optimum results, sprockets for pitch chains up to 150 mm should have between 12 and 21 teeth; those for larger pitch chains should have between 6 and 14 teeth (Link-Belt Industrial Chain Division, 1983).

7.3.2.5 Drive

The drive typically is installed at the head (discharge) end of the conveyor so only the chain's carrying run is under maximum tension.

7.3.2.6 Take-Ups

Take-up devices are used to maintain proper chain tension. Screw take-ups typically are acceptable for most drag conveyors used at wastewater treatment plants. Spring take-ups are useful if shock loads are anticipated. Gravity and catenary take-ups also are available.

7.3.2.7 Head and Tail Sections

Head and tail sections are custom made from steel plate to match the conveyor and installation requirements. They should be designed to support the loads induced by chain tension, which may be substantial. The tail section typically includes the take-up mechanism—unless a catenary take-up method is used, in which case the take-up is in the head section. The end without the take-up should have shaft bearing mounts with slotted holes and jack screws or other mechanisms so operators can align conveyor sprockets accurately.

7.3.2.8 Troughs

Most drag conveyor troughs are fabricated of steel plate and angle iron sections. They also can be constructed of concrete. All troughs should be equipped with replaceable wear bars for the chain or flighting to rest on and slide along.

7.3.3 Other Considerations

Drag conveyors typically cost more to install than other conveyors used for dried solids, but they allow for more flexibility in layouts and configurations then belt or screw conveyors and can handle a much larger spectrum of materials. They also have more capacity per cross-sectional area and can handle higher-impact loads, but are less energy efficient because of the high frictional forces of the solids, flighting, and chain against the trough.

For safety reasons, drag conveyors only should be operated when all covers, enclosures, and other safety appurtenances are in place. Inspection ports and access doors should have a metal screen or welded wire fabric that prohibits operators from inserting arms, legs, or other appendages into moving conveyors.

Most drag conveyors are designed with top covers that are bolted every 100 to 200 mm down the length of the conveyor. While this may complicate maintenance efforts, O&M personnel should be aware that such covers significantly increase the conveyor's structural strength. Tightening only one or two bolts per section when replacing covers may cause result in conveyor buckling and catastrophic failure.

Dried solids are so abrasive that it may be best to avoid chain lubrication. Using conventional lubrication on the drag chain actually may accelerate wear by adhering abrasive particles to the chain, where they then act as a lapping or grinding compound.

For drag conveyors, the most important maintenance consideration is routinely checking and adjusting chain tension. Improper chain tension—whether too much or too little—significantly shortens the lives of the chain, sprockets, and bearings. Other maintenance checks include bearing lubrication, wear bar thicknesses, chain lubrication (if applicable), bolt torques, and alignment of head and tail shafts.

7.4 Bucket Elevators

A bucket elevator is a simple, dependable device for vertically transporting dry materials. It consists of a series of buckets mounted on a belt or chain within a housing. The buckets are filled with material at the base of the unit and discharge it at the top. They are available in a wide range of capacities [10 to 350 kg/s (4 to 140 ton/hr)] and are totally enclosed to prevent dust and odors from escaping.

There are three types of bucket elevators: centrifugal-discharge, positive-discharge, and continuous. Centrifugal-discharge elevators are the most common and are best suited for handling fine, free-flowing materials that can be dug from the elevator boot at the base of the unit. These units have the fewest buckets, which are mounted on either a chain or belt. The buckets are easily loaded and travel rapidly enough [up to 90 m/min (300 ft/min)] to discharge material via centrifugal force as they pass around the head pulley or sprocket.

Positive-discharge elevators are designed for sticky materials that tend to pack. They are similar to centrifugal-discharge units, except that their buckets are mounted on two strands of chain, are large and closely spaced, and are snubbed back under the head sprocket to invert them for positive discharge. (As the snub sprockets engage the chain, the slight impact helps free materials from the buckets.) The buckets also move slower [35 m/min (120 ft/min)] than those in centrifugal-discharge units.

Continuous elevators are recommended for sluggish, aerated, and friable materials—applications in which product degradation is a concern. They have closely spaced buckets mounted on either belts or chains that travel at 38 m/min (125 ft/min). The buckets often are direct-loaded and are designed so the fronts and extended sides form a chute as they pass around the head pulley or sprocket. Gravity allows the material to flow gently out of the buckets and down the chute (formed by the preceding bucket) into the discharge spout.

7.4.1 *Bucket Elevator Applications*

At wastewater treatment plants, bucket elevators are used to transport dried materials vertically within solids-processing units and to load dried biosolids products into trucks or railcars.

For most dried materials, centrifugal-discharge elevators are generally acceptable and the most cost-effective option. However, if the dried biosolids are intended for beneficial reuse, then dust content and degradation are concerns, so designers should consider continuous bucket elevators.

7.4.2 *Bucket Elevator Design and Operation Considerations*

Before selecting a bucket elevator, designers should determine the material's characteristics (abrasiveness, flowability, etc.); its maximum lump sizes; density of the bulk material; capacity needed; and transport height.

Manufacturers sell standard-sized units, and tables are available to help designers size bucket elevators that will convey materials vertically up to 30 m. The tables provide elevator dimensions, bucket sizes, and energy requirements.

Bucket elevators vary in casing thickness, bucket type and thickness, belt or chain quality, and drive equipment.

The casing is constructed of either heavy-gauge steel sections or steel plate and angle iron that are continuously welded for the full length of the unit. The steel is either mild or galvanized. Galvanizing may be preferred over painted casings because it provides corrosion protection for the casing interior. (Most bucket elevator casings are too small for the interior to be painted properly.) Casings can be up to 7 mm thick. A heavier casing may be recommended if the elevator will be outside and exposed to harsh weather conditions. Casings also can be made dust tight. A split, removable hood is recommended for ease of service and maintenance at the top end of the unit.

Buckets are available in a variety of styles, and designers should ask manufacturers for recommendations. Centrifugal elevators typically have malleable iron buckets, which are appropriate for heavy-duty abrasive applications (e.g., dried solids). Ductile iron or steel buckets also can be used if desired. Continuous elevators typically have steel buckets. Polymer and nylon buckets are also available; they resist corrosion and promote discharge of sticky materials. In addition, buckets can be perforated to handle dusty materials. The perforations allow air to be released from the buckets during loading and improve material discharge by eliminating blowing.

Buckets can be mounted on single or dual strands of chains. Class "C" combination chains, which have alternating cast-iron black links, are sufficient for normal applications, according to CEMA. Class "S" chains are stronger and wear less quickly; they are recommended for great heights (up to 45 m) or when transporting abrasive materials. Rubber-covered belts are acceptable for most applications involving belt-mounted buckets. Such belts can be made of impregnated canvas or fabric.

Bucket elevators typically are driven from the head shaft and have take-up bearings in the boot. A shaft-mounted gear reducer with a V-belt drive is recommended for economy and versatility. Another option is a gear motor connected to the elevator head shaft via a chain drive; it is supported on a bracket mounted to the elevator casing. Backstops prevent backward rotation when the elevator stops under load; they may be added to either the head shaft or countershaft. The tail shaft should include a zero-speed switch to indicate motor or conveyor problems, or an overloaded elevator. Drive guards also are required for safety reasons.

Bucket elevators typically use less energy than other types of vertical conveyors. As a general guideline, design engineers can estimate an elevator's electricity requirements as follows:

$$\text{Power needed (kW)} = \text{Capacity (tonne/h)} \times \text{Conveyance height (m)}/103.$$

If the bucket elevator is more than 10 m tall, engineers should include guy wires or structural steel members to provide lateral bracing. Also, designers should include a service platform so operators can inspect and maintain the head terminal and drive more easily. The platform should be accessed by a ladder with a safety cage. Likewise, design engineers should provide a clean-out door in the boot—especially in continuous elevators—so operators can periodically remove any material that has accumulated in the base. In centrifugal units, the casing corners may fill with material that should be removed regularly.

In applications involving dust, designers should ventilate bucket elevators. The ventilation system should be designed in accordance with guidelines established by the American Conference of Governmental Industrial Hygienists in *Industrial Ventilation* (1982). These guidelines require an exhaust point at the top of the elevator and a second one at the bottom if the elevator is more than 10 m high. They also recommend a flow of 30 $m^3/m^2{\cdot}min$ of casing, with a minimum duct velocity of 18 m/s.

Bucket elevators should be designed with explosion relief or suppression mechanisms because organic dusts—including the fine material generated during solids drying—could explode under certain conditions. Explosion relief directs such forces through expendable panels (rupture plates) in the elevator casing and then into the room or outside the facility. Engineers should be extremely careful when designing explosion vents because the force of these explosions could hurt or kill operators who are next to the equipment when a deflagration occurs.

7.5 Pneumatic Conveyors

Pneumatic conveyors use air to move material through a pipeline. There are two types of these conveyors: dilute-phase and dense-phase.

Dilute-phase conveyors have low material-to-air ratios [less than 5:1 (5 kg material/kg air)]; they use a large volume of air to move a small amount of material. Rotary airlocks feed material into the pipeline, and positive-displacement blowers or fans supply low-pressure air [less than 100 kPa (14.5 psi)]. The air velocity typically ranges from 20 to 40 m/s (65 to 130 ft/sec)—high enough to suspend the material in the air stream. The systems then use either positive or negative pressure to push or pull the material through the pipeline.

Dense-phase conveyors have high material-to-air ratios (up to 100 kg material/kg air). A pressure tank and a high-pressure air compressor can provide 350 to 700 kPa (50 to 100 psig) of air, which typically moves at velocities less than 2.5 m/s (8 ft/sec). In these systems, material enters the pressure tank or transporter via gravity and settles in the pipeline. Once a certain volume has settled, the transporter inlet valve closes, the vessel is pressurized using compressed air, and the material flows out of the vessel into the pipeline. As the pressure increases, the material forms a plug that the air pushes to its destination.

There are two types of dense-phase systems: conventional batch and full-line. In a conventional batch system, air is introduced at the pressure vessel with enough force to transport a batch of material in the pressure vessel to the final receiving bin. The pipe is completely purged before the next cycle begins.

A full-line system introduces air in both the pressure vessel and via low-pressure air-booster fittings spaced along the length of the pipeline. The air is introduced at the lowest possible velocity. Once the pressure vessel is emptied, any material left in the pipeline remains there until the next batch is conveyed. The booster fittings serve to move that material when the next batch begins.

7.5.1 Pneumatic Conveyor Applications

Pneumatic conveyors have been used to transport dried biosolids and grit in continuous processes, load railcars or trucks, and help collect and remove fugitive dust.

Dilute-phase conveyors may be used to transport lime, sawdust, and other chemicals typically used at wastewater treatment plants. They are ideal for transporting nonabrasive, powdered materials over short distances. Although successfully used to convey a wide variety of bulk solids, dilute-phase conveyors may not be suitable for dried solids meant for beneficial reuse because they can be abrasive and degrade when exposed to high-velocity air. Also, dilute-phase conveyors have low capital costs, but their energy costs can be quite high because of the large air requirements.

Both conventional-batch and full-line, dense-phase conveyors are appropriate for transporting dried biosolids. Conventional batch conveyors are preferred for short distances (less than 35 m). Full-line conveyors are preferred for longer distances and easily degraded material, such as dried biosolids pellets intended for beneficial reuse.

7.5.2 Pneumatic Conveyor Design and Operation Considerations

Pneumatic conveyor design depends on such parameters as material bulk density, capacity or flowrate, and equivalent pipeline length.

When designing dilute-phase pneumatic conveyors, engineers should be aware that much information is available in manufacturer's brochures, data sheets, and monographs. They can use these resources to determine the optimum pipeline diameter and air volume based on recommended solids ratios and system pressure drop [less than 70 kPa (10 psi)]. Then they can determine the blower or fan size needed based on the calculated conveying air volume.

Pressure systems typically are used when flowrates exceed 9 000 kg/h (2.0×10^4 lb/h). Vacuum systems are used when flowrates are less than 7 000 kg/h and the equivalent length is less than 300 m.

The rotary airlock is sized based on desired flowrate. Carbon steel construction is acceptable for most materials encountered in a wastewater treatment plant.

The system will need a dust-collection device to clean the conveying air before it is exhausted. This device, which may be a cyclone separator or fabric filter, should be sized to handle the air flows determined in accordance with manufacturer's recommendations.

When designing dense-phase pneumatic conveyors, engineers will need more input from manufacturers because there are no readily available monographs and such designs often are considered more an art than a science. Given a set of design parameters, the system manufacturer should be able to provide the optimum pipeline diameter for the conveying air volume and pressure required. The manufacturer also should recommend a standard transporter size (typically corresponding to 5 to 15 cycles per hour) and spacing requirements for air-booster fittings (typically every 1.5 to 6 m).

Transporter size depends on conveying distance; transport-cycle frequency lessens as conveying distances exceed 150 m (500 ft). Also, the transporter should include a 60-degree hopper at the bottom to encourage dried solids to flow out of the vessel.

If continuous conveying is required (e.g., in solids treatment processes), design engineers should provide a dedicated air compressor, an air receiver, and a compressed air dryer.

The pressure vessel typically is made of carbon steel. The pipeline should be constructed of either Schedule 40 carbon steel or galvanized steel.

When designing dense-phase conveyors, engineers should place vertical runs as early as possible and avoid using back-to-back bends. Flat-back elbows also may be considered when transporting heavy-duty, abrasive material. Designers also should be aware that full-line dense-phase conveyors have less pipeline wear because they function at the lowest velocity.

In fact, because dried biosolids are abrasive, all pipeline bends in dilute-phase or dense-phase systems should be long-radius, sweeping bends. Also, status lights and pressure indicators should be included to help plant staff monitor operations.

The pneumatic conveyor's blower, fan, or air compressor typically only requires routine maintenance. However, the exhaust air is odiferous and may require further treatment before discharge. The air also will contain some dust that must be removed before discharge. So, design engineers should equip the pressure vessel with a vent line connected to a dust-collection device and provide for odor control.

Because of the energy requirements, pneumatic conveyors are one of the least efficient methods to transport dried biosolids and other dry granular materials. However, a properly designed and operated pneumatic conveyor will have fewer O&M requirements than mechanical conveyors. Because the conveyor is totally enclosed, housekeeping and odor control are simpler, but problems are more difficult to identify. The conveyor's operating costs are high, but it has a small footprint, easily retrofits into existing facilities, and can handle long distances and multiple discharge points.

8.0 REFERENCES

American Conference of Governmental Industrial Hygienists (1982) *Industrial* Ventilation, 17th Ed.; American Conference of Governmental Industrial Hygienists: Lansing, Michigan.

American Society of Civil Engineers (2000) *Conveyance of Wastewater Treatment Plant Residuals;* American Society of Civil Engineers: Reston, Virginia.

Barbachem, M. J.; Pyne, J. C. (1995) Pipeline Hydraulics of Dewatered Non-Newtonian Cakes. *Proceedings of the 68th Annual Water Environment Federation Technical Exposition and Conference,* Miami Beach, Florida, Oct 21–25; Water Environment Federation: Alexandria, Virginia; pp. 41–49.

Bassett, D. J.; Howell, R. D.; Haug, R. T. (1991) Hydraulic Properties Evaluation for Sludge Cake Pumping. *Proceedings of the 64th Annual Water Environment Federation Technical Exposition and Conference;* Toronto, Ontario, Oct 7–10; Water Environment Federation: Alexandria, Virginia.

Battistoni, P. (1997) Pretreatment, Measurement Execution Procedure and Waste Characteristics in the Rheology of Sewage Sludges and the Digested Organic Fraction of Municipal Solid Wastes. *Wat. Sci. Tech.,* **36,** 33.

Bechtel, T. B. (2003) Laminar Pipeline Flow of Wastewater Sludge: Computational Fluid Dynamics Approach. *J. Hydr. Eng.,* **129** (2), 153.

Bechtel, T. B. (2005) A Computational Technique for Turbulent Flow of Wastewater Sludge. *Water Environ. Res.,* **77,** 417.

Borrowman, D. (1985) *Wastewater Sludge Characteristics and Pumping Application Guide;* WEMCO Pump Co.: Sacramento, California.

Bourke, J. D. (1992) *Pumping Abrasive Slurries with Progressing Cavity Pumps;* Moyno Industrial Products: Springfield, Ohio.

Bourke, J. D. (1997) *Handling High Solids Content Non-Newtonian Fluids;* Moyno Industrial Products: Springfield, Ohio.

Brar, S. K.; Verma, M.; Tyagi, R. D.; Valéro, J. R.; Surampalli, R. Y. (2005) Sludge Based *Bacillus huringiensis* Biopesticides: Viscosity Impacts. *Water Res.,* **39,** 3001.

Carthew, G. A.; Goehring, C. A.; Van Teylingen, J. E. (1983) Development of Dynamic Head Loss Criteria of Raw Sludge Pumping. *J. Water Pollut. Control Fed.,* **55,** 472.

Chilton, R. A.; Stainsby, R. (1998) Pressure Loss Equations for Laminar and Turbulent Non-Newtonian Pipe Flow. *J. Hydr. Eng.,* **124** (5), 522.

Conveyor Equipment Manufacturers Association (1979) *Belt Conveyors for Bulk Materials,* 2nd ed.; CBI Publishing Co.: Boston, Massachusetts.

Crow, H.; Cortopassi, R. (1994) *Schwing KSP 17V(K) Pump Demonstration from July, 1994 through August 3, 1994;* Schwing America Inc., Environ. Div.: Danbury, Connecticut.

Dillon, M. L. (2007) Comparing PD Pump Designs for Transferring Dewatered Sludge Cake. *Pumps & Systems,* September, 84.

Doty, D. (2005) New Progressing Cavity Pump Developments in Sludge Transfer. *World Pumps,* October, 24.

El-Mashad, H. M.; van Loon, W. K. P.; Zeeman, G.; Bot, G. P. A. (2005) Rheological Properties of Dairy Cattle Manure. *Bioresource Technol.*, **96,** 531.

Florida Department of Environmental Protection (2008), *Chapter 62-640: Biosolids*, Draft Version; Florida Department of Environmental Protection: Tallahassee, Florida.

Guibaud, G.; Dollet, P.; Tixier, N.; Dagot, C.; Baudu, M. (2004) Characterisation of the Evolution of Activated Sludges Using Rheological Measurements. *Process Biochem.*, **39,** 1803.

Health and Safety Executive (2005) *Control of Health and Safety Risks at Sewage Sludge Drying Plants;* HSE 847/9; Health and Safety Executive: London, United Kingdom.

Hentz, L.; Cassel, A.; Conley, S. (2000) The Effects of Liquid Sludge Storage on Biosolids Odor Emissions. Proceedings of 14th Annual Residuals and Biosolids Management Conference; Boston, Massachusetts, Feb 27–29; Water Environment Federation: Alexandria, Virginia.

Honey, H. C.; Pretorius, W. A. (2000) Laminar Flow Pipe Hydraulics of Pseudoplastic-Thixotropic Sewage Sludges. *Water SA*, **26,** 19.

Jones, H. (1993) *Solving Sludge Handling Problems with Progressive Cavity Pumps*. Robbins & Myers Inc., Fluid Handling Group: Springfield, Ohio.

Laera, G.; Giordano, C.; Pollice, A.; Saturno, D.; Mininni, G. (2007) Membrane Bioreactor Sludge Rheology at Different Solid Retention Times. *Water Res.*, **41,** 4197.

Levine, L. (1986) *Coming to Grips with Rheology;* Viscous Products.

Levine, L. (1987) *An Introduction to the Measurements of Viscosity;* Viscous Products.

Link-Belt Industrial Chain Division (1983) *Link-Belt Chains and Sprockets for Drives, Conveyors and Elevators;* Link-Belt Industrial Chain Division: Homer City, Pa.

List, E. J.; Hannoun, I. A.; Chiang, W.-L. (1998) Simulation of Sludge Pumping. *Water Environ. Res.*, **70,** 197.

Lue-Hing, C.; Zenz, D.; Tata, P.; Kuchenrither, R.; Malina, J.; Sawyer, B. (1998) *Municipal Sewage Sludge Management: A Reference Text on Processing, Utilization and Disposal;* Technomic Publishing Co. Inc.: Lancaster, Pennsylvania.

Lottman, S. (2008) Personal communication; Siemens: Berlin, Germany.

Metcalf and Eddy, Inc. (2003) *Wastewater Engineering: Treatment and Reuse;* Tchobanoglous, G.; Burton, F. L.; Stensel, H. D., Eds; 4th ed.; McGraw-Hill Inc.: New York, New York.

Mori, M.; Isaac, J.; Seyssiecq, I.; Roche, N. (2008) Effect of Measuring Geometries and of Exocellular Polymeric Substances on the Rheological Behaviour of Sewage Sludge. *Chem. Eng. Res. Des.*, **86,** 554.

Mulbarger, M. C.; Copas, S. R.; Kordic, J. R.; Cash, F. M. (1981) Pipeline Friction Losses for Wastewater Sludges. *J. Water Pollut. Control Fed.*, **53,** 1303.

Mulbarger, M. C. (1997) Selected Notions about Sludges in Motion, and Movers. Paper presented at Central States Water Environment Association Education Seminar, Madison, Wisconsin.

Murakami, H.; Katayama, H.; Matsuura, H. (2001) Pipe Friction Head Loss in Transportation of High-Concentration Sludge for Centralized Solids Treatment. *Water Environ. Res.*, **73,** 558.

National Fire Protection Association (1995) *Report on Comments A2007 – NFPA 820 Standard for Fire Protection in Wastewater Treatment and Collection Facilities;* National Fire Protection Association: Quincy, Massachusetts.

Novak, J.; Adams, G.; Chen, Y.-C.; ; Erdal, Z.; Forbes, R. H, Jr.; Glindemann, D.; Hargreaves, J. R.; Hentz, L.; Higgins, M. J.; Murthy, S. N.; Witherspoon, J.; Card, T. (2004) Odor Generation Patterns from Anaerobically Digested Biosolids. *Proceedings of the Joint WEF/A&WMA Odors and Air Emissions Conference;* Bellevue, Washington, Apr 18–24; Water Environment Federation: Alexandria, Virginia.

Pilehvari, A. A.; Serth, R. W. (2005) Generalized Hydraulic Calculation Method Using Rational Polynomial Model. *J. Energy. Res. Technol.*, **127,** 15.

Radney J. (2008) Personal communication; Cerlic USA: Atlanta, Georgia.

Ram Pumps Take the Drudgery out of Sludge Transfer (1999). *World Pumps*, Feb, 18–19.

Sanks, R. L.; Tchobanoglous, G.; Bosserman, B. E., II; Jones, G. M., Eds. (1998) *Pumping Station Design;* 2nd ed.; Butterworth-Heinemann: Boston, Massachusetts.

Setterwall, F. (1972) Discussion/Communication on Pumping Sludge Long Distances. *J. Water Pollut. Control Fed.*, **44** (1), 648.

Spaar, A. (1972) Pumping Sludge Long Distances. *J. Water Pollut. Control Fed.*, **43** (1), 702.

Spinosa, L.; Lotito V. (2003) A Simple Method for Evaluating Sludge Yield Stress. *Adv. Environ. Res.*, **7,** 655.

Spinosa, L.; Vesilind, P. A. (2001, reprinted 2007) *Sludge into Biosolids—Processing, Disposal, Utilization;* IWA Publishing: London, United Kingdom.

U.S. Army Corps of Engineers (1984) *Engineering and Design—Domestic Wastewater Treatment Mobilization Construction;* EM-1110-3-172; U.S. Army Corps of Engineers: Washington, D.C.

U.S. Environmental Protection Agency (1979) *Process Design Manual, Sludge Treatment and Disposal;* EPA-625/-79-011; U.S. Environmental Protection Agency, Munic. Environ. Res. Lab.: Cincinnati, Ohio.

U.S. Environmental Protection Agency (1982) *Handbook: Identification and Correction of Typical Design Deficiencies at Municipal Wastewater Treatment Facilities;* EPA-625/6-82-007; U.S. Environmental Protection Agency, Munic. Environ. Res. Lab.: Cincinnati, Ohio.

U.S. Environmental Protection Agency (1983) *Process Design Manual—Land Application of Municipal Sludge;* EPA-625/1-83-016; U.S. Environmental Protection Agency: Washington, D.C.

U.S. Environmental Protection Agency (1995) *Process Design Manual—Land Application of Sewage Sludge and Domestic Septage;* EPA-625/R-95/001; U.S. Environmental Protection Agency: Washington, D.C.

U.S. Environmental Protection Agency (2000) *Guide to Field Storage of Biosolids and Other Organic By-Products Used in Agriculture and for Soil Resource Management;* EPA/832-B-00-007; U.S. Environmental Protection Agency: Washington, D.C.

Wagner, R. L. (1990) *Sludge Digester Heating;* Alfa-Laval Thermal Co.: Ventura, California.

Wanstrom, C. (2008) Personal communication. Schwing Bioset: Somerset, Wisconsin.

Water Environment Federation (2004) *Control of Odors and Emissions from Wastewater Treatment Plants;* WEF Manual of Practice No. 25; McGraw-Hill: New York.

Water Environment Research Foundation (2008) *Identifying and Controlling Odor in Municipal Wastewater Environment Phase 3: Biosolids Processing Modifications for Cake Odor Reduction;* Water Environment Research Foundation: Alexandria, Virginia.

Chapter 22

Chemical Conditioning

1.0 INTRODUCTION

Conditioning does not reduce the water content of solids; it alters the physical properties of solids to facilitate the release of water during thickening and dewatering. Mechanical thickening and dewatering techniques would not be economical for a utility without solids conditioning beforehand.

Conditioning is a chemical or thermal treatment used to improve the efficiency of thickening or dewatering processes. Chemical conditioning processes use inorganic chemicals, organic polymers, or both to improve solids' thickening and dewatering characteristics. Physical conditioning techniques (e.g., thermal conditioning) use heat to condition and stabilize solids. (For more information on thermal conditioning, see Chapter 26.)

This chapter discusses chemical conditioning, and includes theory and design considerations. Some thickening and most dewatering of wastewater residuals particularly those containing solids from biological treatment processes (e.g., fixed-film and suspended growth activated solids treatment systems) typically are not practical without some type of conditioning. This conditioning step can take the form of either a chemical or physical process.

Conditioning can significantly reduce the moisture content of the solids, depending on the thickening or dewatering process used, the characteristics of the incoming solids, and the method for thickening or dewatering. Typically, chemical conditioning can increase the residuals' solids content from about 1% up to between 15 and 30%. Conditioning not only removes water but also increases the thickening or dewatering rate significantly by adjusting the chemical and physical properties of solids.

In the United States, the most typically used inorganic chemicals for conditioning solids are iron salts (e.g., ferric chloride, and lime). In Great Britain, the typical practice is ferrous sulfate with lime.

The use of organic polyelectrolytes (polymers) in municipal wastewater treatment plants was introduced during the 1960s and was rapidly adopted for both thickening and dewatering processes. The primary advantage of polymers is that they do not significantly increase solids production: Every kilogram of inorganic chemicals added during conditioning will produce a kilogram of extra solids. Inorganic chemicals (e.g., iron salts and lime) can increase the final product by 20 to 30% (dry solids), and fly ash can increase the final product by 50 to 100% (dry solids).

The conditioning method must be compatible with the proposed thickening or dewatering method. For example, centrifuges use pressure to compact the solids, whereas belt filter presses permit the water to pass through the void spaces; therefore, a single type of conditioning agent cannot be expected to be useful for all applications (WEF, 2003).

2.0 FACTORS AFFECTING CONDITIONING

The type and dosage of conditioning agent needed depends on the residuals' characteristics, solids handling and processing before and after conditioning, and the mixing process after agent addition.

2.1 Residuals Characteristics

Most of the physical and chemical transformations associated with additions of organic or inorganic chemicals are not well understood on a fundamental level (WPCF, 1982). However, wastewater treatment professionals have identified several residuals' characteristics that affect conditioning requirements (U.S. EPA, 1979d; WPCF, 1983). These characteristics include:

- The source of residuals,
- Solids concentration,
- Alkalinity and pH,
- Biocolloids and biopolymer production,

- Particle size and distribution,
- Degree of hydration,
- Particle surface charge, and
- Volatile suspended solids (VSS) content.

2.1.1 *Source of Residuals*

To some extent, the conditioning method depends on the type of solids that must be treated. Municipal primary, secondary, combined, and digested solids are a good indicator of the range of probable doses of conditioning agents that are required for subsequent solids treatment. Cationic polymers work best on primary, secondary, and digested solids, while anionic polymers may be effective with inorganic solids.

An examination of published data for a variety of thickening and dewatering devices suggests that primary solids require lower doses than secondary solids do, and that fixed-film secondary solids require lower doses than secondary solids do (U.S. EPA, 1979d). Depending on the thickening or dewatering method used, aerobically and anaerobically digested solids require conditioning doses comparable to those for secondary solids. Similarly, combined solids (primary and secondary solids) have properties that are closer to those of secondary solids, although they are affected by the respective composition of each type. More importantly, the characteristics of solids from the same source vary from plant to plant and also can vary seasonally, so the conditioner dose depends on the specific conditioning agent used and the goal (thickening or dewatering solids).

Chemical solids are solids that have been mixed with an inorganic conditioning agent (e.g., the addition of aluminum or iron salts and lime to precipitate phosphorus or improve suspended solids removal). These types of solids are more variable than traditional primary and secondary solids, and so are hard to categorize with respect to dose, and their conditioning requirements are often qualitatively different from those for primary and secondary solids. For example, adding lime to mixed-liquor suspended solids before secondary clarification may improve suspended solids removal; however, the resulting solids may require an anionic polymer, not the cationic one typically used for primary or secondary solids, because the positively charged calcium has neutralized some of the negative surface charge. While charge neutralization is a fundamental part of the process, an equally significant piece is the interconnection of solids particles via the polymer chain.

2.1.2 *Solids Concentration*

In many applications, conditioning neutralizes the colloidal surface charge by adsorbing oppositely charged organic polymers or inorganic complexes. The residuals' solids concentration will affect the dosage and dispersal of the conditioning agent. Therefore, for a given particle size distribution, increasing the suspended solids concentration increases the required coagulant dose for effective surface coverage (on a volumetric basis).

The suspended solids concentration also affects two additional aspects of conditioning agents. First, the process is less susceptible to overdosing at higher solids concentrations. Second, the solids and the conditioning agent are more difficult to mix at higher solids concentrations.

2.1.3 *Alkalinity and pH*

When inorganic conditioning agents are used, alkalinity and pH are the most important chemical parameters affecting conditioning. Coagulation occurs when coagulant interacts with the surface of solids' colloids, and nature of the charged surface and the coagulant charge are both pH-dependent. When inorganic conditioners are used, the solids' pH determines which chemical species are present.

Iron and aluminum salts behave like acids when added to water (i.e., these conditioners reduce pH), so the pH of the conditioning process will depend on the solids' alkalinity and the dosage of iron or aluminum salts. This pH, in turn, determines the predominant coagulant species and the nature of the charged colloidal surface. The high alkalinity typically associated with anaerobically digested solids is one reason for the higher coagulant doses required. Low-molecular-weight coagulants tend to be more effective over a broader pH range than inorganic conditioning agents.

2.1.4 *Biocolloids and Biopolymers*

Although more commonly measured by researchers than by design engineers, these fundamental parameters provide some insight into the conditioning process. The biopolymers in activated solids flocs seem to affect the physico-chemical properties of flocs (e.g., floc density, floc particle size, specific surface area, charge density, bound water content, and hydrophobicity).

Other studies have shown that cations can affect bioflocculation and change the settling and dewatering properties of activated sludge flocs (Eriksson and Alm, 1991; Bruus et al., 1992; Higgins and Novak, 1997a, b). Divalent cations bridge across negatively charged biopolymers to form a dense, compact floc structure. Monovalent cations tend to prevent proper flocculation by forming a much weaker structure. As a result, divalent cations promote bioflocculation and produce subsequent improvements in settling and dewatering properties. Monovalent cations tend to degrade settling and dewatering properties. It seems that settling and dewatering properties are further improved when the two divalent cations are added to the feed rather than superficially added to the settling tank (Higgins and Novak, 1997a).

A series of laboratory-scale studies were conducted using waste activated sludge (WAS) to gain insight into the floc-destruction mechanisms that account for changes in solids conditioning and dewatering properties after anaerobic or aerobic digestion. The data indicated that biopolymer was released from solids under both anaerobic and aerobic conditions, but much more was released under anaerobic conditions. In particular, four to five times more protein was released into solution under anaerobic conditions than under aerobic conditions (Novak et al., 2003). Both the dewatering rate (as characterized by the specific resistance to filtration) and the polymer dose depend directly on the amount of biopolymer (protein + polysaccharide) in solution.

2.1.5 *Particle Size and Distribution*

Particle size distributions affect the total particle surface area and the porosity of cakes formed from these particles. These properties affect required coagulant doses and dewaterability. Several researchers (Karr and Keinath, 1978; Novak et al., 1988; Sorensen and Sorensen, 1997) studied the effect of particle size on dewaterability and concluded that particle size was one of the most important parameters in determining dewaterability. Smaller particles (colloidal and supracolloidal) can blind filters and solids cakes (Novak et al., 1988; Sorensen and Sorensen, 1997) and deter the release of water from the solids cake. Also, another study has suggested that an increase in floc density improves dewatering properties via a decrease in bound water associated with the flocs (Kolda, 1995). These studies concluded that dewaterability improvements often associated with other factors (e.g., pH, mixing, biological degradation, and conditioning) all could be explained by the effects of these factors on particle size distributions.

2.1.6 *Degree of Hydration*

Excessive bound water has been suggested as the cause of dewatering difficulties. The percent of bound water associated with the floc also indicates the maximum dryness that can be achieved in the solids cake by mechanical means (Robinson, 1989). Additionally, Vesilind (1979) reviewed the work of several investigators on water distribution in activated sludge. This water was described as free water, floc water, capillary water, and bound (particle) water. These categories are defined based on the amount of centrifugal acceleration required to release a given portion of water. Vesilind suggested that the water distribution in a given solids could determine the applicability of a specific thickening or dewatering operation.

2.1.7 *Particle Surface Charge*

Solids particles (e.g., subcolloidal and macromolecular constituents) typically have a negative surface charge, so they tend to repulse each other. The resulting spaces between these constituents are occupied by cations and water. If the charge can be eliminated, thickening or dewatering improves. This is why chemical conditioners are positively charged, or become positively charged when added to water.

In most cases, polymer conditioning is optimal when the charge is neutralized, so measuring charge can be useful in laboratory comparisons of polymers and doses. A streaming current detector (zeta meter) can be used to measure this charge. It also can be used to monitor or control polymer dose real-time in dewatering processes (Dentel et al., 1995).

Shear during mixing or dewatering tends to open up new negative surfaces in the biocolloids, thereby undoing the charge effects of cationic polymers. So, increases in mixing shear increase the required polymer dose (Dentel, 2001).

2.1.8 *Wastewater Cations*

Numerous studies have suggested that cations interact with the negatively charged biopolymers in activated solids to change the structure of the floc (Higgins, 1995; Bruus et al., 1992; Eriksson and Alm, 1991; Novak and Haugan, 1979; Tezuka, 1969). One study indicated that monovalent cations tend to deteriorate settling and dewatering characteristics, while divalent cations tend to improve them (Higgins, 1995). The effects of charge density on activated sludge properties could decrease the polymer dose needed to condition secondary solids.

Researchers have studied influent concentrations of cations (e.g., aluminum, ammonium, calcium, iron, magnesium, potassium, and sodium) extensively. They postulated that cations play a critical role in bioflocculation. Cations have been found to influence the thickening and dewatering characteristics of biological solids. For example, high concentrations of sodium typically resulted in poor dewatering; however, if the floc contained enough aluminum and iron concentrations, it typically offset the deleterious effects of sodium (Park et al., 2006). The data associated with aluminum further revealed that WAS with low aluminum levels contained high concentrations of soluble and colloidal biopolymer (protein + polysaccharide), resulting in a high effluent COD concentrations, a need for larger doses of conditioning chemical, and poor solids dewatering properties (Park et al., 2006). Studies have shown that iron may contribute to floc strength, and it seems that the reduction and solubilization of iron during anaerobic digestion may be a reason why digested solids dewater poorly. Also, the presence of proteins in solution contributes to poor dewatering and larger conditioning chemical doses (Novak et al., 2001).

2.1.9 *Rheology*

Many studies have focused on the rheological properties of wastewater solids in an attempt to correlate solids properties with chemical conditioning requirements (Ormeci et al., 2004). For example, yield strength and viscosity have been used to optimize chemical conditioning. Another study demonstrated that mixing considerably affected the rheological characteristics of conditioned solids (Abu-Orf and Dentel, 1999). Another showed that solids conditioning could be improved by monitoring centrate or filtrate viscosity (Bache and Dentel, 2000). In another study, both laboratory- and full-scale testing showed that the network strength of the sludge could be used to optimize chemical conditioning and achieve drier solids (Abu-Orf and Ormeci, 2005). In general, the type of conditioner (e.g., polymers or fly ash) and wastewater solids (e.g., chemical, WAS, or biosolids) involved will determine which rheological parameters should be used.

2.2 Handling and Processing Conditions Before Conditioning

The efficiency of any conditioning process depends to a large degree on the solids' chemical and physical characteristics (e.g., origin, solids concentration, inorganic content, chemistry, storage time, and mixing) before conditioning. The solids' physical characteristics are a function of the physical stresses they were exposed to before conditioning. For example, any process that damages the flocculant nature of solids particles typically either increases chemical conditioning demand or reduces performance in the final treatment stage. The extent of mixing and shear stress before and after conditioning can significantly affect conditioning efficiency and, ultimately, solids treatment performance.

2.2.1 Storage

There are two types of storage for liquid residuals: long-term and short-term. Long-term storage may occur in stabilization processes with long detention times (e.g., aerobic and anaerobic digestion; see Chapter 25) or in specially designed tanks (see Chapter 21). Short-term storage may occur in wastewater treatment process (e.g., increasing solids inventory) or in smaller, specially designed tanks. Storage helps smooth out fluctuations in solids production, make the solids feed rate more uniform. It also provides a place to keep solids during equipment downtime. However, long-term storage has been reported to negatively affect the dewaterability of solids.

Unstabilized solids that have been stored for long periods typically require more conditioning chemicals than fresh solids because the degree of hydration and percentage of fine particles increased. Also, storing activated sludge increased sludge's specific resistance to filtration and, subsequently, conditioning requirements (Karr and Keinath, 1978). Storing aerobically or anaerobically stabilized solids for extended periods typically lowers temperature significantly and can change pH and alkalinity.

Temperature drops typically increase conditioning requirements. However, if the temperature decrease is small, the negative effect on conditioning may be more than offset by an increase in solids concentration.

2.2.2 *Pumping*

Pumping subjects solids to shear forces; the level of shear depends on the type of pump and the flow rate. Solids particles are fragile, and pumping typically causes some of them to fragment. Researchers have shown that the major demand for chemical conditioning is associated with the fraction of particles in the colloidal and supracolloidal range (Karr and Keinath, 1978; Roberts and Olsson, 1975), so any process that reduces particle size will increase conditioning chemical requirements.

Conditioned solids should not be pumped, because pumping introduces shear forces that tend to break down flocs. If required, however, then the pump should be designed to minimize shear.

2.2.3 *Mixing*

During conditioning, the solids and added chemicals must be mixed enough to ensure that the chemical is evenly dispersed throughout the solids. However, the mixer must not break the floc once it has formed. Design engineers should optimize the mixing time with these two goals in mind. Mixing requirements depend on the thickening or dewatering method used. In-line mixers typically are used with most modern thickening and dewatering units. A separate mixing and flocculation tank is provided with some older thickening and dewatering devices.

For many municipal solids, intense mixing (a mean velocity gradient in the range of 1 200 to 1 500 s^{-1}) should be followed by much gentler agitation (a mean velocity gradient less than 200 s^{-1}), so fine particles can flocculate into particle aggregates that settle or can be readily filtered. Anaerobically digested solids need mean velocity gradients up to 12 000 s^{-1}. Once the solids and conditioning chemical have been mixed thoroughly, a hydraulic retention time (HRT) of 15 to 45 seconds in the pipeline or flocculation tank will complete the flocculation process before solids enter the thickening or dewatering system.

2.2.4 *Solids Concentration*

The residuals' solids concentration can significant affect conditioning system performance and cost. In most cases, as the influent solids concentration increases, the conditioning cost decreases to a certain level (WPCF, 1980). However, it becomes increasingly difficult to evenly mix coagulant in residuals containing 4% solids (or more) before further dewatering.

2.2.5 *Stabilized and Unstabilized Solids*

Polymer requirements also are affected by the type of solids to be conditioned. In fact, this may have the most effect on the quantity of chemical needed. Solids that are difficult to dewater require the largest doses of chemicals, typically yield a wetter cake, and result in poorer-quality sidestreams (filtrate, centrate, etc.). The following types of solids are listed in increasing order (approximately) of conditioning chemical requirements (Metcalf and Eddy, 2003):

- Untreated (raw) primary solids,
- Untreated mixed primary and trickling filter solids,
- Untreated mixed primary solids and WAS,
- Anaerobically digested primary solids,
- Anaerobically digested mixed primary solids and WAS,
- Untreated WAS, and
- Aerobically digested primary solids.

Digestion changes solids' chemical and physical characteristics, increasing alkalinity while reducing mass. However, stabilized solids typically are more difficult to dewater than unstabilized solids. Anaerobically digested solids contain considerably more colloidal and supracolloidal solids than primary solids or activated sludge does (Karr and Keinath, 1978). Aerobic digestion detains solids for 30 days or more, greatly reducing the dewatering characteristics of the resulting biosolids. Digested solids typically have higher specific resistance values (Karr and Keinath, 1978), which, in turn, mean higher chemical conditioner doses to achieve a specified dewatered solids concentration.

Solids with inorganic contents in the range of 15 to 35% (e.g., biological solids) typically have cationic charge-neutralization requirements. Digestion also produces solids with cationic charge-neutralization requirements, although the inorganic solids content may increase to between 30 and 50%. Occasionally, lime-stabilized or chemically treated solids contain higher levels of inorganic solids, which respond better to an anionic or non-ionic polymer. As a general rule, residuals containing less than 50% inorganic solids have a cationic charge demand, while those containing more than 50% inorganic solids have an anionic or non-ionic charge demand (regardless of whether they are accompanied by a pH shift).

Whenever possible, raw, undigested, or unprocessed solids should remain separate from biological or chemically treated solids until just before dewatering. This is especially true if biological solids are generated at a plant that practices biological nutrient removal. Septic conditions cause bacteria to release their bound phosphorus to the filtrate, which then is recycled back to the influent. If such biological solids will be dewatered with primary solids, they should not be mixed until just before they enter the thickening and dewatering device.

2.3 Purpose of Conditioning: Thickening and Dewatering

The fundamental objective of conditioning is to cause fine solids to aggregate via coagulation with inorganic or organic coagulants, flocculation with organic polymer, or both (IWPC, 1981). It should improve the efficiency of thickening, dewatering, and other subsequent treatment processes. Also, conditioning is a significant item in a solids-management O&M budget, so it is desirable to select the most cost-effective method that produces acceptable liquid and solid output streams.

To be effective, the conditioning method must be compatible with the proposed methods of solids thickening, dewatering, and ultimate use or disposal. For example, belt filter presses, gravity belt thickeners, and rotary drum thickeners perform better when the solids are a uniform floc size that increases the voids between particles, thus allowing free water to filter more rapidly through the porous belt or drum. Polymer conditioning is the easiest way to produce such a floc. Other dewatering methods (e.g., pressure filtration and sand bed filtration) performed well when the solids were conditioned via the addition of chemical solids (organic and inorganic) or bulking materials.

Design engineers should take all of the subsequent solids treatment processes into account. For example, if the dewatered solids will be sent to a composting system or thermal dryer, then the conditioning and dewatering systems must produce a cake with maximum solids content. However, in concept, the conditioning and dewatering systems should not reduce the fraction of volatile solids; use an exotic, expensive polymer; or add inorganic chemicals that dramatically increase the volume of material to be dried or composted.

3.0 ULTIMATE DISPOSAL OR USE OF BIOSOLIDS

Title 40 of the *Code of Federal Regulations* addresses the use and disposal of solids generated during the treatment of domestic wastewater. Wastewater solids disposed in municipal landfills or used as landfill cover material must comply with the requirements of 40 CFR 253, as well as state and local requirements. For example, it is becoming increasingly common for local or state authorities to require that residuals contain 35 to 40% solids before they can be codisposed with municipal solid waste. Landfilled solids may have to meet certain levels of biological stability, soil engineering properties, or both. A sufficiently high dose of lime can both stabilize and condition solids, while a dose of lime, fly ash, or other bulking materials can improve a cake's mechanical properties. The solids' mechanical strength can be measured by a slump test (similar to that used for concrete).

On the other hand, if biosolids will be land-applied or otherwise beneficially used, then the material must meet the pollutant concentrations, pathogen-reduction requirements, and vector-attraction reduction standards outlined in 40 CFR 503. Because of continuing public concerns about biosolids use, many farmers will only accept Class A solids, which in turn affects overall conditioning and treatment options. Also, some farmers only accept solids that were treated with specific conditioners. In addition, biosolids characteristics and site conditions (e.g., groundwater and soils) may limit the use of certain conditioners and treatment methods. For example, certain crops are better cultivated in acidic soils, and land-applying lime-treated biosolids to such fields would not help the overall agricultural operation.

4.0 TYPES OF CHEMICAL CONDITIONING

Solids can be conditioned via a number of methods (e.g., chemical, heat, and freeze–thaw). The most popular is chemical conditioning (e.g., polymers, inorganic chemicals, or both). Heat treatment and freeze–thaw conditioning have been used to a limited extent, but their use has declined in recent years because they were problematic.

In the mid-20th century, wastewater treatment professionals used various inorganic chemicals and natural organics to condition solids. The most common inorganic chemicals used were lime and iron salts. When synthetic organic polymers were introduced in the late 1960s, they quickly were adopted for solids conditioning because they did not significantly increase the amount of solids to be thickened and dewatered.

4.1 Inorganic Chemicals

Inorganic chemical conditioning is principally associated with plate-and-frame filter presses, although they also have been used for belt filter presses. The chemicals typically used are ferric chloride and lime. Compared to polymers, larger doses of inorganic chemicals are required to condition solids, and this affects the volume of solids to be managed. For example, adding iron salts and lime can increase the solids mass (and volume) by as much as 20 to 40% (WEF, 2003).

Lime and liquid ferric chloride are the two most widely used inorganic conditioning agents for recessed-chamber filter presses. These conditioning agents are readily available and can condition a wide range of solids. In addition, the resulting biosolids are suitable for land-application or composting.

Less commonly used inorganic coagulants include liquid ferrous sulfate, anhydrous ferric chloride, aluminum sulfate, and aluminum chloride. Other inorganic materials (e.g., fly ash, power plant ash, cement kiln dust, pulverized coal, diatomaceous earth, bentonite clay, and sawdust) have been used to improve dewatering, increase cake solids, and in some cases, reduce the required dosage of other conditioning agents.

In addition to increasing the volume of solids to be managed, inorganic chemical conditioners reduce the solids' heat value. However, cake combustibility depends on the ratio of water to dry volatile solids, not the level of chemical precipitates, in the cake.

4.1.1 Lime and its Characteristics

Some plants use lime to control pH and improve settling in wastewater treatment processes, as well as condition and stabilize solids. Lime is commercially available in two main dry forms:

- Pebble quicklime (CaO) and
- Powdered hydrated lime [$Ca(OH)_2$]

In either form, lime is caustic, tends to produce dust, and tends to precipitate when slurried, forming a calcium carbonate scale on conveyance equipment.

As a conditioner, lime typically is used to raise the pH, which was lowered by ferric chloride addition. It also forms calcium carbonate and calcium hydroxide precipitates, which improve dewatering by acting like a bulking agent, increasing porosity while resisting compression. Some dissolved calcium hydroxide also is available at high pH levels.

Quicklime is typically 85 to 95% pure; it typically is called *calcined lime* because it is manufactured by burning crushed limestone (calcium carbonate) in high-temperature kilns to drive off carbon dioxide, leaving calcium oxide (quicklime). It typically is purchased in pebble form to minimize dust problems during handling. However, it rarely is applied in dry form, except to stabilize solids. Instead, it typically is mixed with water and converted to the more reactive hydrated form (calcium hydroxide) before application. This hydration reaction (typically called *slaking*) emits heat as part of the reaction:

$$CaO + H_2O \rightarrow Ca(OH)_2 + \text{Heat} \tag{22.1}$$

The quicklime pebbles rupture during slaking, splitting into microparticles of hydrated lime, which have a large total surface area and are highly reactive. A high-grade quicklime produces a quick-slaking, highly reactive, calcium hydroxide slurry, while a low-grade quicklime produces a slow-slaking, less reactive slurry (see Table 22.1). Low-grade quicklime requires critical water control to maximize slaking efficiency and minimize calcium hydroxide particle size.

TABLE 22.1 Relative slaking ability of quicklimes (NLA, 1982).

CaO content, %	Degree of burn	Slaking ability
High	Soft	Very quick
	Normal	Medium
	Over to hard	Medium to slow
Medium	Soft	Quick to medium
	Normal	Medium
	Over to hard	Slow
Low	Soft	Quick to medium
	Normal	Medium
	Over to hard	Slow to very slow

Quicklime must be stored under controlled conditions, because prolonged contact with carbon dioxide in moist air causes quicklime to *air slake*, cake, and become less reactive. Likewise, air or excessively hard water (alkalinity more than 180 mg/L as calcium carbonate) in a hydrate slurry encourages carbonate scaling, which eventually can lead to plugging problems in conveyance pumps and piping.

For quality control, quicklime should be highly reactive, quick-slaking, and able to disintegrate without producing objectionable amounts of dissolved or unslaked products. Medium-slaking limes are not preferred. Low-slaking and run-of-kiln quicklimes are unacceptable.

Hydrated lime is a powdered form of calcium hydroxide; its composition and characteristics depend on the quality of its parent quicklime (see Table 22.2). Hydrated lime typically costs 30% more than quicklime with the same calcium oxide content because of its higher production and transportation costs. However, at small plants where daily requirements for lime are intermittent or minimal, hydrated lime often is preferred because it does not require slaking. The storage and mixing operations are relatively simple (e.g., typically a dedicated storage area and minimal manual labor). Hydrated lime is more stable than quicklime, so storage precautions are satisfied more easily. However, because of its dusting characteristics, handling is more difficult.

4.1.2 *Ferric Salts*

Both ferric chloride and ferric sulfate react with the bicarbonate alkalinity in solids to form ferric hydroxide precipitates. The precipitate can lead to both charge neutralization and floc aggregation. The chemical reaction may be written as follows:

$$Fe + 3H_2O \rightarrow Fe(OH)_3 + 3H \tag{22.2}$$

$$2FeCl_3 + 3Ca(HCO_3)_2 \rightarrow 2Fe(OH)_3 + 3CaCl_2 + 6CO_2 \tag{22.3}$$

TABLE 22.2 Characteristics of quicklime and hydrated lime (Wang et al., 2007).*

Material	Available forms	Containers and requirements	Appearances and properties	Weight	Commercial strength	Solubility in water
Quicklime	Pebble, 6–19 mm	80–100 lb moisture-proof bags, barrels, and container cars. Store dry, maximum 60 days in tight container and 3 months in moisture-proof bag.	White (light grey to tan). Lumps to powder. Unstable caustic irritant slakes to hydroxide slurry evolving heat. Saturated volume pH approximately 12.5.	3.4–4.7 kg/m^3 sp gr: 3.2–3.4	70–96% CaO	Reacts to form $Ca(OH)_2$; 1 lb of quicklime will form 1.16 to 1.32 lb of $Ca(OH)_2$ with 2–12% grit, depending on the purity.
Hydrated lime	Powder, <200 mesh	50-lb bags, 100-lb barrels, and container cars. Store dry, maximum 1 year	White. Powder free of lumps. Caustic dust irritant absorbs H_2O and CO_2 to form $Ca(HCO_3)_2$. Saturated volume pH approximately 12.4.	1.6–2.5 kg/m^3 sp gr: 2.3–2.4	82–93% $Ca(OH)_2$ 62–74% CaO	10 lb/1 000 gal at 70°F 5.6 lb/1 000 gal at 175°F

* gal × 3.785 = L and lb × 0.4536 = kg.

The acid formed during the reaction caused the pH to drop to 6.0. Adding lime raises the pH as high as pH 8.5, thus allowing the ferric chloride reaction to be more efficient in forming hydroxides. Lime also reacts with bicarbonate to form calcium carbonate, a granular structure that provides the porosity needed to increase the water-removal rate during pressure filtration. This chemical reaction is as follows:

$$Ca(OH)_2 + Ca(HCO_3)_2 \rightarrow 2CaCO_3^- + 2H_2O \tag{22.4}$$

Depending on the type of solids involved, the ferric chloride dosage ranges from 2 to 10% (dry solids basis), and lime dosages range from 5 to 40% (dry solids basis). Activated sludge requires high ferric chloride dosages, anaerobically digested solids require mid-range dosages, and fresh raw primary solids require low dosages (see Table 22.3).

Typically used to flocculate solids, ferric chloride is sold as an orange-brown liquid—containing between 30 and 35% (by weight) ferric chloride. At 30°C (86°F) and a specific gravity of 1.39, a 30% ferric chloride solution typically contains 1.46 kg (3.24 lb) of ferric chloride. Liquid ferric chloride is corrosive, so it must be handled and stored properly. In colder climates, for example, the shipping strength is reduced to prevent a crystalline hydrate from forming on cold rail cars.

TABLE 22.3 Typical dosages of ferric chloride and lime for dewatering wastewater solids (U.S. EPA, 1979).

Application	Sludge type	Ferric chloride, g/kg[a]	Lime, g/kg[a]
Vacuum filter	Raw primary	20–40	70–90
	Raw WAS	60–90	0–140
	Raw (primary + TF[b])	20–40	80–110
	Raw (primary + WAS)	22–60	80–140
	Raw (primary + WAS + septic)	25–40	110–140
	Raw (primary + WAS + lime)	15–25	None
	Anaerobically digested primary	30–45	90–120
	Anaerobically digested (primary + TF)	40–60	110–160
	Anaerobically digested (primary + WAS)	30–60	140–190
Recessed-chamber filter press	Raw primary	40–60	100–130
	Raw WAS	60–90	180–230
	Anaerobically digested (primary + WAS)	40–90	100–270
	WAS + TF	40–60	270–360
	Anaerobically digested WAS	70	360
	Raw primary + TF + WAS	75	180

[a] All values shown are mass of either $FeCl_3$ or CaO per unit mass of dry solids pumped to the dewatering unit.
[b] Trickling filter.

Ferric-chloride coagulation is pH-sensitive; it works best above pH 6. Below pH 6, floc formation is weak and dewaterability is sometimes poor. So, lime is used to adjust the pH to optimize ferric chloride use and solids dewatering. There may be situations in which ferric chloride is effective at a pH less than 6, but these exceptions depend on solids type and cake-dryness requirements.

Liquid ferric sulfate typically is sold as a reddish-brown liquid—water containing 50 to 60% of ferric sulfate. It is a cationic coagulant and flocculant typically used with another conditioning agent (e.g., lime or polymers). It has been reported that using ferric sulfate before solids thickening or dewatering will reduce the amount of polymer needed and improve the filtrate or centrate quality. However, its use as a solids conditioner is limited; it primarily is used in water and wastewater treatment to remove turbidity, color, suspended solids and phosphorus.

Ferrous sulfate (also called *copperas*) is similar to ferric chloride in terms of handling, storage, and stoichiometry; however, its use as a solids conditioner has been limited in the United States. Ferrous sulfate ($FeSO_4 + 7H_2O$) is available in granular form in bags, barrels, and bulk. The product has a bulk density of about 1 000 to 1 100 kg/m^3 (62 to 66 lb/cu ft). Dry ferrous sulfate will begin to cake when stored at temperatures above 20°C (68°F) and will further oxidize and hydrate in moist and humid conditions. Ferrous sulfate should be stored in a dry area, and care should be taken to control dust, which can stain and also irritates skin, eyes, and the respiratory tract. Ferrous sulfate forms an acidic solution, so manufacturer precautions should be followed when storing, feeding, and transporting the material. Ferrous sulfate in granular (dry) form may be fed using gravimetric or volumetric feeding equipment; it also may be fed as a solution.

The effectiveness of ferric coagulation depends on pH and alkalinity. A lower pH favors the formation of positively charged hydroxoiron (III) complexes; a higher pH favors the solid species $Fe(OH)_3(s)$ (see Figure 22.1). Because hydroxoiron (III) complexes are effective coagulants, a lower pH should produce better results (see Figure 22.2). In a study conducted by Tenney et al. (1970), ferric iron was most effective between pH 5 and 8, which is near the pH of maximum precipitation (shown in Figure 22.1).

Alkalinity is important in ferric solids conditioning because it controls solids' pH during conditioning. The ferric ion functions as an acid, lowering pH, while alkalinity maintains the existing pH. For a given solids, the pH decreases as ferric doses increase.

4.1.3 Ferric Salts with Lime

Precipitation of $Fe(OH)_3$ can neutralize charge and lead to effective aggregation and filtration in the pH range 6 to 8. Practically, however, the majority of wastewater solids cannot be adequately conditioned unless lime is added after the ferric salt (Christensen and Stulc, 1979). The iron neutralizes and precipitates organic constituents, but the lime

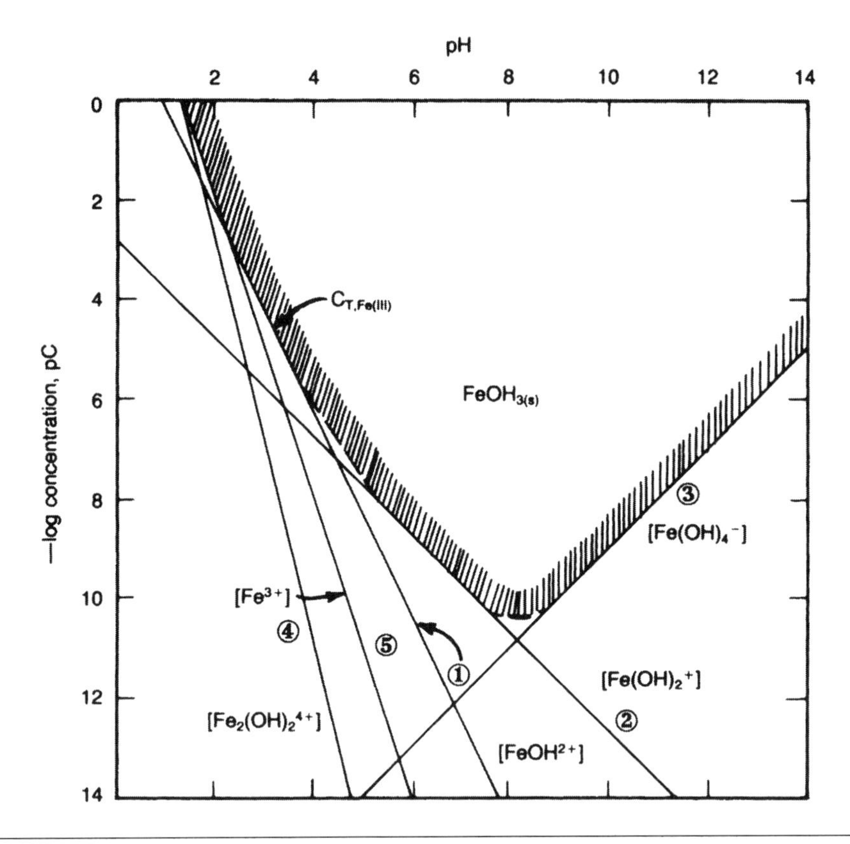

FIGURE 22.1 Equilibrium concentrations of hydroxoiron (III) complexes in a solution in contact with freshly precipitated $Fe(OH)_{3(s)}$ at 25°C (Snoeyink and Jenkins, 1980; reprinted with permission from Wiley & Sons, Inc.).

creates a much more rigid framework of calcium carbonate, which provides a rigid shell around the organic material (Denneux-Mustin et al., 2001). Full-scale filtration involves much more pressure than that typically applied in laboratory tests (e.g., Figure 22.2), which is why lime is required at full scale. The key ingredients are a pH of 11 to 12, a high calcium ion concentration (10^{-2} M), and the presence of solid ferric species (Christensen and Stulc, 1979). Conditioning with ferric salts and lime is not practiced in centrifugation because the solids cannot resist the imposed shear stresses but corrode and abrade metallic surfaces. Ferric should be added before lime at a separate addition point because adding ferric and lime to thickened solids together in the same tank (or in close proximity) adversely affects ferric conditioning (Christensen and Stulc, 1979; Webb, 1974). When conditioning with both ferric and lime, it typically takes two to four times more quicklime than ferric chloride to reach a pH between 11 and 12.

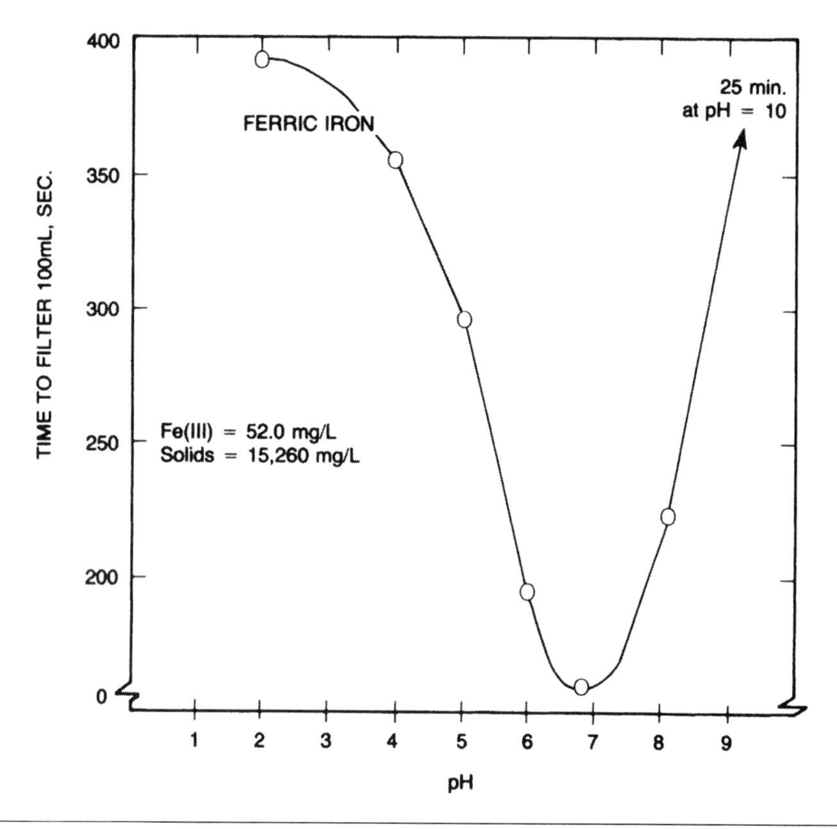

FIGURE 22.2 Effectiveness of ferric iron as a function of pH (Tenney et al., 1970).

The choice of ferric salt for conditioning is more significant when the ferric salt is followed by lime (see Table 22.4). Ferric sulfate followed by lime deteriorates more rapidly and produces a poorer result than ferric chloride followed by lime. The deterioration of solids dewaterability seems to be associated with the formation of insoluble calcium sulfate.

4.1.4 Aluminum Salts

Aluminum salts typically are not used for solids conditioning in the United States, although they have been used at some facilities with limited degrees of success. New age coagulants [e.g., polymerized aluminum chloride (PAC) and aluminum chlorydrate (ACH)] that are widely used in the water treatment industry also are being used in the wastewater treatment industry for phosphorus removal and, to a limited degree, solids conditioning. While aluminum salts are not widely used as a solids conditioner in the

TABLE 22.4 Comparison of iron conditioners used with and without lime (Lewis and Gutschick, 1988; reprinted with permission).

Sludge[a] total solids, %	Iron conditioner	Iron dose, % Fe	CST[b] after iron, seconds	Lime dose, % CaO	Specific resistance after line, Tm/kg[c]
5.5	$FeSO_4 \cdot 7H_2O$	1.72	208	15	1.40
5.5	$FeCl_2 \cdot 4H_2O$	1.72	157	15	0.79
5.5	$Fe_2(SO_4)_3 \cdot 6H_2O$	1.72	41	15	0.50
5.5	$FeCl_3 \cdot 6H_2O$	1.72	26	15	0.26
5.5	$FeSO_4 \cdot 7H_2O$	3.44	180	30	0.60
5.5	$FeCl_2 \cdot 4H_2O$	3.44	139	30	0.29
5.5	$Fe_2Cl_4 \cdot 6H_2O$	3.44	27	30	0.23
5.5	$FeCl_3 \cdot 6H_2O$	3.44	19	30	0.12
7.0	$FeSO_4 \cdot 7H_2O$	3.44	480	20	1.1
7.0	$FeCl_2 \cdot 4H_2O$	3.44	—	20	0.56
7.0	$Fe_2(SO_4)_3 \cdot 6H_2O$	3.44	117	20	0.53
7.0	$FeCl_3 \cdot 6H_2O$	3.44	58	20	0.18

[a] Mixture of raw primary sludge and WAS.
[b] Capillary suction time.
[c] Terameters/kg (Tm/kg) = 10^{12} m/kg

United States, aluminum chlorohydrates have been a popular conditioner in Great Britain for some time.

The primary differences between aluminum and ferric chemistry are the relative solubility of aluminum above pH 7 and the relative insolubility of ferric above pH 7. The practical significance is that ferric hydroxide is relatively insoluble at the highest pH values used in ferric and lime conditioning (i.e., pH 12 to 12.5), while aluminum hydroxide is quite soluble above pH 10. Aluminum salts, therefore, are unlikely to be effective with the same lime doses often used with iron salts.

4.1.5 Process Design Considerations

The following subsections describe the use of inorganic coagulants for thickening and dewatering. However, because the inorganic conditioners discussed in this chapter increase the total solids to be managed by about 20 to 40%, their use in thickening and dewatering applications is limited. Therefore, only one thickening and two dewatering applications that may use inorganic chemicals are discussed.

4.1.5.1 Conditioning for Gravity Thickening

Gravity thickening characteristics depend on the concentration and flocculant nature of the solids being thickened. In many cases, conditioning agents are not used; it depends on the type of solids being thickened. While polymers are the first choice if chemical conditioning is required, alum and ferric salts—with or without lime—also could be used (see Table 22.5). Lime also could be important for odor control because it acts as a bactericide on fresh solids before land application.

The primary mechanism when using these inorganic chemicals is coagulation and flocculation. Efficient flocculation increases solids loading rates, improves solids capture, improves supernatant clarity, and may increase underflow concentrations from conventional gravity solids thickeners. Inorganic conditioning agents also increase the dry solids volume by about 20 to 30%. When designing any thickener, engineers should determine, whenever possible, appropriate coagulants and their dosage rates by using bench-scale tests to evaluate the effectiveness of the conditioning agents during thickening operations.

4.1.5.2 Conditioning for Recessed Plate Filter Press Dewatering

Recessed-plate filter presses are one of the oldest dewatering devices and can produce the highest cake solids concentration of any mechanical dewatering equipment (Kemp, 1997). They are used more often in industrial applications than in municipal wastewater treatment plants. Unless the inorganic content of the feed solids is high, conditioning chemicals are required for successful filter press dewatering (Kemp, 1997). Plate-and-frame filter presses used to rely on lime and ferric chloride for conditioning. While these chemicals typically produced a dewatered cake containing more than 40% solids, they increased the mass to be stored, transported, and used or disposed. Lime also is associated with ammonia releases, which must be considered in overall facility design—especially ventilation and odor-control requirements.

To produce a low-moisture cake via a recessed-plate filter press, biological solids first must be conditioned with lime and ferric chloride, polymers, or a polymer combined with either inorganic chemical. (Using only polymer tends to decreases the units' performance.) Proper conditioning to a specific resistance of about 1×10^{12} m/kg or less

TABLE 22.5 Typical chemical dosages for gravity sludge thickening (WEF, 1996).

	Nature of solids/dosage of chemical			
	Raw		Anaerobically digested	
Solids	**$FeCl_3$ (mg/L)**	**CaO (mg/L)**	**$FeCl_3$ (mg/L)**	**CaO (mg/L)**
Primary	1–2	6–8	1.5–3.5	6–10
Primary + trickling filter	2–3	6–8	1.5–3.5	6–10
Primary + WAS	1.5–2.5	7–9	1.5–4	6–12
WAS	4–6	No data	No data	No data

WAS = waste activated sludge.

typically is required for good dewatering. The specific-resistance test should be used to determine whether the solids are properly conditioned and evaluate and improve various combinations of conditioning chemicals. This test is reliable but time-consuming. If a quick field test is needed, the Buchner funnel test can be conducted onsite; if 200 mL can be dewatered in 100 seconds or less, then the solids should be filterable.

Thin, poorly conditioned solids will approach design pressure rapidly (typically in 5 to 10 minutes); at this point, the sticky, hard-to-handle cake will not release, and the filtrate is inconsistent and of poor quality. However, thin cake continues to dewater if pressed for an extended period, yielding a good discharge from the filters. A properly conditioned solids builds up pressure slowly, which indicates little resistance to filtration (U.S. EPA, 1979c). The most efficient use of chemicals in the dewatering process occurs when thickening before dewatering has been optimized.

Most researchers have found that the pumping, storage, and application methods for conditioned solids significantly affect the amount of conditioning chemical needed and the performance of the dewatering process. Conditioned solids may be excessively agitated in either the conditioning tank or the surge tank. The conditioning tank should provide good mixing without shearing the floc. The surge tank should equalize conditioned solids within a maximum detention time of 30 minutes before feeding them to the press. Engineers also should design the solids feed pump to minimize floc shearing.

The chemicals typically used to condition solids for a plate-and-frame filter press unit are lime and ferric, either alone or with fly ash or polymers. Typically, the required dosage is about a 3:1 ratio [70 to 150 g/kg (140 to 300 lb/ton) of lime and 20 to 50 g/kg (40 to 100 lb/ton) of ferric chloride]. Ferric sulfate may be substituted for ferric chloride; however, it typically requires a higher dosage. Actual dosage requirements for inorganic conditioners depend on the ratio of secondary solids to primary solids, as well as the percentage of solids fed to the dewatering process. Cassel and Johnson (1978) have shown that once the ratio of secondary to primary solids is more than 1:1, secondary solids become the controlling factor in dewaterability. Because the percentage of feed solids directly affects dewatering, it is important to perfect the thickening process.

Table 22.6 summarizes the performance of various types of domestic wastewater solids conditioned with 10 to 30% lime and 5 to 7.5% ferric chloride (dry weight basis). While the dewatered cake is a drier cake containing upwards of 45% dry solids, when lime and ferric chloride are used as conditioners, then a significant portion (15 to 40%) of cake solids will offset the weight reduction of high water removal efficiency.

4.1.5.3 Conditioning for Belt Filter Press Dewatering

Inorganic chemicals typically are not used to condition solids before dewatering in a belt filter press, nor is it recommended because of chemical deposits that can "blind"

TABLE 22.6 Recessed-plate filter press dewatering experience (U.S. EPA, 2000).

Type of wastewater solids	Feed total solids (%)	Typical cycle time (hours)	Dewatered cake total solids (%)
Primary + WAS	3–8	2–2.5	45–50
Primary + WAS+ Trickling Filter	6–8	1.5–3	35–50
Primary + WAS + $FeCl_3$	5–8	3–4	40–45
Primary + WAS + $FeCl_3$ (digested)	6–8	3	40
Tertiary with lime	8	1.5	55
Tertiary with alum	4–6	6	36

WAS = waste activated sludge.

the belt, as well as excessive wear on the rollers and belt (reducing the equipment's overall life expectancy). So, the amount of information available on using inorganic conditioners with belt filter presses is limited. Alum is sometimes used, and other inorganic chemicals (e.g., lime) may be important for chemical stabilization before land application.

As with other dewatering processes, the optimal dose depends on feed solids concentration and type, mixing intensity, and mixing time. The limited information available indicates that the chemical dosage required varies directly with the ratio of secondary to primary solids. At a secondary-to-primary ratio of 1:1, an approximate dosage of 5% for ferric chloride and 15% for lime can be expected. Doubling the secondary-to-primary ratio to 2:1 could double the required lime and ferric chloride dosages.

While using inorganic coagulants may be advantageous when routinely dewatering a widely varying solids, there will be a significant increase in solids to be disposed of and, therefore, increased hauling, handling, and use or disposal costs to consider. Also, design engineers should consider ventilating the belt filter press rooms because of the strong ammonia odor that may result from lime addition.

4.2 Organic Polymers

The organic chemicals used to condition solids are primarily long-chain, water-soluble, synthetic organic polymers. Polyacrylamide, the most widely used polymer, is formed by the polymerization of a monomer acrylamide. Polyacrylamide is non-ionic. To carry a negative or positive electrical charge in aqueous solution, the polyacrylamide must be combined with anionic or cationic monomers. Because most solids carry a negative charge, cationic polyacrylamide copolymers typically are the polymers most used to condition biological solids. Polymers are further categorized by the following characteristics: molecular weight (varies from 0.5 to 18 million), charge density (varies from

0 to 100%), active solids levels (varies from 2 to 100), and form (e.g., dry, liquid or solution, emulsion, or gel).

High-molecular-weight, long-chain polymers are highly viscous in liquid form, extremely fragile, and difficult to mix into aqueous solution. Unmixed polymers in a diluted solution look like fish eyes. As the polymer's molecular weight increases, so does the difficulty in mixing and diluting it.

Unlike the inorganic chemicals discussed earlier, polymers have become attractive because they do not appreciably add to the volume of solids to be used or disposed of. Nor do they lower the fuel value of thickened or dewatered solids. Also, polymers are safer and easier to handle, and result in easier maintenance than inorganic chemicals, which require frequent cleaning of equipment, typically via acid baths. However, polymers are not completely stable, can plasticize at high temperatures, and are slippery when spilled on floors.

4.2.1 *Properties of Organic Polymers*

Polymers are classified by the polymer compound's charge (e.g., anionic, non-ionic, or cationic), molecular weight, and form (when received). A combination of the polymer molecule's charge and molecular weight is useful in product identification.

4.2.1.1 Polymer Charge

To some extent, the chemical reactions for polymers and inorganic chemicals are similar (e.g., they neutralize surface charges and bridge particles). Neutralizing a particle's negative electrical charge via the polymer's positive charge reduces the electrostatic repulsion between particles and, therefore, encourages aggregation. In polymer bridging, a long-chain polymer molecule attaches itself, via adsorption, to two or more particles at once. Flocs formed by particle bridging tend to resist shear more than flocs formed by charge neutralization.

Charge is developed by ionizable organic constituents distributed throughout the polymer molecule. Measuring the charge of a specific polymer under field conditions is nearly impossible, so its relative charge (sometimes called the *application charge*) can be used to measure its charge capability.

For anionic and non-ionic polymers, the application charge does not change significantly because the usual levels of dissolved materials present do not overcome the ionic equilibrium among anionic-charged particles in the solids. For cationic polymers, charge neutralization brought about by the influence of water alkalinity counter-ions depletes the cationic-charged species of the polymer. This effect typically causes deteriorating charge levels over time. Some polymers seem to be more charge-stable than others; however, polymer-charge stability typically is a manufacturer trade secret.

Most polymer manufacturers use the phrase *relative charge* to describe the measured titratable charge level of their products under specified test conditions. So, comparative charge levels among different manufacturers may be practically meaningless, and users should be wary of claims relating to charge in applications without onsite testing under controlled conditions. Charge is not the sole governing criterion that determines a polymer's effectiveness in a given application.

4.2.1.2 Polymer Molecular Weight

Conditioners typically can be categorized as low, intermediate, and high molecular weight. A polymer's *molecular weight* is a rough indication of the length of polymer chain that holds the charged sites apart. It also affects other product attributes (e.g., solubility, viscosity, and charge density in aqueous solution).

Although there are exceptions, lower-molecular-weight products tend to be more soluble, less viscous, and have higher charge density in water. Low-molecular-weight polymers often are called *primary coagulants*, a term typically reserved for products ranging from 2.0×10^4 to 1.0×10^5 (Kemmer and McCallion, 1979). These water-soluble products typically are marketed in concentrations of 30 to 50%. They have low viscosities (close to the viscosity of water) and can be easily diluted and mixed with water at the application point. These polymers are useful for clarification applications where there are many small dispersed particles to be destabilized and settled. They are typically in oily waste and biological waste treatment applications where low concentrations of solids are being treated. They also are sometimes used as the first part of a two-polymer program in which high-charge density is required to break the suspension.

Intermediate-molecular-weight products are available as solutions and in dry and liquid- emulsion forms. It is difficult to generalize about the entire class of intermediate-molecular-weight products; however, most require wetting (e.g., mixing activation to disperse the polymer) and aging to develop full-product activity in application. Solutions of intermediate-molecular-weight products are typically more viscous than lower-molecular-weight products. In fact, product handling of feeding characteristics typically limits commercial solutions of these products to 1% (dry solids basis) or less. Consequently, supplemental dilution water is typically needed to improve polymer disbursement in the solids being conditioned.

Intermediate-molecular-weight products are common in thickening and dewatering systems treating wastewater solids, especially those with high concentrations of secondary solids. Virtually all charge variations are available in the intermediate molecular weight range.

High-molecular-weight polymers can be cationic, anionic, and non-ionic, and are available as liquid viscous solutions, emulsions, or dry powder. Their molecular weights vary from 2×10^6 to more than 12×10^6. Solubility and viscosity considerations typically dictate the solution concentrations available. Product solutions are made up at 0.25 to 1.0% solids concentration and allowed to age for several hours before further dilution at the application point.

4.2.2 *Polymer Cross-Linkage*

A relatively recent development in polymer formulation is the use of controlled degrees of cross-linkage. Such polymers can be highly branched, rather than linear, and are called *structured polymers*. Larger doses of structured polymers may be needed to reach an optimum performance, but the resulting floc is stronger (Dentel, 2001). High-shear dewatering applications (e.g., centrifuges and some recessed-plate filter presses) can benefit from such flocculants. Some suppliers use such terms as XL, FS, FL, and FLX to indicate the cross-linked forms (Dentel et al., 2000a).

4.2.3 *Polymer Forms and Structure*

Polymers are available in two physical forms: dry and liquid. Dry polymers can be delivered in a microbead or gel powder form, while liquid polymers can be delivered as a solution or emulsion. All dry and liquid polymers can be prepared with three charge types—cationic, anionic, and non-ionic—and can be purchased in a wide array of molecular weights, charge densities, and active solids levels. The form, charge, and activity level of the polymer can greatly affect their reactivity with solids.

A polymer's "activity" relates to the percent of the molecular weight that is available to react with and flocculate solids particles; it can greatly vary with the form of the polymer. The polymer dosing criteria are stated in grams of active polymer per kilograms of dry solids. This method allows polymer types with different activity levels to be compared on an equivalent basis. For example, a polymer with an activity of 9% will require 10 times more grams of bulk polymer than a similar polymer with an activity of 90%.

4.2.2.1 Dry Polymers

Dry polymers can have an active solids level as high as 94 to 100%. The shelf life of dry polymers is typically 2 years. Storage areas that are susceptible to wet and humid conditions should be avoided, because dry polymers will tend to cake and deteriorate.

Most dry polymers are difficult to dissolve. To make up a working solution, an eductor is used as a pre-wetting device to disperse polymers in water. The solution is slowly mixed in a mixing tank until the dry polymer particles are dissolved, and then aged in accordance with manufacturer recommendations. Aging time typically ranges from 30 minutes to 2 hours. Aging allows polymer particles to "unfold" into long chains. However, once the dry polymer is diluted and converted into a solution, it is only stable for about 24 hours.

The quality of the water used to dissolve dry polymer particles is important. Hard water (greater than 120 mg/L as calcium carbonate) or water containing more than 0.5 mg/L of free chlorine can cause the solution to deteriorate within a few hours.

4.2.2.2 Emulsion Polymers

Emulsions are dispersions of polymer particles in a hydrocarbon oil or light mineral oil. Surface active agents typically are applied to prevent the polymer–oil phase from separating from the water phase. Provisions must be made to mix the bulk storage tank regularly to prevent the oil and water from separating. With emulsions, it is possible to achieve a high molecular weight and maintain an active solids level of 30 to 50% without producing a solution that has a high viscosity. The approximate viscosity of emulsion polymer in its as-delivered state ranges from 300 to 5 000 cP. The shelf life of emulsion polymers is typically 6 months to 1 year. The initial breaking of the emulsion and aging are critical for optimum performance, which can be accomplished with a static mixer, high-speed mixer, or wet dispersal unit.

Emulsion polymers can have higher molecular weights and higher charges than dry polymers without the operating problems. The primary disadvantages of emulsion polymers are the potential for oil and water separation and the higher cost per volume of active material.

One concern about these types of polymers is the adverse environmental impacts of the surfactants used in them. Such surfactants include alkylphenoethoxylates, which decompose to nonylphenol, a known endocrine disruptor. Recent developments in emulsion polymer manufacturing have been to abandon the use of mineral oils and surfactants for a new class of water-soluble emulsions. The process essentially involves dissolving the polymers in an aqueous salt of ammonium sulfate. A low-molecular-weight dispersant polymer is added to prevent aggregation of polymer chains.

Additionally, some of the typically used copolymers are susceptible to chemical hydrolysis at high pHs. If dewatered solids will be stabilized using an alkaline chemical (e.g., kiln dust, lime, etc.), odor problems could occur from the generation of trimethylamine, which has a "fishy" odor (Chang et al., 2005).

4.2.2.3 Mannich Polymers

A Mannich polymer typically contains 3 to 8% active polymer; it is produced by using a formaldehyde catalyst to promote the chemical reaction to create the organic compound. Because vapors from formaldehyde pose a safety hazard and can be carcinogenic, Mannich polymers should be stored carefully and only used in well-ventilated areas. Mannich polymers are viscous (from 50 000 to more than 150 000 counts/s), difficult to pump, and have a relatively short shelf life. However, they can be effective and economical for large treatment plants, depending on the shipment cost.

4.2.4 *Polymer Dosage*

Various polymers can enhance the performance of thickening or dewatering processes. The dose needed depends on the specific process used and the solids or biosolids to be thickened (see Table 22.7).

Most dewatering processes (except recessed-chamber filter presses) also require polymer addition (see Table 22.8). (Recessed-chamber filter presses typically use ferric chloride and lime as solids conditioners, either alone or with fly ash or polymers, and typically produce a slightly thinner cake with polymer conditioning than with ferric chloride and lime conditioning.) Centrifuges and belt filter presses cannot achieve optimum dewatering performance without polymer addition. Both applications require polymers with high positive charge and high molecular weight to produce a strong and durable floc.

4.2.5 *Application of Polymers*

Because of the wide range of polymers now available, the performance of almost any conditioning or dewatering process can be enhanced by their use. Depending on the application, polymers may improve unit throughput, solids capture, filtrate quality, thickened or dewatered solids, or a combination of these parameters.

TABLE 22.7 Typical dosages of polymer for thickening wastewater solids (U.S. EPA, 1979).

Application	Sludge type	Polymer dosage, g/kg
Gravity thickening	Raw primary	2–4
	Raw (primary + WAS + TF*)	0.8
	Raw WAS	4.3–5.6
Dissolved air flotation	WAS (oxygen)	5.4
	WAS	0–14
	P + TF	0–3
	P + WAS	0–14
Solid-bowl centrifuge	Raw WAS	0–3.6
	Anaerobically digested WAS	2–7.2
Rotary drum	WAS	6.8
Gravity belt	Digested secondary	5

* Trickling filter.

TABLE 22.8 Typical dosages of polymer for dewatering solids (U.S. EPA, 1979; WPCF, 1983).

Application	Sludge type	Polymer dosage, g/kg
Belt filter press	Raw (primary + WAS)	2–5
	WAS	4–9
	Anaerobically digested (primary + WAS)	6–10
	Anaerobically digested primary	4–7
	Raw primary	2–3
	Raw (primary + TF*)	3–6
Solid-bowl centrifuge	Raw primary	0.5–2.3
	Anaerobically digested primary	2.7–5
	Raw WAS	5–10
	Anaerobically digested WAS	1.4–2.7
	Raw (primary + WAS)	2–7
	Anaerobically digested (primary + WAS + TF)	5.4–6.8
Vacuum filter	Raw primary	1–5
	Raw WAS	6.8–14
	Raw (primary + WAS)	5–8.6
	Anaerobically digested primary	6–13
	Anaerobically digested (primary + WAS)	1.4–7.7
Recessed chamber filter press	Raw (primary + WAS)	2–2.7

* Trickling filter.

As with inorganic chemical conditioning, proper organic-chemical conditioning centers on three basic requirements:

- Correct dosage of polymer,
- Correct mixing procedures, and
- Continuous observation of results and response to those observations.

Adhering to these requirements is more critical to conditioning performance when using polymers than when using inorganic chemicals. The polymers perform under a narrower range of operating conditions than inorganic agents do, so they are more sensitive to dosage and mixing. Although this sensitivity requires more operator attention, it can promote efficiency because the dewatering process will not work if conditioning is not closely controlled.

4.2.5.1 Dosage

The correct chemical dosage is critical to proper operation. Chemical conditioning tests [e.g., Buchner funnel or capillary suction time (CST)] should be conducted frequently to determine conditioning requirements. Maintaining the correct dosage requires knowledge and control of the solids stream and chemical feed(s). Continuous metering equipment should be used and the solids content monitored to determine the mass flow. Chemical mass feed rates also should be continuously monitored and controlled to maintain desired dosages. Flow-measuring devices should be used to determine the chemical volume being fed. The consistency of chemical concentration also should be monitored and maintained throughout the process. Although feed-solution concentrations can be measured using total solids or viscosity measurements, an accurate flow-metering system on the feed solution makeup system is the preferred option.

Polymers should be used at specific solution strengths based on the manufacturer's recommendation and the results of any laboratory tests performed. Dilute solutions may be required because of the polymer's chemical activity or to allow good contact of relatively small chemical quantities with large solids volumes. In some cases, poor quality dilution water (e.g., secondary effluent) affects polymer solution activity. Although this is not a common concern, high-quality water typically should be used to prepare feed solutions.

4.2.5.2 Mixing Procedure

Proper mixing is critical to conditioning. It has two primary components: intensity and duration. Getting a viscous material thoroughly dispersed in the solids is of utmost importance when using polymers. Supplemental dilution water is used to reduce the polymer's viscosity, and high-intensity, short-duration mixing is needed to disperse the polymer. This high-intensity mixing is often accomplished with a polymer-injection ring and an adjustable check-valve device that delivers a high shearing action. The flocculation phase requires 15 to 45 seconds of gentle agitation to allow the chemical reaction between the polymer and solids to occur.

Mixing duration typically is on the order of 15 to 60 seconds, and can be accomplished a number of ways (e.g., in the pipeline to the equipment or in a separate flocculation tank). For example, multiple addition points located at HRTs of 15, 30, and 45 seconds from the inlet to the thickening or dewatering equipment. At each addition point in the solids feed pipe, a flexible coupling and a polymer pipe drawoff point should be provided to allow the insertion of a polymer-injection ring and an in-line, high-intensity mixing unit (see Figure 22.3). This arrangement provides the most flexibility in allowing operators to fine-tune the process feed rate.

Another method of providing HRT is to provide a flocculating tank directly upstream of the dewatering unit. This tank provides 15 to 30 seconds of flocculating time based on the design loading rate. A disadvantage of the flocculant tank is that it can create dead zones, which could result in improper conditioning.

FIGURE 22.3 Polymer injection ring and inline high-intensity mixing unit (courtesy of the City of Orlando, Florida).

4.2.5.3 Process Monitoring and Control

To ensure that thickening or dewatering performance is optimal, both processes should be monitored and frequently tested. The thickened or dewatered cake's total solids and released water should be analyzed at least once per shift to monitor solids loading and polymer performance. Although one sample per shift is sufficient, taking composite samples each shift would allow an operator to detect any operational changes and potential equipment problems.

Polymer monitoring includes performance and quality control checks. Polymer use should be recorded during each shift via drawdown readings in bulk or solution-storage tanks, timers on polymer-transfer pumps, or flow meters on polymer feed lines. To confirm solution strengths, total solids tests should be conducted regularly. From these data, product performance or process variations may be detected.

There has been much technological development on a variety of automated process sensors, controllers, and related software for managing thickening and dewatering operations to optimize performance and polymer use. These systems typically include a solids probe in the drain line from the dewatering unit's filtrate line and a controller on the solids feed pump and polymer feed equipment. Typically, when the solids probe detects a sudden increase in filtrate solids concentration, the controller first tries to increase the polymer flow rate. If this does not clear the filtrate after a specific time period, the controller sends a signal to decrease the solids feed rate. These alternate steps are repeated until the filtrate clears. Other control systems monitor the viscosity or particle charge in filtrate. The particle charge measurement informs the control system whether to increase or decrease the polymer dose (Dentel et al., 2000a). The primary benefit of such automated systems is that they tend to smooth out variations in unit operations and decrease polymer use by 20 to 50%. Systems that control dewatering based on filtrate or centrate properties typically will not provide optimal cake solids (Abu-Orf and Dentel, 1999). So, if cake dryness significantly affects handling costs, a control system should not be based solely on polymer savings.

Several publications provide detailed information on automating thickening and dewatering operations that use polymers as conditioning agents (Gillette and Scott, 2001; Pramanik, et al., 2002; WERF, 1995, 2001). For example, liquid stream current monitors have been studied and show promise to continuously optimize a facility's chemical conditioning requirements. One study concluded that streaming current detectors (SCD) using different conditioning agents to dewater undigested solids and biosolids are suitable for monitoring and optimizing chemical conditioning requirements (Abu-Orf and Dentel, 1997).

4.3 Process Design Considerations for Thickening and Dewatering

Below are summaries of design considerations for several types of thickening and dewatering processes. Typical polymer dosages for thickening and dewatering applications are provided in Tables 22.9 and 22.10, respectively.

4.3.1 Conditioning for Gravity Thickening

Conventional gravity thickening typically does not require the use of organic polymers. That said, using these chemicals increases solids and hydraulic loading rates by a factor of two to four and improves solids capture. However, they have minimal effect on the resultant underflow solids concentration. Also, using polymer increases the overall cost of gravity thickening, so it only should be used to prevent operating problems caused by solids carryover.

TABLE 22.9 Polymer dosages associated with various solids thickening processes.

Facility name	Type of sludge	Influent feed solids (%)	Method of thickening	Polymer dosage (lb/ton)	Thickened solids (%)	Method of stabilization
Water Reclamation Facility, Bend, Oregon	Secondary (100%)	0.4–0.6	Gravity belt thickening	8–10	4–5	Anaerobic digestion
Irwin Creek WWTF Charlotte, North Carolina	Primary (60%) Secondary (40%)	3.2 0.5–1.0	Conventional gravity thickening	0	4–5	Anaerobic digestion
City of Greeley WPCF Greeley, Colorado	Secondary (100%)	0.6–0.8	Centrifuge	1.7–2.5	4.5–6.5	Anaerobic digestion
JEA Buckman Street WWTP, Jacksonville, Florida	Primary (60%) Secondary (40%)	1–3	Gravity belt thickening	9–12	3–5	Anaerobic digestion
Greenfield Road WRP Mesa, Gilbert and Queen Creek, Arizona	Primary and secondary	1–1.25	Centrifuge	0.5	5	Anaerobic digestion
Iron Bridge Water Reclamation Facility, Orlando, Florida	Secondary (100%)	0.4–0.6	Gravity belt thickening	7–9	2–3	Lime stabilization
Northwest Water Reclamation Facility, Orange County Utilities, Orlando, Florida	Secondary (100%)	0.7–0.8	Conventional gravity thickening	0	1–2	Contract lime stabilization
Water Conserv I, Orlando, Florida	Secondary (100%)	0.7–1.2	Gravity belt thickening	10–12	4–5	Transported to larger facility for further stabilization
Water Conserv II, Orlando, Florida	Secondary (100%)	0.4–0.8	Gravity belt thickening	10–12	2–4	Anaerobic digestion
Orange County Utilities, South Water Reclamation Facility, Orlando, Florida	Secondary (100%)	0.4–1.25	Gravity belt thickening	2.4–6.5	3.1–6.4	Anaerobic digestion
91st Avenue Wastewater, Treatment Plant, Phoenix, Arizona	Secondary (100%)	1.0–1.5	Centrifuge	2.0–3.0	5.7–6.2	Anaerobic digestion
McAlpine Creek WWTF, Pineville, North Carolina	Primary (50%) Secondary (50%)	1.0	Conventional gravity thickening	0	3–5	Anaerobic digestion
			Centrifuge	3–5	0.9–1	
Roger Road Wastewater Reclamation Facility	Secondary	0.2–0.4	Gravity belt thickening	9–10	5–6	Anaerobic digestion
Pima County, Tucson, Arizona	Primary	0.4–0.6	Conventional gravity thickening	0	4–5	Anaerobic digestion
Columbia Boulevard WWTP, Portland, Oregon	Secondary (100%)	0.5–1.0	Gravity belt thickening	5–9	4–6	Anaerobic digestion

WWTF = wastewater treatment facility and WWTP = wastewater treatment plant.

TABLE 22.10 Polymer dosages associated with various solids dewatering processes.

Facility name/location	Type of sludge	Type of stabilization	Influent feed solids (%)	Method of dewatering	Polymer dosage (g/kg)	Dewatered cake solids (%)
Water Reclamation Facility Bend, Oregon	Primary (55%) Secondary (45%)	Anaerobic digestion	1.8–2.1	Belt filter press	3.75–6	12–15
Irwin Creek WWTF Charlotte, North Carolina	Primary (60%) Secondary (40%)	Anaerobic digestion	1.4	Belt press	3.5–4	18.75
City of Greeley WPCF Greeley, Colorado	Primary (60%) Secondary (40%)	Anaerobic digestion	1.5–2.0	Centrifuge	5–8	19–22
Buckman Street WWTP, Jacksonville, Florida	Primary (60%) Secondary (40%)	Anaerobic digestion	2–4	Centrifuge	3.75–6.25	19–22
Greenfield Road WRP Mesa, Gilbert and Queen Creek, Arizona	Primary and Secondary	Anaerobic digestion	2.75–3.0	Centrifuge	5.75	22–23
Water Conserv II Orlando, Florida	Secondary	Anaerobic digestion	2.0–2.5	Belt filter presses	3.5–4	12
Eastern Water Reclamation Facility Orange County Utilities Orlando, Florida	Secondary (100%)	Contract lime stabilization	<1	Belt filter presses (three belt)	3–3.5	16–17
Iron Bridge Water Reclamation Facility Orlando, Florida	Secondary	Lime stabilization	2.0–3.0	Belt filter presses	3.5–4	17
Iron Bridge Water Reclamation Facility Orlando, Florida	Secondary	Lime stabilization	0.4–0.6	Belt filter presses (three belt)	3.5–4	17
Northwest Water Reclamation Facility Orange County Utilities Orlando, Florida	Secondary (100%)	Contract lime stabilization	1–2	Belt filter presses	3.75–6.25	14–16
South Water Reclamation Facility Orange County Utilities Orlando, Florida	Secondary	Anaerobic digestion	2.3–3.7	Belt filter presses	0.5–1.1	9.3–19.7

(*continued*)

TABLE 22.10 Polymer dosages associated with various solids dewatering processes (*continued*).

Facility name/location	Type of sludge	Type of stabilization	Influent feed solids (%)	Method of dewatering	Polymer dosage (g/kg)	Dewatered cake solids (%)
McAlpine Creek WWTP Pineville, North Carolina	ORC reported a 1:1 ratio of primary to secondary solids	Anaerobic digestion	2.4	Centrifuge	3.75	20
Columbia Boulevard WWTP Portland, Oregon	Freshly digested (65–75%) thickened WAS (20%) Primary (80%) and previously digested, lagoon stabilized thickened WAS (25–35%)	Anaerobic digestion	1.5–2.0	Belt filter presses	3.75–5	19–22
Thomas P. Smith Water Reclamation Facility Tallahassee, Florida	Secondary (100%)	Anaerobic digestion	2.81–4.39	Screw presses	3.75–6.75	13–20

WAS = waste activated sludge; WPCF = water pollution control facility; WRP = water reclamation plant; WWTF = wastewater treatment facility; and WWTP = wastewater treatment plant.

Bench tests and other laboratory or field investigations should be performed to test the relative effectiveness of flocculating aids (either alone or in combinations). Also, care should be taken during feeding and mixing to prevent the overfeeding or poor mixing that causes "islands" to form (WPCF, 1980).

Using about 2 to 4.5 g of active polymer/kg of dry solids (4 to 9 lb/ton) can produce a solids-loading rate of about 22 to 34 kg/m^2·d (4.5 to 7.0 lb/d/sq ft) when thickening primary solids. Adding about 4.5 to 6.0 g/kg (9.5 to 12.5 lb/ton) of polymer can increase the thickener's loading rate to 12 to 16 kg/m^2·d (2.4 to 3.2 lb/d/sq ft) when treating WAS (Ettlich et al., 1978; U.S. EPA, 1978c, 1979d; WPCF, 1980).

4.3.2 Conditioning for Dissolved Air Flotation Thickening

Chemical conditioning is unnecessary for dissolved air flotation (DAF) thickening if low hydraulic and solids-loading rates are used. However, if high loading rates are required, or compaction is poor and the sludge volume index is high, chemical conditioning improves solids capture and can increase the float solids concentration. Although the increase in float solids is typically small (in the range of 0.5%), polymers may be required for WAS if a 4% float solids concentration is to be achieved. Float solids can routinely be 6% or higher when co-thickening mixtures of primary solids and WAS. Unless problems exist, solids capture without polymers is typically about 95%. With polymers, solids capture can increase to 97 or 98%, thereby improving subnatant quality and lessening the effect of recycle solids on plant performance. Also, with polymer addition, it is possible to as much as double the solids loading rate. [A typical rate is 10 kg/m^2·h (2 lb/sq ft/hr).]

Typically, a cationic polymer with a moderate charge and high molecular weight is used; however, lower-charge cationic polymers are starting to show better performance. Typical dosages are from between 2 and 5 g/kg (4 and 10 lb/ton) up to 7.5 g/kg (15 lb/ton) of dry solids.

A common problem when conditioning solids for DAF thickening is improper mixing of conditioner and solids. To mitigate this problem, a more dilute polymer solution (0.25 to 0.5%) should be used, or the flocculant should be mixed with pressurized recycle before contacting the solids (Ettlich et al., 1978; U.S. EPA, 1978c; WPCF, 1980).

4.3.3 Conditioning for Centrifugal Thickening

A solid-bowl conveyor centrifuge has been used to thicken a wide variety of solids. Centrifugal thickening typically does not require polymer addition when treating biological and aerobically stabilized solids. Well-digested solids, however, have little natural flocculating tendency, and require polymer additions to achieve acceptable solids recovery levels. So, engineers should make provisions for polymer addition in the initial design, even if chemical conditioning is not planned. The design should be flexible enough to allow conditioning chemical to be added at one of several points in the influent piping.

Dry or liquid high-molecular-weight cationic polymers are effective thickeners. When dry polymers are used, a 0.05 to 0.1% feed solution is used, while liquid polymers can range in concentration up to 0.5% on an active basis. It is important that a solids capture of at least 95% is obtained to prevent recycling filamentous bacteria and fines to the wastewater treatment process. Waste activated sludge produces a weak floc that tends to shear inside the centrifuge; a dose of up to 4 g/kg (8 lb/ton) of polymer can be used to formulate a tougher floc. Aerobically and anaerobically digested solids have little natural floc and, therefore, require about 4 to 8 g/kg (8 to 16 lb/ton) of polymer.

4.3.4 *Conditioning for Gravity Belt Thickening*

Gravity thickening works well with many types of solids. Difficult-to-thicken solids only require minor modifications of polymer dosages and solids loading rates to keep the effluent solids concentration and percent solids capture high. Gravity thickeners have been used to treat solids containing as little as 0.4% solids or as much as 10% solids with polymer addition. Polymer dosages range from between 1.5 and 3 g/kg (3 and 6 lb/ton) (dry weight basis) for raw primary solids up to between 4 and 6 g/kg (8 and 12 lb/ton) for anaerobically stabilized solids. In all cases, solids capture remained above 95%.

4.3.5 *Conditioning for Rotary Drum Thickening*

Rotary drum thickeners work much like gravity belt thickeners: in both systems, a moving, porous media retains conditioned solids while allowing free water to drain through. A polymer is injected into the feed line and mixed with incoming solids before entering the flocculation tank. Once inside the tank, the mixture is exposed to a low-shear rotary mixer to ensure maximum flocculant development. Conditioned solids then flow onto the distribution tray, where they are directed onto the rotating drum. The free water passes through the openings in the drum, while captured solids remains on the drum surface for further dewatering. Radial flights inside the drum slowly transport thickening solids toward the discharge end of the drum. Thickened solids exit the unit and fall through a discharge chute into a storage tank, pump hopper, or other suitable receiving device.

Drum speed, mixer speed, and spray water cycling is adjustable to ensure maximum performance with minimal polymer and water use. Polymer requirements are about 10 to 20% greater than those associated with gravity belt thickeners. Rotary drum thickeners are suited for high-fiber solids, as well as raw and digested solids with a significant fraction of primary solids. Their success with municipal WAS is variable and depends on solids characteristics. Residuals typically can be thickened to 5 to 7% total solids (in some cases, more than 10% total solids) with up to 99% capture of feed solids at polymer dosages ranging from 4 to 6 g/kg (8 to 12 lb/ton) (dry weight basis).

4.3.6 *Conditioning for Centrifugal Dewatering*

Polymers have been used with solid-bowl conveyor centrifuges to increase machine throughput without lowering cake dryness, to improve solids recovery, or both. Typically, a moderate-to-high charge, high-molecular-weight cationic polymer is used. Pilot studies are needed to determine the correct conditioning agent and dosage. Designs should include facilities for feeding both dry and liquid polymers. Polymer use typically increases solids capture; however, too much polymer can lead to a wetter cake because more fines are captured. Therefore, the relationship between recycled solids and cake dryness determines the dosage of polymer to be used.

4.3.7 *Conditioning for Belt Filter Press Dewatering*

The performance of belt filter presses depends on proper conditioning, and organic polymers traditionally are used. A properly conditioned product has a 95 to 98% solids recovery rate. However, the quantity of polymer required for proper conditioning varies widely; it depends on solids type, solids concentration, and ash content (typically, less polymer is needed when the ash content is high). For example, primary solids require a polymer dose of 3.5 to 5 g/kg (7 to 10 lb/ton), anaerobically digested solids require a dose of 7 g/kg (14 lb/ton on a dry weight basis), and autothermal thermophilic aerobically digested solids require a dose between 18 to 23 g/kg (37 to 47 lb/ton).

Insufficient conditioning causes inadequate dewatering in the initial sections of the press, which, in turn, can cause solids to extrude from the press section, overflow in the drainage section, or blind the belt. Over-conditioning can cause belt blinding and over-flocculation, which causes solids to drain too fast and mound on the belt, resulting in poor dewatering. The goal is to remove as much water as possible in the gravity section of the press. Over-flocculation may be mitigated by using deflection plates to even out the mounds before pressing begins, or by selecting a belt filter press with an extended gravity table.

Because of the shearing action between belts, Novak and Haugan (1980) have suggested using turbulent mixing when adding polymers for conditioning before dewatering. The best dosage and overall system performance depend on solids concentration, mixing intensity, and mixing time.

4.3.8 *Conditioning for Screw Press Dewatering*

The screw press is a simple, slow-moving mechanical device that gradually compresses conditioned, thickened solids as they move through the unit. Dewatering is continuous;

it begins with gravity drainage at the inlet end of the screw and then dewatered the result of increasing pressure at the end of the unit. Proper screw design is critical, because different solids require different polymer dosages, screw speeds, and configurations to maintain a desired dewatered cake concentration and solids capture rate.

Proper solids conditioning is essential to produce a consistent dewatered cake. Slower operations will produce a dryer cake but also will reduce solids throughput. Therefore, it is important that a relationship between polymer dosage, solids throughput, and cake dryness be established. Depending on the influent solids characteristics, the polymer dosage may range from 8 to 12 g/kg (16 to 24 lb/ton) to produce a cake containing between 12 and 25% dry solids and a solids capture rate of 90 to 95%.

4.3.9 *Conditioning for Rotary Press Dewatering*

The rotary press is sometimes confused with the screw press when, in fact, it operates quite differently. Another misconception is that the dewatering channels on the rotary press somehow incorporate a converging or narrowing channel. Its operating principle is relatively simple. After dosing with polymer to promote flocculation, solids are pumped into a hollow cavity between porous screens. Free water (filtrate) passes through the screens, and a cake begins to form inside the cavity. The screens constantly, slowly rotate and are can "grip" the dry cake (via frictional force) near the outlet, extruding it continuously through a pressure-controlled port. A septum separates the inlet side of the cavity from the outlet.

The polymer dosage depends on the type of solids to be dewatered. Work performed at the Daniels and Plum Island facilities in Charleston, South Carolina, indicated that 4.5 to 6 g/kg (9 to 12 lb/ton) of polymer was needed for a mixture of raw primary and secondary solids, resulting in an average dewatered cake concentration of 25%. However, in St. Petersburg, Florida, polymer dosages ranging from 15 to 18 g/kg (31 to 37 lb/ton) were required to dewater aerobically digested secondary solids to an average of 15% solids. In both cases, solids capture was more than 95%.

4.3.10 *Conditioning for Drying Beds*

Conditioning solids before sending them to drying beds is not widely practiced. In fact, the "Ten State Standards" (Great Lakes, 2004) and other design guidelines do not consider using conditioning chemicals in this application, even though they may significantly reduce drying time and, therefore, the bed area required. Relatively small polymer doses (as little as 50 mg/L) can considerably improve the drainage capabilities of properly digested solids by flocculating smaller particles. Flocculation speeds up the drainage period of the dewatering cycle and maintains a porous cake that is more readily susceptible to evaporation.

Studies have indicated that conditioning significantly increases the loading rate for digested primary solids and WAS; the unconditioned solids-loading rate was 73 kg/m^2·a (15 lb/sq ft/yr), while the conditioned solids-loading rate was 270 kg/m^2·a (55 lb/sq ft/yr). A well-conditioned product will dry in about one-third the time (approximately 10 to 15 days) required for unconditioned solids. Such performance improvements typically result from a dosage of about 15 to 23 g/kg (31 to 46 lb/ton) of a cationic polymer with a moderately high or high charge and a high molecular weight.

The conditioning system must be designed to avoid rupturing the conditioned floc during transport to the drying beds. Rupturing typically occurs during pumping; it can be overcome by locating the flocculation tank close to the drying beds, and allowing conditioned solids to flow by gravity from the flocculating chamber. Excessive holding increases the percentage of fines, which will impair flocculation and dewatering.

Both wedge-wire and vacuum-assisted drying beds use polymers to coagulate fines and promote rapid cake formation. The polymer is injected into the solids in the inlet line or in a flocculation tank next to the bed. Typical doses are between 1.5 and 3 g/kg (3 and 6 lb/ton).

5.0 CHEMICAL STORAGE AND FEED EQUIPMENT

Critical to any thickening or dewatering design is the decision of what chemicals are to be used, how they are shipped and stored, and what type of feed equipment should be used. For details on designing chemical conditioner-handling facilities, see Chapter 9, which includes a discussion on sizing the various unit operations and processes, as well as the necessary appurtenances. Because many of the chemicals are corrosive and available in various forms (e.g., liquid, dry, and gel), design engineers need to pay special attention to the design of chemical storage, feeding, piping, and control systems. For example, dry conditioners typically are converted to solution or slurry form before being introduced to solids. Liquid chemicals typically are delivered in a concentrated form and must be diluted before being mixed with solids. Other issues that must be considered when designing these systems are local building codes and the need to maintain operations during natural disasters (e.g., earthquakes, floods, hurricanes, and tornados).

As noted in Chapter 9, the sizing of storage facilities begins with an investigation of the chemicals to be used and their dosage requirements. Many conditions must be evaluated to determine the appropriate range of feed rates, which determine the feed-equipment capacities for each chemical. However, most facilities are limited by subsequent thickening or dewatering equipment capabilities, number of shifts (operating times), and desired final product.

5.1 Inorganic Chemicals

Because ferric chloride and lime have different chemical characteristics, they require different storage, pumping, piping, and handling procedures. The most important consideration when designing facilities for both chemicals is providing enough flexibility to accommodate variations in solids characteristics.

5.1.1 *Ferric Chloride*

Ferric chloride is corrosive and can be delivered in either liquid or dry form. Liquid ferric chloride is dark brown and has a shipping weight of 1.3 to 1.5 kg/L (11.2 to 12.4 lb/gal) for a 35 to 45% solution. Dry ferric chloride shipments should be stored in a dry room. Once opened, the chemical immediately should be used or mixed with water and stored in solution.

Storage tanks for ferric chloride typically are made of fiberglass, rubber- or plastic-lined steel, polypropylene, or spiral-wound extruded high-density polyethylene. Storage tanks must be insulated and, if holding a 45% solution, heated when the ambient temperature is expected to fall below 16°C (60°F).

Liquid ferric chloride feed equipment includes transfer pumps, day tanks, and metering pumps (see Figure 22.4). Rubber- or plastic-lined, self-priming centrifugal transfer pumps are used to convey bulk solution from storage tanks to day tanks. Double-diaphragm metering pumps are used to control the chemical feed rate at the application points. Chemical feed rates typically are paced according to solids feed rates. Dilution water should not be added because of the potential for hydrolysis.

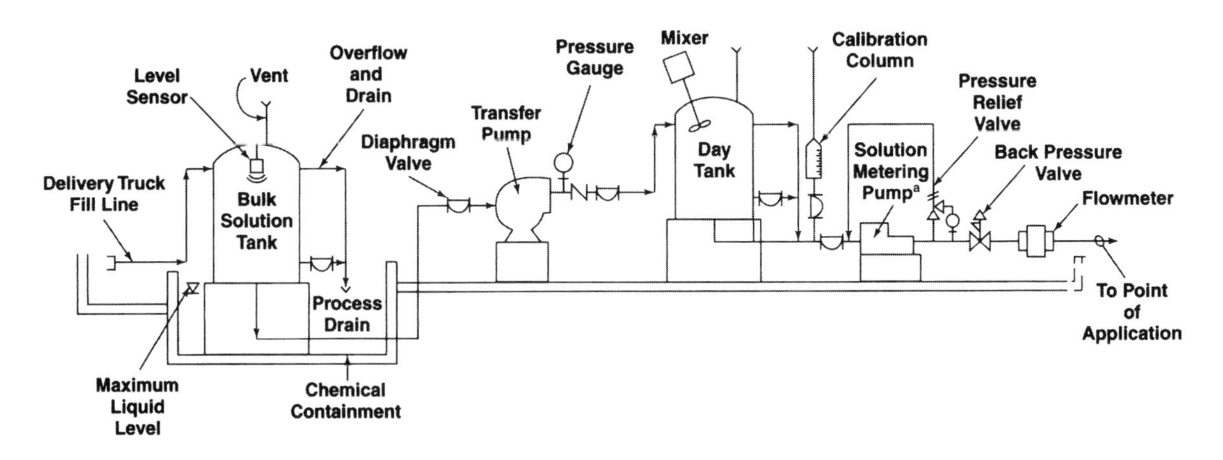

FIGURE 22.4 Simplified polymer-solution feed system (aprovide pressure relief on the discharge side of all positive-displacement polymer pumps).

Aboveground piping and valves typically are made of polyvinyl chloride (PVC), and rubber- or plastic-lined steel is used for buried applications.

The choice of feed bulk solution (30 to 45%) or diluted solution (20%) typically depends on total ferric chloride use and the expected ambient temperature. If this temperature is below the bulk solution's freezing temperature, feed facilities should be insulated and heat traced. Diluting a bulk solution may lower its freezing temperature below the lowest expected ambient temperature (thereby avoiding insulation and heat tracing), but it increases the size of day tanks, piping, valves, and feed pumps.

5.1.2 Lime

Large facilities use pebble quicklime, while small ones use hydrated lime. For lime-application rates in excess of 1 800 to 2 700 kg/d (2 to 3 ton/d), bulk quicklime is typically more economical than hydrated lime. Bagged lime requires a waterproof, well-ventilated storage building; bulk lime requires watertight and airtight storage bins. Bagged lime should be stored on pallets in a dry place for no longer than 60 days. Bulk lime can be pneumatically transferred in bins or conveyed to the bins via conventional bucket elevators or screw conveyors.

5.1.2.1 Lime Silos

Quicklime bins typically have a 55- to 60-deg slope to the bin outlet; hydrated lime bins have a 60- to 66-deg slope. Tall, slender structures with a height-to-diameter ratio (H:D) ratio of 4:2.5 are preferred. The design volume should be based on the average bulk density of the chemical, with an allowance for 50 to 100% extra capacity beyond that required to accommodate a typical delivery. Quicklime and hydrated limes are abrasive, but not corrosive, so steel or concrete bins can be used. It is imperative that the storage bins be airtight and watertight to prevent the effect of air slaking.

Hydrated lime bins should be equipped with bin agitation and a non-flooding rotary feeder at the bin outlet.

Vibrators, air pads, or both should be used on hoppers and silos to maintain lime flow to the outlet. However, design engineers should consider the type of material being handled and select vibrators with caution; the worst possible situation occurs when fine material (e.g., hydrated lime) is overvibrated and packed. For example, electromagnetic vibrators are more suitable for quicklime primarily because they tend to pack hydrated lime. Vibrators can be used for hydrated lime, however, if the unit is operated intermittently (e.g., a system that produces a vibration pulse every 1 to 2 seconds several times a minute). When transporting pebble lime, a vibrator may be operated continuously during discharge. If selected, a vibrator typically is bolted directly to the conical hopper face, one-fourth of the distance (or less) from the discharge point to the cone top.

Air jets and pulsating air pads typically are used to fluidize light materials (e.g., hydrated lime). Some of the best results are obtained by operating jets or pads periodically. Air activation is not recommended for quicklime because any moisture in the air causes air slaking.

Numerous other devices are available. The most popular is the "live" bin bottom, which operates continuously during discharge, using gyrating forces or upward thrusting baffles in the hopper to eliminate bridging and rat-holing. Less sophisticated devices include double-ended cones supported centrally in the hopper, rotating chains or paddles, and horizontal rods run from wall to wall.

Other required appurtenances include air-relief valves, access hatches, and a dust-collector mechanism.

5.1.2.2 Lime Feed System

A typical lime storage and feed system is illustrated in Figure 22.5. Bulk quicklime typically is fed to a slaking device, where oxides are converted to hydroxides, producing a paste or slurry that is further diluted before being piped or pumped to the application points. There are several manufacturers of suitable dry feeders; the choice depends on plant capacity and the degree of accuracy desired. For example, if a gravimetric-type feeder and slaker combination is indicated, larger plants can use pebble lime because gravimetric feeders are the most accurate (0.5 to 1% of set rate). This translates into cost reductions in large operations. Large, medium, and small plants also will find volumetric feeders satisfactory, with their accuracy range of ±1 to 5%.

Paste and detention slakers both operate at elevated temperatures, with or without auxiliary heaters, because of the exothermic reactions between quicklime and water. Both include a feeder, a water flow-control valve, temperature controls, a grit-removal device, a dilution chamber, and a final reaction vessel. All slakers require an integral water vapor and dust collector to maintain a slight vacuum in the slaker and discharge clean air to prevent damage to the feeders. A paste slaker has a water-to-lime ratio of 2:1 and a 5-minute slaking time at 1 040°C (1900°F). A detention slaker has a water-to-lime ratio of 3:1 to 4:1 and a 10-minute slaking time at 870°C (1 600°F). The slurry can reach 28% (by weight). For the sake of stabilization, the slurry should remain in a holding tank for 2 hours. Hydrated lime is already slaked and requires only enough water to form slurry. Typically, a 6% slurry is kept in wetting or dissolving tanks for 5 minutes.

A reasonably pure lime slurry is not corrosive and is relatively easy to keep in suspension, provided that it has been stabilized once all chemical reactions between the water and quicklime were completed. The suggested method for transferring slurry is via gravity and open trough, as long as the slurry is stabilized. If piping and transfer pumping cannot be avoided, the feed loop should be designed with a minimum velocity

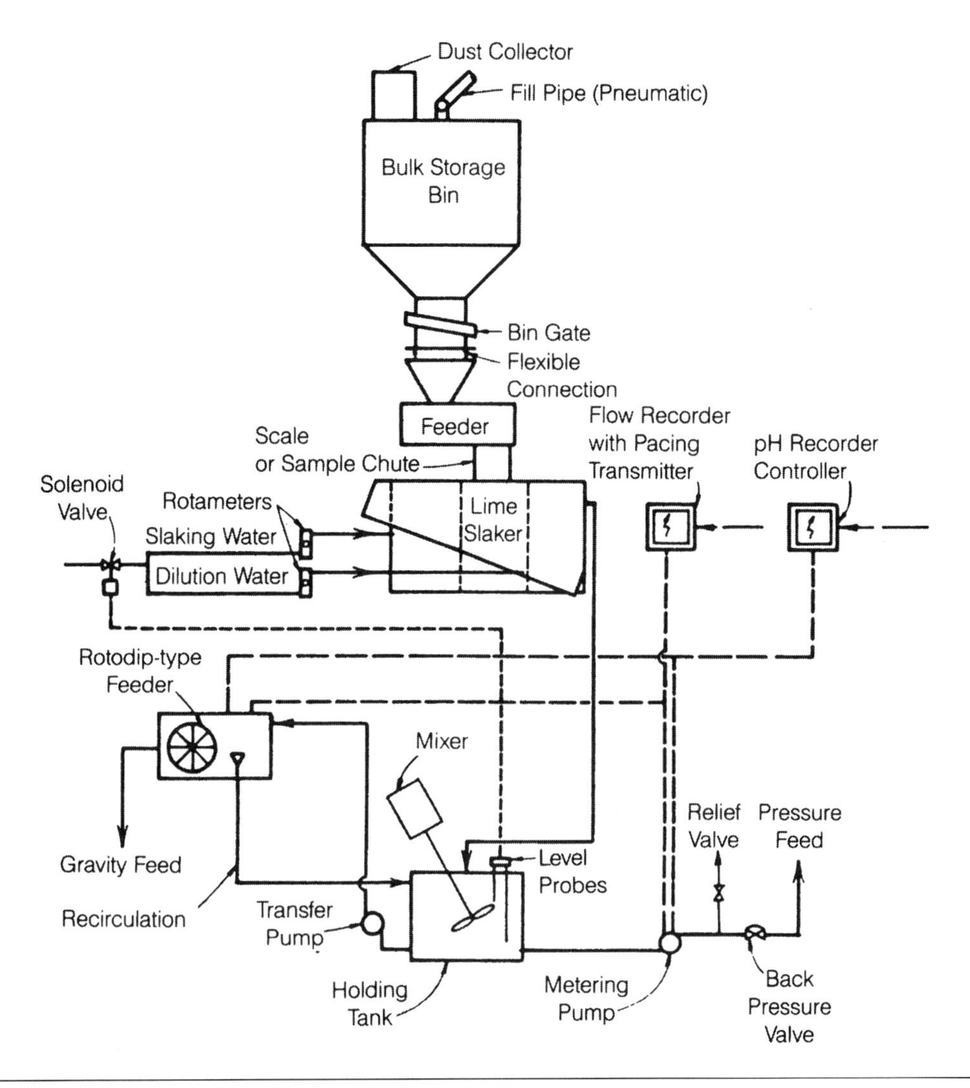

FIGURE 22.5 Typical lime feed system (vapor remover not shown).

of 0.9 to 1.5 m/s (3 to 5 ft/sec). Pinch valves are preferable to ball-and-plug valves. For a short transfer distance with a velocity less than 0.9 m/s (3 ft/sec), a flexible fire hose can be used. In general, feed piping should be at least 50 mm (2 in.) in diameter and have minimal turns and bends.

Slurry pumps typically fall within two categories: centrifugal and positive-displacement. Centrifugal pumps typically are used for low-head transfer or recirculation. Replaceable liners and semi-open impellers are desired. The pump layout should provide for easy dismantling for cleanout and repairs, and should not include water-flushed seals because they tend to scale.

Positive-displacement pumps should be used when the slurry flow must be metered or positively controlled. However, because of the abrasive nature of lime slurry, these pumps are subject to excessive wear and replacement (e.g., pistons and tubing). Turbine pumps and eductors should be avoided because of scaling problems that occur in the pipelines. The lime feed rate can be either controlled by pH or paced with the incoming solids flow.

5.2 Organic Polymers

The feed system needed to mix, store, and feed polymers depends on the type of polymer to be delivered (e.g., dry or liquid). Many facilities feed commercial-strength liquid polymer direct from shipping containers or storage tanks, or else manually prepared dry polymer solutions from batch mixing tanks. The relatively high cost of chemical conditioning requires maximum activation with minimum waste; pre-engineered feed systems can accomplish both goals. Ideally, the system should be able to handle both dry/emulsion and solution polymers. Also, if both thickening and dewatering will be performed, design engineers should consider a system that can prepare and deliver two products concurrently.

5.2.1 *Dry Polymer Feeders*

In the United States, dry polymers are supplied in bags that should be stored in a dry, cool, low-humidity area and used in proper rotation. A bulk storage time of 15 to 30 days is adequate for dry polymers. Some dust is produced when bags of dry polymer are emptied, so polymer-makeup areas should be well ventilated.

Batch-mixing and solution feed equipment consists of a dry storage hopper, dispenser and conveyor (pneumatic or hydraulic), dust collector, mix tank and agitator, aging tank, flow control valves and polymer metering feed pumps (see Figures 22.6 and 22.7). The system can be semiautomatic or fully automatic. The dry polymer can be dispensed either by hand or via a volumetric dry feeder (e.g., screw or vibrator) to a wetting jet (eductor). The polymer then is sent to a mixing (aging) tank that produces a working solution (stock solution) in 30 minutes to 2 hours. Metering pumps dispense the polymer to the solids stream. In most cases, the solution is further diluted with secondary dilution water and mixed in a static mixer to produce polymer concentrations as low as 0.01%.

Polymer feeders should be flexible enough to accommodate any type and grade of polymer. The aging tank's mixer should be variable-speed, with a maximum speed exceeding 500 rpm. The metering pump should be positive-displacement with a variable-

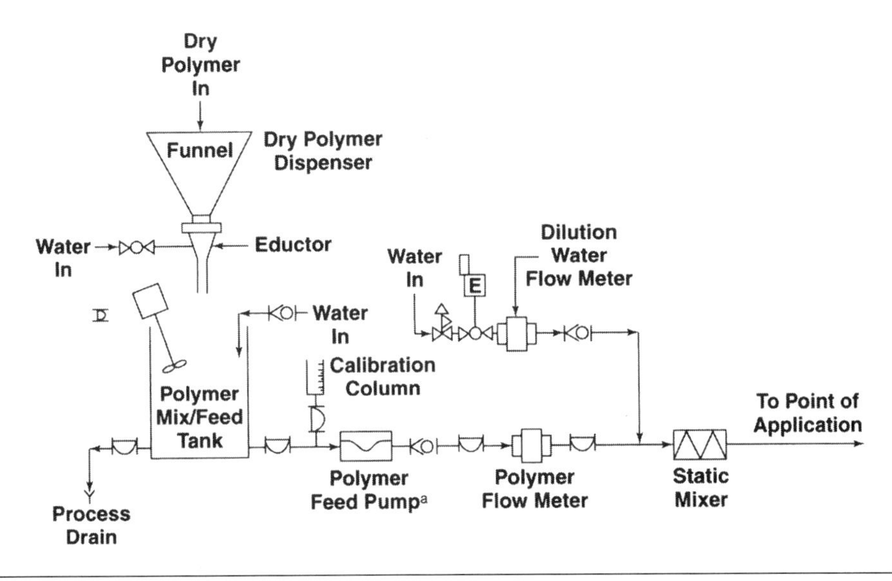

FIGURE 22.6 Typical dry polymer batch-makeup system (aprovide pressure relief on the discharge side of all positive-displacement polymer pumps).

FIGURE 22.7 Dry polymer feed system (City of Knoxville Wastewater Treatment Plant; courtesy of VeloDyne–Velocity Dynamics, Inc.).

speed controller. In general, diaphragm pumps are used for applications of about 380 L/h (100 gal/h) and less. Progressing-cavity or gear pumps are used in applications greater than 380 L/h. The speed controller can be adjusted manually or set to automatically change in response to solids flow variations. The dilution water should have a flow meter and a control valve for adjustment.

Tanks, piping, and valves should be constructed of PVC or fiberglass. Any metal parts that contact polymer solution should be constructed of stainless steel. Floors, platforms, and steps should be be provided with anti-slip patterns to prevent hazardous working conditions.

5.2.2 *Liquid Polymer Feeders*

Liquid polymers should be stored in a heated building or in heat-traced tanks. If it is stored in a building, harmful fumes and unpleasant odors can occur, so the building should be well ventilated.

The only difference between liquid and dry polymer-feed systems is the equipment used to blend polymers with water to prepare a working solution (see Figure 22.8). Solution preparation typically is a hand-batching operation in which the mixing and aging tank is manually filled with water and polymer. Variable-speed metering pumps may control the dose of liquid polymer to the aging tank.

Compact polymer-blending units can automatically mix and dilute polymers and deliver the resulting solution to the application point (see Figures 22.9 and 22.10). These pre-engineered equipment packages include a flow metering pump, valves (e.g., check,

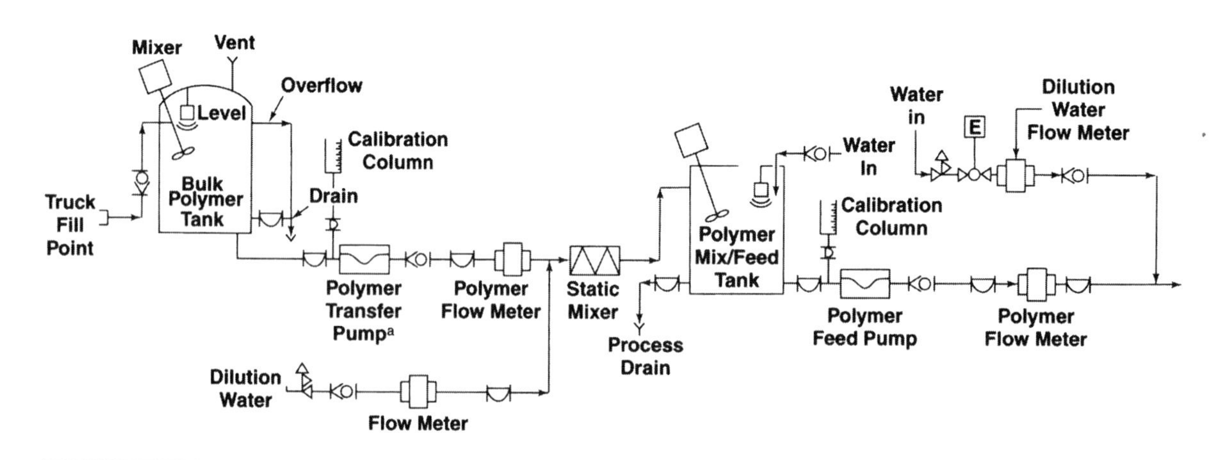

FIGURE 22.8 Typical liquid polymer batch-makeup system (aprovide pressure relief on the discharge side of all positive-displacement polymer pumps).

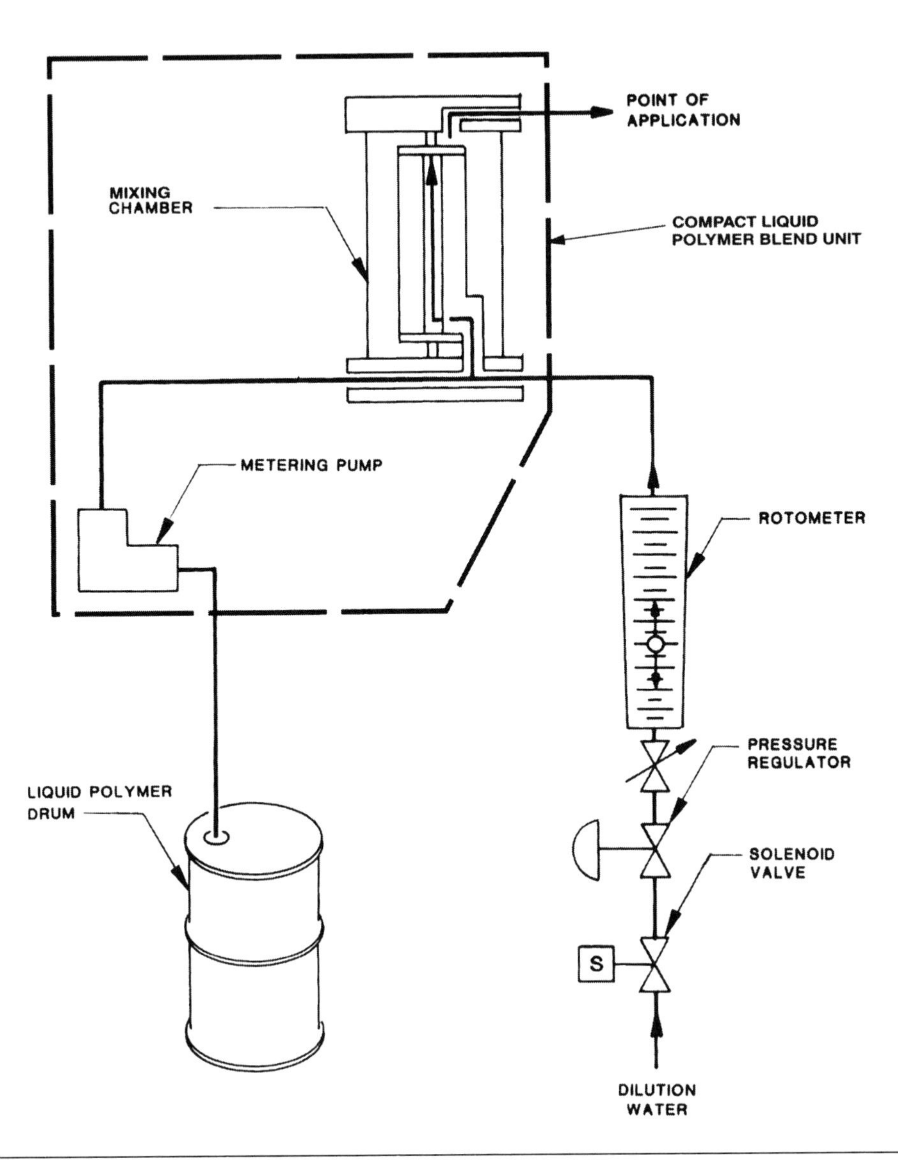

FIGURE 22.9 Compact blending system for liquid polymers.

pressure-relief, and back-pressure), a dilution-water flow-control valve, an integral mixing chamber, and instrumentation and controls. They use a high-shear mixing energy zone rather than a conventional aging tank. However, there is some question as to whether this zone fully activates the polymer. Some plants report more polymer efficiency when an aging tank is provided.

FIGURE 22.10 Compact polymer feed unit (City of Winter Haven Wastewater Treatment Plant No. 3, courtesy of PolyBlend).

5.2.3 Emulsion Polymers

Emulsion polymers consist of a high-molecular-weight polymer concentrated in a hydrocarbon solvent (oil) dispersed in water. This form allows a manufacturer to provide a high-solids organic polymer in liquid form without high-solution viscosity or limited solubility. Anionic, non-ionic, and cationic polymers are available in this form.

The storage and handling facilities for emulsion polymers are similar to those for liquid polymers. Except for the solution-preparation area, the feed system is also similar. The critical issues are aging and the initial breaking of the emulsion. Emulsion polymers must be *activated*—dispersed in water—before they are used. Activation is a two-step process. The first step, called *inversion*, involves a brief period of strong mixing to disperse the oil (continuous phase) in water (dissolving phase). The second step is a quiescent aging period, which allows the flocculant to become fully active. Anionic latex polymers require 3 to 15 minutes of aging to be completely active. Non-ionic latex polymers typically require up to 20 to 30 minutes (even longer in colder water). Some cationic latex polymers only need a few minutes to be fully active, while others need as much as 30 minutes.

It is possible to invert latex emulsion flocculants in a batch makeup system. A measured amount of neat polymer [about 20 kg (40 lb)] is dissolved in makeup water [about 1 800 L (480 gal)] in the vortex of an agitated tank. Inversion by this method takes 30 to 60 minutes to complete, so a separate aging tank is recommended (see Figures 22.11

FIGURE 22.11 Polymer storage and transfer pumps (courtesy of the City of Orlando, Florida, Conserv II).

and 22.12). Typical makeup concentrations for anionic, non-ionic, and cationic polymers are 0.5%, 1.0%, and 0.5 to 2.0% (as neat product), respectively.

Neat emulsion-polymer piping can "cake up" with dried polymer when not in use. To minimize this problem, piping should be at least 30 mm (1.25 in.) in diameter, sloped away from the polymer feed system, and include appropriately placed diaphragm or ball valves to isolate sections, as well as appropriately placed unions and blanked-off tees. In addition, a light-to-moderate machine oil [e.g., Society of Automotive Engineers (SAE) 10W-30] should be used to flush the polymer makeup system and piping whenever the system is taken out of service for more than a week. The oil can be fed via the system's calibration cylinder.

Storage tanks for emulsion flocculants should be designed with vents and breather tubes outdoors to keep fumes and vapors from being vented inside. A dehydration cell is recommended in humid environments. Some means of agitation (e.g., mechanical

FIGURE 22.12 Polymer mix and aging system (courtesy of the City of Orlando, Florida, Conserv II).

mixers or recirculation pumps) to maintain product homogeneity is also advisable, because emulsion flocculants tend to separate into oil and water.

Design engineers should avoid components made of most natural and synthetic rubber elastomers, brass, mild steel, aluminum, and plastics that soften in petroleum solvents. Positive-displacement, rotary gear, or progressing-cavity pumps typically are used to feed emulsion flocculant solutions (see Figure 22.13). Positive-displacement pumps should have low-level alarm and shutoff controls to avoid running dry and damaging feed equipment.

5.3 Safety

Most of the chemicals used as conditioning agents can cause eye burns, skin irrigation, and possibly serious burns. Appropriate safety equipment [e.g., personal protective equipment (safety glasses, filter mask, rubber gloves, boots, aprons, etc.); safety showers; water hoses; and eyewash stations] should be clearly marked and easily accessible in

FIGURE 22.13 Polymer feed pumps (courtesy of the City of Orlando, Florida, Conserv II).

the unloading, storage, and feeding locations. Other safety provisions include a dust-collection system at dry-chemical handling points (e.g., a dry pickup vacuum around feeders and slakers).

Dry chemical bags should be stored in clean, dry places to avoid picking up moisture. (The intense heat generated if quicklime accidentally contacts water could ignite flammable materials nearby.)

A vital slaker safety measure is a thermostatic valve to prevent overheating and possible explosion. This danger can occur if the controlled water supply fails while the lime feed continues, thereby allowing lime to overheat and produce excessive steam. A safety valve delivers a supply of cold water as soon as the maximum safe temperature is exceeded.

Design engineers should avoid using one conveyor or bin to handle both quicklime and other coagulants containing water of crystallization (e.g., copperas, alum, and ferric sulfate). Quicklime could withdraw the crystallization water and generate enough heat to cause a fire. When lime mixes with alum in an enclosed bin, the intense heat (greater than 590°C) generated during the reaction may release enough hydrogen to cause an explosion. Any facilities that must be alternately used should be cleaned thoroughly between applications.

6.0 DOSE OPTIMIZATION FOR ORGANIC CONDITIONERS

Selecting the right dosage of a chemical conditioner is critical to optimum performance. Dosage affects not only cake dryness but also the solids capture rate and solids disposal costs. Dosage is determined based on pilot-plant tests, bench tests, and on-line tests. The dosage should be re-evaluated periodically because solids characteristics can change.

6.1 Cost-Effectiveness of Chemical Conditioner and Dosage

Economic factors often are a consideration when selecting a chemical conditioner and dosage. Vendors typically are willing to conduct the testing and using their expertise to set up the tests (e.g., chemicals tested, dosage ranges, and injection locations) could significantly reduce plant personnel's workload. Once testing conditions are established, however, the vendor's involvement should end; all actual performance testing should be done by plant personnel.

When analyzing the cost-effectiveness of a polymer-enhanced dewatering technology, for example, investigators should begin by establishing minimum performance standards (e.g., a specified cake solids, feed rate, and solids capture rate) for the dewatering unit involved. Polymers that cannot meet these standards should be eliminated from further consideration.

Then investigators should calculate a recycle-reduction credit for polymers whose solids capture rates exceed the minimum standard, because it can cost as much to reprocess recycled solids as it does to process influent solids the first time through the liquid treatment process. This reduction credit is the product of the recycled solids volume multiplied by the reprocessing cost.

Naturally, investigators need to estimate the costs associated with re-processing recycled solids, as well as the anticipated biosolids use or disposal method (hauling, landfilling, incineration, land-application, etc.). Such costs typically depend on the percentage of solids in residuals, and investigators can develop a cost curve illustrating this relationship (i.e., solids management cost per kilogram of dry solids as a function of solids percentage). A good record of the O&M and energy costs for solids management is critical for this step.

Investigators then should conduct onsite prequalification tests, using identical operating conditions and solids feed characteristics for all polymers. First, they should adjust the operating conditions, polymer application rate, and dilution water feed rate to obtain the best performance for each polymer. Because it is difficult to maintain constant solids feed conditions from day-to-day, each polymer should be tested against a "standard" polymer. If the performance of the standard polymer changes during the test, a ratio can be developed to correct the performance of the polymer being tested. Second, investigators should analyze how various doses of each polymer affect cake solids, throughput, and filtrate quality. These tests should range from smallest dose that has any effect to those that clearly overdose the solids (i.e., produce a complete dosage curve). Third, investigators should determine the minimum polymer dosage that produces acceptable conditions (e.g., the driest cake with the best filtrate quality).

Investigators then should analyze test results to determine the lowest net operating condition for each polymer. Any dosage that results in an acceptable solids recovery rate and cake dryness should be used for the cost-effectiveness analysis.

Next, investigators should give each vendor the performance data for their specific products to obtain unit prices for the polymers meeting the minimum standards. The treatment plant's polymer cost is the product of polymer dosage multiplied by polymer unit price. Also, any special equipment needed to apply a particular polymer should be added to the polymer cost.

The net cost of the optimum dosage is calculated as follows:

$$\text{Net cost} = \text{CP} + \text{DC} - \text{RC} \tag{22.5}$$

Where

CP = cost of polymer,
DC = disposal costs, and
RC = reduction credit.

When the annual net cost and polymer dosage are tabulated for all of the tested polymers, the one with the lowest annual net cost is the most cost-effective polymer type and dosage.

This procedure can be modified to fit any thickening or dewatering process or any condition.

6.2 TESTS FOR SELECTING CONDITIONING AGENTS AND DOSAGES

Conditioning agents are critical to the optimum performance of any thickening and dewatering processes. The choice of conditioning agent and dosage affects solids capture, product dryness, and use or disposal costs. Bench-, pilot-, or full-scale conditioning tests typically are used to determine the best method for conditioning solids. Also, the dosage should be re-evaluated periodically because changes in other wastewater treatment processes may influence conditioning requirements.

Numerous laboratory tests are available to determine the effectiveness of conditioning agents in thickening and dewatering processes. Test objectives include

- Evaluating various conditioning and dewatering chemicals to determine which provides the best dewaterability;
- Developing design criteria for pilot- or full-scale dewatering processes;
- Comparing and evaluating different conditioning techniques; and
- Using different conditioning techniques to control the dewatering process.

For the results to be useful, a representative solids sample must be tested. The sample must be fresh (i.e., tested within 24 hours of collection) because storage can affect solids properties and result in erroneous conditioning data. If the sample must be stored or shipped before testing, an acceptable preservative should be used. The conditioning agents also must be fresh (i.e., storing a diluted polymer sample too long can decrease its activity).

For a detailed explanation of each test and the procedures used to select the most cost-effective conditioning agents, see *Operation of Municipal Wastewater Treatment Plants* (WEF, 2007).

7.0 DESIGN EXAMPLE

A belt filter press system used to dewater anaerobically digested solids operates under the following conditions:

- Two belt filter presses with an effective belt width of 2 m (one unit as a standby);
- Operation is 5 d/week, 7 h/d;
- Peak weekly solids production is 110 m^3/d (0.001 27 m^3/s);
- Total solids concentration of the belt filter press feed is 35 000 mg/L;
- Specific gravity of the solids feed is 1.03; and
- Polymer solution is 0.2% and is added before the belt filter press at a rate of 25 L/min.

Calculate the polymer dosage requirements.

7.1 Step 1: Calculate the peak weekly solids to be dewatered

$$\begin{aligned}\text{Wet solids} &= 110\ \text{m}^3/\text{d} \times 7\ \text{d/week} \times 1000\ \text{kg/m}^3 \times 1.03 \\ &= 793\ 100\ \text{kg/week}\end{aligned}$$

$$\begin{aligned}\text{Dry solids} &= 793\ 100\ \text{kg/week} \times 0.035 \\ &= 27\ 760\ \text{kg/week} \\ &= (27\ 760\ \text{kg/week})/(5\ \text{d/week}) \\ &= 5560\ \text{kg/d} \\ &= (5560\ \text{kg/d})/(7\ \text{h/d}) \\ &= 795\ \text{kg/h}\end{aligned}$$

7.2 Step 2: Determine whether solids loading and hydraulic loading rates are within operating parameters

$$\begin{aligned}\text{Solids loading} &= (795\ \text{kg/h})/2\ \text{m} \\ &= 398\ \text{kg/h·m (determined to be within acceptable range of 230 to 455 kg/h·m)}\end{aligned}$$

$$\begin{aligned}\text{Hydraulic loading} &= 110\ \text{m}^3/\text{d} \times (1\ \text{d}/1440\ \text{min}) \times 1000\ \text{L/m}^3 \\ &= 76\ \text{L/min} \\ &= 76\ \text{L/min} \times (7\ \text{d}/5\ \text{d}) \times (24\ \text{h}/7\ \text{h})/2\ \text{m} \\ &= 183\ \text{L/min·m (determined to be within acceptable range of 150 to 190 L/min·m)}\end{aligned}$$

7.3 Step 3: Calculate the polymer dosage

$$\begin{aligned}\text{Dosage} &= 25\ \text{L/min} \times 60\ \text{min/h}\\ &= 1\,500\ \text{L/h}\\ &= (1\,500\ \text{L/h} \times 0.002 \times 1\ \text{kg/L})/(2\ \text{m} \times 398\ \text{kg/h·m}) \times (1\ \text{tonne}/1\,000\ \text{kg})\\ &= 3.77\ \text{kg/tonne}\end{aligned}$$

8.0 REFERENCES

Abu-Orf, M. M.; Dentel, S. K. (1997) Polymer Dose Assessment Using the Streaming Current Detector. *J. Water Environ. Res.*, **69** (6), 1075–1084.

Abu-Orf, M. M.; Dentel, S. K. (1999) Rheology as Tool for Polymer Dose Assessment and Control. *J. Environ. Eng.*, **125** (12), 1133–1141.

Abu-Orf, M. M.; Ormeci, B. (2005) Measuring Sludge Network Strength Using Rheology and Relation to Dewaterability, Filtration, and Thickening—Laboratory and Full-Scale Experiments. *J. Environ. Eng.*, **131** (8), 1139–1146.

Bache, D. H.; Dentel, S. K. (2000) Viscous Behaviour of Sludge Centrate in Response to Chemical Conditioning. *Water Res.*, **34** (1), 354–358.

Bruus, J. H.; Nielsen, P. H.; Keiding K. (1992) On the Stability of Activated Sludge Flocs with Implication to Dewatering. *Water Res.*, **26**, 1597–1604.

Cassel, A. F.; Johnson, B. P. (1978) Evaluation of Filter Presses to Produce High-Solids Solids Cake. *J. New Eng. Water Pollut. Control Assoc.*, **12**, 137.

Chang, J. S.; Abu-Orf, M. M.; Dentel, S. K. (2005) Alkylamine Odors from Degradation of Flocculant Polymers in Sludges. *Water Res.*, **39**, 3369–3375.

Christensen, G. L.; Stulc, D. A. (1979) Chemical Reactions Affecting Filterability in Iron–Lime Sludge Conditioning. *J. Water Pollut. Control Fed.*, **51**, 2499.

Dentel, S. K.; Abu-Orf, M. M.; Griskowitz, N. J. (1995) *Polymer Characterization and Control in Biosolids Management*; Publication D43007; Water Environment Research Foundation: Alexandria, Va.

Dentel, S. K.; Gucciardi, B. M.; Griskowitz, N. J.; Chang, L.; Raudenbush, D. L.; Arican, B. (2000a) Chemistry, Function, and Fate of Acrylamide-Based Polymers. In *Chemical Water and Wastewater Treatment VI*; Hahn, H. H.; Odegaard, H.; Hoffmann, E., Eds.; Springer Verlag: Berlin, Germany. pp. 35–44.

Dentel, S. K.; Abu-Orf, M. M.; Walker, C. A. (2000b) Optimization of Slurry Flocculation and Dewatering Based on Electrokinetic and Rheological Phenomena. *Chem. Eng. J.*, **80** (1–3), 65–72.

Dentel, S. K. (2001) Conditioning. In Sludge *into Biosolids*; Spinosa, L.; P.A. Vesilind, P. A., Eds; IWA Publishing: London.

Eriksson, L.; Alm, B. (1991) Study of Bioflocculation Mechanisms by Observing Effects of a Complexing Agent on Activated Sludge Properties. *Water Sci. Technol.*, **24**, 21–28.

Ettlich, W. F.; Hinrichs, D. J.; Lineck, T. S. (1978) *Operations Manual:* Sludge *Handling and Conditioning*; EPA-68/01-4424; U.S. Environmental Protection Agency: Washington, D.C.

Gillette, R. A.; Scott, J. D. (2001) Dewatering System Automation: Dream or Reality? *Water Environ. Technol.*, **13** (5), 44–50.

Great Lakes Upper Mississippi River Board of State Sanitary Engineering Health Education Services Inc. (2003) *Recommended Standards for Wastewater Facilities*; Great Lakes Upper Mississippi River Board of State Sanitary Engineering Health Education Services Inc.: Albany, New York.

Higgins M. J. (1995) The Roles and Interactions of Metal Salts, Proteins, and Polysaccharides in the Settling and Dewatering of Activated Sludge. Ph.D. dissertation, Virginia Polytechnic Institute and State University, Blacksburg, Virginia.

Higgins M. J.; Novak J. T. (1997a) The Effect of Cations on the Settling and Dewatering of Activated Sludge: Laboratory Results. *J. Water Environ. Res.*, **69**, 215–224.

Higgins M. J.; Novak J. T. (1997b) Dewatering and Settling of Activated Sludges: The Case for Using Cation Analysis. *J. Water Environ. Res.*, **69**, 225–232.

IWPC (1981) *Sewage* Sludge *II: Conditioning, Dewatering and Thermal Drying*; Manual of British Practice in Water Pollution Control; IWPC: Maidstone, Kent, G.B.

Karr, P. R.; Keinath, T. M. (1978) Influence of Particle Size on Sludge Dewaterability. *J. Water Pollut. Control Fed.*, **50**, 1911.

Kemmer, F. N.; McCallion, J. (1979) *The NALCO Water Handbook*; McGraw–Hill: New York.

Kemp, J. S. (1997) Just the Facts on Dewatering Systems: A Review of the Features of Three Mechanical Dewatering Technologies. *Water Environ. Technol.*, **9** (12), 47–55.

Kolda, B. C. (1995) Impact of Polymer Type, Dosage, and Mixing Regime and Sludge Type on Sludge Floc Properties. Master's thesis, Virginia Polytechnic Institute and State University, Blacksburg, Virginia.

Lewis, C. J.; Gutschick, K. A. (1988) *Lime in Municipal Sludge Processing*; National Lime Association: Washington, D.C.

Metcalf and Eddy, Inc. (2003) *Wastewater Engineering: Collection, Treatment, Disposal;* McGraw-Hill: New York.

Mysels, K. J. (1951) *Introduction to Colloid Chemistry*; Interscience Publishers: New York.

National Lime Association (1982) *Lime Handling, Application, and Storage in Treatment Processes*, 4th ed.; Bulletin 213; National Lime Association: Arlington, Virginia.

Novak, J. T.; Haugan, B. E. (1979) Chemical Conditioning of Activated Sludge. *J. Environ. Eng.*, **105**, EE5, 993.

Novak, J. T.; Haugan, B. E. (1980) Mechanisms and Methods for Polymer Conditioning of Activated Sludge. *J. Water Pollut. Control Fed.*, **52**, 2571.

Novak J. T.; Goodman G. L.; Pariroo, A.; Huang, J. C. (1988) The Blinding of Sludges during Filtration. *J. Water Pollut. Control Fed.*, **60**, 206–214.

Novak, J. T.; Miller, C. D.; Murthy, S. N. (2001) Floc Structure and the Role of Cations. *Water Sci. Technol.*, **44** (10), 209–213.

Novak, J. T.; Sadler, M. E.; Murthy, S. N. (2003) Mechanisms of Floc Destruction During Anaerobic and Aerobic Digestion and the Effect on Conditioning and Dewatering of Biosolids. *Water Res.*, **37**, 3236.

Ormeci, B.; Cho, K.; Abu-Orf, M. M. (2004) Development of a Laboratory Protocol to Measure Network Strength of Sludges Using Torque Rheometry. *J. Residuals Sci. Technol.*, **1** (1), 35–44.

Park, C.; Muller, C. D.; Abu-Orf, M. M.; Novak, J. T. (2006) The Effect of Wastewater Cations on Activated Sludge Characteristics: Effects of Aluminum and Iron in Floc. *Water Environ. Res.*, **78**, 31–40.

Pramanik, A.; LaMontagne, P.; Brady, P. (2002) Automation Improvements: Installing an Integrated Control System Can Improve Sludge Dewatering Performance and Cut Costs. *Water Environ. Technol.*, **14** (10), 46–50.

Roberts, K.; Olsson, O. (1975) The Influence of Colloidal Particles on the Dewatering of Activated Sludge with Polyelectrolyte. *Environ. Sci. Technol.*, **9**, 945.

Robinson, J. K. (1989) The Role of Bound Water Content in Designing Sludge Dewatering Characteristics. Master's thesis, Virginia Polytechnic Institute and State University, Blacksburg, Virginia.

Snoeyink, V. L.; Jenkins, D. (1980) *Water Chemistry*; Wiley and Sons: New York.

Sorensen, B. L.; Sorensen, P. B. (1997) Applying Cake Filtration Theory to Membrane Filtration Data. *Water Res.*, **31** (3), 665–670.

Tenney, M. W.; Echelberger, W. F., Jr.; Coffey, J. J.; McAloon, T. J. (1970) Chemical Conditioning of Biological Sludges for Vacuum Filtration. *J. Water Pollut. Control Fed.*, **42,** R1.

Tezuka, Y. (1969) Cation-Dependent Flocculation in *Flavobacterium* Species Predominant in Activated Sludge. *Appl. Microbiol.*, **17**, 222.

U.S. Environmental Protection Agency (1978a) *Innovative and Alternative Technology Assessment Manual*; EPA-430/9-78-009; U.S. Environmental Protection Agency, Office of Water Program Operations: Washington, D.C.

U.S. Environmental Protection Agency (1978b) *Operations Manual for Sludge Handling and Conditioning*; EPA-430/9-78-002; U.S. Environmental Protection Agency: Washington, D.C.

U.S. Environmental Protection Agency (1978c) *Sludge Treatment and Disposal, Sludge Treatment, Vol. 1*; EPA-625/4-78-012; U.S. Environmental Protection Agency: Cincinnati, Ohio.

U.S. Environmental Protection Agency (1979a) *Chemical Aids Manual for Wastewater Treatment Facilities*; EPA-430/9-79-018; U.S. Environmental Protection Agency: Washington, DC.

U.S. Environmental Protection Agency (1979b) *Chemical Primary Sludge Thickening and Dewatering*; EPA-600/20-79-055; U.S. Environmental Protection Agency, Municipal Environmental Research Laboratory, Office of Research and Development: Cincinnati, Ohio.

U.S. Environmental Protection Agency (1979c) Evaluation of Dewatering Devices for Producing High-Solids Sludge Cake; EPA-600/2-79-123; U.S. Environmental Protection Agency, Water Resources Management Administration, Municipal Environmental Research Laboratory: Cincinnati, Ohio.

U.S. Environmental Protection Agency (1979d) Process Design Manual for Sludge Treatment and Disposal; EPA-625/1-79-011; U.S. Environmental Protection Agency, Municipal Environmental Research Laboratory, Office of Research and Development: Cincinnati, Ohio.

U.S. Environmental Protection Agency (1979e) *Review of Techniques for Treatment and Disposal of Phosphorus-Laden Chemical Sludges*; EPA-600/2-79-083; U.S. Environmental Protection Agency, Municipal Environmental Research Laboratory, Office of Research and Development: Cincinnati, Ohio.

U.S. Environmental Protection Agency (2000) *Biosolids Technology Fact Sheet Recessed-Plate Filter Press*; EPA-832/F-00-058; U.S. Environmental Protection Agency, Office of Water: Washington, D.C., Sep.

Vesilind, P. A. (1979) *Treatment and Disposal of Wastewater Sludges*; Ann Arbor Science Publishers: Ann Arbor, Michigan.

Wang, L. K.; Pereira, N. C.; Hung, Y. T. (2007) *Handbook of Environmental Engineering Biosolids Treatment Processes*, 6th ed.; Humana Press: Totowa, New Jersey.

Water Environment Federation (2003) *Wastewater Treatment Plant Design*; IWA Publishing: London.

Water Environment Federation (2007) *Operation of Municipal Wastewater Treatment Plants*, 6th ed.; Manual of Practice No. 11; McGraw-Hill: New York.

Water Pollution Control Federation (1980) *Sludge Thickening*; Manual of Practice No. FD-1; Water Pollution Control Federation: Washington, D.C.

Water Pollution Control Federation (1982) *An Analysis of Research Needs Concerning the Treatment, Utilization, and Disposal of Wastewater Treatment Plant Sludges*; Water Pollution Control Federation: Washington, D.C.

Water Pollution Control Federation (1983) *Sludge Dewatering*; Manual of Practice No. 20; Water Pollution Control Federation: Washington, D.C.

Webb, L. J. (1974) A Study of Conditioning Sewage Sludges with Lime. *J. Water Pollut. Control Fed.*, **73**, 192.

Chapter 23

Solids Thickening

1.0 INTRODUCTION

Wastewater treatment plants typically use thickening processes to make primary solids or a combination of primary and waste activated solids (combined solids) more concentrated.Thickening reduces the volumetric loading and increases the efficiency of subsequent solids-processing steps. Initially, most treatment plants used gravity-based thickening processes; now, solids-flotation, centrifugal, gravity-belt, and rotary drum thickening processes are widely accepted. These methods differ significantly in process configuration; degree of thickening provided; and chemical, energy, and labor requirements.

Liquid sidestreams from thickening processes often are recycled to the wastewater treatment train upstream of primary clarifiers. When recycling sidestreams, design engineers should assess their effects on the liquid treatment process because their flow, solids, and ammonia loadings can be significant.

This chapter primarily describes thickening processes, presents related design information, and offers a general comparison of these processes. It also includes a brief discussion of cothickening and its advantages and disadvantages. For cost information, see other references [e.g., *Process Design Manual for Sludge Treatment and Disposal* (U.S. EPA, 1979), *Sludge Thickening* (WPCF, 1980), and *Handbook of Estimating Sludge Management Costs* (U.S. EPA, 1985)].

2.0 GRAVITY THICKENER

Gravity thickeners function much like settling tanks: solids settle via gravity and compact on the bottom, while water flows up over weirs (see Figure 23.1). They also provide some solids equalization and storage, which may be beneficial to downstream operations.

Gravity thickeners work best on primary and lime sludge, but also are effective on primary sludge combined with trickling filter solids, primary and activated sludge, anaerobically digested solids, and to a lesser degree, activated sludge. Primary and lime

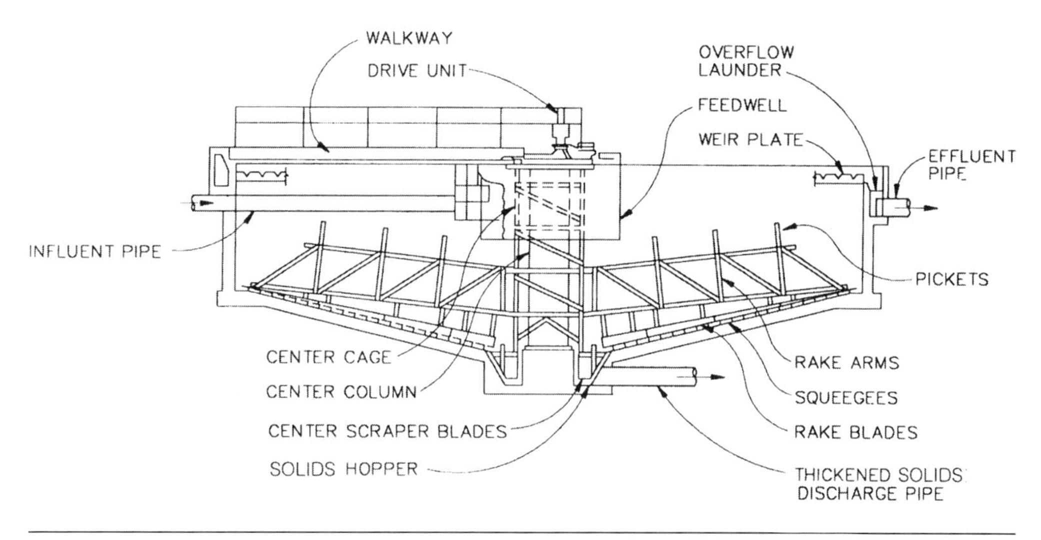

FIGURE 23.1 Example of a gravity thickener.

sludge typically settle quickly and achieve a high underflow concentration without chemical conditioning. Biological solids—particularly waste activated sludge (WAS)—typically have lower capture rates and underflow solids concentrations.

The most common gravity-thickener design is a circular tank with a side water depth of 3 to 4 m (10 to 13 ft). Such tanks typically range from to 21 to 24 m (70 to 80 ft) in diameter. Larger-diameter tanks increase solids detention time, which can cause anoxic and anaerobic activity that leads to gasification and flotation problems. The tank floor typically has a slope between 2:12 and 3:12 (steeper than that of a standard settling tank). The steep slope allows for minimum solids detention while maximizing solids depth over the withdrawal pipe in the center of the floor. It also reduces raking transport problems.

Combination clarifier–gravity thickener units are typically circular sedimentation tanks with a deeper center section that functions as a gravity thickener. Combined units are seldom rectangular because of difficulties associated with solids removal.

See Table 23.1 for typical operating results for gravity thickeners at various overflow rates.

2.1 Evaluation and Scale-Up Procedures

Experience has shown that solids-thickening characteristics vary considerably—not only among various types of solids, but also among samples of one type of solids taken from different locations. These variations can be caused by a wide range of factors (e.g.,

TABLE 23.1 Reported operating results for gravity thickeners at various overflow rates (U.S. EPA, 1979).[a]

Location	Type of solids[b]	Influent solids concentration (percent solids)	Hydraulic loading rate ($L/m^2 \cdot h$)	Mass loading rate ($kg/m^2 \cdot h$)	Thickened solids concentration (percent solids)	Overflow suspended solids, (mg/L)
Port Huron, Michigan	P + WAS	0.6	330	1.67	4.7	2 500
Sheboygan, Wisconsin	P + TF	0.3	760	2.35	8.6	400
	P + (TF + Al)	0.5	780	3.58	7.8	2 000
Grand Rapids, Michigan	WAS	1.2	180	2.06	5.6	140
Lakewood, Ohio	P + (WAS + Al)	0.3	1 050	2.94	5.6	1 400

[a] Values shown are average values only. For example, at Port Huron, Michigan, the hydraulic loading varies between 300 and 400 $L/m^2 \cdot h$, the thickened solids in the underflow varies between 4.0 and 6.0% solids, and the suspended solids in the overflow ranges from 100 to 10 000 mg/L.
[b] Al = alum solids; P = primary solids; TF = trickling filter solids; and WAS = waste activated sludge.

physical properties of solids particles, type and volume of industrial wastes treated, wastewater treatment processes used and their operating conditions, and solids-handling practices before thickening). So, engineers should design a thickening process based on criteria developed via a specific test program. If solids are not readily available for testing, engineers should design the process using performance data available from a similar thickening operation.

The two main parameters in gravity-thickener tank design are depth and area. Engineers can calculate depth based on solids volume and storage requirements; it is not controlled by the type of sludge being thickened. Tank area, on the other hand, depends greatly on solids type; it typically is determined via one of four methods: existing data, batch-settling tests, bench-scale testing, or pilot-scale testing. (For information on depth requirements, clarification function, and other design considerations, see Section 2.2.)

2.1.1 Determining Area Based on Existing Data

Engineers can use empirical data from similar applications to determine the area of a gravity thickener. However, two plants using the same upstream processes can produce solids with very different characteristics, so using empirical data may not always provide the desired results.

Table 23.2 presents typical surface-area design criteria for various types of solids. [The mass loading rate is the quantity of solids allowable per unit area of thickener per

TABLE 23.2 Typical surface area design criteria for gravity thickeners (U.S. EPA, 1979).

Type of solids	Influent Solids Concentration (% solids)	Expected Under Concentration Flow (% solids)[a]	Mass loading rate ($kg/m^2 \cdot hr$)[b]
Separate solids			
Primary (PRI)	2–7	5–10	4–6
Trickling filter (TF)	1–4	3–6	1.5–2
Rotating biological contractor (RBC)	1–3.5	2–5	1.5–2
Waste Activated Sludge (WAS)			
WAS-air	0.5–1.5	2–3	0.5–1.5
WAS-oxygen	0.5–1.5	2–3	0.5–1.5
WAS-extended aeration	0.2–1.0	2–3	1.0–1.5
Anaerobically digested sludge from primary	8	12	5
Thermally Conditioned Solids			
PRI only	3–6	12–15	8–10.5
PRI+WAS	3–6	8–15	6–9
WAS only	0.5–1.5	6–10	5–6
Tertiary Solids			
High lime	3–4.5	12–15	5–12.5
Low lime	3–4.5	10–12	2–6.5
Alum	—	—	—
Iron	0.5–1.5	3–4	0.5–2
Mixed Solids			
PRI+WAS	0.5–1.5	4–6	1–3
	2.5–4.0	4–7	1.5–3.5
PRI+TF	2–6	5–9	2.5–4
PRI+RBC	2–6	5–9	2–3.5
PRI+iron	2	4	1.5
PRI+low lime	5	7	4
PRI+high lime	7.5	12	5
PRI+(WAS+iron)	1.5	3	1.5
PRI+(WAS+alum)	0.2–0.4	4.5–6.5	2.5–3.5
(PRI+iron)+TF	0.4–0.6	6.5–8.5	3–4
(PRI+alum)+WAS	1.8	3.6	1.5
WAS+TF	0.5–2.5	2–4	0.5–1.5
Anaerobically digested PRI+WAS	4	8	3
Anaerobically digested PRI+(WAS+iron)	4	6	3

[a] Data on supernatant characteristics is covered later in this chapter.
[b] This term typically is given in $kg/m^2/day$. Because wasting to the thickener is not always continuous, it is more realistically to use $kg/m^2/hr$.

unit time ($kg/m^2{\cdot}h$) to achieve the indicated underflow solids concentration.] This table can be used to determine the gravity thickener area by dividing the actual solids loading rate by the mass loading rate associated with the type of solids and desired underflow concentration. That said, design engineers should carefully evaluate site-specific conditions, particularly with respect to the quantity of wastes treated.

2.1.2 Determining Area Based on Batch Settling Tests

Another method for determining the area of a gravity thickener is the solids flux theory. It requires that design engineers determine the relationship between settling flux and solids concentration. This relationship is based on batch-settling test results and the premise that a suspension's settling rate is solely a function of solids concentration. Because this premise is not true for wastewater solids with high solids concentrations, the method is not completely valid, but it may give satisfactory results if batch-settling conditions resemble those in a full-scale continuous thickener.

To develop the relationship between settling flux and solids concentration, engineers perform batch-settling tests at various solids concentrations. For each concentration, they plot the depth of the solids–liquid interface and the time required for it to develop. Once enough data have been collected, engineers can plot a subsidence curve (see Figure 23.2).

Engineers may use the graphical method of Yoshioka et al. (1957) to determine the area needed to accomplish a desired degree of thickening (see Figure 23.3). They draw an operating line as a tangent to the settling flux curve. The intercept on this line's abscissa is the underflow solids concentration, and the intercept on the ordinate is the *limiting solids flux* (G_t)—the maximum solids flux that can be transported to the bottom of the thickener. Engineers then calculate the required thickener area as follows:

$$A = \frac{c_0 Q_0}{G_t} \tag{23.1}$$

Where

A = thickener area (m^2),
Q_0 = influent flow (m^3/d),
c_0 = influent solids concentration (kg/m^3), and
G_t = limiting solids flux ($kg/m^2{\cdot}d$).

In this procedure, thickener operation is assumed to be strictly one-dimensional (i.e., solids are distributed uniformly and horizontally at the feed level, and thickened-underflow removal produces equal downward velocities throughout the tank). How-

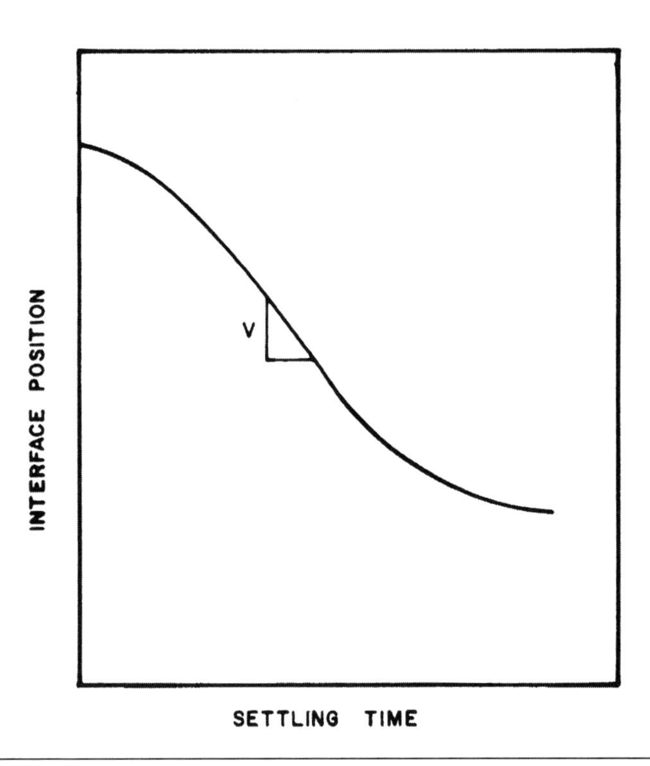

FIGURE 23.2 Example of a subsidence curve for a liquid–solids interface.

ever, full-scale thickeners typically cannot meet these conditions,because of the relatively small feed-well and central withdrawal of thickened solids. No data are available on the effects of non-uniform solids distribution and removal, although these factors should be considered when sizing a thickener. Design methods based on a single-batch settling test are available in the literature (Talmage and Fitch, 1955; Wilhelm and Naide, 1979).

2.1.3 Determining Area Based on Bench-Scale Testing

William and Naide (1979) developed a useful method when using bench-scale studies to help design a gravity thickener. It has three basic steps:

- Compute the settling velocity based on settling curves taken at several feed solids concentrations (at least three).
- Obtain the constants a and b using the following equation:

$$V = aC^{-b} \tag{23.2}$$

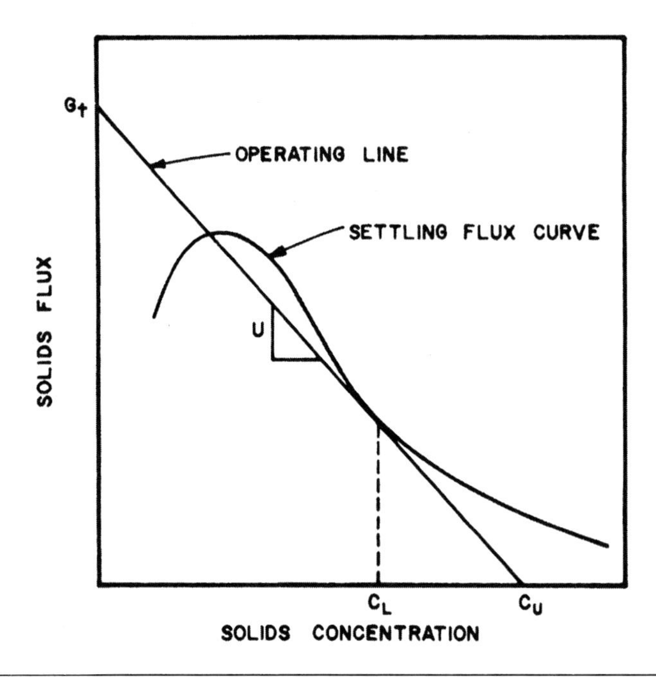

FIGURE 23.3 An example of the graphical method of Yoshioka et al. (1957).

Where

V = settling velocity (m/d),
C = solids concentration (kg/m^3), and
a,b = constants.

The constant a is a measure of the relative ease of settling; it is a function of particle size and shape, liquid and solid densities, liquid viscosity, and attractive or repulsive forces between particles. The exponent b is calculated from the slope of the line. It typically is constant over a certain range of concentrations, but gradually increases as particle-to-particle contact increases.

- For each straight line on a log-log plot of velocity versus concentration, calculate the unit area as follows:

$$UA = \frac{[(b-1)/b]^{b-1}(C_u)^{b-1}}{ab} \tag{23.3}$$

Where

UA = unit area (m^2/kg·d) and
C_u = the underflow's solids concentration (kg/m^3).

Also, bench tests have been developed to evaluate the significance of flocculating agents during thickening. Coagulants (e.g., alum, ferric salts, or organic polyelectrolytes) can enhance flocculating characteristics and reduce the required settling area. Polymers may double the solids concentration in a given unit area.

Once suitable flocculants have been selected, engineers can conduct additional testing in a 1- to 2-L cylinder to determine the underflow concentration that can be achieved. In this test, they should add a relatively dilute concentration of polymer (less than 1 000 mg/L) to solids. Then they should insert picket rakes in the cylinder and continue thickening for a standard time (1 to 24 hours, depending on the flocculant's effectiveness). The ultimate density achieved will be a fair but conservative measure of what to expect in a full-scale unit. For more accuracy during scale-up, engineers should use a test cylinder whose depth is closer to that of the full-scale unit.

2.1.4 *Determining Area Based on Pilot-Scale Testing*

If a treatment plant can provide enough solids for pilot-testing, engineers can obtain reliable design data for a thickener by operating a continuous pilot unit. When sizing the pilot unit, engineers should consider the availability of test solids and the means for withdrawing thickened solids at the low flowrates required. If at all practicable, the unit should be at least 2 m (6 ft) in diameter, have a side water depth of at least 2 m (6 ft), and a bottom slope ratio of 70 mm: 3 010 mm (2.75 in.: 12 in.) (vertical distance to tank radius). It also should be equipped with a feed well and a mechanism for directing solids to the withdrawal point on the tank bottom.

Engineers should conduct pilot-scale tests at several solids loading rates to determine the effect on required solids-withdrawal rates and resulting underflow solids concentrations. During each test run, engineers should ensure that the thickener is operated under steady-state, fully loaded conditions. A thickener is operating at steady state when the solids feed and withdrawal rates are equal and do not change the unit's solids inventory. It is fully loaded when it has a solids blanket but does not lose solids in the overflow. Attaining such conditions is difficult and time consuming. One approach is to start with a slightly overloaded thickener, gradually increase the solids-withdrawal rate until the overflow is solids-free, and then maintain these conditions to stabilize the blanket level. Ideally, the blanket level should be constant under all solids loading conditions. (Engineers can conduct a separate study at a convenient solids loading rate to identify any effects blanket depth may have on thickener performance.)

Once steady-state, fully loaded conditions have been maintained for a certain period of time as determined by site conditions, but typically 0.5 to 2 hours of hydraulic retention time, engineers should monitor thickener performance by measuring the following parameters at convenient intervals:

- Solids feed rate (as determined from feed flowrate and solids concentration),
- Underflow's volumetric rate and solids concentration,
- Overflow's volumetric rate and suspended solids concentration, and
- Concentration profile of thickener at the end of the run.

2.2 Process Design Considerations and Criteria

When designing gravity thickeners, important factors include

- Solids' source and characteristics,
- Nature and extent of flocculation (including flocculation induced by chemical additives),
- Concentration of suspended solids in overflow and effect of recycling fines on plant performance,
- Solids loading,
- Solids retention time in thickening zone or blanket,
- Blanket depth,
- Hydraulic retention time and surface loading rate,
- Solids withdrawal rate,
- Tank shape (including bottom slope),
- Physical arrangement of feed well and inlet pipe, and
- Arrangement of withdrawal pipe and local velocities around the piping.

2.2.1 Loading Rate

The critical design parameter for gravity thickening is the loading rate in terms of weight of total solids per unit area per unit time. Design loadings are determined via one of the methods given in Section 2.2. A thickener's capacity (allowable solids loading rate) typically is expressed in kilograms per square meter per day. For specific feed solids, the capacity is primarily a function of removal rate and desired underflow solids concentration. To increase the underflow concentration, the solids removal and loading rates both must be reduced. For any given feed solids, engineers can establish an operating range, with capacity expressed as a function of underflow solids concentration.

2.2.2 Overflow Rate

The second most important parameter when designing gravity thickeners is the thickener overflow rate. The maximum overflow rate for primary solids is typically 15.5 to

31.0 $m^3/m^2 \cdot d$ (380 to 760 gal/d/sq ft); the maximum for secondary solids is typically 4 to 8 $m^3/m^2 \cdot d$ (100 to 200 gal/d/sq ft). If the hydraulic loading is too high, solids carryover can be excessive. If hydraulic loading is too low, detention times lengthen and septic conditions (floating sludge and odors) can occur. When thickening primary solids, design engineers often select a feed pumping rate that will maintain a desired overflow rate. They also add a dilution-water supply (e.g., plant effluent) to maintain aerobic conditions and may add chlorine, potassium permanganate, or hydrogen peroxide (typically via the dilution-water supply) to control odor and septicity.

2.2.3 *Inlet*

Engineers should design the thickener inlet to minimize turbulence in the feed well. Most circular thickeners at domestic wastewater treatment plants use bottom-feed inlets to a center feed well; the feed flows vertically and then laterally with low turbulence. Most industrial and some domestic wastewater treatment plants use other configurations (e.g., overhead feed). Tangential entries or opposing tangential feed entries via a T connection are preferred to a system that directs feed straight down. A horizontal feed entry just under the liquid surface that is directed toward the center of the feed well typically will produce satisfactory results. Design engineers should avoid air entrainment in the feed entry to reduce froth formation on the thickener surface.

2.2.4 *Pickets*

Gravity thickening mechanisms often include pickets to help release water from solids (see Figure 23.1). Pickets typically are constructed of 0.6- to 2-m-high (2- to 6-ft-high) angle irons or pipes spaced 150 to 460 mm (6 to 18 in.) apart. The design depends on the type of solids being handled. The rake provides the necessary agitation in the lower part of the tank; however, if the rake only consists of one pipe arm (or similar construction), pickets can improve thickening performance.

For maximum benefit, pickets should be operated in dense solids zones. They should not be used in thickeners treating WAS from pulp and paper plants or fibrous wastes from other systems. Fibrous material tends to collect on pickets, eventually causing the entire mass to rotate in the thickener. Nor should pickets be used when thickening thermally conditioned solids because they will increase torque unnecessarily.

That said, reports of their effectiveness have been varied. Many carefully performed studies have produced contradictory results. Ettelt and Kennedy (1966), Voshel (1966), Sparr and Grippi (1969), and Dick and Ewing (1967) all indicate that using some device (e.g., pickets) to stir the solids blanket improves thickening performance. Dick and Ewing (1967) found that pickets seemed to help destroy the solids' macrostructure in static areas of the thickener, thereby permitting subsidence and consolidation to continue. Others found that when solids produced enough gas to prevent subsidence, pickets provided a channel for gas release, thereby enhancing thickening. On the other hand, Vesilind (1968) and Jordan and Scherer (1970) reported that mixing was not beneficial; in fact, pickets actually could hinder thickening (Vesilind, 1968). Likewise, if the thickener mechanism provided enough agitation on its own, pickets may have been redundant. Therefore, application of pickets on gravity thickening mechanisms should be determined on a case-by-case basis. Bench and pilot testing and more extensive research on similar applications are recommended.

2.2.5 *Drive Mechanisms*

The drive mechanisms for gravity thickeners are heavier than those for primary settling processes. Lifting devices typically have been unnecessary when treating municipal solids, but in certain instances—particularly when handling lime or heat-treated solids—hinged lift mechanisms are used so the scraper arms lift when the torque exceeds a preset limit. The machine continues to operate with the rakes lifted [up to 0.3 to 1 m (1 to 3 ft) above the bottom] until torque drops. However, most of a thickener's severe loads (e.g., those caused by island formation in highly viscous solids) actually prevent the self-lifting raking arm from functioning properly, making them unreliable in this application. Cables and other lifting mechanisms also have been used. These mechanisms can be automatic or manual.

In some cases, it may be more desirable to simply provide an oversized machine if intermittent, extraordinary loads are expected.

2.2.6 *Skimmers and Scrapers*

Gravity thickeners typically require skimmers and baffling to remove scum and other floating material. However, baffles, skimmers, and scrapers are vulnerable to earthquake forces—particularly sloshing liquid. To overcome such forces, engineers should design enough torque in both the rake structure and in the gearing and motor used to drive the unit. The torque rate should be high enough to provide sufficient driving force to get the mechanism out of trouble when necessary. However, operating continuously at torques greater than the mechanism's rated capacity greatly shortens the operating life of gears and bearings. So, a thickener's normal operating torque should not exceed 10% of the rated (maximum) torque value.

Design engineers typically calculate the torque for a typical thickener as follows (Boyle, 1978):

$$T = Kd^2 \tag{23.4}$$

Where

T = torque (kg·m),
K = constant (kg/m), and
d = diameter of the thickener (m).

K is a function of the material being thickened and is application-specific (see Table 23.3).

Skimmer and scraper speeds depend on the thickener's diameter. Peripheral velocities typically are kept between 4.6 and 6 m/min (15 and 20 ft/min), which is substantially greater than the velocities in clarifiers.

2.2.7 *Underflow Piping*

Underflow piping is a critical design element for gravity thickeners. Headlosses are high, so the underflow suction line should be as short as possible. Line velocities of 0.6 to 1.5 m/s (2 to 5 ft/sec) are typical. For operation and maintenance (O&M) purposes, the underflow pump should be beside the thickener and below the thickener's water level to ensure that the suction is flooded.

In addition to being as short as possible, the underflow suction piping between the thickener discharge cone and pump inlet should have adequate access points for cleanout. Design engineers also should include access for snaking from the pumps to the solids-well. If excessive fouling or plugging is anticipated, especially with lime solids, dual withdrawal lines are necessary so normal operations can continue while the plugged line is being cleaned.

TABLE 23.3 Design criteria for gravity thickeners.

Sludge	Diameter, m	K, kg/m	Pickets	Overflow rate, $m^3/m^2 \cdot d$
Raw primary	3–24	11	No	33
Raw primary and waste activated	3–21	7	Yes	33
Waste activated	3–15	4	Yes	33
Primary and waste activated heat treated	3–18	15	No	16
$CaCO_3$	3–30	22	No	41
Metal hydroxides	3–21	15	Yes	16
Pulp and paper sludges	3–30	22–30	No	33
Heavy storage duty	3–30	30–45	—	Varies with application

Gravity thickeners are often a significant source of odor in a treatment facility. They be covered and provided with odor-control measures. For example, adding chlorine, hydrogen peroxide, or other chemicals to thickener influent can control odors. Aerating thickener feed also may be beneficial. In addition, design engineers should provide enough dilution water to avoid the thick, aging, anaerobic solids that can lead to septicity and upset in thickeners.

2.2.8 *Rectangular Thickener Considerations*

The most common problems with rectangular thickeners are rat-holes and machine breakage. (A *rat-hole* is a conical hole in the solids that is as deep as the solids bed.) For these reasons, circular designs are more common.

When sizing rectangular units, design engineers use many of the same principles and criteria as for circular units. However, two additional factors should be considered: mechanism strength and inventory at the withdrawal point. The design should include a mechanism to move solids laterally or transversely to the withdrawal point(s). This machine should be strong enough to handle the added load that results from heaping solids near the withdrawal point(s).

Meanwhile, the hopper should be deep enough to prevent rat-holes from forming.

2.3 Operational Considerations Related to Design

2.3.1 *Feed Solids Source and Characteristics*

The source and characteristics of feed solids greatly influence gravity thickener design (and applicability). Depending on temperature, primary solids can be retained in the thickener for 2 to 4 days before upset conditions develop. However, a solids retention time (SRT) of 1 to 2 days is best.

Waste activated sludge settles slowly and resists compaction, significantly reducing mass-loading rates. It also tends to stratify because the continued biological activity produces gas, which creates a flotation effect.

The following precautionary measures apply when considering using a gravity thickener to treat activated sludge (WPCF, 1980):

- In climates where wastewater temperatures exceed 20°C, gravity thickening should be avoided unless the activated sludge's SRT exceeds 20 days;
- Thickener inventory should be less than 18 hours to reduce the undesirable effects of continued biological activity;
- Thickener diameter should be 10.7 to 13.7 m (35 to 45 ft) or less; and
- Solids should be wasted directly from the aeration basin to the thickener.

2.3.2 *Polymer*

Although seldom practiced, a polymer can be added to gravity thickeners to improve solids capture. Synthetic polyelectrolytes work better than inorganic coagulants (e.g., alum and ferric chloride) in this application because they do not yield metal hydroxides that add to the solids volume.

2.3.3 *Underflow Withdrawal*

Thickener operations are most effective if underflow is withdrawn continuously. If intermittent withdrawal is necessary, a time-controlled system will allow operators to achieve efficient thickening performance. Pumping should be frequent and brief rather than for longer periods only once or twice per shift. Frequent pumping minimizes the solids blanket variation required to maintain a suitable average underflow concentration.

The effect of compacting solids depth can be significant but is poorly understood. A certain minimum depth [typically about 1 to 2 m (3 to 6 ft)] is required to achieve a desired thickened underflow. Deeper solids can increase the underflow concentration and, to a lesser extent, the capacity of a given thickening area.

2.4 Ancillary Equipment/Controls

Apart from feed and underflow pumps, the significant ancillary equipment includes blanket-depth indicators, flow indicators, and solids-density monitors. Maintaining a balance between solids input and output is essential to good overall operation. For example, a suspended solids monitor on the overflow may be useful if recycled fines cause special problems. Multiple or variable-speed pumps are desirable to control inventory and remove solids at the correct rate.

2.5 Design Example

Design engineers need to size a circular thickener for a wastewater treatment plant with a primary influent solids loading rate of 22 680. Using Table 23.2, they should select the higher solids loading rate to allow for operation with one unit out of service.

So, they select a loading rate of 6 kg/m^2·hour.

Solids loading of primary sludge (dry weight) = 22 680 kg/d
Number of operating units = 1 unit
Design loading rate (6 kg/m^2·hour) = 147 kg/m^2·d

Equation used:

$$\text{Surface area} = \frac{\text{Solid Loading}}{\text{Design Loading Rate}} \tag{23.5}$$

$$\text{Radius} = \sqrt{\text{Surface Area} * \frac{1}{\pi}} \tag{23.6}$$

Surface area = 155 m^2	Use Equation 23.5
Radius = 7 m	Use Equation 23.6
Diameter = 14 m	

Assumption:

(1) Thickening facility operates continuously.
(2) Typically, two tanks operate simultaneously, however, this calculation allows for operations with one out-of-service.

Note: Thickener diameter is based on solids loading per unit from manufacturer.

3.0 DISSOLVED AIR FLOTATION THICKENER

In a dissolved air flotation (DAF) thickener, solids and liquid are separated via the introduction of fine gas bubbles (typically air) to the liquid phase. The bubbles attach to solids particles, making them buoyant. They then rise to the liquid surface, where a skimmer collects them. This process typically is used to thicken WAS, aerobically digested solids, and contact-stabilized, modified activated, or extended-aeration solids without primary settling. It typically is not used for primary or trickling filter solids because gravity settling is more economical. However, it can effectively cothicken (settle and consolidate) primary sludge and WAS. (The advantages and disadvantages of cothickening are discussed later in this section.)

The main components of a DAF thickener are the pressurization system and DAF tank (see Figure 23.4). The pressurization system has a recycle pressurization pump, an air compressor, an air saturation tank, and a pressure-release valve. The dissolved air flotation tanks are either rectangular or circular and are equipped with both surface skimmers and bottom solids-removal mechanisms. The surface skimmers remove floating solids from the tank surface to maintain a constant float blanket depth. The bottom mechanism removes the heavier solids that settle on the tank floor.

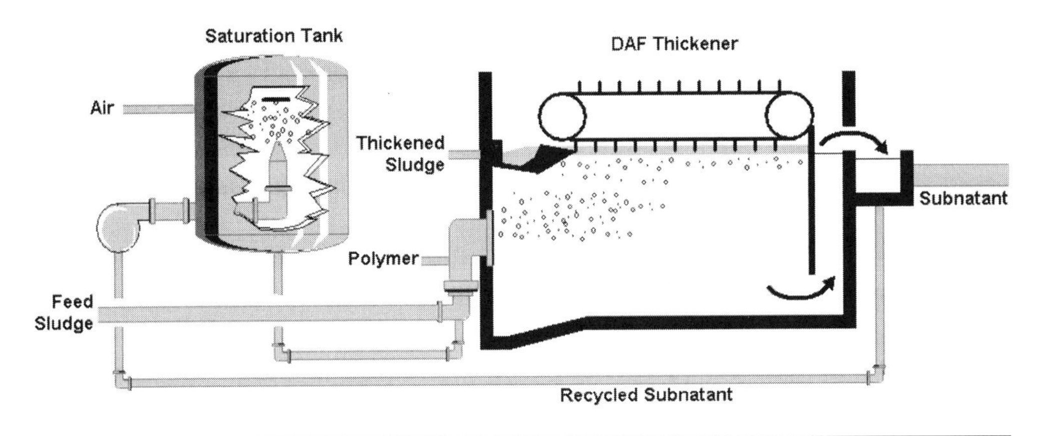

FIGURE 23.4 Schematic of a dissolved air flotation thickener.

Dissolved air flotation tanks also have baffles and an overflow weir. Clarified effluent passes under either an end baffle (rectangular units) or a peripheral baffle (circular units) and then over the weir to an effluent launder. The weir controls the liquid level in the tank with respect to the float collection box and helps regulate capacity and performance. The float collection box collects the particles that rapidly float to the water surface with supersaturated air bubbles, leaving clarified water behind.

Flotation can be used to either clarify liquids or concentrate solids. The quality of the liquid effluent (*subnatant*) is the primary performance factor in clarification applications (e.g., refinery, meat-packing, meat-rendering, and other oily wastewaters). The concentration of floating solids is the main performance criteria in concentration applications (e.g., waste solids of biological, mining, and metallurgical processes).

3.1 Evaluation and Scale-Up Procedures

Dissolved air flotation has been widely used to thicken waste biological solids since the mid-1960s. Engineers typically can size flotation equipment based on design experience. However, bench- or pilot-scale performance investigations can provide valuable information, such as

- Thickened solids concentration, solids recovery rates, and chemical needs;
- DAF designs that can satisfy performance requirements; and
- The causes of poor or suboptimum DAF performance.

Before conducting any bench- or pilot-scale tests, however, engineers should collect a representative sample of the solids to be thickened. Then they should determine its suspended solids content, volatile solids content, and sludge volume index (SVI) for WAS.

3.1.1 Bench-Scale Evaluations

Bench-scale tests provide insight into the thickening characteristics of specific solids. Manufacturers have designed and built bench-scale units that are available for such evaluations. They also have scale-up criteria for their own equipment that enable engineers to predict full-scale operations with reasonable accuracy.

A typical bench-scale unit consists of a pressurization chamber, a flotation chamber, a pressure-release valve, and ancillary equipment. The test typically is conducted as follows:

- Engineers introduce a sample of the fluid (typically clarified liquid) to the pressurization chamber (a full-scale unit typically uses subnatant).
- They then adjust an air-bleed valve to allow compressed air to bubble through the liquid. After a suitable pressurization period (typically 10 minutes), they close the air-bleed valve.
- Engineers place a measured sample of the material to be floated (e.g., waste biological solids) in the flotation chamber.
- They then open a pressure-relief valve, and pressurized fluid enters the flotation chamber and is distributed about the space. Engineers close the pressure-relief valve when the total volume in this chamber reaches a predetermined level. The material is permitted to float in the chamber for a suitable period (typically 10 minutes).
- Engineers then collect samples of subnatant and floating material.

To identify the optimum value, engineers should perform enough tests to determine system performance at several air-to-solids ratios. Air-to-solids ratios can be varied by changing the solids concentration or volume of sample to be floated. Further tests may be required to assess the efficacy of chemical conditioning and the effects of feed-solids concentration.

Bench-scale tests are especially useful for predicting float-solids content and solids capture, as well as for evaluating the effects of chemical flotation aids on float-solids and solids capture. However, they are seldom used to establish design loading rates because of uncertainties in scale-up.

One option for using bench-scale test data to size full-scale DAF units involves applying batch or limiting flux methods to the interface height-versus-time data obtained during flotation (Wood, 1970). Engineers must develop separate flux curves for each air-to-solids level of interest. As with gravity thickening, scale-up uncertainties have limited application of this procedure. Also, engineers have accumulated experience in designing DAF system to thicken WAS and developed other means to thicken the solids.

3.1.2 *Pilot Flotation Units*

The flotation performance at a given installation depends on the interaction of many factors. In most situations, pilot-scale flotation units are the best way to identify this performance. Results obtained from pilot and field equipment are analogous when the devices are geometrically, kinematically, and dynamically similar. However, complete similarity is seldom achieved because of innate physical differences between pilot- and full-scale equipment, so the goal is to be as similar as practical.

Two different-sized systems are *geometrically similar* if they are proportional in all corresponding dimensions (e.g., the length, width, and depth of the flotation unit). They are *dynamically similar* when the ratios of all corresponding forces are equal. They are *kinematically similar* if velocities at corresponding points have the same ratio. Kinematic similarity is approached when geometrically similar pilot- and full-scale equipment have identical hydraulic loading rates, and when the pilot-scale unit's pressure-relief valve is a properly scaled-down version of the valve on the full-scale unit.

Ideally, both units should treat the same feed material, create the same size gas bubbles, and operate at the same pressure. Also, the loading rate and air-to-solids ratio used during pilot-scale tests must be applicable to full-size equipment.

When scaled-up, pilot-testing data can only reveal the full-scale unit's probable performance because pilot units are not completely similar to full-scale units. Equipment manufacturers have scale-up information specific to their own equipment.

3.2 Process Design Considerations and Criteria

Before feed solids entering a DAF tank, they typically are mixed with a recycled flow. The recycled flow is pressurized up to 520 kPa (75 psi) and added at a rate that depends on feed-solids concentration. It typically is DAF tank effluent, although a backup source is advisable in case poor DAF performance leads to an effluent containing high levels of suspended solids. Recycled flow first is pumped to an air-saturation tank, where compressed air dissolves into the flow. When returned to the DAF tank (whose surface is at atmospheric pressure), the pressure release creates the air bubbles used for flotation. These bubbles typically range from 10 to 100 μm in diameter.

The air combines with solids particles and floats, forming a blanket on the DAF tank surface that typically is 150 to 300 mm (6 to 12 in.) thick. Meanwhile, clarified effluent flows under the tank baffle and over the effluent weir. A properly designed and operated DAF thickener typically captures between 94 and 99% of suspended solids.

Other DAF pressurization systems do not use recycled flow; instead, they pump all or part of feed solids through an air-saturation tank and then into the DAF tank. Such systems are inadvisable for wastewater treatment applications because they subject solids to high-shear conditions and the solids can clog various pressurization-system components.

Polymers can enhance DAF performance by significantly increasing applicable solids-loading rates and solids capture; and they also can increase the concentration of floating-solids concentrations to some degree. If used, a polymer typically is introduced at the point where feed solids and recycle flow are mixed. For the best results, design engineers should introduce polymer to the recycle flow just as the bubbles are being formed (before it is mixed with feed solids). Good mixing (enough to ensure chemical dispersion while minimizing shearing forces) will provide the best solids–air bubble aggregates.

Table 23.4 presents operating data from selected DAF thickener installations. Numerous factors affect DAF process performance, including

- Type and characteristics of feed solids,
- Hydraulic loading rate,
- Solids loading rate,
- Air-to-solids ratio,
- Chemical conditioning,
- Operating policy,
- Float-solids concentration, and
- Effluent clarity.

TABLE 23.4 Typical operating data for dissolved air flotation thickeners.

Location	Activated-sludge type	Feed solids concentration, mg/L	Solids loading rate, kg/m²·h	Float concentration, %	Polymer dosage, g active polymer/kg of solids	Solids capture, %
Green Bay, Wisconsin	Contact stabilization	4 000	1.5	3–4	None	80–85
San Francisco, California Southeast Plant	High-purity oxygen	6 000	3.4	3.7	1.6	98.5
Salem, Oregon	High-purity oxygen	14 800–20 300	19.5	5	48–59	95+
Milwaukee, Wisconsin South Shore Plant	Conventional	5 000	4.9	3.2	1.5–2.5	90–95
Tri-Cities, Oregon	Conventional	11 300	—	3.9	1.5–2.5	98
Arlington, Virginia	Conventional	10 000	8.5	2.6	1.5–2	95+
Kenosha, Wisconsin	Conventional	8 600	5.4	4.3	None	99+

These factors often act synergistically to produce a net positive or negative effect on DAF performance. Isolating each factor's effect is often difficult, but Bratby and Marais (1975a) have proposed a model to predict DAF performance as a function of various conditions.

3.2.1 *Type of Solids*

Dissolved air flotation can thicken a variety of solids, including conventional WAS, solids from extended aeration and aerobic digestion, pure-oxygen activated sludge, and solids from dual biological processes (trickling filter plus activated-sludge). The performance characteristics of each type of solids are difficult to document because site-specific conditions (e.g., type of process, SRT, and SVI in the aeration basin) affect DAF performance more than flotation-equipment adjustments (e.g., air-to-solids ratio) do. Gulas et al., (1978) and Wood and Dick (1975) discuss the effects of some plant operating parameters on DAF performance in considerable detail.

3.2.2 *Mixed-Liquor Sludge Volume Index*

One of the solids characteristics that affect DAF performance is an activated sludge's mixed-liquor SVI. The floating-solids concentration typically decreases as SVI increases. To produce a 4% floating solids concentration with nominal polymer doses, SVI should be less than 200 mL/g. If SVI is low, solids are compacting well, and a broad band of floating solids exists, then other factors clearly are influencing DAF performance. At higher values, SVI has a deleterious effect on floating solids. Large doses of polymer typically are required when thickening WAS from systems with excessively high SVI.

3.2.3 *Hydraulic Loading Rate*

The *hydraulic loading rate* is the sum of the feed and recycle flowrates divided by the net available flotation area. Engineers typically design DAF thickeners for hydraulic loading rates of 30 to 120 $m^3/m^2 \cdot d$ (0.5 to 2 gpm/sq ft), with a suggested maximum daily hydraulic loading of 120 $m^3/m^2 \cdot d$ if no conditioning chemicals are used. If the hourly hydraulic loading rate exceeds 5 $m^3/m^2 \cdot h$, the added turbulence may prevent a stable float blanket from forming and reduce the attainable floating-solids concentration. Also, fewer solids may be captured because increased turbulence forces the flow regime to convert from plug flow to mixed flow. A polymer flotation aid typically is required to maintain satisfactory performance when hourly hydraulic loading rates are greater than 5 $m^3/m^2 \cdot h$.

3.2.4 *Solids Loading Rate*

The solids loading rate for a DAF thickener typically is denoted in terms of solids weight per hour per effective flotation area (see Table 23.5). Without chemical conditioning, the

TABLE 23.5 Percent suspended solids captured when using dissolved air flotation to thicken WAS (U.S. EPA, 1974; Komline, 1976).

	Solids loading rate ($kg/m^2 \cdot h$)	
Type of solids	**No chemical addition**	**Optimal chemical addition**
Primary only	4–6	up to 12
Waste activated sludge (WAS*)		
Air	2	up to 10
Oxygen	3–4	up to 11
Trickling filter	3–4	up to 10
Primary + WAS (air)	3–6	up to 10
Primary + trickling filter	4–6	up to 12

* WAS = waste activated sludge.

loading rates for DAF processes thickening WAS range from about 2 to 5 $kg/m^2 \cdot h$ (0.4 to 1 lb/hr·sq ft); this produces a thickened underflow of 3 to 5% total solids (Ashman, 1976; Burfitt, 1975; Jones, 1968; Mulbarger and Huffman, 1970; Reay and Ratcliff, 1975; U.S. EPA, 1974; Walzer, 1978). With polymer, the solid loading rate typically can be increased 50 to 100%; producing a thickened underflow that contains up to 0.5 to 1% more solids.

Operating difficulties may arise when the solids loading rate exceeds about 10 $kg/m^2 \cdot h$ (2.0 lb/hr·sq ft). These difficulties typically are caused by coincidental operation at excessive hydraulic loading rates and by float-removal difficulties. Even when the hydraulic loading rate can be kept below 120 $m^3/m^2 \cdot d$ (2 gpm/sq ft), operating at solids loading rates more than 10 $kg/m^2 \cdot h$ can cause float-removal difficulties. The extra floating material created at high solids loading rates necessitates continuous, often rapid, skimming.

Faster skimming, however, can disturb the float blanket and lead to a subnatant with unacceptable solids levels. In these circumstances, a polymer flotation aid can increase the solids' rise rate and float-blanket consolidation rate, thereby alleviating some of the operating difficulties. Although stressed conditions (e.g., mechanical breakdown, excessive solids wastage, or adverse solids characteristics) may make it necessary to operate in this manner periodically, the flotation system should not be designed on this basis.

3.2.5 *Feed-Solids Concentration*

Feed-solids concentration affects DAF processes in two ways. As in sedimentation processes, feed-solids concentration directly affects the floating solids' characteristics in terms of initial and—to a lesser extent—hindered rise rate. Within the normal range of feed-solids concentration (5 000 to 10 000 mg/L), more dilute feed solids result in more rapid initial and hindered rise rates. However, this phenomenon only has a minor effect on DAF sizing and performance because the solids blanket's hindered rise rate and compression rate govern design and performance for most thickening applications.

Feed-solids concentration also indirectly affects DAF via resulting changes in operating conditions. For example, if the feed flowrate, recycle flow, pressure, and skimmer operations remain constant, then increasing feed-solids concentration decreases the air-to-solids ratio. Changes in feed-solids concentration also changes float-blanket inventory and depth. Float skimmer speed may need adjustments when the operating strategy involves maintaining a specific float-blanket depth or range of depths.

3.2.6 Air-to-Solids Ratio

The *air-to-solids ratio*—the ratio (by weight) of air available for flotation to the floatable solids in the feed stream—is the most important factor affecting DAF performance. Reported ratios range from 0.01 : 1 to 0.4 : 1 (U.S. EPA, 1979); at most municipal wastewater treatment plants, adequate flotation occurs at ratios of 0.02 : 1 to 0.06 : 1. Design engineers size pressurization systems based on many variables (e.g., design solids loading, pressurization-system efficiency, system pressure, liquid temperature, and dissolved solids concentration). Pressurization-system efficiencies vary among manufacturers and system configurations; they can range from as low as 50% up to more than 90%. The U.S. Environmental Protection Agency (1979) provides detailed information is available on designing, specifying, and testing pressurization systems.

Because the solids blanket in a DAF thickener contains a considerable amount of entrained air, design engineers should use positive-displacement or centrifugal pumps that do not air bind, and consider suction conditions. Initially, the density of skimmed solids is about 700 kg/m^3 (6 lb/gal). After they are held for a few hours, the air escapes and solids return to normal densities.

Up to a point, solids blankets increase as air-to-solids ratios increase; then, further increases in air-to-solids ratios result in little or no increase in floating solids (Gehr and Henry, 1978; Gulas et al., 1978; Maddock, 1976; Mulbarger and Huffman, 1970; Turner, 1975). The solids blanket typically is maximized when air-to-solids ratio are between 2 and 4%.

There are several explanations for this wide range. First, the optimum air-to-solids ratio is related to the type of feed solids and its characteristics. For example, activated sludges with low SVIs require lower air-to-solids ratios than those with high SVIs.

Second, evaluating the effects of air-to-solids ratios is difficult because other DAF operating conditions (e.g., blanket depth) can vary as air-to-solids ratio changes. So, the effect of a change in air-to-solids ratio is often masked by other changes.

Third, differences among the DAF systems reasearched (e.g., the pressurization system's air-dissolving efficiency, gas-bubble size distribution, and feed-recycle mixing methods) undoubtedly are responsible for some of the differences in optimum air-to-solids ratios.

Although the optimum air-to-solids ratio probably is related to solids type and characteristics, lower air-to-solids ratios seem to be required to maximize the performance of systems that operate at a high air-dissolving efficiency, produce optimum air-bubble size distribution, and correctly contact the feed solids and minute air bubbles at the proper time.

3.2.7 *Float-Blanket Depth*

The floating solids produced during DAF must be removed from the tank. This solids-removal system typically consists of a variable-speed float skimmer and a beach arrangement. The volume of floating solids that must be removed during each skimmer pass depends on the solids loading rate, the chemical dose rate, and the consistency of floating solids.

A blanket of waste biological solids consists of two sections: one above the nominal water level and one below it. When evaluating a DAF system, Bratby and Marais (1975b) found that its ratio of float depth above the surface to float depth below the surface was 0.2 : 1 when the air-to-solids ratio was 0.02 : 1. They stated that the optimal ratio of above-surface and below-surface solids will differ according to the type of feed solids involved.

The concentration of solids on the surface of the solids blanket is always greater than the average concentration of solids within the blanket. Bratby and Marais (1975b) also suggested that DAF thickening occurs as water drains from the section above water to that below it. Maddock (1976) found that the solids concentration at the blanket surface was nearly twice that at the blanket–subnatant interface.

Blanket skimmers are designed and operated to maximize float drainage time by incrementally removing only the top (driest) portion of the blanket and preventing the blanket from expanding to the point where solids exit the system in the subnatant. The optimal float depth varies from installation to installation. A float depth of 300 to 600 mm (1 to 2 ft) is almost always sufficient to maximize floating-solids content.

3.2.8 *Polymer Addition*

Chemical conditioning can enhance DAF performance. Conditioning agents can improve clarification or increase the floating solids concentration. Design engineers should determine the amount of conditioning agent required, the point of addition (in the feed stream or recycle stream), and the intermixing method for each installation. Bench- or pilot-scale tests are the most effective method of determining the optimal chemical-conditioning scheme for a particular installation.

Typical polymer doses range from 2 to 5 g dry polymer/kg dry feed solids (4 to 10 lb/ton). Adding polymer typically affects solids capture more than floating-solids content. For example, adding dry polymer at a dose of 2 to 5 g/kg dry solids typically increases floating solids content by up to 0.5%.

If design engineers use the lower ranges of hydraulic and solids loadings, well-designed and -operated DAF thickeners typically do not need polymer. Maintaining proper design and operating conditions results in stable operations and satisfactory solids capture and floating-solids concentration. Routine additions of polymer should only be considered for designs with extreme loading conditions or when solids are expected to have poor compaction characteristics (i.e., high SVI).

Without polymer addition, a properly sized DAF unit typically will recover more than 90% of solids. High loadings or adverse solids conditions can cut solids recovery to 75 to 90%. Polymer-aided recovery can exceed 95%.

Under normal operations, solids recycled from the DAF unit will not damage the treatment system but rather increase WAS. However, if solids or hydraulic loading already is excessive, recycled solids pose an additional burden on the system. Under these conditions, polymers should be used to maximize solids capture from the DAF unit.

3.2.9 *Floating Solids Concentration*

As with any thickening process, flotation performance strongly depends on the type and characteristics of the solids being thickened. Although municipal wastewater treatment plants typically use DAF to thicken WAS, they also have used it to thicken raw primary solids, trickling filter humus, and various combinations of these.

The floating solids concentration that a DAF treating WAS can obtain is influenced by various factors, the most important of which are innate solids characteristics (i.e., SVI), solids loading rate, air-to-solids ratio, and polymer application. Test results demonstrate that the floating solids concentration typically decreases as solids loading rates increase (see Figure 23.5). They also indicate (with few exceptions) that polymers must be used to achieve the higher loading rates. Although high loadings of 15 to 29 $kg/m^2{\cdot}h$ (3 to 6 lb/hr·sq ft) can be achieved, these results are neither typical of the average plant nor a relevant basis for new designs. In some cases, a lot of expensive chemicals are necessary to achieve a loading level in excess of 10 $kg/m^2{\cdot}h$ (2 lb/hr·sq ft).

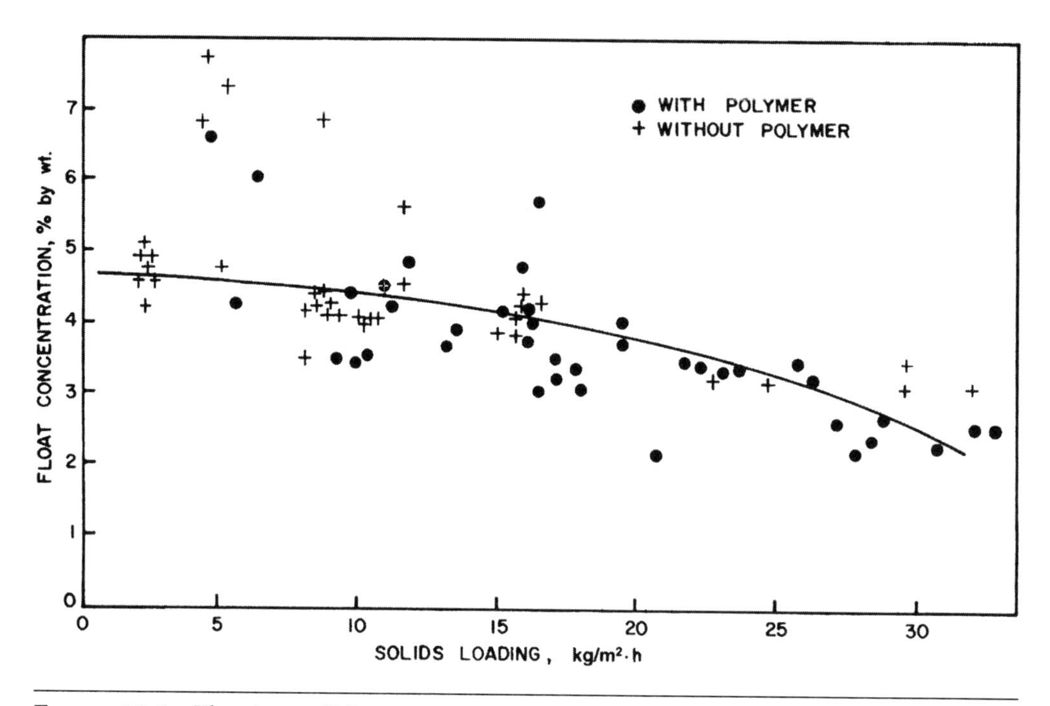

FIGURE 23.5 Floating-solids concentration versus solids loading rates (Noland and Dickerson, 1978).

The curve in Figure 23.5 does not indicate the effect of polymers on floating solids concentration. Polymers can improve poor float concentration up to 1.0% (2.0 to 3.0% TSS), but their effect lessens as the concentration of untreated floating solids increases.

Dissolved air flotation thickeners typically are designed for floating-solids concentrations of 3.5 to 4.0% total solids—a reasonable goal based on the data presented (see Figure 23.5) and other published information (U.S. EPA, 1974; Wanielista and Eckenfelder, 1978). However, DAF performance, like other solids processing equipment performance, is influenced by factors beyond the design engineers' control. Therefore, designers should anticipate variations in float-solids concentration when sizing downstream unit operations.

3.2.10 Solids Capture

Overall solids capture measures how efficiently a DAF unit recovers solids at a fixed set of operating conditions. The solids-capture calculation is based on a material balance about the DAF unit (Mulbarger and Huffman, 1970). The flows of interest include feed solids, subnatant, and floating solids. Overall solids capture is defined as:

$$R = \frac{TS_P(TS_F - TSS_S)}{TS_F(TS_P - TSS_S)} * 100 \qquad (23.7)$$

Where

R = percentage recovery;
TS_P = total solids concentration in thickened material (% by weight);
TS_F = total solids concentration in influent (% by weight); and
TSS_S = total suspended solids concentration in subnatant (% by weight).

Published results of suspended solids capture at numerous DAF plants indicate that they can capture at least 95% of solids without using polymer (see Table 23.6). With polymer, they typically capture at least 97 to 98%. When recycling subnatant, most thickening operations need to capture at least 95% of solids to minimize adverse effects on other treatment processes.

3.2.11 *Solubilization Efficiency*

The most cumbersome procedure associated with DAF equipment is determining the solubilization efficiency of the dissolving tank in the pressurization system. *Solubilization efficiency* is the ratio of the amount of air (oxygen and nitrogen) actually dissolved in the tank to the amount that theoretically could be dissolved in the tank under existing conditions. Data collection and calculation procedures have been developed and published (APHA et al., 1976; Leininger and Wall, 1974; U.S. EPA and ASCE, 1979).

A variety of pressurization systems are available from DAF manufacturers, and their air-dissolving efficiency ranges from 65 to 85%, according to Leininger and Wall (1974). Dissolving efficiency is important because the air-to-solids ratio is critical to DAF

TABLE 23.6 Design solids loading rate for and are of dissolved air flotation (DAF) systems.*

	Solids loading—winter operation			
Plant size, m^3/d	**kg/wk**	**kg/operating day (5 day/wk)**	**kg/operating hour**	**Area, m^2**
4 000	2 130	430	108	12.5
20 000	10 640	2 130	532	62.0
40 000	21 280	4 300	716	84.0
200 000	106 400	21 280	1 480	173.0

* Summer operation at 0.65 to 0.8 of winter conditions; DAF loading at 8.6 $kg/m^2 \cdot h$.

performance. Design engineers must use rigorous test procedures to accurately determine the amount of air available for flotation. For example, procedures that do not distinguish between dissolved and undissolved (free) air (e.g., those based on conventional air mass-balance calculations) will not provide accurate results.

The flotation system should be designed based on winter solids production, or else the equipment may have to operate 25 to 50% longer to handle the increased solids, which are also more difficult to thicken as temperature decreases. Solids loadings to DAF systems for various plant sizes are shown in Table 23.7.

3.3 Mechanical Features

3.3.1 *Typical Flotation Systems*

The number and configuration of DAF thickeners to be installed at a wastewater treatment plant depends on the plant size, method of operation, the quantity of solids to be thickened under both average and peak conditions, and the degree of operating flexibility desired.

3.3.2 *Rectangular Versus Circular*

Dissolved air flotation tanks can be rectangular or circular. Both rectangular and circular units have been used in wastewater treatment plants ranging from 3800 to more than 380 000 m^3/d (1 to more than 100 mgd).

Standard rectangular flotators vary from 9 to 167 m^2 (100 to 1 800 sq ft). Length-to-width ratios typically are 3:1 to 4:1. Rectangular tanks typically are used in smaller applications (e.g., inside buildings where circular units would not fit as well). Their surface skimmers can be closely spaced and designed to skim the entire surface. The solids collector on the bottom typically has a separate drive, so it can be operated independently of the skimmer. The liquid level can be adjusted more easily because of the straight-end weir configuration.

Circular units typically range from 29 to 130 m^2 (300 to 1 400 sq ft). They often are used when land availability is not a constraint. Also, their structural requirements and mechanical equipment cost less than those of a rectangular unit (U.S. EPA, 1979).

3.3.3 *Materials of Construction*

Dissolved air flotation tanks can be constructed of concrete or steel. Typically, larger units are made of concrete, while rectangular units up to 41.8 m^2 (450 sq ft) [2.4 to 3 m (8 to 10 ft) wide] and circular units up to 9 m^2 (100 sq ft) are made of steel. The size of steel DAF units is limited by structural and shipping considerations; they typically are completely assembled and only require a concrete foundation pad, piping, and wiring

TABLE 23.7 Performance averages when using a dissolved air flotation thickener to cothicken solids (1994–1995) (Butler et al., 1997).

Location	Loading		Polymer	% Suspended solids
	$kg/m^2 \cdot h$	$m^3/m^2 \cdot h$		
Kenosha, Wis.	53	—	No	99+
Chicago, Ill.	28.3	—	No	99+
Amarillo, Texas	14.2	—	No	92+
East Fitchburg, Mass.	42.5–85.0	—	No	99.5
Xenia, Ohio	42.5	—	No	99.5
Eugene, Ore.	35.4	—	No	90+
Bernardsville, N.J.	62.3	2.9	No	94.5
Morristown, N.J.	48.1	1.1	No	97.0
Bay Park, N.Y.	36.8	0.7	No	94.0
East Wenatchee, Wash.	42.5–0.8	2.9	No	97.3
Incline Vill., Nev.	31.2	2.5	No	96.4
Fairfax, Va.	17	3.2	No	93.0
Plano, Texas	36.8	5.8	No	95.0
San Pablo, Calif.	17	2.9	No	99.0
Richmond, Calif.	42.5	3.6	No	98.9
Tenneco, Texas	5.7	2.2	No	98.8
Adolf Coors, Golden, Colo.	104.8	—	Yes	99.4
Springdale, Ark.	70.8	—	Yes	99+
Biddeford, Maine	90.6	—	Yes	99+
East Fitchburg, Mass.	85	—	Yes	98.3
Athol. Mass.	90.6	—	Yes	99.4
Somerset, Mass.	99.1	—	Yes	99+
Dartmouth, Mass.	90.6	—	Yes	98
The Dalles, Ore.	68	—	Yes	99.3
Denver, Colo.	104.8	—	Yes	99
Amarillo, Texas	65.1	—	Yes	99.2
Warren, Mich.	59.5	—	Yes	99+
Atlanta, Ga.	135.9	—	Yes	99.7
Chicago, Ill.	70.8	—	Yes	99+
Abington, Pa.	82	2.9	Yes	96.2
Hatboro, Pa.	83.5	1.8	Yes	96.0
Omaha, Neb.	87.8	1.8	Yes	99.4
Bellview, Ill.	107.6	1.1	Yes	98.7
Indianapolis, Ind.	59.5	3.6	Yes	95.0
Frankenmuth, Mich.	184.1	3.2	Yes	99.1
Oakmonth, Pa.	85	2.5	Yes	98.7
Columbus, Ohio	93.5	2.5	Yes	99.5
Levittown, Pa.	82.1	2.5	Yes	99.4
Bay Park, N.Y.	138.8	2.9	Yes	99.6
Nashville, Tenn.	144.4	1.4	Yes	99.6
East Wenatchee, Wash.	45.3	3.2	Yes	98.6
Plano, Texas	53.8	5	Yes	98.8
Richmond, Calif.	42.5	3.6	Yes	98.0
San Pablo, Calif.	31.2	2.9	Yes	98.6

hookups. Steel tank systems have higher equipment costs but avoid field-installation costs (e.g., structural, labor, and equipment components). That said, concrete tanks typically are more economical for a large installation requiring multiple or large tanks (U.S. EPA, 1979).

3.3.4 *Location*

The capacity of DAF units is an order of magnitude greater than that of gravity thickeners, so their space requirements typically are low. At large wastewater treatment plants with an influent 5-day biochemical oxygen demand (BOD_5) of 150 to 200 mg/L, a DAF process with polymer addition needs 0.37 to 0.5 $\times 10^{-3}$ $m^2/m^3 \cdot d^{-1}$ (15 to 20 sq ft/mgd); without polymer addition, the process needs 0.7 to 1.0 $\times 10^{-3}$ $m^2/m^3 \cdot d^{-1}$ (30 to 40 sq ft/mgd). At small wastewater treatment plants with the same BOD_5, a DAF process with polymer addition needs 0.5 to 0.7 $\times 10^{-3}$ $m^2/m^3 \cdot d^{-1}$ (20 to 30 sq ft/mgd); without polymer addition, the process needs 1.0 to 1.5 $\times 10^{-3}$ $m^2/m^3 \cdot d^{-1}$ (40 to 60 sq ft/mgd). Because they do not need much space, DAF thickeners often are located inside buildings. This is especially desirable in locales where odor control is required, or cold or wet climates could adversely affect a DAF unit's mechanical performance.

3.3.5 *Skimmers and Rakes*

Dissolved air flotation tanks are equipped with surface skimmers and floor rakes. Surface skimmers remove floating solids from the tank to maintain a constant solids-blanket depth. They can be controlled manually or automatically. The most common method is manual control of skimmer speed based on site-specific operating conditions. A more preferable arrangement is the use of automatic timers to control skimmer operation so the solids blanket remains 300 to 500 mm (12 to 18 in.) deep. This approach maximizes both floating-solids concentration and solids drainage before removal. Design engineers can use skimmer on–off cycles of variable durations to maximize floating-solids detention time while maintaining a stable blanket. They also should use variable-speed skimmers [up to about 7.6 m/min (25 ft/min)] to maximize operating flexibility and should time the skimmer cycle so the skimmer's maximum speed is 300 mm/min (1 ft/min).

Floor rakes remove heavier solids that settle to the tank bottom. Such deposits should be minimal if the treatment plant's grit-removal facilities are effective, but design engineers still should make provisions to remove this material. Engineers should design floor rakes and surface skimmers as separate systems. Continuously operated floor rakes sometimes reduce solids-capture efficiency because they increase turbulence and mixing of the subnatant inventory. Providing a separate drive system for the floor rakes allows operators to operate them only as required.

3.3.6 *Overflow Weir*

Dissolved air flotation units also are baffled and equipped with an overflow weir. The weir controls the liquid level in the flotation tank with respect to the float collection box, thereby regulating the unit's capacity and performance. To maximize capacity and performance under widely fluctuating conditions, the overflow weir should be adjustable.

3.3.7 *Pressurization System*

Pressurization systems dissolve gas (typically air) into the liquid used during DAF. The theoretical principles of pressurization systems are well known and have been discussed by several researchers (e.g., Vesilind, 1974b; Bratby and Marais, 1975a, 1975b, and 1976; Speece et al., 1975).

Historically, three methods have been used to provide gas bubbles for a DAF system: total, partial, and recycle pressurization flow schemes (see Figures 23.6, 23.7, and 23.8). The total-pressurization flow scheme pressurizes the entire wastestream entering the DAF unit; it is only practical for small flowrates, oily liquids, or other situations where turbulence in the pressurization systems will not degrade solids enough to impair DAF performance. This approach should not be used when the influent contains flocculated solids because the turbulence in the tank and pressure-relief valve would destroy flocs. This approach also should not be used when the influent contains abrasive or large solids, which can wear eductors and clog pumps; recycle pressurization should be used instead.

Partial pressurization systems pressurize a fraction of influent; how much depends on the air-to-solids ratio needed for optimal performance. This flow scheme typically is only practical for small rates of nonflocculated oily wastewaters. Its limitations are the same as those for total pressurization.

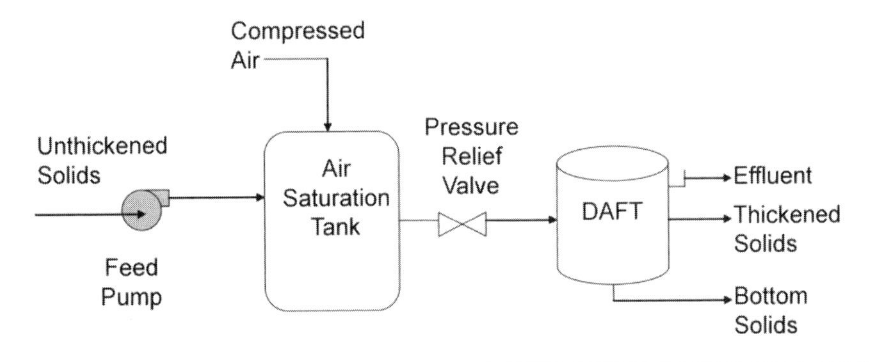

FIGURE 23.6 A dissolved air flotation thickener using a total-pressurization-of-solids flow scheme to produce gas bubbles.

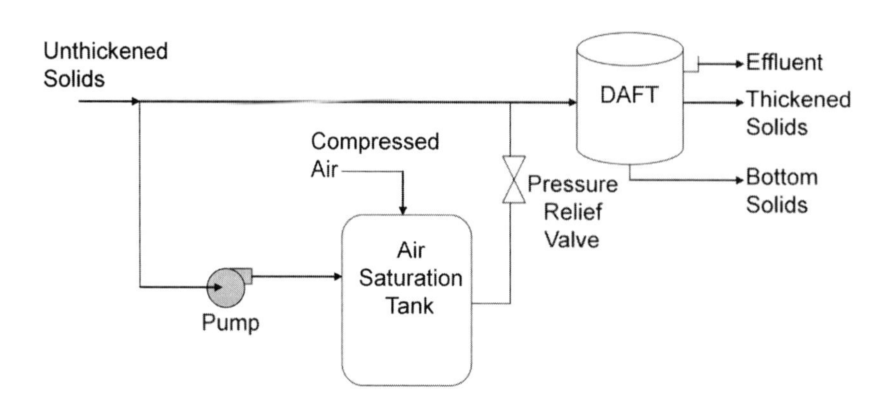

FIGURE 23.7 A dissolved air flotation thickener using a partial-pressurization-of-solids flow scheme to produce gas bubbles.

Most DAF units thickening WAS use recycle pressurization systems, in which some of the subnatant is pressurized. Influent solids do not pass through the pressurization system but are mixed with the pressurized recycle stream before entering the DAF unit.

The pressurization system consists of a recycle pressurization pump, an air compressor, an air-saturation tank, and a pressure-relief valve. Most systems operate at

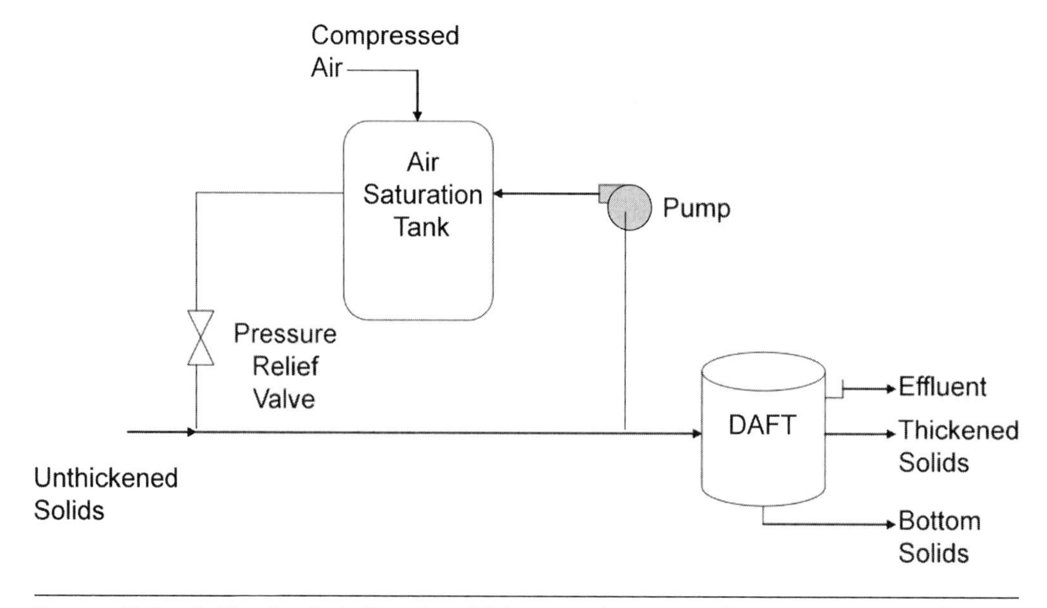

FIGURE 23.8 A dissolved air flotation thickener using a recycle-pressurization-of-solids flow scheme to produce gas bubbles.

280 to 480 kPa (40 to 70 psi). About 40 to 90% (depending on system design) of the oxygen and nitrogen in the air entering the tank is dissolved in the liquid. As dissolved air is released from solution, operators use the pressure-relief valve to control pressure loss and evenly distribute flow.

Recycle pressurization systems are used in large DAF applications and when the influent contains flocculated (typically biological) solids. Most systems include auxiliary recirculation flow (e.g., plant effluent) to start up the process. Because the system is complex and consists of numerous valves and fittings, staff training programs are essential to ensure proper operation.

3.3.8 Pressurization Tanks

Each pressurization system includes one or more of the following: a pressurization pump, pressurization tank, air compressor, airflow control panel, recycle-flow indicator, pressure-release valve, other valves, piping, and pressure gauges. The primary component is the pressurization tank.

This tank is designed to dissolve air efficiently into the pressurized recycle liquid. It provides the liquid residence time and the mass-transfer surface (in some cases, internal structures) necessary to permit air to dissolve in liquid. If the air is injected upstream of the pressurization tank, the tank also may be designed to separate undissolved air from the recycle stream. If the tank has internal structures designed to create liquid mass-transfer surface (e.g., trays, packing, and nozzles), they must be designed to be nonclogging. The recycle stream typically contains 100 to 200 mg/L of biological solids; it can contain 3 000 mg/L or more during upsets. These solids will clog most traditional mass-transfer packing surfaces.

The pressurization tank should be next to the flotator and pressurization pump(s) to minimize piping requirements and headloss via interconnecting piping. Any pressure loss downstream of the pressurization tank tends to release dissolved air from solution. Released air can enter the flotator as entrained air bubbles and create disruptive turbulence in the inlet section.

Pressurization tanks typically are constructed in accordance with the American Society of Mechanical Engineers (ASME) code for unfired pressure vessels with a working pressure of 700 kPa (100 psi); however, they typically are tested hydrostatically to 1 300 kPa (150 psi). The society's design includes a corrosion allowance whose magnitude depends on the specific constituents anticipated in the recycle stream. If more corrosion protection is required, a layer of epoxy coating is applied to the tank's internal surfaces. Stainless steel vessels also can be used.

The pressurization tank typically has steel legs or other support systems, a drainage opening, an access manhole for inspection and maintenance, a liquid-level sight glass, a pressure gauge protected by a diaphragm element, and a pressure-relief safety valve. It also may have one or more air-inlet connections, an air-release valve, and a liquid-level control valve.

3.4 Ancillary Equipment and Controls

In addition to the thickening tank, a complete DAF system includes a number of appurtenances (e.g., solids and conditioning-chemical feeding equipment; a pressurization pump, air compressor, and other pressurization-system equipment; and various control elements).

3.4.1 Pipes, Valves, and Instruments

Typical recycle-flow DAF systems have numerous valves and fittings (e.g., interconnecting pipe and pipe fittings, liquid and gas flow-control valves, gas and liquid flowrate indicators, and a level-control valve). All must be properly designed to ensure proper DAF operations.

Pressurization-system components must be spaced as closely as possible to reduce costs and minimize pressure loss and air release in the pipes. Liquid-recirculation piping typically is sized to produce a liquid velocity of 0.9 to 1.5 m/s (3 to 5 ft/sec) and manufactured with Schedule 40 or 80 carbon steel or coal-tar, epoxy-coated, carbon steel materials.

Design engineers should specify traditional piping practice, including the installation of eccentric reducers and expanders on the suction and discharge sides of the recirculation pump. Isolation valves (e.g., ball or gate valves) with a maximum open passage and minimum pressure drop in the full-open position, should be installed on the influent and effluent side of the flotation vessel (feed, float discharge, and subnatant); recirculation pump; and pressurization tank.

Air-supply piping should include oil and moisture traps, a pressure-regulating valve, a rotameter with appropriate temperature and pressure gauges, isolation valves about the rotameter, a rotameter bypass line and valve, and a check valve next to the air-injection port in the pressurization tank or recirculation piping. Isolation valves should be either ball or gate valves. The pressurization tank's air-supply piping should include a solenoid valve that is wired to shut off process air when the pressurization pump is off.

The air-supply line also should include a pressure-regulating valve, which typically is set to discharge air at 70 kPa (10 psi) above the air-absorption tank's pressure to ensure a constant airflow despite small fluctuations in tank pressure. The airflow rotameter should be direct reading in standard volumetric units and equipped with a stainless steel float and a safety shield. Operators should use a needle valve downstream of the rotameter to control the airflow rate. All valves in the air-supply system and interconnecting air piping should be made of stainless steel.

If the pressurization system is designed to accommodate a variety of flowrates, a recycle-liquid flow indicator and control valve can be useful. The flow indicators should be able to handle solids-laden streams. Venturi and vortex-shedding indicators work well in this application. The flow indicator and control valve should be installed in the pump discharge piping upstream of the pressurization tank. Ball, eccentric-plug, and diaphragm valves are effective flow-control valves that also can serve as isolation valves for pump discharge.

All pressurization systems use a pressure-relief valve, which typically is located next to the flotator in the pressurization tank's discharge line. The valve reduces recycle-liquid pressure to atmospheric conditions; the air dissolved under pressure is precipitated at the valve in the form of microscopic air bubbles. These air bubbles contact the solids to be floated. Sometimes the pressure-relief valve can be used to control recycle-liquid flows. Design engineers should consult pressurization-system manufacturers in each instance.

Operators can use a float-controlled air bleed-off valve to maintain the liquid level in the pressurization tank. It typically bleeds off a small amount of excess air. If the water level rises, the float closes the bleed port so the air will force the liquid level back down, after which the air bleed resumes. If an alarm circuit is used to indicate a high water level, a float switch can be wired to an air-bleed solenoid valve that bleeds off excess air.

The flotator feed line and subnatant recirculation piping should include provisions for feed and subnatant sampling. Polymer-addition taps should be installed in both feed and subnatant lines and should be far enough upstream of the discharge point to allow for thorough mixing. The proper location is site-specific. Drain plugs should be installed in all low points in feed and subnatant piping. Cleanout pipe "T"s should be used rather than elbows so that operators can remove any debris that becomes lodged in process piping.

3.4.2 Pumps and Compressors

Design engineers can use positive-displacement, diaphragm, piston, or progressing cavity pumps to feed solids to DAF thickeners, although centrifugal pumps have been preferred. The pumps should possess variable capacity and an operating range wide enough to accommodate expected variations in solids-production and thickening requirements, as well as variations in feed-solids characteristics. They also should be equipped with a flow totalizer or monitor so operators can maintain records of the amount of solids thickened and control DAF operations. Each flotator should have its own pump.

Dissolved air flotation processes often have polymer systems, which include mixing and storage tanks and chemical feed pumps. The systems can be purchased as a unit from polymer suppliers or designed by engineers. Either way, they should use variable-capacity, positive-displacement pumps so operators can accurately control the amount of polymer used. Each flotation thickener should have its own chemical pump.

A key element of any DAF system is the pressurization pump, which feeds enough liquid into the pressurization tank to ensure that the flotation tank will receive the desired amount of dissolved air. Open-impeller, centrifugal pumps typically are used for this purpose. Single- and two-stage pressurization pumps also have been used. Most currently operating DAF thickeners use single-stage pressurization systems. Two-stage pumps reportedly provide more air-dissolving efficiency than single-stage ones. If using two-stage pumps, compressed air is delivered to the suction end of the second stage.

For system flexibility, design engineers should use pressurization pumps with a relatively steep head-capacity curve. Operators can adjust pump flow by throttling the pump isolation valve between the pressurization pump and air-dissolving tank. Throttling typically can control the discharge from pumps with steep head-capacity curves. This is not true for centrifugal pumps with a flat curve; in this case, using a throttling valve could induce pump surging.

Pressurization pumps typically use single-speed motors. The use of two-speed motors or adjustable sheaves for variable head and flow capability depends on several factors (e.g., number of flotators, operating method, quantity of solids to be thickened under both average and peak conditions, and degree of flexibility desired). Although initial costs are higher, variable-speed pressurization pumps can lower power costs and enhance flexibility.

A variety of air compressors (e.g., reciprocating piston, rotary vane, and screw) can be used to provide air for the DAF process. Some wastewater treatment plants use central compressors to meet DAF air requirements as well as other needs in the plant.

Most flotation systems have their own air compressors. Reciprocating piston-type units are the most common and typically are sized to deliver at least twice the maximum air theoretically required for saturation so the compressor can operate in an unloaded condition about 50% of the time.

In addition to the compressor, a pressure reservoir, air filter, oil trap, pressure regulator, and airflow meter are required.

3.5 Cothickening Primary and Secondary Solids

In cothickening processes, settled solids from primary and secondary clarifiers are mixed together and then thickened. Cothickening used to be rare; however, recent pilot testing and full-scale operations have indicated the benefits of cothickening over thickening primary and secondary solids separately (Butler et al., 1997):

- Ability to increase DAF thickener solids loading rate (could double the solids loading rate per surface area compared to separate thickening);
- Ability to reduce soluble BOD and chemical oxygen demand (COD) as much as 80% and 60%, respectively;
- Lower present-worth and operating costs; and
- Significantly reduced secondary BOD loading while reducing grit because of thickened-solids recycling.

Table 23.8 shows DAF thickener operations and performance averages for 1994 and 1995 at a wastewater treatment plant.

TABLE 23.8 Typical solid loading rates for a dissolved air flotation thickener (U.S. EPA, 1979).

	Averages	
	1994	**1995**
Mixed Sludge Feed		
Flow with recycle, m^3/min	12.4	12.5
Total solids with recycle, mg/L	5 100	5 492
Primary : secondary solids ratio	1.20	1.22
Primary sludge, total solids, mg/L	7 189	7 540
Waste activated sludge TSS, mg/L	5 878	4 467
DAFT Operations		
Solids loading rate, $kg/m^2/day$	108.4	110.8
Hydraulic loading rate, $m^3/m^2/hr$	3.33	4.02
Air-to-solids ratio, kg/kg	0.05	0.06
Polymer dose, kg/dry metric ton	15.9	16.1
Recycle ratio, Qr/Qfeed	2.3	3.3
Dissolution pressure, kPa	400	483
DAFT Performance		
Solid capture, %		
Float solids only	84.2	81.1
Float solids + bottom solids	96.8	95.4
Thickened float volume, m^3/min	.84	0.74
Thickened float solids, % TS	5.94	6.18
Thickened overflow,[a] m^3/min	9.28	10.45
Thickened overflow TSS, mg/L	235	296
Bottom sludge, m^3/day	1.97	0.736
Bottom sludge total solids, % TS	0.55	1.75

[a] Thickener overflow refers to the clarified subnate.

Cothickening typically is used at small wastewater treatment facilities and must include polymer to improve clarification and thickening. Design engineers should ensure that primary solids and WAS are mixed thoroughly because variable concentrations of mixed solids can cause operating problems and result in poor thickening performance. They also make it difficult to maintain consistent polymer doses in the thickening process.

Torpey (1954) was the first to recognize that treatment plant staff had to manage several factors for stable cothickening operations. His research established that:

- Primary solids and WAS should not contain more than 3500 mg/L of solids,
- A dilution-to-primary liquid ratio of 8:1 will keep the thickener fresh, and
- Operators must maintain specific loading rates for primary and secondary solids.

Diluting influent can cause poor cothickening. Cothickener influent typically is 18% of plant flow, of which nearly 98% is recycled dilution water. If wastewater temperatures never exceed 15 to 20°C, a dilution-to-primary liquid ratio of 4:1 to 6:1 is satisfactory; however, higher temperatures require more dilution. The key to successful cothickening is optimizing the volume of fresh dilution liquid required to minimize gasification and biological activity.

3.6 Design Example

A municipal wastewater treatment plant is planning on using DAF thickening for WAS. It has primary clarification and complete-mix aeration basins.

- WAS has a solids concentration of 0.8 to 1.0%.
- The maximum design WAS production is 3 000 lb/day.
- The treatment plant will be staffed 5 days per week, 8 hours per day, but the DAFT thickening process will be operated 24 hours per day, 7 days per week.
- WAS flow from the secondary clarifier will be continuous.
- It has been decided that a circular cast-in-place concrete DAF tank will be constructed.
- It has been decided that the DAF process will use recycle pressurization.
- It has been decided that polymer will be added only if performance is unsatisfactory.

Maximum daily WAS production = 1 360 kg/d
Solid loading rate = 2 kg/m2/hr (Table 23.8—select a conservative loading rate (WAS air)
Efficiency factor = 85% (Efficiency of 85% is selected because polymer is not used)
Operation hours per day = 24 hrs
WAS minimum solids concentration = 0.8%
Minimum solids concentration based on SLR = 4%

Pressurized Recycle Flow Rate

The design pressurized flow should be based on the maximum gross solids load (EPA, 1979). In this example, a recycle rate of 100% will be assumed. Therefore, the recycle flow rate is 37 gal/min.

Maximum net hourly load = 57 kg/hr	Use Equation 23.8
Maximum surface area required = 29 m^2	Use Equation 23.9
Required diameter for circular DAF tank = 6 m	
Maximum gross solids load = 67 kg/hr	Use Equation 23.10
Maximum feed flow rate = 37 gal/min	Use Equation 23.11
Thickened solids flow rate = 6 gal/min	Use Equation 23.12
Subnatant (total) flow rate = 74 gal/min	Use Equation 23.13
Hydraulic surface loading rate = 2.6 gal/min^2	Use Equation 23.14

Note: Per Section 3.2.3, DAFT typically is designed for a hydraulic loading rate between 5 and 21 gal/min/m (1 gal = 3.78 kg).

Number of Units

Ideally, design engineers should provide two DAF units so one can be taken out of service and the process still will have adequate capacity of WAS thickening.

4.0 CENTRIFUGE

Centrifugal thickening is analogous to gravity thickening except that centrifuges can apply a force 500 to 3000 times that of gravity. The centrifugal force causes suspended solids particles to migrate through the liquid toward or away from the centrifuge's rotation axis, depending on the difference between the liquid's and solids' densities. The increased settling velocity and short particle-settling distance accounts for a centrifuge's comparatively high capacity.

Centrifuges have been used to thicken waste solids since the early 1920s (WPCF, 1969). Solid-bowl conveyor centrifuges are the most widely used in this application. Variables affecting centrifuge thickening are grouped into three basic categories: performance, process, and design. Performance is measured by the thickened solids concentration and the suspended solids recovery in the centrate. The recovery is calculated as the thickened dry solids as a percentage of feed dry solids. Using the commonly measured solids concentrations, recovery is calculated as follows:

$$R = \frac{TSS_P(TSS_F\ TSS_C)}{TSS_F(TSS_P\ funcTSS_C)} \times 100 \qquad (23.15)$$

Where

R = percent recovery,
TSS_P = total suspended solids concentration in thickened solids (% by weight),
TSS_F = total suspended solids concentration in influent (% by weight), and
TSS_C = total suspended solids concentration in centrate (% by weight).

Process variables that affect thickening include feed flowrate, the centrifuge's rotational speed, differential speed of the conveyor relative to the bowl, pond depth, chemical use, and the physicochemical properties of the liquid and suspended solids (e.g., particle size and shape, particle density, temperature, and liquid viscosity). These variables are the tools that wastewater treatment plant operators must use to optimize centrifuge performance.

4.1 Operating Principle

A centrifuge's main components are the bowl and the scroll. The bowl is mounted horizontally and turns rapidly to create the centrifugal force. The scroll is mounted inside the bowl and conveys solids from one end of the bowl to the other.

The bowl consists of a cylindrical section and a conical section (see Figure 23.9). Both typically are cast stainless steel but also can be made of rolled stainless steel plate. The sections are bolted together in the factory, machined, and balanced at a high speed.

The scroll consists of a stainless steel screw conveyor mounted on a hollow shaft (see Figure 23.10). It can either be an open design mounted to the shaft via spokes or a closed design mounted directly to the shaft. The entire scroll is mounted inside the bowl and can turn independently.

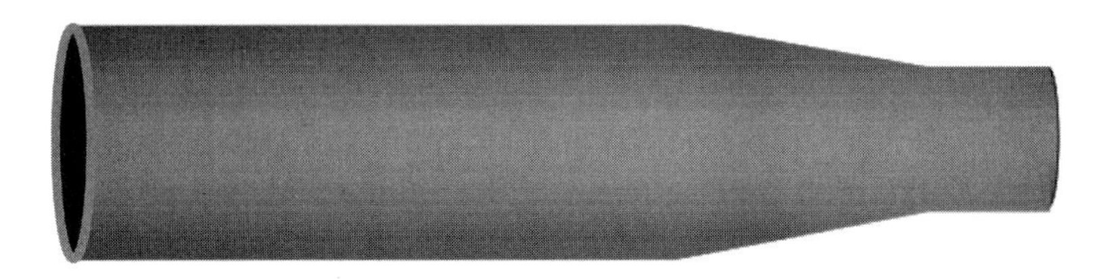

FIGURE 23.9 The geometry of a centrifuge bowl.

To thicken solids, the bowl and scroll typically operate at more than 1500 rpm; the scroll rotates just a few rpm faster (or slower) than the bowl to create a differential speed. Feed solids and polymer are injected into the scroll's hollow shaft and discharged into the spinning bowl. The bowl's centrifugal force causes solids to settle along the bowl wall. The scroll's screw conveyor moves the solids up the conical section of the bowl and discharges them. Meanwhile, the liquid is discharged at the opposite end of the bowl via openings in the end plate (see Figure 23.11).

4.2 Physical Features

Each centrifuge has unique physical features that can affect throughput, capture efficiency, polymer dose, cake solids concentration, and power.

4.2.1 Bowl Geometry

The bowl is one of the centrifuge's most critical features. Bowl geometry significantly affects throughput, capture efficiency, and cake solids concentration.

The bowl consists of two major sections: the cylinder and the cone (see Figure 23.12). The critical dimensions that manufacturers use to describe a particular centrifuge are bowl diameter (D1), bowl length (L1), discharge diameter (D2) and beach angle (A). Each dimension influences centrifuge performance.

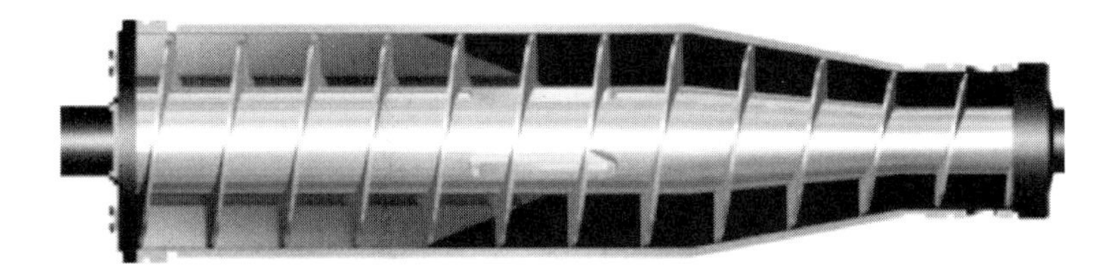

FIGURE 23.10 An example of a centrifuge scroll (courtesy of GEA Westfalia Separator, Inc.).

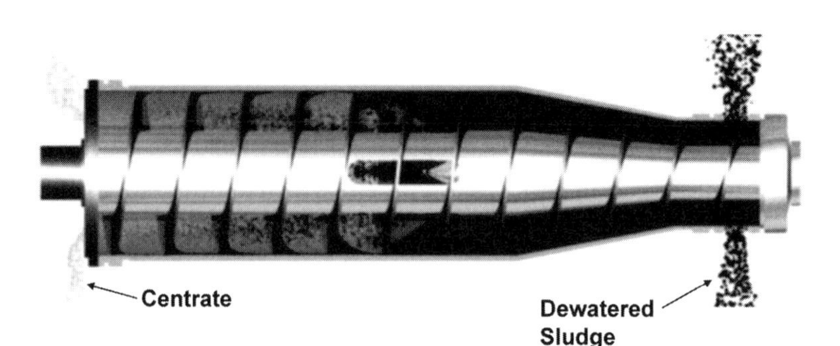

FIGURE 23.11 An example of a centrifuge thickening solids (courtesy of GEA Westfalia Separator, Inc.).

- Together, bowl diameter and bowl speed dictate the centrifugal force at the bowl wall. At a given bowl speed, centrifugal force at the bowl wall increases as bowl diameter increases.
- The discharge diameter dictates the pond depth of solids in the centrifuge. This is associated with the maximum volume of solids that the centrifuge can hold. At a given bowl diameter, pond depth and maximum solids volume decrease as discharge diameter increases.
- Thickened solids are conveyed up the beach (conical section of the centrifuge) and then discharged. Manufacturers have found that a 15- to 20-deg beach angle is optimum for thickening centrifuges.

4.2.1.1 Bowl Volume

Table 23.9 presents the data used to calculate the total volume of a centrifuge bowl. However, the centrifuge's *usable volume*—total volume minus the air space associated

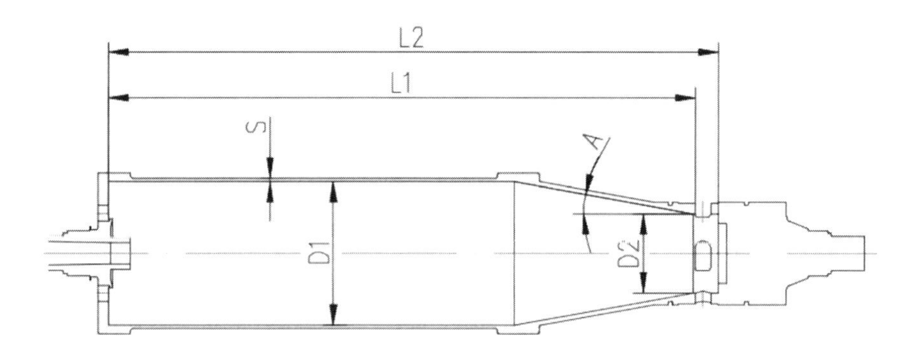

FIGURE 23.12 The critical dimensions of a centrifuge bowl.

TABLE 23.9 Characteristics of various centrifuge bowls.

Manufacturer	Bowl diameter (mm)	Bowl circumference (m)	Bowl length (m)	Cone length (mm)	Cone angle (deg)	Cylinder length (m)	Cylinder volume (m^3)	Discharge diameter (mm)	Usable bowl volume (m^3)	Unusable bowl volume (m^3)	Bowl speed (sec/rev)	Normal operating speed (rpm)	Angular velocity (m/s)	Centripetal acceleration (m/s^2)	G-force at bowl wall (m^3)	G-force at bowl wall (gal)
A	740	2.32	3.05	450	20	2.6	1.1	410	0.8	0.3	0.024	2500	97	25 500	2 600	600 000
B	690	2.15	2.92	470	15	2.5	0.9	430	0.5	0.4	0.023	2 600	93	25 000	2 600	380 000
C	740	2.32	3.1	490	20	2.6	1.1	380	0.8	0.3	0.026	2 300	89	22 000	2 200	500 000
D	760	2.39	3.07	380	20	2.8	1.3	480	0.8	0.5	0.027	2 200	89	21 000	2 100	500 000

with the discharge diameter (the space that the scroll occupies)—is the actual volume of solids that a centrifuge can hold (see Figure 23.13).

4.2.1.2 Cylinder Volume

Thickening centrifuges use the conical section of the bowl to convey solids to the discharge point. So, the only part of the bowl separating solids from liquids is the cylindrical portion.

4.2.2 Scroll Geometry

The scroll conveys solids along the bowl and up the beach, where they are discharged. This scroll may be open or closed.

4.2.2.1 Open Scroll

An open scroll consists of a steel ribbon flight attached to a scroll shaft by spokes. Manufacturers that use this type of scroll claim that it reduces turbulence in the bowl because the centrate does not have to travel around the flight and agitate solids along the bowl wall. This could reduce polymer use and improve capture efficiency, according to the manufacturers.

4.2.2.2 Closed Scroll

A closed scroll consists of a flight directly attached to the scroll shaft. Manufacturers that use this type of scroll claim that it allows solids inventory to be built up higher than the open scroll does. (The open scroll permits more cake compression, potentially producing higher solids.) One manufacturer has used closed scrolls successfully in several dewatering applications.

4.2.3 Scroll Configuration

The scroll can be configured to lead or lag (see Figure 23.14). A leading scroll runs slightly faster than the bowl, while a lagging one runs slightly slower.

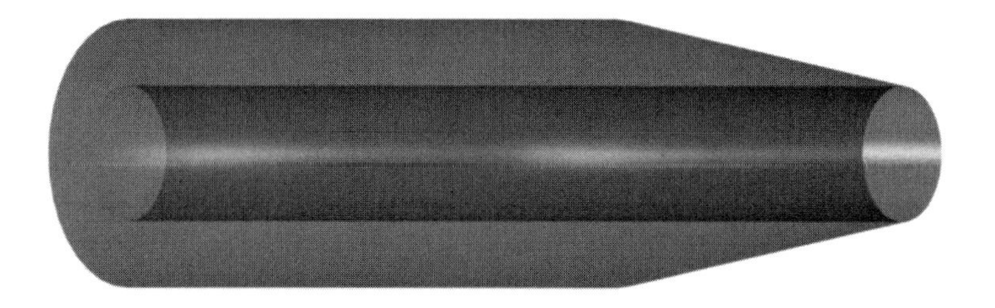

FIGURE 23.13 The usable volume in a centrifuge bowl.

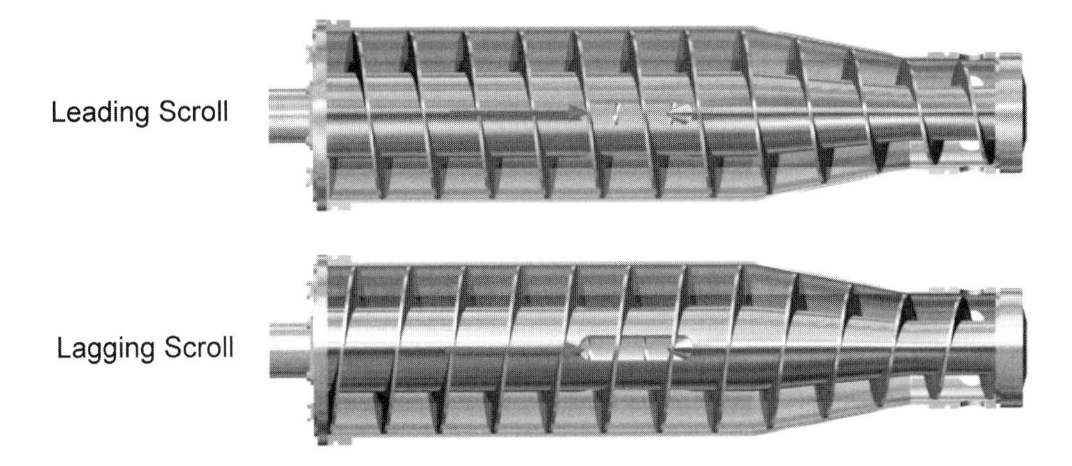

FIGURE 23.14 Possible scroll configurations (courtesy of GEA Westfalia Separator, Inc.).

4.2.4 *Scroll Drive Features*

The scroll drive (back drive) system turns the scroll relative to the bowl creating a differential speed that can range from 1 to 15 rpm. At higher differential speeds, thickened solids are removed from the centrifuge more rapidly and have lower solids concentrations. At lower differential speeds, thickened solids are removed from the centrifuge more slowly and have higher solids concentrations.

4.2.5 *Differential Speed Adjustment*

Thickening centrifuges operate at low differential speeds to keep solids in the bowl as long as possible. Differential speed is a fine adjustment that operators use to achieve the highest cake solids possible while providing the required throughput and capture efficiency, and minimizing polymer dose. They make fine adjustments via a separate scroll drive operating system that converts a high motor speed to a slow scroll speed. The finer the scroll speed adjustment, the more control operators have to optimize the centrifuge.

4.3 Evaluation and Scale-Up Procedures

There are relatively few centrifuge manufacturers and they typically guard their evaluation and scale-up technologies, so basic, widely accepted design criteria for centrifugal thickeners do not exist. Instead, design engineers typically rely on experience, laboratory testing, and pilot tests to estimate centrifuge performance. Feed solids rarely are available for experimentation, so designers must make judgments based on past experiences with similar solids under identical conditions. If solids are readily available, however, designers should make every effort to base their designs on pilot-test results.

Design engineers typically use bench-scale tests to determine whether centrifugation is feasible, choose a chemical conditioner, and select an appropriate dose of that chemical. They use pilot tests to generate centrifugal design data because equipment manufacturers are reluctant to guarantee performance without such data.

Pilot tests should be conducted on a full-scale centrifuge whose design and proportions are similar to the commercial unit being considered. Many manufacturers have test units available on a rental or trial basis; several of them have truck-, van-, or skid-mounted units that can move readily from site to site. Operating a pilot unit through a broad range of machine and process variables allows design engineers to assess the effects of normal variations in feed-solids flow and quality. In particular, they should evaluate how the following parameters affect centrifuge performance:

- Hydraulic feed rate, including polymer solution flow;
- Thickened solids (cake) discharge rate;
- Polymer dose rate;
- Clarification area;
- Pool depth and volume;
- Solids retention time;
- Conveyor differential speeds;
- Centrifugal force (bowl speeds);
- Percentage solids recovery; and
- Cake concentration.

Testing results typically provide enough accurate information for designers to select and size the most economical full-scale equipment. That said, the final selection of a solid-bowl, scroll-type centrifuge typically is a compromise between two operations that are intrinsic to successful centrifuge operations: solids separation (which is a function of clarification area) and solids consolidation and removal (via the screw conveyor). If these cannot be balanced, then centrate clarity and/or solids concentration will deteriorate.

4.3.1 Theoretical Capacity Factors

Researchers have developed certain theoretical equations for use in scaling up pilot data to the full-scale commercial unit. [Complete derivations of these scale-up factors can be found in literature by Perry and Chilton (1963), Vesilind (1974a and 1974b), and Purchas (1977).]

The two most important criteria for successful centrifuge operations are solids separation (hydraulic or clarification capacity) and solids removal (cake-conveying capacity). A solid-bowl centrifuge's hydraulic capacity (Σ) is determined as follows:

$$\sum = 2\Pi\, l \frac{\omega^2}{100\ g}(0.75\, r_2^2 + 0.25\, r_1^2) \qquad (23.16)$$

Where

Σ = theoretical hydraulic capacity (cm^2);
l = centrifuge bowl's effective clarifying length (cm);
ω = centrifuge bowl's angular velocity (rad/s);
g = acceleration from gravity (m/s^2);
r_1 = radius from centrifuge centerline to the liquid surface in the centrifuge bowl (cm); and
r_2 = radius from centrifuge centerline to the inside wall of the centrifuge bowl (cm).

Below is a simpler way to calculate Σ (applicable only to solid-bowl centrifuges) (Vesilind, 1974b):

$$\sum = \frac{V\,\omega^2}{g\ \ln(r_2/r_1)} \qquad (23.17)$$

where V = centrifuge pool volume (cm^3).

When scaling up solids-handling (cake-conveying) capacity, the assumption is that if two geometrically similar (but different-sized) machines have the same ratio of solids discharge rate to theoretical solids-handling capacity, then their performance will be similar for a given solids feed (Vesilind, 1974a). The scale-up relationship is as follows:

$$\frac{Q_{SP}}{\beta_p} = \frac{Q_{SF}}{\beta f} \qquad (23.18)$$

Where

Q_{SP} = the pilot-scale unit's solids discharge rate (m^3);
Q_{SF} = the full-scale unit's solids discharge rate (m^3);
β_p = the pilot-scale unit's solids-handling capacity factor (m^3/h); and
β_f = the full-scale unit's solids-handling capacity factor (m^3/h).

The best approach is to develop full-scale requirements based on both capacity and solids-loading considerations. The limiting criterion would govern machine selection.

There are limitations with the theoretical relationships just cited. Neither Σ nor β takes into consideration interactions between the clarification and cake-storage zones. Certain wastewater solids (e.g., WAS) are thyrotrophic and may have difficulty moving up the conical section of the bowl before discharge. Also, the full-scale unit may not be able to achieve the theoretical solids depth (cake pile) because of the theoretical nature of cake solids.

Design engineers should consult with centrifuge manufacturers to properly identify limiting design factors and develop full-scale requirements based on both hydraulic and solids loading considerations. Some limitations may be overcome by altering the centrifuge design (e.g., bowl or conveyor speed, conveyor pitch, number of conveyor leads, or pool depth) to provide a more conducive environment for solids separation and removal.

4.4 Process Design Conditions and Criteria

Centrifuge manufacturers offer designs with substantially different features. Table 23.10 lists the major design and operating variables that influence the operation of a horizontal solid-bowl centrifuge. These variables are discussed at length in the literature (U.S. EPA, 1979; WPCF, 1980). A desirable characteristic of the centrifuge is that its performance—as measured by thickened solids and solids capture—can be adjusted to desired values by modifying control variables [e.g., feed flowrate, bowl and conveyor differential speed, conditioning-chemical (polymer) use, and pool depth].

Table 23.11 indicates how a horizontal solid-bowl centrifuge's capabilities relate to basic rotating assembly size and operating speed. (Specific design recommendations

TABLE 23.10 Factors affecting centrifugal thickening.

Basic machine design parameters	Adjustable machine and operational features	Solids characteristics
Flow geometry	Bowl speed	Particle and floc size
Countercurrent	Bowl and conveyor differential speed	Particle density
Cocurrent	Pool depth and volume	Consistency
Internal baffling	Feed rate	Viscosity
Bowl/conveyor geometry	Hydraulic loading	Temperature
Diameter	Solids loading	SVI
Length	Flocculant use	Volatile solids
Conical Angle		Solids retention time
Pitch and lead		Septicity
Maximum pool depth		Floc deterioration
Solids and flocculant feed points		
Maximum operating speed		

TABLE 23.11 Operating results reported for horizontal solid-bowl centrifuges.

Location	Activated-sludge type	Feed solids concentration, mg/L	SVI (% VSS)	Feed flow rate, L/min	Thickened solids concentration, %	Solids capture, %	Polymer use, g active polymer/dry kg of solids	Machine size (bowl diameter × length), mm	Bowl speed, r/min	Centrifuge configuration
Atlantic City, New Jersey	Conventional	3 000	100 (60)	1 230	10	95	2.5	740 × 2 340	2 600	Countercurrent
Los Angeles, California, Hyperion	Conventional	4 8000–6 000	110–190	2 300–3 000	3.7–5.7	88–91	None	1 100 × 4 190	1 600	Cocurrent
					3.6–6.0	77–96	0.2–2.2	1 100 × 4 190	1 600	Cocurrent
	Conventional	4 800–6 000	110–190	2 300–3 000	1.9–7.9	47, 89	None	1 100 × 3 600	1 995	Countercurrent
					1.7–8.2	57–97	0.4–1.4	1 100 × 3 600	1 995	Countercurrent
Oakland, California, East Bay MUD	High-purity oxygen	5 000	250–400	4 200	7	66	6	1 000 × 3 600	1 995	Countercurrent
Naples, Florida	Conventional	10 000–15 000	70–80	380	6	90–92	None	740 × 3 050	2 000	Countercurrent
Milwaukee, Wisconsin, Jones Island	Conventional	6 000–8 000	80–150 (75)	1 100–1 900	3–5.5	92–93	—	—	1 000	Cocurrent
Littleton, Colorado	Conventional	6 000–8 000	100–300	570–1 100	6–9	88–95	3–3.5	740 × 2 340	2 300	Countercurrent
Lakeview, Ontario (Canada) (PM 75 000)	Conventional	7 560	80–120	840	4.7	77	None	740 × 2 340	2 300	Countercurrent
Lakeview, Ontario (Canada) (XM-706)	Conventional	7 120	80–120	1 350	6.1	65	None	740 × 3 050	2 600	Countercurrent

are omitted because anticipated performance ranges vary widely due to design differences and solids characteristics.)

Sometimes polymer addition can increase a centrifuge's hydraulic loading while maintaining its solids-capture and -thickening characteristics. Polymer use typically can improve solids-capture efficiencies to between 90% and more than 95%.

Following are significant design considerations for centrifuge thickeners:

- Provide effective wastewater degritting and screening or grinding. If wastewater screening or grinding is inadequate, feed solids should be sent through grinders before entering the centrifuge to avoid plugging problems.
- Use a feed source with a relatively uniform consistency (a mixed storage or blend tank is often appropriate) and feed it to the centrifuge via an adjustable-rate pump with positive flowrate control.
- Consider handling thickened solids via one of the following methods: direct discharge to a collection well followed by transport using a positive-displacement pump, direct discharge to an open-throat progressing cavity pump, or discharge to a screw conveyor.
- Consider recycling centrate to either primary or secondary treatment processes and providing the ability to vent and/or suppress foam in the centrate piping.
- Consider structural aspects (e.g., static and dynamic loadings from the centrifuge, vibration isolation, and provision of an overhead hoist for equipment maintenance). In regions subject to earthquakes, provide snubbers (to isolate vibrations), piping flexibility, and connections to auxiliary equipment.
- Provide water to flush the centrifuge during equipment shutdowns.
- Consider whether a heated water supply will be needed to periodically flush grease buildup.
- Ensure that the centrifuge is vented properly and consider whether odor controls are needed.
- When thickening anaerobically digested solids, consider the potential for struvite (ammonium magnesium phosphate) to form.
- Pay attention to polymer feed-system design.

4.4.1 *Process Design Criteria*

Engineers typically define process design criteria by establishing which solids characteristics are significant in a given application and determining how they affect process performance. Unfortunately, specific design criteria are not possible for centrifugal thickeners because of all the variations in both solids characteristics and centrifuge designs.

Solid-bowl conveyor centrifuges are versatile; they can be used to thicken a variety of wastestreams. Most municipal wastewater treatment plants use them to thicken WAS.

One important design parameter is hydraulic loadings. Hydraulic loadings to the centrifuge control the liquid-phase residence time, which typically ranges from 30 to 60 seconds in thickening applications.

Solids concentration is another important factor; it determines the specific solids load applied to the centrifuge (kilograms of dry solids per day) and the thickened solids (wet-cake) output volume (cubic meters per day). These measurements help centrifuge designers determine the machine's parameters. They also help operators adjust machine and process variables (e.g., conveyor differential and feed rates) to balance load demands.

The density of the feed material's solids and liquid fractions also is important, and operators have little control over this characteristic. Because activated sludge and mixed liquor typically have similar low floc densities. Operators frequently need to add chemical conditioners to increase the effective density of the aggregate floc, thereby increasing sedimentation or centrifugal settling rates.

The size and distribution of particles in the feed solids significantly affect the centrifuge's thickening and dewatering performance, but these characteristics are difficult to measure accurately. For the most part, operators simply measure solids concentration rather than particle size or density. One concern—particularly with WAS—is that these naturally well-flocculated materials consist of small particles loosely bound together in one aggregate floc. This naturally occurring aggregate often breaks up easily, particularly under the high shearing forces in a centrifuge. So, polymers may be necessary to make the floc aggregate more coherent.

4.5 Mechanical Features

A basic solid-bowl conveyor centrifuge has the following main components: base, case, bowl, conveyor, feed pipe, main bearings, gear unit, and back drive.

4.5.1 Motor Type and Size

Most centrifuge manufacturers provide two motors and two VFDs with their systems: a scroll drive motor and VFD, and a main drive motor and VFD. The AC motors receive current directly from the VFDs. The scroll drive motor speeds up or slows down the scroll; it does not seem to affect the total connected horsepower of either VFD. However, when the scroll drive motor is used to speed up the scroll, the size requirement is typically larger than when it is used to slow down the scroll. Conversely, when the scroll drive motor is used to speed up the scroll, the main drive motor size requirement is typically smaller than when the scroll drive motor is used to slow down the scroll.

4.5.2 *Base*

The base provides a solid foundation on which to mount and support the centrifuge's main components. Vibration isolators between the base and the machine foundation reduce the transmission of centrifuge vibrations.

4.5.3 *Case*

The case completely encloses the rotating assembly; it serves as a guard and noise dampener. [A solid-bowl centrifuge's noise typically ranges from 80 to 90 dbA at 0.9 m (3 ft).] The case also contains and directs cake solids and centrate as they are discharged from the rotating assembly.

4.5.4 *Bowl*

A solid-bowl centrifuge's bowl typically resembles a cylinder or cone. Proportions vary, depending on manufacturer. Bowl diameters range from 0.23 to 1.38 m (9 to 54 in.), and the bowl length-to-diameter ratio ranges from 2.5:1 to 4:1. Centrifuge capacity typically ranges from 40 to 3 000 L/min (10 to 800 gpm).

Bowls used for wastewater treatment applications typically are made of carbon steel or 300 series stainless steel with strips or grooves on the inside that retain a protective layer of solids. Sometimes the bowl has a stainless steel or ceramic liner.

4.5.5 *Conveyor*

A helix or screw-conveyor assembly consists of a central core or hub, a feed compartment and feed ports lined with abrasion-resistant ceramic or tungsten carbide. The helical flights leading surfaces and blade tips are coated with abrasion-resistant materials. Modern solid-bowl centrifuges use replaceable conveyor segments that are made of ceramic or tungsten carbide. The entire assembly fits concentrically into the centrifuge bowl. Conveyor speed is controlled by the gear unit and back-drive assembly. Flocculent aids are added either to the feed compartment or via a separate injection port in the machine.

4.5.6 *Feed Pipe*

The feed pipe is removable, and design engineers should determine the length of the pipe. They typically provide two polymer-feed locations: into the feed pipe and directly into the centrifuge. Engineers should make sure the feed pipe has a flexible inlet connection (rather than a valve or fitting) so the isolator can properly protect against vibrations.

4.5.7 *Bearings*

Depending on machine size and speed, three types of main bearings—ball, spherical, and cylindrical—support the entire rotating assembly. The bearings are lubricated by grease, a static oil bath, or an external circulating-oil system and typically have an L10 life of 100 000 hours.

4.5.8 *Back-drive*

The gear unit and back-drive assembly allow the bowl and the conveyor to maintain different speeds. It typically consists of a planetary or cyclo gear and a mechanical, hydraulic, or electrical back-drive. The gear unit typically is lubricated by either an oil-bath or grease-lubrication system.

4.5.9 *Abrasion Areas*

Centrifuges have a number of sensitive abrasion areas (e.g., the bowl's interior wall, the conveyor blades, the feed compartment, the feed ports, and the solids discharge area). Such areas typically are protected by various hard-facing materials (e.g., sintered tungsten carbide or ceramic). Modern techniques have increased conveyor lives to between 10 000 and 20 000 hours.

4.5.10 *Vibration*

Every centrifuge vibrates to some degree. To dampen vibrations transmitted to the foundation or piping, the centrifuge base should be supported on vibration isolators and directly connected piping should have flexible connectors.

4.5.11 *Electrical Controls*

A solid-bowl centrifuge's control circuitry typically is designed to protect the centrifuge from malfunctions (e.g., torque overload, loss of oil pressure, excessive vibration, high oil temperature, and motor overload). Also, design engineers should provide interlocks that shut down feed solids and initiate a water-flush sequence during centrifuge shutdowns.

4.6 Ancillary Equipment

4.6.1 *Pumps*

Design engineers typically prefer to use progressive-cavity pumps for small to medium centrifuges because the positive displacement and steady pumping rate allow for both effective metering and close control of the solids-feeding rate. Progressive-cavity drives should be variable speed with at least a fivefold range [e.g., 80 to 400 L/min (20 to 100 gpm)] so that operators can meet solids-loading goals regardless of variations in feed solids concentrations.

If centrifugal pumps are selected, design engineers should be aware that changes in solids consistency will affect the pumping rate. Therefore, it is important to choose appropriate flow meters and controllers to maintain centrifuge loadings.

It is a good design practice to use flow meters with either type of pump, particularly when other measurement controls (e.g., density sensors) will be used to maintain a relatively constant solids load to the centrifuge.

4.6.2 *Thickened Solids Transportation*

Centrifuges discharge a thickened cake containing between 3 and 15% solids (by weight), depending on the application. This cake is highly viscous and often thixotropic. When discharged from the centrifuge's directional chute, the cake can be

- Directly discharged to a collection well for subsequent pumping,
- Directly discharged to an open-throat progressing cavity pump, or
- Discharged to a horizontal screw conveyor, which carries it to a sump or open-throat pump.

The last alternative works well when the centrifuge is used for both thickening and dewatering. A reversing screw conveyor can direct thickened solids in one direction and dewatered cake in the opposite direction.

4.7 Performance-Control Systems

Maximizing cake solids, maintaining reasonable capture efficiency, and minimizing polymer requirements are difficult tasks to accomplish without instrumentation and controls. Centrifuges used to be controlled by an automatic torque feature, which changed the scroll's speed relative to the bowl speed to maintain a predetermined torque setpoint. This helped maintain consistent cake solids but did not address capture efficiency and polymer dose. Capture efficiency was maintained by manually changing the polymer dose or torque setpoint until the centrate was clear. This system worked well, but required significant operator attention.

The current state-of-the-art control system uses a combination of sensors and software to control differential speed, bowl speed, pond depth, and polymer dose.

4.7.1 *Feed-Forward System*

A feed-forward system uses a sensor to track the feed solids concentration and a flow monitor to track the flowrate. This information is used to determine the solids loading rate. Then, the system automatically adjusts the polymer dose to catch an operator-specified dose per pound of solids processed. It also automatically adjusts the torque (differential speed) based on historical information to produce the desired cake solids concentration. If the polymer dose cannot be increased sufficiently to provide visably clean centrate, the system adjusts the torque setpoint until the centrate clears.

4.7.2 *Feed-Backwards System*

The feed-backwards system uses a sensor to measure the centrate's solids concentration and uses this information to adjust the polymer dose until the desired centrate solids concentration is achieved. It also automatically adjusts the torque to produce a consistent cake solids concentration. If the desired centrate quality cannot be achieved regardless of polymer dose, the system automatically reduces the torque until the centrate clears.

4.8 Variable-Speed Bowl and Scroll

All centrifuge manufacturers now offer variable-frequency drives on both the main and scroll drives. The drives can be changed while the centrifuges are operating (i.e., both differential and bowl speed can be adjusted). This relatively new feature allows operators to reduce bowl speed and differential speed simultaneously until the desired solids concentration is obtained. Also, operating at lower bowl speeds can reduce power consumption and may reduce both wear and polymer requirements.

4.9 Pond-Depth Adjustments

The characteristics of co-mingled primary solids and WAS may tend to change throughout the day. Solids concentration, for example, could change from 1.0% to as high as 4.5%. Such significant changes in solids concentration may require changes in centrifuge pond depth to maintain the desired solids-capture efficiency and thickened-solids concentrations at a minimum polymer dose.

Manually changing pond depth is a trial-and-error process that involves shutting the unit down, unbolting and adjusting the centrate weirs, starting the centrifuge back up, collecting and analyzing thickened-solids and centrate samples, and repeating as necessary until the desired results are obtained. Manufacturers now offer a variable pond-depth feature that allows operators to adjust pond depth while the centrifuge is operating (see Figure 23.15). This system changes pond depth by restricting centrate flow via a motor-operated plate.

4.10 Chemical Conditioning

Polymers characteristics vary widely, so design engineers should consult manufacturers about their properties, as well as preparation and hauling techniques. For information on polymer mixing and feeding systems, see Chapter 22.

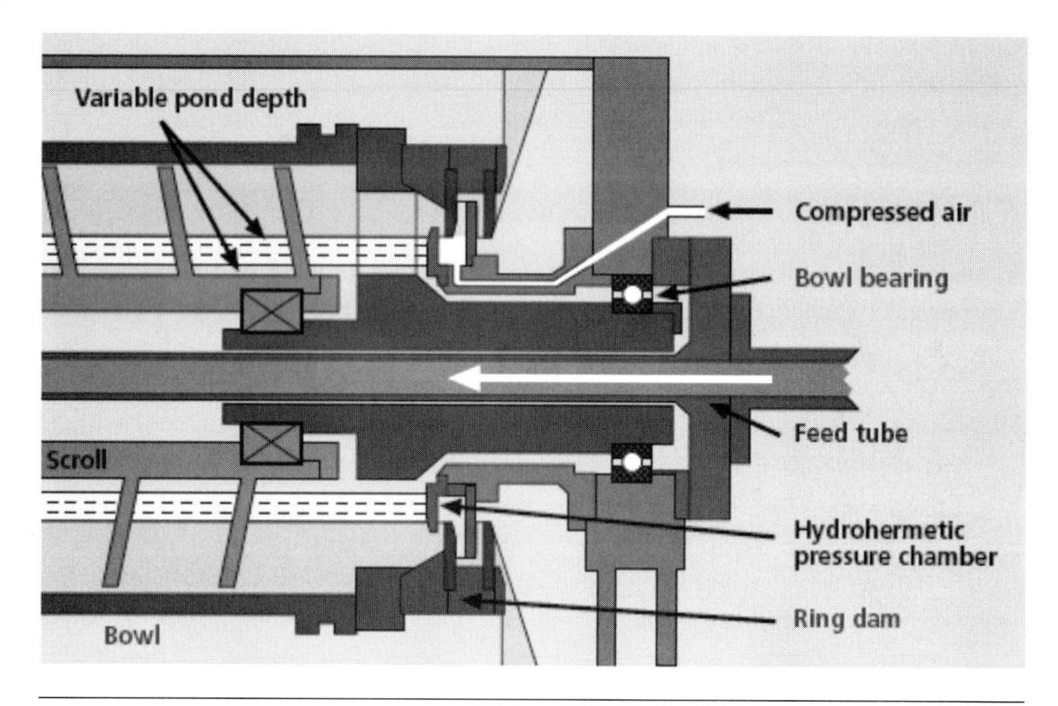

FIGURE 23.15 Mechanisms for varying pond depth in a centrifuge.

4.11 Design Example

Suppose design engineers have calculated that a maximum of 10 883 kg/d of WAS (dry weight) must be thickened. The following operating criteria applies:

- Thickening facility operates 7.5 hr/d
- WAS solids concentrations is 0.5% solids
- New facility with no pilot testing
- Design for three operating units and one standby unit

$$\text{Maximum daily WAS production} = 10\,883 \text{ kg/d}$$
$$\text{No. of operating units} = 3 \text{ units}$$
$$\text{Hours operating per day} = 7.5 \text{ hours}$$
$$\text{WAS minimum solids concentration} = 0.5\%$$
$$\text{WAS specific gravity} = 1$$

Equations used:

$$\text{Maximum net hourly load} = \frac{\text{Maximum Daily WAS Load}}{\text{No. of Operating hours per day}} \tag{23.19}$$

$$\text{Maximum net hourly load per unit} = \frac{\text{Maximun Net Hourly Load}}{\text{No. of Units}} \tag{23.20}$$

$$\begin{array}{c}\text{Volume per operating}\\ \text{minute per unit}\end{array} = \frac{\text{Net Hourly Load per Unit}}{\begin{array}{c}\text{WAS Solids Conc.} * \text{Specific Gravity (WAS)} *\\ \text{No. of units}\end{array}} \tag{23.21}$$

$$\text{Bowl Circumference (m)} = (\text{bowl diameter (m)}) \times \text{pi} \tag{23.22}$$

$$\text{Bowl Speed}\left(\frac{\text{sec}}{\text{rev}}\right) = \frac{\text{No. of seconds /min}}{\text{Bowl Speed}\left(\frac{\text{rev}}{\text{min}}\right)} \tag{23.23}$$

$$\text{Angular Velocity}\left(\frac{m}{\text{sec}}\right) = \frac{\text{Bowl Circumference (m)}}{\text{Bowl Speed}\left(\frac{\text{sec}}{\text{rev}}\right)} \tag{23.24}$$

$$\text{Centripital Acceleration}\left(\frac{\text{m}}{\text{sec}^2}\right) = \frac{\left(\text{Angular Velocity}\left(\frac{\text{m}}{\text{sec}}\right)\right)^2}{\text{Bowl Radius (m)}} \tag{23.25}$$

$$\text{G} - \text{force @ bowl wall} = \frac{\text{Centrifugal Acceleration}\left(\frac{\text{m}}{\text{sec}^2}\right)}{\text{Acceleration due to gravity}\left(\frac{\text{m}}{\text{sec}^2}\right)} \tag{23.26}$$

$$\text{Cylinder Length (m)} = \text{Bowl Length (m)} - \text{Cone Length (m)} \tag{23.27}$$

$$\text{Cylinder Volume (m}^3) = pi \times \left(\frac{\text{(Bowl Diameter (m))}}{2}\right)^2 \times \text{Cylinder Length (m)} \tag{23.28}$$

$$\text{Unusable Cylinder Volume (m}^3) = \text{pi} \times \left(\frac{\text{Discharge Diameter (m)}}{2}\right) \times \text{Cylinder Length (m)} \tag{23.29}$$

$$\text{Usable Volume of Bowl Cylinder (m}^3) = \text{Cylinder Volume (m}^3) - \text{Unusable Cylinder Volume (m}^3) \tag{23.30}$$

$$\text{Usable Volume of Bowl Cylinder (gal)} = \text{Usable Volume of Bowl Cylinder (m}^3) * \frac{\text{No. of gal}}{\text{m}^3} \tag{23.31}$$

$$\text{GVolume} = \text{GForce @ Bowl Wall} \times \text{Usable Volume of Bowl Cylinder (gal)} \tag{23.32}$$

Maximum net hourly load	= 1 451 kg/hr	Use Equation 23.19
Maximum hourly load per unit	= 483 m^2	Use Equation 23.20
Volume per operating minute	= 427 gal/min/unit	Use Equation 23.21

Contact several manufacturers. Give them information about the application, along with the desired mass and flow rate criteria for each unit. Request specific references for the size unit recommended.

Refer to Table 23.9.

5.0 GRAVITY BELT THICKENER

Introduced in 1980, a *gravity belt thickener* is a belt filter press with a modified upper gravity drainage zone that allows water to drain through the moving, fabric-mesh belt while coagulating and flocculating solids (see Figure 23.16). It originally was designed to be a dewatering pretreatment method, but subsequent improvements have made it more suitable for solids thickening. Gravity belt thickeners currently are used to treat aerobically or anaerobically digested solids, alum and lime solids, primary solids, WAS, and blended solids that initially contain between 0.4% and 8% solids. They typically capture more than 95% of solids when 1.5 to 5 g/kg (3 to 10 lb/ton) (dry weight) of polymer is used to concentrate the material and avoid excessive solids losses. When treating municipal WAS and biosolids, gravity belt thickeners can produce a material containing 6% solids.

Gravity belt thickeners are gaining popularity because of their efficient space requirements, low power use, and moderate capital costs. Improvements in both throughput and polymer use are ongoing, and will make this process even more cost-effective.

Experience has shown that gravity belt thickeners work well with many types of wastewater solids and are less affected by plant operating problems than many other

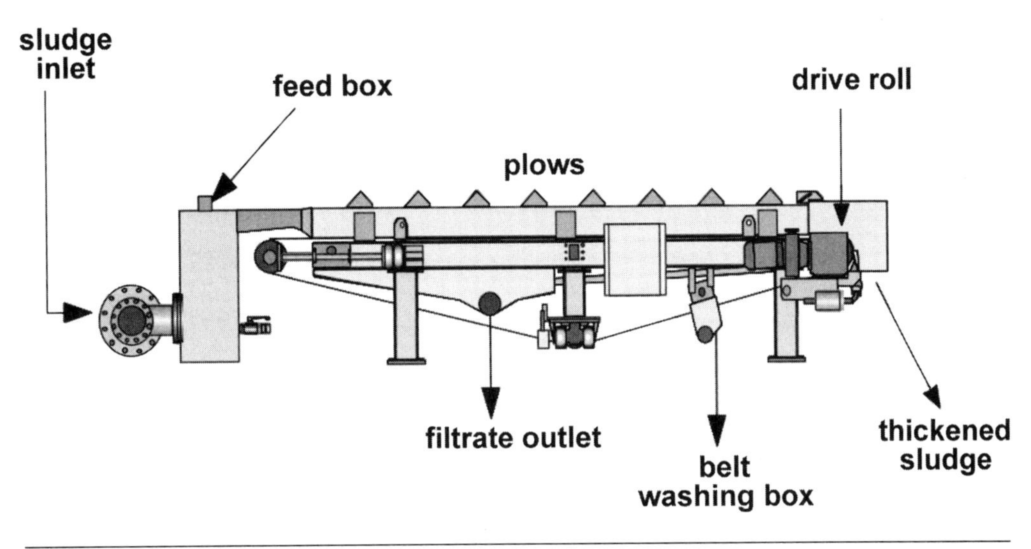

FIGURE 23.16 Schematic of a gravity belt thickener (courtesy of Seimens Water Technologies).

thickening processes. They typically handle even difficult-to-thicken solids via minor modifications to polymer dose, hydraulic loading rates, and solids loading rates.

If design engineers intend to use a gravity belt thickener as part of a cothickening process, then they should ensure that primary solids and WAS are mixed properly before the material is loaded onto the belts. This will eliminate belt blinding caused by grease and debris in the primary solids.

5.1 Evaluation and Scale-Up Procedures

In the past, engineers have tested gravity belt thickener performance to verify design parameters before installation. Over the years, they tested this process on many types of solids (e.g., WAS, anaerobically digested solids, primary solids, trickling filter solids, aerobically digested solids, and pure-oxygen activated sludge) using a trailer-mounted, 1-m gravity belt thickener that came with all necessary ancillary equipment. Results showed a good correlation between pilot- and full-scale operations.

The process' flexibility and overall good performance have made pilot testing unnecessary for most wastewater solids applications. Design engineers now primarily use pilot tests to compare the cost-effectiveness of various manufacturers' machines, to evaluate an unusual thickening application, or to demonstrate the machine's performance to gain acceptance. Laboratory testing typically consists of sending a solids sample to the manufacturer for jar and free-drainage tests to determine polymer type and dose.

As with other wastewater thickening processes, performance is solids-specific. However, performance and design criteria can be predicted based on similar full-scale installations (see Table 23.12).

5.2 Process Design Considerations and Criteria

The primary design components of a gravity belt thickener are

- Feed solids pumps and feed flow control,
- Polymer system and feed control,
- Gravity belt thickener,
- Belt washwater supply,
- Thickened-solids pumps, and
- Odor control.

Before designing these components, engineers need to determine potential operating modes. For example, the process may need to thicken 7 days' worth of solids in a shorter timeframe. Common operating modes include continuous thickening 7 days per week, continuous thickening 5 days per week, one-shift thickening 7 days per week, and one-shift thickening 5 days per week.

All components must be sized to handle both minimum and maximum potential solids-feed rates. Other important design considerations and criteria are noted in the following subsections. (For design information related to ancillary equipment and controls, see Section 5.4.)

TABLE 23.12 Typical performance of gravity belt thickeners (reprinted with permission from Ashbrook Simon-Hartley, Houston, Texas).

Type of biosolids	Initial concentration, %	Solids loading, kg/m·h	Polymer dosage, g/kg	Final concentration, %
Primary	2–5	900–1 400	1.5–3	8–12
Secondary	0.4–1.5	300–540	3–5	4–6
(50% P)(50% S)	1–2.5	700–1 100	2–4	6–8
Anaerobic (50% P)/(50% S)	2–5	600–790	3–5	5–7
Anaerobic (100% S)	1.5–3.5	500–700	4–6	5–7
Aerobic (100% S)	1–2.5	500–700	3–5	5–6

5.2.1 *Unit Sizing*

Engineers can design gravity belt thickeners based on pilot-test results and manufacturers' flow and solids loading capacity criteria (see Table 23.13). They could rely on manufacturer criteria for most municipal WAS, anaerobically digested solids, aerobically digested solids, drinking water solids, and pulp and paper (recycled paper) solids because of the abundance of historical operating data available. However, some testing is recommended to verify that a given solids can be thickened at typical polymer doses.

In lieu of pilot-test data, engineers could use a conservative design value of 800 L/m·min (200 gpm/m) for the hydraulic loading rate. As feed flowrates increase, more operator attention is needed to maintain stable operations.

Operating experience demonstrates that some gravity belt thickeners can treat WAS with 0.6 to 1.5% solids at a hydraulic loading rate of up to 1 500 L/m·min (400 gpm/m) and a solids loading rate of up to 500 kg/m·h (1 100 lb/hr/m). They also can treat digested solids with 2 to 4% solids at a hydraulic loading rate of up to 1 100 L/m·min (300 gpm/m) and a solids loading rate of up to 770 kg/m·h (1 700 lb/hr/m). In both cases, they can produce a thickened material containing 4 to 7% solids. Solids capture typically ranges from 90 to 98%.

Optimizing operations and increasing the polymer dose sometimes can produce a thickened material containing 10% solids; however, it can be difficult to pump and treat in downstream processes, so gravity belt thickeners typically are designed for a maximum of 5 to 7% thickened solids. Also, engineers must design thickened-solids and filtrate-conveyance systems to handle a range of possible conditions.

Pilot testing is recommended for atypical solids or plants receiving atypical industrial contributions. Such tests allow engineers to determine flow and solids-loading capabilities, solids capture, and required polymer dose. Typically the only difference between pilot- and full-scale models is belt width, so testing results are determined on a per-metre belt-width basis to allow for a directly proportional scale up.

Table 23.13 Typical hydraulic loading ranges for gravity belt thickeners (MacConnell et al., 1989).

Belt size (effective dewatering width), m	Hydraulic loading range,* L/min
1.0	400–950
1.5	570–1 420
2.0	760–1 900
3.0	1 100–2 800

* Assumes 0.5 to 1.0% feed solids for municipal sludges. Variations in sludge density, belt porosity, polymer reaction rate, and belt speed will act to increase or decrease the rates of flow for any given size belt.

As part of pilot testing, design engineers should select a suitable polymer. Polymers typically are screened via bench-scale tests and selected based on pilot-test performance.

5.2.2 *Other Design Considerations*

When designing gravity belt thickeners, engineers also need to consider mixing, flocculation, belt speed, plow, and discharge design. This information relates to items often supplied with the thickener and so affect the choice of manufacturer and equipment model.

Other details to consider include mechanical durability, corrosion resistance, availability and cost of replacement parts, and service assistance provided by the manufacturer. To determine mechanical durability and corrosion resistance, engineers should evaluate a manufacturer's engineering drawings and specifications, as well as interview O&M staff at other installations. To evaluate service-assistance and parts issues, they should survey O&M staff at other installations.

5.2.3 *Mixing Design*

The process design should provide for adequate mixing of polymer and solids followed by enough flocculation time before solids are discharged onto the belt. Poorly flocculated particles will blind the belt, hinder thickening, and result in poor solids capture. The polymer must be injected immediately upstream of the mixing device. Mixing must be intense enough to provide good contact without breaking the floc apart or shearing the polymer. Adjustable mixers (e.g., an adjustable orifice mixer) are recommended so mixing conditions can be optimized. The mixer should be installed next to a feed tank so operators can visually evaluate the floc immediately after adjustments. The feed tank typically provides enough time for flocculation before solids enter the gravity belt thickener.

Once the mixer is properly adjusted, it typically will not require adjustments to react to small changes in solids characteristics or feed rates. Larger changes, however, will require mixer adjustments to minimize polymer use and improve thickening and solids capture.

5.2.4 *Flocculation Design*

After mixing, the polymer-conditioned solids need time to agglomerate into larger floc. Operating experience has shown that the conditioned solids need at least 30 seconds to flocculate before entering the gravity belt thickener. So, design engineers should provide for 30 seconds of contact time under peak-flow conditions.

Flocculation time can occur in the piping or in a feed tank, and turbulence should be avoided. One manufacturer suggests that the piping downstream of the mixing point include no more than three 90-degree elbows or similar fittings. Although the significance of such fittings is not well documented, design engineers may need to adhere to such manufacturer recommendations to enforce performance guarantees. It typically is difficult to detain flow in the feed piping for 30 seconds without using more than three elbows, so manufacturers offer a feed tank for this purpose. Design engineers should size this feed tank to provide about 30 seconds of detention time at the peak feed rate.

Whenever the gravity belt thickener will be shut down for more than a few hours, the feed tank should be drained to avoid excessive odors. If the drain valve is inaccessible, it should automated.

5.2.5 *Belt Speed Design*

Operators need to be able to adjust belt speed so they can control thickening and maximize solids capture. Gravity belt thickeners typically perform well within a certain range of belt speeds, depending on manufacturer's specification. Below this range, the solids feed rate must be limited to avoid flooding the belt. Above this range (without other performance improvements), solids capture degrades because more residuals are being washed off the belt in the wash station. With proper polymer conditioning, however, a faster-moving belt can accommodate higher mass loadings that can offset the solids loss due to higher speed. Design engineers should perform a mass balance to confirm this. They also should survey O&M staff at existing installations about the performance records of the manufacturers being considered.

If operators can adjust the belt speed, they can maintain operations at the slowest speed that will accommodate the feed rate without the possibility of flooding. Belt speed can be changed mechanically or electrically; manufacturers often provide a mechanical adjustment mechanism as standard equipment and offer a variable-frequency drive as an option. If solids characteristics and feed rates are expected to be fairly constant, the mechanical adjustment mechanism may be sufficient. If frequent belt-speed adjustments are anticipated, adjustable-frequency drives can be used. Also, the thickener's control panel should include a potentiometer so that operators can adjust belt speed and monitor the result more easily.

5.2.6 *Plow Design*

The clearance between the plows and the belt must be large enough for the belt seam, which typically protrudes above the rest of the belt, and yet small enough for the plows to clear drainage pathways right down to the belt. It should be adjustable to allow for proper installation, plow wear over its lifetime, and manufacturing variations among replacement belts. If the plows press too hard on the belt, excessive wear will reduce belt life.

5.2.7 *Discharge Design*

Some gravity belt thickeners have an adjustable ramp over which solids must flow before discharging from the belt. The ramp can act as a quasi-dam, causing the solids to thicken further as they roll up and over it. A steeper ramp typically results in thicker solids. Some installations, however, may require the exit ramp to have little or no angle to avoid flooding the belt.

Rather than a ramp, some manufacturers use a wedge under which solids must squeeze before discharging from the belt. Side-by-side pilot tests have indicated that, under identical operating conditions, the ramp design can produce slightly thicker solids. Put another way, the ramp design would need less polymer to produce a thickened material with the same solids content as that produced by the wedge design. Design engineers would need to conduct side-by-side pilot tests of the two systems to quantify differences in polymer use and solids thickening for a given solids stream and compare them to differences in capital costs.

5.3 Mechanical Features

5.3.1 *Solids Polymer Injection and Mixer*

Good polymer distribution optimizes polymer use and improves thickening, so immediately after being added, polymer should be well mixed with the feed solids. Manufacturers of gravity belt thickeners often provide polymer-injection and -mixing devices. A device with multiple injection points around the solids feed pipe probably will provide better distribution. One option is a polymer manifold; it connects to the end of the polymer-feed pipe and has feed tubes that connect to an injection ring on the solids feed pipe. Design engineers should avoid injection points that extend into the solids path, because they may cause fibrous or stringy material to entangle and plug the pipe. Some manufacturers provide manifolds with clear, flexible feed tubes that make installation easier and let operators see whether polymer is flowing to each point on the injection ring. If one of the tubes is plugged, operators sometimes can clear it by using tubing clamps on one or more open lines, which increases the pressure and scouring velocity in the plugged line. To keep the tubing transparent, it occasionally must be removed and cleaned or replaced.

The polymer mixer should be immediately downstream of the injection point. Design engineers should select adjustable mixers (e.g., an adjustable orifice mixer) that allow mixing optimization both during initial operations and whenever solids feed rates or characteristics change. Adjustable orifice mixers can include an adjustable counterweight lever that changes the orifice size so that operators can adjust the mixer according to solids variations to optimize mixing.

5.3.2 *Flocculation Tank and Feed Distribution*

Unless the feed line offers proper flocculation conditions, a flocculation tank is required. Polymer-conditioned solids flow into the bottom of this tank so they have enough time to flocculate before they overflow the tank onto the filter belt. As the solids leave the feed tank, they should be distributed across the entire working width of the belt. Some manufacturers use a feed chute with wedges for this purpose.

Also, the tank should be equipped with a drain valve so it can be emptied when feed is discontinued to the gravity belt thickener.

5.3.3 *Frame*

A gravity belt thickener has a frame that supports and holds its components, except for the polymer-injection ring, polymer–solids mixer, flocculation tank, and tracking and tensioning power unit. The frame typically is galvanized or otherwise coated with a durable corrosion-resistant surface; it can be constructed of stainless steel. Some manufacturers offer gravity belt thickeners constructed of stainless steel plate that encloses the typically open areas. This provides for better odor control.

5.3.4 *Gravity Drainage Area*

The feed chute distributes solids onto a continuously moving, horizontal filter belt, which retains solids but lets free water pass through. The belt is made of a porous, woven mesh and is seamed to form a continuous loop between the feed chute and discharge point. A series of adjustable plows along the belt ensure that solids are distributed evenly across the mesh, turn solids over to promote water separation, and create solids-free areas that enable free water to drain through the belt. Adjustable retention plates with seals prevent solids from spilling off the sides of the belt. A drainage grid underneath the belt supports it and enables filtrate to drain into collection trays below.

5.3.5 *Discharge Area*

The back end of the unit often has a ramp or wedge, depending on the manufacturer. If the unit has a ramp, its leading edge contacts the top of the belt. Solids roll as they move onto the ramp, squeezing out a little more water. The angle of the ramp also acts as a partial dam, increasing product depth and, therefore, retention time on the belt—further thickening the solids before discharge. The ramp angle is adjustable to optimize thickening. The ramp also can be rotated out of the way if it hinders thickening operations.

If the unit has a wedge, its leading edge has the widest clearance from the belt. Solids pass under the wedge and are squeezed as the clearance between the wedge and belt decreases. This removes a little more water just before discharge.

Thickened solids may be discharged to a wet well, open-top pump, or another conveyor. Meanwhile, once the belt clears the ramp or wedge, it moves past a scraper blade, which separates remaining solids from the belt. Such solids are discharged to the same location as the rest of the thickened material.

5.3.6 *Belt Washing*

After being scraped, the belt passes through a wash station, which removes embedded particles from that portion of the belt before it begins another thickening cycle. The wash station consists of a washwater supply pipe, spray nozzles, and a housing to contain spray. The station typically is constructed of stainless steel.

5.3.7 *Filtrate and Wash Water*

Filtrate is the free water that passes through the belt. It collects in trays below the belt and then discharges to drain pipes, which convey it to a large filtrate header or to a large floor drain that leads directly to a sump.

Used washwater (from the wash station) also is discharged to the sump or filtrate header. Another option is to collect washwater and recycle it to the thickener feed tank. This reduces the load on downstream treatment processes and improves solids capture because the solids in washwater are combined with the incoming floc.

5.3.8 *Gravity Belt Thickener Drive, Tracking, and Tensioning*

Gravity belt thickeners require multiple rollers to drive, steer, adjust the tension, and guide the belt. The drive roller pulls the belt through the machine. The roller-drive motor typically has either mechanically or electrically adjustable speed controls to vary belt speed. The steering roller maintains proper alignment of the belt in response to sensing devices on the gravity belt thickener. The tensioning roller pulls on the belt to create the belt tension required for traction between the belt and belt-drive roller. The guide roller directs the belt through the wash station.

A gravity belt thickener requires a power unit that provides pressure to the belt-tracking and -tensioning systems. Some manufacturers use recirculating hydraulic fluid for this purpose, while others use compressed air. Unlike the belt-drive motor, which is connected to the gravity belt thickener, the power units may be remotely located.

5.4 Ancillary Equipment/Controls

A gravity belt thickener's ancillary components primarily include the following:

- Feed pumps and feed flow control,
- Polymer system and feed control,

- Belt washwater supply,
- Thickened-solids pumps, and
- Odor control.

All components must be sized to handle both maximum and minimum potential feed rates. Ideally, all controls should be near the thickener in a place where operators can see the top of the gravity belt while making process adjustments. Also, design engineers should include an emergency stop cord along the unit so operators can stop the thickener, solids feed pump, and polymer feed pump for safety reasons.

5.4.1 *Feed Pumps and Feed Flow Control*

A gravity belt thickener can operate over a large range of feed rates, but each feed rate requires its own polymer dose and belt speed. Feed-rate changes also may require adjustments in discharge-ramp angle, polymer dilution water, and the position of the solids–polymer mixer. So, design engineers should provide a flow meter (e.g., an electromagnetic flow meter) and an adjustable speed pump or a flow-control valve so operators can maintain a constant solids flowrate.

Screw-induced centrifugal pumps work well when enough suction head is available and the discharge head is not too high. These pumps require less maintenance than positive-displacement pumps because of the low contact between the slurry and screw impeller. Recessed-impeller centrifugal pumps are used when the slurry is abrasive but enough suction head is available and discharge head is not too high. Positive-displacement pumps (e.g., progressing cavity pumps) and certain rotary lobe pumps are good choices when pumping slurries with higher friction losses. They provide good suction draw and can pump against higher heads. Design engineers can set up a control loop that uses a flow meter to track solids feed and adjusts pump speed to maintain a setpoint feed rate. This arrangement requires that each gravity belt thickener has a dedicated pump.

Another feed-control strategy involves flow-control valves and centrifugal pumps. It can be used to split flow among multiple thickeners. Design engineers can set up a control loop in which feed-pump speed is adjusted to maintain a setpoint pressure in a header, which serves as a manifold with branches to each thickener. Each branch requires a flow-control valve and flow meter. Alternatively, design engineers can set up a control loop in which the flow-control valve is adjusted to maintain a setpoint feed rate (as monitored by the flow meter). This alternative does not require each thickener to have a dedicated pump, so design engineers can reduce the number of pumps involved. They also may put a flow-control valve downstream of a dedicated constant-speed pump, cutting costs by eliminating the need for an adjustable-speed drive.

However, this option requires a pump that can operate at the intended thickener-feed rates and restrict the range of acceptable feed rates. Also, the variable-orifice mixer may have wide swings in pressure drops (depending on the type of solids involved), making stable thickener operations problematic.

5.4.2 *Polymer System and Feed Control*

Adding polymer to the feed solids is essential to successful thickening when using a gravity belt thickener. It promotes flocculation of solids and release of free water. Without polymer, the belt would blind (because of fine solids filling belt pores) and flood (because of poor water release). Design engineers should test various polymers to determine which is the most effective. Cationic (positively charged) polymers typically are chosen because wastewater solids often are negatively charged. However, if the solids contain significant amounts of aluminum or ferric salts (which impart a positive charge), then an anionic (negatively charged) polymer may be better.

Design engineers also need to determine the appropriate polymer dose—the minimum and maximum amounts that can be added to get good results. Typically, solids capture and concentration increase as polymer doses rise above the minimum effective level. They eventually level off until the polymer dose exceeds the maximum effective level, when solids capture and concentration can be reduced. Excess polymer can blind belt pores and create a floc more susceptible to breakup.

If gravity belt thickeners will be used to thicken both WAS and digested solids, then design engineers typically select one polymer that is suitable for both feed solids because this is more cost-effective than adding two chemical-feed systems. They typically can find a polymer that is effective for both solids, although it may not be the optimal choice for either material.

Adding higher doses of a less expensive polymer can be cheaper than using lower doses of a polymer with a high charge strength or molecular weight. So, design engineers should specify performance requirements based on dollars of polymer per thousand kilograms of solids rather than on grams of polymer per kilogram of solids. During equipment performance tests, manufacturers should choose the polymer so they have full control of and responsibility for test results. Afterward, however, design engineers can invite chemical companies to test other polymers.

Polymer is made in batches; it often is made at higher concentrations (up to 1%) to reduce the size and number of batches required. The appropriate concentration depends on use (to avoid excessive storage and keep up with demand). The batch should be diluted downstream of the polymer feed tank before being added to solids, and this dilution rate depends on polymer feed rate and concentration. The polymer's concentration can affect both solids thickening and polymer efficiency, so design engineers should make both the polymer-batching and -dilution systems flexible so operators can adapt them as needed.

5.4.3 *Belt Washwater Supply*

Before each portion of the belt begins another thickening cycle, it should be washed to remove embedded solids and excess polymer. Belt wash stations typically are installed on the belt's return loop and use about 80 L/min (20 gpm) of water per metre of belt. Washwater flow and pressure recommendations depend on the manufacturer. Pressure recommendations typically range from 517 to 586 kPa (75 to 85 psi), although some manufacturers recommend pressures of 760 to 830 kPa (110 to 120 psi). This pressure is created by nozzle losses, minor losses caused by fittings, pipe-friction losses, and any elevation differences between the nozzles and the water source. The actual pressure at the wash station typically depends on nozzle losses (i.e., nozzles create the backpressure). So, design engineers need an accurate curve of nozzle losses versus flow (typically available from the nozzle manufacturer) to design the washwater-supply system properly.

Washwater use depends on thickener size (typically defined by belt width), number of thickener units in operation, and time of operation. If water use is insignificant, a plant's potable water supply can be used. Manufacturers typically provide the control valve, and a booster pump can compensate for inadequate plant water pressure. If water use is significant, plant effluent can be used. Design engineers will have to add pumps and automatic strainers to remove larger suspended solids if the effluent's suspended solids concentration is less than 50 mg/L. Alternatively, some manufacturers offer the option of using the gravity belt thickener's filtrate (which also requires pumps and automatic strainers). The controls for washwater pumps should be interfaced with those for the gravity belt thickener and water-supply valve. Design engineers also should select automatic strainers for uninterrupted operation during cleaning. The strainers typically are either cleaned continuously or based on a timer and/or pressure loss.

5.4.4 *Thickened-Solids Pumps*

Headlosses are high when thickened solids flow through a pipe. The amount of headloss depends on pipe size and type, flow velocity, type of solids, and solids concentration. Engineers have used various curves, models, and "general rule" multipliers to estimate headlosses. The estimating procedures developed for certain solids (e.g., paper-stock solids) are fairly reliable, but those for municipal solids are less standardized. Headloss can become more difficult to estimate as piping distance and solids concentration increase. So, unless available data indicate otherwise, design engineers should use conservative models and worst-case assumptions.

Because of the high headlosses, design engineers should use positive-displacement pumps (e.g., progressing cavity pumps, certain rotary lobe pumps, and air-operated diaphragm pumps) to transport gravity belt-thickened solids. These pumps provide good suction draw and can pump against higher heads. Engineers also should design suction pipes to be straight and as short as possible. If thickened solids must be pumped long distances, then special provisions (e.g., multiple-stage progressing cavity pumps) probably will be necessary.

5.4.5 Odor Control

Odors from gravity belt thickeners are not unique but may be stronger because of turbulence. The most common odors are related to reduced sulfur compounds. Ammonia odors also can occur; they typically are related to anaerobically digested biosolids. So, high ventilation rates are necessary to provide an acceptable working environment. A minimum of 15 fresh air changes per hour is recommended. The exhaust rate should be slightly faster than the fresh air supply rate to maintain a slight vacuum in the room, which will prevent odor problems in adjacent areas.

The odorous air should be treated before discharge to prevent odor nuisances in areas near the treatment plant. Odor-treatment options include packed-tower scrubbers, mist scrubbers, activated-carbon beds, and biofilters. Hypochlorite and sodium hydroxide typically are used to scrub sulfur-related odors, while sulfuric acid solutions can be used to scrub ammonia. Another options is adding an odor-control compound (e.g., hydrogen peroxide) to the feed solids, but the compound first should be tested to determine whether it interferes with polymer efficiency and solids thickening.

5.5 Design Example

Suppose a wastewater treatment plant plans to install a gravity belt thickener to treat a mix of primary solids and WAS. Design engineers have calculated that the plant produces 2 200 L/min of combined solids on a continuous basis. The following operating criteria apply:

- thickening facility operates 8 hr/d, 5 days/week,
- WAS solids concentrations is 0.5 to 1.0% solids,
- the gravity belt thickeners have an effective thickening width of 2 m, and
- the design should allow for one unit undergoing maintenance and one in standby.

Using the criteria in Table 23.15 to select a hydraulic loading rate based on one unit out of service and one unit undergoing maintenance, design engineers choose a rate of 800 L/m·min.

Plant's hydraulic loading rate = 2 200 L/min
Effective gravity belt thickening width = 2 m
Design loading rate = 800 L/m·min

Equations used:

$$\text{Hydraulic loading per week} = \text{Hydraulic Loading} * \frac{\text{No. of min}}{\text{week}} \tag{23.33}$$

$$\text{Weekly loading capacity per belt} = \frac{\text{Design Loading Rate} * \text{Belt Width} * \text{No. of operating min}}{\text{week}} \tag{23.34}$$

$$\text{Belts needed to meet hydraulic requirement} = \frac{\text{Hydraulic Loading per Week}}{\text{Weekly Loading Capacity}} \tag{23.35}$$

Hydraulic loading per week = 22 176 000 L/week — Use Equation 23.33
Weekly loading capacity per belt = 3 840 000 L/belt/week — Use Equation 23.34
Belts needed to meet hydraulic requirement = 6 belts — Use Equation 23.35

Six gravity belt thickeners must be operational at all times. Therefore, the systems will need eight thickeners if one will be on standby and one is undergoing maintenance.

6.0 ROTARY DRUM THICKENER

A rotary drum thickener (also called a *rotary screen thickener*) basically consists of an internally fed rotary drum, an integral internal screw, and a variable- or constant-speed drive (see Figure 23.17). Both gravity belt and rotary drum thickeners allow free water to drain through a moving, porous media while retaining flocculated solids. In rotary drum thickeners, the rotating drum imparts centrifugal force to separate liquids and solids, while the internal screw transports thickened solids or screenings out of the drum.

Rotary drum thickeners are most applicable to thickening WAS at small- to medium-sized wastewater treatment plants because the largest unit has a capacity of about 1 100 L/min (300 gpm). Its advantages include efficient space requirements, low power use, moderate capital costs, and ease of enclosure, which improves housekeeping and odor control.

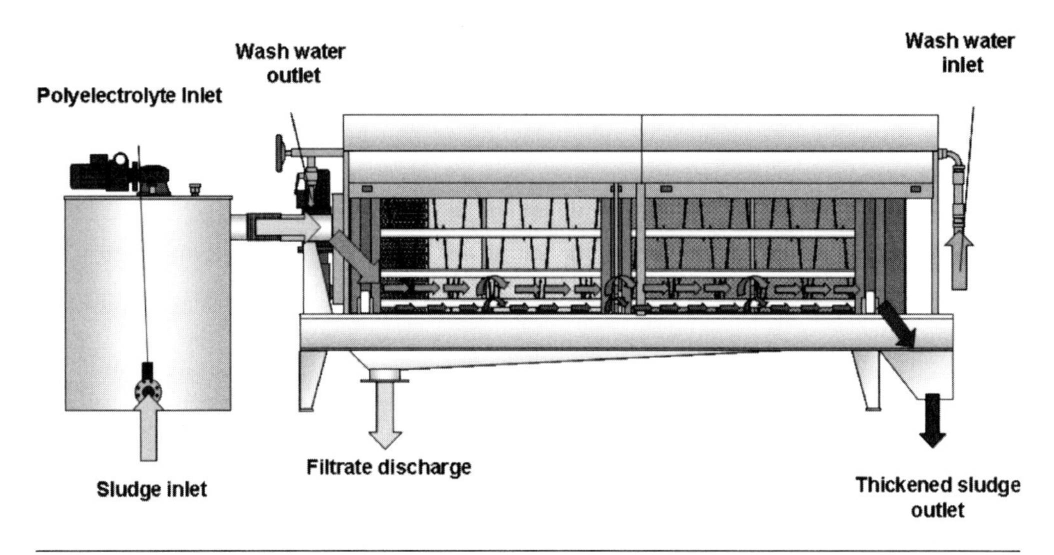

FIGURE 23.17 Schematic of a rotary drum thickener (courtesy of Seimens Water Technologies).

6.1 Evaluation and Scale-Up Procedures

Rotary drum thickeners work well on many types of wastewater solids. They have been used to remove screenings from raw wastewater or primary solids. They are well suited to thickening high-fiber solids (e.g., those found in the pulp and paper industry) and raw or digested solids containing a significant fraction of primary solids. They also can pretreat solids before belt filter press dewatering. Their success with municipal WAS is variable and depends on actual solids characteristics.

Performance and sizing tests can be done by manufacturers at their factories or by design engineers using pilot-scale units brought to the treatment plant. Design engineers often conduct pilot tests to compare units from various manufacturers or to compare rotary drum thickeners to other thickening devices (e.g., gravity belt thickeners). They also perform pilot tests to select an appropriate polymer and its dose. Adding large amounts of polymer can be a concern in rotary drum thickening because of floc sensitivity and potential for shearing.

As with other thickening processes, performance is solids-specific. However, expected performance and design criteria can be approximated using data from similar full-scale installations (see Table 23.14).

6.2 Process Design Conditions and Criteria

The primary components to be designed in a rotary drum thickener are:

TABLE 23.14 Performance data for rotary drum thickeners.

Drum size, m	Influent flow rate, L/min	Influent solids concentration, %	Effluent solids concentration, %
0.6	Average = 160	Average = 1	Average = 6
	Range = na	Range = 1–2	Range = 5.5–6
1.5	Average = 900	Average = 0.9	Average = 5.5
	Range = 490–1 100	Range = 0.001–1.5	Range = 4–9

- Solids feed pumps and feed flow control,
- Polymer system and feed control,
- Rotary drum thickener,
- Screen washwater supply,
- Thickened-solids pumps, and
- Odor control.

The feed-solids pumps, polymer system, and thickened-solids pumps are similar to those described for gravity belt thickeners. The screen washwater supply can be treatment plant effluent; manufacturers recommend 230 L/min (60 gpm) at 150 kPa (22 psi).

Odor control is greatly simplified by the enclosures that rotary drum manufacturers provide. Odors can be drawn off these enclosures for treatment in appropriate odor-control units.

Equipment sizes and configurations vary among manufacturers. The following four variables can be used to make the most of rotary drum thickener operation: sludge feed rate, polymer feed rate, pool depth, and drum speed.

6.2.1 Solids Feed Rate

Operators can maximize throughput and optimize solids concentration by varying the solids feed rate (within the rotary drum thickener's range of capacities). Before deciding on rotary drum thickener capacity, design engineers should consider the treatment plant's solids production rate and determine whether it would be more cost-effective for the thickener to run continuously or intermittently. Most small- and medium-sized plants operate their thickeners for one shift per day, 5 days per week. Design engineers also must size all components to handle both maximum and minimum potential solids feed rates.

6.2.2 Polymer Feed Rate

The polymer feed system must be sized to handle both maximum and minimum feed rates, and it must be flexible enough to allow operators to vary the dose to meet flocculation requirements. Both under- and overdosing can degrade thickener performance. Gravity-belt-thickener feed rates typically also apply to rotary drum thickeners (see Section 5.4.2). For polymer feed rate calculations, see Section 6.5.

6.2.3 Pool Depth

Operators can control pool depth by adjusting the inclination of the drum. The drum can be angled from a horizontal position to about 6 deg above horizontal. Steeper angles produce drier solids but reduce drum capacity; shallower angles increase capacity but produce wetter solids. The preferred drum angle is solids-specific but typically is between 1 and 3 deg above horizontal.

6.2.4 Drum Speed

Operators can change drum speed in response to changes in feed solids concentrations and flowrate to maintain the desired thickened-solids concentration. When treating WAS, rotary drum thickeners capture between 90 and 99% solids and produce a thickened material containing between 4 to 9% total solids. Typical performance ranges for other types of solids are shown in Table 23.15.

The need for polymer conditioning makes rotary drum thickening a cost-intensive process. However, this is less of a concern at small plants where short-term or seasonal operations are feasible.

Rotary drum thickeners occupy less space than DAF thickeners, centrifuges, or gravity belt thickeners that treat the same amount of solids, so they can be installed inside smaller buildings than any other thickening alternative. The unit size required to thicken a given amount of solids depends on machine capacity, solids characteristics, and polymer dose.

Rotary drum thickeners and gravity belt thickeners have similar solids capture rates and thickened solids percentages. The need for polymer conditioning and neces-

TABLE 23.15 Typical performance ranges for rotary drum thickeners.

Type of solids	Feed, % TS	Water removed, %	Thickened solids, %	Solids recovery, %
Primary	3.0–6.0	40–75	7–9	93–98
WAS	0.5–1.0	70–90	4–9	93–99
Primary and WAS	2.0–4.0	50	5–9	93–98
Aerobically digested	0.8–2.0	70–80	4–6	90–98
Anaerobically digested	2.5–5.0	50	5–9	90–98
Paper fibers	4.0–8.0	50–60	9–15	87–99

sary operator attention are cost considerations related to O&M. Rotary drum thickeners offer the flexibility of varying processing performance with solids and polymer feed rate control and drum speed adjustment.

6.3 Mechanical Features

A rotary drum thickener consists of an internally fed rotary drum with an internal screw, lubricated trunnion wheels, a variable- or constant-speed drive, an inlet pipe, a filtrate-collection trough, a discharge chute, and an optional rotating-brush drum cleaner.

6.4 Ancillary Equipment and Controls

Ancillary equipment includes feed pumps, thickened-solids pumps, and a polymer-mixing and feed system. These items are similar to those described for gravity belt thickeners in Section 5.4.

6.5 Design Example

Suppose a wastewater treatment plant is installing a rotary drum thickener treat its primary solids and WAS. Design engineers calculated that the treatment plant produces 2 200 L/min of combined solids on a continuous basis. The following operating criteria apply:

- Thickening facility operates 8 hours per day, 5 days per week,
- WAS solids concentrations is 0.5 to 1% solids,
- Rotary drum thickeners have an effective drum size of 1.5 m, and
- The design should allow for one unit undergoing maintenance and one unit in standby.

Using the criteria in Table 23.16 to select a higher hydraulic loading rate based on one unit out of service and one undergoing maintenance, design engineers select a design loading rate of 1 100 L/m·min.

Plant's hydraulic loading rate = 2 200 L/min
Effective drum size = 1.5 m
Design loading rate = 1 100 L/m·min
Plant's hydraulic loading rate = 2 200 L/min
Effective thickening width of each screen drum = 1.5 m
Design loading rate = 1 100 L/m·min

TABLE 23.16 The advantages and disadvantages of various thickening technologies.

Method	Advantages	Disadvantages
Gravity	Simple Low operating cost Low operator attention required Ideal for dense rapidly settling sludges such as primary and lime Provides a degree of storage as well as thickening Conditioning chemicals not typically required Minimal power consumption	Odor potential Erratic for WAS Thickened solids concentration limited for WAS High space requirements for WAS Floating solids
Dissolved air flotation	Effective for WAS Will work without conditioning chemicals at reduced loadings Relatively simple equipment components	Relatively high power consumption Thickening solids concentration limited Odor potential Space requirements compared to other mechanical methods Moderate operator attention requirements Building corrosion potential, if enclosed Requires polymer for high solids capture or increased loading
Centrifuge	Space requirements Control capability for process performance Effective for WAS Contained process minimizes housekeeping and odor considerations Will work without conditioning chemicals High thickened concentrations available	Relatively high capital cost and power consumption Sophisticated maintenance requirements Best suited for continuous operation Moderate operator attention requirements
Gravity belt thickener	Space requirements Control capability for process performance Relatively low capital cost Relatively lower power consumption High solids capture with minimum polymer High thickened concentrations available	Housekeeping Polymer dependent Moderate operator attention requirements Odor potential Building corrosion potential, if enclosed
Rotary drum thickener	Space requirements Low capital cost Relatively low power consumption High solids capture can be easily enclosed	Polymer dependent Sensitivity to polymer type Housekeeping Moderate operator attention requirements Odor potential if not enclosed

Equations used:

$$\text{Plant's hydraulic loading per week} = \text{Hydraulic Loading} * \frac{\text{No. of min}}{\text{week}} \tag{23.36}$$

$$\text{Weekly loading capacity per belt} = \frac{\text{Design Loading Rate} * \text{Screen width} * \text{No. of operating min}}{\text{week}} \tag{23.37}$$

$$\text{Belts needed to meet hydraulic requirement} = \frac{\text{Hydraulic Loading per Week}}{\text{Weekly Loading Capacity}} \tag{23.38}$$

Hydraulic loading per week	= 22 176 000 L/week	Use Equation 23.36
Weekly loading capacity per belt	= 3 960 000 L/belt/week	Use Equation 23.37
Drums needed to meet hydraulic requirement	= 6 screen drums	Use Equation 23.38

To meet the design loading rate, six drums must be operational at all times. So, the final system must have eight drums in case one is on standby and one is undergoing maintenance.

Polymer Addition Calculation:

Suppose design engineers must calculate the total volume of polymer needed for an influent flow of 3 000 gal/min. They are using a low- to medium-viscosity liquid polymer and the following conditions apply:

- influent solids concentration is 0.001 to 1.5% (Table 23.16),
- effluent solids concentrations is 4 to 7% (Table 23.16),
- polymer flow rate should be adjusted based on primary and secondary flow and turbidity,
- the polymer's specific weight is 1 kg/L.

Using criteria in Table 23.7, they select a typical polymer dose for rotary drum thickeners (6.8 mg/L)

Influent flow rate = 3 000 gal/min
Influent solids concentration = 1.5 mg/L
Effluent solids concentration = 7 mg/L
Polymer dose = 6.8 mg/L

Equations used:

$$\text{Volume flow of settled materials} = \frac{\text{Influent Flow Rate} * \text{Influent Solids Conc.}}{\text{Effluent Solids Conc.}} \quad (23.39)$$

$$\text{Volume to be treated by polymer} = \text{Influcent Flow Rate} - \text{Volume Flow of Settled Materials} \quad (23.40)$$

$$\text{Volume of polymer flow required} = \frac{\text{Required Dosage of Polymer} * \text{Volume to be treated}}{\text{Specific Weight of Polymer}} \quad (23.41)$$

Volume flow of settled materials = 643 gal/min	Use Equation 23.39
Volume to be treated by polymer = 2 357 gal/min	Use Equation 23.40
Volume of polymer flow required = 0.016 gal/min	Use Equation 23.41

7.0 COMPARISON OF THICKENING METHODS

A model comparing the cost-effectiveness of various thickening processes rarely applies to all situations because many factors that govern the final decision may be site-specific and more qualitative than quantitative. Such factors include sensitivity to upset, the benefits of achieving the highest possible solids concentration, the quality of operation required, installation size, compatibility with existing thickeners, the effect of downstream processing methods, and various personal preferences based on experience. Design engineers should consider all of the thickening alternatives in this chapter (see Table 23.16).

8.0 REFERENCES

American Public Health Association; American Water Works Association; Water Environment Federation (1976) *Standard Methods for the Examination of Water and Wastewater*, 14th ed.; American Public Health Association: Washington, D.C.

Ashman, P. S. (1976) Operational Experiences of Activated Sludge Thickening by Dissolved Air Flotation at the Aycliffe Sewage Treatment Works. Paper presented at the Conference on Flotation in Water and Waste Treatment; Felixstowe, Suffolk, Great Britain.

Boyle, W. H. (1978) Ensuring Clarity and Accuracy in Torque Determinations. *Water Sew. Works*, **125** (3), 76.

Bratby, J.; Marais, G. V. R. (1975a) Dissolved Air (Pressure) Flotation, An Evaluation of the Interrelationships Between Process Variables and Their Optimization for Design. *Water SA*, **1**, 57.

Bratby, J.; Marais, G. V. R. (1975b) Saturation Performance in Dissolved Air (Pressure) Flotation. *Water Res.* (G.B.), **9**, 929.

Bratby, J.; Marais, G. V. R. (1976) A Guide for the Design of Dissolved Air (Pressure) Flotation Systems for Activated Sludge Systems. *Water SA*, **2**, 87.

Burfitt, M. L. (1975) The Performance of Full-Scale Sludge Flotation Plant. *Water Pollut. Control* (G.B.), **74**, 474.

Butler, R. C.; Finger, R. E; Pitts, J. F.; Strutynski, B. (1997) Advantages of Cothickening Primary and Secondary Sludges in Dissolved Air Flotation Thickeners. *Water Environ. Res.*, **3**, 69.

Dick, R. I.; Ewing, B. B. (1967) Evaluation of Activated Sludge Thickening Theories. *J. Sanit. Eng.*, **93** (EE4), 9.

Ettelt, G. A.; Kennedy, T. J. (1966) Research and Operational Experience in Sludge Dewatering at Chicago. *J. Water Pollut. Control Fed.*, **38**, 248.

Gehr, R.; Henry, J. G. (1978) Measuring and Predicting Flotation Performance. *J. Water Pollut. Control Fed.*, **50**, 203.

Gulas, V.; et al., (1978) Factors Affecting the Design of Dissolved Air Flotation Systems. *J. Water Pollut. Control Fed.*, **50**, 1835.

Jones, W. H. (1968) Sizing and Application of Dissolved Air Flotation Thickeners. *Water Sew. Works*, **115**, R-177.

Jordan, V. J., Jr.; Scherer, C. H. (1970) Gravity Thickening Techniques at a Water Reclamation Plant [part I]. *J. Water Pollut. Control Fed.*, **42**, 180.

Komline, T. R. (1976) Sludge Thickening by Dissolved Air Flotation in the USA. Paper presented at the Conference on Flotation in Water and Waste Treatment; Felixstowe, Suffolk, Great Britain.

Leininger, K. V.; Wall, D. J. (1974) Available Air Measurements Applied to Flotation Thickener Evaluations. *Highlights/Deeds Data*, **11**, D1.

MacConnell, G. S.; et al. (1989) Full Scale Testing of Centrifuges in Comparison with DAF Units for WAS Thickening. Paper presented at the 62nd Annual Water Pollution Control Federation Technical Exposition and Conference; San Francisco, California, Oct 15–19; Water Pollution Control Federation: Alexandria, Virginia.

Maddock, J. E. L. (1976) Research Experience in the Thickening of Activated Sludge by Dissolved Air Flotation. Paper presented at the Conference on Flotation in Water and Waste Treatment; Felixstowe, Suffolk, Great Britain.

Mulbarger, M. C.; Huffman, D. D. (1970) Mixed Liquor Solids Separation by Flotation. *J. Sanit. Eng*, **96** (SA4), 861.

Noland, R. F.; Dickerson, R. B. (1978) *Thickening of Sludge*, Vol. 1; EPA-625/4-78-012; U.S. EPA Technology Transfer Seminar on Sludge Treatment and Disposal; U.S. Environmental Protection Agency: Washington, D.C.

Perry, R. H.; Chilton, C. H. (1963) *Chemical Engineer's Handbook*, 4th ed.; McGraw-Hill: New York, N.Y.

Purchas, D. B. (1977) *Solid/Liquid Separation Equipment Scale-up*; Uplands Press Ltd.: Croydon, U.K.

Reay, D.; Ratcliff, G. A. (1975) Experimental Testing on the Hydrodynamic Collision Model of Fine Particle Flotation. *Can. J. Chem. Eng.*, **53**, 481.

Sparr, A. E.; Grippi, V. (1969) Gravity Thickeners for Activated Sludge. *J. Water Pollut. Control Fed.*, **41**, 1886.

Speece, R. C.; et al. (1975) Application of a Lower Energy Pressurized Gas Transfer System to Dissolved Air Flotation and Oxygen Transfer. *Proceedings of the 30th Purdue Industrial Waste Conference*; West Lafayette, Indiana; 465.

Talmage, W. P.; Fitch, E. B. (1955) Determining Thickener Unit Areas. *Ind. Eng. Chem., Fundam.*, **47**, 38.

Torpey, W. N. (1954) Concentration of Combined Primary and Activated Sludges in Separate Thickening Tanks. *Proc. Am. Soc. Civ. Eng.*, **80**, 443.

Turner, M. T. (1975) The Use of Dissolved Air Flotation for the Thickening of Waste Activated Sludge. *Effluent Water Treat. J.* (G.B.), **15** (5), 243.

U.S. Environmental Protection Agency (1974) *Process Design Manual for Sludge Treatment and Disposal*; EPA-625/1-74-006; U.S. Environmental Protection Agency, Office of Technology Transfer: Cincinnati, Ohio.

U.S. Environmental Protection Agency (1979) *Process Design Manual for Sludge Treatment and Disposal*; EPA-625/1-79-011; U.S. Environmental Protection Agency, Municipal Environmental Research Laboratory, Office of Research and Development: Cincinnati, Ohio.

U.S. Environmental Protection Agency (1985) *Handbook of Estimating Sludge Management Costs*; EPA-625/6-85-010; U.S. Environmental Protection Agency, Water Engineering Research Laboratory: Lancaster, Pa.

U.S. Environmental Protection Agency; American Society of Civil Engineers (1979) *Proceedings of the Workshop Towards Developing an Oxygen Transfer Standard*; EPA-600/9-78-021; U.S. Environmental Protection Agency: Washington, D.C.

Vesilind, P. A. (1968) The Influence of Stirring in the Thickening of Biological Sludge. Ph.D. thesis, University of North Carolina, Chapel Hill.

Vesilind, P. A. (1974a) Scale-Up of Solid Bowl Centrifuge Performance. *J. Environ. Eng.*, **100**, 479.

Vesilind, P. A. (1974b) *Treatment and Disposal of Wastewater Sludges*; Ann Arbor Science Publishers Inc.: Ann Arbor, Michigan.

Voshel, D. (1966) Sludge Handling at Grand Rapids, Michigan, Wastewater Treatment Plant. *J. Water Pollut. Control Fed.*, **38**, 1506.

Walzer, J. G. (1978) Design Criteria for Dissolved Air Flotation. *Pollut. Eng.*, **10**, 46.

Wanielista, M. P.; Eckenfelder, W. W. (1978) *Advances in Water and Wastewater Treatment, Biological Nutrient Removal*. Ann Arbor Science Publishers Inc.: Ann Arbor, Michigan.

Water Pollution Control Federation (1969) *Sludge Dewatering*; Manual of Practice No. 20; Water Pollution Control Federation: Washington, D.C.

Water Pollution Control Federation (1980) *Sludge Thickening*; Manual of Practice No. FD-1; Water Pollution Control Federation: Washington, D.C.

Wilhelm, J. H.; Naide, Y. (1979) Sizing and Operating Continuous Thickeners. Paper presented at the American Institute of Mechanical Engineers Meeting; New Orleans, Louisiana.

Wood, R. F. (1970) The Effect of Sludge Characteristics upon the Flotation of Bulked Activated Sludge. Ph.D. thesis, University of Illinois, Urbana–Champaign.

Wood, R. F.; Dick, R. I. (1975) Factors Influencing Batch Flotation Tests. *J. Water Pollut. Control Fed.*, **45**, 304.

Yoshioka, N.; et al. (1957) Continuous Thickening of Homogeneous Flocculated Slurries. *Chem. Eng.* (Jpn.), **21**, 66.

9.0 SUGGESTED READINGS

Albertson, O. E.; Vaughn, D. R. (1971) Handling of Solid Wastes. *Chem. Eng. Prog.*, **67** (9), 49.

Ashbrook-Simon-Hartley (1992) *Aquabelt Operations & Maintenance Manual*; Ashbrook-Simon-Hartley: Houston, Texas.

Coe, H. S.; Clevenger, G. H. (1916) Methods for Determining the Capacities of Slime Settling Tanks. *Trans. Am. Inst. Min. Eng.*, **55**, 356.

Dick, R. I. (1970) Thickening. In *Advances in Water Quality Improvement, Physical and Chemical Processes*; Gloyna, E. F., Eckenfelder, W. W., Jr., Eds.; University of Texas Press: Austin.

Dick, R. I. (1972a) Gravity Thickening of Waste Sludges. *Proc. Filtr. Soc., Filtr. Sep.*, **9**, 177.

Dick, R. I. (1972b) Thickening. In *Water Quality Engineering: New Concepts and Developments*; Thackson E. L., Eckenfelder, W. W., Jr., Eds.; Jenkins Publishing Co.: New York.

Eckenfelder, W. W., Jr. (1970) *Water Quality Engineering for Practicing Engineers*; Barnes & Noble: New York.

Fitch, B. (1966) A Mechanism of Sedimentation. *Ind. Eng. Chem., Fundam.*, **5**, 129.

Fitch, B. (1974) *Unresolved Problems in Thickener Design and Theory*; Dorr-Oliver Inc.: Stamford, Connecticut.

Fletcher, N. H. (1959) Size Effect in Heterogeneous Nucleation. *J. Chem. Phys.*, **29**, 572.

Flint, L. R.; Howarth, W. J. (1971) The Collision Efficiency of Small Particles with Special Air Bubbles. *Chem. Eng. Sci.* (G.B.), **26**, 1155.

George, D. B.; Keinath, T. M. (1978) Dynamics of Continuous Thickening. *J. Water Pollut. Control Fed.*, **50**, 2561.

Hassett, N. J. (1958) Design and Operation of Continuous Thickener [parts I, II, and III]. *Ind. Chem.*, **34/116/169**, 489.

Javaheri, A. R. (1971) Continuous Thickening of Non-ideal Suspensions. Ph.D. thesis, University of Illinois, Urbana–Champaign.

Kos, P. (1977) Gravity Thickening of Water Treatment Plant Sludges. *J. Am. Water Works Assoc.*, **69**, 272.

Kynch, G. J. (1952) A Theory of Sedimentation. *Trans. Faraday Soc.* (G.B.), **48**, 166.

Shin, B. S.; Dick, R. I. (1975) Effect of Permeability and Compressibility of Flocculent Suspensions on Thickening. *Prog. Water Technol.*, **7**, 137.

Tarrer, A. R.; et al. (1974) A Model for Continuous Thickening. *Ind. Eng. Chem., Process Des. Dev.*, **13**, 341.

Vaughn, D. R.; Reitwiesner, G. A. (1972) Disk-Nozzle Centrifuges for Sludge Thickening. *J. Water Pollut. Control Fed.*, **44** (9), 1789.

Vesilind, P. A. (1968) Design of Thickeners from Batch Tests. *Water Sew. Works*, **115**, 9.

Chapter 24

Dewatering

1.0 INTRODUCTION

1.1 Objectives of Dewatering

Dewatering is the process of removing water from solids to reduce its volume and produce a material suitable for further processing, beneficial use, or disposal. The difference between thickening and dewatering is the nature of the product they produce. Thickening processes concentrate solids but generate a product that behaves and flows like a liquid. Dewatering processes produce a cake that behaves like a semi-solid or solid material. The objective of solids dewatering is to reduce the volume of material and prepare the solids for further processing, beneficial use, or disposal. Reducing the volume cuts subsequent solids-management costs.

Dewatering systems frequently require a relatively large capital investment and a substantial share of a facility's annual budget for operation and maintenance (O&M). To design cost-effective dewatering facilities, engineers need to systematically and holistically analyze a wide array of dewatering options, solids characteristics, and site-specific variables (e.g., other treatment processes and sidestreams). The information in this chapter will help design engineers make wise choices in the selection and design of dewatering facilities.

1.2 Key Process Performance Parameters

All dewatering processes generate two products: a solids cake and a liquid stream that consists of the water removed from the cake and some residual solids (see Figure 24.1). The liquid stream, which has many names (e.g., supernatant, decant, underdrainage, filtrate, centrate, and evaporated moisture), often is recycled to the head of the wastewater treatment plant.

Dewatering process performance is measured by two primary parameters: the cake solids content and solids capture rate. The *cake solids content* is a measure of cake dryness; it is the XXX of the weight of total dry solids in the cake divided by the total

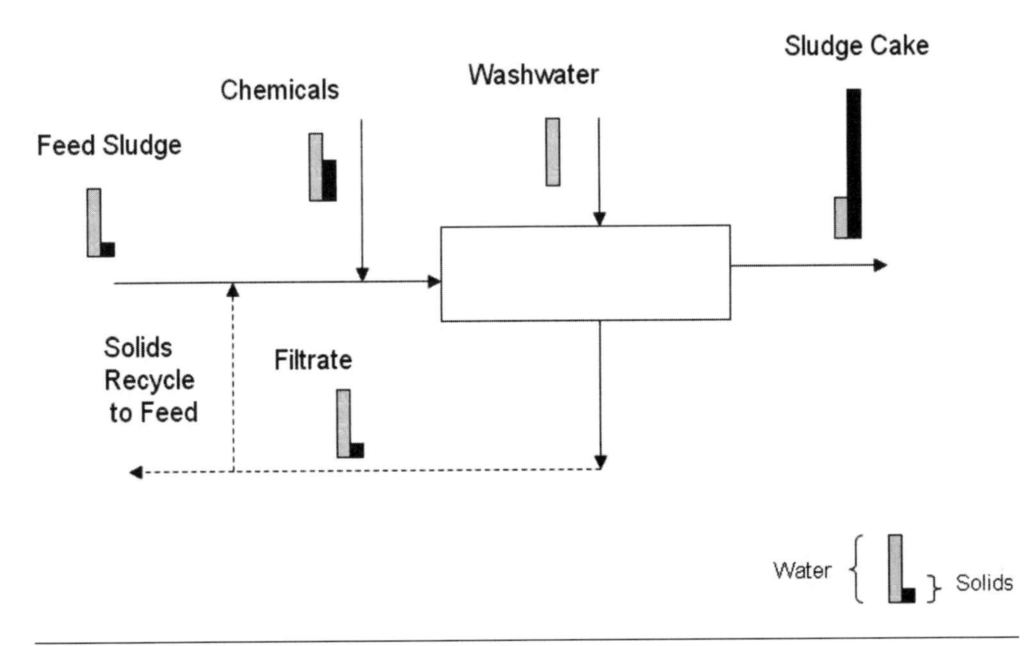

FIGURE 24.1 Solids balance for a dewatering system.

weight of the cake (solids plus water), expressed as a percentage. Because the amount of dissolved solids is tiny compared to the amount of suspended solids in dewatered cake, a cake's total solids and total suspended solids (TSS) are essentially the same. The solids content of dewatered cake typically ranges from around 12% to as high as 50%.

The *solids capture rate* is a measure of the percentage of suspended solids in the residuals fed to the dewatering process that remain in the dewatered cake. If a significant amount of chemical is being added to condition the solids, then it should be included in the feed solids when computing the solids capture rate. The effect of any washwater addition also needs to be included in the computation. The best way to calculate capture efficiency is by performing a solids and water balance around the process (see Figure 24.1). The solids capture rate for a belt press that recycles both its washwater and filtrate can be expressed by the following equations, which are derived from a solids mass balance around the belt press:

$$E = \frac{Cc(Ci - rCf)}{Ci(Cc - Cf)} \tag{24.1}$$

$$r = (Qi + Qw)/Qi \tag{24.2}$$

Where

E = solids capture (%),
r = recycle (%),
Cc = cake solids content (mg/L),
Ci = feed solids content (mg/L),
Cf = filtrate solids content (mg/L),
Qi = feed flow (L/m),
Qw = washwater flowrate (L/m),
Qf = filtrate flowrate (L/m), and
Qc = cake flowrate (L/m).

A dewatering process' solids capture rate typically exceeds 90%, and can be as high as 99%. Uncaptured solids are rcycled with the supernatant and may become a recirculating load. Once the cake solids content and solids capture rate have been calculated, design engineers can use the solids balance around the dewatering device to determine the solids concentration in the recycle stream.

When determining the solids loading to the dewatering device, the effect of recycled solids should be taken into account. For example, if the centrate or filtrate is recycled to the head of the plant, the solids in the filtrate will be removed during the wastewater treatment process, increasing the solids generated. In other words, the recycled solids will increase the total solids loading to the dewatering device. Therefore, a dewatering device with a capture efficiency of 90% will experience a solids load that is about 111% of the unadjusted solids-generation rate. At steady-state conditions, the mass of solids leaving the dewatering device should be equal to the solids generated during wastewater treatment, assuming that none of the recycled solids were destroyed but rather added to the solids being managed.

A third important parameter is the cake's *bulk density*, which is the weight of the cake divided by its volume. The bulk density takes into account the air voids that can be present as the cake solids concentration increases and the product takes on soil-like or granular properties. The bulk density will be change as the cake is handled and conveyed. For example, a cake's bulk density will be higher as it leaves a dewatering device than it is when the cake is being trucked offsite. (Some cake compaction and water release may occur as the cake is conveyed or loaded onto trucks, thereby changing its bulk density.) Bulk density is useful when determining how much volume dewatered solids will fill in a plate-and-frame press, when sizing a cake conveyance system, and when determining the number of trucks needed to transport dewatered cake offsite.

1.3 Overview of Dewatering Technologies

It is difficult to make a generalized comparison of dewatering processes. For example, while a centrifuge may handle one type of solids better than a belt press, the reverse may be true when treating another type of solids. The physical and chemical characteristics of solids can affect the choice of dewatering process. Belt filter presses, for example, heavily rely on in-line polymer feeding, which may not work well with solids that have a high pH or contain a lot of salt. Using metal salts instead of polymers may increase equipment and storage requirements, thereby losing some of the advantages of belt filter presses. Also, proper conditioning is important for belt filter-press performance, so feedstock that varies considerably from day to day can create operational problems.

Centrifuges also are affected by solids variations, but they can be adjusted to achieve different combinations of cake solids and solids capture. Polymer, which is necessary to coagulate centrate colloids, typically is used to improve the solids capture rate.

Fixed-volume filter presses using ferric chloride and lime conditioners are less sensitive to physical and chemical variations in solids, but often require large (expensive) quantities of chemicals. Design engineers should evaluate the use of polymers with filter presses, where feasible.

Performance is affected not only by the type of solids and the nature of the water it contains, but also by how the solids are conditioned and the specific design of the dewatering equipment. Dewatering-equipment performance can be optimized by adjusting process variables and testing different types and concentrations of conditioning chemicals. Full-scale pilot testing is desirable so systems can be optimized and then compared.

The selection and design of a dewatering process can be based on performance observed at similar installations, on bench-scale testing (sometimes), or on full-scale pilot testing.

Because solids vary from plant to plant, the most accurate prediction of how a dewatering process will perform is best determined via full-scale pilot testing. Pilot testing will allow optimization of chemical conditioning. Different types of polymers can be tested and the relationship between dosage, hydraulic and solids loading, solids capture efficiency, and cake solids can be ascertained. In many cases, side-by-side pilot testing of different types of dewatering equipment from multiple manufacturers enables the project team to select a process that minimizes capital and O&M costs while meeting process objectives. However, this is not possible if the wastewater treatment process is being changed, and a representative solids sample is not available for testing.

1.4 Effect of Solids Characteristics

The ease with which different solids dewater varies widely. A waste activated sludge (WAS), for example, is difficult to dewater, while well-digested primary solids dewaters more easily. Much of this variability has to do with the solids constituents involved and how water bonds to them. In waste activated sludge, for example, much of the water is difficult to remove because it is attached to bacterial cells or tied up chemically in the cell structures.

Dewatering performance can even vary when treating the same type of solids, depending on how the material was managed beforehand. It depends on the solids' chemical and physical characteristics and the influence of chemical conditioning, which in turn depends on the solids' salt concentration, the solution's pH, or the character of the organic matter present.

The solids' pH can adversely affect dewatering. Studies by Novak and Haugan (1979) have shown that activated sludge dewaters poorly at pH 8 and higher; it dewaters best at about pH 3. Ferric chloride lowers the pH to a more desirable level, while polymers raise the pH above 9. Only lime conditioning optimally performs at high pH because mixing dense, porous calcium carbonate with solids provides a matrix that promotes rapid water removal.

Organic matter (e.g., organic acid and exocellular biopolymers) can play an important role in dewatering. Most of the naturally occurring organics that are of significance in dewatering are anionic biopolymers. These materials typically are removed from wastewater via charge neutralization or adsorption to solid surfaces. Polymers are ineffective at removing these organics; in fact, they can combine with the organics to create a material with poorer dewatering characteristics. Organic molecules typically are present in wastewater that is undergoing biological activity. For example, primary solids should contain low levels of organic biopolymer during winter, when biological activity is low. Increased biological growth during summer can generate enough biopolymer to greatly alter process performance. Industrial organics also may alter solids properties. They typically have the same effect as naturally occurring organics, but may be present at higher concentrations and may not coagulate as easily, particularly if they adsorb poorly.

A key to evaluating the effect of these organics on dewatering is to compare the relative benefits of anionic and cationic polymers in a laboratory. Cationic polymers typically work well and can improve dewatering characteristics via either charge neutralization or molecular bridging. Anionic polymers perform well if the solids contain little biopolymer (e.g., when the solids are not biologically active). If anionic polymers worsen dewatering performance, excess anionic biopolymer probably is present. If neither polymer works well, the cause may be excess salts, high pH, or organic matter.

Some consider the particle size of the feed solids to be the single most important factor influencing solids dewaterability (U.S. EPA, 1979; Oerke, 1981). As the average particle size decreases (which may result from excessive mixing and shear), the surface volume ratio increases exponentially (Heukelekian and Weisberg, 1958). Adding such particles needs to be considered as feed solids when the cake solids and solids capture are determined. Increased surface area results in greater hydration, higher chemical demand and increased resistance to dewatering (Oerke, 1981).

Several physical parameters (e.g., temperature) also can influence dewatering. Temperature can influence biological activity and thereby alter solids properties. Lower temperatures slow chemical reactions, affecting the performance of both metal ion conditioners and polymers. Typically, reactions will be less complete and chemical doses may need to be increased when temperatures drop, although some of these reactions may be offset by a decrease in biological activity.

The presence of particles (e.g., coal or another precoating agent) in solids can enhance dewatering. Such particles can serve as adsorption surfaces for organic biopolymers and thus allow alum solids to serve as a chemical conditioning agent.

A higher solids concentration in the feed solids has been shown to directly increase the cake solids that can be obtained in most mechanical dewatering processes. For example, a 1985 survey of more than 100 municipal belt-press installations showed a strong correlation between cake solids, feed solids, and the percentage of WAS in the feed (Koch et al., 1988). Similar results have been reported for centrifuges and plate-and-frame filter presses (Koch et al., 1989).

Other studies have shown that feed solids concentration significantly affects filter-press performance (Oerke, 1981). Increasing the feed solids concentration produced equivalent cake solids but at much lower chemical conditioner doses and increased both process yield and solids loading during full-scale and pilot-scale testing at the South Shore Wastewater Treatment Plant in Milwaukee, Wisconsin, (completed as part of the Milwaukee Metropolitan Sewerage District's Water Pollution Abatement Program). This was also the most common and significant conclusion that Pietila and Joubert (1979) and Morris (1965) reached when analyzing the dewaterability of solids using several pieces of dewatering equipment.

While residuals with high solids concentrations often produce dewatered cake with higher solids contents, there are some exceptions. This is not true when the gel point of the feed material has been reached, and when large amounts of liquid are required to provide good interaction between the solids and conditioning chemicals (e.g., polymers).

1.5 Pretreatment

While most solids only require conditioning before being dewatered, some pretreatment systems have been found to be useful to protect equipment, improve cake solids, or enhance the quality of the dewatered cake product. For example, grinders or macerators can be incorporated into the pumping and feed system to reduce the size of the solids entering the equipment. Grinders can prevent entry of elongated or jagged pieces of material that could tear the expensive filter belt cloth or clog centrifuge nozzles. Grinders should be added to the suction side of the belt-press feed pump, even if other grinders will be installed throughout the treatment plant. Although grinders typically are considered high-maintenance items themselves, they help protect the belt cloth and prolong its life.

In other applications, in-line screening devices have been placed behind the dewatering device to remove large particles from dewatered cake. Screens typically have been used when the dewatered solids will be further processed into a Class A biosolids product (e.g., a pelletized fertilizer). The screen removes plastic and debris and produces a more uniform pelletized product.

Other pretreatment processes have been used ahead of the dewatering process to improve dewaterability. These processes use elevated temperatures or pressures, ultrasonics, high-shear devices (e.g., ball mills), or chemicals (acids or bases) to break down WAS, reduce chemical conditioning requirements, or produce a high-solids cake.

Some stabilization processes combine thermal conditions with anaerobic digestion to improve the dewatering characteristics of solids (see Chapter 25). There also has been limited experience with adding bulking materials (e.g., wood chips or ash) to increase the cake's fiber content and improve its dewaterability or bulking properties. The addition of bulking materials has been largely limited to filtration devices (e.g., belt presses and plate-and-frame presses).

1.6 Chemical Conditioning

All mechanical dewatering methods benefit from some form of chemical conditioning. Although many physical conditioning methods have been used for dewatering, the most effective method has been chemical conditioning (Genter, 1934; Lecey, 1980; Sharman, 1967; Tenney and Stumm, 1965). Chemical conditioning improves cake solids and the solids capture efficiency that can be achieved. The chemicals used can be either inorganic or organic compounds. The most common types of chemicals used are inorganic salts (e.g., lime and ferric chloride) and organic polymers.

The inorganic chemicals typically used are metal salts (e.g., ferric chloride, aluminum chloride, or ferrous sulfate). Their activity is pH-dependent, so pH may need to be controlled for the full use of the chemical to be obtained. The organic chemicals used are high-molecular-weight, water-soluble organic polymers. Past economic evaluations indicate that organic polymers have taken a major role in dewatering because they are cost-effective and have better maintenance, performance, and safety records than inorganic chemicals.

Optimizing the polymer dosage can improve centrifuge and belt press operations (Lecey, 1980). The ideal test of a chemical conditioner is actual application of the product to the particular solids being dewatered.

Optimum polymer addition points depend on the chemical and the solids involved. Polymers may require long reaction times. Trial and error typically can determine whether chemicals should be added inside the centrifuge, to the feed pipe of the centrifuge, or even further upstream (e.g., at the suction of the feed pump). To determine the suitability of individual polymers, bench-scale tests should be used. Jar tests and capillary suction time (CST) measurements should be used to classify the types of polymer required. Typically, a strong floc is best. Other bench tests (e.g., piston filter presses and batch centrifuges) can provide indications of likely cake solids content. These are likely to be more convenient than side-by-side dewatering trials.

For plate-and-frame filter presses, ferric chloride and lime (with or without ash) typically are used to improve solids particle size distribution, while ash may reduce the conditioned-solids compressibility. While the use of ferric chloride and lime significantly increases the amount of inert material in the dewatered solids, it improves compressibility and enhances cake release. There has been some success with using polymer to condition solids for plate-and-frame presses, but care must to taken to ensure that a sticky cake is not formed that is difficult to remove from the press. A sticky cake often is the result of incomplete mixing or overdosing polymer.

1.7 Effect of Recycle Streams

The water removed from solids during dewatering typically is returned to the treatment plant headworks for processing. This liquid wastestream contains constituents that can affect wastewater treatment processes and add to the plant's influent loading [e.g., biochemical oxygen demand (BOD), ammonia, and phosphorus]. The degree of effect will depend on solids capture efficiency and the nature of the wastewater and solids treatment processes before dewatering. For example, if the treatment plant provides biological phosphorus removal and anaerobic digestion, then soluble phosphate may be released. This phosphate can increase the influent phosphate loading to the plant or cause magnesium-based scale (e.g., struvite) to precipitate in the pipeline or pumping system that conveys recycled filtrate or centrate. The recycled centrate or filtrate also can be a source of additional nitrogen loading to the plant as a result of the ammonia released during anaerobic digestion. (For a more detailed discussion of the effect and treatment of recycle streams, see Chapters 11 and 25.)

1.8 Odor Control

Although all dewatering processes separate liquid from solids, offgases and odors also may be released. The likelihood of odor production depends on how the solids were processed and how long they were held and stored before dewatering. If the solids are stored in unaerated tanks, they most likely will begin to anaerobically digest and release odorous compounds. In some dewatering devices (e.g., centrifuges), the odors will be contained in the equipment; it will only be necessary to control odors from the dewatered cake and any offgases from the equipment. In other devices (e.g., belt presses) it may be necessary to install a hood over the equipment to provide negative pressure to ventilate the dewatering process and collect the air for odor control. (For details on odor containment and control, see Chapter 7.)

1.9 Pilot Testing

Many types of solids can be dewatered successfully by various types of dewatering equipment. The reliability of a dewatering unit, however, only can be evaluated by pilot-testing the intended wastestream using trailer-mounted test units.

Field tests involving more than one machine or type of dewatering device should be run concurrently. Side-by-side operations will alleviate concerns that operating conditions or solids characteristics changed between tests. Several belt filter press manufacturers have mobile trailer-mounted pilot units that can be brought to a plant site for testing. Most of these units are small production machines and will provide performance comparable to larger models. Centrifuge manufacturers and plate-and-frame press manufacturers also can provide pilot-scale trailer-mounted units. Data to be collected as part of the pilot test include hydraulic- and solids-loading rates, polymer type and use, percent solids, and capture efficiency. While small production units can provide an indication of the expected performance of larger units, care must be exercised when scaling up the data to the larger unit. Design engineers should consult with the equipment manufacturer to develop full-scale criteria. Where possible, testing of a full-scale unit is preferable.

As an alternative to a full-scale pilot test, many manufacturers have their own in-house testing equipment that can be used to predict performance. A sample of the material to be dewatered can be sent to the manufacturer, who will provide design criteria based on their tests. These tests have been shown to provide adequate projections of equipment performance for most applications. Care needs to be taken to ensure that the solids do not change during transport by getting too hot or being subjected to excessive vibration. Tests should be performed as soon as possible to ensure that the solids do not change properties as they age.

When evaluating the performance of a dewatering device, the quantity and quality of the filtrate and backwash, and their effects on the wastewater treatment system, should be considered. While a pilot test provides good data, they typically are operated with consistent feed conditions and much operator attention. Designs based on pilot-test data should account for variable feed and should not require constant operator attention to produce acceptable performance.

1.10 Design Example

The following spreadsheet design example outlines the key steps in selecting and designing dewatering equipment (see Figure 24.2). The example takes the solids generation rate and applies peaking factors and the number of operating hours to determine the design solids and hydraulic loading rates. The chemical dosage, percent solids capture, and cake solids typically are obtained from pilot-testing or from the performance of similar units treating similar solids. The design example illustrates how a spreadsheet can be used to track the solids and water balance across the units and take into account the recycle streams. Finally the hydraulic and solids loading are used to pick the size and number of dewatering devices. Manufacturers should be contacted to determine the acceptable hydraulic and solids loading for a specific piece of dewatering equipment. Some flexibility should be provided to allow for the use of different types of conditioning chemicals. (The design of the feed pumps and chemical handling system is covered in other chapters.)

Hourly loading adjusted to reflect number of shifts and actual operating hour of unit and time for cleanup between shifts

Average Sludge Generation Rate	16 kg/hr
Assumed Peaking Factor	1.5
Design Sludge Generation Rate	24 kg/hr
Hours/Week	168
Operating Hours/Week	40
Design Dewatering Rate	101 kg/hr

Sludge Feed

Dry solids Feed	101 kg/hr	
Per Cent Solids	2.00%	
Wet sludge feed rate	5050kg/hr	
Water in feed	4949kg/hr	
Volumetric Loading	5050l/hr	(assuming density of water 1000kg/l)

Device Performance and Chemical Dosage based upon pilot testing of specific sludge

Belt Press and Centrifuge		Plate and Frame Press	
Per Cent Capture	92.00%	Per Cent Capture	97.00%
Per Cent Solids	20.00%	Per Cent Solids	45.00%
Added Chemicals	1.00%	Added Chemicals	35.00%

Assume all recycle returned to plants, removed in primaries and returned to dewatering

At steady state solids leaving dewatering must equal solids generated

	Belt Press and Centrifuge	Plate and Frame Press
Dry solids generated by wastewater treatment	101.0kg/hr	101.0kg/hr
Chemical solids added	1.0 kg/hr	35.4kg/hr
Dry Solids Leaving dewatering	102.0kg/hr	136.4kg/hr
Wet Cake Leaving Device	510.1kg/hr	303.0kg/hr
Water in Cake	408.0kg/hr	166.7kg/hr
Assumed bulk density	900.0kg/cubic meter	1200.0kg/cubic meter
Cake Volume	1.0 cubic meters/hr	0.1 cubic meters/hr
Dry Solids entering device based upon capture	110.9kg/hr	140.6kg/hr
Wet Solids entering device base upon % feed solids	5544.0kg/hr	7028.4kg/hr
Water entering device	5433.1kg/hr	6887.8kg/hr
Recycle Dry Solids	9.9 kg/hr	4.2 kg/hr
If belt cleaning washwater is being added to unit it must be inclued in recylce water (assumed zero for example)		
Recycle Water	5025.1kg/hr	6721.1kg/hr
Recycle Per Cent Solids	0.2%	0.1%
Recycle Volume	5.0 l/hr	6.7 l/hr

Number of Belt Presses and Centrifuges sized on basis of hydraulic and solids loading per unit from manufacturer
Number of Plate and Frame Presses sized based upon size and number of plates, final pressed cake volume and cycle time

Additional units added based upon assumed redundancy and spares required

FIGURE 24.2 Data on belt filter press, centrifuge, and plate-and-frame press for design example.

1.10.1 *Input Parameters*

Because the primary function of dewatering is to produce a dry cake, the most important input parameter is the TSS concentration of the feed entering the dewatering devices. The primary performance parameters are cake solids content and solids capture. The input parameters that have been assumed for the design example are listed below:

- Solids generation rate = 16 dry kg/h,
- Feed solids concentration = 2%,
- Peaking factor = 1.5, and
- Dewatering operations = 40 h/week.

1.10.2 Assumptions

It is assumed that the actual solids loading to the dewatering equipment will be increased above the solids generation rate to account for chemical addition and the device's solids capture efficiency. The assumption implies that any solids in the recycle steam that are returned to the wastewater treatment process will be removed from the wastewater and dewatered again, and thus contribute to the solids production and solids loading of the dewatering process. If a solids stabilization process (e.g., anaerobic digestion) precedes dewatering, some of the recycle solids may be destroyed and this assumption may be overly conservative. A more thorough mass balance or simulation around the entire wastewater treatment process can be used to provide a less conservative estimate of the solids loading to the dewatering process.

The design example is intended to demonstrate the effect of cake solids, capture efficiency and chemical dosage on total cake production. A process that uses a high chemical dose and produces a dry cake (recessed plate-and-frame filter press) was compared with a process that uses a lower chemical dose but produce a wetter cake (belt press or centrifuge) The assumptions for the two dewatering process are listed below:

Belt press or centrifuge

- Solids capture efficiency = 92%
- Cake solids = 20%
- Chemical conditioning = lime and ferric chloride
- Solids added from chemical conditioning = 1%
- Cake bulk density = 1200 kg/m^3

Plate and frame filter press

- Solids capture efficiency = 97%
- Cake solids = 45%
- Chemical conditioning = polymer
- Solids added from chemical conditioning = 35%
- Cake bulk density = 900 kg/m^3

1.10.3 Calculations

The loading rate to the dewatering equipment first is calculated by applying the peaking factor and hours of operation to the solids generation rate. The feed solids concentration is used to compute the water and wet solids loading rate from the solids loading rate expressed in dry kg/hr.

Next, the loading rate is increased to account for the chemical addition by adding the extra solids added by chemicals. Finally, the loading rate is divided by the solids capture efficiency to obtain the effective loading rate to the dewatering equipment. As mentioned previously, this is a conservative assumption that assumes all the solids in the filtrate or centrate will be removed in the liquid processing train and returned to the dewatering device, thereby increasing the solids loading. This assumption sets the solids leaving the dewatering device equal to the amount of solids generated in the wastewater treatment process.

1.10.4 Output

Using the assumed loading rates, solids capture efficiency, cake solids, and bulk density, the day and wet weight and volume of the cake produced by the dewatering device can be calculated. The volume and solids concentration of the centrate and filtrate can be computed by performing a mass balance around the dewatering equipment. Even though the plate-and-frame press will produce more cake on a dry solids basis, the volume of the cake is ten times less than that produced by the belt press or centrifuge because of the dried cake and higher bulk density (see Figure 24.2).

The number of dewatering units then is selected based on the processing rates of each dewatering machine and the amount of redundancy required on a case-specific basis. If it is desired to have a standby unit, smaller-capacity units may be preferable. Having a larger number of units with lower throughput also allows operators to take some out of service during periods of lower solids production. Another design approach is to use fewer units (or even one unit) and respond to changes in solids production by changing the number of hours in operation.

Most dewatering units are rated based on hydraulic loading rates, although they may become solids loading limited at high feed solids concentrations. Centrifuge manufacturers typically offer units having different nameplate hydraulic processing rates. Belt press manufacturers offers units with different belt widths to accommodate different hydraulic loading rates. Manufacturers of plate-and-frame presses offer different plate sizes and can vary the number of plates to accommodate different hydraulic loading rates. Ancillary facilities (e.g., feed pumps, chemical conditioning, washwater, odor control, conveyors, and cake storage) are then sized and coordinated with the number of dewatering units selected.

2.0 CENTRIFUGES

This chapter will only discuss solid bowl centrifuges because basket centrifuges and disc nozzle decanter centrifuges are no longer typically used to dewater municipal wastewater solids.

2.1 Introduction

Centrifuges work on the basis of sedimentation, much like clarifiers and thickeners. However, centrifuges rotate quite rapidly and subject process solids to an acceleration rate between 1 500 and 3 000 times that of the earth's gravity. Centrifuges are used for both thickening and dewatering residuals, although the internal design dimensions for thickening are quite different than for dewatering.

Centrifuges are relatively simple to operate. Operators set the conveyor torque to control cake dryness, and control centrate quality by changing the polymer dosage. Centrifuges require less operator attention, are easier to automate, and often produce drier cake than other common dewatering equipment. They also typically are energy-intensive, noisy, vibrate, and may vent smelly air to the environment.

2.2 Process Design Conditions and Criteria

2.2.1 *Mechanical Features*

In modern centrifuges, all wetted parts are made of stainless steel, because it is difficult to maintain tolerances with rusting carbon steel (see Figure 24.3). The main structural parts are typically cast stainless steel. Centrifugal casting produces better quality than static casting, and is preferred for high-stress parts.

There are distinct differences in the mechanical design of small and large centrifuges. Typically, smaller centrifuges are of lower quality, have shorter bearing lives, less rigorous hard surfacing, and lighter construction, on the assumption that they will not be subjected to rigorous duty. Centrifuges are designed to meet the same vibration standard (6- to 7-mm/sec velocity at their maximum rated speed). The least costly way for manufacturers to meet the standard is to load dead weight into the stationary frame, followed by lowering the speed. However, strengthening the frame is more common. With this in mind, engineers should specify a maximum static/dynamic weight (S/DW) ratio for centrifuges. This is calculated as:

$$\text{Static Dynamic Weight Ratio} = \frac{\begin{array}{c}\text{Total static weight on the four}\\ \text{main vibration isolators}\end{array}}{\text{Weight of the rotor filled with water}} \qquad (24.3)$$

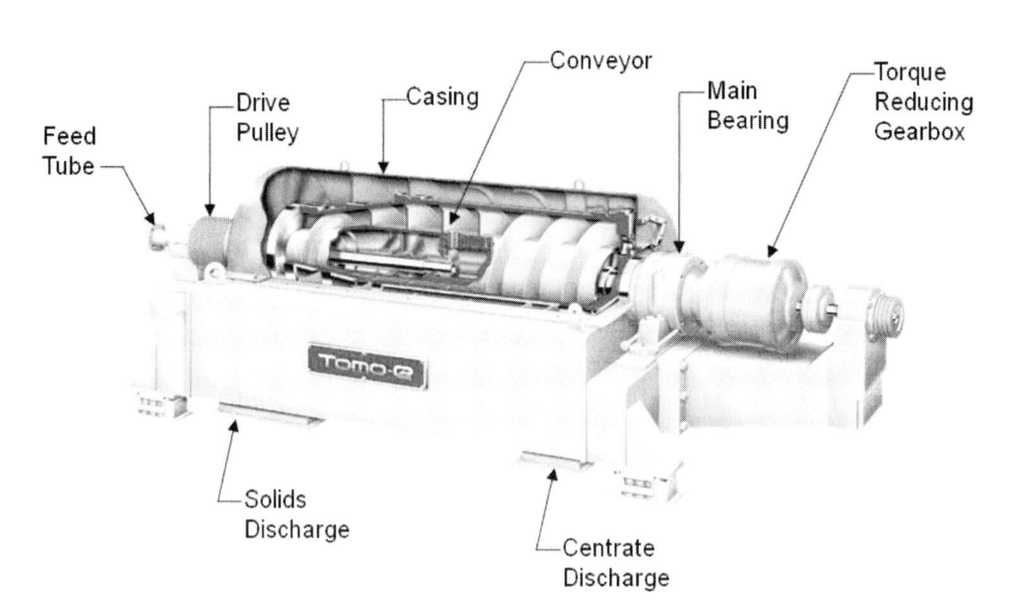

FIGURE 24.3 Cutaway view of a centrifuge.

Where

S/DW ratio $\leq$ 2 is excellent;

S/DW ratio = 2 to 3 is fair (bearing life will be shortened); and

S/DW ratio = 3 to 4 is poor.

Smaller centrifuges with higher S/DW ratios are acceptable if they only operate a few days per week.

The expected bearing life is difficult to evaluate. Centrifuges have main bearings that carry the rotor (high-speed), conveyor bearings (low-speed), and a thrust bearing that carries the axial load of the centrifuge. Bearings have a calculated B10 life, which is a theoretical life that predicts failure due to cycles and loading. However, the B10 life assumes no problems with lubrication, contamination, or alignment—all common causes of bearing failure—so it is not a useful predictor of quality. The bearing choices are spherical rollers, roller bearings, and ball bearings. Bearing choice is part of an engineering design, so there is not an obvious ranking of the various types. Spherical roller bearings typically are excellent for the main bearings. Conveyor bearings are such slow-speed bearings that they are not critical. The largest centrifuges need circulating oil to cool high-speed bearings. On smaller centrifuges, either grease or oil lubrication is used. Designs with an air space between the main bearings and the centrifuge casing are less troublesome than designs in which the bearing housings are attached directly to the casing.

Conveyor bearings are slow-speed bearings that typically fail via contamination. The best design is one that allows the bearing to be purged with grease, thus periodically removing the inevitable contamination. Sealed bearings are inexpensive, but because all seals fail, they have a limited life.

All centrifuges have a torque reducer; the options are a planetary gearbox, a cyclo gearbox or a hydraulic motor. It is impossible to accurately predict the quality of a gear reducer using the data made available by vendors. Failure typically occurs as a result of long-term use. Of two identical centrifuges, the unit running at 30% of rated torque should outlast the one running at 60% of rated load. A quality torque reducer should last 20 000 to 40 000 hours without mechanical problems or rebuild, and the best designs last 50 000 to 100 000 hours or more. Asking reference installations about their experience and what the typical torque load is *as a percentage of the allowable load* is the only practical way to predict repair costs. Much like automobile manufacturers, centrifuge manufacturers have a series of centrifuge models, each of a different diameter and length, and each model is available with a variety of options. The same centrifuge typically is sold for all applications within water and wastewater treatment plants.

All dewatering centrifuges typically require polymer, and the following guidelines are suggested. Design engineers should supply equipment that can use at least two forms of polymer, one of which is dry polymer. They also should supply sufficient aging tank capacity to ensure a minimum of 40 minutes aging time of 0.2% active concentration at 150% of the expected polymer use. Making up a concentrated solution and then diluting it typically is not a good option, unless further aging storage is available. The best polymer addition point depends on the solids, the polymer, and the dewatering objectives. Four addition points are suggested: inside the centrifuge, immediately before the centrifuge, 8 m (25 ft) ahead of the centrifuge, and 16 m (50 ft) ahead of the centrifuge. The best way to accomplish this is to install a manifold in a convenient location (see Figure 24.4). To change the addition point, operators only need to open one valve and close another. While presses typically include more devices to mix the polymer with the solids, there is no evidence that they benefit centrifuges, which provide their own mixing. Periodically, operators need to evaluate polymers. Load cells under the polymer storage tanks make it easy to measure polymer consumption without climbing on the tanks to do a drawdown test. Engineers should design the process piping so another polymer can be tested easily. For emulsions, the distance from the emulsion pump to the makeup unit should be small, so a large manifold need not be filled. Fittings should be provided so an alternate emulsion can be fitted up without difficulty.

Process instrumentation is also important. Magnetic flow meters on the feed and the polymer solution are recommended. Owners also will need convenient sample taps

FIGURE 24.4 A polymer-addition manifold.

for the feed entering centrifuge, feed before polymer addition, polymer as delivered, polymer solution after aging, centrate, and dewatered cake.

Ideally, the sample taps should be close to the main run of a line to minimize the wastage to get a fresh sample. A floor drain under the sample tap should be provided for operators to dispose of samples. Sample sinks need a provision for water to backflush sample runs. Also the trap and drain under the sink need to be oversized to handle the flows. Because centrate quality is the most important operator observation, it is important to make it easy to sample. The best practice is to have a representative centrate sample flow continuously into a drain easily visible to operators as they walk by. The best layout is to put the centrifuges on the floor above the operator's station, because it is much easier to get samples on the floor below the centrifuge, and typically quieter as well. It is important when troubleshooting the system to be able to sample both the as-delivered polymer and the diluted solution going to the centrifuge.

Flow meters are also important. With the advent of variable-frequency drives on pumps, there is an inclination to assume that flow is always proportional to the pump speed. This is often true when the pump is new, and rarely in extended use. However, it is difficult to troubleshoot a system if the flowrates are not known with reasonable accuracy.

2.2.1.1 Installation Considerations

The centrifuge typically is designed to vibrate, and all electrical and process connections must use flexible connections with enough flex to minimize the load on the piping. If either solids or liquid backs up the chutes and reaches the bowl, it will cause enough drag to shut the centrifuge down on high motor amps. Centrate is sometimes a problem because there can be excessive amounts of foam. Avoid any traps that will hold back foam, and provide a minimum 7 m (20 ft) drop before entering a horizontal, or near horizontal run.

Solids conveying can be difficult. Placing the centrifuges on the upper floor means that the solids will be transported horizontally or downward. Most conveying systems have problems when they have to elevate the solids. Belt conveyors are less common because of odor and housekeeping issues. Cake pumps are expensive to purchase and maintain, which leaves screw conveyors as the most common conveyance devices. Inclined screw conveyors work well, but vertical ones are troublesome. All centrifuges spill liquid out the solids chute on startup, and occasionally thereafter. The most successful way to handle this is with a diverter gate. When the torque is low, the gate diverts everything to the drain. When the torque reaches some to-be-determined value, the gate opens and the solids drop into the conveying system

Venting is a problem in many installations. Centrifuges suck air in near the center line, and blow it out down the centrate and the cake chute. Restricting the flow of air by buttoning up the centrate chute, or especially the cake chute, results in air carrying solids into the casing and eroding the bowl severely enough to require replacing the bowl. Both chutes should be vented to the same pressure, and provisions made to flush the vent lines as well. Given the potential costs of bad venting, the centrifuge vendor should formally sign off on the vent design.

2.2.2 *Hydraulic Loading Rates*

Centrifuges typically are purchased with a process guarantee (e.g., feed rate, solids loading, cake dryness, polymer dosage, and centrate quality). Ideally, having an older piece of equipment as a benchmark helps because the guarantee then can be expressed as "matching the performance of the existing dewatering equipment". The major money issues that affect the cost and performance of a centrifuge installation include cake dryness, polymer dosage, feed rate, and centrate quality. Of the four money issues, typically one can be increased at the expense of another. As a result, there are few absolute limits on loading rates. Vendors may elect to bid a small centrifuge against the competition's larger one, in the hope of achieving a sale at lower price. The result is that capacities may be unrealistic for continuous duty, and owners may not get the centrifuge capacity they need. Because centrifuge manufacturers do not add a service factor, engineers should base the process penalties at a feed rate 30% higher than that needed for the design.

2.2.3 *Solids Loading Rates*

There are a number of limitations on scale-up. As the hydraulic or solids loading rate increases, performance decreases. More polymer typically helps, but there is a limitation. Engineers should design an installation to ⅔ of the vendor's maximum capacity. Feed solids concentration is important because the solids and the polymer should be completely mixed for the chemical reaction between them to occur. If the solids (or polymer) is viscous, it can be diluted to allow adequate chemical contact. Secondary solids much thicker than 2%, and mixed solids much more concentrated than 4% can be problems.

2.2.4 *Bowl Speed*

The centrifuge rotates to create a centrifugal force, which drives the solid separation. Traditionally the acceleration is referred to as "g force". The acceleration of the earth's gravity is 9.8 m/s^2 (32 ft/sec^2), typically referred to as "1 g", and centrifuge manufacturers use this as a unit of acceleration. Most centrifuges in service operate between 1 500 to 3 000 g. In practical terms, if a centrifuge operating at 3 000 g had a 1-lb (0.45-kg) imbalance on the bowl wall, the centrifuge would shake with a 1.260-kg (3 000-lb) force. The relationship between bowl speed and *g* force is calculated as:

$$g = k \times RPM^2 \times \text{diameter} \tag{24.4}$$

Where

RPM = the rotational speed [revolutions per minute (rpm)],
k = 0.000 000 56 when the diameter is measured in millimeters or 0.000 014 2 when the diameter is measured in inches.
diameter = the inside diameter of the bowl.

The "best" operating g force is application-specific, and higher is not always better. Higher g forces require better materials, better manufacturing techniques, and therefore, costs more money to build. As a result, the g force at which a centrifuge can meet its vibration and noise specification is an excellent measure of the quality of the mechanical design, even if the owner chooses to operate it at a lower speed. Specifying a g force and the noise and vibration level at that speed is one of the best ways to ensure the quality of the centrifuge.

2.2.5 *Pool Depth*

As with all sedimentation devices, deeper ponds provide more force to compress solids (see Figure 24.5). *Pool depth* is the depth of the liquid relative to the centrifuge bowl. Because larger centrifuges intrinsically have deeper ponds than smaller ones, then if all else is equal, larger centrifuges will produce drier cake than smaller ones. Most centrifuge manufacturers have reduced the solids discharge diameter to make the pond deeper and to reduce power consumption. Some manufacturers have moved the dam openings inward by a like amount. This limits the differential head pressure and, therefore, the cake dryness that the centrifuge can achieve. Limiting the cake dryness limits the torque, allowing smaller torque reducers and thinner flight material without bending flights. A good centrifuge design should have

- The capability to provide a 25-mm negative pond and
- A conveyor hub whose largest diameter is at least 50 mm smaller than the solids discharge diameter.

One unfortunate consequence of negative ponds is that all dewatering centrifuges discharge feed material or sloppy cake during startup until a seal is established (typically in 5 to 15 minutes). Design engineers should take that into account.

2.2.6 *Structural Support*

From a design point of view, smaller centrifuges can be skid-mounted and have minimal foundation requirements. Larger ones weigh 13 to 18 metric ton or more, and require more substantial design effort. Plants in earthquake zones have more problems because the rotating mass tends to remain in place, while the building under the centrifuge

FIGURE 24.5 Difference in pool depth between similar centrifuges with different diameter scrolls.

moves. Centrifuge manufacturers provide installation drawings giving the relevant dimensions, static and dynamic loadings, and the process requirements. Most large installations rely on particularly thick concrete floors to absorb vibration and deaden noise. Avoid metal gratings because they tend to rattle. Sooner or later, the centrifuges must be removed for service. Traveling bridge cranes with adjustable vertical and horizontal speed controls are required for assembly and disassembly of the scroll.

2.2.7 *Safety*

All centrifuges should be protected by interlocks for torque, motor amperage, and vibration. Formerly vibration switches were used, but now vibration readouts should be specified. The controls should clearly state the vibration level for an alarm, and for a shutdown. No centrifuge cover could contain the pieces in the event of a catastrophic failure. Such failures are extraordinarily rare.

2.2.8 *Capture Efficiency*

As with all sedimentation processes, centrifuges separate solids based on particle size and density. Capture should be targeted at about 95% to minimize recycle. Centrifuges can achieve 99+% capture, but at considerable risk of underusing the centrifuge volume.

2.2.9 *Area/Building Requirements*

While it may be expensive, placing the centrifuge high in the building results in a better installation. Then, because there is nothing to be seen on the centrifuge floor, put the control room on the floor below the centrifuge. This reduces the operator's need for noise protection by allowing the operator to sample the centrate and cake to make control adjustments from the control room, rather then the noisy centrifuge room.

2.3 Ancillary Equipment and Controls

All modern centrifuges have two modes of control: differential and torque. Centrifuge manufacturers typically hire a panel shop to design and build the controls. Most controls use a variable frequency drive (VFD) main drive and a VFD backdrive motor. There is little or nothing proprietary about controls, and the brand of motor control and programmable logic controller (PLC) is the customer's choice. For serviceability, it is important to require the supplier to provide an electronic copy of the control program, along with the passwords and software to access and service the controls and drives.

Automation is available from both the centrifuge manufacturers and independent controls engineers. Some suppliers use a black box approach for control, in which the owner cannot alter anything. A better option is to require open software, which anyone can alter to suit changing needs. Automation measures the feed solids and controls the solids loading. Process instrumentation can measure the centrate solids and adjust the polymer to maintain fixed centrate quality. Most use backdrive torque as a means of measuring cake dryness, but there are some instruments that can measure this as well.

In addition, it is very helpful to have the supervisory control and data acquisition (SCADA) system calculate the solids volume and the net disposal cost in real time, and by shift. Before operators can run the dewatering process in a cost-effective way, they have to know what the costs are.

2.3.1 *Feed System*

All processes are challenged by changing feed conditions. It is especially a problem when the feed material is a blend of two or more streams. For example, when the ratio of primary to secondary solids changes, the dewaterability of the solids changes as well, and unless operators react quickly and correctly, the centrifuge will be either severely overdosed or underdosed with polymer. Other sources of variation are unmixed storage tanks, and digesters whose WAS is pumped directly to the dewatering system rather than to a blending tank.

Storing solids can also be a problem. In storage, solids become septic, which makes them harder to dewater. Excessive solids mixing in storage tanks also should be avoided because it tends to reduce the particle size of feed solids, increase the required polymer dose, and reduce dewaterability (Oerke, 1981). The less time solids spend in the bottom of clarifiers or holding tanks, the better.

Manufacturers typically prefer low-shear pumps; they recommend progressing-cavity, rotary-lobe, or double-disc pumps. Variable-speed centrifugal pumps are acceptable, especially for larger capacities, but only with a properly designed feed-piping arrangement. Because the hydraulic residence time inside centrifuges is about 1 second, pulsating, intermittent pumps with check valves never should be used.

2.3.2 *Scroll Tip Linings*

Any solids containing grit causes erosion, and centrifuges have to be designed to control erosion. There are four areas where erosion can be a problem:

- The feed zone, where the feed material is accelerated up to bowl speed;
- The edges of the conveyor flights between the feed zone and the solids discharge;
- The openings through which the dewatered solids leave the rotating assembly; and
- The cover/casing liner, where solids traveling at 300 km/hr (200 mph) come to a dead stop.

There are two issues in wear protection: service life and replacement cost. It typically is more economical if maintenance personnel can replace worn parts themselves, rather than sending the unit to a repair shop.

Urethane- and plasma-applied hardsurfacings work well in feed zones, except when solids flow is concentrated in a small area. The edges of the conveyor flights typically have sintered carbide tiles that are easily replaced without balancing. At the solids discharge end, tungsten carbide is preferred, except for nozzles that protrude beyond the rotating assembly. These are particularly prone to breakage, and a less brittle material often gives a longer life. The casing wear liner should be long enough to cover the flexible boots on the solids chute. Urethane and rubber liners give the longest service life, followed by hard faced steel, and a distant third is stainless steel.

2.3.3 *Scroll/Bowl Differential Speed Controls Drives*

The functionality of controls and drives has not changed significantly in 30 years; the actual components offered by vendors are the result of market prices. Currently, VFD main and back drives are the most common. Assuming the starting current is not too costly, starting systems have the lowest lifetime costs. Controls should have one-button starts, the ability to do a hot restart at any time, and a self-powering feature, so in the event of power failure, the drive goes into breaking mode, and maintains power to the controls as the centrifuge comes to a stop. Without this feature, the centrifuge may become plugged and need service attention before restart. All modern controls operate in differential control and torque control. Most are intuitive, the operator enters a differential or torque setpoint, and the centrifuge holds that figure. Others are more complicated. A good control system should hold torque to ±8%, except when facing severe jumps in feed solids.

Process automation was of dubious benefit in the 1990s, but now all centrifuge manufacturers and several independent automation companies offer automation packages for centrifuges. The most sophisticated automation mimics operators, using sensors to measure centrate solids, and using torque as a measure of cake dryness. Operators set the feed rate and torque, and the system uses the centrate measurement to adjust the polymer rate to maintain a centrate setpoint. More advanced systems also measure the feed solids concentration and feed rate. With this information, they then can flow pace the system, maintaining a constant solids and moisture load to an incinerator, thermal dryer, or other downstream process. Adding the cost of both solids disposal and polymer, the SCADA display can show the dewatering cost in real time. The automation system does not replace operators, but it does "watch" the operation and hold it to setpoint. Operators still must troubleshoot the system, maintain the instruments, and periodically optimize the system. At a minimum, plan for automation in the future by including spool pieces in the piping to allow for the insertion of instruments.

2.3.4 *Dynamic Loads*

A centrifuge imbalance—when multiplied by the g force at which the unit is operating—creates large dynamic loads on the bearings, which are transferred to the building structure. At 3 000 g, a 0.454-kg (1-lb) imbalance results in a 1 263-kg (3 000-lb) eccentric load. Lowering the g force (reducing the bowl speed) lowers the dynamic loading. [Dynamic loadings are in the X–Y direction, not in the Z (axial) direction.] The vibration frequency is the rotational speed of the centrifuge. When the centrifuge is coasting to a stop, it goes through its natural harmonic frequencies, which can result in the centrifuge pounding on the floor through its isolators.

2.3.5 *Vibration/Noise Control*

All centrifuges vibrate and make noise. The manufacturer typically certifies the noise and vibration level of the centrifuge when running dry at the factory. The certification is not useful unless the speed at which the measurement takes place is also specified. It is common to find that many new centrifuges cannot meet the noise and vibration guarantee at their nameplate-rated speeds. For centrifuges with the same S/DW ratio running at the same g force, the amount of vibration and noise are direct indications of overall design quality. It is reasonable to expect the manufacturer to state what peak noise and vibration levels they will have during process-performance tests, at full rated speed, with some penalty for failure. The operating vibration typically is about 1.5 times the factory level. Sound deadening in the room or covers over the centrifuges is expensive, and the covers are especially awkward. The best solution is to close up the centrifuge room and place the controls on the floor below. This makes it easier to sample the cake and centrate, and greatly reduces the noise level. Purchasing larger, higher-speed centrifuges than necessary for the process load and operating them at slower speeds is also an effective way to reduce sound and vibration levels.

2.3.6 *Cake Discharge*

All modern centrifuges spill feed solids out of the cake discharge during startup, and occasionally may do so during operation, as well. Common solutions to the problem are a sliding diverter gate, which deflects most of the slop to drain, or an inclined screw conveyor that reverses to drain when there is slop coming out. (Otherwise, the screw conveyor would send the slop and poorly dewatered solids to the solids handling system.)

2.3.7 *Chemical Conditioning Requirements*

All centrifuges require polymers to dewater wastewater solids. Polymer is one of the four major costs, and is part of the economic balance. Dryer cake requires more polymer than wetter cake. Polymers come in three forms: dry, emulsions, and solutions. Smaller installations may not be able to justify a polymer system that can use more than one form of polymer. Larger installations need to be able to handle both dry polymer and emulsions to foster more competition and minimize dewatering costs.

Inorganic chemicals used for purposes other than dewatering may still affect dewatering. For example, ferric chloride reduces polymer demand, while alum and lime increase polymer demand. Ferric chloride is both high in chlorides and acidic. When diluted by solids, neither is a problem, but an interlock should be added to ensure that ferric is added only when the feed solids are entering the centrifuge. Peroxides and permanganate are used in such small quantities that they have no discernable effect on dewatering. Most inorganic chemicals cannot justify their cost based on a corresponding reduction in polymer demand, but may have other benefits.

2.3.8 *Energy Requirements*

Energy consumption is precisely known and typically guaranteed by the manufacturer. Power is proportional to the square of speed, so bowl speed is a critical factor. Power also is proportional to the radius of discharge and the feed rate. The efficiency of the drives and gear reducers is only a minor factor. Design engineers should focus on the energy consumption for the centrifuge system at the rated speed and loading, because this determines the power usage.

2.3.9 *Washwater Requirements*

Large volumes of washwater are not useful. If the normal flow of feed solids [e.g., 23 m^3/s (100 gpm) of feed material] does not clear a blockage, then a like amount of flushing water will not do it either. Roughly 20% of the rated feed rate is plenty. The feed tube is an open pipe, so the flush water pressure is nominal. For maintenance purposes, and to prevent erosion on the outside of the bowl, most manufactures offer a spray header and nozzles as an option. The surface velocity of the centrifuge is more than 320 km/h (200 mph), so the difference in velocity between the bowl and low-pressure water is more than enough to keep the area clean. In rare cases where the plant water has high chloride levels, letting the flush water evaporate can concentrate the chlorides to the point where corrosion is a problem. In those cases, a short final flush with potable water is advised before shutting down.

3.0 BELT PRESSES

3.1 Introduction

Belt filter presses continuously dewater solids using two or three moving belts and a series of rollers. The filter belt separates water from solids via gravity drainage and compression. Belt filter press machines evolved from paper-making applications to dewatering municipal wastewater solids. The belt filter press was introduced to North America in the 1970s as a lower-energy alternative to centrifuges and vacuum-filter equipment. Belt presses are used throughout the United States and are available from more than a dozen manufacturers.

Compared to other mechanical dewatering devices, belt presses still have the lowest energy consumption per volume of solids dewatered. However, energy consumption is not the only factor to consider when selecting, sizing, and designing a dewatering system.

3.2 Process Design Conditions and Criteria

The main design elements for belt filter presses include cake solids and solids capture, hydraulic and solids throughput capacity, solids and polymer feed, belt washing, filtrate and dewatered-cake conveyance, equipment access and layout, and odor control.

The desired use or disposal option for the dewatered solids (and the solids characteristics needed for that option) also must be considered.

Belt press performance data indicate significant variations in the dewaterability of different types of solids or biosolids. Although the press typically can produce a dewatered cake containing 18 to 25% solids when treating a typical combination of primary and secondary solids, many plants produce a cake containing 15 to 18% solids when dewatering anaerobically digested material. The solids capture rate (total solids recovery, including washwater solids) ranges from 85 to 95%.

Feed cake with higher solids concentrations directly increase the dewatered cake solids that most mechanical dewatering processes can obtain. For example, a 1985 survey of more than 100 municipal belt-press installations showed a strong correlation between cake solids, feed solids, and the percentage of WAS in the feed (Koch et al., 1988). Table 24.1 summarizes the results of this survey; it depicts a linear regression of the cake solids concentration verses the feed solids concentration for different blends of primary and secondary solids. While improvements in belt press technology since 1989 may have improved belt press performance somewhat, the relative trend for the effect of feed solids and solids blend remain unchanged.

In recent years, new press designs have been developed (with more rollers or a separate gravity-drainage or gravity-thickening deck) to produce higher dewatered-cake solids concentrations. While these presses have improved dewatering significantly for some applications, improvements have been only marginal for others.

The best method for evaluating belt filter press performance on a specific material is to dewater the solids using a full-size pilot-scale test unit. Several belt press manufacturers have mobile trailer-mounted pilot units that can be rented for testing. Most pilot units are small production machines that can perform comparably to larger models when the size reduction is belt width, not belt length or solids path. Data to be collected during the pilot test include hydraulic- and solids-loading rates, polymer type and use, percent solids in both feed cake and dewatered cake, and percent of solids capture (total solids recovery).

TABLE 23.1 A comparison of the solids concentrations in feed solids and belt filter-pressed cake (%).

Percent primary sludge (%)	Percent secondary sludge (%)	Percent cake solids (%)
0–10	90–100	X/(0.044 + 0.0426X)
10–40	60–90	X/(0.0297 + 0.0402X)
40–60	40–60	X/(0.059 + 0.0307X)
60–80	20–40	X/(0.062 + 0.0306X)
80–100	0–20	X/(0.071 + 0.0266X)

As an alternative to a full-scale pilot test, many manufacturers have in-house testing equipment that can be used to predict belt press performance. A sample of the material to be dewatered can be sent to the manufacturer, who will provide design criteria based on their tests. These tests have been shown to provide adequate projections of belt filter press performance for smaller applications. Whenever possible, polymer dosages and feed rates should be specifically optimized for the characteristics at the facility. Specific resistance is used to determine the filtration characteristics of solids and to determine the optimum coagulation requirements. A pilot- or bench-scale filter test unit is used to conduct these tests. These tests typically indicate that material with a higher WAS proportion require larger polymer doses.

When evaluating belt filter-press performance, the quantity and quality of the filtrate and belt washwater, and their effects on the wastewater treatment system should be considered. Typically, the recycle flow's BOD varies from 150 to 300 mg/L, and its TSS varies from 600 to 1 100 mg/L.

3.2.1 *Mechanical Features*

Each belt press manufacturer produces machines with slightly different mechanical features and operating characteristics. Presses are available in widths ranging from about 0.5 to 3.5 m. Most municipal presses use 1- to 2-m belt widths. The main components of a belt filter press include feed equipment and piping frame, belts, belt-tracking and -tensioning systems, belt wash system, rollers and bearings, cake-discharge blades, chutes, cake conveyance, drive system, belt-speed control, and chemical conditioning and flocculation (see Figure 24.6).

3.2.2 *Hydraulic Loading Rates*

The throughput capacity of a belt press is the primary design criterion when sizing a belt press system. Throughput capacity typically is considered to be either hydraulically limited or solids limited, depending on feed solids concentration. Belt presses have a maximum liquid or solids-loading capability for a given unit of width that can be attained only when solids are conditioned correctly.

Nominal design hydraulic loading rates for a belt press range from 3 to 4 L/s/m of belt width (15 to 22 gpm/ft of belt width). The maximum hydraulic loading limit it typically 6 to 9 L/s/m of belt width (30 to 45 gpm/ft).

3.2.3 *Solids Loading Rates*

Solids characteristics, origin, and degree of stabilization all significantly affect belt press loading and obtainable dewatering performance. Dilute solids (0.5 to 1.0% total solids) require more gravity drainage, more polymer, and a longer dewatering time than more

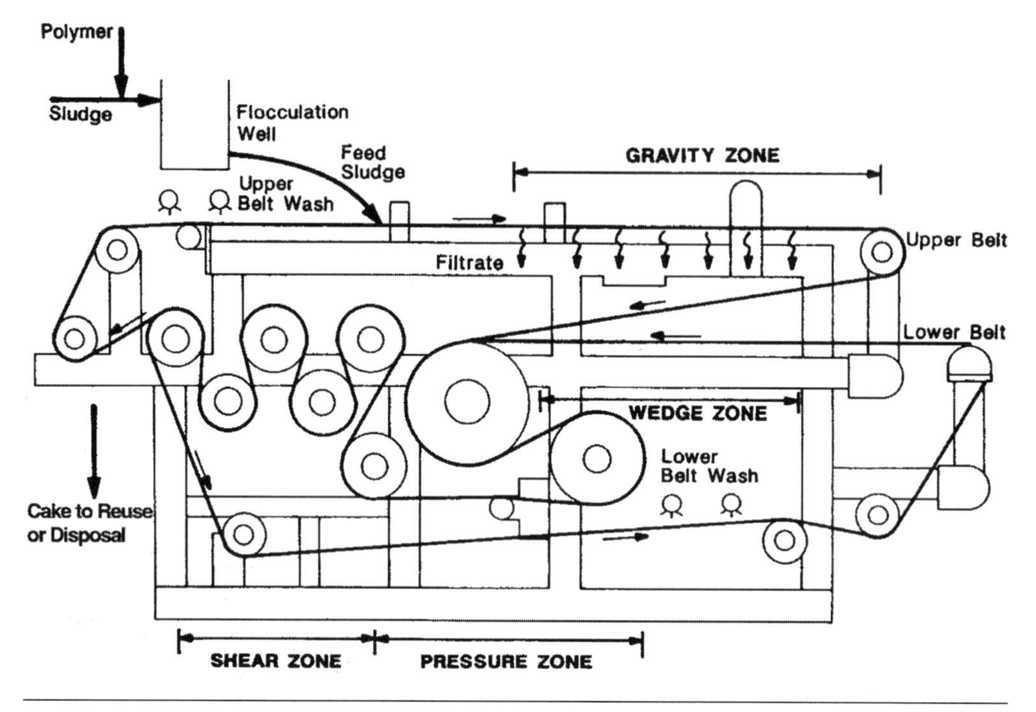

FIGURE 24.6 Schematic of a belt filter press.

concentrated material. Most manufacturers indicate that solids throughput and possibly the percentage of cake dryness increase as the percentage of solids in the feed increases. In addition, the type of process used for stabilization has a direct bearing on the maximum solids content achievable. Although manufacturers' estimates differ regarding the amount of dewatering that can be achieved, they typically agree that anaerobically digested solids are easier to dewater than aerobical digested solids. Typically, digested solids with lower volatile solids contents produce thicker dewatered cakes. In addition, as the primary-to-secondary solids ratio increases, dewatering becomes easier and the cake solids concentration increases.

Other variables that affect the maximum solids loading rate include the degree of stabilization, amount and type of fiber in the solids, shear strength of the solids, type of chemical conditioning, belt type, and maximum pressure applied to the solids. Typical solids loading rates range from 150 to 300 kg/m·h (100 to 200 lb/ft/hr) (dry solids basis). A typical maximum solids loading rate is 450 kg/m·h (300 lb/ft/hr). Recommended loading limits vary per machine manufacturer and should be verified when performing sizing calculations and comparing units.

3.2.4 *Gravity Drainage Zone*

Belt presses dewater solids in three stages: chemical conditioning, gravity drainage to a non-fluid consistency, and compaction in a pressure and shear zone. Dewatering operations begin when polymer-flocculated solids enter the gravity drainage zone. The conditioned solids are evenly applied to the gravity feed belt via a distribution system, which typically is provided. The continuous, porous belt provides a large surface area through which gravity drains free water. Filtrate from the gravity zone is collected and piped to a drainage system.

The gravity drainage zone is a flat or slightly inclined belt that is unique to each manufacturer. The effectiveness of this zone is a function of solids type, chemical conditioning, belt fabric, and detention time. Two-belt machines have a continuous flow path for the solids (i.e., the belt where gravity drainage occurs carries the material directly into the pressure zone). Three-belt machines have a single-belt gravity drainage zone (similar to a gravity belt thickener) mounted above, and discharging to, a two-belt pressure zone. Some systems use a rotary screen for gravity drainage rather than a single-belt gravity zone. This provides a larger gravity zone that would likely benefit more dilute solids. The three-belt machine also allows for different hydraulic loading rates and belt speeds for the gravity-drain and dewatering sections of the press. Equipment manufacturers select the length of the gravity drainage zone based on the inlet solids concentration and the relative drainage rate of the conditioned solids. Gravity-zone lengths typically range from 2 to 4 m (6.6 to 13 ft).

3.2.5 *Pressure Zones*

After gravity thickening, solids move into the pressure zones. Typically, a low-pressure zone is followed by a high-pressure zone. In the high-pressure zone, forces are exerted on the solids by the movement of the upper and lower belt relative to each other as they move over and under rollers with decreasing diameters. Some machines have an extended roller section that provides higher pressure, and sustains that pressure for a longer time.

The low-pressure zone is the area where the two belts first come together with the gravity drained solids between them. This can either be a "wedge zone" where the solids are sandwiched between the two belts or a large-diameter drum screen. The low-pressure zone provides enough dewatering to form a cake that can withstand the additional pressure and shear in the high-pressure zone without extruding out the edges of the belts. A common mistake is applying too much pressure before enough water has been removed from the flocs.

Pressures continue to increase as the solids pass through the wedge zone and enter the high-pressure (drum-pressure) stage of the belt filter press. The belt tension squeezes the cake sandwich as the belts proceed around several drums or rollers (of varying diameters) to maximize shearing action. As solids moves through the press, increasingly smaller-diameter rollers progressively increase the pressure. Average pressures applied typically are 35 to 105 kPa (5 to 15 psi), although they can range up to 210 kPa (30 psi) depending on the size and arrangement of the rollers. Arbitrarily increasing belt tension to increase cake solids may reduce belt life and solids capture, and embed more solids in the belt.

3.2.6 *Frame*

The structural frame of the belt filter press is the skeleton of the unit; it typically is constructed of steel. All belt filter press components are supported and attached to this frame. Belt presses typically operate in a wet, corrosive environment. The selection and specification of the frame material and coatings is the key to the installation's long-term durability. Frames can be made of coated carbon steel or stainless steel; the most common coating for carbon steel frames is hot-dipped galvanizing. Depending on the site, another epoxy or baked-on enamel coating could be considered. Stainless steel frames provide corrosion resistance and require no coating maintenance. The framing structural steel can be channels, I-beams, or tubing. (However, tubing can be difficult to protect from internal corrosion.)

The frame's structural integrity is important to ensure that the rollers are supported and function properly. The frame should be designed (specified) to accommodate operating and static loads with a factor of safety not less than 5, so the machine can operate without deflection, deformation, or vibration. Seismic design of conduit and piping connections, and anchorage of the frame are important considerations.

Access to the belt press building, room, or area needs to be considered when specifying the frame. The specification should include whether the machine can be installed in one piece. If the installation needs to be in pieces because of limited access, the size and/or weight of the largest piece needs to be defined. Dismantling the frame and rebuilding it in the field can affect the frame's critical protective coating. Lifting lugs should be specified to facilitate placement or removal of the units. An overhead crane, hoist, or portable lifting device that is sized to handle the largest equipment component should be included in the building design.

The structural design of the frame should include platforms or walkways so an operator can observe the gravity portion of the belt press and perform routine maintenance. Structural members of the walkways must be clear of the rollers and bearings.

Also, the layout of the belt press needs to provide enough clear space between units to remove individual rollers.

3.2.7 *Rollers*

Rollers support the porous cloth belts and provide tension, shear, and compression throughout the pressure stages of the belt press. Rollers can be made of a variety of materials, including stainless steel. Corrosion and structural considerations are important. The most common coating systems include rubber for the drive rollers and thermoplastic nylon for the others. Roller deflection at the rated belt tension of at least 8.75 kN/m (50 lb/in.) should be limited to 1 mm (0.05 in.) at roller midspan. Belt tension should be based on at least 5.4 kPa/cm of belt width (200 lb/in. of belt width), and drive tension should be calculated based on a belt speed of at least 4.6 m/min (15 ft/min).

Some manufacturers use perforated stainless steel rollers in the initial pressure stages to enhance drainage.

3.2.8 *Belts*

Most belt presses have two operating belts, but there are three-belt units available that provide a separate gravity thickening section before the high-pressure dewatering section. Belts are made of woven synthetic fibers, typically monofilament polyester. Nylon belts are available but typically are used for specific applications (e.g., high-pH solids or abrasive slurries). Both seamed and seamless belts are available. Seamed belts have either stainless steel clipper-type seams or zipper-type seams; they tend to wear quickly at the seam because of a high degree of discontinuity and stress concentration at that point. The raised metal seam also causes wear on the rollers and the doctor blade (i.e., a belt scraper). Zipper-type seams have a lower profile and provide less discontinuity than clipper seams, and have a longer life. Seamless belts are continuously woven, endless belts that have a longer service life than any other belt type. However, seamless belts are more costly and difficult to change out. Several manufacturers market belt presses that accommodate seamless belts. Available in various materials and weave combinations, belts should be evaluated relative to the expected solids characteristics, solids capture required, and durability.

3.2.9 *Bearings*

Bearings are an important part of the belt press. Many manufacturers mount the bearings directly on the structural mainframe so they are accessible for maintenance and service on the exterior of the units. These bearings typically are pillow-block construction and should be rated for at least an L-10 life of 300 000 hours based on forces and loads (e.g., belt tension, roller mass, and drive torque loads). Bearings should be double- or triple-sealed to prevent contamination and wear resulting from press washdown and solids penetration. Bearings should be self-aligning. A split-housing type of bearing is necessary if ready access is unavailable outside of the mainframe. A centralized lubrication system is an option offered by some manufacturers.

3.2.10 Safety

Personnel safety must be fully considered and incorporated into the design. The design must provide for and facilitate maintenance, provide safety stops and trip wires around the belt press and any cake conveyors, convenient and safe equipment access, drainage and spill containment, non-slip walkways and floors, sufficient lighting, noise reduction, ventilation, and odor control.

System interlocks should be provided to stop the solids and polymer feed pumps when the press is shut down.

3.2.11 Press Enclosures

Because of the open nature of a belt press, there is a significant potential for odors and sprays. Workers in the belt press areas can be exposed to aerosols from the belt-wash spray nozzles, as well as pathogens and hazardous gases (e.g., hydrogen sulfide). One alternative for containing odors is installing a ventilation hood above the belt press, as well as enclosures that surround the machine. Ventilation hoods reduce the amount of foul air to be treated, compared with presses in an open room. However, they can restrict lifting-equipment access.

Some manufacturers offer enclosed belt presses. While more expensive, enclosing the units better contains odors, reduces odor-handling volumes, and better contains sprays and spills. However, the enclosed system is more susceptible to moisture and chemical corrosivity, which must be considered when establishing ventilation rates and selecting materials and coatings. Enclosures also limit visual and physical access to the machine and to the solids being processed.

In large installations, another option is to house the belt presses in a separate room. This helps reduce ventilation requirements and improve the overall building environment.

3.2.12 Capture Efficiency

The solids capture rate (total solids recovery, including washwater solids) ranges from 85 to 95%. Capture efficiency is affected by the hydraulic loading rate, the solids loading rate, the nature of the solids being dewatered, the mesh size of the belt, and chemical conditioning. It is important test polymers to determine the best type of polymer, as well as the relationship between polymer dosage, solids capture efficiency, and cake solids. Capture efficiency will decline if the solids are not properly conditioned.

3.2.13 Area/Building Requirements

Because of the continuous backwash and the potential for occasional solids spillage from the belts, belt presses typically are enclosed with containment walls or grating to capture the water running off the press. This also permits the units to be hosed off when they are taken out of service. This typically involves installing the belt press several feet above the floor. Access to the belt press for lubrication of bearings, and inspection of bearings, belts, and rollers typically is provided by placing elevated metal platforms and walkways around the presses. These walkways need to allow room for removing belts and rollers and to not interfere with the moving parts of the press.

Sufficient space should be provided between adjacent units to allow for removal of the rollers. Overhead cranes typically are provided to facilitate roller removal, although portable cranes and hoists also can be used.

Typically, a belt press dewatering room will be at least three or four times larger than the footprint of the presses to accommodate all of these the requirements.

3.3 Ancillary Equipment and Controls

3.3.1 *Controls and Drives*

A control panel (typically custom-designed for the site) is necessary to control the belt presses and their ancillary systems. The panel should provide for automatic, semiautomatic, and manual starting or stopping of the system's components. Sequencing relays or programmable controllers can be provided, as well as electrical and safety interlocks. Critical alarms should be annunciated at the panel and at a central location, and a systemwide emergency power shutdown should be provided. The controls should be in a dry area within sight of the belt press but away or protected from the potentially corrosive atmosphere and spray from equipment washdown. Control panels should meet National Electrical Manufacturers' Association (NEMA) 4X standards to protect components from the moist, corrosive environment.

The controls for each part of the dewatering system should be interconnected to ensure that system operations are coordinated. Solids feed, polymer feed, and belt press and conveyor startup and shutdown must be properly sequenced for either automatic or manual operation. Polymer feed should keep pace with the solids feed rate. The dewatering equipment should automatically shut down for a belt-drive failure, conditioning-system failure, belt misalignment, insufficient belt tension, loss of pneumatic or hydraulic system pressure, low belt-washwater pressure, emergency stop (trip wire), and stoppage of the cake-conveyance system.

3.3.2 *Feed System*

A belt press feed system typically grinds, pumps, pipes, conditions, and flocculates solids before distributing them onto the press. Adjustable-flow-rate pumps (typically progressing-cavity pumps or gear pumps) are used to feed solids to the belt filter press.

3.3.2.1 Feed Pumps

Feed pumps run continuously while the belt press is operating. To match solids production rates and to adjust or optimize press performance, the pumps should have variable-speed drives. Because of the residuals' high solids concentration, potential variability in feed solids characteristics, and the desire to pump at a known or selected rate, positive-displacement pumps are recommended. Centrifugal pumps are inadvisable because of their potential to damage floc formation and the difficulty maintaining a constant feed rate when using a variable orifice mixer. As a good practice, one pump per press should be provided for uniform loading to each press. For multiple-press installations, interconnecting piping and valving are needed for redundancy and reliability. Feed controls typically are incorporated into the main belt press panel.

3.3.2.2 Feed Piping

As with other solids-handling systems, smooth-lined pipe (e.g., glass-lined ductile iron or steel) can be used for the dewatering system's piping. Pressures, velocities, and plugging all require consideration. Velocities should be maintained at 1 m/s (3 ft/sec) or higher to prevent solids deposition and clogging problems. Cleanouts and flushing connections are needed at bends and tees.

Piping systems should include multiple locations for polymer injection so operators can vary the detention time between polymer addition and dewatering, as needed, for best results. Ideally, polymer-injection locations should be spaced at 15-second intervals along the piping system.

3.3.2.3 Conditioning System

The upstream feed-piping system should include several taps or spool pieces (e.g., injectors and/or mixing equipment). The contact time between conditoner and solids affects dewatering performance. If the feed piping is too short to provide adequate mixing and flocculation, design engineers should consider adding a flocculation tank ahead of the belt press.

Variable-output positive-displacement pumps are recommended for chemical metering. Pump output can be either manually or automatically adjusted via speed controllers or stroke-length positioners. For automated systems, the chemical-pump control would be integrated with the belt press control panel.

While polymer is the most common conditioner used with belt filter presses, other chemicals (e.g., ferric chloride) have been used. Lime also has been added for stabilization before dewatering, which affects press performance. These alternate designs often require nonstandard press components (e.g., special material for the belts) and should be reviewed with press manufacturers.

3.3.3 *Belt Speed*

Compressive and shear forces are exerted as the solids passes between the belts and wind through the belt press. Belt speed directly relates to SRT in various sections of the press, the dryness of cake solids, and throughput. Belt speed should adjustable at the belt press control panel.

3.3.4 *Belt Tracking*

The belt-tracking system maintains proper belt alignment by keeping the belts centered on the rollers. It uses sensing arms connected to a limit switch to sense movement in the belt position. A continuously adjustable roller senses the shift and automatically adjusts the belt to compensate. This roller is connected to a pneumatic, hydraulic, or electrically operated response system. An automatic, continuous modulating control must be an integral part of the system.

3.3.5 *Tensioning*

Belt-tension adjustments can be one of the operators' process-control variables. During operations, belt tension is maintained and controlled either pneumatically, mechanically, or hydraulically. Increasing belt tension will increase dewatering pressure. Several manufacturers offer separate control systems for the upper and lower belts so the tension of each can be adjusted independently. An automatic adjustment system, similar to the one for the tracking system, is necessary. A pressure gage (or similar device) is recommended to indicate belt tension. The belt-tensioning system should be able to accommodate at least a 3% increase in belt length. The system should adjust to maintain the desired belt tension as the belt stretches under normal use and wear. (Note that belt life decreases as belt tension increases.)

Systems with exposed gearing are a safety hazard; those that do not act continuously will jar the belts each time they start up. Retrofitting continuous tensioning and tracking systems deserves consideration, although it typically is difficult and costly. For new facilities, specifications should require continuous-acting systems with ready access for easy maintenance and properly covered gearing to minimize potential safety hazards.

3.3.6 *Belt Cleaning System*

The belt cleaning system includes the discharge blade and the belt wash system.

3.3.6.1 Discharge (Doctor) Blade

Often called a *doctor blade*, the discharge (scraper) blade is typically a knife edge constructed of ultrahigh-molecular-weight plastic. It typically is located at the outlet end of the high-pressure section to scrape or peel dewatered solids from the belt into the cake disposal or conveyance system. Worn or poorly adjusted blades reduce belt life and deteriorates the belt seam. A blade-tension system can adjust the pressure exerted by this blade against the belt, as well as the angle at which the blade touches the belt. The blade-tension system's components should be made of corrosion-resistant material (e.g., polycarbonate) and inspected frequently.

Doctor blades are considered a wear item and should be removable for easy replacement.

3.3.6.2 Belt-Wash System

After cake is discharged, the part of the belt that was in contact with solids should be washed before returning to the pressing zones. This belt washing system consists of piping, nozzles, drip pans, and spray-containment shields. A belt-wash station typically is provided for each belt. The belt-wash pipe and nozzle, housed in either a stainless steel or fiberglass enclosure, provides a high-pressure water spray to cleanse the belt of any dried or residual solids, grease, polymer, or other material that blinds the mesh. Self-cleaning nozzles are suggested; however, most manufacturers provide a manual cleaning feature that includes a handwheel-operated brush internally mounted in the nozzle header pipe. Spray piping and nozzles shall be adequately braced and pressure-rated to withstand the pressure transients caused by sudden valve closures.

Also, the gravity dewatering section, the pressure dewatering section, and each belt-wash area need a drainage system to collection and transport filtrate and washwater. Drainage pans, shield, and piping should be designed to confine spray and splashed liquids and should discharge to a sump or floor-drainage system directly below the unit. Drainage connections should be self-venting to prevent overflow. Drainage capacity must be sufficient to allow for washdown of the unit. If possible, the drainage piping should be hard-piped to the floor drainage system to minimize turbulence, and thereby reduce odors.

When the drainage system is sized, both filtrate and washwater flows must be included. One 2-m belt press, for example, can discharge between 450 and 950 L/min (120 and 250 gpm) of drainage flow (filtrate and washwater). This recycle flow typically is routed back to the headworks or primary clarifiers. The concentration of TSS in combined filtrate and washwater typically ranges from 400 to 800 mg/L. Provisions for sampling washwater should be provided in the design.

3.3.7 Washwater Requirements

A reasonably clean washwater supply is needed to ensure adequate belt cleaning, especially when dewatering secondary WAS and scum, which tends to rapidly clog the belt.

This water supply, which amounts to 50 to 100% of the solids flowrate to the machine, typically is pressurized up to 700 kPa (100 psi). Sometimes, a booster pump is needed. Belt washwater can be potable water, secondary effluent, or even recycled filtrate water, although a clean supply is preferable.

3.3.8 *Energy Requirements*

A belt filter press' energy requirements are relatively low compared to some other types of dewatering equipment. Typically, a 2.5-m-wide belt press will require around 7 kW (10 hp) per machine. The power required for the polymer conditioning system, conveyors, and any hoist to remove the rollers is installation-specific and must be added to the total energy requirements. Building ventilation and odor-control energy requirements also need to be taken into account when determining the total energy needed for the belt press installation.

4.0 RECESSED-PLATE FILTER PRESSES

4.1 Introduction

Although pressure filter presses have successfully dewatered solids in several wastewater treatment facilities since the mid-1800s, it was not until 1970 that filter presses received widespread consideration and use in the United States. Since 1970, the filter press has evolved from a labor-intensive batch process to a partially automated operation. The reduction in overall labor requirements has made the filter press a more practical solids dewatering option.

The main advantage of a pressure filter press system is that it typically produces cakes that are drier than those produced by other dewatering equipment. If the cake solids content must be more than 35%, filter presses can be a cost-effective dewatering option. Filter presses also can adapt to a wide range of solids characteristics, are acceptably reliable, and have energy requirements to comparable vacuum-filter dewatering systems, and produce a high-quality filtrate that lowers recycle stream treatment requirements.

The main disadvantages of filter presses are their high capital cost, relatively high O&M costs, the substantial quantities of treatment chemicals required, and the periodic adherence of cake to the filter medium, which must be manually removed. It also requires significant amounts of energy to pressurize the units. Typical energy requirements are on the order of 0.04 to 0.07 kWh per kilogram of dry solids processed.

During the mid-1800s, filter presses were successfully used in England to dewater wastewater solids with and without chemical pretreatment. A few U.S. cities used pressure filter presses in the early to mid-1900s. Until the 1960s, the essential mechanical features of pressure filter presses remained virtually unchanged.

Pressure filter presses did not receive widespread consideration and use in the United States before 1970 because of the high labor requirements involved and the lack of need for high cake solids. However, improvements in mechanization and automation (e.g., automatic plate shifting, cake discharge, and washing) reduced the overall labor requirement. Moreover, the filter press' range of capacities substantially increased (i.e., fewer presses are required for larger facilities), so they have become more cost-effective.

As of 1998, there were fewer operating pressure filter presses in the United States than other dewatering devices (e.g., centrifuges and belt presses). Most of the pressure filter press installations in the United States are semi-mechanized and use a fixed-volume chamber. The mechanized and automated systems typically are being replaced with more reliable manual systems, although other fully mechanized and automated filter presses continue to be developed and marketed.

In terms of both capital and O&M costs, the pressure filter press system typically remains more expensive than other dewatering alternatives; however, when disposal requirements dictate drier cakes, pressure filter presses often have been proven cost-effective because of the lower use and disposal costs associated with drier cakes. Moreover, many landfills have adopted more stringent criteria for the moisture content of solids cakes; often, dewatered cake is required to contain more than 35% solids before it can be landfilled. Other dewatering devices cannot reliably and routinely meet this requirement. Pressure filter presses also are cost-effective when the dewatered cake must be incinerated. Often, the drier filter press cake (which increases the ratio of volatile matter to water content) enables autogenous combustion in incinerators, thus reducing the need for other fossil fuels (e.g., natural gas or fuel oil).

Pressure filtration uses a positive pressure differential to separate suspended solids from a liquid slurry. Recessed-chamber filter presses are operated as a batch process. Solids pumped to the filter press under pressure ranging from 700 to 2 100 kPa (100 to 300 psi) force the liquid through a filter medium, leaving a concentrated solids cake trapped between the filter cloths that cover the recessed plates. The filtrate drains into internal conduits and collects at the end of the press for discharge. Then the plates separate, and the cake drops via gravity into a conveyor, collection hopper, or truck.

When the operating pressures are 1600 kPa (225 psi) or higher, the unit's pressures typically are expressed in terms of atmospheres (bars). A machine rated at 1 600 kPa (225 psi) would be called a 15-bar unit. Similarly, a pressure filter press rated at 2 100 kPa (300 psi) is called a 20-bar filter press.

Filter press dewatering is both a constant-rate and constant-pressure process because of the boundary conditions set by the type of equipment used and the complex, often unpredictable interrelationships of the process variables. In this process, the beginning of the cycle uses a constant filtration rate up to the maximum pumping head available, and then switches over to constant-pressure filtration until the rate diminishes to a predetermined low level.

The typical filtration cycle is characterized by temporal variations in flowrate, pressure, and solids loading. Design capacities and controls further define these relationships by limiting the maximum pressure under which the system will operate, and the maximum and minimum flowrates based on the design limitations of the feed system selected. Figure 24.7 illustrates the typical relationships between feed rate and pressure during a filter press cycle. The portion of the cycle in which wastewater solids are actually being applied to the filter is called the *form cycle* (P_c). During the form cycle, resistance to filtration remains relatively low and constant, until enough solids collect on the media to fill the pressure chambers. The form cycle is characterized by high, constant feed rates and relatively low pressure.

As solids accumulate and resistance to filtration builds, the flowrate declines and pressure increases. Solids accumulate at a relatively high but steadily declining rate until the cake experiences a significant change in porosity, which severely restricts the

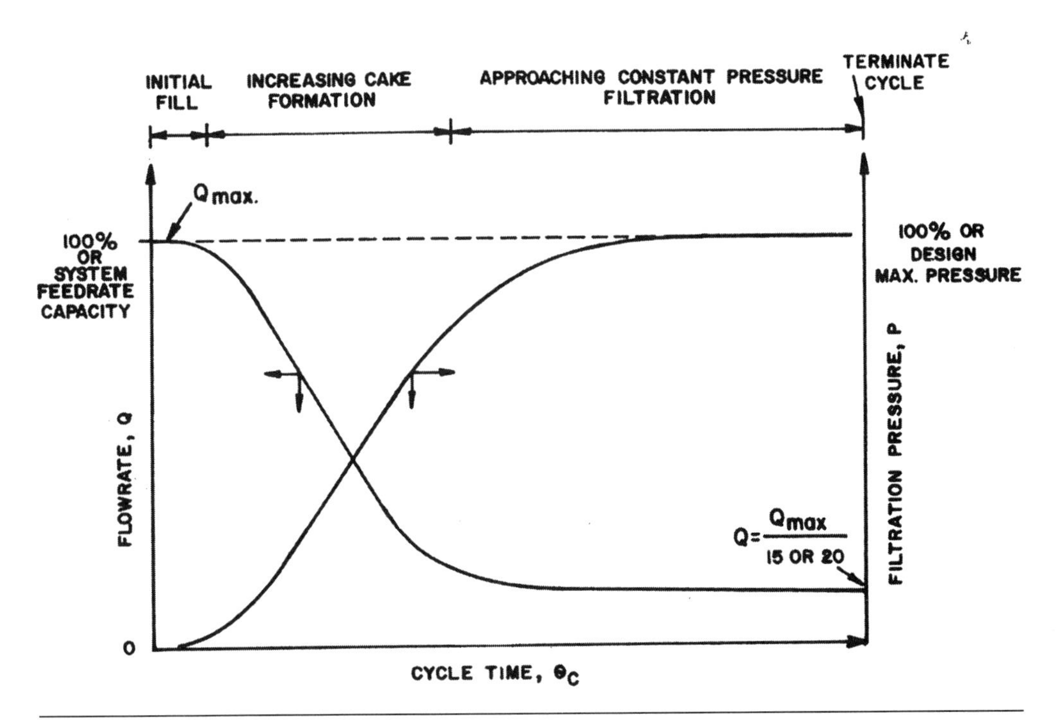

FIGURE 24.7 Filter-press cycle relationships.

amount of flow discharged as filtrate. Thereafter, because of the increased resistance to flow through the cake, pressure will continue to increase (but more slowly) while the flowrate will continue to decrease (also more slowly). The system's pressure will continue to increase until the setpoint pressure is reached; thereafter, system pressure remains relatively constant, while the flowrate continues to decline. Meanwhile, water will still trickle through the cake and solids will continue accumulating in proportion to the flowrate (if the particle concentration remains constant).

Pressure filtration is affected by several factors (e.g., particle size, specific gravity, and particle concentration). The effect of particle size on filtration can best be illustrated using several examples discussed by Thomas (1971). If the particles were the same size, the resulting cake would be loosely packed and relatively unstable (like a stack of marbles), particularly if the cycle incurred a large pressure drop. If the particles were relatively flat (platelike), the resulting cake would resemble a relatively impervious envelope with a highly fluid, moist center. However, wastewater solids consist of a wide variety of particle sizes and shapes that, under certain conditions, help keep an open matrix of particles, thus promoting free filtration. Ideally, the voids between larger particles can be filled with smaller particles, and adequate flow channels will exist between individual particles and throughout the entire physical structure to promote and maintain free filtration. For most wastewater solids, the range of particles must be altered via chemical conditioning before entering the filter press. Creating an open matrix of biological solids is difficult; biological solids are gelatinous, leaving relatively small void spaces for filtrate even when they are well conditioned. If the feed solids consist of a large fraction of biological solids, lime may be needed to ensure that the solids matrix will have adequate flow channels.

If low concentrations of solids particles with a wide range of specific gravities are pumped into the filter, they may settle in the lower chambers of the press, resulting in poor cake formation and unbalanced cake pressures. As solids concentration increases, the viscous drag created among particles inhibits coarser solids from settling out, unless there is a significant difference in specific gravity. This effect becomes less important when the feed consists of fine particles.

Particle concentration in the solids has a significant effect on the filtration cycle time. A feed with a higher solids concentration will increase cake yield and decrease cycle time.

4.2 Process Design Conditions and Criteria

Performance criteria and process design conditions are critical to a pressure filter press. The principal design elements include cycle time, operating pressure, number of plates, feed method, type of feed system, layout and access, type of press, mechanical features, and safety.

4.2.1 Cycle Time

The total cycle time is governed by solids characteristics, desired cake solids concentration, and the relationship between feed rate and pressure. In general, a longer cycle time will result in a drier cake, but after a while, the amount of water being removed from the cake is negligible. Figure 24.7 illustrates the typical relationship between feed rate (which is equal to the water-removal rate) and pressure in a filter press cycle. The filtrate discharge rate and filtrate volume for each filter cycle are valuable control parameters in filter press operations, so a flow-measurement system with a recorder to plot filtrate flow versus time and a totalizer to sum filtrate volume should be included with each filter press. The shape of the filtrate flow curve indicates the feed solids' dewatering characteristics and will allow operators to note any changes in the filtrate curve over multiple filter runs that might suggest chemical dose adjustments or filter media blinding. The filtrate flow curve also can be used to indicate when to terminate the filter cycle. For specific solids at constant conditioning-chemical dosages, a desired cake solids concentration can be calculated as follows:

$$S_T = S_I + \frac{V_T}{V_F}(S_F - S_I) \tag{24.5}$$

Where

S_T = cake solids concentration (% dry solids);
S_I = initial feed-solids concentration (% dry solids);
S_F = reference cake solids concentration (% dry solids);
V_T = cumulative filtrate volume at S_T (L); and
V_F = cumulative filtrate volume at S_F (L).

Using a family of cumulative filtrate curves, operators can terminate a particular cycle when its flowrate matches the proportion of the initial feed rate corresponding to the preselected cake solids concentration.

Alternatively, for a given feed solids concentration, filter press chamber or cake volume, and desired cake solids concentration, operators can use the equation to estimate the required filtrate volume (solids feed volume) and terminate the filter cycle when the filtrate totalizer reaches the calculated value.

Because of the wide range in filtrate flowrates, it is important to have a flow-measuring device that has a reasonable degree of accuracy over the full range—particularly at low flows. Parshall flumes and V-notch weirs have good characteristics for measuring a wide range of flows and are well suited for filter press use. Bubbler tubes, ultrasonic sensors, and capacitance probes have been used to accurately read the levels produced in the Parshall flume or V-notch weir for conversion to the respective flowrate. Depending on the filtrate characteristics and the filtrate piping configuration, foam can be generated and occasionally can cause difficulty in obtaining accurate level readings. Generally, foam is caused when the filtrate from the press goes through severe turbulence. Bubbler tubes are not affected by foam and produce reliable results. Ultrasonic sensors are affected by foam, however, producing inaccurate level readings.

Filtering digested solids typically produces appreciable concentrations of ammonia in the filtrate. To avoid accumulating high concentrations of ammonia in the work space, the filtrate should not be open to atmosphere.

4.2.2 *Operating Pressure*

Pressure is the driving force for filtration in these systems. Filter press systems typically are designed to operate at 700, 1 600, or 2 100 kPa, (100, 225, or 300 psi) without many variations in between. Researchers cite successful applications at all operating pressures; however, increases in filtration pressure can simultaneously increase cake resistance if the solids are compressible. Excessively high pressures can inhibit the process by tightly packing the solids, thereby reducing cake porosity and increasing resistance. Higher cake resistance reduces filtration flowrates. Experience at several installations has demonstrated that close attention to filter media selection and proper solids conditioning will often overcome these problems.

Difficult-to-dewater industrial sludges sometimes respond well to increased dewatering pressure. Increasing pressure on highly compressible solids often can decrease the rate of compression.

4.2.3 *Number of Plates*

The number of plates in a filter press can affect the process' overall efficiency of. When well-conditioned solids are filtered in a press with a large number of plates, solids distribution throughout the filter chambers may be poor. In this situation, chambers nearest the entry points begin filling and start filtering, while chambers toward the center or end of the press have not yet received solids. As a result, unequal pressures develop throughout the length of the press during the filter cycle, producing cakes of randomly poor quality that often do not meet design criteria. The unbalanced forces created by poorly formed cakes can warp and eventually break plates (those made of plastic, not ductile iron).

4.2.4 *Type of Feed System*

The feed system must deliver conditioned, flocculated feed solids to the filter presses under various flow and pressure requirements. There are two ways to feed a pressure filter press, and the system should be capable of both. In the first (more typical) method, the feed system will complete the initial fill cycle by achieving an initial system pressure of 70 to 140 kPa (10 to 20 psi) within 5 to 15 minutes to minimize uneven cake formation. This can be done using separate fast-fill pumps, or running two feed pumps to one press during the initial fill cycle.

As the cake forms, the resistance to filtration increases, requiring higher pressures to feed the press. During this period, the feed system should provide a relatively constant high-solids feed rate at continually increasing pressure until the maximum design pressure is reached. Then, the solids feed rate decreases to maintain a constant system pressure.

The second method, although slower, achieves the same result. A lower flowrate is used to fill the press (typically less than half of the feed pump's capability). When pressure starts building to about half of the operating pressure, the feed pumps are ramped up to full flow and then are controlled by pressure (similar to the first method). This method has been used with coarse cloths to prevent blinding at the high initial flows used in the first method. Whichever method works best for the solids should be used.

4.2.5 *Capture Efficiency*

Recessed-plate filter presses typically can achieve capture efficiencies from 95% to more than 99%, depending on the nature of the solids. These capture efficiencies do not include the water used to wash the plates at the end of a filtration recycle, which also contributes solids to the recycle stream from the dewatering operation.

4.2.6 *Area/Building Requirements*

The design of filter press dewatering facilities demands a careful, methodical approach and attention to detail because of the size and weight of the filter press equipment and the numerous filter press support systems (see Figures 24.8 through 24.10). The size of the filter press room is dictated not only by the size of the filter press itself, but by the clear space around the filter press necessary to facilitate cake release, plate removal, and routine maintenance operations. Generally, at least 1 to 2 m (4 to 6 ft) of clearance is required at the ends of the filter presses; however, a typical clearance of 2 to 2.5 m (6 to 8 ft) between filter presses is desirable. Height clearance must be sufficient for removal of plates via a bridge crane. Some filter press installations with a sidebar design use the bridge crane to remove each plate to make cloth removal and replacement easier; this critical maintenance procedure typically is an annual event.

Design engineers should consider how the filter presses will be installed in and removed from the building, even though larger filter presses have equipment lives of more than 20 years. In addition, provisions for installing future filter presses typically

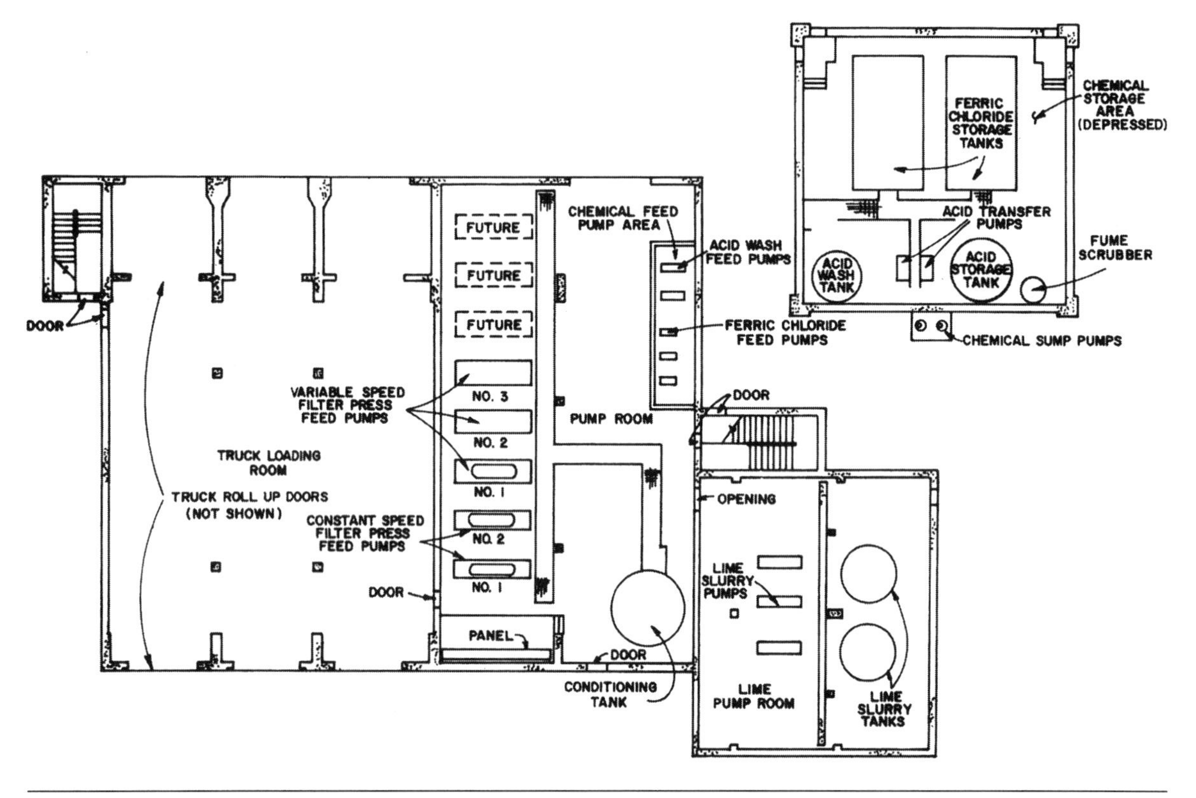

FIGURE 24.8 Floor plan of a filter-press building (ground level).

are included in building designs. The building should have openings large enough to allow major filter press components (e.g., the fixed end, moving end, and plate support bars) to be passed through. It also needs an overhead bridge crane rated to lift the heaviest individual filter press component for maintenance. The bridge crane is also used to lift filter press plates for removal, replacement, or inspections. With larger filter presses, it is impractical to size the bridge crane for use in installing new presses. The headstand alone would require design capacities of 8 000 to 45 000 kg (20 to 50 tons), which would be rarely used.

Operators may need an elevated platform next to one side of the filter press to assist cake release and inspect equipment. The press floor would be an appropriate platform if the press itself is not elevated above this floor. Sufficient floor area near the filter presses should be provided for storage of spare filter plates (not required with ductile iron plates), filter cloths, and other spare parts.

Clearance for truck-loading facilities (where applicable) should be amply sized for a wide range of possible vehicles. A minimum vertical clearance of 4 m (12 ft) is recom-

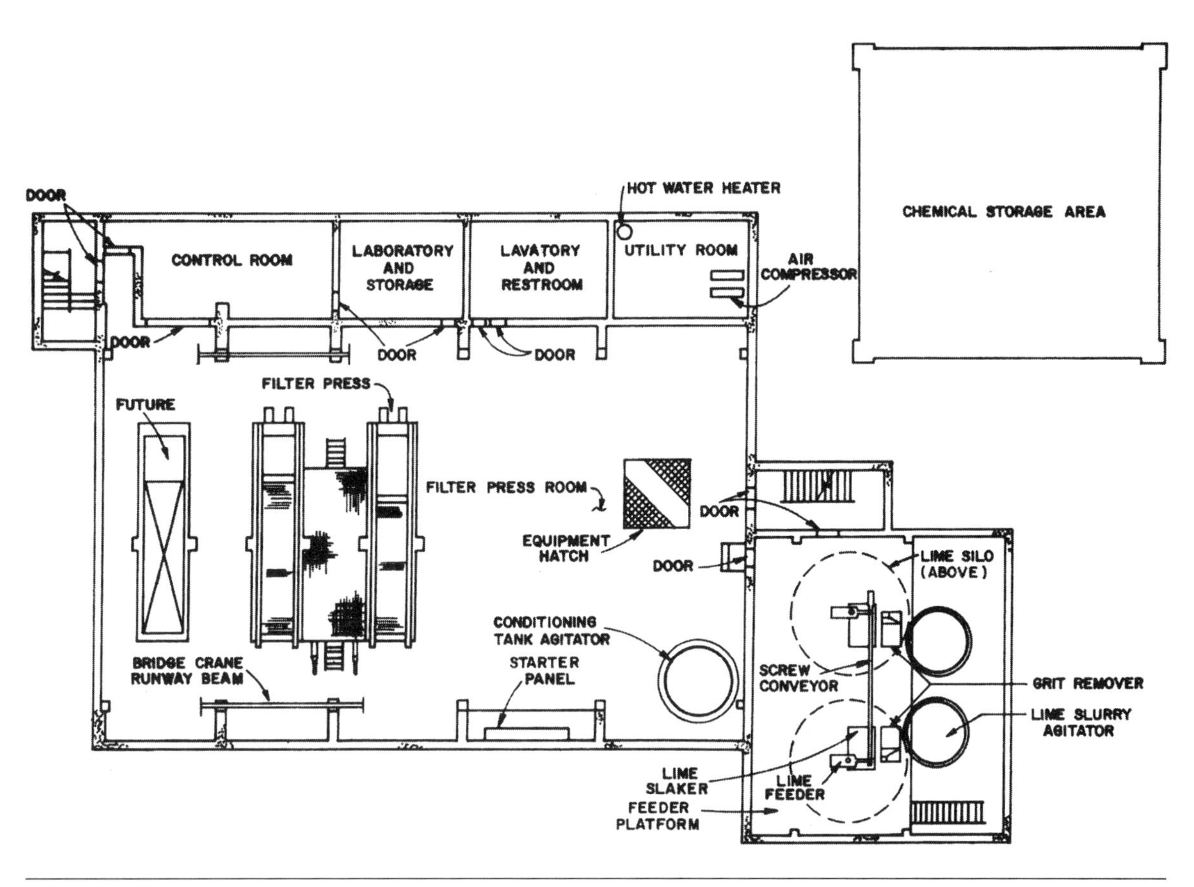

FIGURE 24.9 Floor plan of a filter-press building (second level).

mended. Driveways in truck-loading areas that permit trucks to drive through in either direction are preferable to one-way driveways that require trucks to back in or out.

To a large extent, the heating requirements for filter press buildings depend on site conditions. At a minimum, all building areas should be prevented from freezing, localized heaters should be provided where work activities are concentrated, and control rooms should be designed to meet office-environment conditions. If rubber-coated, steel filter plates are used, the filter-press and plate-storage areas must be kept above 4°C (40°F) to avoid damage to the rubber-covered plate because of thermal contraction.

Design engineers should recognize that cake-conveying systems consistently pose a housekeeping problem for filter press installations. Anywhere cake transfers onto a conveyor is an opportunity for cake material to bounce, roll, splatter, or cling in the immediate area. The return runs of a conveyor continually release cake material that was not removed at the discharge point. Cake breakers disperse cake particles in any

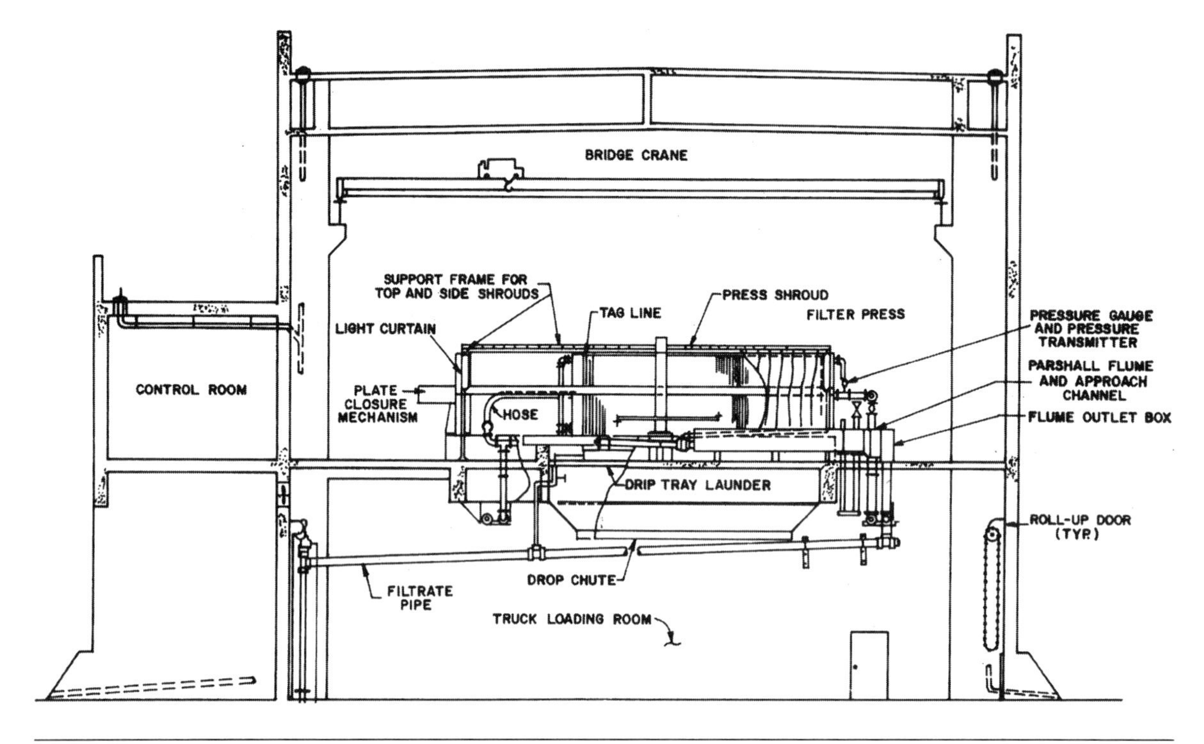

FIGURE 24.10 Sectional view of a filter-press building.

direction through the smallest openings or cracks. Design engineers can take several precautions to minimize the housekeeping problems associated with cake-conveying systems. They can reduce the number of cake-transfer points, as well as the drop distance at the remaining transfer points. They can add flexible discharge chutes at each transfer point to contain the cake. Skirt boards can be installed on belt conveyors to help keep cake on the conveyor.

Under the conveyors, the design can include drip troughs that are wider than the belt to collect any spillage. Such troughs should be U- or V-shaped to facilitate washdown and drainage. These troughs contain and convey all drainage from the filter press to the building drain system and prevent these liquids from discharging onto the conveyors or trucks below. This liquid is released at the end of each filter press cycle from the filtrate ports, from filter-cloth cleaning operations, from leakage between plates during filtration, and during general equipment washdown for housekeeping. Drip troughs are particularly important when the dewatered dake will be incinerated because excess water is detrimental to incinerator operation. Drip trays are the hinged single-leaf or double-leaf type that are sloped to one or both sides for drainage to a launder trough parallel to the length of the filter press. If filter presses discharge filter cake directly to outdoor loading areas, the drip trays also serve as a barrier to the outdoors. They can open up or down, and there are advantages and disadvantages to each type. Drip trays are an essential housekeeping feature for filter press.

Cake-breaking cables or bars typically are provided beneath the filter presses to break up the filter cakes as they drop from the plate chambers, making them easier to manage. The cables or bars typically are spaced 300 to 600 mm (12 to 24 in.) apart and aligned parallel to the length of the filter press.

Blowout curtains are a highly desirable housekeeping feature for filter press operations. Filter press blowouts occur when the plates do not seal properly. During blowouts, a high-pressure stream of solids can be emitted in any direction. Although blowouts are undesirable and every precaution should be taken to prevent them, they inevitably occur. Blowout curtains are positioned over the top and sides of the filter press and can be mounted on a frame supported from the filter press. The top curtain or canopy typically is fixed in place, and the side curtains should be designed to slide to the side to provide ready access to the filter press plates at the end of a filter press cycle. Large, high-pressure filter presses typically have considerably fewer plate blowouts.

4.2.7 *Type of Press*

Two types of filter presses typically are used to dewater wastewater solids. The most common is the fixed-volume, recessed-chamber filter press. The other is the variable-volume, recessed-chamber filter press (also called the *diaphragm filter press*).

4.2.7.1 Fixed-Volume Press

The fixed-volume, recessed-chamber filter press basically consists of a number of plates that are rigidly held in a frame to ensure alignment. These plates are pressed together either hydraulically or electromechanically between a fixed end and a moving end (see Figure 24.11). The plates have a drainage surface, drainage ports to discharge filtrate, and a relatively large centralized port for solids feed (see Figure 24.12). A filter cloth covers the drainage surface of each plate and provides a filter medium. A closing device presses and holds the plates closely together while feed solids are pumped into the press through the inlet port at pressures of 700 to 2 100 kPa (100 to 300 psi). The filter medium captures suspended solids and permits filtrate to drain through the plate drainage channels. A backing cloth (underdrainage), typically made of a rigid cloth or polyvinyl chloride (PVC), sometimes is used to keep the cloth separated from the drainage channels or pipes and the drainage ports during the high-pressure cycle. The backing cloth is the size of the recessed portion of the plate and is held in place by pins.

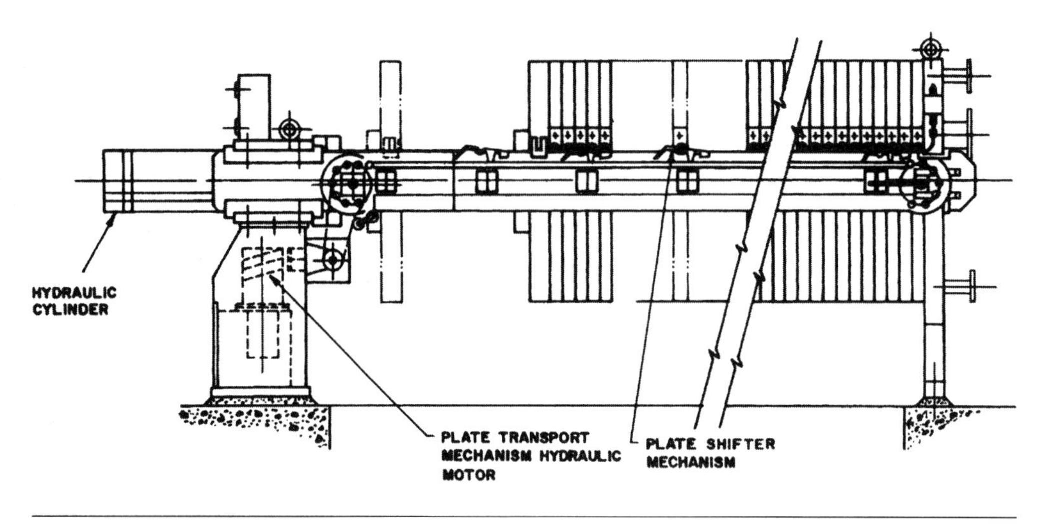

FIGURE 24.11 Filter press elevation drawing.

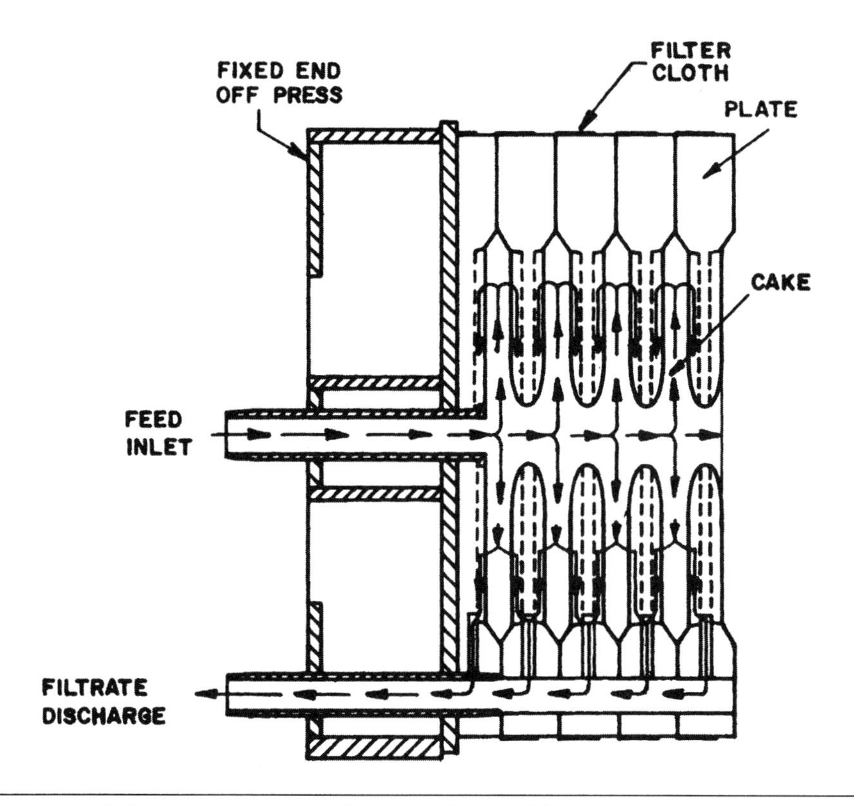

FIGURE 24.12 Schematic of a typical recessed-plate filter press.

Solids collect in the chambers until a practical low feed rate limit is reached (typically 5 to 7% of the initial flowrate) and the filter cycle is terminated. Then, the filter-press feed pump is stopped, the individual plates are shifted, and cakes are discharged.

An example of a typical filter press system is shown in Figure 24.13.

4.2.7.2 Variable-Volume Press

The variable-volume press includes a flexible membrane across the face of the recessed plate. The initial stage of a dewatering cycle is the same in both systems. However, once the plate chamber is filled and the filter-cake formation has started, the membrane is pressurized [between 600 and 1 000 kPa (85 and 150 psi)] with compressed air or water, thereby compressing the filter cake within the plate chamber. Typically, the squeeze pressure is kept relatively low and water is used as the pressurizing media for safety reasons. Diaphragm units typically are operated at a filtration pressure of 700 kPa (100 psi). The physical compression (squeezing) of the filter cake increases the dewatering rate and shortens the cycle time. The results are higher production rates and more flexibility in achieving a desired level of cake dryness.

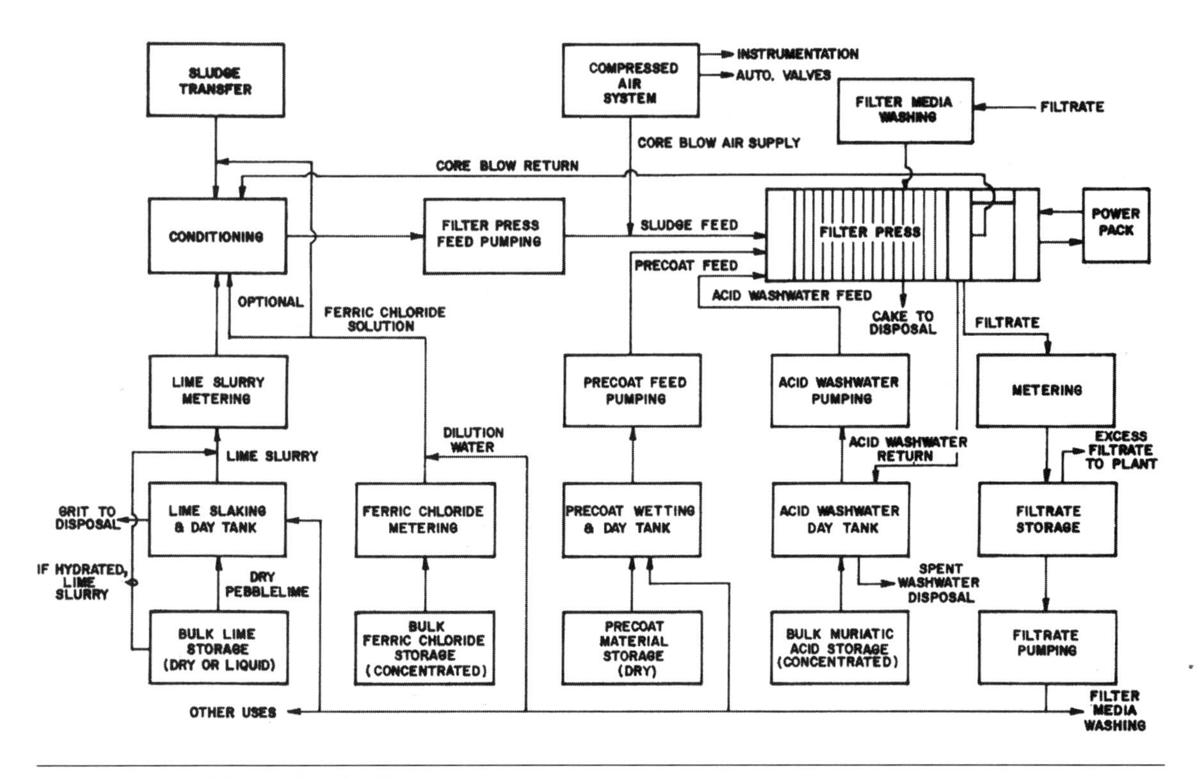

FIGURE 24.13 Schematic of a filter-press system.

The variable-volume press significantly differs from the fixed-volume press in that the volumetric capacity is typically less, cakes are much thinner, and the press is highly automated.

A recent variation of the variable-volume pressure filter press pumps hot water through the plate itself and pulls a vacuum on the chamber where the solids are being dewatered. Thermodynamics show that water evaporates at lower temperatures under a vacuum. This press takes advantage of this principle to obtain a dry cake. The filter press is operated identically to a conventional variable-volume press, including the squeeze. Hot water is pumped through the plates to warm the solids to 60°C (140°F) or higher. During the squeeze, a vacuum is pulled on the solids chamber to 135 kPa (4 in. mercury) absolute. The length of time the solids are kept hot and under a vacuum determines how dry will be the material.

4.2.8 Mechanical Features

Both fixed-volume and variable-volume filter presses can be reliable when proper attention is given to O&M activities. The main operational difficulty encountered in pressure filter installations is inconsistent separation of the cake from the filter media. This problem may indicate the need to wash the filter media or increase conditioner dosages.

The main mechanical components of filter press equipment include the structural frame, filter press plates, diaphragms, filter cloths, and plate shifters. Various options are available for each component. In many cases, individual manufacturers provide only one option for a particular component. Design engineers should carefully evaluate all requirements of a contract that may either specify or exclude a specific option.

4.2.8.1 Structural Frame

The structural frame of the filter press has a fixed end, a moving end, and either plate-support bars on each side (sidebar type), or one or more beams on the top (overhead type). The fixed end anchors one end of the filter press and plate-support bars. The moving end anchors the opposite end of the filter press and plate-support bars, as well as housing the closing mechanism. The plate support (side or top) span the fixed and moving ends, carrying the filter press plates and the shifting mechanism that moves individual plates for cake discharge. Larger filter presses use intermediate supports midway between the fixed end and the moving end to provide more rigidity and strength for the plate-support bars.

The sidebar support bar supports each side of the plate at a point above the plate center. The overhead support bar suspends each plate at the top center of the plate. The sidebar option allows easier removal of individual plates from the structural frame because they can be lifted directly out of the structural frame. The overhead option allows easier access to, and observation of, individual plates mounted on the structural frame. Because the overhead option supports each plate from one point rather than two, it simplifies plate closing and shifting operations.

The type of plates used is critical to the structural frame design. Ductile iron plates weigh several hundred pounds each, and these deadweight forces must be supported. A filter press using polypropylene plates will be much lighter.

The structural load imposed on buildings by filter presses is substantial (see Sections 4.2.8.5 and 4.3.1.3). Some filter presses use the building structure to provide the support to close the plate stack and maintain sufficient pressure to keep the system closed during high-pressure operations. Other systems are designed to withstand all forces internally so only loads in the vertical direction are imparted to the building structure. It is desirable that all horizontal loads be self-contained within the structural frame. Improper alignment of filter presses can warp the structural frame and twist anchor bolts. The manufacturer should verify that the filter press is properly installed and aligned before it is placed in service.

4.2.8.2 Filter Press Plates

The filter chambers can be precoated with porous materials (e.g., fly ash) to serve as a filter body for fine solids particles and to promote release of the finished cake from the filter cloth.

Filter press plates are available in several types of construction, dimensions, and materials. Recessed plates are used almost exclusively in solids dewatering applications. Plate-and-frame construction is used in some industrial applications but is not practical for municipal solids dewatering. Recessed plates are fabricated with a constant recess depth and area formed in both sides of the plate. The filter cake is formed in the volume or chamber formed by the combined recessed area of two adjacent plates. Additional interior supports (called *stay-bosses*) have the same overall thickness as the plate perimeter to prevent deflection of the plate in the recessed area. The number and size of stay-bosses, which have a truncated cone shape, primarily are a function of the dimensions and structural material of the plate. The face of the plates is machined to close tolerances, and the stay-bosses are similarly machined. Uneven surfaces will cause filter-cloth wear, plate-stack shifting during press closure, and frequent blowouts.

Filter press plates can be round, square, or rectangular, and range in size from 0.5 to 2.6 m (1.6 to 8.5 ft). The plates typically are constructed with a top-center or center feed port and filtrate ports at the corners of the recessed areas. The surface of the plates at the perimeter and stay-bosses is flat, to seal when the plates are closed. The surface of the recessed area of ductile iron plates is typically constructed with rows of drainage channels. The recessed area of plastic plates typically is constructed with cylindrical pipes. The channels or pipes provide support for the filter cloths, and the gaps between them provide paths for filtrate to drain to the filtrate ports.

Filter press plates can be made of epoxy-coated steel, rubber-covered steel, cast iron, ductile iron, and polypropylene. Epoxy-coated steel plates offer low initial cost with good strength and moderate weight, but are susceptible to corrosion if the coating is not maintained. Rubber-covered steel plates offer moderate initial cost with good strength and moderate mass. The rubber covering, which is molded to the plate, offers excellent chemical and corrosion resistance, provided that its integrity is maintained. Pinholes and delaminations in the rubber covering can pose serious corrosion problems. Cast-iron (not readily available) and ductile iron plates offer superior strength with reasonable chemical and corrosion resistance. However, they have the highest initial cost and weigh considerably more than steel plates. Polypropylene plates are a relatively recent introduction; they offer the lowest initial cost, excellent chemical and corrosion resistance, and are lightweight for easier handling. Although the inherent strength of polypropylene is less than steel or ductile iron, the plates are thicker and have more or larger stay-bosses to compensate. The effect of this compensation is less significant for smaller plates than for larger ones.

Selecting plate material is difficult; each material has had successful and unsuccessful installations. Design engineers should consider two factors during the economic evaluation of plate material: mass and strength. Because the mass and strength of each plate material are interrelated, design engineers must consider the tradeoff between more mass–less strength or less mass–greater strength.

The mass of plates affects not only the ease of handling for inspecting, cleaning, and changing filter cloths, but also the cost and mass of the filter press. It also affects the costs related to the structural requirements for the building housing the filter press. In some instances, these considerations may not be relevant because it may be desirable to design the filter press and building for heavier plates than were initially installed. This design approach allows the flexibility to install heavier plates at a later date.

Plate strength is critical. Filter presses operating between 700 to 2 100 kPa (100 to 300 psi) can impart tremendous forces on plates if feed solids are distributed unequally, causing voids on one side of the plate. Unequal distribution can cause plate deflection and deformation, blowouts, and filter-cloth wear. This effect is magnified as plate size increases. Plates constructed of lower-strength material tend to be thicker and have more stay-boss area, which reduces the volume for filter-cake formation. So, a longer plate stack will be needed to dewater a given volume of filter cake. The longer stack will require a larger structural frame and more building space.

4.2.8.3 Diaphragms

A variable-volume press uses a flexible membrane to apply more pressure to the dewatered cake. Toward the end of the filtration cycle, the membrane is inflated with air or water to apply extra pressure on the cake. The membrane is deflated before the plates are separated. While the membrane can increase cake solids concentration, it adds more complexity to the process. The membranes are typically made of rubber or a synthetic plastic and are subject to wear.

The variable-volume recessed plate is a recent development in filter press plate construction. The diaphragms can either be incorporated with the plate (less expensive initially) or removable (more expensive, but a diaphragm failure will not cause the plate to be replaced). In addition, the diaphragm may be on one side of the plate or both sides. When replacing fixed-volume plates with variable-volume plates, it may be desirable to use a mixed stack of plates with the diaphragms on one side only. This reduces the overall stack length compared to those with diaphragms on both sides of the plates.

4.2.8.4 Filter Cloths

Filter media require routine cleaning via high-pressure water spray, closed-circuit acid wash, or both. Although many operational factors can affect the performance of filter cloths, the initial selection of filter-cloth media influences subsequent filter-cloth performance. Important factors to consider when selecting filter-cloth media are durability, cake release, blinding, and chemical resistance. The durability of filter cloths is affected by the media material and construction. Cake release from filter cloths is influenced by the media weave and cleanliness.

Filter cloths are available in a variety of combinations of material, weave, and air permeability. The most widely used material for solids dewatering is polypropylene with nylon; saran is also used to a certain degree. The filter media typically is fabricated with monofilaments into a plain or twill weave. The media strands may have different monofilament diameters in the warp and waft to achieve particular filter-cloth characteristics. The air permeability of filter cloths is a measure of the openness of the weave as determined by airflow through a unit area of media at a given pressure drop [e.g., 1800 $m^3/m^2{\cdot}h$ at 0.1 kPa (100 cu ft/min/sq ft at 0.5 in. water)]. Although the permeability changes during use because solids impregnate the filter cloth, swelling the material and distorting the weave, it does serve as a useful parameter for the initial selection of filter media. A low-permeability rating }{ss than 900 $m^3/m^2{\cdot}h$ (50 cu ft/min/sq ft)] will yield high solids capture, but has a greater tendency for media blinding, poorer cake release, and cleaning difficulty. Medium permeability ratings [900 to 5500 $m^3/m^2{\cdot}h$ (50 to 300 cu ft/min/sq ft)] yield good solids capture without excessive cloth blinding and provide good cake release. High-permeability ratings [greater than 5500 $m^3/m^2{\cdot}h$

[300 cu ft/min/sq ft)] provide advantages when treating difficult-to-dewater solids, where cloth blinding and cake release are critical.

It should be noted that the air-permeability rating has minimal overall effect on solids capture in filter press applications. After the initial stage of operation during a filter-press cycle, solids buildup begins and the filter cake itself serves as the filtration media. As a result, although a low-permeability cloth will result in relatively higher solids capture in the initial stage of a filter press cycle, once cake formation has started, the efficiency of solids capture is independent of the filter media.

Filter cloths are sometimes reinforced at the stay-bosses and plate perimeter to improve wear resistance. Such reinforcement can consist of a double layer of filter-cloth material, impregnation of the media with a coating, or insertion of a different material. Care must be exercised to ensure that the thickness of the material at the stay-bosses and the perimeter remain the same if the filter cloth is altered. If the thickness of the material at the stay-boss is different from that at the perimeter, the closing mechanism will impart unequal forces on the plates, causing blowouts or plate deformation.

The proper attachment of filter cloths to the plates is critical to filter-press performance. The filter cloths must drape across the face of the plate without creases and must remain in place through multiple filter-press cycles. This is particularly important if the cloth has been reinforced at the stay-bosses and perimeter, and the cloths must be exactly aligned. A sewn tube, which connects two filter cloths, is inserted through the feed port of each plate. The tube should be impregnated with a waterproof coating to prevent prefiltration from occurring in the feed port. Rubber feed tubes are also available. The filter cloths are secured with grommets around the perimeter and fastened with ties to the filter cloth on the opposite face of the plate. It is desirable to use a sewn loop at the top edge of the filter cloth through which a rod is inserted to support the filter cloth. This method provides uniform support across the top of the filter cloth rather than point supports at the grommeted openings, which may promote creases in the cloth.

The choice of filter media may be the most important equipment variable affecting cake quality and release, filtrate quality, and filter yield. Tightly woven cloths will improve initial filtrate quality, but typically extend filter cycles and result in difficulties in cake release during discharge. Open-weave media typically facilitate cake release during cake discharge but extend the initial cake-formation time and reduce initial filtrate quality. Moreover, although multifilament-fabric media typically improve filtrate quality by entrapping solids within the multiple filaments, they are susceptible to blinding and have a tendency to result in poor cake release during discharge. Monofilament-type construction facilitates cake releases during discharge and typically is easier to clean and to maintain. Larger plates (2 m × 2 m) require special consideration of cloth to prevent stretching.

4.2.8.5 Plate Shifters

At the end of the press cycle, a plate-shifting mechanism (plate shifters) moves each filter-press plate one by one to release the filter cakes. The shifting mechanism is housed in the plate-support bars and operates via an endless chain or a reciprocating bar. Pawls attached to the chain or bar automatically engage the plate at the end of the plate stack and slide it along the plate-support bar 0.6 to 1.0 m (2 to 3 ft). As each successive plate is separated from the end of the plate stack, the filter cake in the corresponding chamber is freed and drops from the filter plate. Typically, reciprocating-bar shifters are used on sidebar filter presses, and the endless-chain shifter is used on overhead-beam filter presses. At least one manufacturer, however, uses a reciprocating-bar shifter on an overhead-beam filter press.

4.2.9 Safety

With filter presses, the paramount safety consideration is preventing inadvertent plate shifter or moving-head operation while an operator is physically between the plates assisting cake discharge. The safety device typically used in most filter press installations is an electric light curtain on both sides of the filter press. The light curtain consists of a number of vertically stacked photoelectric (or infrared) cells to guard one side of the filter press. The light curtain is automatically activated when the closing or plate-shifting mechanisms are engaged. If an operator interrupts the light beam (is between photoelectric cells) during filter press opening or closing or during a plate-shifting cycle, controls will temporarily stop the mechanism until the light beam is restored to protect any foreign object—including parts of a worker's body—from being caught between the plates. In addition, a tag line along the operating side of the filter press enables operators to stop the plate shifter manually and then resume operation at their discretion.

Other safety concerns include those typically associated with mechanical and electrical equipment (e.g., pumps, tankage, and high-pressure piping and valving) and those associated with chemical storage and handling (WEF, 1994), protection from overpressurization, and adequate ventilation.

The filter press building must be ventilated for operator comfort, odor reduction, and fume protection. Solids conditioning is the greatest source of odor. In particular, when digested solids are conditioned by lime and ferric chloride, significant amounts of ammonia are released as pH rises in the conditioning tank and filter press. The fumes are most noticeable when the press is opened for cake discharge. Design considerations must include covering and ventilating the conditioning tank and providing a system to increase ventilation around the filter press.

4.3 Ancillary Equipment and Controls

Filter presses typically have a number of auxiliary systems (e.g., feed system, solids conditioning, filtrate management, cake handling, and cleaning and housekeeping) to ensure safe and successful performance. The difficulty of designing and controlling a pressure filter press system is the number of ancillary systems that must be coordinated and operated for successful system performance. In fact, the ancillary equipment sometimes requires more space and effort than the press itself.

4.3.1 Feed System

Filter press feed systems must deliver conditioned, flocculated feed solids to filter presses under varying flow and pressure requirements. Feed system components include precoat, rapid fill, pressurization, and cake removal.

4.3.1.1 Precoat System

Solids with a high biological content or industrial sludges that are difficult to dewater often tend to stick to the filtration media. The precoat system aids cake release from the filtration media and protects filtration media from premature blinding. The precoat material can be fly ash, incinerator ash, diatomaceous earth, cement-kiln dust, buffing dust, coal, or coke fines. A thin layer of this material is deposited over the entire filtration surface before each filtration cycle begins.

There are two types of precoat systems: dry-material feeding and wet-material feeding. The dry-material system is used at larger installations, particularly those that operate continuously (approaching 24 hours per day). In this system, a precoat pump draws clear water from a filtrate-storage tank (or other reasonably clean source), circulates it through the filter press, and returns it to the tank. Once the filter press is full of water (all air evacuated), a predetermined amount of precoat material is transferred from a storage hopper to the closed precoat tank. The recirculating water stream is then diverted through the precoat tank and, aided by a baffle arrangement inside the tank, forces the slurry of precoat material out of the tank and deposits it on the filter medium in the filter. To ensure uniform and even precoating, the clean water must be circulated through the filter at high rates. The entire precoat cycle should last between 3 and 5 minutes before each filtration cycle. Precoat material requirements range from 0.2 to 0.5 kg/m^2 (5 to 10 lb/100 sq ft) of filter area; 0.4 kg/m^2 (7.5 lb/100 sq ft) is typical.

The wet-material feed system typically is used at smaller filter press installations with enough space to store the dry precoat material onsite. This system chiefly consists of a precoat preparation tank into which water is metered and the proper amount of dry precoat material is added. An agitator keeps the precoat material in suspension. The precoat-material pump circulates the material from the bottom outlet and discharges it back to the material preparation tank. The filtration cycle is similar to that of the dry-material feed system. At the beginning of the precoat cycle, water from a filtrate-storage tank is pumped through the filter press and returned to the tank. After the filter is completely filled (all air expelled), the precoat-material pump injects the precoat slurry into the piping on the suction side of the precoat pump, and the precoat material is uniformly and evenly distributed throughout the filter.

4.3.1.2 Rapid Fill

Two rapid-fill methods have been developed for filter press systems. The first method uses one pump or a combination of pumps with variable-speed drives are operated to achieve the required flow and pressure characteristics. The second method uses a combination of pumps and pressure tanks to achieve the required flow and pressure characteristics.

The first method typically uses one variable-speed feed pump for each filter press. Automated controls are used to vary the pump speed (maximum flow until system pressure is reached, and decreasing flow to maintain system pressure). For large filter presses where the initial flow requirements are high and the available turndown of the variable-speed pump is too limited to operate from the minimum flow to the maximum flow, a second pump (either constant- or variable-speed) is incorporated to operate parallel to the first pump to achieve the initial high-flow requirements. The second pump is controlled to drop out when flow requirements drop within the first pump's capacity range.

The second method typically uses one feed pump and one pressure tank for each filter press. At the start of the filter press cycle, the pressure tank is filled with feed solids and pressurized with air. To initiate the cycle, an automatic valve is opened to release the feed solids in the pressure tank into the filter press. This method achieves a rapid initial fill because the working volume of the pressure tank is designed to exceed the solids feed volume required for the initial fill. When the solids level in the pressure tank drops, the solids feed pump starts and runs until the working volume in the pressure tank is replenished. Air controls associated with the pressure tank operate to add or release air to or from the pressure tank to maintain the desired system pressure. The automatic outlet valve from the pressure tank closes at the end of the filter cycle to terminate the solids feed to the filter press.

Both methods have been used successfully at many installations. Design engineers should note that the first method requires less building space (floor area and room height) than the second. However, the second method provides more rapid, positive initial fill. If pressure tanks are used, design engineers should take precautions to properly handle the air released from the pressure tank because of possible odor problems.

4.3.1.3 *Pressurization*

Plate pressurization is either hydraulic or electromechanical. It closes the filter-press plate pack and maintains the necessary force to hold the plates closed during a filter press cycle. The hydraulic system consists of one or more hydraulic rams and a hydraulic power pack. Some manufacturers use an air-over-hydraulic system, but these are not common. The electromechanical system consists of a single or twin screw and an electric gear motor. Either system can be equipped with automatic controls to maintain a constant closing force throughout the filter press cycle. This feature is desirable because the closing force required may vary as the solids feed pressure increases, filter cloths and plates compress, and materials of construction expand or contract as the temperature changes.

Smaller filter presses typically only need one piston, which is mounted against a rigid support at the opposite end of the unit. This is a push-to-close unit, and it imposes a large load on the building structure. One piston should not be used on a large plate stack, because the stack can shift if the plate faces are uneven, which can happen if some solids are left behind.

Larger presses typically used a pull-to-close design, in which two or four rods extend from the front to the tail stand, and hydraulic cylinders "pull" the tail stand toward the front. In this design, only vertical loads are imposed on the supporting structure.

4.3.1.4 *Cake Removal*

Filter press cake is a thixotropic material that can change from relatively firm, discrete pieces to a gelatinous, homogenous mass if the material is allowed to settle and compact over time; this condition typically occurs in any cake-storage facility. Consequently, the characteristics of stored cake rarely are the same as the characteristics of fresh cake. Storage bins that slope to a relatively small opening have been susceptible to bridging across the outlet opening, thus preventing the release of cake. Storage bins with vertical sidewalls and helical screws in the bin bottom were rendered useless at one installation because the distance between the outside edges of the screws was sufficient to allow the cake to mass and bridge over the screws. The screws "tunneled" through the mass and could not remove any cake. Therefore, cake-storage bins should be designed with steep sidewalls (vertical-to-horizontal slope greater than 5:1) and true "live bottoms" operating over the full width and length of the bin [e.g., chain-and-flight mechanisms or gauged helical screws with a minimum clearance of ±25 mm (1 in.) between the outside edges of the screw flights]. The live-bottom mechanisms also should be provided with variable-speed drive capability to control loading to the next solids management process. Cake-handling requirements depend on the ultimate use or disposal method. For example, if trucks will haul the cake offsite, the simplest procedure is to drop the cakes directly into the truck.

If the cake will be incinerated, then two approaches are common. The first approach is to provide storage capacity beneath each filter press and meter the cake onto a conveying system that leads to the incinerator. A second approach provides intermediate storage between the filter presses and the incinerator.

The capability for core blow is an optional feature for filter press applications. The filter press "core" is the annular space through the press formed by the feed ports of the plates. At the end of a filter press cycle, the core is filled with liquid residuals that have not been dewatered. When the plates are shifted to drop the filter cakes, the residuals in the core also are discharged. Although the amount of feed solids in the core is minor and has minimal effect on cake solids content, these residuals tend to run down the face of the filter cloths, creating blind spots in the cloths that may result in non-uniform cake formation. A "core blow" uses compressed air to force the residuals out of the feed ports feed solids and back to the solids conditioning tank. Design engineers should consider the duration, flowrate, and air pressure required to provide such core blows. They also should carefully weigh the cost of the equipment, piping, and building space required for this option.

4.3.1.5 Washing System

Filter media periodically are washed with water or acid when the filter press is shut down. As cakes degrade during normal operations, operators will be able to determine when washing is required. The washing cycle depends on solids characteristics, the conditioning system, and filter-cloth weave. Some facilities wash with water after 20 cycles and wash with acid after 100 cycles. If solids only are conditioned with polymer, not lime, then acid washing is not required.

Filter media washing is an essential feature of good filter press operations. Washing removes the following:

- Residual cake left over after normal cake discharge;
- Liquid residuals from the feed-port core that were not dewatered and dribbled down the face of the filter media (if the core blow is not used);
- Solids and grease impregnated in the filter media; and
- Scale and solids buildup behind the filter media on the filter-plate drainage surface.

These materials must be removed to avoid blinding the filter media and to maintain atmospheric pressure between the filter medium and filtrate discharge. Backpressure reduces the effect of the applied pressure on the filtration rate.

Design engineers typically provide facilities for both water spray wash and acid wash. The water spray wash often is used to wash the surface of the filter media to remove accumulated solids. The acid wash is used periodically to remove impregnated solids and scale buildup, which accumulate more slowly and are not readily removed by the water spray wash.

The least expensive and most typically used water spray-wash method is a portable spray-wash unit, which consists of a hydraulic reservoir, high-pressure wash pump, and a portable lance to direct the spray water. Operators direct the high-pressure spray [up to 13 800 kPa (2 000 psi)] wherever buildup is observed. In addition to being labor-intensive and physically demanding, this method is tedious when cleaning filter presses with several large plates.

Filter press manufacturers have developed an automatic water spray-wash system as an option to their equipment package. It consists of controls, which automatically shift plates, and an overhead spray-wash mechanism that washes the entire filter media surface. High-pressure water-booster pumps typically are supplied to provide a satisfactory surface wash pressure. Although more expensive and complicated than the portable spray-wash system, the automatic system provides more thorough, efficient, and frequent media washing with less labor.

4.3.2 *Acid Cleaning*

The acid-wash method cleans filter media *in situ*. A dilute solution of hydrochloric acid is pumped into an empty filter press with the plate pack in the closed position. The acid is either circulated through the plate chambers or detained in the plate chambers to clean the filter cloths. The acid-wash system typically includes a bulk-acid storage tank, acid-transfer pump, dilution appurtenances, dilute-acid-storage tank, acid-wash pump, and associated valves and piping.

Acid for the procedure typically is furnished in carboy containers, tank trucks, or tank car shipments as 32% hydrochloric acid (muriatic acid) solution. A cleaning solution strength of about 5% is recommended, although specific experience may warrant slightly higher concentrations (up to a maximum of 10%).

4.3.3 *Chemical Conditioning Requirements*

Solids conditioning involves adding lime and ferric chloride, polymer, or polymer combined with either inorganic compound to the solids before filtration. The goal is to produce a low-moisture cake. Most existing filter press installations in the United States use lime and ferric chloride for conditioning. At these installations, the solids conditioning system typically has lime slakers to produce a slurry, lime-transfer pumps, lime slurry equipment, ferric chloride equipment, and a conditioning tank. The lime and ferric chloride are added in the conditioning tank on a batch basis. For installations that use only polymer, the solids conditioning system tends to be simpler because polymer is added in line rather than on a batch basis. However, the polymer must be fed on a cycle that matches the flowrate in the solids feed pumps, so instrumentation and control are essential for a polymer system to work.

Recently, several installations have begun using only polymer for solids conditioning because their experience has shown that a small decrease in performance is offset by lower chemical costs, reduced ammonia odors, and smaller volumes of dewatered cake. One problem with only using polymer is cake release from the cloth during the discharge cycle. Several installations in Europe use ferric chloride to enhance cake release; however, combining ferric chloride with polymer leads to severe corrosion of metals in the piping and press. This does not happened when lime and ferric chloride are used together for conditioning because the lime neutralizes the corrosive ferric chloride. New installations that intend to use polymer and ferric chloride should line all metallic surfaces with rubber.

The solids conditioning system typically consists of solids-transfer pumps, lime slurry equipment, ferric chloride equipment, and a conditioning tank. The solids-transfer pumps convey feed solids to the conditioning tank, where lime and ferric chloride are added. Ideally, the pumps withdraw solids from a holding or storage tank equipped with a mixing system. Such tanks allow operators to maintain an inventory of solids rather than scheduling solids wasting, thereby allowing solids dewatering to be independent of the biological treatment process. The mixing system for the solids holding tank ensures that the solids feed concentration is uniform, prevents radical changes in chemical dosage requirements, and minimizes the possibility of chemical overdosing or underdosing.

The solids conditioning tank should include a mechanical mixer to thoroughly and gently mix solids with lime and ferric chloride to develop a flocculated solids. To ensure adequate mixing and floc formation, the minimum retention time in the conditioning tank should be 5 to 10 minutes. A longer retention time (about 20 to 30 minutes) may be desirable to ensure that the lime completely reacts with the solids, minimizing lime scale in the solids feed piping and in the filter media. However, longer retention times than needed to develop a good solids floc tend to promote floc deterioration and breakdown as the solids floc ages. These considerations are further complicated for installations with multiple filter presses and one conditioning tank. The retention time then is affected by the number of filter presses in operation and the solids feed rates required to each filter press at various stages of the filter cycle.

The solids conditioning system must be designed to match the filter presses' feed requirements. Two methods have been frequently used to achieve this design goal. The first method uses variable-speed solids-transfer pumps, which are designed to maintain a nearly constant level in the conditioning tank by matching the tank's inflow and outflow rates. To maintain a constant chemical dosage, the lime slurry and ferric chloride feed pumps also must be variable-speed to proportionately match the rate of the variable-speed solids-transfer pumps. This method ensures a nearly constant retention time in the conditioning tank but requires a wider range in pump capacity, the addition of variable-speed drives and controls, and more operational complexity.

The second method uses constant-speed solids-transfer pumps and varying levels in the conditioning tank. Solids are intermittently pumped to the conditioning tank at a constant rate, and the lime and ferric feed pumps operate at constant, but manually adjustable, rates parallel to the solids-transfer pumps. By varying retention time in the conditioning tank, this method increases conditioning-tank capacity, but requires less-complex controls. Recognizing the inherent difficulties with chemical systems, this operational mode is most desirable.

The solids-transfer pump capacity and the solids conditioning tank volume must be carefully sized to ensure that the filter press feed requirements are fully met. Correspondingly, the lime slurry and ferric chloride feed pumps also must be carefully sized to meet the ranges of solids feed flow, solids feed concentration, and chemical feed concentration and provide the appropriate chemical doses.

5.0 DRYING BEDS AND LAGOONS

5.1 Introduction

Drying beds and lagoons are the oldest solids dewatering methods. Both use a combination of drainage, evaporation, and time to dewater solids. Both also require considerable area compared to other dewatering methods.

Drying beds have been used for more than 100 years. If well designed and properly operated, they are less sensitive to influent solids concentration and can produce a drier product than most mechanical devices (see Table 24.2). Particularly suited to small facilities in the southwestern United States, they can be used successfully in wastewater treatment plants of all sizes and in widely varying climates.

Compared to mechanical dewatering, drying beds are a less-automated process that requires more land. The high capital and operating costs of mechanical systems have caused designers to take a second look at drying beds when adequate land is available and environmental conditions are acceptable. This trend has been accompanied by

TABLE 23.2 Advantages and disadvantages of using drying beds.

Advantages	Disadvantages
Where elaborate lining and leachate control is not necessary and where land is available, capital cost is low for small plants	Lack of rational design approach for sound economic analysis
Low requirement for operator attention and skill	Large land requirement
Low electric power consumption	Stabilized sludge requirement
Low sensitivity to sludge variability	Impact of climatic effects on design
Low chemical consumption	High visibility to general public
High dry cake solids content	Labor-intensive sludge removal
	Permitting and groundwater contamination concerns
	Fuel and equipment costs for bed cleaning systems
	Real or perceived odor and visual nuisances

growing concerns about groundwater contamination. Regulations prohibit unlined drying beds in many areas. The additional costs of bed lining and groundwater quality monitoring of bed systems may make mechanical dewatering more cost-effective for all but small plants. Drying beds also contribute to higher wet weather flows to the treatment plant because they consist of relatively large areas that drain to the plant. Drying beds may, however, be a useful backup to mechanical dewatering methods.

Drying beds and lagoons may be problematic if concerns over odors are high. The large open areas where solids are drying can cause offsite odors. Odor control only possible when drying beds are enclosed.

5.2 Sand Drying Beds

Sand drying beds are the oldest, most widely used drying bed method. The beds typically consist of a bed of sand underlain with a gravel layer and perforated drain piping. The beds are contained by concrete walls around the perimeter. Sand drying beds today also may be lined to prevent liquid from seeping to the groundwater. Drying beds also may be enclosed to prevent water from re-wetting dewatered cake.

Solids on sand beds primarily are dewatered via drainage and evaporation. Removing water from the solids via drainage is a two-step process. Initially, the water is drained into the sand and removed by the underdrains. This step, which typically lasts a few days, continues until the sand becomes clogged with fine particles or all the free water has drained away. Once a solids supernatant layer has formed, decanting removes surface water. This step is especially important for removing rainwater, which slows the drying process if it is allowed to accumulate on the surface. Decanting also may be useful for removing free water released by chemical treatment. Solids drying in beds also can be enhanced by using auger-mixing vehicles in paved beds. Water remaining after initial drainage and decanting is removed via evaporation.

5.2.1 *Process Design Considerations and Criteria*

The operation of a sand drying bed depends on

- Solids concentration;
- Depths of solids applied;
- Loss of water via the underdrain system;
- Conditioning and digestion (degree and type) provided;
- Evaporation rate (which is affected by many environmental factors);
- Type of removal method used; and
- Solids use or disposal method used.

All of these site-specific considerations determine the optimum solids loading, area requirements, and other design criteria for a given bed.

5.2.1.1 Area Requirements

The per-capita area criteria typically used to size sand drying beds are shown in Table 24.3. These criteria are based largely on empirical studies of primary solids conducted in the early 1900s by Imhoff and Fair (1940), who recommended a range from 0.1 to 0.3 m^2/cap (1.0 to 3.0 sq ft/cap), depending on the type and solids concentration of the residuals applied to the bed. Other important factors (e.g., the applied solids depth and number of yearly applications) also were considered in this pioneering work. In attempting to uniformly apply these criteria, however, many design engineers fail to adequately consider the basic parameters of the original work. Changes in the characteristics and quantity of solids produced per person make current use of these criteria highly questionable.

As a result of today's stringent effluent standards, these criteria may no longer suffice for sizing drying beds. The greater quantity of solids produced as a result of lower effluent suspended solids, chemical reactions, garbage grinders, and advanced treatment processes requires larger drying areas. Area requirements are also greater for the thinner combined solids (typically, 2.5 to 4% rather than 7%) that are prevalent today. British experience indicates that a minimum of 0.35 to 0.50 m^2/cap (3.5 to 5.5 sq ft/cap) is necessary because of these changes in solids characteristics.

5.2.1.2 Solids Loading Criteria

Accepted solids-loading criteria for sand drying beds are based on empirical data. Typical requirements vary from 50 to 125 $kg/m^2{\cdot}a$ (10 to 25 lb/yr/sq ft) for open beds and

TABLE 23.3 A summary of recognized, published sand bed sizing criteria for anaerobically digested, unconditioned solids.

	Uncovered beds		
Initial sludge source	Area, m^2/cap	Solids loading, $kg/m^2 \cdot a$	Covered bed area,* m^2/cap
Primary			
Imhoff and Fair (1940)	0.09	134	
Rolan (1980)	0.09–0.14		0.07–0.09
Walski (1976)			
N45°N latitude	0.12		0.09
Between 40 and 45°N	0.1		
S40°N latitude	0.07		0.05
Primary plus chemicals			
Imhoff and Fair (1940)	0.02	110	
Rolan (1980)	0.18–0.21		0.09–0.12
Walski (1976)			
N45°N latitude	0.23		0.173
Between 40 and 45°N	0.18		0.139
S40°N latitude	0.14		0.104
Primary plus low rate trickling filter			
Quon and Johnson (1966)	0.15	110	
Imhoff and Fair (1940)	0.15	110	
Rolan (1980)	0.12–0.16		0.09–0.12
N45°N latitude	0.173		0.145
Between 40 and 45°N	0.139		0.116
S40°N latitude	0.104		0.086
Primary plus waste activated sludge			
Quon and Johnson (1966)	0.28	73	
Imhoff and Fair (1940)	0.28	73	
Rolan (1980)	0.16–0.23		0.12–0.14
Walski (1976)			
N45°N latitude	0.202		0.156
Between 40 and 45°N	0.162		0.125
S40°N latitude	0.122		0.094
Randall and Koch (1969)	0.32–0.51	35–59	

* Only area loading rates available for covered beds.

60 to 200 $kg/m^2 \cdot a$ (12 to 40 lb/yr/sq ft) for enclosed beds (see Table 24.3). Sizing sand drying beds based on solids-loading criteria is a better approach than sizing them based on per-capita area requirements. Furthermore, because the total quantity of solids produced daily by the overall treatment process can be predicted accurately, the risk of

error is minimal. The best criteria would take into consideration climatic conditions (e.g., temperature, wind velocity, humidity, and precipitation).

Several models have attempted to mathematically describe the complex relationships involved in properly functioning sand drying beds. Although early models were strictly empirical, they have been used extensively to size drying beds. The advantage of using mathematical models is that they take local weather conditions (e.g., the amount of rainfall received) into consideration.

Using a rational engineering design approach rather than empirical studies, Rolan (1980) developed a series of equations to determine not only the design criteria for sand drying beds, but their optimal operation. Rolan found that the optimum application depth for a given percentage of dry solids is a function of the desired dry cake thickness, dry solids content, evaporation rate, and number of applications per year. The cost of removing solids (labor, equipment, and sand replenishment) primarily depends on the number of applications per year rather than the volume of solids.

Walski (1976) has developed a similar mathematical model to account for major solids drying mechanisms. However, Walski indicates that the area required is relatively independent of the depth of the solids applied over the range of bed operation. Because neither Rolan nor Walski account for the removal of chemically bound water, these models are not valid where such water must be removed. Nevertheless, their studies show that many design standards in use do not adequately address the environmental and mechanical factors involved in operating sand drying beds and, therefore, result in inadequate designs.

When applying any models, design engineers must recognize that cleaning a bed in an actual plant can depend on many variables other than cake dryness. For example, operators at a large plant in Albuquerque, New Mexico, adopted a policy that no beds would be poured during the hottest month of the year because experience had shown that odors were greatest then, even though bed productivity was high. Figure 24.14 illustrates the effect of evaporation rate on bed loading at various percentages of dry solids applied.

5.2.1.3 Chemical Conditioning

In some cases, new drying bed designs may need to include chemical conditioning to offset unpredictable weather conditions and variable solids characteristics. Conditioning also can help improve the solids drying capacities of existing beds.

Polymers are the primary chemicals used for conditioning. Evaluating their effectiveness and economy is often difficult because of their large variety. Nonetheless, the CST meter can be used for comparative evaluations of both polymer type and dosage. Optimum polymer dosages should be determined with care, because polymer's effec-

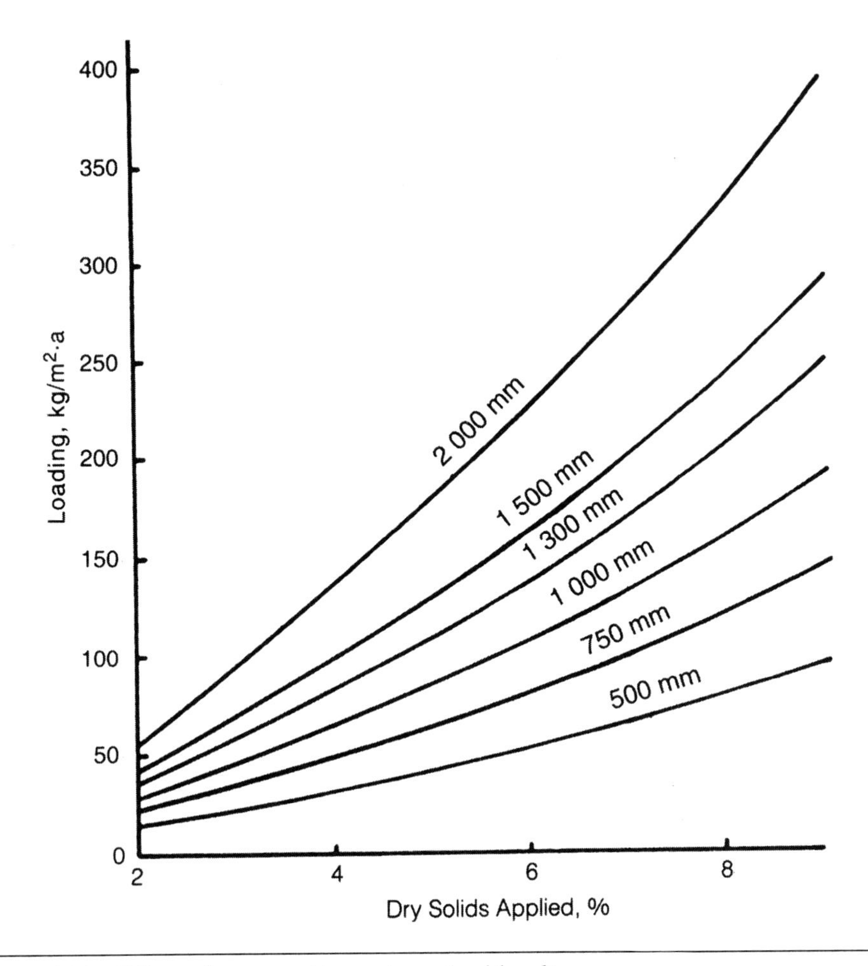

FIGURE 24.14 Effect of evaporation rate on bed loading.

tiveness can be hampered by both underdosing and overdosing. The net and gross bed loadings for chemically treated and untreated beds should be compared in laboratory tests and under actual field conditions. Blinding of the sand can result if chemical usage is excessive.

If the design for the drying bed system includes polymer addition, at least three points of addition are required for optimum effectiveness: one near the suction side of the pump, one at the pump discharge, and one near the discharge point of each bed. Variable-output positive-displacement pumps typically convey the chemical. Where possible, provisions also should be made for recirculating polymer-treated solids to allow dosage optimization with a CST meter before discharging the initial solids to the bed. Such a procedure will help prevent blinding of the sand-bed surface by poorly treated solids.

5.2.1.4 Design Criteria

Several parameters are important in determining what size a drying bed must be and how it will perform.

5.2.1.4.1 Drying Time

The total drying time required depends on the desired final moisture content and relates to the solids-removal method and subsequent use. Ultimate bed sizing is a function of evaporation, application depth, and applied solids concentration.

The time required to achieve a liftable cake depends more on the initial solids content and percentage of total water drained than on the initial drainage rate. This is particularly significant from a dewatering standpoint, because with most solids, the time required for moisture to evaporate is considerably longer than that required for it to drain. Therefore, the total time that solids must remain on the bed is controlled by the amount of water that must be removed by evaporation, which, in turn, is determined primarily by the amount removed by drainage and decanting. The percentage of drainable water strongly depends on initial solids concentration (see Table 24.4).

Quon and Johnson (1966) demonstrated that the percentage of drainable water in aerobically digested activated solids often considerably exceeds that reported for anaerobically digested solids.

5.2.1.4.2 Effect of Digestion

Dried, digested solids on drying beds contain many small cracks that allow for more surface exposure to the drying air, greater drainage of water, and easier passage of rainwater directly to the underlying sand-bed drains compared to typical raw solids (see Table 24.5).

According to Randall and Koch (1969), the dewatering properties of aerobically digested activated solids are closely related to oxygen-use characteristics. Solids obtained from digesters where the dissolved oxygen concentration remained less than 1 mg/L, dewater poorly. Drainage and drying properties are improved by extending the solids retention time (SRT). Some of the solids studied reach a point, however, at which additional digestion does more harm than good.

Digestion also increases friability of air-dried cake, making it easier to remove from sand beds and land-apply (mix with soil). It also minimizes odor problems and reduces grease buildup in soil.

Another advantage of digestion is the destruction of pathogens. The U.S. Environmental Protection Agency (U.S. EPA) guidelines for disinfecting solids suggest that

TABLE 23.4 Total water drained from aerobically digested sludge when dewatered in a sand drying bed.

Sludge solids concentration, g/L	Water loss for applied sludge depth, %		
	100 mm	150 mm	200 mm
7.55			85.7
14.6	77.6		79.3
17.3	86.4	92.2	79.4
18.6	85.4	85.5	80.3
19.5	85.5		74.5
20.45	80.0	85.0	78.0
21.3			73.6
24.1	78.3		71.0
25.2	74.0		72.0
25.4	72.8	77.8	71.0
28.75	77.7	80.0	73.4
28.8	86.3	85.0	70.5
29.2	78.7	82.1	70.5
29.7	77.5	70.7	70.3
34.0			76.3
38.0	69.0	71.8	67.2
Average	**79.2**	**80.0**	**73.4**

anaerobic digestion followed by dewatering on sand drying beds may destroy enough pathogens to allow the unrestricted use of the dried cake, assuming that the end use is not restricted by concentrations of heavy metals or other regulated parameters. Neither anaerobic nor aerobic digestion alone destroys pathogens as effectively as either would in combination with dewatering on sand drying beds. Reimers et al. (1981) noted that the inactivation of viable parasite eggs in raw solids increases as moisture content decreases.

TABLE 23.5 The effect of digestion on sand bed dewatering.

Removal method	Raw sludge removed, %	Digested sludge removed	
		Poor drainage, %	Good drainage, %
Drainage	48–52	28	72
Decantation	4–9	22	2
Evaporation	43–44	50	27

5.2.1.4.3 Effect of Application Depth

Quon and Johnson (1966) reported that the depth of applied solids affects the drainage rate; they concluded that the depth should not exceed 200 mm (8 in.). Haseltine (1951) reported an optimum depth of 230 mm (9 in.), depending on drying time and removal method; his suggested application depths ranged from 200 to 400 mm (8 to 16 in.). The applied depth should result in an optimum loading of 10 to 15 kg/m^2 (2 to 3 lb/sq ft).

Randall and Koch (1969) found that for a given solids concentration and depth, the solids drainage rate was constant after 8 hours. In addition, a typical applied solids depth of 200 mm had been reduced to a total depth of less than 25 mm (1 in.) when it was ready to be removed from the bed. The thickness of the dried cake primarily is a function of the solids concentration and the depth applied (see Figure 24.15).

The solids concentration of removed cake ranges from 44.5 to 95.5% dry solids, with the higher moisture content typically corresponding to a higher initial solids concentration.

According to Coackley and Allos (1962), drying occurs at a constant rate until a critical moisture content is reached; then, it proceeds at a declining rate. In general, the lower the required final moisture content is, the longer the drying time will be.

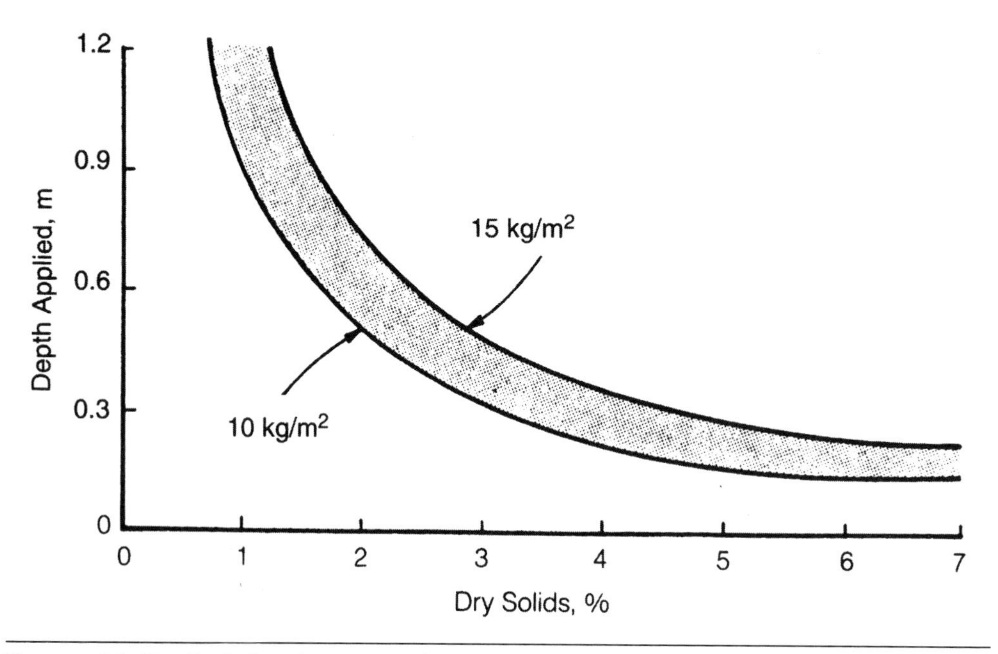

FIGURE 24.15 Bed depth required to obtain optimum loading at various solids concentrations.

5.2.1.4.4 Climatic Effects

Regional climatic conditions greatly affect dewatering on drying beds. The drying time is shorter in regions with more sunshine, less rainfall, and low humidity. Southern localities, where the summers are longer, and arid regions, where humidity is low, are more favorable than northern localities for drying bed use. Higher rainfall and higher humidity in many southern areas, however, can adversely affect drying time. Natural freezing in northern climates also has been reported to improve dewaterability, but it can deactivate a bed for the winter. The prevalence and velocity of wind also affect evaporation rates. So, climatic conditions may warrant some modifications of design criteria. For example, storage of liquid residuals should be included if drying beds may be unavailable for extended periods because of climatic conditions.

5.2.1.4.5 Sidestream Treatment

The only sidestreams from a sand drying bed operation are underdrainage liquor and surface decanting. Little is known about the characteristics of these sidestreams, which typically are not treated separately, but rather are returned to the plant headworks. The high strength and intermittent flow of these sidestreams can adversely affect the performance of some small plants because, although typically low in suspended solids, sidestreams can contain large quantities of soluble BOD and nutrients. Sidestreams also will include collected rainwater if the beds are uncovered. Essentially, all the rain that falls on a solids drying bed will be returned to the plant for treatment.

5.2.2 *Structural Elements of Conventional Beds*

Each drying bed typically is designed to hold, in one or more sections, the full volume of solids removed from a digester or aerobic reactor at one drawing. Structural elements of the bed include the sidewalls, underdrains, gravel and sand layers, partitions, decanters, solids distribution channel, runway and ramps, and possibly bed enclosures (see Figure 24.16).

5.2.2.1 Sidewalls

Construction above the sand surface should include an embankment (vertical wall) with above-sand freeboard of 0.5 to 0.9 m (20 to 36 in.). Walls can be constructed of the following materials: earth sodded with grass, wooden planks (preferably treated to prevent rotting), concrete planks, and reinforced concrete or concrete blocks that are set on an edge around the extremities of the sand surface and extended to the underdrain gravel to help prevent weed and grass encroachment.

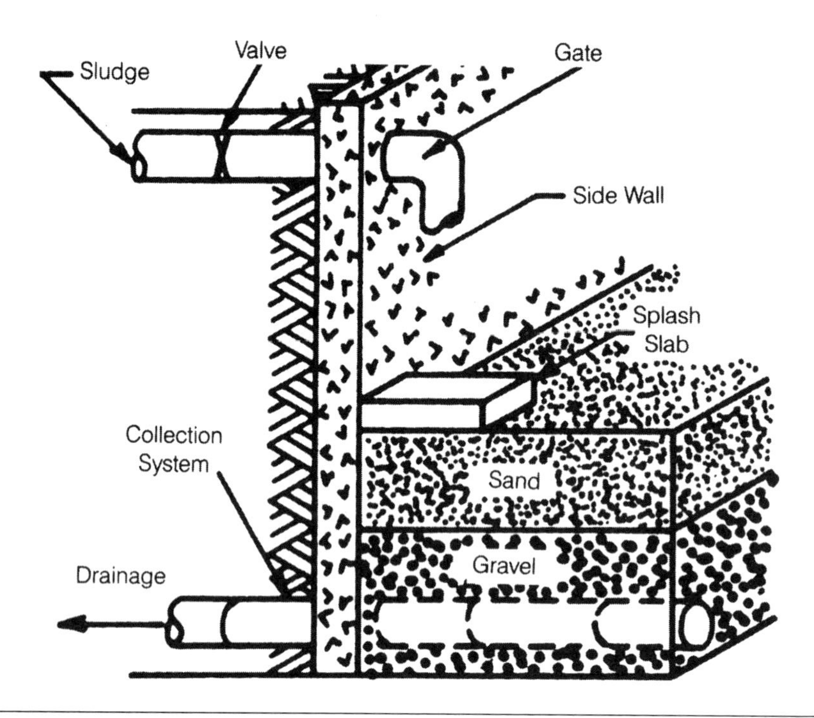

FIGURE 24.16 Schematic of a typical sand drying bed.

5.2.2.2 Underdrains

Underdrains, which typically are constructed of perforated plastic pipe or vitrified clay tile, are sloped toward a main collection pipe (outlet drain). The main underdrain pipes should be no less than 100 mm (4 in.) in diameter and have a minimum slope of 1%. Spacing should range from 2.5 to 6 m (8 to 20 ft) and should take into account the type of solids-removal vehicles to be used to avoid damage to the underdrain.

Lateral tiles feeding into the main underdrain should be spaced from 2.5 to 3 m (8 to 10 ft) apart, with the shorter distance preferred. If infiltration would endanger groundwater, then the earth floor should be sealed with an impervious membrane system approved by local regulators. The area around the drain tiles should be backfilled with coarse gravel; disturbing or breaking the tiles should be avoided. Heavy equipment should be excluded from the bed after the underdrains are laid, unless the bed is designed to accommodate such heavy loads.

5.2.2.3 Gravel Layers

Gravel layers are graded to an overall depth of 200 to 460 mm (8 to 18 in.), with the relatively coarser materials at the bottom. The gravel particles range from 3 to 25 mm (0.1 to 1.0 in.) in diameter.

5.2.2.4 Sand Layer

Sand depth varies from 200 to 460 mm (9 to 18 in.). However, a minimum depth of 300 mm (12 in.) is suggested to secure a good effluent and reduce the frequency of sand replacement caused by cleaning-related losses. A good-quality sand has the following characteristics: particles that are clean, hard, durable, and free from clay, loam, dust, or other foreign matter; a uniformity coefficient that is not more than 4.0, but preferably less than 3.5; and an effective sand-grain size between 0.3 and 0.75 mm (0.01 and 0.03 in.).

In some instances, pea gravel and anthracite coal, crushed to an effective size of about 0.4 mm (0.02 in.), can be used instead of sand. Gradations toward water filter sand should be avoided because this media affords poor traction, and wheeled cake-removal vehicles might become bogged down.

5.2.2.5 Partitions

For manual removal of solids in smaller plants, the drying bed typically is divided into sections about 7.5-m (25-ft) wide. Some mechanical removal methods have wider areas; the width should be designed to accommodate the removal method used (e.g., multiples of loader bucket width and span of vacuum-removal system). Beds have been constructed as long as 30 to 60 m (100 to 200 ft). If polymer use is anticipated, however, the bed length should not exceed 15 to 25 m (50 to 75 ft) to avoid solids-distribution problems. The angle of repose for many polymer-treated solids can be as flat as 1:120, but the angle can be much greater; therefore, unevenly distributed solids can cause inefficient use of the drying bed area. Provisions for flooding the bed with plant water before introducing solids have been used to aid in distribution at some plants. In this approach, the bed drain valves are closed, the bed is flooded with water, liquid residuals are applied, and then the drain valves are opened. Preflooding increases the rate of initial water removal because it adds to the hydraulic head by essentially creating a vacuum under the bed when the drain valves are opened. This vacuum will hold until air begins to leak through the bed into the underdrain system.

The partitions may be earth embankments (where land is plentiful) or walls constructed of concrete block, reinforced concrete, or planks and supporting grooved posts. The posts can be made of wood, although the preferable material is reinforced concrete planks fitted into grooves in reinforced concrete posts. If used, partition planks should extend about 80 to 100 mm (3 to 4 in.) below the top of the sand surface, and the posts should extend 0.6 to 0.9 m (2 to 3 ft) below the bottom of the gravel. The placement of partitions and other structural elements should be designed to accommodate mechanical removal equipment, if it is used. For example, including at least one solid, vertical wall in each bed against which a wheeled front-end loader can push will speed bed cleaning.

5.2.2.6 Decanters

A method for either continuously or intermittently decanting supernatant can be provided on the perimeter of the bed (see Figure 24.17). Decanters can be particularly useful for relatively dilute secondary and polymer-treated solids and for removing rainwater. Properly performed, decanting also can reduce drying time significantly.

5.2.2.7 Solids Distribution Channel

Liquid residuals can be applied to the sand bed sections via a closed conduit or a pressurized pipeline with valved outlets at each sand-bed section, or via an open channel with side openings controlled by sluice gates or hand slide gates. The open channel is easier to clean after each use. With either type, a concrete splash slab 130-mm (5-in.) thick and 0.9-m (3-ft) square is necessary to receive falling solids and prevent erosion of the sand surface.

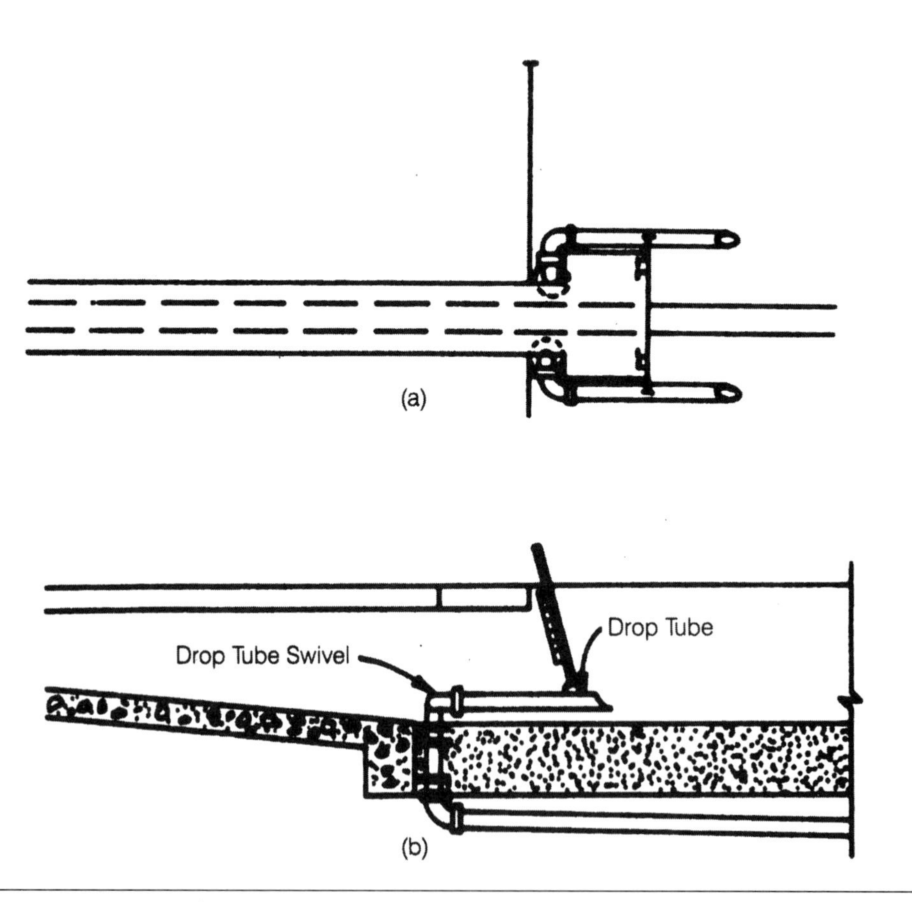

Figure 24.17 Typical decant piping: (a) plan and (b) elevation.

If a pressurized pipeline with valved outlets is used, a 90° elbow should direct the solids trajectory against the splash slab at all pumping rates. Piping and valves should be protected from freezing because draining completely after each bed is filled would be impractical.

Preferably, the distribution channel runs between two series of 7.5-m (25-ft) wide sand-bed sections. This bed width typically suffices for manual removal of cake; however, another width may better accommodate mechanical removal equipment.

5.2.2.8 Runway and Ramps

To remove cake via truck, concrete runways are needed along the central axis of each bed section (see Figure 24.18). Concrete runway slabs are narrow to minimize compaction of the sand filter surface; they are formed to help keep truck wheels on the two strips. In addition to reducing sand compaction and protecting the underdrain system from damage, multiple runway strips reduce the loss of sand and provide a good gauge for sand replacement.

If ramps are included in the entrance design and mechanical removal devices are used, design engineers should consider full-width ramps to avoid problems involving corner access and maneuvering of equipment on the sand.

FIGURE 24.18 A drying bed containing runways and ramp.

5.2.2.9 Enclosures for Covered Beds

Covered-bed enclosures provide a roof over the sand beds. Most of the previously mentioned features of open beds also apply to enclosed beds. Because of snow loads, continuous-slope shed roofs are best for northern latitudes.

Drying beds can be covered with a durable fiberglass-reinforced plastic that is available in various colors (see Figure 24.19). Glass or polyester glass-fiber roofs, which cover the top of the drying bed but leave the sides open to the atmosphere, protect the drying product from precipitation but provide little temperature control. Completely enclosed drying beds, on the other hand, permit more cake withdrawals per year in most climates because of better temperature control. Enclosed beds typically require less area than open beds. Favorable weather conditions, however, allow open beds to evaporate cake

FIGURE 24.19 A covered drying bed with a fiberglass enclosure: (a) interior view and (b) exterior view.

moisture faster than enclosed ones. Consequently, a combination of open and enclosed beds can achieve the most effective use of bed drying facilities.

Most manufacturers have developed standard dimensions for width, length, truss spacing, and other details for bed covers. For interior wood and metal work, paints (e.g., a coal-tar-epoxy bitumastic coating) must resist moisture and hydrogen sulfide. Application of paints and protective covers should conform with manufacturers' recommendations.

Older enclosure designs often include only one row of side sash that opens out, diverting the air across the top of the enclosures rather than across the surface of the solids. Newer enclosure designs typically have two rows of side sash, with the top row opening out and the bottom row opening in. Mechanical ventilation of enclosed beds is suggested in humid climates. Ventilation requirements should ensure that enclosures will not be confined spaces under applicable codes.

5.3 Other Types of Drying Beds in Use

Other types of drying beds currently in use include proprietary systems and mechanically-assisted solar drying systems.

5.3.1 *Polymer-Assisted Filter Bed*

One manufacturer provides a proprietary solids drying system, consisting of a sand layer over specialized drainage panels, that may more accurately be called *polymer-assisted filter beds* (see Figure 24.20). Physically, the filter beds look similar to conventional beds, but have a much quicker turnover time. The key to this process is an upstream polymer activation system designed to properly mix polymer with feed solids and then properly distribute the mixture across the surface of the filter bed.

Sizing a polymer-assisted filter bed system is similar to sizing conventional solids drying beds, but should be confirmed with the manufacturer. Solids are applied to the system much as they would be on a conventional drying bed, but they first must be conditioned with polymer. The underdrain system is designed to provide a siphoning effect that provides some vacuum assistance to the drying process. The plant water system (see Figure 24.20) is used to apply plant water to the underdrain system, which is key to the siphoning effect. Together, polymer conditioning and the siphoning effect allow solids to dewater more rapidly than a conventional drying bed, according to the manufacturer. A specially designed articulated vehicle typically is used to remove dewatered solids from the filter bed.

This system also can be used as part of an U.S. EPA-approved Class A biosolids process, in which the drying beds would be used to dewater biosolids to about 40% solids. The dewatered solids then would be removed, placed in windrows, and turned each day using the air drying (back-blending) method. The manufacturer claims that, depending on

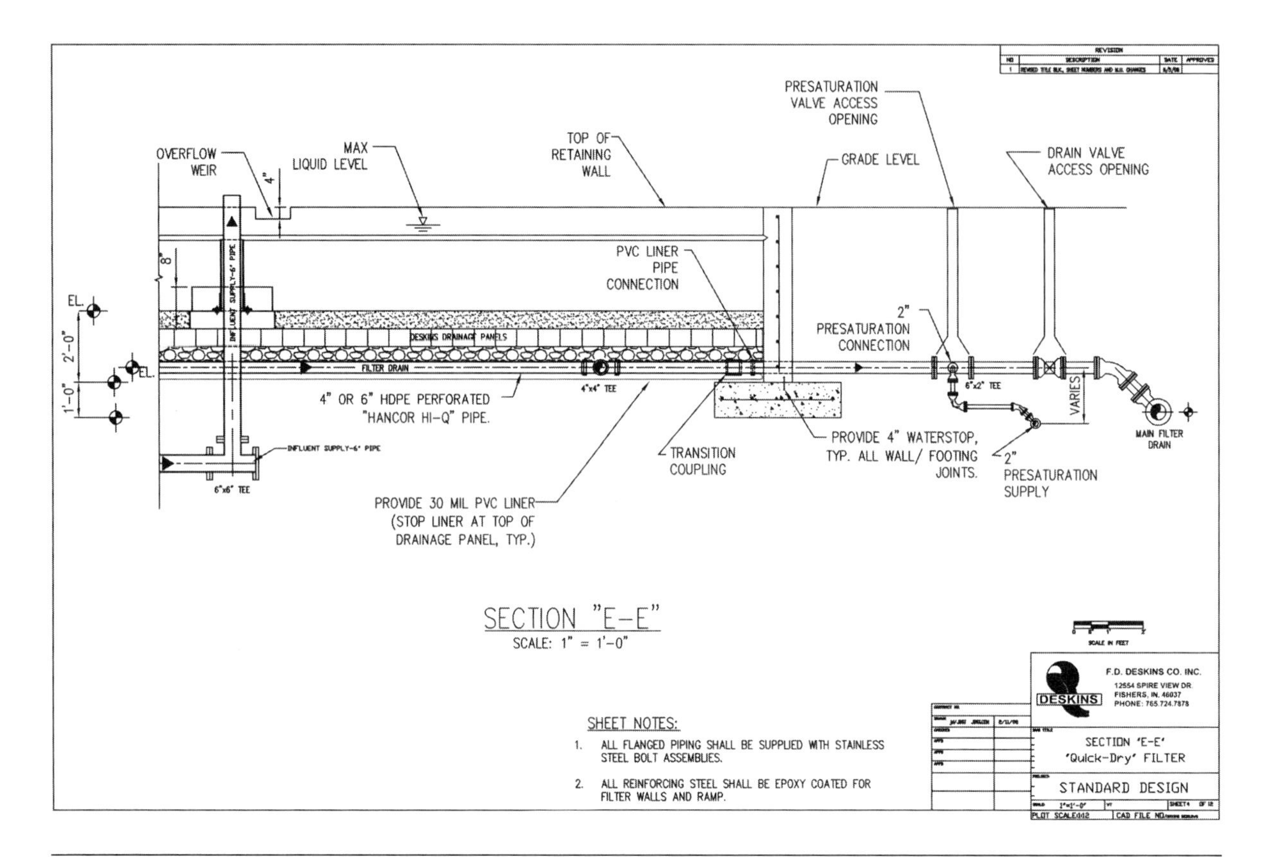

FIGURE 24.20 Schematic of a typical polymer-assisted filter bed system (courtesy of F.D. Deskins Co., Inc.).

weather conditions, liquid biosolids can be converted to dried, Class A biosolids in about 15 to 21 days. However, the final product must be tested in a laboratory to meet the pathogen reduction requirements because this is not approved as a PFRP process.

5.3.2 Mechanically Assisted Solar Drying System

One manufacturer provides a solar drying system consisting of a series of drying beds covered by a translucent, climatically controlled chamber. Sensors monitor the atmosphere in the chamber and control air louvers and ventilation fans to optimize drying conditions. The monitors also control a mobile, electric "mole" that plows the solids cake during the drying cycle up to 10 times per day. The manufacturer claims that systems can be designed to produce dried cake containing 50 to 90% solids. The advantages of this system are protection from rain, elevated temperatures because of the "greenhouse effect" in the chamber, potentially drier cake solids, and shorter drying time. This system is proprietary and information on sizing and design should be obtained from the manufacturer.

5.4 Other Types of Drying Beds Not Frequently Used

Other types of drying beds include paved drying beds, wedge-wire drying beds, and vacuum-assisted drying beds. Because few plants use these methods and some equipment is no longer available, only limited information on their performance is available.

5.4.1 *Paved Drying Beds*

Paved drying beds were constructed with concrete, asphalt, or soil cement liners to help front-end loaders more easily remove cake and mix solids to speed up drying. A series of tests conducted by Randall and Koch (1969) indicated that drying beds with sand bottoms perform better than beds with impervious bottoms.

5.4.2 *Wedge-Wire Drying Beds*

Wedge-wire drying bed systems were used in the United States starting in the early 1970s. In a wedge-wire drying bed, slurry is spread onto a horizontal, relatively open drainage media in a way that yields a clean filtrate and provides a reasonable drainage rate. The cake typically is removed relatively wet (8 to 12% dry solids), which may complicate use or disposal.

5.4.3 *Vacuum-Assisted Drying Beds*

The principal components of vacuum-assisted drying beds are

- A bottom ground slab with reinforced concrete;
- A several-millimeter-thick layer of stabilized aggregate that supports the rigid multimedia filter top (this space is also the vacuum chamber and is connected to a vacuum pump); and
- A rigid multimedia filter top, which is placed on the aggregate.

Liquid residuals are spread onto the filter surface by gravity flow at a rate of 570 min (150 gpm) and to a depth of 300 to 750 mm (12 to 30 in.). Polymer is injected into the solids in the inlet line. Filtrate drains through the multimedia filter into the aggregate layer and then to a sump. From the sump, a level-actuated submersible pump returns filtrate back to the plant. After solids are applied and allowed to gravity-drain for about 1 hour, the vacuum system is started and maintains a vacuum of 34 to 84 kPa (10 to 25 in. Hg) in the sump and under the media plates.

Under favorable weather conditions, this system can dewater a dilute, aerobically digested solids to a 14% solids concentration in 24 hours. The dewatered solids can be lifted from the bed by mechanical equipment. It will further dewater to an about 18% solids concentration in 48 hours. No manufacturers currently supply vacuum-assisted drying bed systems.

5.5 Reed Beds

5.5.1 *Introduction*

The use of reed beds to treat stabilized solids from secondary wastewater treatment plants has been successful in Indiana, Wisconsin, New York, Pennsylvania, and Maine. This method was developed by the Max-Planck-Society of Germany in the 1960s and has been recognized by U.S. EPA as an alternative and innovative system (Riggle, 1991).

The system combines the action of conventional drying beds with the effects of aquatic plants on water-bearing substrates. While conventional drying beds are used to drain more than 50% of the water content from solids, the resulting residue must be hauled away for further treatment or disposal at designated sites. When the drying beds are constructed in a specific manner and then planted with reeds of the genus *Phragmites communis*, further desiccation results from the demand for water by these plants. To satisfy this demand, the plants continually extend their root system into the solids deposits. This extended root system establishes a rich population of microflora that feed on the organic content of the solids. This microflora is also partly kept aerobic by the action of the plants. Degradation by the microflora is so effective that eventually up to 97% of solids are converted into carbon dioxide and water, with a corresponding volume reduction. The beneficial result is that these planted drying beds can reportedly be operated for up to 10 years before the accumulated residues have to be removed. This represents a considerable monetary savings.

5.5.2 *Design Considerations*

The reed-bed treatment system typically consists of a composition of rectangular, parallel basins with concrete sidewalls. The bottom of each bed is lined and provided with two underdrains. In addition, a 230-mm (9-in.) layer of 19-mm (0.75-in.) washed river gravel is topped with a 102-mm (4-in.) layer of filter sand. The reeds are planted in the gravel, with 11 plants per square meter (1 plant per square foot) of filter area. A freeboard of 1.0 to 1.5 m (3.5 to 5 ft) is often provided, depending on storage design requirements. Basins are cyclically loaded. In each cycle, the first basin is loaded over a 24-hour period and then allowed to absorb the loading over a 1-week resting period before the cycle is repeated (Banks and Davis, 1983).

Figure 24.21 shows a typical reed drying bed system. Hydraulic design loadings for residuals containing 3 to 4% solids are 0.004 2 $m^3/m^2{\cdot}h$ (2.5 gal/d/sq ft) or 35 $m^3/m^2{\cdot}a$ (86 gal/yr/sq ft). At this loading rate, about 1.0 m (3.5 ft) of product with 70% moisture will accumulate over 10 years. When solids are removed for disposal, the top layer of sand also will be removed and must be replaced. Generally, the root system remaining

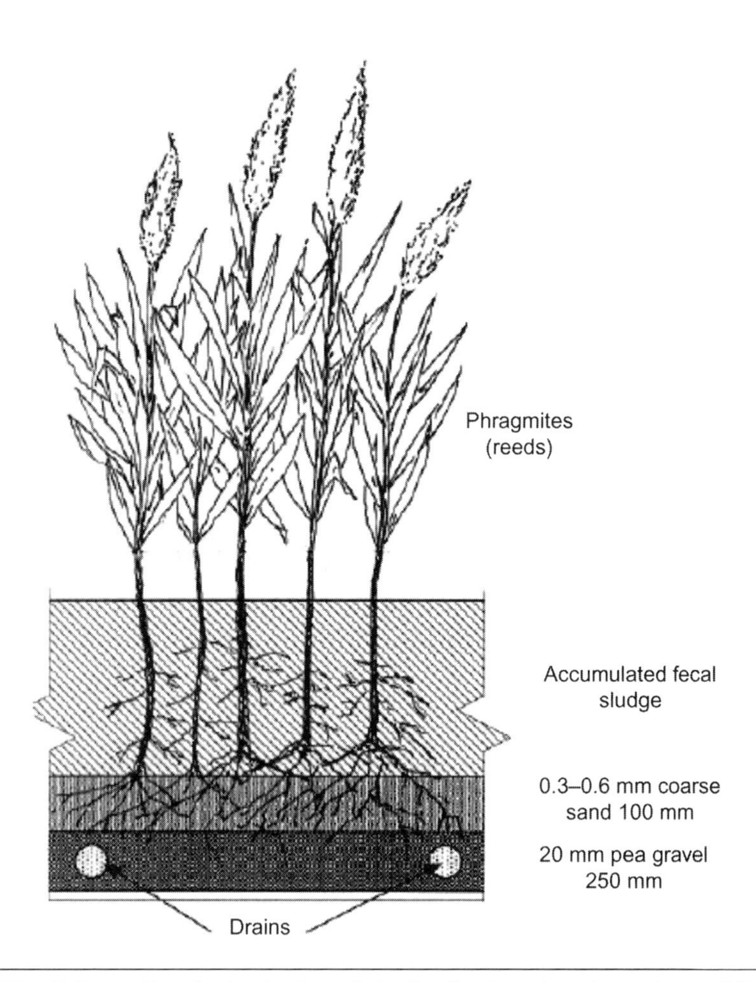

FIGURE 24.21 Schematic of a typical reed drying bed system (courtesy of Crolla et al., 2007).

in the gravel bed will allow the reed plants to regenerate without the need for replanting (Banks and Davis, 1983).

These systems are most applicable in climates where winter temperatures ensure at least one prolonged frost. The planted reeds are harvested in the fall once they have become dormant.

The use of reed-bed systems is a reasonable application for wastewater treatment facilities with less than 5.7 to 7.6 ML/d (1.5 to 2.0 mgd) capacity. The primary benefits of this system in smaller plants are the reduced O&M requirements and the concentration of disposal handling to once every 10 years.

5.6 Lagoons

Lagoons, which are natural or artificial earth basins, can be used for both solids drying and storage. A few communities have used some form of lagoon system, reportedly with favorable results, although the use of lagoons typically starts as a temporary expedient to handle solids volumes in excess of the original plant design. Climatic conditions have a decided effect on the functioning of a lagoon, with warmer, arid climates producing the best results.

Lagoon operations typically involve the following processes:

- Pumping liquid in the lagoon for several months or more. The pumped solids typically are stabilized before application to minimize odor problems.
- Continuously or intermittently decanting supernatant from the lagoon surface and returning it to the wastewater treatment plant.
- Removing the dewatered material with some type of mechanical removal equipment.
- Repeating the cycle.

5.6.1 Environmental Considerations

The location of natural or artificial dewatering lagoons must be considered before selecting lagoons as a treatment method. The proposed site should be sufficiently removed from dwellings and other areas where odors would produce problems. Because they go through a series of wet and dry conditions, large lagoons can produce nuisance odors.

The use of deep lagoons to dry raw solids has resulted in severe odor problems, but the use of shallow lagoons for drying typically has not produced odors more intense than those sometimes experienced with conventional sand drying beds. However, odors produced from lagoons used to store solids can be more of a problem because wet treated municipal solids retain a higher moisture content far longer than solids treated on conventional sand drying beds.

Odor problems with lagoons have been described by Zablatzky and Peterson (1968). Both the intensity and type of odors produced from wastewater lagoons have varied greatly, depending on the condition of the solids and the depth of the lagoon. Odors can vary from a gas- and tar-like odor produced by a well-digested material to the putrid odors produced by decomposing raw solids.

The possibility of polluting groundwater or nearby surface waters should be investigated thoroughly. If the subsurface soil is permeable, the potential for groundwater contamination exists. Clay and/or membrane liners and underdrain systems can minimize this potential.

Baxter and Martin (1982) found that the application depth also can affect groundwater contamination. Their results indicate that a 250-mm (10-in.) application of liquid residuals to earthen drying lagoons can lead to significant quantities of polluted water moving toward the groundwater. However, applications of 0.6 to 1.0 m (2 to 3 ft) of liquid solids lead to a rapid sealing effect by producing an impermeable layer of solids that prevents contaminants from moving toward groundwater.

Finally, all lagoon areas should be fenced to keep animals and other trespassers out and to prevent vandalism and potential liability problems.

5.6.2 *Storage Lagoons*

Storage lagoons can be 1.5 m (5 ft) deep or more and are primarily designed for storage rather than drying. To allow for cleaning, the dikes should be constructed to be about 3 m (10 ft) wide across the top to accommodate trucks and other mechanical equipment used for solids removal or for maintenance. The side slopes of the dikes should be a maximum of 3:1 (horizontal-to-vertical) to provide a slope surface that can be mowed by mechanical equipment. Residuals in storage lagoons typically do not dry or condense enough to permit removal by anything but a dragline, so the maximum width of lagoons must be less than twice the length of a dragline boom or other equipment to be used.

The design volume of lagoons depends on intended use and length of storage. Generally, solids lagoons are designed to provide for the emptying of one or more digesters and can be sized in reference to digester volume and frequency of emptying. Local climate and the proposed disposal of supernatant liquor and rainwater drawoffs influence the length of time between the initial placing of solids and their eventual concentration to a degree that permits mechanical removal.

Properly designed drawoff piping should be provided for the removal of supernatant liquid and rainwater from storage lagoons. The drawoff piping should be arranged to discharge these liquids to the wastewater treatment plant's headworks. Removing these liquids prolongs the life of the lagoon, helps prevent insect breeding, and hastens dewatering.

Nuisances that can result from storage lagoons depend on the type and condition of the solids placed in the lagoon; they can be reduced by chemicals to control or mask odors and prevent insect breeding. If poorly digested wastewater solids are added to a storage lagoon, it may not be possible to control odors. If this situation is suspected during design, one solution to the problem is adequately isolate the lagoon site. Another solution entails putting clean water on top of the lagoon to provide an aerobic layer exposed to the atmosphere.

5.6.3 *Drying Lagoons*

Drying lagoons are used for dewatering when sufficient economical land is available (Vesilind, 1979). Lagoons are similar to drying beds; however, solids are placed at depths three to four times greater than in a drying bed. Generally, solids are allowed to dewater and dry to a predetermined solids concentration before removal; this process may require 1 to 3 years.

Dewatering in lagoons occurs in two ways: evaporation and transpiration. Studies by Jeffrey (1959, 1960) indicate that evaporation is the most important dewatering factor.

Drying lagoons should be shallower than storage lagoons, with about 0.6 to 1.2 m (2 to 4 ft) of dike provided above the bottom of the lagoon. Drying lagoons typically do not include an underdrain system because most of the drying is accomplished by decanting supernatant liquor and by evaporation. However, groundwater pollution is a potential problem. The depth of solids in drying lagoons should not exceed 400 mm (15 in.) after excess supernatant liquor has been removed.

Dikes should be a shape and size that permits maintenance and mowing. The hydraulic loading against the dikes and the possibility of dike leakage are not great because the depth of liquid seldom exceeds about 0.5 m. Dikes should be constructed of compacted material to provide stability on the slopes, and the bottom of the lagoons should be level or have a small slope away from the liquid-residuals inlet opening. The dikes also should be large enough to allow trucks and front-end loaders to enter the lagoons for cleaning and to permit easy mowing with mechanical equipment.

The outlet into the lagoon, supernatant drawoff lines, and other piping should be 0.3 m (1 ft) below the original bottom of the lagoon. Supernatant liquor and rainwater drawoff points should be provided, and the drawn-off liquid should be returned to the treatment plant for further processing. In addition, surrounding areas should be graded to divert surface water around and away from the lagoons.

Wet solids typically will not dry enough to be removed with a fork except in an arid climate or when an extremely long, warm dry spell has occurred. The concentrated cake typically can be removed with a front-end loader.

The potential odor problem from lagoons used to dewater well-digested solids is about the same as it is for sand drying beds. If supernatant liquor and rainwater are removed promptly from the solids surface so the cake is exposed to oxygen in the air and can rapidly dry, there should be minimal odors.

The actual depth and area requirements for drying lagoons depend on several factors (e.g., precipitation, evaporation, type of solids, volume, and solids concentration. Solids-loading criteria specify 35 to 38 $kg/m^3 \cdot a$ (2.2 to 2.4 lb/yr/cu ft) of capacity

(Zacharias and Pietila, 1977). The area provided for drying lagoons varies from 0.1 m^2/cap (1 sq ft/cap) for primary digested solids in an arid climate, to as high as 0.3 to 0.4 m^2/cap (3 to 4 sq ft/cap) for activated sludge plants in areas where the annual rainfall is about 900 mm (36 in.).

6.0 ROTARY PRESSES

6.1 Introduction

Rotary presses are a relatively new technology that can achieve cake solids and solid capture performance similar to belt presses and centrifuges (see Table 24.6) (Crosswell et al., 2004). Rotary press and rotary fan press dewatering technology relies on gravity, friction, and pressure differential to dewater solids. There are currently three major manufacturers of dewatering equipment in this category.

Figure 24.22 illustrates the principal components of a rotary press. Solids are dosed with polymer and fed into a channel bound by screens on each side. The channel curves with the circumference of the unit, making a 180° turn from inlet to outlet. Free water passes through the screens, which move in continuous, slow, concentric motion. The motion of the screens creates a "gripping" effect toward the end of the channel, where cake accumulates against the outlet gate, and the motion of the screens squeezes out more water. The cake is continuously released through the pressure-controlled outlet.

TABLE 24.6 Advantages and disadvantages of rotary presses and rotary fan presses.

Advantages	Disadvantages
• Uses less energy than centrifuges or belt filter presses	• May be more dependent on polymer performance than centrifuges or belt filter presses
• Small footprint	• Low throughput compared to other mechanical dewatering processes
• Odors contained	• Screen clogging potential
• Low shear	• Need for heavy rated overhead crane to lift and maintain channels
• Minimal moving parts	• High capital cost
• Minimal building requirements	
• Minimal start-up and shutdown time	
• Uses less wash water than belt filter presses	
• Low vibration	
• Low noise	
• Modular design	

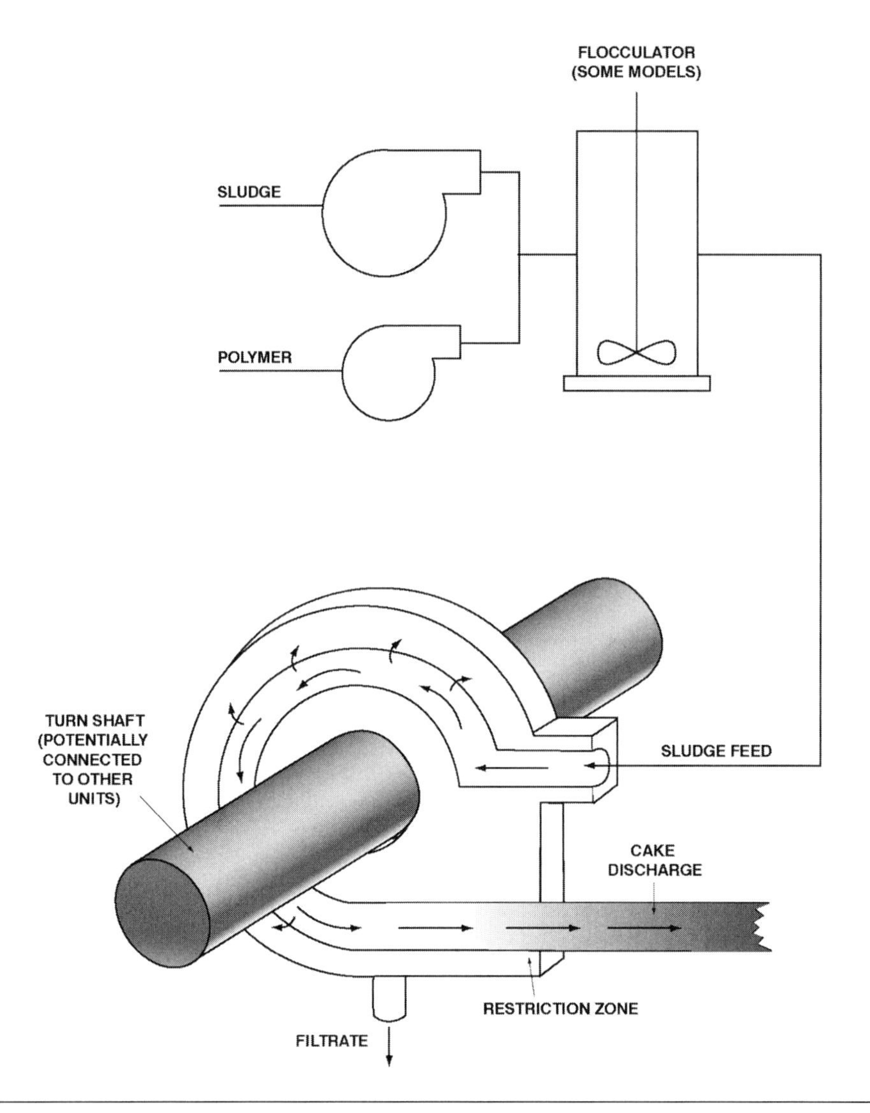

FIGURE 24.22 Schematic of a rotary press system.

6.2 Process Design Conditions and Criteria

6.2.1 Mechanical Features

The major elements of a rotary press are the polymer feed and mixing system, parallel filtering screens, a circular channel between the screens, the rotation shaft, and a pressure-controlled outlet. The key differences between the rotary press and rotary fan press are the screens, drive mechanism, and pressure differential. In the rotary press, the screens consist of two layers of perforated stainless steel, with each layer having different sieve size. The rotary fan press' screens consist of fabricated wedgewire with small openings and linear gaps. The rotary press drive configuration allows up to six rotary press channels to be operated on a single drive. Each channel has bearings, and the combined unit has an outboard bearing cantilevered on one end. The rotary fan press drive configuration uses a maximum of two rotary press channels on a single drive with isolated bearings in a sealed gearbox.

The entry zones of rotary presses and rotary fan presses function much like the gravity phase of belt press dewatering. Free water "falls" through the filtering screen pores and is collected in a filtrate channel. Pressure builds gradually as the solids travel toward the machine outlet. Because the outlet controls the pressure at which cake can be released, cake solids accumulate against the outlet and are further dewatered via friction from the continuous motion of the screens. In the rotary press, the friction generated between the screens and the cake plug translates into mechanical pressure that deflects the cake away from the center and forces it sideways against the restricted outlet. In the rotary fan press, frictional force also is imparted in the outlet zone to dewater solids, but the mechanical pressure is not generated to the same magnitude. In both designs, water is released via friction and is collected in the filtrate channel along with water released by gravity in the entry zone.

A key feature of both rotary press and rotary fan press dewatering technology is their slow rotational speed. Typical installations use speeds of 1 to 3 revolutions per minute (rpm). This provides low vibration, low shear, and low noise.

6.2.2 *Structural Elements*

Rotary press or rotary fan press dewatering equipment can be mounted on concrete flooring, a concrete pad, or a metal skid. The structural housing for rotating equipment is built into the unit. A hopper, conveyor, or additional cake tubing must be supplied at the cake discharge outlet. The flowpath through the units differs between the rotary press and rotary fan press. Rotary presses are fed from the top and discharge on the bottom, while rotary fan presses are fed on the bottom and discharge from the top.

6.2.3 *Hydraulic Loading Rate*

The hydraulic loading rate is a function of the equipment's size and number of channels. The technology is modular, and the hydraulic loading rate of single-drive units ranges from 0.5 to 15 L/s (7 to 250 gpm), although a maximum hydraulic loading rate of 3 L/s (50 gpm) per channel is typical. Rotary presses provide better performance on residuals with higher fiber content (e.g., primary solids).

6.2.4 *Solids Loading Rate*

Because solids capture is a function of the adjustable back pressure, the solids loading rate varies with the hydraulic loading rate. At higher solids concentrations, residuals will accumulate in the outlet zone, form cake, and extrude more quickly

6.2.5 *Operational Control*

Operators can control the performance of the rotary press or rotary fan press by changing polymer type and dosage, feed rate, feed pressure, wheel speed, and outlet pressure. Both types of rotary press require minimal supervision and can be unattended between startup and shutdown.

6.2.6 *Capture Efficiency*

Capture rates depend on solids type and polymer use but can exceed 95% solids capture. Performance depends largely on solids consistency. However, there is limited data at this time comparing performance to other dewatering technologies. The performance characteristics of current installations surveyed for this manual are summarized in Table 24.7; some additional data from another recent study (Crosswell et al., 2004) is shown in Table 24.8.

6.2.7 *Area/Building Requirements*

Building requirements are minimal because rotary presses and rotary fan presses are enclosed and have small footprints. The dewatering area for the presses themselves can be as small as 9.3 m^2 (100 sq ft) for a small system, depending on the model size and number of channels.

6.3 Ancillary Equipment and Controls

6.3.1 *Chemical Conditioning Requirements*

Chemical conditioning is mandatory to attain design performance in rotary press or rotary fan press dewatering. Polymer feed systems can be supplied by the manufacturer or can be procured independently. In both cases, the feed systems typically include a polymer storage tank and metering pump, which feeds the polymer into the mixing or flocculation tank, where it is blended with the solids. Dry or emulsion polymers can be used.

6.3.2 *Energy Requirements*

Rotary press and rotary fan press systems have a connected horsepower of about 3.7 to 15 kW (5 to 20 hp).

TABLE 24.7 Performance characteristics of rotary presses.

Facility	Equipment	Facility size ML/d (mgd)	Solids type	Incoming % TS	Discharge % TS
Raiford, Florida, prison WWTP	Rotary fan press	4.2 (1.1)	Extended aeration	1.6–2.5	19–22
Front Royal, Virginia, municipal WWTP	Rotary fan press	12 (3.3)	Thermophilic anaerobic digestion	3.5	25
Fairfield, California, municipal WWTP	Rotary fan press	59 (15.5)	Conventional anaerobic digestion	2	17
Lafayette, Tennessee, municipal WWTP	Rotary press	1.9 (0.5)	Conventional anaerobic digestion	1.0–1.5	25
Portland, Maine, municipal WWTP	Rotary press	75 (19.8)	Thickened PS/WAS	3–6	19–25
Hampton, New Hampshire, municipal WWTP	Rotary press	9.5 (2.5)	Septage/PS/WAS	6–8	26–28
Murfreesboro, Tennessee, municipal WWTP	Rotary press	61 (15)	PS/WAS	0.8–1	12–14
Scarborough, Maine, municipal WWTP	Rotary press	5.3 (1.4)	Thickened PS/WAS	3	28
Ocean City, Maryland*	Rotary press	53 (14)	Conventional aerobic thickened digestion	4.4	35
Aberdeen, Maryland*	Rotary press	7.6 (2)	BNR anaerobic digestion	2.7	23
Salisbury, Maryland*	Rotary press	26 (6.8)	Thickened anaerobic lagoons	2.6	24
Cambridge, Maryland*	Rotary press	31 (8.1)	BNR thickened	2.6	20
Marley-Taylor, Maryland*	Rotary press	23 (6.0)	BNR anaerobic digestion	2.3	19
Broadwater, Maryland*	Rotary press	7.6 (2.0)	BNR thickened	4.5	25
South Central, Maryland*	Rotary press	67 (23)	BNR thickened	3.7	26
SCRWF, Delaware*	Rotary press	23 (6.0)	WAS aerobic digestion	1.9	16.3
Thurmont, Maryland*	Rotary press	3.8 (1.0)	BNR WAS aerobic digestion	1.5	12
La Plata, Maryland*	Rotary press	5.7 (1.5)	BNR WAS aerobic digestion	1.4	16.9
MCI, Maryland*	Rotary press	4.5 (1.2)	BNR WAS thickened	3.3	15.5

* Crosswell et al. (2004).

6.3.3 *Wash Water Requirements*

Rotary presses and rotary fan presses include a self-cleaning system that must run for 5 minutes per day at the end of use to flush all lines and equipment. The system does not require high-pressure water for flushing. Typically, the normal in-plant water source has sufficient pressure, but in some cases, high-pressure booster pumps may be required.

TABLE 24.8 Full-scale screw press data.

Facility	Facility information/ process[a]	Solids type	Screw press type	Average feed solids (%)	Average cake solids (%)	Solids capture (%)	Typical feed rate (L/m)	Throughput per press (dkg/d)	Polymer dose (gm/dry kg)	Operating schedule	Labor required	Manual cleaning schedule
1[b]	3 ML/d WWTP: OD (SRT= 18–21 d), SC, SH (SRT=3–7 d)	Secondary sludge	Horizontal	4–6	20–45		57–114		11.5	24 h/7 d	8 h/d	
2	7.6 ML/d WWTP: OD (SRT=20 d), SC, SH (SRT=20 d)	Secondary sludge	Horizontal	2–3	15–20		106			16 h/d	1–2 h/week	
3[b]	1.1 ML/d WWTP	Secondary sludge	Horizontal	1.5	30		14–16			24 h/d, 5 d/week	40–50 h/week	Once per week
4[c]		Secondary sludge	Horizontal	1.2	16		13			70–100 h/week		2–3 times per week
5[c]		Secondary sludge	Horizontal	1.2	16		13			150 h/week		Once a day
6	3.8 ML/d WWTP: AS (SRT=24 d), SC, SH (SRT=7–10 d)	Secondary sludge	Inclined	0.07	N/A[d]		38			8 h/d, 5 d/week		
7	17 ML/d WWTP: OD (SRT=2–3 d), SC	Secondary sludge	Inclined	0.5–1	12–17	92	up to 280		7.5–10	24 h/7 d	2 h/d	Once per week
8	7.7 ML/d WWTP: OD, SC, SH	Secondary sludge	Inclined	3.5–4	18–20		up to 150		7.5	8 h/d, 7 d/week		
9	17.8 ML/d WWTP: PC, RBC, SC	Combined primary/ secondary sludge	Inclined	2.5	18–30		up to 150	0.0033		2 d/week, 6–8 h/d		Twice per month
10	17 ML/d WWTP: RBC, SC, SH (SRT=2–4 d)	Combined primary/ secondary sludge (40%/60%)	Inclined	1.1–1.2	30–40		300–340	0.0066		8 hours Monday, 2–4 hours Tuesday–Friday		Once a month

11[c]	.9 ML/d WWTP: PC, TF, SC, AD	Aerobically digested secondary sludge	Horizontal	1.2	18		26–38			24 h/7 d	8 h/d	Once per week
12	16 ML/d WWTP: PC, AB, SC	Primary/ secondary (60%/40%) digested sludge combination	Horizontal	3–4	>18	95	190–378	0.0055	9–11	5 h/d, 5 d/week		Once every 6 months
13	31 ML/d WWTP: BOD removal (SRT=20 d), SH (SRT=1 d)	Aerobically digested secondary sludge	Horizontal	1	15–19	90	95–397	0.0066	8.2	16 h/d Monday–Friday, 8 h/d Saturday–Sunday	Minimal	Once per week
14	9.9 ML/d WWTP: OD (SRT=10 d), SC, AD (SRT=1.5 d)	Aerobically digested secondary sludge	Inclined	0.9	12–14		170			7 h/d, 5 d/week	Operator checks every 1–2 hours	Once per week
15	1.0 ML/d WWTP: AB, SBR, AD (SRT=40 d) sludge	Aerobically digested secondary	Inclined	N/A	22–25		30–34			8 h/d		Once per week
16	6.1 ML/d WWTP	Aerobically digested secondary sludge	Horizontal	0.75–1.0	21–28		22–64		5 d/week	8–12 h/d,	1–2 checks operator by per hour	Once per month
17[c]	83 ML/d	Anaero-bically/ aerobically digested primary/ secondary sludge	Horizontal	3.2	17–21		114					

(*continued*)

TABLE 24.8 Full-scale screw press data (*continued*).

Facility	Facility information/ process[a]	Solids type	Screw press type	Average feed solids (%)	Average cake solids (%)	Solids capture (%)	Typical feed rate (L/m)	Throughput per press (dkg/d)	Polymer dose (gm/ dry kg)	Operating schedule	Labor required	Manual cleaning schedule
18	23 ML/d WWTP: PC, OD (SRT=20 d), SC, and (SRT=30–60 d)	Anaerobically digested primary/secondary sludge	Inclined	3–3.5	13–16		90–227			12:00 A.M. to 8–9 P.M. (3 hours downtime/day)	4–6 h/d	
19	3.7 ML/d WWTP: PC, AB (SRT=3.5 d), SC, CCT, DAF, AND (SRT=41 d)	Anaerobically digested primary/secondary (55%/45%) sludge	Horizontal	1	20–24	High	26–30			4 d/week, 10 h/d		Once per month
20[c]		Anaerobically digested primary/secondary sludge	Horizontal	2–4	25		156–312					
21[c]		Anaerobically digested primary/secondary sludge	Horizontal	2	22		163					
22	7.6 ML/d WWTP: PC, AB (SRT=15 d), RBC, SC, SF, UV, AD (SRT=20 d)	Aerobically digested primary/secondary sludge	Inclined	2	19–24		106–170			24/7	1 h/d	

[a] PC=primary clarifiers, OD=oxidation ditch, AS=activated sludge, TF=trickling filter, SC=secondary clarifiers, AD=aerobic digestion, AND=anaerobic digestion, SH=sludge holding DAF=dissolved air flotation, CCT=chlorine contact tank, SF=sand filter, SRT=solids retention time of liquid or solids treatment process

[b] Class A installations with heat/lime process.

[c] Information included for these facilities was provided by the screw press manufacturer.

[d] Facility had only been in service for two weeks at time of survey. Cake solids percentage had not yet been tested.

7.0 SCREW PRESSES

7.1 Introduction

Although screw presses have been in existence because the 1960s, the technology originally was used solely in industrial applications (e.g., pulp and paper mills and food processing plants). In the municipal market, the application is relatively new in the United States, with most facilities installed after year 2000. The increasing number of installations is still relatively small, compared to traditional solids dewatering technologies.

There are presently two major types of screw presses used in municipal dewatering applications: horizontal and inclined. Inclined screw presses are at angles 15 to 20 degrees from the horizontal. Other areas of difference pertain to solids inlet configuration, screen basket design (wedge wire), basket cleaning from the inside and outside (brushes and rotating wash system), and filtrate water collection. One manufacturer also provides an option in which lime and heat are added to the screw press, which then both dewaters solids and reduces pathogens to produce biosolids that potentially meet the Class A standards in 40 CFR 503.

The major elements of a screw press dewatering system are the solids feed pump, polymer makeup and feed system, polymer injection and mixing device (injection ring and mixing valve), flocculation vessel with mixer, solids inlet headbox or pipe, screw drive mechanism, shafted screw enclosed within a screen, a rectangular or circular cross-section enclosure compartment, and an outlet for dewatered cake (see Figure 24.23). Some horizontal screw press systems (e.g., the combined dewatering and pasteurization process) include a rotary screen thickener before the screw press, which may be desirable for reducing the hydraulic load to the screw press given certain feed solids characteristics in conventional applications.

A screw press is a simple, slow moving device that achieves continuous dewatering (see Figures 24.24 and 24.25). Polymer is combined with solids in flocculation vessels upstream of the screw press to enhance the solids' dewatering characteristics. Screw presses dewater solids first by gravity drainage at the inlet section of the screw and then by squeezing free water out of the solids as they are conveyed to the discharge end of the screw under gradually increasing pressure and friction. The increased pressure to compress the solids is generated by progressively reducing the available cross-sectional area for the solids. The released water is allowed to escape through perforated screens surrounding the screw while the solids are retained inside the press. The liquid forced out through the screens is collected and conveyed from the press, and the dewatered solids are dropped through the screw's discharge outlet at the end of the press. Screw speed and configuration, as well as screen size and orientation, can be tailored for each dewatering application.

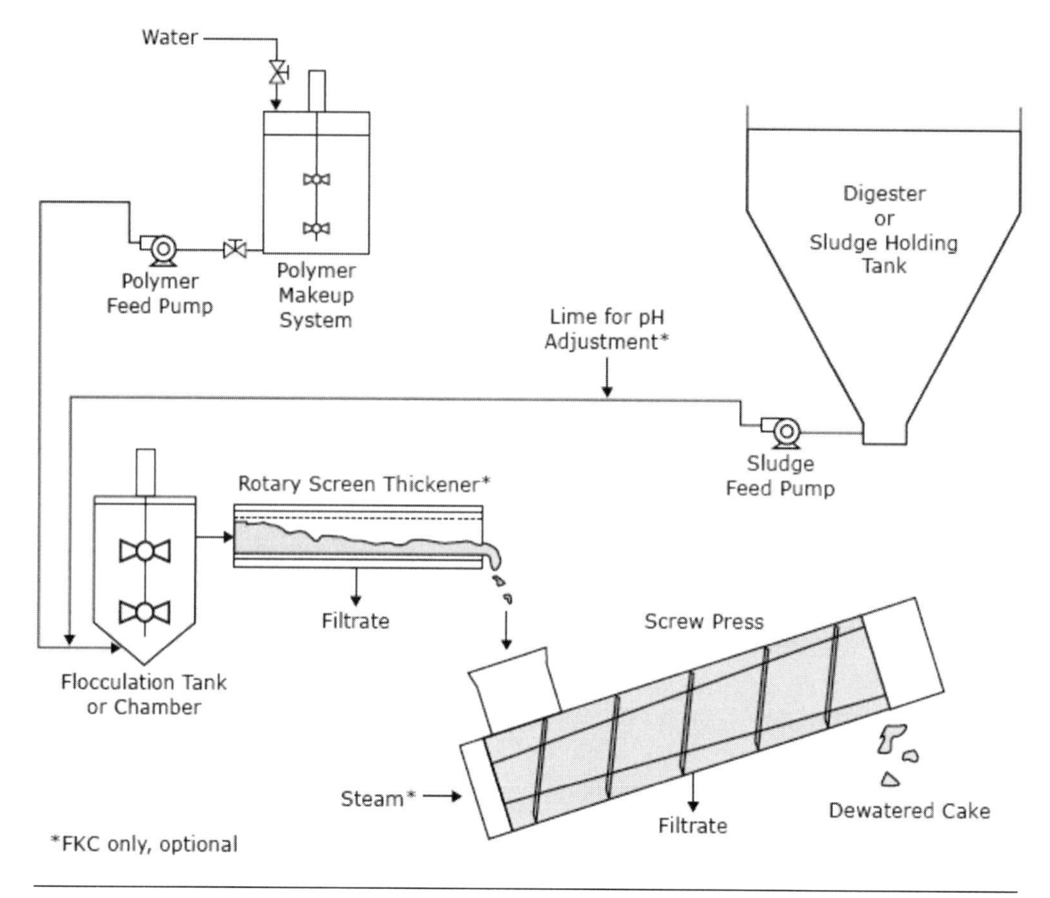

FIGURE 24.23 Schematic of screw press dewatering system.

Solids are combined with polymer and pumped into the flocculation vessel. After flocculation, solids are transferred to the screw press. In the horizontal screw press configuration, solids are fed by gravity from the flocculation tank into the screw press headbox. If a rotary screen thickener is used, solids flow from the flocculation tank to the rotary screen thickener and then to the screw press headbox. Solids then flow from the headbox into the inlet of the screw press.

In the inclined configuration, solids are pumped through the polymer injection and mixing device into the flocculation chamber and then enter the inlet pipe to the screw press. The polymer injection and mixing device is designed to intensely mix solids and polymer (which determines the strength); the flocculation tank is designed to provide the reaction time needed to create the appropriate floc size.

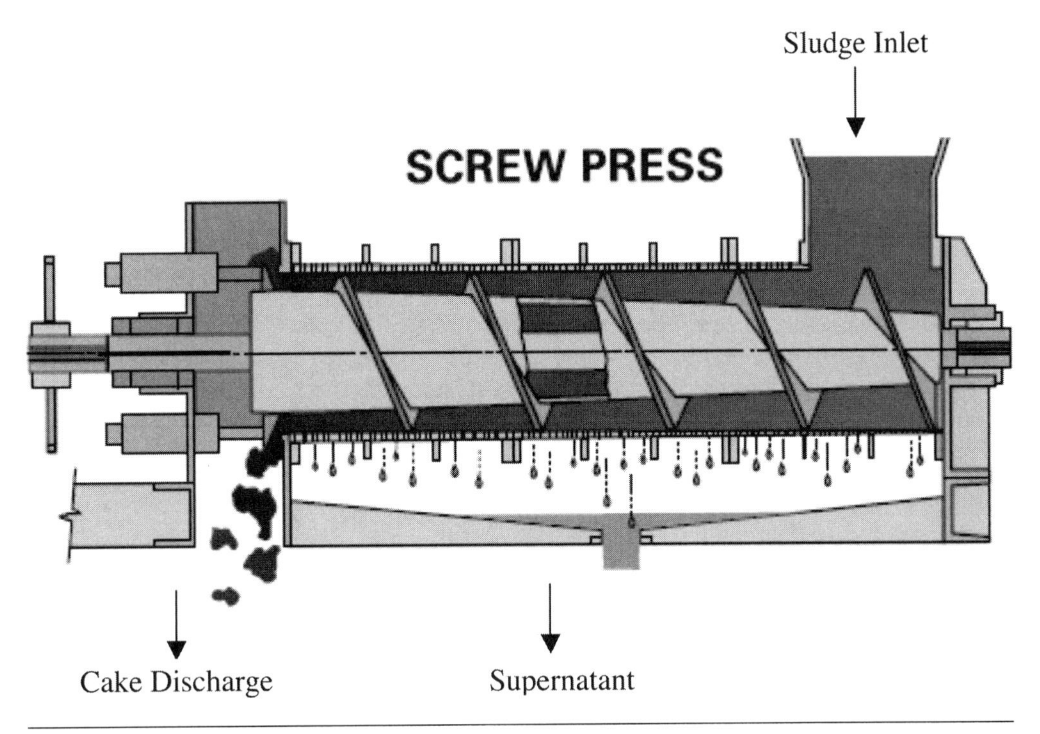

FIGURE 24.24 Cutaway view of a horizontal screw press (courtesy of FKC Co., Ltd., Port Angeles, WA).

A slow rotating screw conveys flocculated solids through a wedge wire drum. Solids compaction increases gradually by reducing the pitch of the flights, increasing the center shaft diameter, and reducing the flight diameter. The inclined screw press includes a pneumatic (or manually) adjusted dewatering cone at the discharge end of the screw press. It can be adjusted to provide an opening between 0.95 and 1.9 cm (⅜ and ¾ in.) and used to regulate the solids pressure, which in turn provides a balance between increasing cake dryness and an associated decrease in pressate clarity (solids capture). The pressure of the solids in the inclined screw press is typically between 34 and 152 kPa (5 and 22 psi), with a max. of 276 kPa (40 psi), depending on the type of unit and solids characteristics. In the horizontal screw press, the screen is cleaned from the inside by the rotating screw flight, which has a nominal clearance of 0.5 mm (1/50 in.). In the inclined screw press, the wedge wire screen is cleaned from the inside by brushes fitted on the screw flights to prevent solids from attaching to the inner surface of the screen. The screen is cleaned from the outside via a spray wash system.

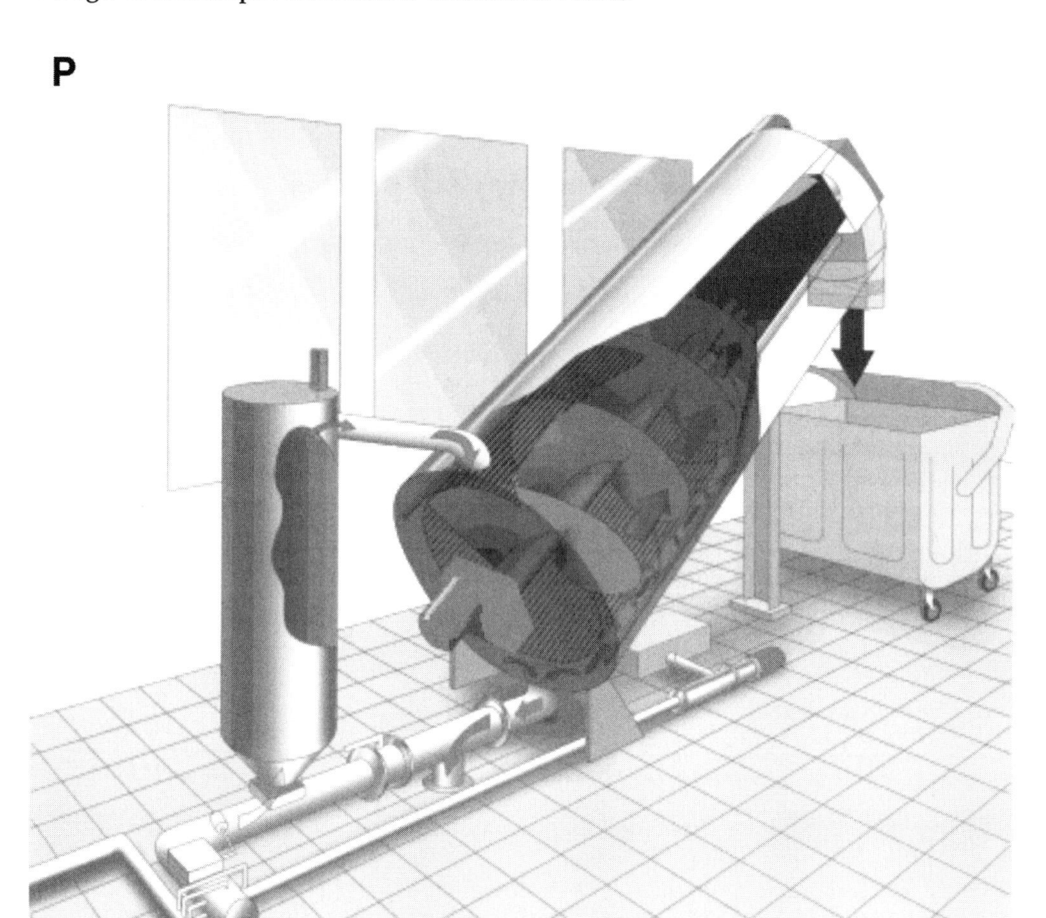

FIGURE 24.25 Cutaway view of an inclined screw press (courtesy of Hans Huber AG [Huber AG] of Germany).

A screw press also is used in a process that combines dewatering and pasteurization. In this patented process, lime is added to solids to raise the pH to 12. The lime-treated solids then are flocculated with polymer and fed to the screw press. Steam is then used to heat the screw press to achieve the pathogen-reduction requirements for Class A biosolids The resulting biosolids typically contain 30 to 50% solids. The process has more odor-control requirements because of the increased temperature and pH. The owner must decide early in the project if a Class A system is, or ever will be, a necessary part of the design because this system's design characteristics are different from those for standard screw presses and it is expensive to upgrade a dewatering screw press to the dewatering–pasteurization screw press. The first dewatering–pasteurization screw press installations started up in 2003, and there is presently little experience with this system in the United States.

Based on a survey of existing installations, solids concentrations in screw press cake vary widely depending on polymer use, solids characteristics, and dewatering application (see Table 24.8). The higher solids concentrations typically are achieved in applications with primary solids or with the combined dewatering–pasteurization process. Typical solids concentrations are 15 to 28% for secondary WAS, and 13 to 40% for a blend of primary and secondary solids. Digested solids typically result in higher concentrations than undigested solids, but require a larger polymer dosage. The dewatering–pasteurization screw press process produces cake containing up to 50% solids because of the added lime. Cake dewatering performance for solids with poor dewatering properties (e.g., in treatment facilities without primary clarifiers) can be significantly lower compared to other dewatering processes (Kabouris et al., 2005).

In addition to extrapolating data from other similar facilities, pilot testing provides useful information for screw press designs. Pilot testing can be used to establish trends and relationships among various performance measures (e.g., hydraulic loading rate, solids loading rate, polymer dose, solids capture, and cake solids concentration). However, because of significant variations in screw design and operating parameters, scaling up pilot information to a full-scale design requires careful evaluation and, potentially, an appropriate full-scale performance-guarantee contract (when pilot-scale units are used). If available, full-scale screw presses should be used in pilot testing to better predict full-scale performance.

Screw presses run continuously at low speeds and do not require close operator supervision; therefore, they are easy to maintain and have low power consumption. The manual cleaning schedule ranges from once per week to once every 30 days. A summary of the advantages and disadvantages of screw press dewatering is given in Table 24.9.

7.2 Process Design Conditions and Criteria

Screw press designs are determined by solids characteristics, proper chemical conditioning, and hydraulic and solids loads, which affect the detention time (speed of the screw). It is essential to select the proper conditioning agent (polymer); flocculation should agglomerate solids into large, strong flocs that release as much water as possible. The released water drains via gravity in the thickening section of the screw press. Screw press performance improves when less water must be removed in the dewatering zone.

Table 23.9 Advantages and disadvantages of screw presses.

Advantages	Disadvantages
Low rotational speed results in low maintenance and noice.	Cake concentration may be relatively low, in particular when there are no primary clarifiers.
Low operating energy consumption.	Larger footprint.
Containment of odors and aerosol, low building corrosion potential	Few manufacturers available and equipment cannot be specified "as-equal." It must be sole-sourced or pre-purchased.
Simple operation with low operator attention.	Requires wash water.
Lower than belt presses wash water demand and wash water pressure requirements.	Relatively few existing installations in the United States.
	Lower solids capture than other dewatering processes in some cases.

7.3 Mechanical Features

Significant variations exist in screw design among manufacturers, and detailed screw design information is proprietary. Primary features include the screw, screen, flocculation system, cleaning, and drive. Stainless steel (either 304 or 316L) typically is used for all wetted screw surfaces.

7.3.1 Screw

Screw press design parameters include the inlet-end screw-flight outside diameter; inlet-end shaft/shell diameter; inlet-end flight pitch or flight-to-flight dimension; screw length; discharge-end screw-flight outside diameter; discharge-end shaft/shell diameter; discharge-end flight pitch or flight-to-flight dimension; combinations of single-helix and double-helix designs. Common designs include constant flight outside-diameter with tapered screw shaft/shell diameter; and varied flight outside-diameter with constant screw shaft/shell diameter.

The inclined screw design (flight thickness, pitch, shaft diameter) is determined by the machine size and screen configuration. The diameter of the screw is 28 to 80 cm (11 to 31.5 in.), and the pitch of the flights typically is 15 to 25 cm (5.9 to 9.8 in.), although up to 40 cm (15.7 in.) is possible. Inclination varies from 15 to 20°. The inclination of the screw press allows the cake conveyor to fit under the discharge so the screw press does not need another support/pedestal. The inclination also facilitates startup, because discharge is above the inlet elevation.

7.3.2 *Screen*

Screw-press screen configurations also vary significantly among manufacturers. Screens can be either perforated stainless steel sheet or wedge wire. The holes used in perforated screens range from 1.0 to 3.0 mm in diameter, depending on solids type and press size. The open area of perforated screens ranges from 2 to 48%, depending on solids type, inlet consistency, and location on the screw press. The drums that support the screens can be either two-piece or one-piece. The advantage of two-piece (split) drums is ease of assembly and disassembly, because the drums (screens) can be removed from the press without removing the screw. Two-piece (split) drums allow an easy changeout of the inner perforated screens to fine-tune press performance, while removing one-piece drums requires disassembling the press.

The inclined screw press is only available with wedge wire screens made of stainless steel [304 or 316 Ti (optional)]. The bar spacing varies from 0.05 to 0.5 mm, depending on screen configuration, solids quality, and pilot-testing results. There are up to three wedge weir sections in a screw press. The open area of the screen sections is 3 to 30%. The support structure for the basket does not affect the open area. The main goal is to ensure a capture rate of at least 95% regardless of what type of solids are dewatered.

There is the possibility of nonstandard press designs for solids with high fiber content (e.g., primary solids). The fiber content allows for bigger openings, which may allow higher loading rates without compromising capture efficiency. Nonstandard screw press designs are always based on pilot tests.

7.3.3 *Cleaning System*

Screw press systems have automatic cleaning systems which involve plant water and spray nozzles. During automated wash cycles, washwater from solenoid valves sprays onto the screw press screen to remove built-up solids. The washwater-system design pressure is 345 kPa (50 psi). The inclined screw press has two cleaning processes. First, the screen is cleaned continuously from the inside via brushes or wipers mounted on the edge of the rotating screw flight. The brushes are made of nylon with stainless steel mounting hardware. This mainly cleans the screen to allow water to drain by gravity (especially in the lower part of the screen) and minimize resistance to water filtration. Clean screens require less dewatering pressure, which improves the solid capture rate. The second cleaning process is an automatic spray wash system, which cleans the screen from the outside. It is comprised of a rotating spray-bar washing system and spray nozzles fed by solenoid valves. The spray-bar system is made of stainless steel piping (304, 316 on request) with flat fan nozzles [made of polyvinylidene difluoride (PVDF)], and the washwater system design pressure is 414 to 517 kPa (60 to 75 psi).

7.3.4 Flocculation System

Flocculation tanks typically are used to mix solids and polymer to condition solids before dewatering. The horizontal screw-press flocculation tank is a vertical cylindrical tank with the solids–polymer feed at the bottom and an overflow at the top. The tank typically is sized for a retention time of 3 to 10 minutes, depending on solids type and inlet consistency. Flocculation tanks are provided with variable-speed agitators to allow operators to optimize mixing energy. Undermixing results in undispersed polymer, while overmixing results in broken floc. If the flocculation tank is undersized, then a non-clog inline static mixer can be used to blend solids and polymer before the mixture enters the flocculation tank.

In the inclined screw-press design, polymer is injected into the solids flow immediately before they enter the mixing device. The mixing valve is equipped with a manually adjustable weight to adjust the mixing energy in response to solids characteristics and to minimize polymer consumption. The flocculation tank typically is sized for a retention time of 0.5 to 1.0 minute and is equipped with a VFD-controlled stirrer to control floc formation.

7.4 Structural Elements and Building Requirements

Structural elements for screw presses include a concrete base (pad) or elevated support system for horizontal screw-press installations, a screw-press support system for inclined screw presses, flocculation vessel, screw press, and support for the rotary screen thickener, if present. Depending on screw-press installation configuration, a stair and landing system or movable ladder may be required to gain access to all parts of the screw press.

Because the fully enclosed design contains odors and minimizes operating noise, screw presses can be installed outdoors in mild climates (see Figure 24.26). If installed indoors, there must be adequate room around the perimeter of the press to accommodate normal maintenance and manual washdown (see Figures 24.27 and 24.28). Adequate overhead space also must be considered for maintenance, as well as lifting equipment. Lifting equipment can be either fixed (e.g., a crane rail centered above the press) or mobile equipment. If mobile equipment will be used, then equipment access needs to be considered.

Because screw presses are fully enclosed, ventilation requirements are minimal. If odors are a concern, screw-press covers can be fitted with ventilation connections. Such connections should have flexible ductwork that can be removed by operators without tools (e.g., a flex hose slipped over a pipe stub, held in place by a hose clamp with a thumbscrew). Typical ventilation airflow rate requirements range from 340 to 680 m^3/hr

FIGURE 24.26 Outdoor installation of a horizontal screw press.

(200 to 400 cu ft/min) for direct connection to the screw press. The screw press housing, screw-press discharge, and/or cake conveyor system also can be attached to a ventilation system for odor control.

Following are other design and layout considerations for screw presses:

- Provide curb around the screw-press area to protect other areas from spills and washdown water.
- Provide access to all parts of screw press via platforms or movable stepladders.
- Provide a hoist or crane for the installation, removal, and/or repair of screw-press components.
- Provide a plant water connection for the automatic washdown system.
- Provide connections to the plant's odor-control system if odor ducts off of the screw press are desired. The dewatered solids conveying system also should be connected to the odor-control system.

FIGURE 24.27 Indoor installation of a horizontal screw press.

- Provide adequate space between screw press units for maintenance and manual cleaning operations.
- Storage tanks from which solids are being fed to the screw press should have enough mixing to ensure a near-constant feed solids concentration during dewatering.
- Primary and secondary solids should be blended in an upstream storage tank; mixing solids in the pipe feeding the dewatering system is not recommended.

7.5 Hydraulic and Solids Loading Rates

In addition to solids loading, solids type, and desired discharge dryness, the hydraulic loading rate is a factor that should be considered in screw press sizing. Higher hydraulic loading rates typically require larger-diameter presses or a coarser screen. An important factor affecting the hydraulic loading rate is solids conditioning, because the screw press can only attain maximum capacity with the optimal polymer type and application. Optimal operation occurs when flocculated solids have sufficient time to fully gravity drain

FIGURE 24.28 Indoor installation of an inclined screw press.

before being conveyed beyond the inlet end of the screw press. Typical hydraulic loading rates for a horizontal screw press range from 3.8 to 2 081 L/min (1 to 550 gpm), depending on the screw press model. Typical hydraulic loading rates for an inclined screw press are between 18.9 and 227 L/min (5 and 60 gpm), depending on the model.

The solids loading rate for screw press dewatering varies, depending on solids characteristics, screw size, and rotational speed. The solids loading rate capacity for a horizontal screw press ranges from 0.91 to 703 kg/h (2 to 1550 lb/hr), depending on screw press model. The typical solids loading rate for an inclined screw press is 22.7 to 295 kg/h (50 to 650 lb/hr), depending on screw press model. The inclined screw press design is mainly controlled by solids loading, but hydraulic loading is an important factor if the residuals have a low solids content.

Screw presses can operate under a wide range of load rates just by changing the screw speed. The design involves balancing an increased solids or hydraulic loading rate with increasing screw speed and higher cake solids concentrations (which improve at slower screw speeds). Screw press operations are determined by solids flow (hydraulic or solids loading) and auger speed (retention time of conditioned solids in dewatering system). Cake solids will improve as auger speed decreases, provided that solids flow is constant. The speed can be reduced until incoming flow exceeds the amount of water that drains by gravity in the lower part of the screw press. Then, solids will start backing up.

If screw speed is constant, then the screw press can be operated with a variable solids-feed flowrate as follows. Because cake solids will improve as solids flow increases, the flowrate can be increased until the incoming flow exceeds the amount of water that drains by gravity at the front of the screw press and solids start to back up. The flowrate then is decreased a little to keep the system in a steady state.

The screw press will operate at its maximum capacity as long as the water drainage is not limited by high solids load and the volume of the auger is filled properly with solids to ensure maximum pressure in the dewatering zone. The loading highly depends on proper conditioning, so it is impossible to predict maximum loading-rate values. Maximum loading rate is plant-specific and must be determined onsite (e.g., during pilot tests or startup).

7.6 Unit Redundancy

Unlike many other dewatering systems, screw presses typically are designed for continuous operation; typical operations range from 5 to 7 days per week. Furthermore, because of the relatively infrequent maintenance requirements, screw press systems often are designed without a redundant (backup) unit. So, in a single-press installation, the plant must have facilities to store solids for several days if the press is off-line. Multiple screw presses may be necessary for larger facilities to provide redundant capacity and increase system reliability. Multiple screw presses also allow for a smaller number of screw presses to be operated during extended periods of lower solids production.

7.7 Rotation Speed

Typical rotation speeds range from 0.1 to 2.0 rpm for horizontal screw presses and from 0.5 to 2.0 rpm for inclined screw presses. In general, an increase in screw rotation speed increases production capacity but decreases cake solids concentration. In a full-scale application, increasing rotational speed from 1 to 1.25 rpm reduced the cake concentration from 23 to 20% (Atherton et al., 2005).

7.8 Ancillary Equipment and Controls

Ancillary equipment needed for screw press dewatering includes a solids feed pump, polymer makeup and feed system, flocculation vessel, and dewatering control system. The screw press manufacturer may provide all of these components as part of a package system.

The dewatering control system is provided by the manufacturer; it includes an operator interface to monitor and adjust screw press operations. The main control panel operates the entire dewatering system [i.e., the solids feed pump, polymer system, flocculation vessel, rotary screen thickener (if applicable), and screw press] as a complete system. The control system is equipped with operating and warning lights for various system monitoring points, audio alarms, emergency shutoff devices, and a display panel. Screw-press dewatering systems typically are operated in an automatic mode.

Screw press performance depends on a consistent inflow of solids. Therefore, positive-displacement or progressing-cavity pumps are highly recommended because their performance is not affected by solids consistency or changing water levels in the storage tanks. The progressing-cavity pumps must be designed properly to minimize the wear and tear caused by grit and other abrasive material.

The polymers used in screw press dewatering are not specially designed for this application; they are widely available from various suppliers. The choice of polymer needs to be determined via jar testing. There are no specific requirements for the polymer makeup system. It needs to be sized properly for the maximum solids load. System controls should allow automatic operation and include a feature to pace polymer feed rate with solids flow. Using an aging tank will maximize polymer efficiency and, therefore, minimize polymer consumption.

The system always needs some operator attention, especially if the residuals' solids concentration is inconsistent. The control system can only adjust the polymer dosing rate when solids flow (hydraulic load) changes; any change in solids loading due to solids concentration must be addressed manually by operators. So, it is essential to operating reliability that the solids concentration is constant (i.e., proper mixing occurs in the storage tank). Solids content could be monitored continuously, but such sensors often require a lot of maintenance and operator attention. Therefore, it may be easier to readjust the dewatering system's settings by changing the "solids content of raw solids" parameter at the operator interface.

For the horizontal screw press, solids pump speed typically is adjusted automatically based on headbox level analog input. The goal is to maintain a constant headbox level.

7.8.1 Chemical Conditioning

Polymer addition promotes particle flocculation and increases the dewatering and solids-capture rates. Jar testing and pilot testing can be used to estimate the type and quantity of polymer necessary for each application, because it may vary significantly depending on solids characteristics. Polymer consumption is affected by multiple parameters (e.g., grit content of the solids, the presence or absence of primary clarifiers, the type of biological treatment, and the type and duration of solids digestion). Polymer doses for screw press systems can range from 3 to 17.5 g of active polymer per kilogram of dry solids (6 to 35 lb of active polymer per dry ton of solids), with a typical range of 6 to 10 g/kg (12 to 20 lb/dry ton). The dewatering–pasteurization process also requires lime addition. Lime dosage typically ranges between 100 and 400 g/kg of dry solids (200 and 800 lb/dry ton of solids). In general, an increase in polymer dose increases cake solids concentration, although other factors (e.g., screw speed) can affect performance as well.

The polymer type and dosage also largely determines the solids capture rate. Solids capture also is affected by the efficiency of polymer injection and mixing, the screen design (size of opening), and the pressure inside the dewatering zone.

7.8.2 *Energy Requirements*

Screw presses have relatively low power requirements. Screw-press motor horsepower ranges from 0.67 to 10 kW (0.5 to 7.5 hp) for horizontal screw presses and from 0.67 to 2.7 kW (0.5 to 2 hp) for inclined screw presses, depending on screw press size. Screw press installations also require a solids feed pump and polymer system, whose power requirements depend on system size. The flocculation tank's mixer motor is typically 2 kW (1.5 hp) or less.

7.8.3 *Washwater and Pressate*

Pressate quantity is a factor of inlet flow, inlet consistency, and washwater flow. No screw press uses continuous washwater; most have automatic, intermittent showers. Typically, a horizontal press' automated wash cycle occurs for 1 minute each hour. The resulting washwater flow volume is 2 to 5% of solids feed rate (on average). Depending on the size of the screw press, the instantaneous flowrate can range from 20 to 120 gpm. The presses also require manual washdown at a frequency ranging from once per week to once per month. Typical washwater volume required for manual washdown is 0.2 to 0.5% of solids feed rate (on average).

The inclined screw press' automatic spray wash system is controlled by a timer and runs intermittently, typically for one revolution (about 60 seconds) every 10 to 15 minutes, but there are installations where the wash cycle only operates once every 30 minutes. Typical washwater demand for an inclined press is 49 to 163 L (13 to 43 gal) per wash cycle per screw press, depending on the screw press size. The actual flowrate is 79 to 132 L/min (21 to 35 gpm). Total washwater volume is 4 to 9% of the processed solids volume (on average), with a maximum of 15%. Water demand depends on solids flocculation performance. Manual spraydown of the flocculation vessel, screw, and screen is done as needed, depending on installation conditions and dewatering schedule.

Manual washdown takes about 1 to 2 hours and is recommended from about once per week to once per month, depending on screw press runtimes.

There are limited data available documenting the full-scale solids capture for screw presses under various field conditions. Solids capture typically ranges from 85 to 97%. Captures of up to 99% have been reported in certain applications, but a maximum of 95% capture is more common. Pressate typically is clear when a screw press dewaters WAS. It is slightly cloudy when primary or digested solids are processed. The particles are typically small, and the solids content varies depending on screw-press status: low solids content during dewatering, and higher solids content during the wash cycle.

8.0 REFERENCES

Atherton P.A.; Steen, R.; Stetson, G.; McGovern, T.; Smith, D. (2005) Innovative Biosolids Dewatering System Proved a Successful Part of the Upgrade to the Old Town, Maine, Water Pollution Control Facility. *Proceedings of the 78th Annual Water Environment Federation Technical Exhibition and Conference* [CD-ROM]; Washington, D.C., Oct 29–Nov 2; Water Environment Federation: Alexandria, Virginia.

Banks, L.; Davis, S. (1983) Desiccation and Treatment of Sewage Sludge and Chemical Slimes with the Aid of Higher Plants. *Proceedings of the 15th National Conference on Municipal and Industrial Sludge Utilization and Disposal*; Atlantic City, New Jersey; Hazardous Materials Control Research Institute: Silver Springs, Maryland.

Baxter, J. C.; Martin, W. J. (1982) Air Drying Liquid Anaerobically Digested Sludge in Earthen Drying Basins. *J. Water Pollut. Control Fed.*, **54**, 16.

Coackley, P.; Allos, R. (1962) The Drying Characteristics of Some Sewage Sludges. *J. Proc., Inst. Sew. Purif.* (G.B.), **6**, 557.

Crolla A.; Goulet, R.; Kinsley, C.; Ho, T., (2007) Septage Treatment Pilot Project, Drying Bed and Reed Bed Filters. Proceedings of the Ontario Onsite Wastewater Association Conference, Mar. 26–28. Ontario Onsite Wastewater Association: Cobourg, Ontario, Canada.

Crosswell, S.; Young, T.; Benner, K. (2004) Performance Testing of Rotary Press Dewatering Unit Under Varying Sludge Feed Conditions. *Proceedings of the 77th Annual Water Environment Federation Technical Exhibition and Conference* [CD-ROM]; New Orleans, La., Oct 2–6; Water Environment Federation: Alexandria, Virginia.

Genter, A. L. (1934) Adsorption and Flocculation as Applied to Sewage Sludges. *Sew. Works J.*, **6**, 689.

Haseltine, T. R. (1951) Measurement of Sludge Drying Bed Performance. *Sew. Ind. Wastes*, **23**, 1065.

Heukelekian, H.; Weisberg E. (1958) Sewage Colloids. *Water Sew. Works*, **105**, 428.

Imhoff, K.; Fair, G. M. (1940) *Sewage Treatment*; Wiley & Sons: New York.

Jeffrey, E. A. (1959) Laboratory Study of Dewatering Rates for Digested Sludge in Lagoons. *Procedings of the 14th Purdue Industrial Waste Conference*; West Lafayette, Indiana; Purdue University: West Lafayette, Indiana.

Jeffrey, E. A. (1960) Dewatering Rates for Digested Sludge in Lagoons. *J. Water Pollut. Control Fed.*, **32**, 1153.

Kabouris J. C.; Gillette, R. A; Jones, T. T.; Bates, B. R. (2005) Evaluation of Belt Filter Presses, Centrifuges, and Screw Presses for Dewatering Digested Activated Sludge at St. Petersburg's Water Reclamation Facilities. *Proceedings of the 78th Annual Water Environment Federation Technical Exhibition and Conference* [CD-ROM]; Washington, D.C., Oct 29–Nov 2; Water Environment Federation: Alexandria, Virginia.

Kinsley, C.; Crolla, A.; Ho, T.; Goulet, R. Septage Treatment Pilot Project: Drying Bed and Reed Bed Filters; Ontario Rural Wastewater Centre, University of Guelph. http://www.oowa.org/conference/2007/SeptageTreatmentPilotProject.pdf (accessed June 2009).

Koch, C. M.; Chao, A; Semon, J. (1988) Belt Filter Press Dewatering of Wastewater Sludge *ASCE J. Environ. Eng.*, **114** (5), 991–1005.

Koch, C. M.; McKinney, D. E.; Fagerstrom, A. A.; Palmer, E. W. (1989) Comparison of

Centrifuge Performance on Oxygen Activated Sludge *Proceedings of the Environmental Engineering Division, American Society of Civil Engineers in cooperation with the University of Texas, at Austin, Civil Engineeering Department;* Austin, Texas, Jul 10–12; American Society of Civil Engineers: New York.

Lecey, R. W. (1980) Polymers Peak at Precise Dosages. *Water Wastes Eng./Ind.*, **17**, 39.

Morris, R. H. (1965) Polymer Conditioned Sludge Filtration. *Water Works Wastes Eng.*, **2**, 68.

Novak, J. T.; Haugan, B. E. (1979) Chemical Conditioning of Activated Sludge. *J. Environ. Eng.*, **105** (EE5), 993.

Oerke, D. W. (1981) Fundamental Factors Influencing Dewaterability of Wastewater Solids. Master's essay, Marquette University, Milwaukee, Wisconsin.

Pietila, K. A.; Joubert, P. J. (1979) Examination of Process Parameters Affecting Sludge Dewatering with a Diaphragm Filter Press. *Proceedings of the 52nd Annual Water Pol-*

lution Control Federation Exposition and Conference; Houston, Texas, Oct 7–12; Water Pollution Control Federation: Washington, D.C.

Quon, J. E.; Johnson, G. E. (1966) Drainage Characteristics of Digested Sludge. *J. Sanit. Eng. Div., Proc. Am. Soc. Civ. Eng.*, **92**, 4762.

Randall, C. W.; Koch, C. T. (1969) Dewatering Characteristics of Aerobically Digested Sludge. *J. Water Pollut. Control Fed.*, **41**, R215.

Reimers, R. S.; et al. (1981) *Parasites in Southern Sludges and Disinfection by Standard Sludge Treatment*; Project Summary; U.S. Environmental Protection Agency: Washington, D.C.

Riggle, D. (1991) Reed Bed System for Sludge *Biocycle*, **32** (12), 64–66.

Rolan, A. T. (1980) Determination of Design Loading for Sand Drying Beds. *J., N.C. Sect., Am. Water Works Assoc., N.C. Water Pollut. Control Assoc.*, **L5**, 25.

Sharman, L. (1967) Polyelectrolyte Conditioning of Sludge. *Water Wastes Eng./Ind.*, **4**, 50.

Tenney, M. W.; Stumm, W. (1965) Chemical Flocculation of Microorganisms in Biological Water Treatment. *J. Water Pollut. Control Fed.*, **37**, 1370.

Thomas, C. M. (1971) The Use of Filter Presses for the Dewatering of Sludges. *J. Water Pollut. Control Fed.*, **43**, 93.

U.S. Environmental Protection Agency (1979) Evaluation of Dewatering Devices for Producing High Sludge Solids Cake. Contract No. 68-03-2455; U.S. Environmental Protection Agency, Office of Research and Development: Cincinnati, Ohio.

Vesilind, P. A. (1974a) Scale-Up of Solid Bowl Centrifuge Performance. *J. Environ. Eng.*, **100**, 479.

Vesilind, P. A. (1979) *Treatment and Disposal of Wastewater Sludges*, revised ed.; Ann Arbor Science Publishers: Ann Arbor, Michigan.

Walski, T. M. (1976) Mathematical Model Simplifies Design of Sludge Drying Beds. *Water Sew. Works*, **123**, 64.

Water Environment Federation (1994) *Safety and Health in Wastewater Systems*, 5th ed.; Manual of Practice No. 1; Water Environment Federation: Alexandria, Virginia.

Water Pollution Control Federation (1983) *Sludge Dewatering*; Manual of Practice No. 20; Water Pollution Control Federation: Washington, D.C.

Zablatzky, H. R.; Peterson, S. A. (1968) Anaerobic Digestion Failures. *J. Water Pollut. Control Fed.*, **40**, 581.

Zacharias, D. R.; Pietila, K. A. (1977) Full-Scale Study of Sludge Process and Land Disposal Utilizing Centrifugation for Dewatering. *Proceedings of the 50th Annual Meeting of the Central States Water Pollution Control Association*; Milwaukee, Wisconsin; Central States Water Pollution Control Association: Milwaukee, Wisconsin.

Chapter 25

Stabilization

1.0 INTRODUCTION

An important component of wastewater treatment plants, stabilization processes treat the solids generated in the liquid treatment train, converting them to a stable (i.e., not readily putrescible) product for use or disposal. They reduce pathogens and odors—provided the solids are properly stabilized and remain stable over time—making the resulting biosolids appealing for beneficial use. The four most common stabilization processes used in the United States today are anaerobic digestion, aerobic digestion, composting, and alkaline stabilization. (Thermal drying also is considered a stabilization process, but it is covered in Chapter 26 because its design is more similar to other thermal processes.)

When designing a stabilization process, engineers should start by evaluating the reasons for stabilization and all the stabilization options that could be easily integrated into the existing wastewater treatment scheme. They also must assess the local market for biosolids before selecting a process. Once a process is chosen, engineers should review all of its aspects (e.g., sidestream management, effectiveness in producing the desired biosolids quality, safety, ease of operation, and ancillary equipment needs) to ensure that the system is well-designed. This includes evaluating all upstream and downstream processes to ensure that they are flexible and reliable enough to consistently produce biosolids with the required characteristics. In other words, design engineers should use a systematic approach that addresses both the economic and noneconomic ramifications of any proposed processing.

Not all wastewater treatment plants stabilize their solids. Those who do, however, typically stabilize them for one or more of the following reasons:

- Aesthetic reasons [e.g., product appearance and odor (putrescibility control)],
- Mass reduction,
- Volume reduction,
- Biogas (renewable energy) production,
- Better dewaterability,
- Reduction of pathogens,
- Vector attraction reduction, and
- Product usefulness and marketability.

Facility planners should consider residuals management when designing or upgrading wastewater treatment plants. They should start by deciding how the solids removed from wastewater will be used or disposed, because this will determine whether and what type of stabilization is needed. For example, if solids will be landfilled with routine cover, federal and many state agencies may not require that the material be stabilized. Stabilization also may not be desirable if the solids will be thermally oxidized, because doing so will reduce the solids' calorific value. On the other hand, if treated solids will be used in agriculture or silviculture, or possibly distributed commercially, the material first must be stabilized to reduce pathogens, odor, and vector attraction. In such cases, planners also should consider whether any public-access or crop restrictions could be involved, because they will affect the choice of stabilization technology.

1.1 Comparison of Processes

The implementation of 40 CFR 503 and the public's growing concern for the environment have increased research into new technologies for beneficially using biosolids. They also have prompted many engineers and municipalities to investigate or design more effective stabilization systems. Tables 25.1 and 25.2 summarize many of the advantages and disadvantages of the principal stabilization processes used today. (Heat drying is not included in these tables because it is evaluated in Chapter 26.)

Anaerobic digestion may be the most widely used solids process discussed in this manual of practice. It produces relatively stable biosolids at a moderate cost, as well as methane gas that can be used to heat digesters, heat or cool buildings, or for cogeneration (e.g., a gas-engine-driven generator with jacket water heat recovery). Compared to other stabilization options, however, it is expensive to build, requires a significant amount of mechanical equipment (particularly if the digester gas is beneficially used); produces a strong ammonia sidestream; needs extra heat to maintain the desired temperature; and can upset because of poor mixing, overloading, lack of temperature control, and heavy metals or other toxic agents in the feed. Such disadvantages are largely resolved via proper design, operations, and pretreatment programs.

Aerobic digestion typically is used at smaller wastewater treatment plants [i.e., those with capacities less than about 19 000 m^3/d (5 mgd)] and those that only produce biological solids or waste activated sludge (WAS). Compared to anaerobic digestion, aerobic digestion is a power-intensive process (because of the power needed for oxygen transfer), but it typically is less expensive to construct and less complex operationally than anaerobic digestion.

Researchers have developed many methods for increasing volatile solids destruction, gas production, and pathogen destruction in aerobic digestion processes. Some use higher temperatures to destroy pathogens and reduce volatile solids, while others use phasing or mechanical disruption to improve performance. Sometimes aerobic digestion is not a separate process; many extended aeration facilities (e.g., oxidation ditches) have a long enough solids retention time (SRT) to provide at least partial digestion via endogenous respiration. However, Part 503 does not permit the volatile solid

TABLE 25.1 Comparison of stabilization processes.

Process	Advantages	Disadvantages
Anaerobic digestion	Good volatile suspended solids destruction (40 to 60%) Net operational cost can be low if gas (methane) is used Broad applicability Biosolids suitable for agricultural use Reduces total sludge mass Low net energy requirements	Requires skilled operators May experience foaming Methane formers are slow growing; hence, "acid digester" sometimes occurs Recovers slowly from upset Supernatant strong in ammonia and phosphorus Cleaning is difficult (scum and grit) Can generate nuisance odors resulting from anaerobic nature of process High initial cost Potential for struvite (mineral deposit) Safety issues concerned with flammable gas
Advanced anaerobic digestion (many process options)	Excellent volatile solid destruction Can produce Class A biosolids using time and temperature based batch operations Can increase gas production Can reduce solid retention time	Requires skilled operators Can be maintenance intensive (see Anaerobic digestion for other disadvantages)
Aerobic digestion	Low initial cost, particularly for small plants Supernatant less objectionable than anaerobic Simple operational control Broad applicability If properly designed, does not generate nuisance odors Reduces total sludge mass	High energy cost Generally lower volatile suspended solids destruction than anaerobic Reduced pH and alkalinity Potential for pathogen spread through aerosol drift Biosolids typically are difficult to dewater by mechanical means Cold temperatures adversely affect performance May experience foaming
Autothermal thermophilic aerobic digestion	Reduced hydraulic retention compared with conventional aerobic digestion Volume reduction Excess heat can be used for building heat Pasteurization of the sludge pathogen reduction	High energy costs Potential of foaming Requires skilled operators Potential for odors Requires 18 to 30% dewatered solids
Composting	High-quality, potentially saleable product suitable for agricultural use Can be combined with other processes Low initial cost (static pile and window)	Requires bulking agent Requires either forced air (power) or turning (labor) Potential for pathogen spread through dust High operational cost: can be power, labor, or chemical intensive, or all three May require significant land area Potential odors
Lime stabilization	Low capital cost Easy operation Good as interim or emergency stabilization method	Biosolids not always appropriate for land application Chemical intensive Overall cost very site specific related to product management costs

(*continued*)

TABLE 25.1 Comparison of stabilization processes (*continued*).

Process	Advantages	Disadvantages
		Volume of biosolids to be managed is increased pH drop after treatment can lead to odors and biological growth in product odorous operations
Advanced alkaline stabilization	Produces a high-quality Class A product Can be started quickly Excellent pathogen reduction	Operator intensive Chemical intensive Potential for odors Volume of biosolids to be managed is increased May require significant land area

reduction (VSR) achieved in aeration tanks to be included as part of the 38% VSR required for biological solids stabilization.

Autothermal thermophilic aerobic digestion (ATAD) is an advanced aerobic digestion process that operates at 50 to 65°C (131 to 149°F). A decade's worth of refinements and lessons learned from early ATAD systems have helped contribute to the success of more recent ATAD systems.

Composting often is used to convert solids into a soil amendment or conditioner. The feedstock can be either raw solids or biosolids, but should contain at least 40% solids. A bulking agent frequently is added to increase solids content, provide carbon for the

TABLE 25.2 Attenuation effect of well-conducted treatment processes on stabilizing wastewater solids.

	Degree of attenuation	
Process	Pathogens	Putrefaction and odor potential[a]
Anaerobic digestion	Fair	Good
Advanced anaerobic digestion	Excellent[b]	Good
Aerobic digestion	Fair	Good
Autothermal thermophilic aerobic digestion	Excellent	Good
Lime stabilization	Good	Good
Advanced lime stabilization	Excellent	Good
Composting	Excellent	Good

[a] In addition to the stabilization process, putrefaction and odor potential also depends on post-processing and storage practices.

[b] For Class A time–temperature processes.

process, improve the material's structural properties, and promote adequate air circulation. Composting typically is a labor-intensive process (e.g., adding bulking agent, turning the material, and recovering the bulking agent). It also can emit odors, especially if the site is poorly designed or operated. In addition, the process may increase the mass of biosolids to be used or disposed, and could spread pathogens via dust from the material.

Lime or alkaline stabilization frequently is used to meet the 40 CFR 503 requirements for Class B biosolids. In some cases, this process can produce a soil amendment or conditioner that meets Class A requirements. Alkaline stabilization typically is less costly and simpler to operate than digestion and composting. However, the resulting biosolids can become unstable if the pH drops after treatment and biological organisms regrow. Also, the lime or alkaline agent often is costly and can significantly increase the mass and cost of solids to be used or disposed. Odors and undesirable working conditions also have been noted at alkaline stabilization facilities.

A number of advanced alkaline stabilization technologies now used in the wastewater treatment field include chemical additives in addition to, or instead of, lime. A few of these processes use chemicals (e.g., cement kiln dust, lime kiln dust, Portland cement, fly ash, and other additives) to meet Class A criteria [processes to further reduce pathogens (PFRPs)]. Other Class A (PFRP) technologies use supplemental heat or other chemicals (e.g., sulfamic acid) to reduce the lime dose needed for pasteurization temperatures.

Alkaline stabilization processes produce a rich, soil-like product containing few pathogens. The biosolids also have a higher pH, which is desirable at farms with acidic soils. However, alkaline stabilization increases the mass of biosolids to be managed, as well as generating strong ammonia and amine odors that may need to be treated. One of the more important parameters for alkaline stabilization is mixing efficiency, which depends on the raw materials used in the process (rheology of dewatered cake and gradation of lime). Improper mixing results in variable biosolids characteristics and odors during storage and land application.

2.0 ANAEROBIC DIGESTION

2.1 Process Development

Anaerobic digestion has been the primary technology used to stabilize wastewater solids for the last 40 years. The major objectives of the technology have historically been as follows:

- Stabilize raw and waste solids,
- Reduce pathogen density,

- Reduce the mass of material via biomethanization, and
- Produce usable biogas as a byproduct.

As sustainable practices have developed in the United States, the role of anaerobic digestion has evolved to include the following objectives:

- Generate a biosolids product with fertilizer value,
- Recover resources by codigesting solids with other organic wastes, and
- Develop power and energy via biogas use in cogeneration facilties.

As the importance of anaerobic digestion has increased in the wastewater industry, a number of new process alternatives, designs, and fundamental understandings have evolved.

Anaerobic digestion is a relatively complex process biochemically, but mechanically it is quite straightforward. It requires both proper design and careful operation. The evolution of the technology has resulted in the need for more understanding of the fundamental aspects of design and process control to ensure that the system operates stably and efficiently.

Drawbacks of anaerobic digestion include the following:

- Handling potentially explosive and corrosive gases,
- A more complex system (biochemically and mechanically), a more complex system than aerobic processes, and
- Completely closed tanks make process monitoring more challenging.

2.2 Process Fundamentals

2.2.1 *Microbiology and Biochemistry*

Anaerobic digestion is driven by a series of syntrophic relationships that convert complex organic matter via a series of intermediate compounds to a variety of low-molecular-weight reduced compounds. The primary products of anaerobic digestion are methane (CH_4), carbon dioxide (CO_2), hydrogen (H_2), hydrogen sulfide (H_2S), ammonia (NH_3), phosphorus (PO_4), and residual organic matter and biomass. Digestion can be viewed as a series of steps in which the waste products of one organism are the substrate for another. Solids destruction is the result of a balanced coupling of a variety of metabolisms.

Figure 25.1 is a simplified flow diagram of the major metabolic processes in anaerobic digestion for converting organic matter to methane and carbon dioxide. Each metabolic pathway represents myriad microorganisms, many of which have yet to be speciated.

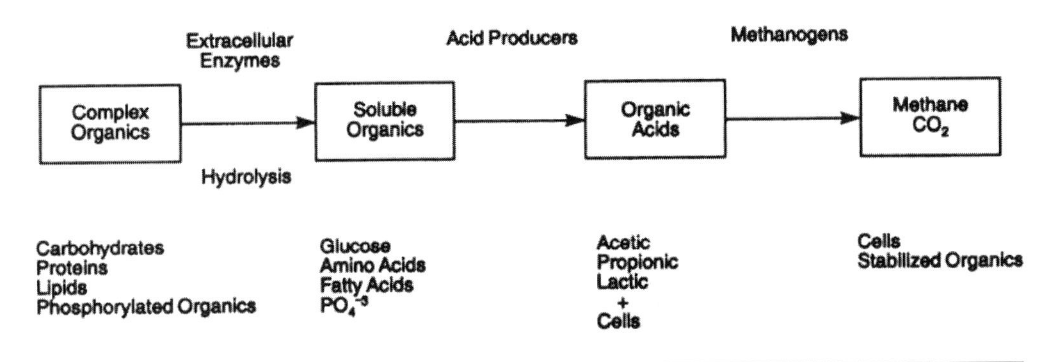

FIGURE 25.1 The major metabolic processes and products of anaerobic digestion.

The microorganisms responsible for digestion are bacteria and archea. Each group provides a unique and indispensable biotransformation. Hydrolysis, acidogenesis, and methanogenesis are the three major metabolic steps in anaerobic digestion. Each step involves several biochemical reactions to convert complex organics to intermediates, such as short-chained organic acids, and final products, such as methane and carbon dioxide.

2.2.2 *Process Rates and Kinetics*

Process rates are impacted by several external factors, including temperature, substrate, interspecies competition, and the presence of toxicants. While the environmental conditions affect the observed process rates, the fundamental limits of a process are regulated by kinetics.

Two kinetic models dominate the fundamental description of process performance: Michaelis–Menten and Monod. Michaelis–Menten kinetics describe the kinetics of enzymatic reactions, which are primarily responsible for the first step in anaerobic digestion: hydrolysis. Monod kinetics describe the process reactions mediated by specific microorganisms (e.g., acetogenesis and methanogenesis).

Table 25.3 provides a small sample of kinetic rates for different microbial populations associated with anaerobic digestion. As with enzymes, the different organisms have significantly different maximum growth rates and half-saturation coefficients. The minimum retention time is set by the slowest-growing organisms in a system.

When designing a system, a sufficient level of conservatism must be built into the design. Depending on substrate characteristics, complex organic matter will hydrolyze at different rates, and the microbial consortia that form in the digester will be a function of initial substrate characteristics. So the observed loading rates and retention time required for one facility will not necessarily translate to another because of differences in substrate characteristics.

Table 25.3 Summary of common microbial populations associated with anaerobic digestion and reported kinetic parameters (Muller, 2006).

Microbial population or metabolic group	μ_{max} (h^{-1})	K_s (mg substrate COD/L)	Source
Amino acid and sugar fermenting bacteria	0.25	20–25	Grady et al. (1999)
Long chain fatty acid oxidation	0.01	800	Grady et al. (1999)
Propionate oxidation	0.0065	250	Gujer and Zehnder (1983)
Propionate oxidation	0.0033	800	Bryers (1984)
Methanosarcina sp.	0.014	300	Grady et al. (1999)
Methanosaete sp.	0.003	30–40	Grady et al. (1999)
H_2 oxidizing methanogens	0.06	0.6	Grady et al. (1999)
Acetoclastic methanogenesis	0.0173	166	McCarty and Smith (1986)

2.2.2.1 *Hydrolysis*

In the first stage (hydrolysis), the proteins, cellulose, lipids, and other complex organics are cleaved into lower-molecular-weight components that can pass through the cell wall for conversion to energy and additional biomass. Hydrolysis is thought to be the rate-limiting step of anaerobic digestion (Pavlostathis and Gossett, 1988, 2004).

Several factors can affect hydrolysis, including the organisms present, growth condition, temperature, particle surface area (Sanders et al., 2000) and solids composition.

2.2.2.2 *Acidogenesis*

In the second stage (acid formation), the products of the first stage are converted to complex soluble organic compounds (e.g., long-chained fatty acids), which in turn are broken down into short-chained organic acids (e.g, acetic, propionic, butyric, and valeric acids). The concentration and relative proportions of these acids can be indicate the overall condition of a digester.

2.2.2.3 *Methanogenesis*

In municipal solids digesters, methanogenesis occurs via two primary metabolic pathways: acetoclasitc methanogenesis and hydrogenotrophic methanogenesis. *Acetoclastic methanogenesis*—methane formed via acetate reduction—is the primary route of methane formation (McHugh et al., 2006). Although *hydrogenotrophic methanogenesis*—methane formed via hydrogen reactions—produces less of the methane in a digester, it plays a critical role in preventing feedback inhibition. McCarty and Smith (1986) reported that when the partial pressure of hydrogen exceeds 5 mPa, fatty-acid hydrolysis becomes thermodynamically unfavorable. Volatile acids then accumulate, reducing pH and souring the digester. Hydrogenotrophic methanogenesis ensures that the system remains in balance.

Another methanogenic population is *methylotrophic methanogens*, which convert simple methylated compounds into methane and a reduced product. These organisms typically consume methyl mercaptan, trimethyl amine, dimethyl disulfide, etc. While not significant in overall digester performance, they play a critical role in controlling organic methylated odorants.

2.2.3 Microbial Ecology

An anaerobic digester can be described as an ecosystem whose environmental conditions are dictated by design engineers, operators, and the composition of the feedstock.

Feed solids composition, SRT, hydraulic retention time (HRT), temperature, and mixing regime all exert selectve pressures on the microbial populations in the digester. Such pressures lead to the proliferation of some species and the recession or absence of others. Selective pressure can be to such a degree that overall digestion capacity can be affected by the relative population of species, such as has been reported for *Methanosarcina spp.* and *Methanoseata spp.* (Conklin et al., 2006).

The relationship among microorganisms in an anaerobic digester can best be described as a syntrophic relationship. The metabolic activity of one population supports another, though not for the mutual benefit of either. Often the waste products of one group of organisms serve as the substrate of another. These relationships result in some distinct control points in the digestion process.

The production of hydrogen from fatty acid metabolism is thermodynamically not a highly favorable reaction. When hydrogen accumulates in the system at partial pressures above 5 mPa, there is a feedback inhibition of the acid oxidation process (McCarty and Smith, 1986, 2004). For the process to continue, as it does under stable digestion conditions, the hydrogenotrophic methanogenic population must be well established and respiring. What is evident from this one example is that a stress or toxin that affects one population may be enough to retard or upset the entire digestion process.

Feed solids characteristics will affect which populations are dominant and can set up conditions where there is competition between a desired population and one that is less desirable. For example, acetate, the substrate from which about 75% of the methane in biogas is generated, is also the preferred substrate of sulfate-reducing bacteria. If the feed solids have high sulfate concentrations, conditions may exist in which the methanogenic population is in direct competition with sulfate-reducing bacteria. When this happens, the biogas' methane content may shrink or overall biogas production can be reduced.

Anaerobic digestion can be adversely affected by loading changes, both quantity and quality. Given the complex nature of the different microbial interactions and the potential for process upset via stress on the weakest population, sufficient care must be taken when changing loading conditions, both quantity and quality.

2.2.4 Feedstock Characteristics

Key objectives of anaerobic digestion are to stabilize raw solids and reduce the mass of the material. Raw, primary, and secondary solids are primarily composed of the following compounds: proteins, polysaccharides, nucleic acids, fatty acids, and lipids. The relative concentrations of these compounds are a direct function of influent wastewater characteristics and the liquid treatment train used.

The overall solids composition will affect digester performance. Park et al. (2003) reported that the metal content—particularly iron and aluminum—can be used as an indicator of the relative digestibility of solids; however, a predictive tool is not available. Muller and Novak (2007) expanded this to the stability of centrifuged biosolids in terms of volatile sulfur-compound release.

2.2.5 Hydraulic and Solids Residence Time

Anaerobic digesters are sized to provide enough residence time in well-mixed reactors to allow significant volatile solids destruction to occur and to prevent slower growing microbe populations from wahing out. Sizing criteria, expressed as SRT or HRT, are defined as follows:

- *Solids retention time*, measured in days, is equal to the mass of solids in the digester (kilograms) divided by the solids removed (kilograms per day), and
- *Hydraulic retention time*, measured in days, is equal to the working volume (liters) divided by the amount of solids removed (liters per day).

Typically, HRT is calculated based on either the solids feeding or removal rate. However, if supernatant is removed from the digester, SRT is calculated based on the solids volume removed. Solids retention time and HRT are equal in digestion systems without recycle and decant.

The solids retention time (or HRT) and the extent of hydrolysis, acid formation, and methane formation during anaerobic digestion are directly related: an increase in SRT increases the extent of each reaction; a decrease in SRT decreases the extent of each reaction (see Figure 25.2). Each reaction has a minimum SRT; if the SRT is shorter, bacteria cannot grow rapidly enough to remain in the digester, the reaction mediated by the bacteria will cease, and the digestion process will fail. Excessively long SRTs would prevent washout, but the extra equipment and infrastructure costs typically are not justified by the marginal increase in process performance.

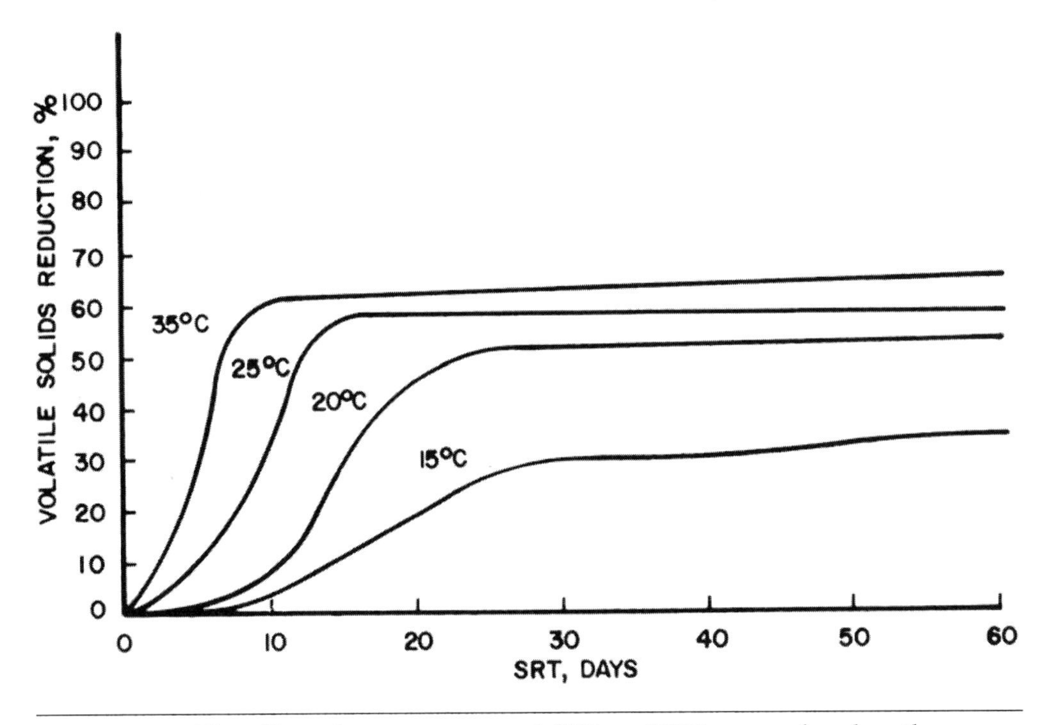

FIGURE 25.2 The effect of temperature and SRT on COD removal and methane production in anaerobic digesters.

Mesophilic anaerobic digestion of typical solids has been characterized via years of experience in operation and design, although design engineers always should consider the fundamental microbiology when designing a digester to optimize it toward maximum efficiency. Integrating process fundamentals into a design becomes increasingly important when the technology has a shorter operational history (e.g., thermophilic and phased systems). Applying inappropriate process parameters can result in improperly sized systems, which perform poorly and result in odors or other adverse consequences.

2.2.6 *Organic Loading Rate and Frequency*

The phrase *volatile solids loading* refers to the mass of volatile solids added to the digester each day divided by the digester's working volume [kg volatile solids/m^3·d (lb volatile solids/d/cu ft)]. Loading criteria typically are based on sustained loading conditions (typically peak month or peak week solids production), with provisions for avoiding excessive loading during shorter periods. A typical design sustained-peak loading rate for mesophilic digesters is 1.9 to 2.5 kg volatile solids/m^3·d (0.12 to 0.16 lb volatile solids/d/cu ft). The upper limit of the volatile solids loading rate typically is determined by the rate at which toxic materials—particularly ammonia—accumulate or methane formers wash out. A limiting value of 3.2 kg volatile solids/m^3·d (0.20 lb volatile solids/d/cu ft) is often used.

Thermophilic systems typically have much higher volatile solids loadings than mesophilic systems because the operating temperature is higher, which increases the growth and metabolic rates. Currently, the limited application of thermophilic systems (compared to mesophilic systems) has not generated the empirical data needed for a recommended operating range. When designing a thermophilic system, engineers should either pilot-test digester configurations to determine loading limits or perform a significant review of currently operating systems.

While design engineers should be conservative when setting an upper loading limit for a design, excessively low volatile solids loading rates can result in designs that are expensive to both construct and operate. Construction is expensive because of the large tank volume needed. Operation can be expensive because gas-production rates may not be sufficient to provide the energy required to maintain the desired operating temperature in the digester. Thickening of solids before digestion may be a cost-effective method for maintaining the design SRT (or HRT) at a low volatile solids loading rate (less than about 1.3 kg volatile solids/m^3·d [0.08 lb volatile solids/d/cu ft]).

The loading frequency can affect operations as well as design. Microorganisms typically prefer to be maintained at a constant metabolic state (steady-state), which is achieved via consistent, constant loading. Constant loading also equalizes gas flow and simplifies gas management. However, it does involve some significant additional design considerations.

To maintain constant loading to the system, the feed and wastage rates need to be balanced, which would mean constant thickening and dewatering, as well as sufficient tankage to equalize flows. A more common alternative is to provide sufficiently large storage tanks before and after digestion so thickening and dewatering can occur during peak staffing periods.

Constant feed is the ideal operation, but one not often achieved because of the cost constraints associated with implementation. Slug loading (semi-continuous feeding) is often used; it involves feeding and wasting solids at set intervals, typically at peak staffing times. This type of operation has been implemented at many utilities but has some process risks (e.g., overloading and foaming).

2.2.7 *Process Stability*

Stable anaerobic-digester operations can provide significant benefits (e.g., consistent solids destruction, pathogen reduction, and biogas generation). Stability is achieved through consistent loading, temperature control, and mixing—all at adequate levels but not exceeding maximum allowable limits.

A well-functioning, stable anaerobic process will exhibit specific digester and biosolids characteristics (see Table 25.4). Unstable digesters are more likely to foam, go acidic, and have microbial populations that are more susceptible to toxins.

Foaming is a common problem for many digesters. It can be caused by several factors (e.g., unstable digester operations, high concentrations of filamentous organisms in raw solids, and surfactants and other agents). Foaming associated with filaments and chemical additives can be remedied via source control. Foaming associated with digester stability is a result of the microorganisms responding to an environmental stresses.

A good design can help promote stable digestion oeprations. For example, boilers and heat exchangers should be sized to meet both the heat demands of raw solids and the shell losses that will occur under the coldest conditions expected. This will ensure that the system's temperature does not drop below the setpoint.

Blending tanks improve process stability by homogenizing raw solids and metering them more constantly to the digester. Minimizing fluctuations in solids strength and loading rate help the microorganisms in the digester maintain a constant metabolic state, which minimizes their stress. Wide, frequent fluctuations in loading can stress the biomass—especially if loadings, substrate, and nutrients are insufficient.

Mixing improves the contact between the biomass and raw solids. Dispersing solids in the digester ensures that its entire volume is used and all of the biomass is engaged in stabilization.

2.2.8 *Temperature*

An anaerobic digester's operating temperature significant affects its observed performance and stability. Temperature affects growth rates (Lawrence and McCarty, 1969;

TABLE 25.4 Typical operating parameters for mesophilic anaerobic digestion of wastewater solids.

Parameter	Value	Units
Volatile solids destruction	45–55	%
pH	6.8–7.2	
Alkalinity	2 500–5 000	mg $CaCO_3$/L
Methane content	60–65	percent by volume
Carbon dioxide content	35–40	percent by volume
Volatile acids	50–300	mg VA/L
Vol. acid : alkalinity ratio	<0.3	mg $CaCO_3$/mg VA
Ammonia	800–2 000	mg N/L

van Lier et al., 1996; Salsali and Parker, 2007); substrate half-saturation constants (Lawrence and McCarty, 1969; van Lier et al., 1996); and microbial diversity (Chen et al., 2005; Wilson et al., 2008a). Lawrence and McCarty (1969) observed that growth rates increased as temperature increased in the mesophilic operation range. Salsali and Parker (2007) evaluated anaerobic digestion performance at 35, 42, and 49°C; they observed volatile solids destruction increased as temperature increased. They did not attempt to derive growth rates from their experiments. Van Lier et al. (1996) evaluated volatile fatty acid degradation by methanogens and suggested that acetate conversion in mesophilic and thermophilic digestion was described by an Arrhenius relationship, suggesting an increase in growth rates as temperature increased. Both Lawrence and McCarty (1969) and van Lier et al. (1996) suggested an increase in substrate (volatile fatty acid) half-saturation constant for acetoclastic methanogenesis with an increase in temperature (i.e., rising temperatures increased residual acetic and propionic acids). Selecting an operating temperature not only affects digester design but also day to day operations. From a design standpoint, the ability to achieve and maintain that temperature is critical to process stability and optimization.

The design operating temperature establishes the minimum SRT (or HRT) required to destroy a given amount of volatile solids (see Figure 25.3). Currently, most anaerobic digesters are designed to operate in the mesophilic temperature range [about 35°C (95°F)]. Some systems have been designed to operate in the thermophilic temperature range [about 55°C (131°F)]. Many new digesters are being designed so they can operate at both thermophilic and mesophilic temperatures, allowing future process flexibility.

Regardless of which temperature is selected, keeping it stable in the digester is of utmost importance. The microorganisms involved (particularly methanogenic populations) are sensitive to temperature changes; fluctuations in temperature can stress the organisms, thereby destabilizing the process. Temperature changes greater than 1°C/d can result in process failure. A good design avoids temperature changes greater than 0.5°C/d.

As a minimum, temperature changes should be held to less than 1°C/d during stable operation. This is a critical consideration when determining feed schedules. However, for start-up of a thermophilic digester, a rapid increase in the temperature from mesophilic to thermophilic conditions has been reported to be an effective means of establishing a stable anaerobic population (Griffin et al., 2000).

Temperature stability not only affects microbial stability but also the classification of the resulting biosolids. Under 40 CFR 503 Alternative 1 (time and temperature), solids must be maintained at a specific temperature above 50°C for a set period of time to achieve Class A status. In this instance, time is separate from HRT because every particle must be treated, requiring a batch held in an isolated tank, unless process equiva-

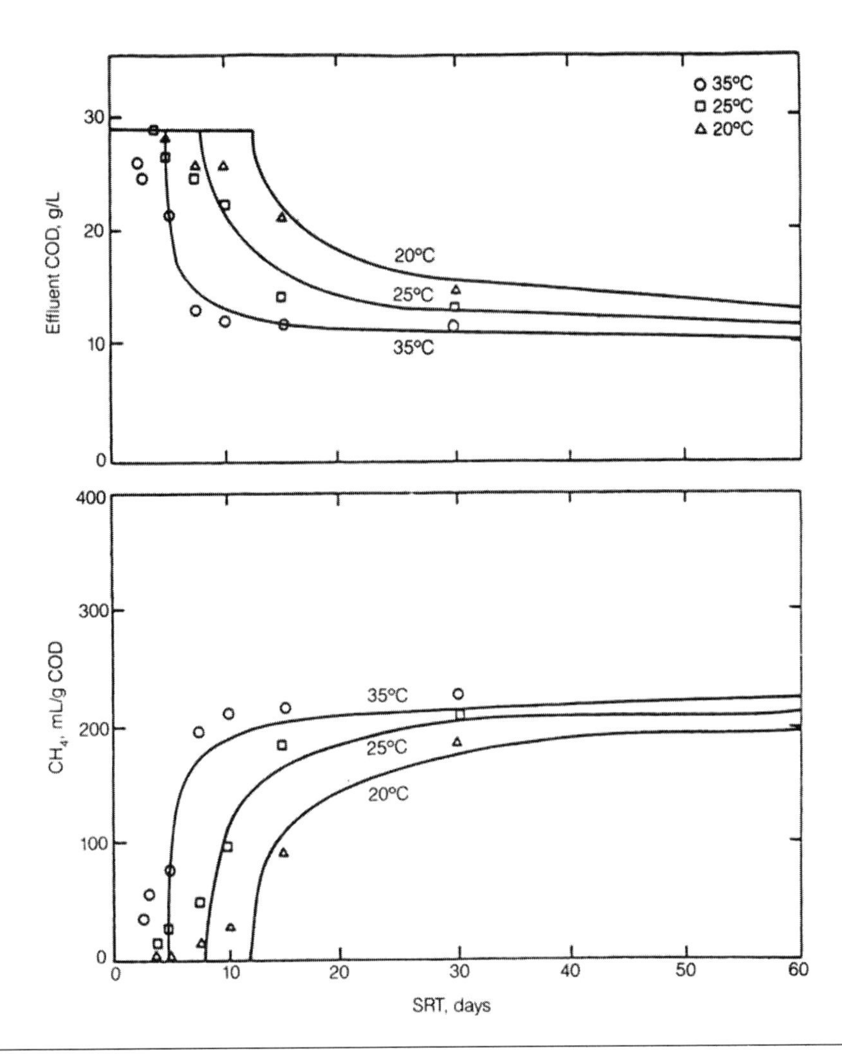

FIGURE 25.3 The effect of SRT and temperature on the rate and extent of VSR during anaerobic digestion (O'Rourke, 1968).

lency has been granted. The effect of temperature on pathogen inactivation has made it one of the core mechanisms for achieving Class A biosolids, so it is important to design the system to maintain temperature (within close tolerances) under varying loads.

2.2.9 *Volatile Fatty Acids, Concentration and Composition*

When designing anaerobic digestion systems, engineers need to understand how volatile acid concentrations affect system design. Volatile fatty acids are the primary intermediates between complex organic matter and methanogenesis. The gross concentration of volatile acids can indicate how complete digestion is, while the composition of the acids can indicate a process upset or disturbance. (High concentrations of propionic and/or butyric acid typically indicate a process upset or disturbance.) Volatile fatty acids also can lead to onsite odors because of fugitive emissions from digesters or dewatering processes.

In many cases, the concentration of volatile fatty acids in a digester is also a function of operating conditions. For example, the residual volatile fatty acid concentration increases when the operating temperature is high (van Lier et al., 1996). Higher ammonia concentrations also can result in higher residual volatile fatty acid concentrations (Nielsen and Angelidaki, 2008). The temperature and ammonia effects represent normal operating conditions as long as pH is not depressed.

As volatile acid concentrations increase, alkalinity is consumed. Once the buffer capacity is consumed, pH will decrease, leading to process upset and failure. Monitoring acid production and concentration can provide evidence of impending upset or recent disturbance, so operators can take remedial actions.

Volatile acid concentrations of 50 to 300 mg/L are considered normal for an anaerobic digester operating at mesophilic temperatures. This is not necessarily true in other anaerobic digestion systems, however; thermophilic systems [e.g., temperature-phased anaerobic digestion (TPAD)] and phased systems (e.g., acid–gas phasing) will have vastly different volatile acid concentrations in their reactors.

2.2.10 *Alkalinity and pH*

Anaerobic bacteria—particularly methane formers—are sensitive to pH. Optimum methane production typically occurs when the pH is maintained between 6.8 and 7.2. Acid forms continuously during digestion and tends to lower pH. However, methane formation also produces alkalinity—primarily carbon dioxide and ammonia, which buffer changes in pH by combining with hydrogen ions.

A reduction in pH (by various causes) promotes more acid formation and inhibits methane formation. As acid production continues, methane and alkalinity formation are further inhibited, possibly leading to process failure. Mixing, heating, and feed-system designs are important in minimizing the potential for such upsets. Design engineers also should include provisions for adding chemicals (e.g., lime, sodium bicarbonate, or sodium carbonate) to neutralize excess acid in an upset digester.

2.2.11 *Toxicity in Digesters*

If concentrations of certain materials (e.g., ammonia, heavy metals, light metal cations, and sulfide) increase sufficiently, they can create unstable conditions in an anaerobic digester (see Tables 25.5 and 25.6). A shock load of such materials in plant influent or a

TABLE 25.5 Concentrations of selected inorganic compounds that inhibit anaerobic processes (Parkin and Owen, 1986).

Substance	Moderately inhibitory concentration, mg/L	Strongly inhibitory concentration, mg/L
Na^+	3 500–5 500	8 000
K^+	2 500–4 500	12 000
Ca^{++}	2 500–4 500	8 000
Mg^{-+}	1 000–1 500	3 000
Ammonia-nitrogen	1 500–3 000	3 000
Sulfide	200	200
Copper (Cu)	—	0.5 (soluble) 50–70 (total)
Chromium VI (Cr)	—	3.0 (soluble) 200–250 (total)
Chromium III	—	180–420 (total)
Nickel (Ni)	—	2.0 (soluble) 30.0 (total)
Zinc (Zn)	—	1.0 (soluble)

TABLE 25.6 Concentrations of select organic chemicals that reduce anaerobic digester activity by 50%.

Compound	Concentration resulting in 50% activity, mM
I-Chloropropane	0.1
Nitrobenzene	0.1
Acrolein	0.2
I-Chloropropane	1.9
Formaldehyde	2.4
Lauric acid	2.6
Ethyl benzene	3.2
Acrylonitrile	4
3-Chlorol-1,2-propandiol	6
Crotonaldehyde	6.5
2-Chloropropionic acid	8
Vinyl acetate	8
Acetaldehyde	10
Ethyl acetate	11
Acrylic acid	12
Catechol	24
Phenol	26
Aniline	26
Resorcinol	29
Propanol	90

sudden change in digester operation (e.g., overfeeding solids or adding excessive chemicals) can create toxic conditions in the digester.

Typically, excess concentrations of such toxicants inhibit methane formation, which typically leads to volatile acid accumulation, pH depression, and digester upset. Depending on the concentration and type of toxicant, the effect can be acute (e.g., instant process failure) or chronic (e.g., depressed performance). Chemicals can be added to control the concentrations of dissolved forms of some toxicants (e.g., using iron salts to control sulfide).

Design engineers typically can only address toxicity by mitigating a known impact. Identifying and monitoring process toxicity (e.g., sampling and analytical techniques and practices) is typically an operational issue and beyond the scope of this text. However, a sound monitoring and control program, and an understanding of toxic agents, can greatly improve the design of mitigation systems.

2.2.12 Volatile Solids and COD

Volatile solids and chemical oxygen demand (COD) are common measures of the substrate entering a digester. *Volatile solids* are the ignitable (550°C) fraction of total solids. They typically are thought of as the organic fraction. Volatile solids measurements typically are used as part of determining overall process performance, regulatory compliance, and for mass-based calculations. One must be careful when using data from a volatile solids test. There are artifacts to the test, because a significant amount of inorganic salts—especially ammonium-based salts—can volatilize in analytical tests, skewing the volatile solids concentration higher and the volatile solids destruction lower (Beall et al., 1998; Wilson et al., 2008b).

Chemical oxygen demand is a measure of the chemically oxidizable material in solids. As with volatile solids, this measure has limitations (e.g. the accounting of non-substrate components). Chemical oxygen demand typically is used in kinetic models (e.g., ADM 1) because all of the parameters are based on COD coefficients (WEF, 2009; Batstone et al., 2002).

Most plants provide volatile solids data rather than COD data in their regulatory reports because of the relative ease of conducting the volatile solids test.

2.2.13 Biogas Production and Characterization

The quality and quantity of digester gas (biogas) produced also can be used to evaluate digester performance. Biogas production is directly related biochemically to the amount of volatile solids destroyed; it often is expressed as volume of gas per unit mass of volatile solids destroyed. The gas-production rate is different for each organic substance in the digester (see Table 25.7). The gas-production rate of fats ranges from about 1.2 to

TABLE 25.7 Gas-production rates from various organic substrates (Buswell and Neave, 1939).

Material	Specified gas production per unit mass destroyed	
	m^3/kg	Methane content, %
Fats	1.2–1.6	62–72
Scum	0.9–1.0	70–75
Grease	1.1	68
Crude fibers	0.8	45–50
Protein	0.7	73

1.6 m^3/kg (20 to 25 cu ft/lb) of volatile solids destroyed; the gas-production rate of proteins and carbohydrates is 0.7 m^3/kg (12 cu ft/lb) of volatile solids destroyed. The gas-production rate of a typical anaerobic digester treating a combination of primary solids and WAS should be about 0.8 to 1 m^3/kg (13 to 18 cu ft/lb) of volatile solids destroyed. The amount of gas produced is a function of temperature, SRT, and volatile solids loading. Specific gas production should be measured until an average value can be obtained and used for monitoring.

The two main constituents of digester gas are methane and carbon dioxide; it also contains trace amounts of nitrogen, hydrogen, and hydrogen sulfide. Performance data from healthy digesters suggest that methane concentrations should be 60 to 70% (by volume) and carbon dioxide concentrations should be 30 to 35% (by volume). Tortorici and Stahl (1977) have published data on typical digester gas characteristics (see Table 25.8).

TABLE 25.8 A survey of the characteristics of biogas from anaerobic digesters (Tortorici and Stahl, 1997).

Constituent	Values for various plants, % by volume[a]							
Methane	42.5	61.0	62.0	67.0	70.0	73.7	75.0	73–75
Carbon dioxide	47.7	32.8	38.0	30.0	30.0	17.7	22.0	21–24
Hydrogen	1.7	3.3	[b]	—	—	2.1	0.2	1–2
Nitrogen	8.1	2.9	[b]	3.0	—	6.5	2.7	1–2
Hydrogen sulfide	—	—	0.15	—	0.01	0.06	0.1	1–1.5
Heat value, Btu/cu ft[c]	459	667	660	624	728	791	716	739–750
Specific gravity (air = 1)	1.04	0.87	0.92	0.86	0.85	0.74	0.78	0.70–0.80

[a] Except as noted.
[b] Trace.
[c] Btu/cu ft × 37.26 = kJ/m^3.

An increase in carbon dioxide levels (percent) often indicates an upset digester. Excessive concentrations of hydrogen sulfide can indicate unbalanced digestion, industrial waste sources, or saltwater infiltration. Hydrogen sulfide may be responsible for odor problems and excessive corrosion in the digester and adjacent piping. Heavy metals can precipitate as metallic sulfide, thereby minimizing hydrogen sulfide concentrations in biogas.

2.2.14 *Pathogens*

Pathogen and pathogen-indicator reduction are major stabilization criteria. The rate and extent of pathogen or pathogen-indicator destruction (inactivation) are process-specific. The degree of pathogen or pathogen-indicator reduction required depends on the biosolids quality desired (Class B or Class A, assuming beneficial use is desired). Design engineers should consult 40 CFR 503 and other applicable regulations for pathogen or pathogen-indicator reduction requirements for anaerobic processes. (For more information on pathogen-reduction regulations, see Chapter 20.)

Design engineers should carefully consider which digestion system to use. While multiple methodologies meet the desired degree of pathogen reduction, each comes with costs and degrees of process complexity.

2.3 Process Options

Process options for anaerobic digestion of wastewater solids have advanced significantly since the early 1990s. This section provides some historical context and discusses process options that are being considered and implemented more frequently in the 21st century (e.g., staged and phased systems, and mesophilic and thermophilic processes).

2.3.1 *Low-Rate Digestion*

Before the 1950s or 1960s, solids were anaerobically digested in "low-rate" systems, which consisted of a cylindrically shaped tank with a flat or domed roof. External mixing typically was not provided, and heating often was limited or non-existent. So, stabilization resulted in stratified conditions in the digester.

Scum often accumulated on the liquid surface. Digester supernatant typically contained high ammonia and phosphorus concentrations; it often was drawn off and recycled to the plant headworks or primary clarifiers. Stabilized solids settled to the tank bottom for removal and further processing.

Low-rate digestion is characterized by intermittent feeding, low organic loading rates, no mixing other than that caused by rising gas bubbles, and detention times of 30 to 60 days. The tanks are large because grit and scum layers accumulate on the bottom and top, respectively, thereby decreasing the effective volume. It may or may not have external heat source to increase the digestion rate. Optimum digestion conditions are not maintained. Relatively few low-rate digestion systems are in service today because technology advancements have made them uneconomical and unattractive.

2.3.2 *High-Rate Digestion (Mesophilic and Thermophilic)*

Wastewater treatment professionals kept tinkering with the low-rate system, making improvements that eventually led to the "high-rate" anaerobic digestion system that was more common in the 1960s (see Figure 25.4). This system is widely used for mesophilic digestion.

2.3.2.1 Process Development

High-rate anaerobic digestion was developed after research demonstrated the benefits of controlling environmental conditions in the digester. High-rate digestion is characterized by supplemental heating and mixing, relatively uniform feed rates, and prethickening of solids (solids feedstock typically should contain 4 to 5% solids, although some recent improvements allow for thicker or thinner solids). These factors

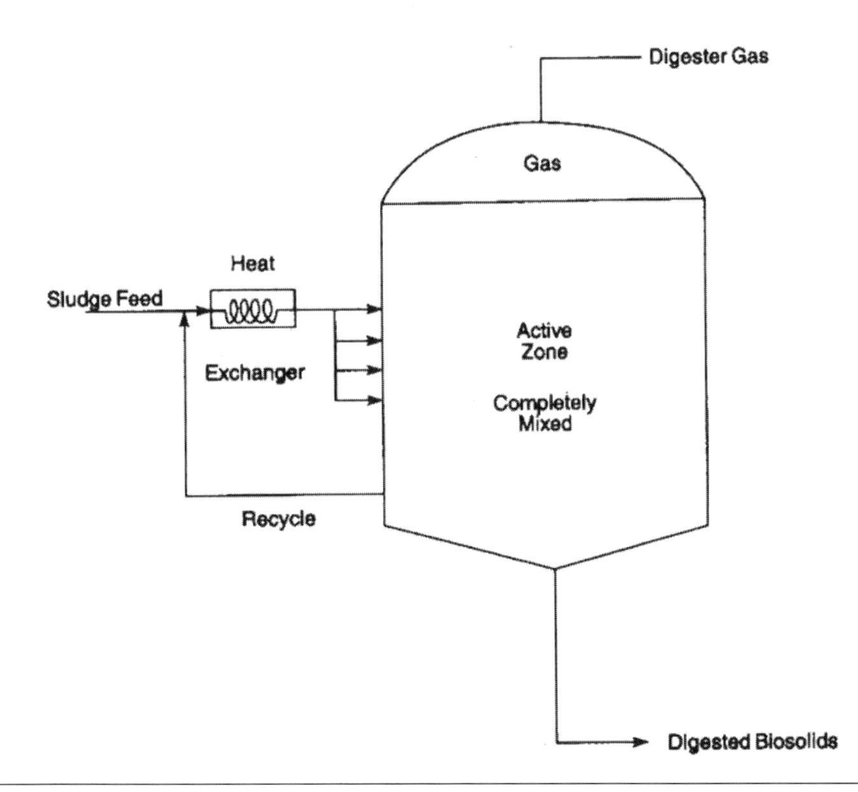

FIGURE 25.4 Simplified flow schematic of high-rate anaerobic digestion.

result in relatively uniform conditions throughout the reactor, leading to lower overall tank volume requirement and increased process stability.

Several heating methods (e.g., steam injection, internal heat exchangers, and external heat exchangers) have been used for anaerobic digesters. External heat exchangers are the most popular because of their flexibility and easily maintained heating surfaces. Internal coils can foul because of caking; fouled coils must be removed, or the digester must be emptied to clean them. Steam injection dilutes the contents of the digester but avoids the heat-exchange equipment.

2.3.2.2 Design Criteria—Mesophilic

Typically, converting from a low-rate system to a high-rate one will increase gas production, solids destruction, and overall process stability. In the 1970s, U.S. EPA extensively evaluated single-stage, high-rate anaerobic digesters operated at mesophilic temperatures with residence times exceeding 15 days; regulators found that the process achieves significant pathogen reduction and solids stability. The agency defined it as a process to significantly reduce pathogens (PSRP) in its 1979 rule (40 CFR 257), and it essentially became a baseline for wastewater solids stabilization.

The basic design criteria for such mesophilic digesters typically are as follows:

- Volatile solids loading rate of 1.9 to 2.5 kg volatile solids/m^3·d (0.12 to 0.16 lb volatile solids/cu ft/d), and a typical limiting value of 3.2 kg volatile solids/m^3·d (0.20 lb volatile solids/cu ft/d);
- SRT of at least 15 days when feeding at peak 15-day or month loads [a 15-day SRT is the minimum allowed under the PSRP (Class B) requirement in Part 503];
- Mesophilic temperatures [35 to 39°C (95 to 102°F); the PSRP (Class B) requirement in Part 503 is at least 35°C];
- Enough mixing to ensure that the temperature is relatively consistent throughout the reactor (and to minimize bottom deposits and surface scum/debris, although this is only partially achieved in many high-rate digesters); and
- Feed cake containing between 4 and 5% solids (historically), although more facilities are aiming for between 5 and 7% solids.

Frequent solids feeding helps maintain steady-state conditions in the digester. Methanogens are sensitive to changes in substrate levels; uniform feeding and multiple feed-point locations in the tank reduce shock loading to these microorganisms. Excessive hydraulic loading should be avoided because it decreases detention time, dilutes the alkalinity needed for buffering capacity, and requires more heat to achieve process goals. Good mixing also is required to disperse feedstock, mix microorganisms with fresh feed, and ensure that the temperature is consistent throughout the reactor.

Improvements in mixing, heating, and solids loading enhanced anaerobic digestion performance. Mixing and heating provide better contact between substrate and microorganisms, increasing stabilization while reducing short-circuiting (less short-circuiting makes pathogen kill more consistent, thereby increasing biosolids stability).

The relative success of high-rate mesophilic digestion has made this process the most common means of solids stabilization in the world. It also is the standard for evaluating future process variations.

2.3.2.3 Design Criteria—Thermophilic

Although most anaerobic digesters are operated at mesophilic temperatures [i.e., 35°C (95°F)], they also can be operated at thermophilic temperatures [typically between 50 and 57°C (122 and 135°F)]. Thermophilic digesters have somewhat different design and performance criteria than those for mesophilic digestion. For example, volatile solids loading rates can be higher and SRTs can be lower (Schafer et al., 2002). Thermophilic digesters also reduce more volatile solids than identical-size mesophilic digesters, as suggested by the Arrhenius relationship. Because the temperature is higher, however, more energy is needed to provide heat. To reduce heating costs, waste heat from cogeneration systems could be used to heat sludge in thermophilic digesters.

Thermophilic digestion has a number of advantages over mesophilic digestion. For example, thermophilically digested solids have better dewatering characteristics, so heating costs may be offset by reduced dewatering costs. Thermophilic digestion systems also destroy more volatile solids (Schafer et al., 2002) and typically produce biosolids containing fewer pathogens. However, Part 503 classifies both mesophilic and thermophilic anaerobic digestion (non-batch, non-phased systems) as PSRPs (Class B processes), so continuous-flow thermophilic digesters do not get regulatory credit for their pathogen-reduction performance. Part 503 also specifies that any process to further reduce pathogens (PFRP) (Class A process) must precede or be concurrent with the vector-attraction reduction process (e.g., mesophilic or thermophilic anaerobic digestion) to allay concerns about post-pasteurization regrowth (Clements, 1982; Keller, 1980). So, to meet Class A requirements, a thermophilic digester may need to be designed to be partly or wholly operated in batch mode, where every particle meets the time and temperature relationship established by U.S. EPA.

According to Part 503, a thermophilic digestion process typically meets Class A requirements by maintaining its temperature at or above 50°C (more typically, 55°C) for a specific period of time in a batch operation. The amount of time is calculated using the formula (U.S. EPA, 1999b) below:

$$D = \frac{50,070,000}{10^{0.14T}} \quad (25.1)$$

Where

D = time (days) and

T = temperature (°C).

This equation can be applied to cake containing less than 7% solids. Based on this equation, one point of compliance would be 55°C for a minimum of 24 hours. Another point of compliance would be 50°C for a minimum of 120 hours. As the thermophilic digestion temperature increases, the batching time (and thus tank volume) required to destroy pathogens or pathogen indicators is considerably reduced.

If a utility chooses to disinfect cake containing more than 7% solids, the time and temperature requirements are determined using the following equation:

$$D = \frac{131{,}700{,}000}{10^{0.14T}} \quad (25.2)$$

Where

D = time (days) and

T = temperature (°C).

So, for a cake containing more than 7% solids, the minimum batching time is 63.1 hours at 55°C. In practice, however, thermophilic digestion reactors are unlikely to be operated at such high solids concentrations because of the higher viscosity involved. Digesters need much more energy to pump and mix high-viscoity solids.

Early full-scale tests at Los Angeles indicated that thermophilic digesters were difficult to operate (Garber, 1982). However, more recent work shows that thermophilic digester operations are reliable when temperatures are constant and good mixing and feeding systems are used (Krugel et al., 1998). It has been suggested that early thermophilic digesters did not have the advantage of current mixing, heating, and temperature-control systems.

In summary, thermophilic digestion destroys more volatile solids and pathogens, and produces more biogas, but can be more costly to implement and operate than mesophilic digestion. It has been claimed that the increased costs can be offset by waste heat recovery and use, as well as lower dewatering costs. Design engineers may wish to perform pilot tests with the actual feedstock before deciding to use thermophilic digestion.

2.3.3 Primary–Secondary Digestion

A primary–secondary digestion system divides fermentation and solids-liquid separation into two separate tanks in series (see Figure 25.5). The first tank is an anaerobic

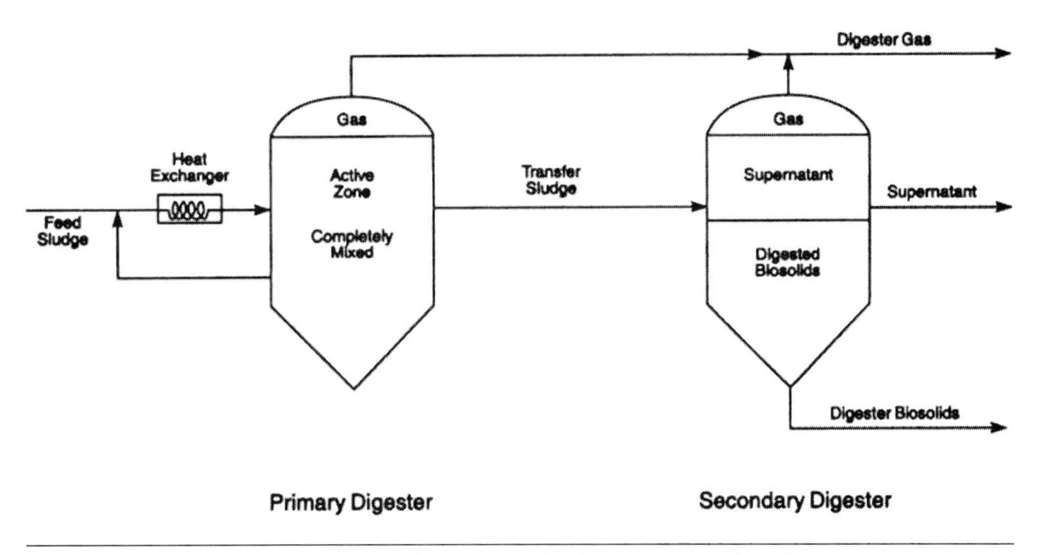

FIGURE 25.5 Process schematic of two-stage high-rate anaerobic digestion.

digester with a varying loading rate, while the second is a solid–liquid separator. The second reactor typically does not have mixing or heating facilities, unless it also is used to provide standby digester capacity. In fact, the second tank may serve several other functions (e.g., providing storage capacity and insurance against process short-circuiting). Primary–secondary digestion typically works well on primary clarifier solids (settled solids), because the second tank typically provides good separation.

However, if biological solids are introduced to this process, the second tank is not likely to separate solids from liquids nearly as well. If the second tank's supernatant contains a high concentration of suspended solids, recirculating it can be detrimental to the liquid treatment train. Poor settling can be caused by incomplete digestion in the primary digester (which generates gases in the secondary digester and causes floating solids) or by fine solids with poor settling characteristics (typically a concern when treating secondary or tertiary solids, including chemical solids). Because of the problem with biological solids, primary–secondary digestion is much less common today.

2.3.4 *Recuperative Thickening*

Recuperative thickening is a process that separates HRT from SRT in an anaerobic digester (see Figure 25.6). This process increases SRT; how much it increases depends on the amount or proportion of thickened solids and the degree of thickening provided. It also returns anaerobic bacteria to the digester to potentially increase biological activity.

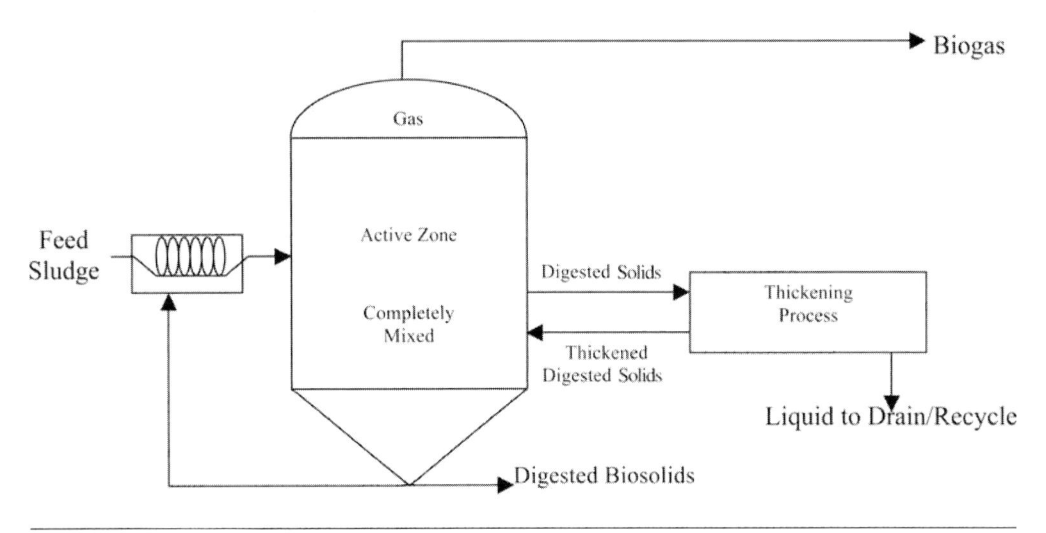

FIGURE 25.6 Simplified schematic of the recuperative thickening process.

Thickening options include centrifuges, gravity belt thickeners, and dissolved gas flotation thickeners [both air flotation (DAF) and anoxic gas flotation have been used]. Dissolved air flotation recuperative thickening was tested at the Spokane, Washington, treatment plant, and the bacteria survived the oxygenation effect (Reynolds et al., 2001).

Recuperative thickening has been used for several reasons (e.g., temporarily increasing digester SRT while some of the plant digestion capacity is off-line for maintenance or construction). It also can be used to delay construction of more digestion tank capacity. Sometimes, existing thickening equipment is used for recuperative thickening purposes.

The disadvantages of recuperative thickening include operational complexity, polymer use, maintenance needs, and costs for an additional process.

2.3.5 *Staged Digestion*

The concept of staged (phased) digestion has been used in various ways over the years (e.g., to stage metabolisms, operating temperatures, and redox conditions), but it is increasingly being recognized for its pathogen-control benefits. Primary–secondary digestion, for example, is basically a two-stage mesophilic process that produces well-stabilized solids because of the relatively long retention times and reduction of solids short-circuiting. Other staged mesophilic digestion options also are being recognized and used.

2.3.5.1 Two-Stage Mesophilic Digestion

Two-stage mesophilic digestion is an extension of single-stage high-rate anaerobic digestion that uses two complete-mix digesters in series (see Figure 25.7). The first stage

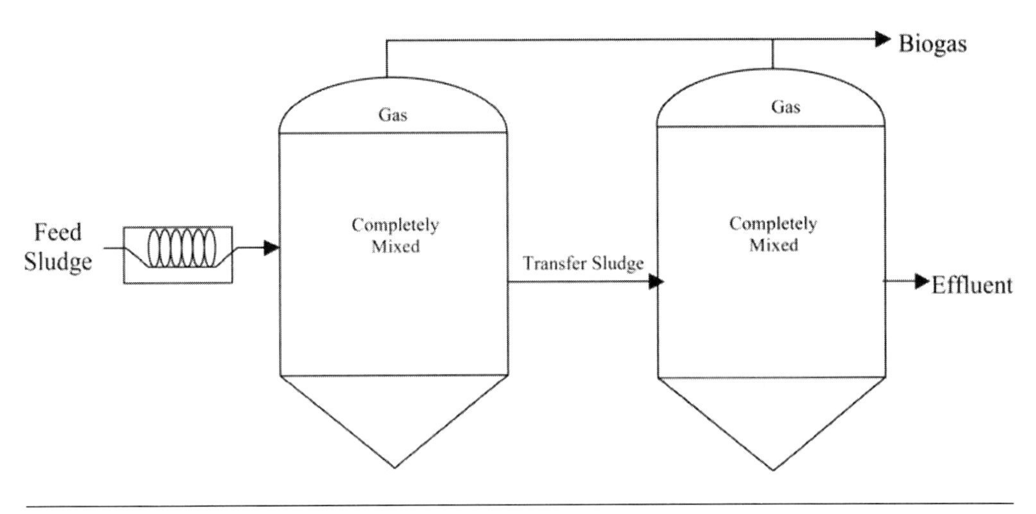

FIGURE 25.7 Schematic of the two-stage mesophilic digestion process.

must be designed to provide reliable mesophilic digestion (e.g., sufficient SRT and reasonable volatile solids loading). Because much of the process considerations are met in the first reactor, the second stage can operate with a relatively shorter SRT. Both stages are heated and mixed. Placing the tanks in series causes the reaction kinetics to behave more like a plug-flow rather than a complete-mixed process, thus reducing short-circuiting and improving process efficiency.

Schafer and Farrell (2000) reported that, compared to single-stage digestion, two-stage mesophilic digestion can

- Improve product stability (because more volatile solids are destroyed) and
- Reduce short-circuiting of raw solids and pathogens.

Zahller et al. (2005) reported that two-stage digestion provided better volatile solids destruction and biogas composition (i.e., more methane content) than a single-stage digester with an equivalent SRT. However, the study also noted that the two-stage system seemed to have less capacity to absorb large variations in loadings than the single-stage system, as measured by additional acetate use capacity.

2.3.5.2 *Multiple-Stage Thermophilic Digestion*

Compared to mesophilic processes, thermophilic digestion offers more gas production, solids reduction, and pathogen destruction. Likewise, two-stage thermophilic digesters are more effective than single-staged systems. The Annacis Island Wastewater Treatment Facility in Vancouver, British Columbia, is an example of multiple-stage thermophilic digestion (see Figures 25.8 and 25.9). Krugel et al. (1998) predicted, based on the equations describing time and temperature relationships for batch systems, that Annacis' process would achieve pathogen and pathogen-indicator reductions equivalent to a Class A batch process, and the agency's monitoring results confirm the predictions. The system also has been reported to show low organic sulfur release from centrifugally dewatered biosolids, as well as no *E. coli* regrowth following dewatering—something not observed in other thermophilic or mesophilic anaerobic processes (Chen et al., 2008).

2.3.6 Temperature-Phased Anaerobic Digestion

Temperature-phased anaerobic digestion uses both thermophilic and mesophilic digestion to improve digestion performance (see Figure 25.10). Such systems are not nearly as common as conventional mesophilic systems. The Western Lake Superior Sanitary District in Duluth, Minnesota, has a TPAD system (see Figures 25.11 and 25.12). Constructed in 2001, this system has four tanks: one that operates as a thermophilic digester, followed by three tanks operating in parallel as mesophilic digesters.

2.3.6.1 Process Development

Researchers in Germany identified the potential advantages of temperature-phased digestion. Anaerobic digesters in Cologne, Germany, have been operated in a tempera-

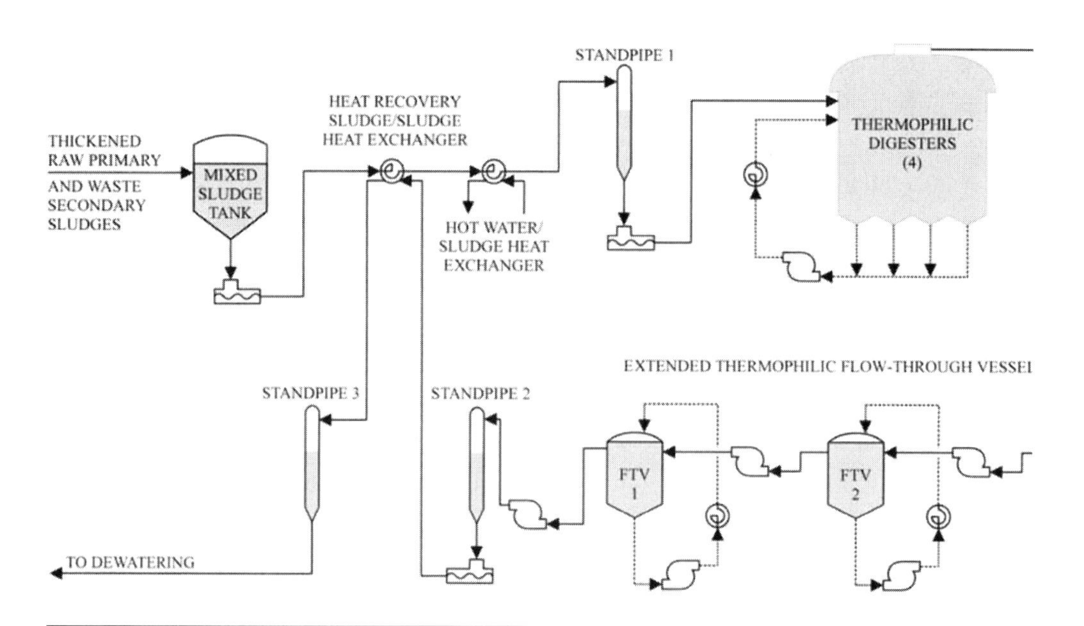

FIGURE 25.8 Schematic of the extended thermophilic digestion process at Annacis Island, Vancouver, British Columbia (Krugel et al., 1998; reprinted from *Water Science and Technology*, with permission from the copyright holders, IWA).

FIGURE 25.9 Photograph of the extended thermophilic digestion process at Annacis Island, Vancouver, British Columbia (courtesy of Brown and Caldwell).

ture-phased mode since August 1993 (Dichtl, 1997). In the United States, Han and Dague (1996) conducted laboratory studies documenting the advantages of temperature-phased anaerobic digestion, and a patent for the TPAD process was issued to Iowa State University in the 1990s based on Dague's work.

The thermophilic digester's greater hydrolysis and biological activity tends to provide more volatile solids destruction and gas production than an all-mesophilic digestion process. The system also reduces the tendency of high-rate mesophilic digesters to foam when treating combined solids (primary solids and WAS) (Han and Dague, 1996). Other advantages include lower coliform counts in digested solids (Han and Dague, 1996) and the potential to meet Class A pathogen criteria under 40 CFR 503.

The TPAD's mesophilic stage provides additional volatile-solids destruction and biogas production, as well as conditions solids for further handling. It also reduces the

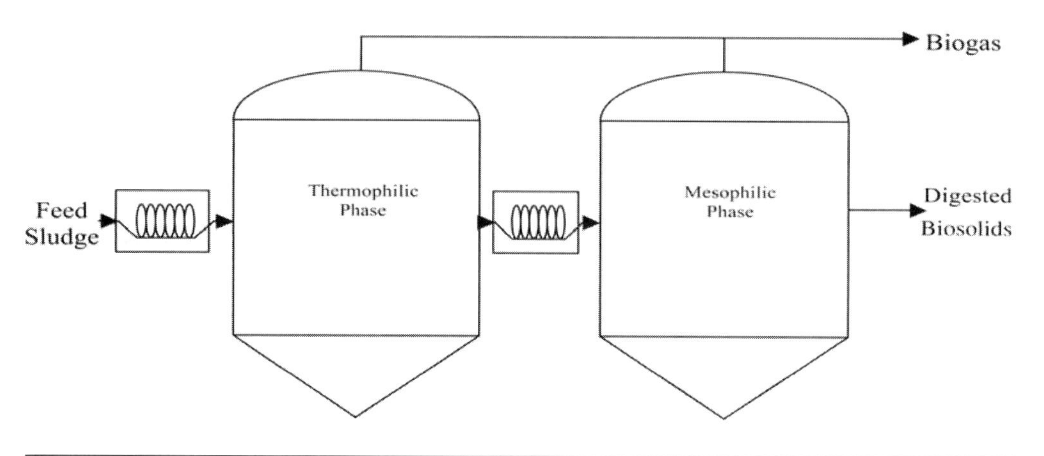

FIGURE 25.10 Schematic for the temperature-phased anaerobic digestion process.

concentration of odorants (mostly fatty acids) that are common to thermophilic digestion, increases operational stability, and produces biosolids with more consistent characteristics. The biosolids also produce higher cake solids content during dewatering than those produced by mesophilic digesters.

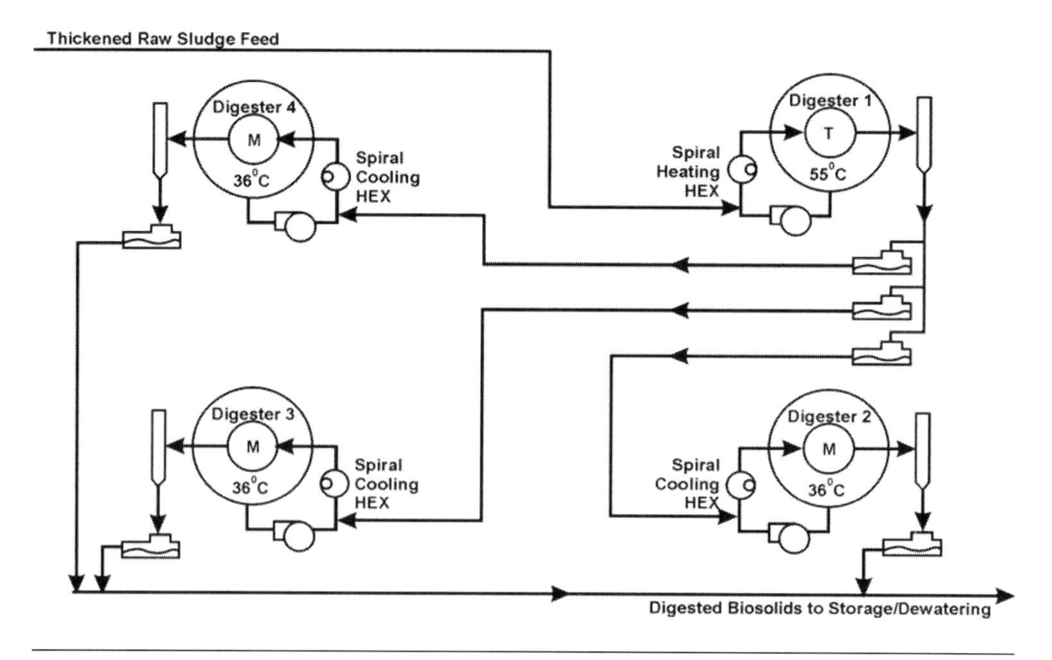

FIGURE 25.11 Process flow schematic of the TPAD installation at Western Lake Superior Sanitation District in Duluth, Minnesota (Krugel et al., 2006).

FIGURE 25.12 Photograph of the TPAD installation at Western Lake Superior Sanitation District in Duluth, Minnesota (courtesy of Brown and Caldwell).

The temperature-phased process is designed to take advantage of thermophilic digestion rates, which are estimated to be four times faster than mesophilic digestion (Dague, 1968). Dague evaluated a system in which the thermophilic stage operated at 55°C (131°F) with a 5-day detention time and the mesophilic stage operated at 35°C (95°F) with a 10-day detention time. Other researchers have tried different residence times for the thermophilic and mesophilic phases and found performance improvements at a variety of residence times for each stage. Few full-scale plants have operated the thermophilic phase at 5 days SRT or less, but research shows that this can be successful.

2.3.6.2 Design Criteria

Design criteria for TPAD vary because performance success has been demonstrated under various situations. However, based on most of the research and full-scale experience to date, design criteria for a typical TPAD system are

- Thermophilic temperatures of 50 to 57°C;
- Thermophilic residence times of 4 to 10 days (existing TPAD systems may have longer SRTs because a large tank was available for the thermophilic stage, or early-year loads were less than design loads);
- Mesophilic temperatures of 35 to 40°C; and
- Mesophilic residence times of 6 to 12 days (again, existing TPAD systems may have longer SRTs because of the tankage used or loads that are less than design loads).

When designing a TPAD system, engineers should choose design criteria based on project objectives, solids feed characteristics and variability, and existing facilities (because most TPAD systems are modifications of existing digestion systems). If there are wide variations in feedstock quantity and characteristics, design engineers may want to use a longer residence time in the first-stage thermophilic reactor—perhaps a 10-day SRT or longer. If an existing tank can be used for thermophilic digestion, but its SRT is only 4 or 5 days, the TPAD system may work well as long as the mesophilic system's SRT is long enough to adequately handle some variable performance from the first-stage thermophilic reactor. Total SRTs of about 15 days (minimum) are considered good design practice for peak 15-day or peak month loads. If the plant needs to ensure that its biosolids meet Class B standards, then a 15-day total SRT (minimum) typically is required. Another Part 503 requirement is that digester temperatures in the mesophilic stage must remain above 35°C (if mesophilic SRT is needed to meet Class B requirements).

2.3.6.3 Performance

The performance of TPAD systems often is measured based on volatile solids reduction (VSR) or biogas production. Schafer et al. (2002) reported significant improvement in VSR at several plants using the TPAD process, compared to the performance of a mesophilic system with a similar SRT. (If the mesophilic and TPAD systems have different SRTs, then direct performance comparisons are more difficult to quantify without additional information.)

Schafer et al. (2002) also reviewed pilot- or demonstration-scale studies and found that TPAD outperformed mesophilic digestion systems when fed the same feedstock and using similar total SRTs. Improvements in VSR are often cited as follows:

- A high-rate mesophilic digester with a 20-day SRT achieved 50% VSR, while
- A TPAD system with a 20-day total SRT and identical or similar feedstock achieved 57% VSR.

2.3.6.4 Heating, Cooling, and Other Design Considerations

Temperature-phased anaerobic digestion can be heated (and cooled) in several ways, with some precautions:

- For energy efficiency, the heat from thermophilic solids can be recycled to heat cold feedstock and partially cool the thermophilic solids. Various arrangements have been used for this heat recycling concept (e.g., solids/solids heat exchangers and solids/water/solids heat exchangers). This approach requires supplemental heat (e.g., heat exchangers or steam addition) to ensure that feedstock reaches thermophilic conditions.
- Solids can be heated to thermophilic temperatures via heat exchangers and/or steam addition without heat recycling. In this case, thermophilic solids typically must be cooled before entering the mesophilic stage, and this heat can be transferred to plant effluent, blown into the atmosphere, or used to heat water for building heating or other purposes.
- In colder seasons, and if the mesophilic stage's SRT is long enough, a purposeful cooling system may not be needed because the mesophilic stage can cool itself via thermal losses to the atmosphere and ground. Design engineers should calculate whether this is possible and if the mesophilic stage may be too hot for reliable performance under certain conditions.

Engineers should ensure that the thermophilic stage is operated at a consistent temperature because its microorganisms are more sensitive to temperature changes, particularly increases. For example, if the design temperature is 55°C, the control system should ensure that this temperature is maintained (close tolerances). Also, a mixing system is required to ensure that all tank contents are close to the measured thermophilic temperature. The temperature tolerances for the mesophilic stage are not as exacting, but must be evaluated carefully for reliable performance.

Other design considerations include protection against odor release from thermophilic digesters. For example, floating covers typically are avoided because of odor release from the annular space. Also, the gas-handling system must be designed to handle the much larger production rate and moisture content of thermophilic biogas. In addition, design engineers may need to evaluate whether an existing mesophilic digester is adequate for thermophilic service. This includes a structural evaluation of the tank, its mechanical systems, piping, and coating/lining systems.

2.3.7 *Two-Phase Anaerobic Digestion*

Two-phase digestion (sometimes called *acid–methane digestion*) separates two major anaerobic reactions—acid formation (acidogenesis) and methane generation (methanogenesis)—to benefit the overall stabilization process (see Figure 25.13). The most practical way to separating phases is via kinetic control, by regulating the detention time and loading rate for each reactor. Increasing loadings to the first-stage digester and reducing SRT (HRT) favors acidogenic organisms because the low pH and retention time are unfavorable to methane formers. In the second stage, a larger digester (or multiple digesters) increases SRT, so methanogens proliferate. High influent concentrations of short-chain fatty acids also promote the growth of methanogens.

2.3.7.1 Process Development

Early work on the process was largely completed by Professor Sam Ghosh (Ghosh et al., 1975, 1987; Lee et al., 1989). Raw solids initially are fed to a reactor with 1- to 2-day SRT, called an *acid-phase digester*. In this reactor, low-pH environment (typically 5.5 to 6.2) is established, suspended organic matter is hydrolyzed, and then low-molecular-weight fatty acids are formed. Methane generation is limited in this phase. This first phase has been tested at both mesophilic and thermophilic temperatures, although few full-scale systems have operated the acid phase at thermophilic temperatures.

Acid-phase biomass then is fed to a second vessel with a 10- to 15-day SRT (called a *methane-phase digester*). This phase also can be operated at mesophilic or thermophilic temperatures. Conditions in this phase are similar to those found in conventional high-rate digesters, which are operated to maintain an optimum environment for methanogenic bacteria.

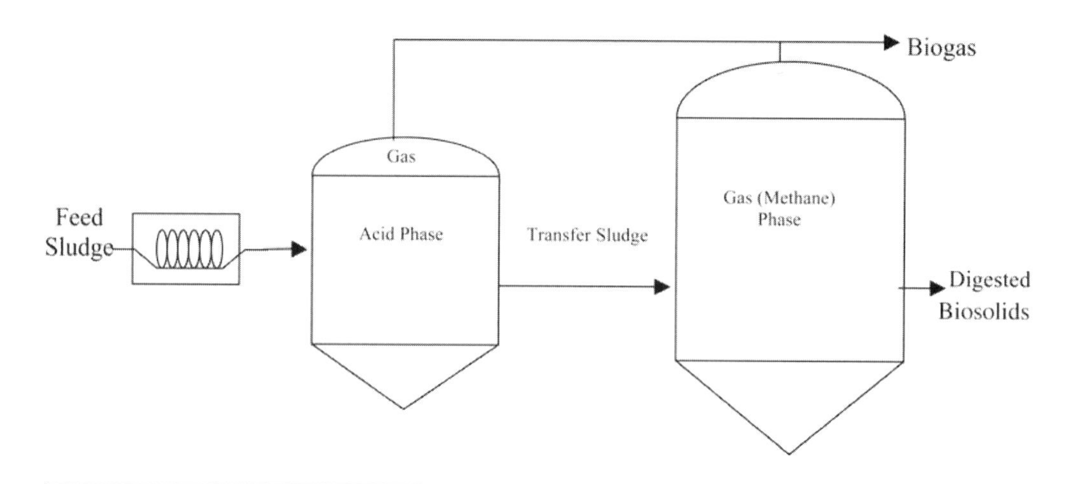

FIGURE 25.13 Schematic of the two-phase anaerobic digestion process.

2.3.7.2 *Design Criteria*

Laboratory- and small-scale work led to the development of larger scale systems for wastewater solids (Ghosh et al., 1991, 1995). Experience indicated that anaerobic digester performance could be improved by optimizing the acid-forming and methane-generation phases separately. Compared to the single-phase systems, two-phase anaerobic digestion systems have higher rates of volatile solids reduction and biogas production, produce biogas that contains more methane, inactivate more pathogens, minimize foam, and are overall more resilient and stable.

The key design criteria are loading rates and retention times. The recommended process design criteria for the acid-phase digester typically are as follows:

- Volatile solids loading rate of 25 to 40 kg volatile solids/m^3/d (1.5 to 2.5 lb volatile solids/cu ft/d),
- Feedstock that contains 5 to 6% solids,
- Solids retention time of 1 to 2 days (at mesophilic temperature),
- Total volatile fatty acid (VFA) concentrations of 7000 to 12 000 mg/L, and
- pH range of 5.5 to 6.2.

The methane reactor in two-phase digestion can be loaded at higher rates than conventional high-rate digestion systems because of the hydrolysis that occurred in the acid-phase reactor. Volatile-solids loading rates for the methane reactor often are similar to those for conventional high-rate mesophilic digestion. Residence times of about 10 days have been tested and promoted by process proponents; however, most full-scale systems have longer residence times (often 15 days or more). The total SRT for the entire two-phase digestion process is rarely less than 15 days, which is required if the resulting biosolids must meet Class B criteria under Part 503.

2.3.7.3 *Performance*

Two-phase digestion performance often has been measured via VSR or biogas production. Schafer et al. (2002) reported that this process significantly improved VSR at the DuPage County, Illinois, facility, but other agencies had seen less improvement. Barnes et al. (2007) reported that Denver, Colorado's two-phase digestion facility had not shown any significant increase in VSR compared to its prior high-rate mesophilic system, but digester foaming was no longer a major problem. The DuPage County facility also had a major reduction in digester foaming (Ghosh et al., 1995). Minimal foam is considered a prime advantage of two-phase digestion. The theory is that the hydrolysis and high concentrations of volatile acids in the acid-phase reactor are a significant factor in breaking down biological solids and other constituents that promote digester foam.

2.3.7.4 Process Variation—Three-Phase Digestion

Three-phase digestion is a variation of two-phase digestion that uses both thermophilic digestion and a third reactor, which may have variable temperatures (see Figure 25.14). This process has been used at the DuPage County, Illinois, facility and at the Inland Empire Utilities Agency in Chino, California. The primary objectives of this approach are to ameliorate the higher VFA levels that can occur in thermophilic digestion and provide another phase of digestion to reduce short-circuiting and allow for more pathogen control. Inland Empire's three-phase digestion process has been reported to produce biosolids that meet Class A requirements under Part 503 (Drury et al., 2002).

2.3.7.5 Process Variation—Enzymic Hydrolysis and Digestion

Enzymatic hydrolysis expands the acid phase of the system into as many as six tanks in series at 42°C. The goal is to shift reactor kinetics away from complete mix to plug flow, which can provide more treatment. Figure 25.15 shows the two enzymatic-hydrolysis configurations tested at the Blackburn Wastewater Treatment Plant in Lancashire, United Kingdom: the standard enzymatic-hydrolysis process and the enhanced enzymatic-hydrolysis process, which includes a pasteurization step at 55°C (Werker et al., 2007). Accoding to proponents, enzymatic hydrolysis proponents increased biogas and solids destruction, and enhanced enzymatic hydrolysis greatly improved pathogen destruction.

2.3.8 Pre-Pasteurization

Pre-pasteurization is a disinfection process developed primarily in Europe to allow biosolids to be applied directly to sensitive croplands or pastures.

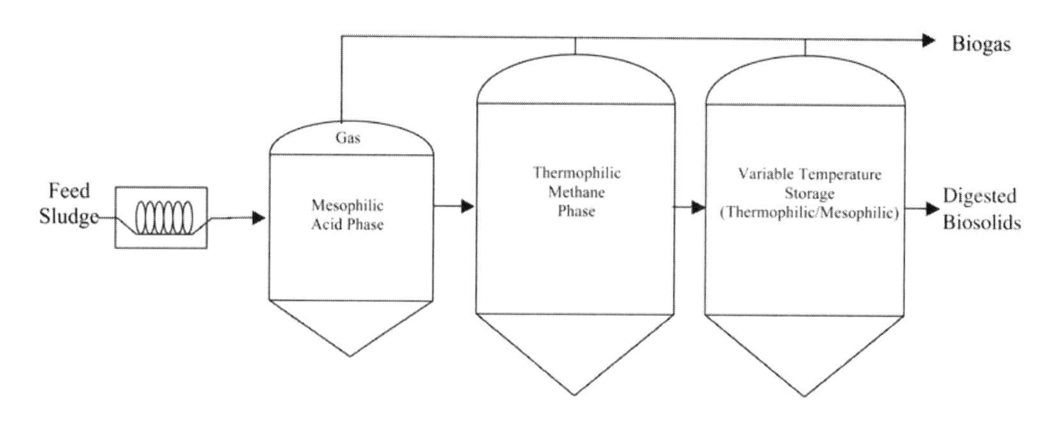

FIGURE 25.14 Schematic of the three-phase (acid–gas phased) anaerobic digestion process.

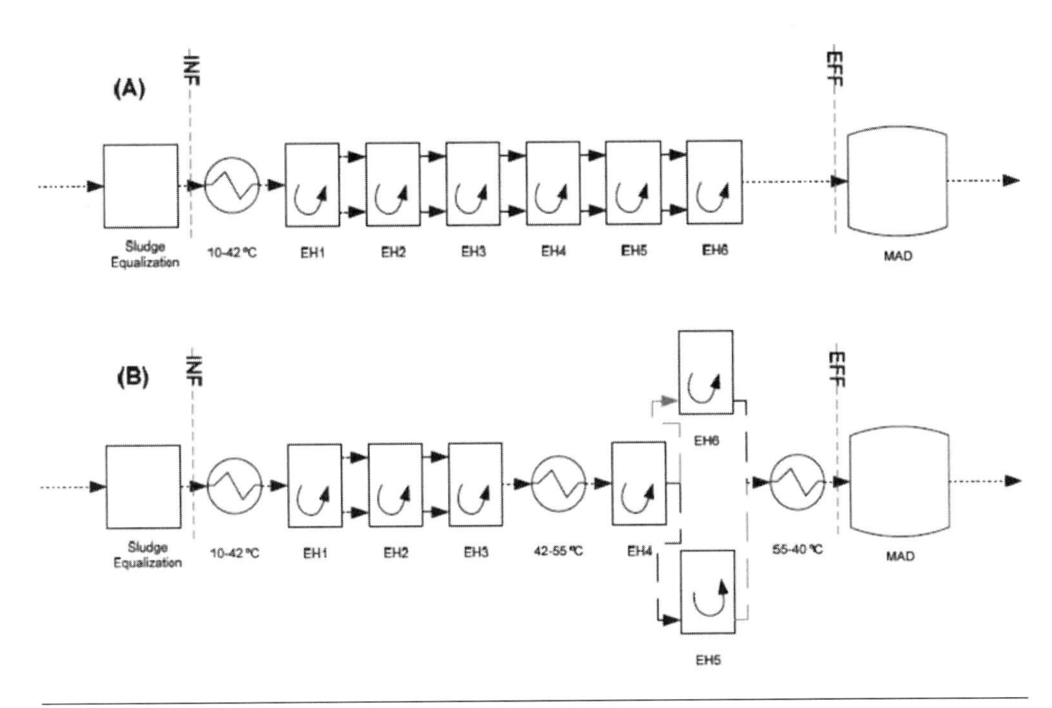

FIGURE 25.15 Flow schematics for two configurations of the enzymatic hydrolysis process: (a) standard operation and (b) enhanced operation (Werker et al., 2007).

2.3.8.1 Process Development

In an early paper on pasteurization, Clements (1982) reported that Switzerland had issued regulations in 1971 requiring biosolids to be hygienized before they could be applied to grazing land. About 70 post-pasteurization facilities were constructed in the following 6 years; they typically processed anaerobically digested biosolids (i.e., biosolids were digested, pasteurized, and then either used or stored). When veterinary scientists investigated the material, they reported that the stored products often contained extremely high densities of pathogens, even though they had contained few enteric bacteria and no *Salmonella* immediately after pasteurization. Further investigation showed that regrowth was due to surviving organisms or to contamination. Investigators concluded that, because the pasteurized material had no remaining vegetative bacteria, any bacteria present later could grow explosively in the absence of competitors.

Keller (1980) presents data on regrowth of bacteria in biosolids from a post-pasteurization process. The data show that, after pasteurization, biosolids contained between 20 and 75 colony-forming units (CFU) of *Enterobacteriaceae* per gram of solids. When it was transported from the treatment facility, however, the material contained between 207 000 and 35 mil. CFU/g. Shortly afterwards, treatment plants modified the process to pasteurize solids before digestion (called *pre-pasteurization*), and the regrowth problem disappeared. The practice has successfully sustained itself over several decades in many countries (mostly in Europe). A few U.S. facilities now use the process.

2.3.8.2 Design Criteria

The main reason to use a pasteurization process is to disinfect (hygienize) solids; it has not been reported to enhance VSR significantly. The pre-pasteurization process meets Part 503's Class A pathogen standards by typically maintaining its temperature above 65°C (more typically 70°C) for a specific period of time in a batch operation. The amount of time is calculated using Equation 25.1, which applies to cake containing less than 7% solids.

Calculations indicate that solids must be maintained at 65°C for about 1 hour. As pasteurization temperature increases, the required time (and, therefore, tank volume) shrinks. Most systems are designed to achieve at least 70°C for 0.5 hour, even though operating at slightly lower temperatures may lower operations and maintenance (O&M) costs.

If a utility chooses to pasteurize cake containing more than 7% solids, the time and temperature requirements are determined using Equation 25.2.

Part 503 specifies that the pasteurization process must precede the vector attraction reduction process (e.g., mesophilic or thermophilic anaerobic digestion) and should be operated in a batch mode to allay regrowth concerns. Hence, post-pasteurization is not allowed under this rule.

2.3.8.3 Pre-Pasteurization Vessel

The U.S. Environmental Protection Agency has indicated that to produce a Class A biosolids that meets the requirements in Alternative 1 under Part 503, every particle of solids should be exposed to a minimum temperature for a minimum time. So, design engineers should avoid using completely mixed systems or systems with potential for back-mixing or short-circuiting as pre-pasteurization tanks. Most vendor-supplied systems are batch tanks; only one vendor supplies a plug-flow tank for pre-pasteurization. Design engineers should consult U.S. EPA staff or other pertinent regulators before using a non-batch system.

Batch pre-pasteurization systems are operated in a fill/hold/draw mode, with several batch vessels used to perform each cycle if continuous operation is desired. The vessels should be well mixed to ensure that the monitored temperature reflects the entire contents (i.e., every solids particle meets the time and temperature requirements). If the downstream anaerobic digestion process uses an intermittent feed cycle and upstream storage is adequate, the system needs fewer than three batch vessels for the required fill, hold, and draw cycles.

2.3.8.4 Ancillary Equipment for Pre-Pasteurization

Design engineers should consider three important ancillary features when installing a pre-pasteurization process:

- Solids heating and cooling,
- Solids screening, and
- Temperature monitoring and control.

Because the temperature of pre-pasteurization systems typically is maintained at 70°C, solids must be heated and then cooled. Heat exchangers are the most common method for heating and cooling solids. If desired, design engineers could include a heat-recovery step to use heat from cooling biosolids to preheat raw solids. The recovery step will require a substantial amount of heat-exchange capacity. If heat exchangers are used, the solids may need to be screened before pre-pasteurization—even if fine screens are used in the plant headworks. Screening also helps produce a more aesthetically pleasing product.

Good temperature monitoring and control are required to maintain Class A compliance. It is critical to have a well automated system to both ensure pasteurization and prevent downstream contamination, which would take months to remedy. It should prohibit unpasteurized solids from passing through, and either waste or recirculate material that did not meet the time and temperature requirements. If necessary, standby equipment should be included to maintain time and temperature, because compromising these parameters could contaminate the downstream anaerobic digestion process.

2.3.8.5 Performance

The pre-pasteurization process can meet the *Salmonella* criteria in Part 503. Ward et al. (1999) showed that pre-pasteurized solids resisted regrowth even after they were seeded with *Salmonella*; instead, the organisms died off. Higgins et al. (2008b) also observed that pre-pasteurization effectively destroyed *Salmonella*, even though fecal coliforms regrew (suggesting that it may be necessary to measure the actual pathogen rather than the indicator to ensure Part 503 compliance).

In summary, pre-pasteurization is an effective method for destroying pathogens in solids. Used throughout the world, it is one of the more prevalent solids disinfection processes associated with anaerobic digestion. Several U.S. facilities use this process, including a 204 400-m^3/d (54-mgd) treatment plant operated by the Alexandria Sanitation Authority in Virginia (see Figures 25.16 and 25.17).

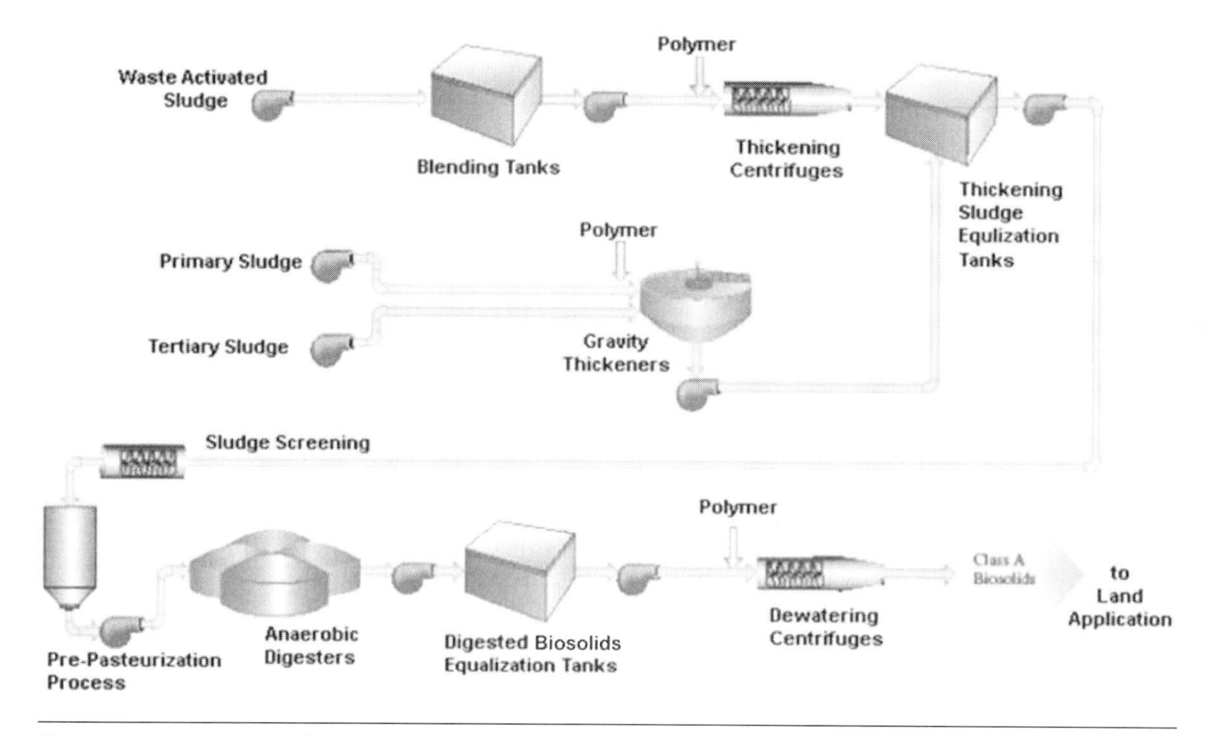

FIGURE 25.16 Process flow diagram of the Alexandria Sanitation Authority in Virginia; it includes the screening and pre-pasteurization vessels (courtesy of Alexandria Sanitation Authority).

2.3.9 *Thermal Hydrolysis*

Thermal hydrolysis is a pre-digestion conditioning process. It treats solids in a batch reaction at elevated temperature and pressure. The process improves the digestability of biological solids, in particular, while reducing the size of digestion tankage and improving dewatering performance.

2.3.9.1 Process Development

Some of the earliest work on using thermal hydrolysis as an anaerobic-digestion pretreatment step was initiated in the United States (Haug et al., 1978; Haug et al., 1983; Stuckey and McCarty, 1984). These early publications (along with Li and Noike, 1992) propose the basic optimized format of the full-scale thermal hydrolysis process used today, which involves temperatures of 150 to 170°C, a 30-minute SRT, and a pressure of about 827 kPa (120 psi). The pressure prevents water from vaporizing out of the solids and reduces overall energy needs.

Thermal hydrolysis was first developed in the United States, but successful implementation occurred in Europe. The first full-scale system was implemented at the Hias

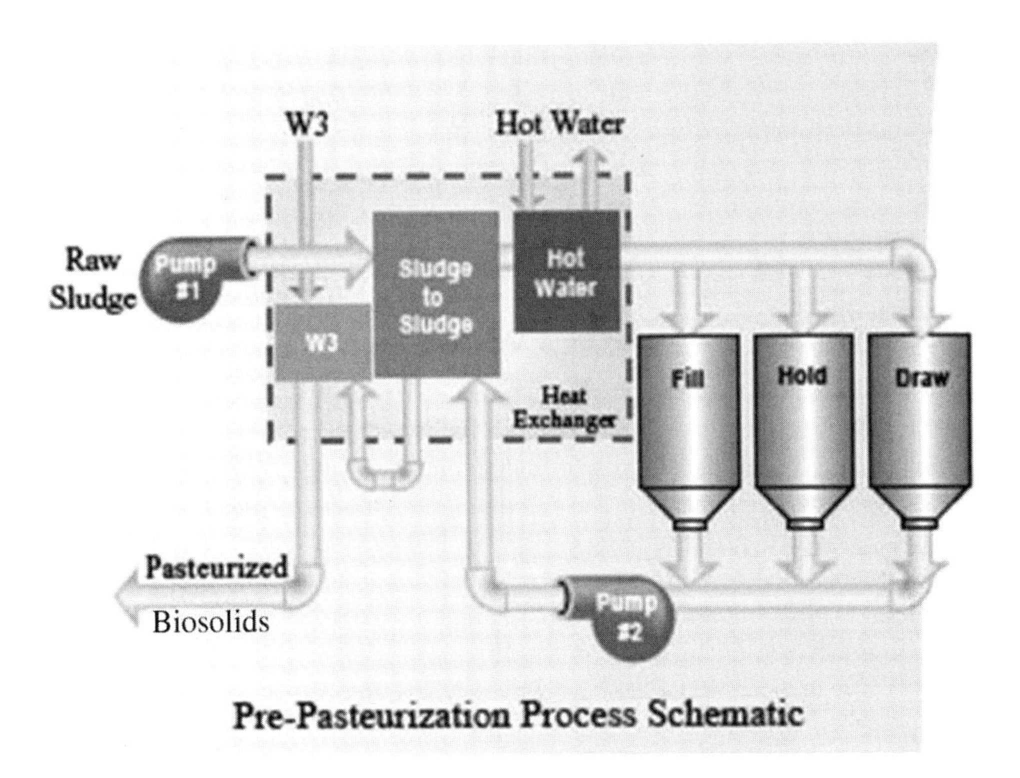

FIGURE 25.17 Krugers Biopasteur™ process at the Alexandria Sanitation Authority (courtesy of Alexandria Sanitation Authority).

Wastewater Treatment Plant in Norway in 1996. The largest plant in operation is in Dublin, Ireland; it serves a population equivalent of 1.6 million. More than 20 large and small systems are currently in operation, mostly in northern Europe.

Another aspect of the full-scale thermal hydrolysis process is the rapid depressurization step. It occurs after the reaction step and is reported to help burst cells, further promoting hydrolysis and disinfection.

When used before anaerobic digestion, thermal hydrolysis achieves one or more of the following:

- Enhances digestion rates and gas production,
- Reduces the size of the anaerobic digestion system,
- Disinfects solids, or
- Prepares solids for thermal processing downstream of anaerobic digestion.

2.3.9.2 Design Criteria—Thermal Hydrolysis Vessels

Thermal hydrolysis vessels are made of Type 316 or better stainless steel and are built to withstand both pressure and vacuum. The vessel configuration depends on the vendor.

Operated at temperatures between 150 and 170°C for 30 minutes at the corresponding vapor pressure, thermal hydrolysis solublizes and hydrolyzes solids (Li and Noike, 1992), and disintegrates biological cells (e.g., bacteria and viruses). According to Li and Noike (1992), maximum solublization occurs at 170°C and the optimal SRT is between 30 and 60 minutes. In practice, a 30-minute SRT optimizes reactor size and delivers a solubilized and hydrolyzed product.

Thermal hydrolysis basically consists of a preheating step, a heating and batch-reaction step, and a rapid depressurization step for further solubilization and rupturing of microbial cells (see Figure 25.18). The preheating step is used to conserve spent heat from the reaction and depressurization steps, as well as produce a favorable energy balance. The system's feedstock is dewatered cake containing 14 to 18% solids. Dewatering considerably improves the heat balance and reduces the volume of the downstream anaerobic digestion process by about 50%.

If a Class A product is desired, design engineers and operators should ensure that every particle of solids in the reactor meets the time and temperature requirements. Solids screening also may be required if the process will use heat exchangers to recover heat.

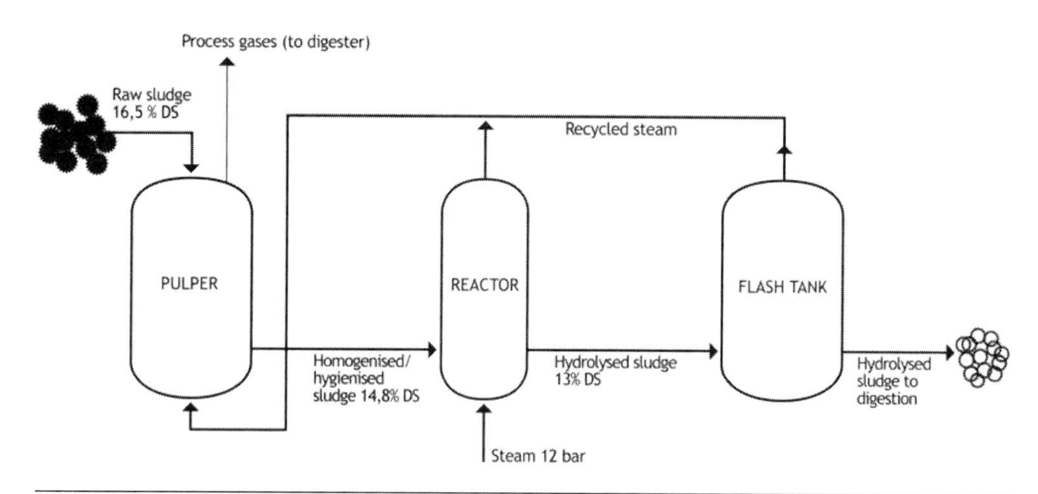

FIGURE 25.18 Schematic of the thermal hydrolysis process (courtesy of Cambi, Norway).

Because the cake contains more than 7% solids, time and temperature requirements are determined using Equation 25.2. Calculations indicate that solids must reach 150°C and stay that hot for only a fraction of a second, so the system's 30-minute SRT far exceeds U.S. EPA's time and temperature requirements. This retention time is used more to optimize hydrolysis and solubilization, and to ensure that the required temperature has diffused to the interior of all solids particles in the solids mass.

2.3.9.3 *Ancillary Equipment for Thermal Hydrolysis*

Design engineers typically consider the following four ancillary features when installing a thermal hydrolysis system: heating and heat-recovery, screening, process control, and odor management.

2.3.9.3.1 Solids Heating and Recovery

After depressurization, thermally hydrolyzed solids are about 100°C and must be cooled before entering the anaerobic digestion process. So, the low-pH solids are combined with higher-pH solids recycled from the anaerobic digester and then cooled in heat exchangers. Plant effluent typically is used for heat exchange. Heat recovery can be performed if desired, but is typically not used to minimize size of heat exchanger and, therefore, maintenance requirements.

2.3.9.3.2 Solids Screening

Because thermal hydrolysis process includes a heat exchange step, biosolids screening is recommended. Screening can considerably improve downstream solids processes and aesthetically enhance the biosolids.

2.3.9.3.3 Temperature and Pressure Monitoring and Control

Monitoring pressure and temperature is critical. Furthermore, the process needs to be installed with pressure-safety and vacuum-breaks valves. The systems must be well automated to ensure disinfection and prevent downstream contamination of anaerobic digesters, which would take months to remedy. A well automated system should prohibit undisinfected solids from passing through, and either waste or recirculate material that does not meet the time–temperature requirements. If necessary, standby equipment or an upstream cake silo buffer should be provided.

2.3.9.3.4 Odor Management

Unlike wet air oxidation, where both the process and final product are odorous, thermal hydrolysis followed by anaerobic digestion does not produce an odorous biosolids. However, the process itself could emit odors, which must be contained and treated. The odorous gases are biodegradable and water soluble, so a convenient treatment method is to use water scrubbers and discharge the water into the downstream anaerobic digester, which will treat the process odors. The valving design for thermal-hydrolysis vessels is critical to minimize vented odors. Tank cleanout prevents odors when tanks need maintenance or inspection work.

2.3.9.3.5 Sidestream Treatment

The return liquor from thermal hydrolysis contains colloidal material that will contribute organic nitrogen, COD, and color to the mainstream process. For example, it will increase the plant effluent's organic nitrogen content by 0.75 to 1.5 mg/L. If the treatment plant has low limits for any of these constituents, then they should be removed before the liquor enters the mainstream process. Treating these constituents with chemical conditioners (e.g., iron or aluminum) in the dewatering step has been proposed by Wilson et al. (2008b).

2.3.9.4 Process Mode Variations

Two full-scale versions of thermal hydrolysis are currently in use (see Figure 25.19). One process mode consists of a pre-heat tank, a reactor tank, and a flash tank (i.e., three tanks in series). The other process mode consists of one tank in which preheating, reaction, and depressurization occur. This mode could be designed with multiple parallel vessels.

2.3.9.5 Anaerobic Digestion Performance

The solubilized and hydrolyzed solids are easier to digest and, therefore, can increase digestion rates or reduce digester SRT. For example, Li and Noike (1992) report that the digester reached a stable maximum methanogen population and degraded most of the substrate in a 5-day SRT, suggesting that this was the minimum digester SRT to prevent washout or process instability. However, their work did not evaluate thermal hydrolysis performance under the high ammonia concentrations observed in highly loaded systems today, so full-scale processes are operated at a minimum 15-day SRT under average conditions. Design engineers can decrease the required anaerobic digester SRT by up to 10 to 25% (compared to conventional high-rate digestion) and expect similar process performance. For example, Wilson et al. (2008b) operated parallel conventional high-rate anaerobic digesters with and without thermal hydrolysis, using solids from the District of Columbia Water and Sewer Authority's Blue Plains facility. Results showed that both digesters had similar volatile solids and COD destruction, but the digester with thermal hydrolysis achieved these results at a 25% shorter digestion SRT (i.e., in 15 days rather than 20).

The solubilized solids from thermal hydrolysis are much less viscous (Kopp and Ewert, 2006). Hydrolized cake contains 10% solids or more (compared to a typical digester feed cake containing 5% solids). Frequently, cake containing about 16% solids

FIGURE 25.19 The thermal hydrolysis process in Cotton Valley, UK, which processes 20 000 tonne/a of combined solids (courtesy of Cambi, Norway).

enters thermal hydrolysis process, where it is heated with steam. The steam both increases the temperature and dilutes the solids. The hydrolyzed cake typically contains between 9 and 12% total solids, which can be adjusted with plant-dilution water, if needed, to maintain fairly constant solids loading to the anaerobic digester. The maximum solids content depends on the concentration of ammonia-nitrogen produced during digestion; ammonia-nitrogen concentration is typically kept below 2 500 mg/L to prevent process inhibition.

The digested and dewatered biosolids contain 6 to 8% more solids (Kopp and Ewert, 2006; Wilson et al., 2008b) than that produced via conventional digestion. So, thermal hydrolysis is attractive when the resulting biosolids will be hauled long distances for land application or will be thermally processed (e.g., heat drying). (Influent cake dryness and process evaporative capacity are major considerations when determining size of a thermal process.) Kopp and Ewert (2006) report that cake solids increased from 25 to 34% when thermal hydrolysis pretreatment was used. Wilson et al. (2008b) had similar results.

2.3.10 *Aerobic Pretreatment*

Aerobic pretreatment of solids has been practiced in North America and Europe for more than a quarter century. It involves adding air or oxygen to solids at thermophilic temperatures as an initial "conditioning" step before anaerobic digestion.

2.3.10.1 Process Development

Aerobic pretreatment developed differently in Europe and North America; the aerobic thermophilic pretreatment (ATP) process is mainly practiced in Europe, while dual digestion is mainly practiced in North America. Both processes are intended to enhance VSR and disinfection. The main difference between the two processes is the method of heating solids. In ATP, the heat used to attain thermophilic temperatures is waste heat from cogeneration of digester gas (not an autothermal process). Aerobic thermophilic pretreatment's SRT mainly depends on both process and Part 503 requirements. It typically is 24 hours or less, depending on the digestion temperature and the results of time–temperature requirements in Equation 25.1 to attain Alternative 1 in Part 503 (EPA, 1999b). The minimum temperature is typically 55°C and the maximum is 65°C (when biological conditioning of raw solids is encouraged to enhance VSR).

In dual digestion, the aerobic step is an autothermal step in which heat generated during microbial aerobic metabolism is used to increase the process' temperature to thermophilic conditions. The temperature range for this process is also between 55 and 65°C. Dual digestion's SRT depends on two factors: U.S. EPA's time and temperature equation, and the time needed to autothermally raise the temperature of raw solids to the required setpoint. In most cases, the time needed to meet the temperature setpoint is greater than that demanded by U.S. EPA's time–temperature equation. Using oxygen rather than air reduces SRT requirements and improves the heat balance. Also, heat recovered from thermophilic solids can be used to help raise the input solids' temperature (see Figure 25.20). All of the North American installations have been at treatment plants that already used high-purity oxygen in their activated solids process. Dual digestion's SRT is about 1 to 2 days, depending on the raw solids' volatile solids content. Feed containing more volatile solids significantly helps the heat balance to achieve autothermal temperatures. Several dual digestion plants were commissioned in the 1980s, and three U.S. treatment plants are operating the process successfully today. The 143 800-m^3/d (38-mgd) Central Treatment Plant in Tacoma, Washington, has used this process for more than a decade (see Figure 25.21), producing and marketing a Class A soil amendment product (called *Tagro*) from the resulting biosolids (Eschborn and Thompson, 2007).

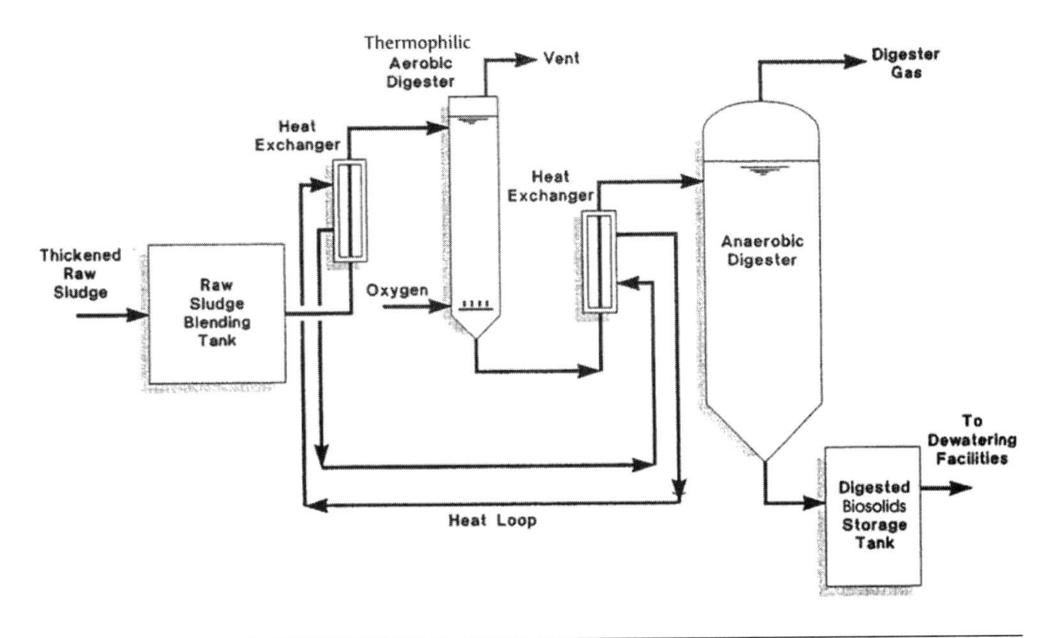

FIGURE 25.20 Simplified schematic of the dual digestion process (courtesy of Brown and Caldwell).

2.3.10.2 Design Criteria

Aerobic pretreatment has two primary goals: solids disinfection and enhanced VSR. Both dual digestion and ATP are designed to operate in the thermophilic temperature range, so the tanks should be well insulated to maintain a favorable heat balance. In addition, solids screening may be required if heat exchangers are used for heat recovery. Under Part 503, processes must meet pathogen reduction requirements to achieve Class A biosolids status. Aerobic pretreatment meets these requirements by typically maintaining a temperature between 55 and 65°C for a specific period of time in a batch (plug-flow) operation. The amount of time is calculated using Equation 25.1 (the equation for solids content less than 7%). Two possible options are 60°C for a minimum of 4.8 hours or 55°C for 24 hours. Hotter temperatures typically reduce time and, therefore, tank volume. In dual digestion, however, the minimum SRT depends more on the time required to achieve the desired autothermal temperature than on U.S. EPA's time–temperature equation.

If a utility chooses to disinfect feedstock containing more than 7% solids, the time and temperature requirements are determined using Equation 25.2.

Aerobic thermophilic pretreatment typically is heated using waste heat from a cogeneration process. The heat balance largely depends on the insulation of the preheat tank and the decision to use heat-recovery heat exchangers.

FIGURE 25.21 Photograph of the dual digestion and oxygen activated sludge processes at the Central Treatment Plant in Tacoma, Washington (courtesy of City of Tacoma, Washington).

Dual digestion is mainly heated autothermally (the mixers introduce some heat). Three important parameters in dual digestion designs are the decay rate, the biological heat of reaction (BHR), and oxygen demand. Gemmell et al. (1999) estimated an average decay rate of $0.087 \pm 0.010\ d^{-1}$ at an average temperature of 37°C. Arant and Boden (2000) estimated a similar decay rate ($0.08\ d^{-1}$ at 35°C). Gemmell et al. (1999) determined a BHR of 16.6 MJ/kg volatile solids destroyed for the dual digestion process at Barrie, Ontario, in initial trials with a 26% VSR. Grady et al. (1999) suggest a BHR value of 18.8 MJ/kg volatile solids destroyed (for a design involving autothermal thermophilic aerobic digesters). Messenger et al. (1993) determined a BHR of 18.6 MJ/kg volatile solids destroyed. Haas (1984) found that BHR ranged from 17.4 to 23.3 MJ/kg volatile solids destroyed at 20 and 10% VSR, respectively, during trials conducted at Hagerstown, Maryland. The oxygen demand depends on the type of solids and the ratio of primary and biological solids. Values in the range of 1.7 kg O_2/kg volatile solids destroyed have been reported by Pitt and Ekama (1996) and Gemmell et al. (1999). This value can easily be determined experimentally and should be tested, because it can vary from plant to plant.

2.3.10.3 Aerobic Vessel Design

To produce a Class A biosolids, the treatment process must meet the time and temperature requirements in Alternative 1 under Part 503, which specify that every particle of solids should be exposed to a minimum temperature for a minimum period of time (U.S. EPA, 1999b). So, complete mixed systems or systems that could back-mix or short-circuit should be avoided. Design engineers should consult EPA staff or other pertinent regulators before using a non-batch plug-flow system. The batch systems should be designed to operate in a fill/hold/draw mode and be well mixed to ensure that every particle is maintained at the required temperature (for the required time) and that the monitored temperature reflects the entire contents of the batch. Typically, if continuous operation is desired, three batch vessels are needed (one for each cycle). However, if the downstream anaerobic digester can handle an intermittent feed cycle and upstream storage is adequate, then fewer batch vessels can be supplied.

2.3.10.4 Ancillary Equipment for Aerobic Pretreatment

Design engineers typically consider three important ancillary features when installing an aerobic pretreatment process:

- Solids heating and recovery (aerobic thermophilic pretreatment) or oxygen system (dual digestion),
- Solids screening, and
- Temperature monitoring and control.

The aerobic vessel for ATP and dual digestion typically is maintained between 55 and 65°C. In the ATP process, solids must be heated and then cooled. The most common method for heating and cooling solids is heat exchangers. In dual digestion, solids are heated autothermally so an external heat source is unnecessary. Both processes must cool treated solids before digestion—unless the anaerobic digester also is operated at thermophilic temperatures. The cooling method could include a heat-recovery step in which the heat transferred from cooling solids is used to preheat raw solids. This step depends on owner, engineer, and vendor preference, because it will require a substantial amount of heat-exchange capacity. If heat exchangers are used, the solids may need to be screened first. Screening also helps produce an aesthetically pleasing biosolids. Good temperature monitoring and control are required to maintain Class A compliance. If necessary, standby equipment should be provided to maintain time and temperature, because compromising these parameters could contaminate the anaerobic digester with inadequately disinfected solids. The solids should be thickened to at least 5% total solids for successful operations; therefore, downstream solids pumps and pipes should be designed to handle thicker solids adequately.

Also, dual digestion will need an oxygen supply (hence, dual digestion typically is installed at facilities that already use high-purity oxygen in their activated sludge processes).

2.3.10.5 Performance

Preconditioning of solids with air is intended to increase the overall VSR. Researchers (Pagilla et al., 1996; Cheunbarn and Pagilla, 1999; Cheunbarn and Pagilla, 2000) have extensively evaluated ATP performance and compared it to mesophilic digestion performance. They confirmed the European full-scale observations of Baier and Zwiefelhofer (1991) that ATP enhances VSR and gas production. Cheunbarn and Pagilla (1999) showed that VSR increased as the ATP's SRT (0.6 to 1.5 days) and temperature (55 to 65°C) increased. In pilot-testing work at Sacramento, Pagilla et al. (1996) compared ATP to conventional mesophilic digestion and determined that ATP enhanced VSR from 53 to 59% for combined solids (primary solids and WAS). They also showed that ATP could meet U.S. EPA's Part 503 requirements for fecal coliform, *Salmonella*, enteric virus, and helminth ova. In addition, they determined that ATP effectively controlled and destroyed *Nocardia* filaments. Finally, they observed that ATP-treated, centrifuged biosolids contained between 32 and 36% total solids, compared to mesophilically digested, centrifuged biosolids, which only contained 30% total solids.

Gemmell et al. (1999) suggested that dual digestion could achieve stable performance when the first high-rate aerobic reactor had an HRT of 1 to 2 days and the second high-rate anaerobic reactor had an HRT of 9 to 12 days. Overall, this retention time was 6 to 10 days shorter than the 20-day SRT typically required for conventional high-rate digestion. Also, Gemmell et al. (2000) suggested that the full-scale dual digestion process at the Barrie Wastewater Treatment Plant achieved 60% VSR. Operators at a 143 800-m^3/d (38-mgd) treatment plant in Tacoma, Washington, have been producing and marketing a biosolids-based soil amendment for more than a decade; they attribute much of its high quality to the dual digestion process (Eschborn and Thompson, 2007).

2.3.11 Lagoon Digestion

It used to be common to digest solids via open lagoons, but anaerobic lagoons typically caused odor problems and were phased out of use. Today, various wastewater treatment agencies use facultative solids lagoons (FSLs), which have an aerobic cap layer to help oxidize and control the odorous decomposition products that rise from the anaerobic activity below. The Sacramento (California) Regional Wastewater Treatment Plant has a 50-ha (125-ac) FSL system of lagoons filled with liquid 4.5 m (15 ft) deep. Considerable research was conducted when this system was developed in the 1970s (Schafer and Wolfenden, 1982).

2.3.11.1 System Performance

Large-scale lagooning is conducted at ambient temperatures, which typically encompass the psychrophilic temperature range. A key feature of such digestion is that more digestion occurs in warmer seasons, and less occurs in colder seasons.

The approach used at Sacramento, Chicago, and other plants is to achieve maximum VSR and solids stabilization via long-term digestion (1 to 5 years). Sacramento achieves almost 60% VSR in its mesophilic digesters, and has documented another 40 to 45% VSR in its FSL (Schafer and Wolfenden, 1982). Such long-term stabilization results in biosolids with relatively little product odor.

Design and operating criteria vary widely for FSLs, but most systems currently are fed either mesophilically digested biosolids or aerobic solids from extended aeration plants. Treatment plant effluent is often used for the cap water layer.

At Chicago and other plants, dredged solids are air dried in warmer-weather months, producing biosolids that contain at least 60% solids. The biosolids are used for land application, land reclamation, landfill cover material, and other beneficial purposes. FSLs have been reported to produce Class A biosolids when batch storage is used to prevent short-circuiting (WERF, 2004).

2.3.11.2 Covered Lagoons for Methane Emission Control

Since the 1990s, concerns about odors and methane emissions from open waste lagoons (mostly animal waste lagoons) have increased. In response, the animal-waste treatment industry has begun covering some lagoons to collect the biogas generated during anaerobic digestion and use it to produce power. Such systems are becoming more common in North America, Australia, and Asia.

For example, the Western treatment plant in Melbourne, Australia, has used an extensive floating cover system on its wastewater ponds for almost a decade; the high-density polyethylene (HDPE) system covers 7.8 ha (19 ac) of anaerobic disgestion ponds (DeGarie et al., 2000). The collected biogas is used to generate more than 2 MW of electricity, which powers other portions of the plant. This system reduces direct methane emissions to the atmosphere (cutting greenhouse gas emissions) and generates renewable power (offsetting carbon emissions from fossil fuel-based generators elsewhere). Memphis, Tennessee, also has covered some of its FSLs since the 1990s to control odor and collect biogas for energy use.

2.3.12 Solids Disintegration Processes

Solids disintegration technologies are designed to increase the rate and extent of anaerobic solids digestion by applying external energy to render solids more bioavailable. These processes typically are applied to WAS because it is considered the most difficult to digest.

The means of energy application is technology-specific, but the reported effects are consistent:

- More biogas production,
- Increased VSR, and
- Reduced mass of solids for disposal.

Table 25.9 lists various disintegration technologies that currently are commercially available. However, only ultrasonics has been used for more than 5 years in a full-scale installation.

2.3.12.1 Ultrasonic Technologies—Process Development

Full-scale implementation of ultrasonic technologies to enhance anaerobic digestion began in Europe in the mid- to late 1990s. The technology was developed as a means of increasing biogas production while reducing the mass of biosolids for disposal; it essentially increases the digester's gasification rate.

Ultrasonics generate transient acoustic cavitation, which improves anaerobic digestion of WAS, in particular. Acoustic cavitation occurs when ultrasonic waves compress and rarefy the liquid. During rarefaction, enough energy may be applied to exceed intermolecular forces, forming a void (cavitation bubble). The bubble's subsequent collapse generates significant amounts of heat (>4000°C), pressure (about 1000 atm) (Christi, 2003), and shear forces from the liquid jets formed (Mason and Lorimer, 1988). Particles near or within a collapsing bubble can be exposed to one or more of these forces, breaking them down to a size of 40 000 Da (Portenlänger and Heusinger, 1997).

TABLE 25.9 Examples of solids disintegration technologies.

Process	Example Manufacturer	Disintegration Method
Ultrasonics	Sonix™, Enpure	Acoustic cavitation
High-pressure homogenization	Microsludge™, Paradigm Environmental	Chemical pretreatment with hydrodynamic cavitation and shear
Pulsed electric field	OpenCel	Electroporation

Pressure, temperature, and shear forces are the primary forces responsible for disintegrating WAS. Chemical transformations also are theoretically possible: The environment in the cavitation bubble could lead to the formation of free radicals, which can interact with various elements in the surrounding fluid. Depending on the ultrasonic probe's frequency, either mechanical or sonochemical forces can dominate. At lower frequencies (around 20 kHz), mechanical forces typically dominate; free radical formation becomes more common at higher frequencies.

The horn's material of construction also affects process performance and longevity. Ultrasonic horns typically are made of aluminum, stainless steel, or titanium. Titanium is by far the most expensive material but has the best acoustic properties and longevity, resisting pitting due to cavitation. Given the harsh operating environment, the design requirement to reduce O&M, and the desire for predictable performance, titanium typically is used in wastewater applications.

Other factors (e.g., solids concentration, particle size, and liquid gas saturation) affect the generation of cavitation bubbles, but optimizing conditions based on these parameters is unlikely because of the effects on digester operation and the inherent variability of wastewater treatment systems.

2.3.12.2 Ultrasonic Technologies—Process Variations

The basic principles of ultrasonics are consistent regardless of the technology applied. The differences between technologies primarily depend on the site and the proprietary configurations of the ultrasonic reactors. At press time, two ultrasonic technologies were commercially available. Both apply ultrasonic energy to thickened WAS before digestion. The primary difference between them is reactor configuration.

The toroidal horn configuration uses a 20-kHz stack transducer attached to a radial horn via a booster (see Figure 25.22). This horn configuration can process 200 m^3/d of WAS at an energy input of 4 to 5 kJ/L. Demonstration projects and permanent installations have reported varying degrees of improvement to the digestion process and ancillary solids-handling processes (i.e., from no improvement to significant increases in VSR, biogas production, and solids content).

The serpentine reactor is configured so a 20-kHz ultrasonic probe is located at each turn (a total of five probes) (see Figures 25.23 and 25.24). The system can process 35 m^3/d (9250 gpd) of thickened WAS containing 5 to 7% total solids. One unit consumes about 3 to 8 kWh/m^3 while processing 25 to 33% of the total thickened WAS flow.

2.3.12.3 Ultrasonic Technologies—Design Considerations

Solids disintegration may enhance anaerobic digestion by making typically refractory material bioavailable. However, most solids disintegration technologies must be coupled with electrical cogeneration or beneficial biogas utilization to help offset operating

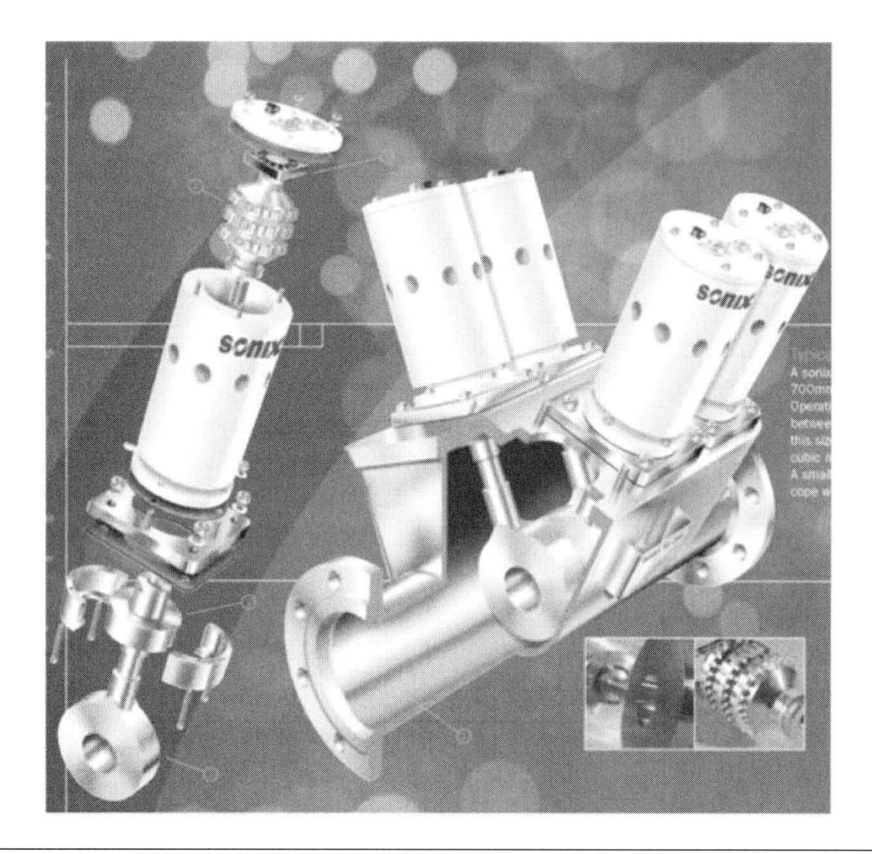

FIGURE 25.22 Schematic of the toroidal (radial) horn system (source is Sonix Brochure, courtesy of Enpure Ltd.).

costs. While the process may increase VSR, the lower hauling and tipping/use costs typically are not sufficient to cover the system's capital and O&M costs.

2.3 Digestion Processing

This section discusses some key processes that can improve digestion performance and biosolids characteristics.

2.4.1 Thickening Before Digestion

Thickening solids before digestion is a means of preserving or expanding digester capacity. There are a variety of technologies for thickening solids (see Chapter 23), and their design and operation should be coordinated with the anaerobic digesters. A thicker cake will preserve volumetric capacity and reduce heating requirements, but it can cause an

FIGURE 25.23 Photograph of a serpentine ultrasonic reactor for digestion enhancement (courtesy of Eimco Water Technologies, LLC, a GLV Company).

organic overload of a digester. Excessively thick solids can increase the sludge viscosity, thus increasing the energy required for digester mixing and sludge pumping.

2.4.2 *Debris Removal*

Debris typically is removed via screening and grit removal at the plant headworks; however, subsequent screening of primary solids, scum, and other digestion feedstocks may be necessary. If debris enters the digester, several of the following process issues could arise:

- Loss of digester volume (because of accumulated debris in the tank),
- Excessive wear on pumps,
- Clogging and additional cleaning of heat exchangers, and
- Ragging and binding of mixing and pumping equipment.

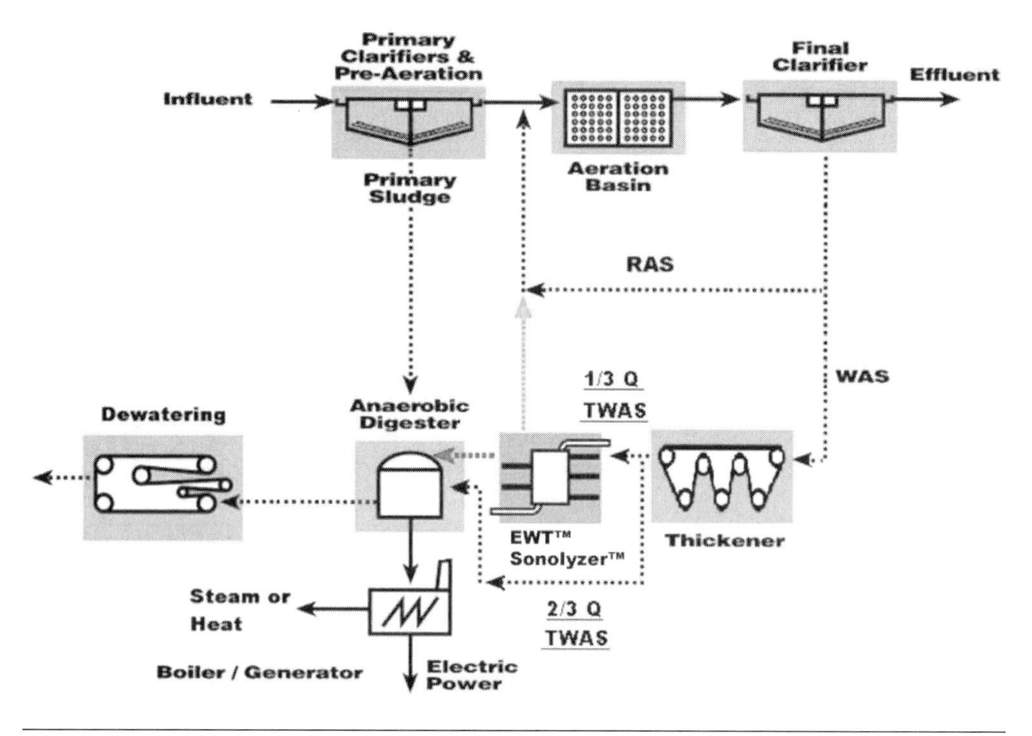

FIGURE 25.24 Schematic of ultrasonic system for digestion enhancement (courtesy of Eimco Water Technologies, LLC, a GLV Company).

Debris also affect biosolids quality. Large quantities of debris (e.g., plastic materials) can degrade the aesthetic qualities of biosolids, making it undesirable or even unfit for beneficial use.

A variety of technologies (e.g., rotary drum screens and strain presses) can remove debris from solids. Screening (straining) has been done on raw solids, raw scum, and unthickened, thickened, and digested solids in slurry form. When evaluating technologies, design engineers should consider the intended application, the material to be processed and other site-specific constraints (e.g., whether the sewer is combined or separated, extent of debris removal from wastewater, whether grinders have already been used, whether stringy material could bind pumps or other equipment, desired use for the biosolids, biosolids aesthetics, and regulatory requirements).

2.4.3 Debris Size Reduction (Reduction in "Identifiables")

Reducing the size of debris protects process equipment from blockages and binding; it also makes debris less identifiable in biosolids. Debris can be reduced by grinding solids before pumping or dewatering them. Inline solids grinders often will be placed in front of recirculation, feed, or wasting pumps to reduce debris size. At press time, some states (e.g., Washington) required identifiable debris to be removed before biosolids could be land-applied or otherwise used beneficially.

2.4.4 Batch and Plug-Flow Systems

Most existing digestion systems can be classified PSRPs, which meet Class B requirements for land application (assuming vector attraction requirements are met). The systems can be upgraded to PFRPs, which produce Class A biosolids.

One upgrade option typically involves higher temperatures (e.g., thermophilic) and batch or plug-flow operations. To meet the time and temperature requirements of Part 503, every particle must be treated at a temperature higher than 50°C for a prescribed period of time. Typically, at least three tanks are required to meet this requirement: one in fill mode, one in hold mode, and one in draw mode (see Figure 25.25).

Another option is using a plug-flow reactor followed by complete mix reactor—both operating at thermophilic temperatures. Developed by the Columbus (Georgia) Water Works, this combination is thought to achieve similar pathogen reductions as the batch system but has fewer control points and less process complexity (Willis et al., 2003). The U.S. Environmental Protection Agency recently determined that this process is a conditional, site-specific, PFRP-equivalent process.

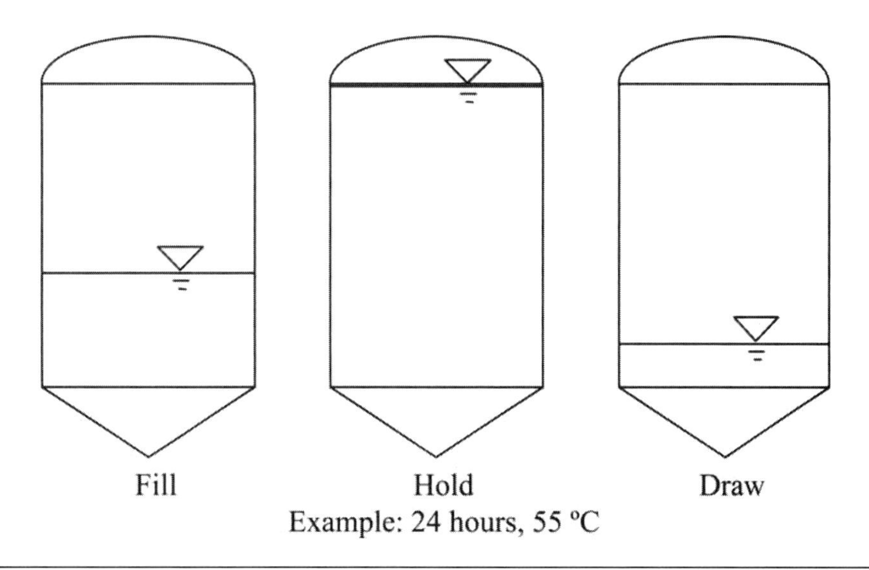

FIGURE 25.25 Basic sequence of batch (plug-flow) operations in a digester.

2.5 Post-Digestion Processing

2.5.1 Process Development

After digestion, solids are considered "stabilized" because the available substrate has largely been depleted and microbial activity has largely been reduced, so the product is far less likely to emit odors, attract vectors, or regrow pathogens. However, because of its biological origin, any perturbation of biosolids characteristics can "destabilize" the material. So, the methods used to store and handle biosolids can be extremely important.

Murthy et al. (2003), Chen et al. (2005), Chen et al. (2006) and Higgins et al. (2006a) evaluated the headspace of bottle-stored anaerobically digested biosolids and found that destabilizing biosolids could increase odor production by increasing the available substrate and decreasing methanogenic activity. Research has shown that a key group of odorants are the volatile organic sulfur compounds (VOSCs), which are mainly methanethiol (or methyl mercaptan), dimethyl sulfide, and dimethyl disulfide. When present in air samples, VOSCs correlate well with odor panel measurements from biosolids (Adams et al., 2004). Higgins et al. (2006b, 2008b) also showed that fecal coliform regrowth was possible from post-digestion solids processing. These authors suggested that biosolids shearing caused both fecal coliform regrowth and odorant production.

Murthy et al. (2002b) and Higgins et al. (2006a, 2008a) suggested that an increase in substrate (e.g., bioavailable protein) will increase volatile solids production. Higgins et al. (2008a) evaluated 10 full-scale mesophilic plants with anaerobic digestion and found that the protein content in biosolids was well correlated with VOSC emissions. Murthy et al. (2003) confirmed this relationship in incubation-bottle headspace experiments (see Figure 25.26).

Another important factor in VOSC production is methanogenic activity. More active methanogens (as indicated by gas production) produced fewer odors (Murthy et al., 2003) because they degraded organic sulfur compounds in the biosolids headspace (Higgins et al., 2006a). These researchers showed that biosolids with inhibited methanogenesis produced substantially more volatile sulfur than the uninhibited control did, suggesting that methanogens play a major role in biosolids deodorization (see Figure 25.27). They suggested that methanogens demethylated volatile organosulfur groups and converted them to inorganic sulfides (which remained in biosolids as a chemical precipitate). In summary, methanogens are seen to be responsible for degrading VOSC odorants; destabilizing methanogens result in a "net" odorant production (Murthy et al., 2003).

Post-digestion processes should minimize conditions that would destabilize biosolids or the population dynamics within the material.

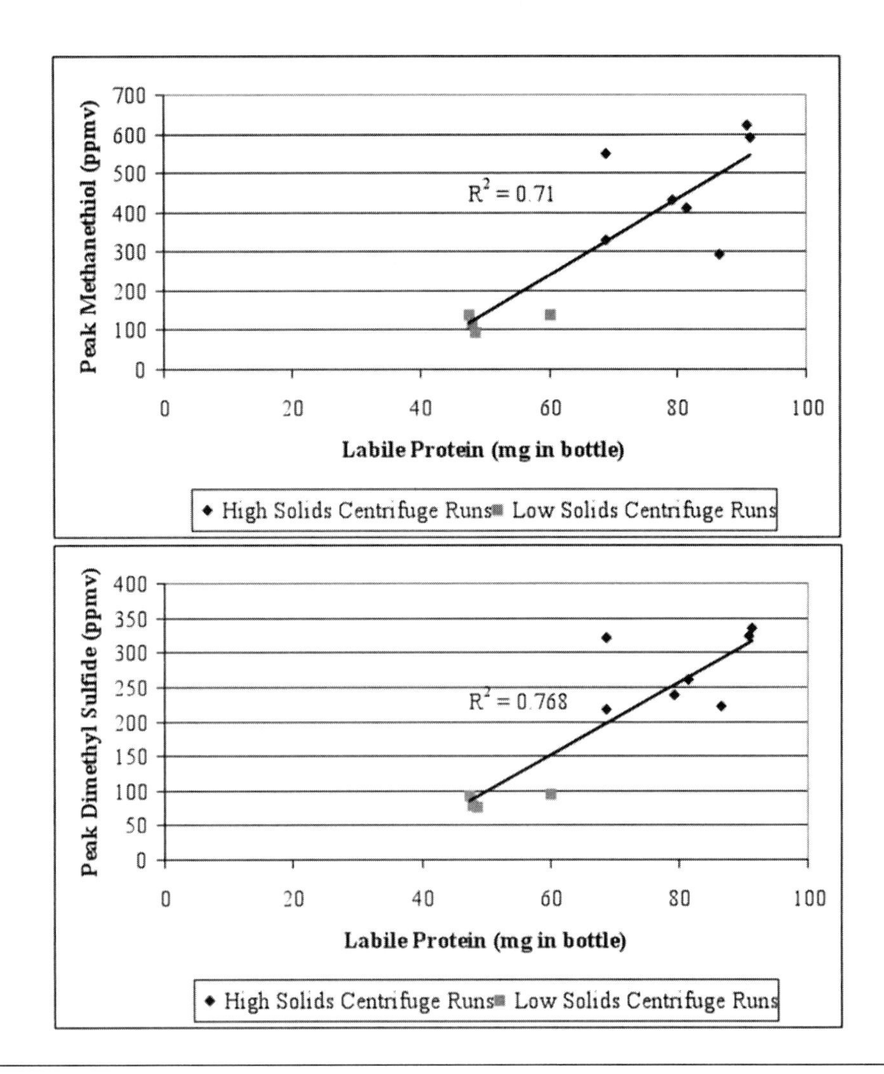

FIGURE 25.26 Relationship between labile protein extraction and emissions of volatile sulfur compounds (Murthy et al., 2003).

2.5.2 *Storage of Biosolids*

Biosolids in liquid and cake form can be stored under different conditions (see Chapter 24). One aspect of storage that affects microbial populations is freeze–thaw, in which stored biosolids alternately freeze and thaw during cold-weather seasons. Freezing biosolids can disrupt cells, thus releasing substrate and inhibiting methanogens. Thawing them could increase biological activity, resulting in odors. Eschborn et al. (2006) showed that the internal temperature of a field storage pile dropped during winter, and

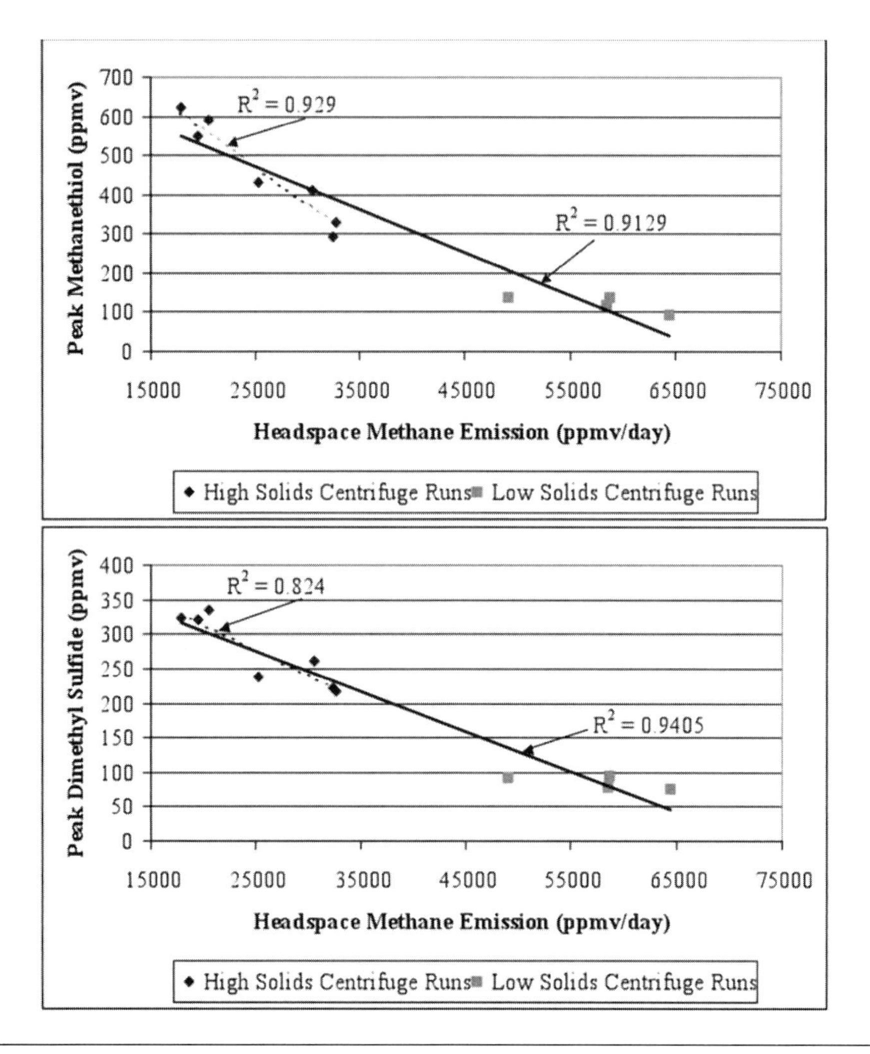

FIGURE 25.27 Relationship between headspace methane emissions and peak emissions of volatile sulfur compounds (Murthy et al., 2003).

the outer layer of the pile froze. As temperatures increased the following spring, odorant production also increased. When designing biosolids storage, engineers should take these factors into account, especially in regions where long-term winter storage is anticipated. Higgins et al. (2003) simulated freeze–thaw conditions in laboratory headspace experiments and showed that odorant production could be substantial when frozen cake samples were thawed (see Figure 25.28). Freezing biosolids led to a delay in methanogen recovery when the material thawed (see Figure 25.29).

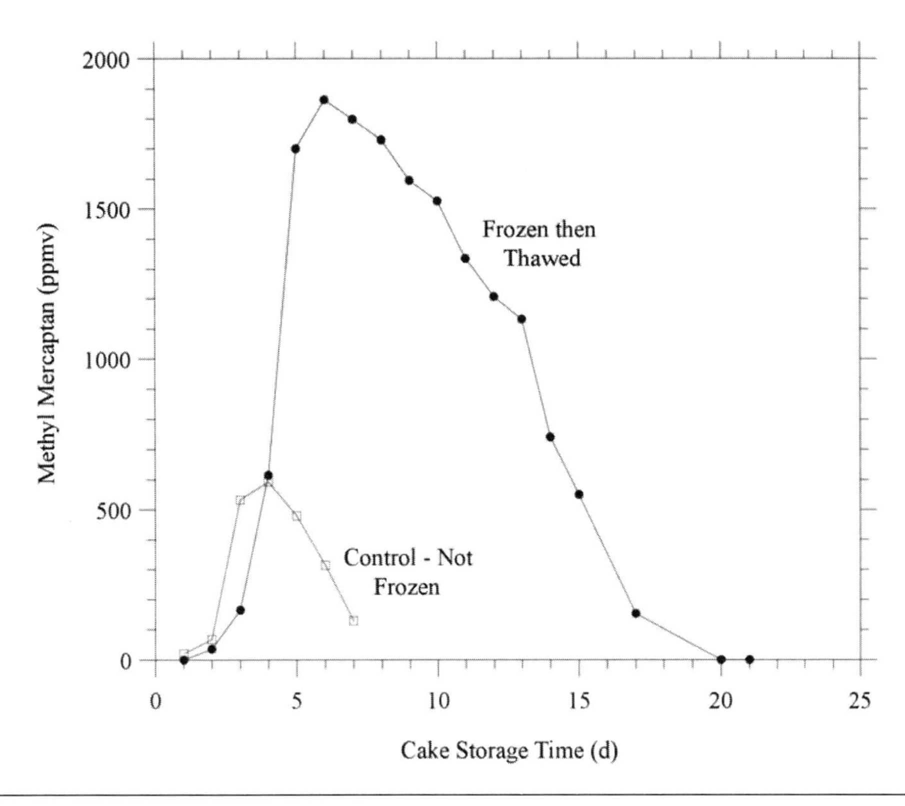

FIGURE 25.28 The effect of freeze–thaw conditions on methyl mercaptan emissions from biosolids (Higgins et al., 2003).

Managing biosolids storage before land application can help reduce nuisance odors. The goal of storage is to allow solids to restabilize once odorant production begins, thus allowing VOSC-associated odors to dissipate. For example, once odorant production begins, it typically peaks about a week or two later—although this depends on temperature (see Figure 25.30) (Higgins et al., 2003). Cooler temperatures increase the time needed for odors to peak and VOSC concentrations to reduce.

When designing storage systems for biosolids, engineers should consider the following:

- If frozen biosolids are stored for several days once thawing begins, odors can dissipate before the material is beneficially used;
- Fresh biosolids emit more odors during land spreading than biosolids that had been in long-term field storage (to get past the peak concentrations of odorants), so proper storage is important to managing odors (Eschborn et al., 2006);

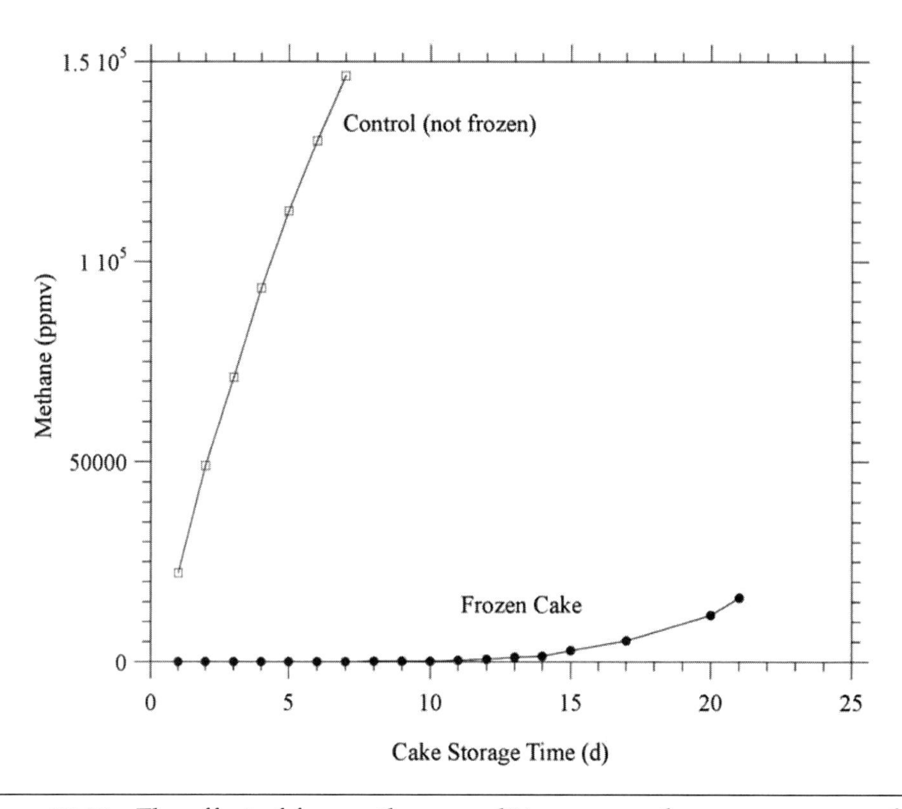

FIGURE 25.29 The effect of freeze–thaw conditions on methanogen recovery from dewatered biosolids(Higgins et al., 2003).

- Mixing old and new biosolids mitigates odorant production by bioaugmenting fresh biosolids with active methanogens that can degrade VOSCs (Chen et al., 2005; Williams et al., 2008); and
- Storage conditions should minimize freezing and perturbations that can destabilize population dynamics in biosolids.

2.5.3 *Cake Conveyance Impacts*

The effect of high-shear conveyance has not been studied in great detail. Murthy et al. (2002a) report that high-shear conveyance methods increase the biosolids' odorant production profile. Headspace experiments involving anaerobically digested biosolids confirm this research (see Figure 25.31). So, when designing solids conveyance systems, engineers should consider the following:

- Keep transport distances as short as possible;
- Use low-shear conveyance methods, if possible;

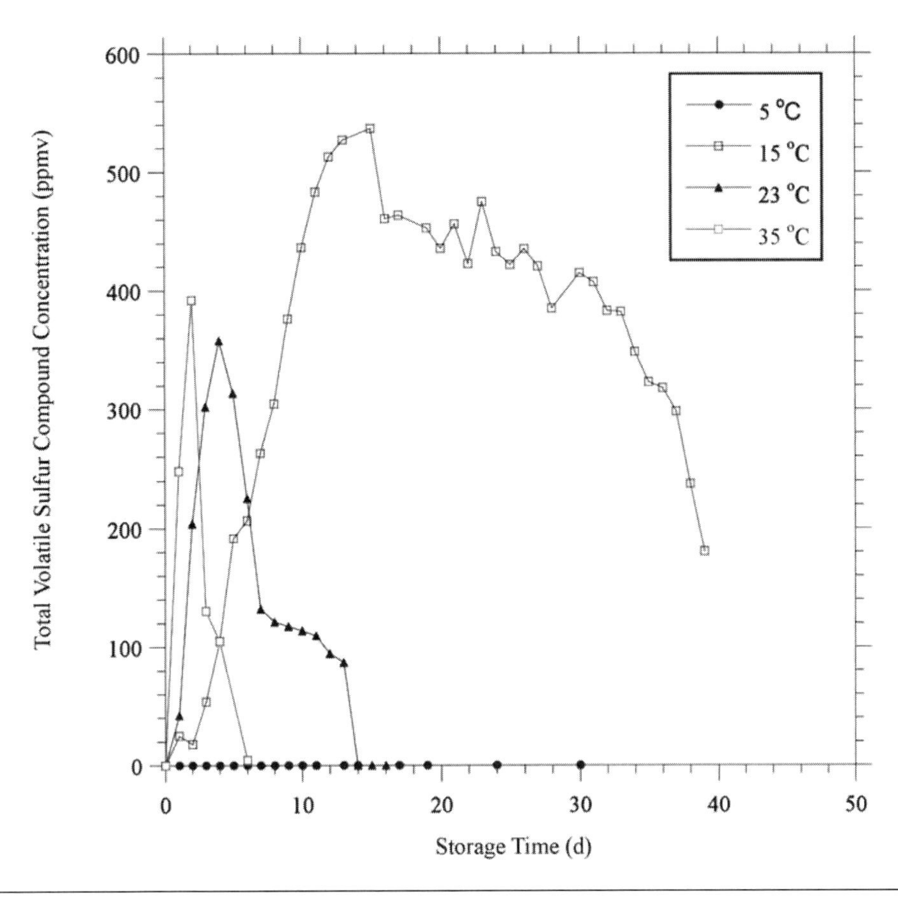

FIGURE 25.30 The effect of incubation temperature on emissions of total volatile sulfur compounds from dewatered biosolids (Higgins et al., 2003).

- Minimize the use of vertical high-shear conveyance; and
- Maintain a top–down design philosophy, if possible (i.e., install dewatering equipment at the top of a building and storage silos at the bottom to minimize conveyance distance and shear).

2.5.4 *Dewatering Impacts*

Murthy et al. (2002a) and Higgins et al. (2002) have suggested that VOSC production is influenced by a combination of factors. They conducted side-by-side tests on three dewatering systems: a high-solids centrifuge, a low-solids centrifuge, and a belt filter press simulator. The centrifuges were adjusted to produce lower cake solids (similar to the belt press device), so researchers could compare cakes from the same plant with

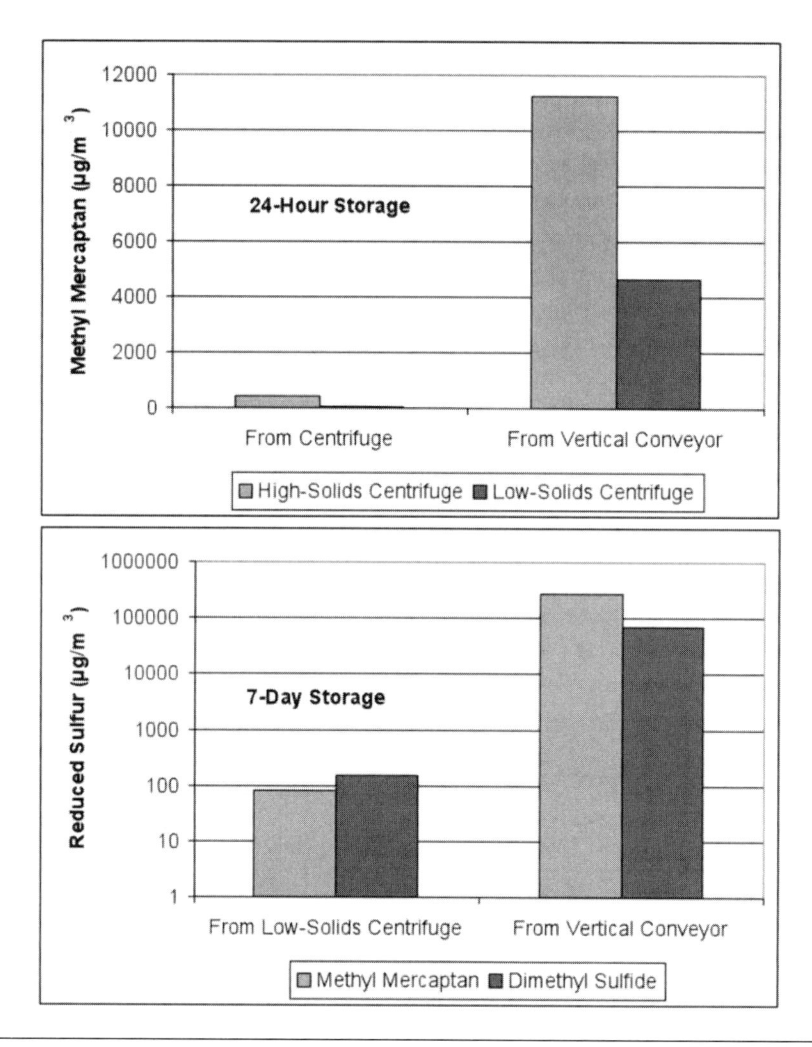

FIGURE 25.31 The effect of a vertical screw conveyor on odorant production in dewatered biosolids (Murthy et al., 2002a).

similar solids content. Results showed that cakes from the high-solids centrifuge produced more odorants than the other two devices (see Figure 25.32). The authors found that solids sheared during centrifugation released both labile protein and inhibited methanogenesis, thus increasing odorant production.

Higgins et al. (2006b) also showed that centrifuging biosolids promoted the regrowth of fecal coliforms. In some cases, this regrowth exceeded fecal coliform concentrations in raw solids. Further work by Higgins et al. (2008b) showed that Salmo-

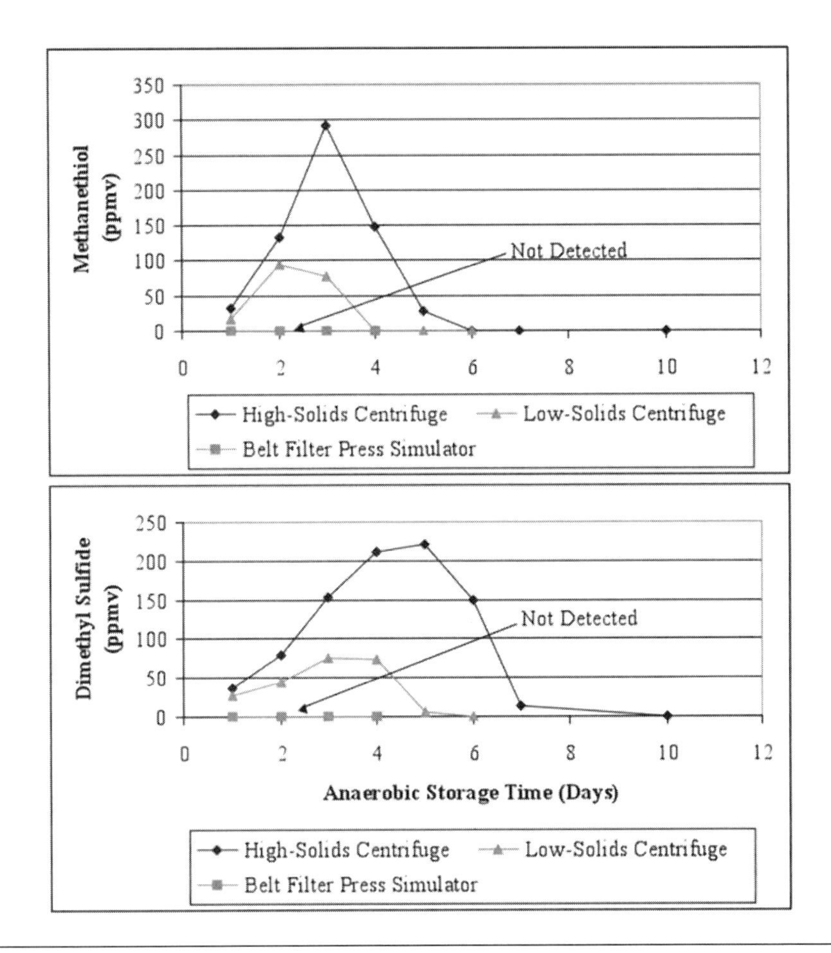

FIGURE 25.32 The effect of dewatering equipment on VSC emissions from biosolids (centrifuge cake contained 26% solids, belt filter press cake contained 25% solids, and detection limit is 1 ppmv). (Murthy et al., 2003).

nella regrew in Class B biosolids during storage. The biosolids had been mesophilically digested and centrifuged. However, Salmonella did not regrow in stored Class A and Class B biosolids that had been thermophically digested and centrifuged.

Design engineers should consider pilot-testing dewatering equipment, and monitoring for possible odorant production and regrowth before selecting a unit.

2.5.5 Digestion Process Impacts

While low-shear dewatering equipment produces fewer odors regardless of the anaerobic digestion process used, this is not the case for high-shear dewatering. More complete digestion (e.g., enhanced or thermophilic digestion) can affect overall odorant production and in some cases reduce it, especially if high-shear post-processing is used. Figures 25.33 and 25.34 show the total VOSC production profile for several full-scale digestion processes. These figures suggest that enhanced digestion processes can produce fewer odors than a conventional mesophilic digestion process.

If high-shear post-digestion processes are being proposed or already exist at a facility, design engineers should consider using enhanced digestion processes to mitigate overall odorant production if land application is proposed. Processes that have been shown to reduce odorant production from high-shear solids processing include thermophilic digestion, temperature-phased anaerobic digestion, and thermal hydrolysis.

2.6 Co-Digestion Processing

Historically, anaerobic digestion was used to minimize solids volume, stabilize solids, and reduce pathogens. As process options evolved, process efficiency improved, and

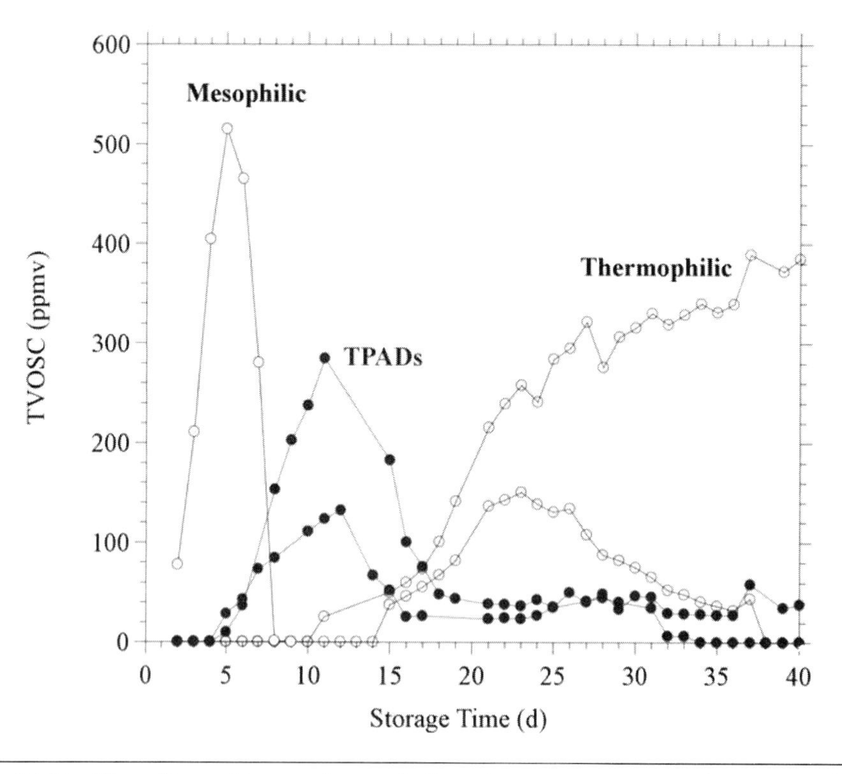

FIGURE 25.33 The effect of two enhanced digestion processes on odorant production (courtesy of Dr. Matthew Higgins, Bucknell University).

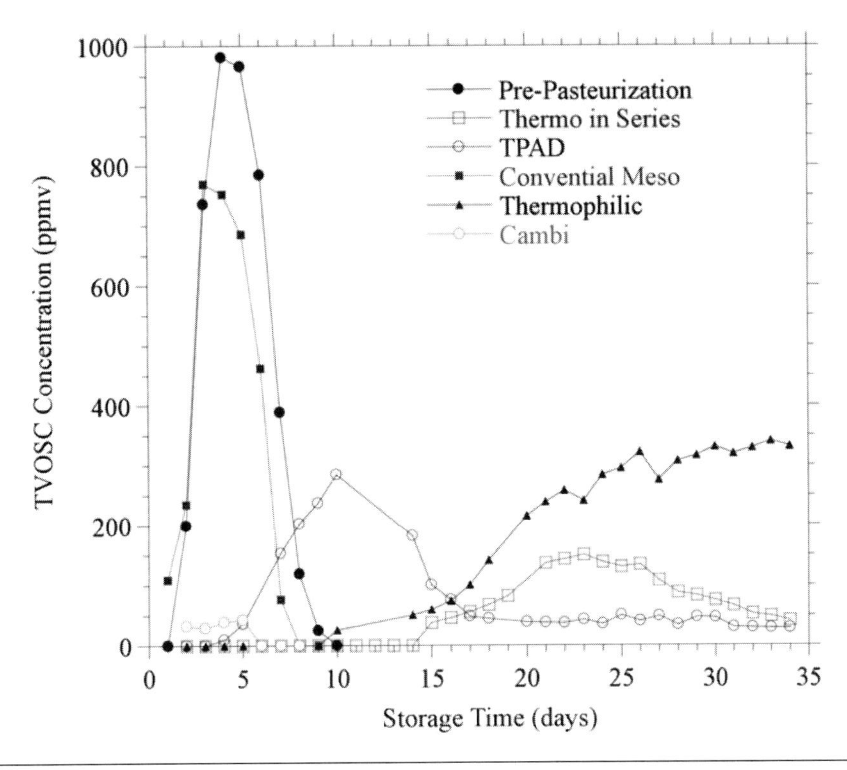

FIGURE 25.34 The effect of five enhanced digestion processes on odorant production (courtesy of Dr. Matthew Higgins, Bucknell University).

the focus on biogas production and energy use grew, interest in anaerobic digestion expanded. Some wastewater utilities would like to further expand biogas production by adding other feedstocks [fats, oil and grease (FOG), food wastes, and organic liquid wastes] to anaeroic digesters.

2.6.1 *Dry Digestion and Wet Digestion*

Co-digestion of wastes can encompass a wide variety of organic feedstocks. Some organic feedstocks are more amenable to "dry digestion", in which feedstocks contain 15 or even 20% solids. Such systems were developed primarily for organic solid waste or bulk-waste materials, and are seen primarily in technologies coming from Europe. Most of these systems originally were applied to the solid waste industry. The digestion vessels are often developed for plug-flow movement (e.g., feedstocks are added to the top of a silo-shaped reactor and move downward over a 20-day digestion period). Mixing systems are often limited, but can include paddles and other systems that work in thick solids. The use of wastewater solids in dry-digestion systems is limited because the cake typically contains less than 10% solids. This may change over time, so readers may wish to gather more information about dry digestion from the Internet or other literature sources.

In this manual of practice, the discussion of anaerobic digestion is devoted to liquid–slurry ("wet digestion") in which feedstocks contain less than 10 to 15% solids (they frequently contain 5 or 6% solids). In wet-digestion systems, feedstocks are pumpable materials and the mixing systems are compatible with slurries that are typically less than 5% solids. Wet-digestion systems typically are operated as complete-mix reactors that minimize deposits on the bottom of the tank and minimize floating layers of solids on the liquid surface. Wastewater utilities in North America, Europe, and elsewhere are not only adding solids to their wet-digestion systems but also a variety of compatible organic feedstocks.

2.6.2 *FOG and Grease Wastes*

Fats, oil, and grease are generated in a variety of locations in an urban environment. To protect the collection system, many municipalities strictly limit FOG discharges, so the material is accumulated at point sources (e.g., waste drums, grease traps, and grease interceptors). While municipalities do not want this material in the collection system, its use in anaerobic digesters can provide major benefits. FOG is an energy-rich substrate which is readily degradable in anaerobic digestion systems.

Digesters with FOG loads as high as 30% of feedstock or volatile solids loading still can maintain stable operations (Schafer et al., 2008). Kester et al. (2008) has reported that some facilities digesting both solids and FOG have both enhanced biogas production and enhanced VSR.

The benefits of FOG addition are well documented, but there are challenges as well. At ambient temperatures, FOG tends to be highly viscous and adheres to metallic and concrete surfaces, making it difficult to introduce to a digester via a conventional hauled-waste receiving station. Also, FOG tends to stratify during transport and storage. Receiving stations used for FOG have a wide variety of tankage. Tanks of FOG used to feed digesters need to be mixed (to obtain consistent feedstock characteristics) and may need to be heated via steam injection, hot water circulation, or heat exchanger. Also, screening and grinding often are required to remove debris (e.g., stones, rags, and metallic objects). Because of its association with the food service industry, FOG can contain other extraneous food debris that must be removed to protect pumps and downstream equipment.

To minimize stratification and eliminate a possible grease layer on the liquid surface in the digester, FOG must be properly mixed with the mass in the tank. Utilities often introduce FOG into a recirculating slurry of digesting mass, so it becomes mixed and diluted quickly. Also, the digester's mixing system should be carefully evaluated to ensure that a surface layer of scum or grease does not develop and cause operational problems.

2.6.3 *Liquid and High-Strength Wastes*

Liquid and slurry wastes are typically easiest for a digestion system to accept because its infrastructure is designed to pump liquids. Some liquid wastes may require screening or grit removal, depending on its composition. These wastes often come from food processing industries (e.g., fruit and vegetable processing), the beverage industry, pharmaceutical industry, and others. While liquid wastes are relatively easy to accept, they can have significant negative effect on digester capacity. Most digesters do not use solids–liquid separation for SRT control, so the added volume of liquid waste takes up a digester's hydraulic capacity. So, high-strength wastes should be used before lower-strength wastes because they produce the most biogas per unit hydraulic capacity consumed. When characterizing liquid wastes, design engineers should ensure that their components will not negatively affect digestion (e.g., promote struvite formation or produce a compound that will affect the liquid treatment train). Also, increasing levels of total dissolved solids or other dissolved constituents can be introduced via digestion and typically will be returned to the liquid treatment train via the post-digestion dewatering system. Such dissolved material may be inert and proceed directly to the outfall or can affect the liquid treatment process.

Liquid or slurry wastes being considered for co-digestion should be carefully evaluated for their compatibility with the digestion process, digestion-feeding system, mechanical systems used in digestion, and biogas management-system capacity. High-protein wastes, for example, could greatly affect foam production in digestion vessels. Sudden or large foam production is often debilitating for digester operations; it has caused digesters to overflow material onto the ground and caused accidents that resulted in major structural damage to tanks and tank covers.

These wastes often need to be carefully metered into the digestion system to prevent overfeeding microbes and causing digester upsets and foaming. Rapid feed of highly digestible waste also can lead to large spikes in gas production that the gas piping cannot accommodate. Instead, it is directly released to the atmosphere via gas-relief valves. In extreme cases, gas production could outpace the gas system's ability to discharge gas (especially if foam is blocking gas-relief valves), leading to gas pressure buildup in the vessel and potential catastrophic tank or cover failure.

2.6.4 *Food Waste Materials*

As with FOG, food scraps or post-consumer food waste materials are a source of renewable energy when anaerobically digested and converted to methane. One of the challenges of food scraps (the organic fraction of municipal solid waste) is removing contaminants to protect the digestion process. Also, the waste itself typically requires preprocessing to make it pumpable and amenable to efficient digestion without compromising biosolids aesthetics. Preprocessing can include screening; manual separation of debris and extraneous materials; removal of metals, aluminum, glass, grit, and plastics; and pulping—depending on the specific situation. Technologies that preprocess food scraps or food-waste materials are being developed primarily in Europe. Gray et al. (2008) describe one of these systems, which was tested at the East Bay Municipal Utility District in Oakland, California.

2.7 Design Considerations

2.7.1 *Design Data and Parameters*

Before designing the anaerobic digestion process, engineers should consider modeling it using available models (Batstone et al., 2002; Jones et al., 2008a). Jones et al. (2008b) recommend using a full-plant model to simulate influent wastewater characteristics and quantify fractions of solids produced to predict anaerobic digester performance. The Water Environment Federation (2009) provides more details on anaerobic digester modeling.

Historically, anaerobic digesters have been designed based on SRT, organic loading rate [volatile suspended solids (VSS) per volume], and volume per capita (see Table 25.10). In the absence of operating data (or estimates of projected plant flows) and cali-

TABLE 25.10 Typical design parameters for low- and high-rate digesters (Burd, 1968).

Parameter	Low rate	High rate
Solids retention time, days	30–60	15–20
Volatile suspended solids loading, lb/cu ft/d (kg/m^3·d)	0.04–0.1	0.12–0.16
	(0.64–1.6)	1.9–1.5
Volume criteria, cu ft/cap (m^3/cap)		
Primary sludge	2–3	1.3–2
	(0.06–0.08)	(0.03–0.06)
Primary sludge + trickling filter sludge	4–5	2.6–3.3
	(0.11–0.14)	(0.07–0.09)
Primary sludge + waste-activated sludge	4–6	2.6–4
	(0.11–0.17)	(0.07–0.11)
Combined primary + waste biological sludge feed concentration, % solids-dry basis	2–4	4–6
Anticipated digester underflow concentration, % solids-dry basis	4–6	4–6

brated system modeling, volume figures per capita can be used to estimate influent volumes at treatment plants. Low-rate digesters typically have organic loading rates of about 0.5 to 1.5 kg VSS/m^3·d (0.04 to 0.1 lb VSS/d/cu ft). High-rate digesters with mixing and heating typically have organic loading rates of 1.9 to 2.5 kg-VSS/m^3·d (0.12 to 0.16 lb/d/cu ft).

Typical SRTs are about 30 to 60 days for low-rate digestion and 15 to 20 days for high-rate digestion at mesophilic temperatures. The *solids retention time* is the ratio of the total mass of solids in the system to the quantity of solids withdrawn per day. In two-stage digestion, the typical SRT above refers to that of the first reactor in the system. For anaerobic digesters with no internal recycle, the SRT equals the HRT. If settled solids are recycled, the SRT would be larger than the HRT. Recycling is characteristic of the anaerobic contact or two-phase digestion process or recuperative thickening.

A minimum SRT is essential in anaerobic digestion; it ensures that the necessary microorganisms are being produced at the same rate as they are wasted daily. It also is different for various constituent groups. For example, lipid-metabolizing bacteria grow most slowly and, therefore, need a longer SRT, while cellulose-metabolizing bacteria require a shorter SRT (see Figure 25.35).

If the SRT is too short, then the microbial population of methanogens will wash out and the system will fail. Lawrence (1971) has published the minimum SRT needed to reduce several specific substances; a function of temperature, these SRTs range from less than 1 day for hydrogen to 4.2 days for wastewater solids (see Table 25.11). Hotter temperatures reduce the SRTs needed for maximum performance because they increase specific gas production (see Figure 25.36).

However, digester SRT is not only a function of system microbiology; the selected use for biosolids also must be considered, especially for Class B biosolids. Both thermophilic and mesophilic anaerobic digestion are considered PSRPs. One method for achieving this status, along with maintaining temperature, mixing and anaerobic conditions, is maintaining a minimum SRT of 15 days. Furthermore, for overall process stability, ease of control, to account for grit and scum accumulation, imperfect mixing, variability in solids production rates and biosolids stability most digesters operate at 15 days.

Later sections will provide equations for process parameter estimation. Given the variability in reactor configurations and solids composition, generic and lumped kinetic parameters may not sufficiently describe system performance. Pilot or full-scale testing in conjunction with using newer more complex process models, with sufficient background data, should be considered when the accuracy of the projected process performance is critical.

Pilot testing, before design, should also be considered when implementing a new process configuration or the raw solids contain unconventional substrates such as

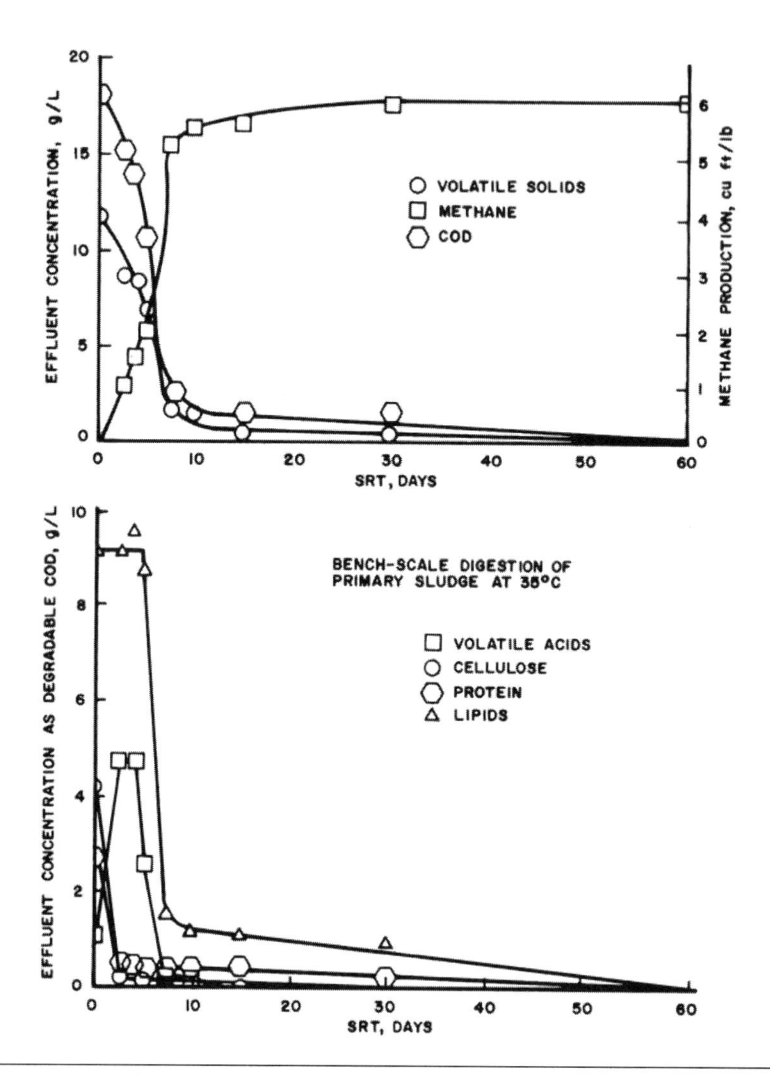

FIGURE 25.35 The effect of SRT on the relative breakdown of degradable waste components and methane production (cu ft/lb × 0.062 43 = m^3/kg).

industrial inputs. The effect of industrial wastes on anaerobic digestion is always questionable until the specific waste is characterized and tested.

2.7.2 Process Design

2.7.2.1 Sizing Criteria

When sizing digesters, the key parameter is SRT. For digestion systems without recycle, there is no difference between SRT and HRT. The volatile solids loading rate is

TABLE 25.11 Minimum values of solids retention time (θ_c) for anaerobic digestion of various substrates (reprinted with permission from Lawrence, A. W. [1971] Application of Process Kinetics to Design of Anaerobic Processes. In *Anaerobic Biological Treatment Processes.* F. G. Pohland [Ed.], Advances in Chemistry Series; American Chemical Society: Washington, D.C., 105. Copyright 1971 American Chemical Society).

Substrate	35°C	30°C	25°C	20°C
Acetic acid	3.1	4.2	4.2	—
Propionic acid	3.2	—	2.8	—
Butyric acid	2.7	—	—	—
Long-chain fatty acid	4.0	—	5.8	7.2
Hydrogen	0.95[a]	—	—	—
Wastewater sludge	4.2[b]	—	7.5[b]	10

[a] For 37°C.
[b] Computed value.

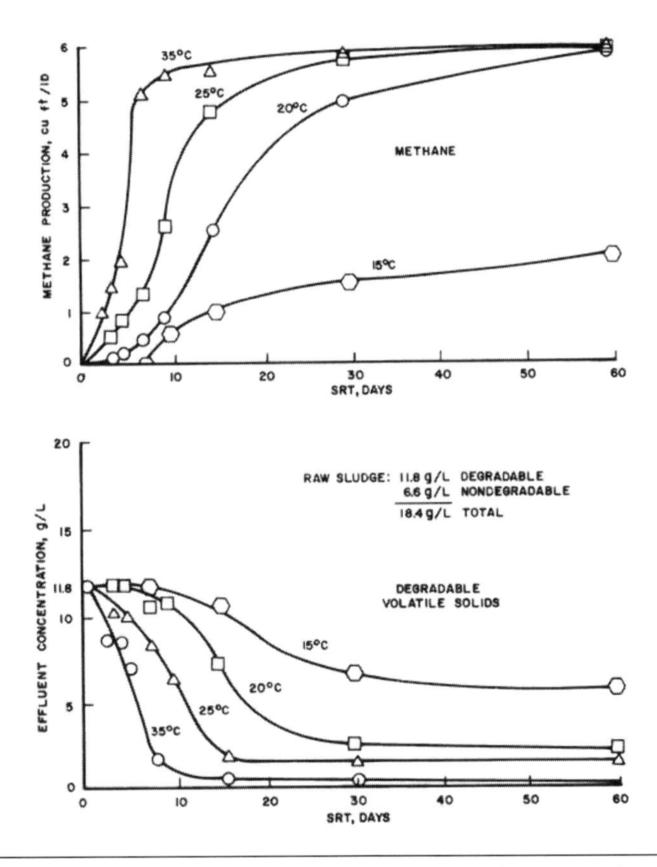

FIGURE 25.36 Effect of temperature and SRT on methane production and VSR. (O'Rourke, 1968).

also frequently used because it is directly related to SRT or HRT. However, SRT is considered the more fundamental parameter. The selection of the design SRT and volatile solids loading rate should consider several factors (e.g., process microbiology, stability, biosolids regulatory requirements, biosolids stability, and industrial inputs).

Presently, the design minimum SRT typically is selected based on experience, general rules, and regulatory requirements. Design engineera should note that a range of solids-production conditions must be considered when developing appropriate SRT design criteria.

Researchers are developing more quantifiable approaches to understanding digestion and its limits. The selection of a design SRT directly affects system kinetics and, consequently, process performance. Both complex models (e.g., ADM-1) and simplified models [e.g., that presented by Parkin and Owen (1986)] have been suggested as a means of estimating or predicting the limiting (minimum) digester SRT, although the data required for its application are limited. The Parkin and Owen approach is based on applying a safety factor to a limiting SRT to establish the design SRT. If the limiting SRT is based on a desired digestion efficiency and the digester approaches complete-mix conditions, the limiting SRT can be estimated as follows:

$$\mathrm{SRT}_{\mathrm{min}} = \frac{YkS_{\mathrm{eff}}}{K_c + S_{\mathrm{eff}}} - b^{-1} \tag{25.3}$$

Where

$\mathrm{SRT}_{\mathrm{min}}$ = limiting SRT for required digester performance;

Y = yield of anaerobic organisms resulting from growth (g VSS/g COD destroyed);

k = maximum specific substrate use rate (g COD/g VSS-d);

S_{eff} = concentration of biodegradable substrate in digested solids (therefore, in digester) (g-COD/L); which is equivalent to $S_o\,(1-e)$, where S_o is the concentration of biodegradable substrate in the feed solids, g COD/L, and e is the digestion efficiency at removing S_o,

K_c = half saturation concentration of biodegradable substrate in feed solids (g COD/L); and

b = endogenous decay coefficient (d^{-1}).

Values for constants in Equation 25.3 have been proposed for typical municipal primary solids within a temperature range of 25 to 35°C (77 to 95°F). The following proposed values (Parkin and Owen, 1986) are based on laboratory experiments (full-scale data are not currently available):

- $k = 6.67$ g COD/g VSS-d (1.035^{T-35});
- $K_c = 1.8$ g COD/L (1.112^{35-T});
- $b = 0.03\ d^{-1}\ (1.035^{T-35})$;
- $Y = 0.04$ g VSS/g COD removed; and
- T = temperature (°C).

Using the calculated value of the limiting SRT for required digester performance, the safety factor for anaerobic digestion then can be calculated as:

$$SF = \frac{SRT_{actual}}{SRT_{min}} \tag{25.4}$$

For example, suppose engineers were designing a new anaerobic digestion system with the help of data from similar existing systems (see Table 25.12) (ASCE, 1983). The data indicate a median average SRT of about 20 days. Assuming the new system would have a digestion efficiency of 90%, a design temperature of 35°C (95°F), and feed solids with a biodegradable COD concentration of 19.6 g/L, they used Equation 25.3 to calculate that the minimum SRT would be 9.2 days (Parkin and Owen, 1986). (Biodegradable COD is assumed to represent the degradable fraction of volatile solids in feed solids

Table 25.12 Solids residence times reported for anaerobic digestion operations in the United substrates (reprinted with permission from ASCE, 1983).

	Percent of facilities in each range	
SRT, days	**Primary sludge only**	**Primary plus secondary**
0–5	0	9
6–10	0	15
11–15	0	9
16–20	11	12
21–25	45	25
26–30	11	3
31–35	11	15
36–40	0	6
41–45	0	0
46–50	22	0
Over 50	0	6
Total number of plants	(12 plants)	(132 plants)

and liquor.) Then, using Equation 25.4 and the median SRT of the existing systems, engineers calculate that the new system's safety factor is 2.2. This safety factor obtained should be used to estimate if short-term or dynamic increase in hydraulic loadings could be accommodated within the process.

Pilot-testing may be required to determine a system's actual limits because using generic and/or lumped kinetic parameters can over- or underestimate SRT requirements, depending on actual operating conditions and the composition of the substrate. For example, if feedstocks contain significant amounts of materials (e.g., lipids) that are more difficult to degrade then typical municipal primary solids, then the constants given for Equation 25.3 will not apply. In such cases, or if consistently high VSR is critical, engineers may need to use higher design SRT values than those listed in Table 25.13. If solids readily degrade (e.g., typical primary solids without biological solids), then engineers may be able to use slightly lower design SRT values than those listed in Table 25.13.

Other limits (e.g., regulatory limits to meet PSRP requirements) need to be considered as well. While a shorter design SRT may reduce tank and ancillary equipment size and cost, the ease of biosolids use or disposal also must be considered. Any initial cost savings may be lost if the biosolids are poor quality, do not meet regulatory standards, or incur significant O&M costs to ensure compliance.

Furthermore, to minimize the likelihood of a digester upset, design engineers should select the design SRT based on a critical operating period (e.g., a high solids-loading period when grit and scum has accumulated, or when a digester is out of service). The choice of critical operating period will depend on plant size, anticipated solids production, and other site-specific factors.

2.7.2.2 *Loading Rates and Frequency*

A digester's loading rate and frequency significantly affects digester performance. Constant loading will produce the most stable operations because the microorganisms will reach and maintain steady-state conditions. Relative loading will affect the process' overall stability because both over- and underloading can impair process performance.

TABLE 25.13 Estimated volatile solids destruction.

Digestion type	Digestion time (days)	Volatile solids destruction (%)
High-rate (mesophilic range)	30	55
	20	50
	15	45
Low-rate	40	50
	30	45
	20	40

The design loading rate should be coupled with the process selected (suggested values can be found in the sections discussing each process).

Furthermore, design engineers also should consider the potential effects of a specific feeding regime. Continuous feeding may necessitate storage tanks or 24-hour thickening and dewatering operations. Slug loading may lead to foaming.

In addition, design engineers should ensure that upstream processes are adequate to meet design loading. Thickening performance could prevent a design loading rate or SRT from being met consistently. Design engineers should take this into account when sizing tanks because it could lead to process limitations that could reduce the expected life of the system.

2.7.2.3 *Solids Blending*

Adding a solids blending tank before the reactor can improve digestion stability (see Figure 25.37). The blend tank homogenizes solids before digestion, particularly when multiple solids streams are being loaded to a digester. Primary solids and secondary solids degrade differently, so without homogenization and flow pacing, the digester can experience a wide fluctuation of organic loads throughout the day, even at constant pumping speeds. (The effective organic loading rate is a function of degradable volatile solids rather than the total volatile solids load.) Blending tanks reduce diurnal loading variability by providing a "wide spot" in the line. The tank absorbs high solids flows, allowing the digester to maintain more consistent operations, rather than peaking with the solids wasting protocols.

Blending tanks typically are mixed by mechanical mixers or pumps. The degree of mixing selected must be balanced by equipment costs and power use. When sizing blending tanks, design engineers should take care not to oversize them because that can promote acid-reactor conditions if the detention time is long enough. Sometimes blending tanks are heated to partially heat solids before they enter the digester. Heating also can exacerbate acid formation in exececessively large blending tanks.

2.7.2.4 *Solids Destruction and Gas Production*

Design engineers can estimate the expected VSR using previous data (40 to 65.5%) or equations relating VSR to detention time (Liptak, 1974). For a standard-rate system,

$$V_d = 30 + \frac{t}{2} \tag{25.5}$$

Where

V_d = VSR (%),

t = time of digestion (days), and

FIGURE 25.37 Thickened solids blending tank at King County's South Wastewater Treatment Plant in Renton, Washington (courtesy of Brown and Caldwell, Seattle, Washington, and King County, South Treatment Plant, Renton, Washington).

For a high-rate digestion system,

$$V_d = 13.7\ln(\theta_d^m) + 18.94 \tag{25.6}$$

Where

V_d = VSR (%), and
θ_d^m = design SRT.

Additional estimates can be made using Table 25.13. The concentration of fixed solids entering the digester will remain constant. Table 25.14 compares the alternative

TABLE 25.14 Comparison of methods for estimating digester volatile solids destruction.

Description	Value	Units	Source or notation
Process parameters for estimated solids destruction			
Raw solids biodegradable COD	19.3	mg-COD/L	Example in section 2.7.2.1. Parkin and Owen (1986)
COD removal efficiency	90	%	Example in section 2.7.2.1. Parkin and Owen (1986)
Digester temperature	35	°C	
Design solids retention time (SRT)	20	Days	
Process type	HR		HR = high rate
Volatile solids destruction based on Equations 25.8 and 25.15			
SRT_{min}	9.2	Days	Equation 25.8, using kinetic parameters in section 2.7.2.1
V_d	49	%	Equation 25.15
Volatile solids destruction based on Table 25.13			
V_d	60	%	
Volatile solids destruction based on Figure 25.38			
V_d	42	%	Solids age × temperature = 700

methods for estimating VSR for a single source; it shows that VSR estimates depend on the method selected, even for the same data set. Design engineers should understand the underlying assumption(s) behind each method before choosing one. If operational data are available, engineers should use them rather than these methods.

A close estimation of the solids (kg/d) that would enter the second-stage digester of a two-stage system is given by the following equation:

$$\text{Solids} = \text{TS} = (A \times \text{Total solids} \times V_d) \tag{25.7}$$

Where

TS = total solids entering digester [kg/d (lb/d)],

A = volatile solids (%), and

V_d = volatile solids destroyed in primary digester (%).

Equation 25.7 can be used to estimate the solids load to a second-stage digester and the degree of thickening (%). However, the secondary digester's volume often is equal to that of the primary digester to allow units to be taken out of service.

Design engineers can estimate the specific gas production at municipal plants by using the relationship of about 0.8 to 1.1 m^3/kg (13 to 18 cu ft/lb) of VSR. Gas production increases as the percentage of FOG in the feedstock increases (as long as adequate SRT and mixing are provided) because FOG is the slowest to metabolize. The total gas volume produced is as follows:

$$G_v = (G_{sgp})\, V_s \tag{25.8}$$

Where

G_v = volume of total gas produced [m^3 (cu ft)];
V_s = VSR [kg (lb)]; and
G_{sgp} = specific gas production, taken as 0.8 to 1.1 m^3/kg VSR (13 to 18 cu ft/lb VSR).

The total amount of methane produced can be estimated from the amount of organic material removed each day:

$$G_m = M_{sgp}\,[\Delta OR - 1.42(\Delta X)] \tag{25.9}$$

Where

G_m = volume of methane produced [m^3/d (cu ft/d)],
M_{sgp} = specific methane production per mass of organic material (COD) removed [m^3/kg COD (cu ft/lb COD)],
ΔOR = organics (COD) removed daily [kg COD/d (lb COD/d)], and
ΔX = biomass produced [kg VSS/d (lb VSS/d)].

Because digester gas is about two-thirds methane, the total digester gas produced is equal to the following:

$$G_T = \frac{G_m}{0.67} \tag{25.10}$$

where G_T = total gas produced [m^3/d (cu ft/d)].

Expected methane concentrations can range from 45 to 75%; typical methane concentrations range from 60 to 75% (by volume). Typical carbon dioxide concentrations range from 25 to 40%, (by volume). Biogas typically includes hydrogen sulfide, but excessively high concentrations should be investigated (e.g., by determining any sources of industrial wastes or saltwater infiltration). The expected heat value of digester gas depends on the biogas' composition.

2.7.3 *Tank Configuration and Shape*

The tank configuration (shape) significantly affects the operating characteristics of anaerobic digestion, as well as the cost of construction and O&M. Typically, digesters are available in three basic shapes: short cylinder ("pancake"), tall cylinder ("silo"), and egg-shaped.

2.7.3.1 *Egg-Shaped Digesters*

Egg-shaped digesters have been in service for more than 40 years in Europe, and since the early 1990s in the United States (Volpe et al., 2004). This shape typically is considered the optimal shape for a digester, providing excellent mixing characteristics, very few dead zones, and good grit suspension.

This digester's shape is optimal for good mixing (see Figures 25.38 and 25.39). The tapered base, with centrally located mixing, is designed to promote the resuspension of

Figure 25.38 Egg-shaped digesters at the Central Valley Water Reclamation Facility in Utah (courtesy of Central Valley Water Reclamation Facility, Salt Lake City, Utah).

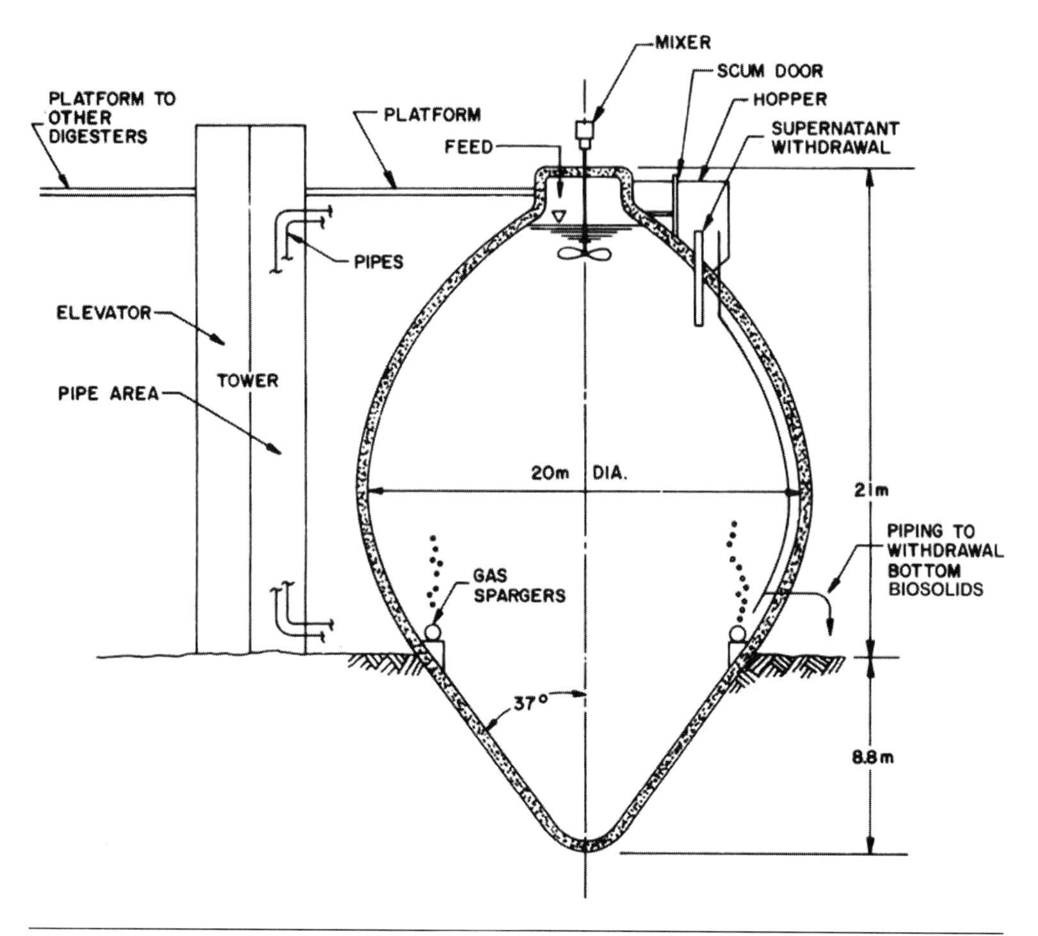

FIGURE 25.39 Schematic of a typical egg-shaped digester.

grit and other heavy materials into the bulk fluid. This minimizes the amount of material retained in the digester, increasing the active fraction and reducing the out-of-service time for cleaning. In a review of egg-shaped digesters, Volpe et al. (2004) reported that some have been in service for 20 years without needing to be cleaned.

The digester's small gas dome provides operational advantages and disadvantages. First, the degree of liquid agitation in the dome due to mixing and the foam-suppression system is typically high, preventing a scum layer from forming; instead, it remains entrained in the liquid phase and can be withdrawn. Depending on tank configuration, however, withdrawal from the small gas dome can be problematic (see Figure 25.40).

The German style of solids withdrawal—bottom withdrawal—can be problematic for systems with solids with a high propensity to foam. In such cases, foam can accumu-

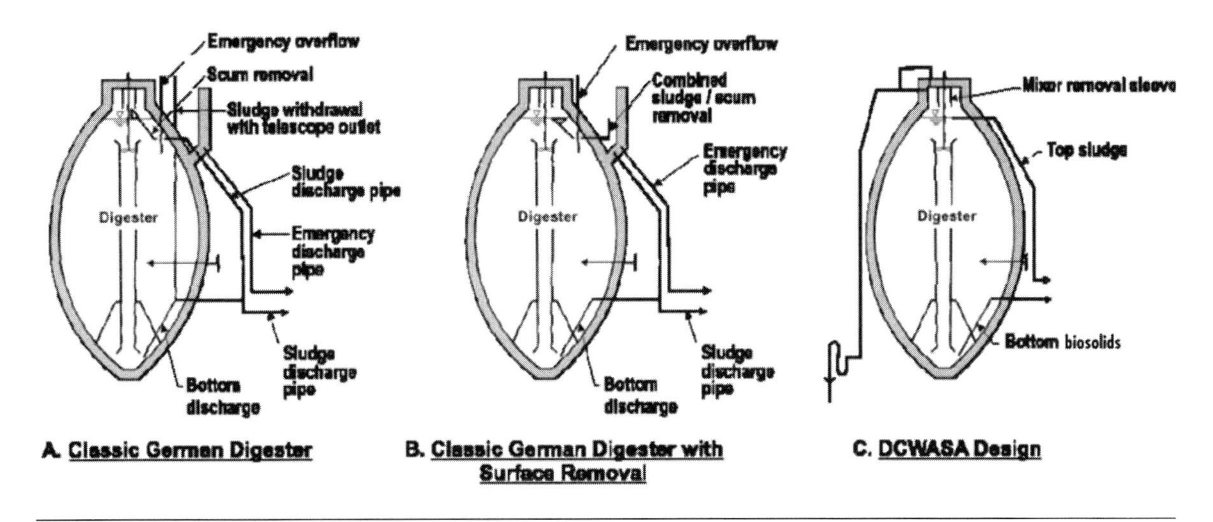

FIGURE 25.40 Solids-withdrawal configurations for egg-shaped digesters (Volpe et al., 2004).

late in the gas dome despite foam suppression. If the foaming event is great enough, the only means of exit is via the gas-handling system. So, both surface and bottom withdrawal should be provided.

Egg-shaped digesters typically are made of either steel or concrete. Both are susceptible to corrosion by the digester contents, so a corrosion-resistant coating typically is applied. The amount and locations are the purview of the manufacturer. Egg-shaped digesters made of steel could use stainless steel in areas prome to corrosion, although this increases the cost. Steel eggs also need a corrosion-resistant coating on the outside of the digester. The insulation system often serves as part of the exterior coating, preventing corrosion by minimizing moisture contact.

Their relatively tall profile makes these digesters visible from a distance but allows for a larger volume in a relatively small footprint. Also, their shape and construction constraints make their construction a specialty. A limited number of firms have the equipment and experience to build them, so egg-shaped digesters typically are more expensive to construct than a similar-sized cylindrical or silo digester. A recent review of egg-shaped digester construction reported only five vendors in the United States who supply them (Volpe et al., 2004).

Volatile solids loading in these digesters ranges from 0.64 to 2.4 kg volatile solids/m^3·d (0.040 to 0.150 lb volatile solids/cu ft/d), with a maximum loading of 1.6 to 2.9 kg volatile solids/m^3·d (0.106 to 0.175 lb volatile solids/cu ft/d). Solids retention time ranges from 15 to 45 days, and VSR ranges from 42 to 65%, depending on the intended use for the resulting biosolids. Gas production ranges from 0.68 to 1.08 m^3/kg volatile solids destroyed (11 to 17.5 cu ft/lb volatile solids destroyed). Gas, jet, mechanical, and pump mixing systems can be used, and mixing energy varies from 3.16 to 13.19 W/m^3 (0.12 to 0.5 hp/1 000 cu ft). Heating systems primarily are heat exchangers, although a few use steam injection.

2.7.3.2 Silo Digesters (Tall Cylinders)

Silo digesters are a newer version of cylindrical (pancake) digesters, with a greater height-to-diameter ratio to gain some of the mixing advantages of egg-shaped digesters without the cost (see Figure 25.41).

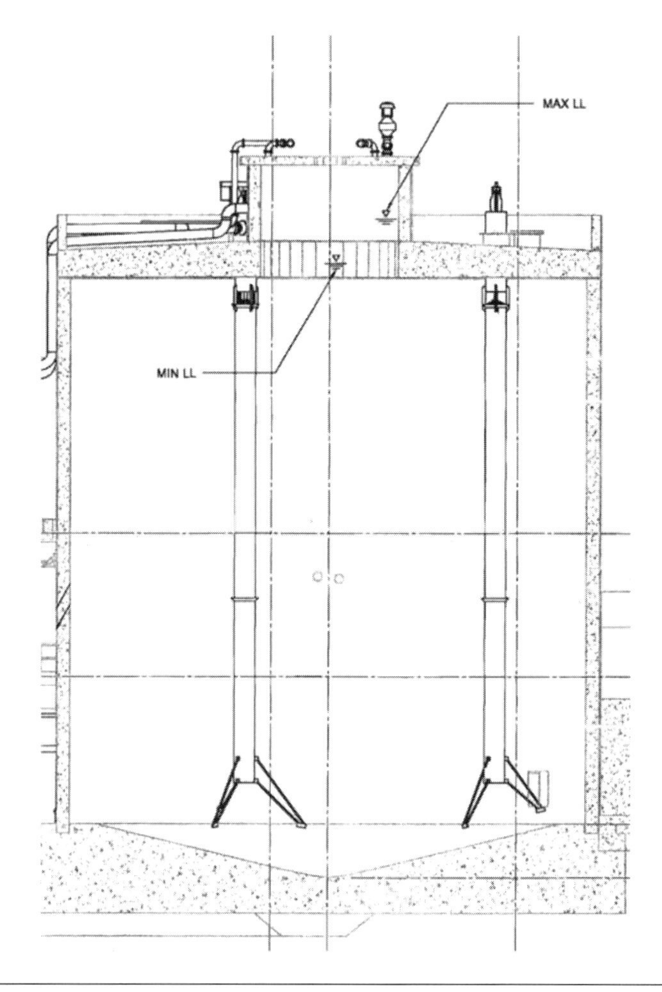

FIGURE 25.41 Cross-section of a typical silo digester with internal draft tube mixers (courtesy of Brown and Caldwell and King County Brightwater WWTP).

Silo digesters can resemble egg-shaped digesters when equipped with a submerged fixed cover (see Section 2.7.4.6). The small liquid–gas interface of the submerged cover is similar to the gas dome of the egg-shaped digester, and provides many of the same advantages. Other cover options include gas-holder, fixed, and Downe's, allowing design engineers more flexibility to customize the tank for a specific use or even multiple uses (something egg-shaped digesters are not well suited for).

Currently, few silo digesters are operating in the United States.

2.7.3.3 Cylindrical Digesters

Cylindridal (pancake) digesters are by far the most common tank design for anaerobic digesters in the United States (see Figures 25.42 and 25.43). Their relatively low height-

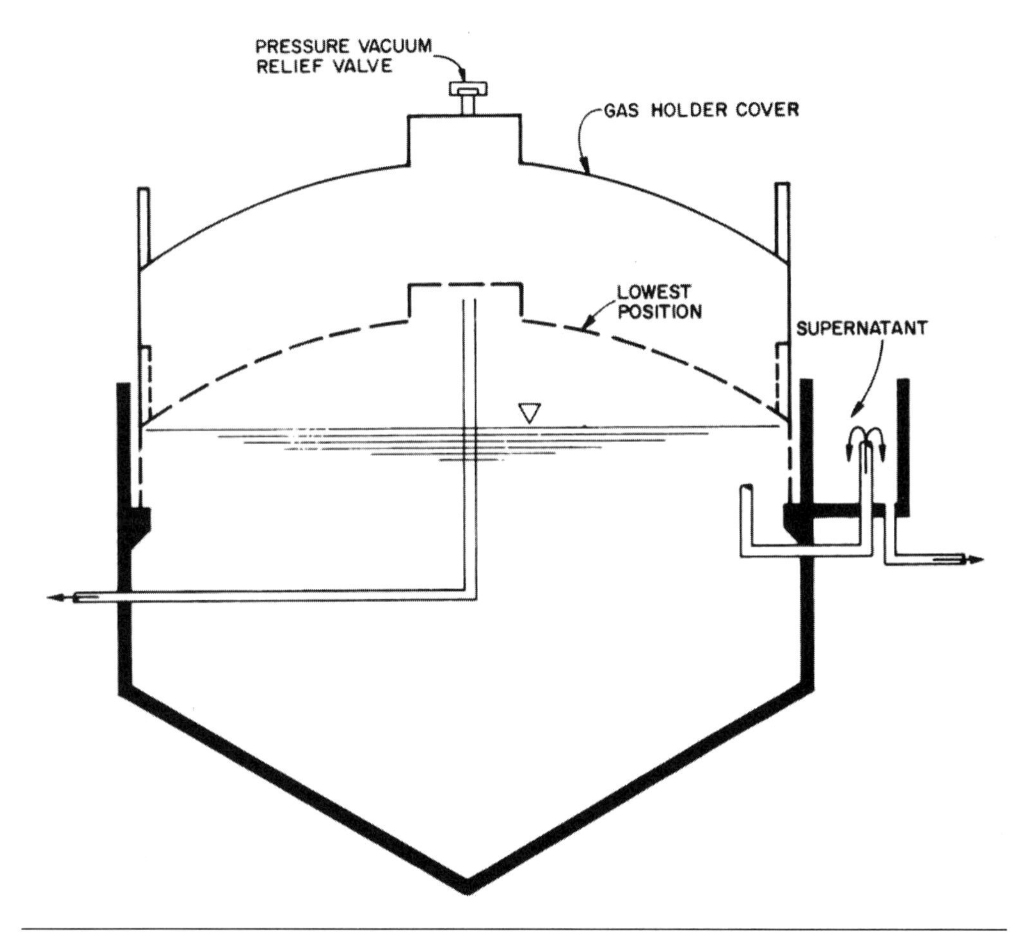

FIGURE 25.42 Schematic of a typical pancake (short cylinder) digester with a gas-holder cover (covers are interchangeable).

FIGURE 25.43 The anaerobic digesters at King County's South Wastewater Treatment Plant in Renton, Washington (courtesy of Brown and Caldwell and King County, Washington).

to-diameter ratio makes them easier and less expensive to construct. However, this tank does not have the process benefits realized by egg-shaped and silo digesters.

Pancake digesters are more prone to dead zones and poor mixing regimes, resulting in lower VSR and more grit deposition. So, cleanings are needed more often, increasing O&M costs.

However, pancake digesters can be fitted with a variety of cover configurations, which can allow for significant process flexibility and multiple roles (e.g., gas storage and variable liquid level). Egg-shaped digesters are only suited for service as anaerobic reactors.

These tanks typically are made of reinforced concrete, with sidewall depths ranging from 6 to 14 m (20 to 45 ft) and diameters ranging from 8 to 40 m (25 to 125 ft) (U.S. EPA, 1979). Conical bottoms are preferred for cleaning purposes, with slopes varying between 1:3 and 1:6 (WPCF, 1987). Slopes greater than 1:3, although desirable for grit removal, are difficult to construct and difficult to stand on while cleaning. Conical bottoms also minimize grit accumulation; instead providing for relatively continuous grit removal. The floors can have either one central withdrawal pipe or be divided into wedges, each with its own withdrawal pipe (waffle-bottom digester). The latter are more costly to construct but may reduce cleaning costs and frequency.

Where necessary, cylindrical digesters have been insulated using brick veneer and an air space, earth fill, polystyrene plastic, fiber glass, or insulation board.

2.7.4 Digester Cover Type-Shape

2.7.4.1 Fixed-Cover Digester

Standard fixed covers can be constructed of either concrete or steel (see Figure 25.44). They provide the maximum solids–gas interface, which makes foam and scum control

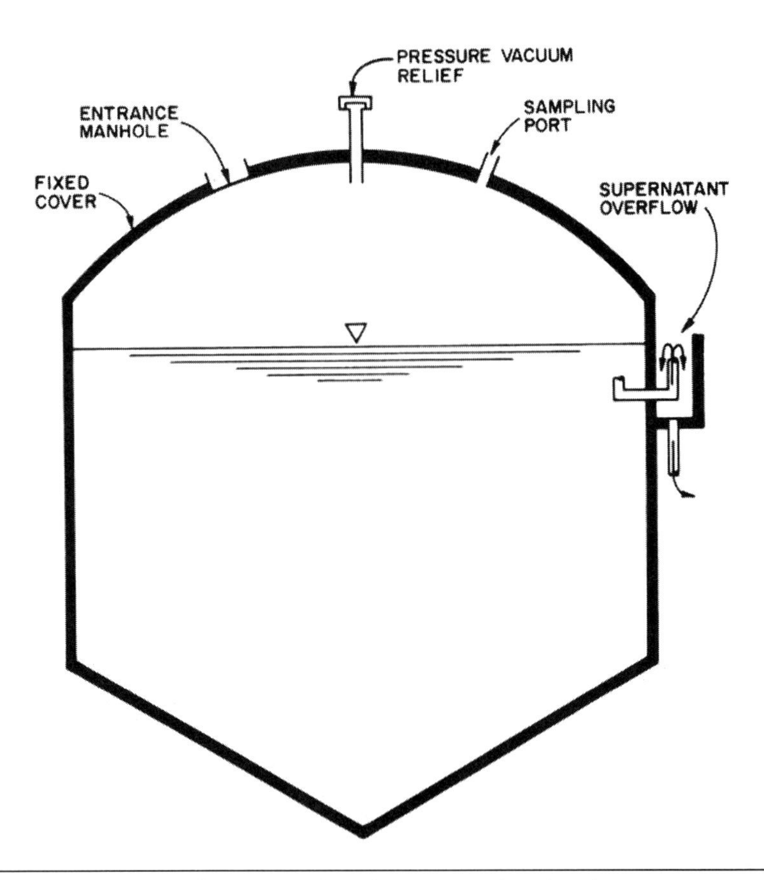

FIGURE 25.44 Schematic of an anaerobic digester with a fixed cover.

difficult. (Depending on the location of the draw-off point, foam and scum can be trapped in the digester.) The fixed cover minimizes fugitive odors, which can be released from the annular space between any floating cover and the digester wall. The cover is also more stable when the disgester is filled and emptied.

These covers do not require the ballasting that floating or gas holder covers do, decreasing the complexity of construction and engineering. While less complicated to construct, these covers can allow a tank to become hydraulically overpressurized.

2.7.4.2 Floating-Cover Digester

A floating cover protects the digester from hydraulic overloading because it rides up and down with the digester liquid level (see Figure 25.45). Floating covers require ballasting to ensure that they are balanced because piping and equipment on the cover will make it float unevenly on the liquid surface. Design engineers should take care to prevent the cover from dropping to the corbels and losing its natural gas-seal characteristics.

Ballasting is only one of the special requirements involved when designing a floating cover. All connections between the cover and the digester wall must be flexible enough to allow full movement of the cover without stressing any of the equipment (e.g., gas pipes, foam suppression pipes, or stairways on and off the cover). Also, guides

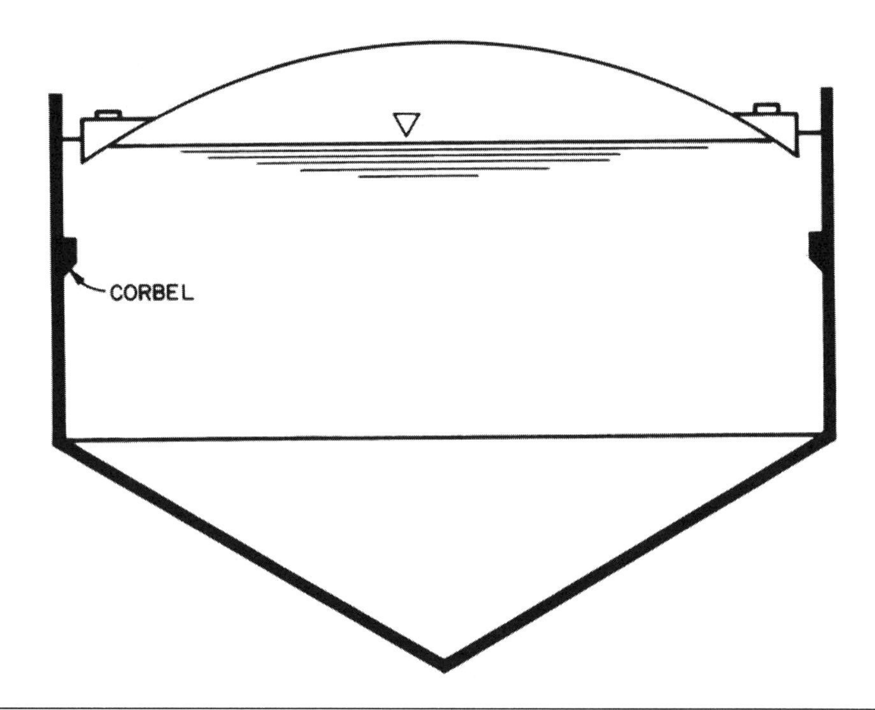

FIGURE 25.45 Schematic of an anaerobic digester with a floating cover.

are needed to ensure that the cover moves up and down without impinging on the walls and to prevent wind torsion (see Figure 25.46).

Floating covers typically are made of steel, which is not a good insulator. So, they typically are coated or covered with some type of insulation (e.g., spray-on polyurethane foam with a protective coating or a modular insulation system). Without insulation, shell losses would be significant—particularly in cold environments—increasing heating requirements.

2.7.4.3 Downe's Floating Cover

Downe's floating cover uses an attic system to create an air gap between the cover and the solids in the digester (see Figure 25.47). It sits down on the liquid to a greater extent than a traditional floating cover, creating a small gas–liquid interface in the gas dome that reduces corrosion on the cover and provides a smaller area for foam to accumulate. A foam-suppression spray typically is installed in the dome to reduce the opportunity for foam to enter the digester gas lateral. However, without surface withdrawal, foam can accumulate if the spray cannot entrain the foam sufficiently into the bulk liquid.

As with all floating covers, hydraulic overpressurization of the tank is reduced because solids can escape through the annular space; however, odors can be emitted as

(a)

(b)

FIGURE 25.46 Additional equipment required for a floating-cover digester: (a) cover guides and roller and (b) flexible gas piping (courtesy of Brown and Caldwell and the Budd Inlet WWTP, Olympia, Washington, LOTT Alliance).

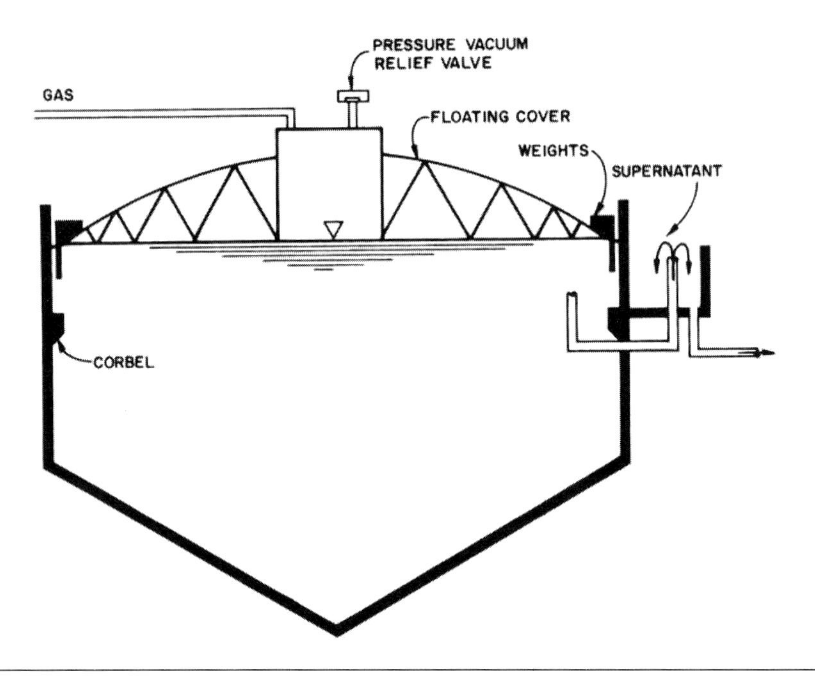

FIGURE 25.47 Schematic of a typical Downe's floating cover.

well. As with all floating covers, mixing choices are limited when using a Downe's cover. Also, as with other steel covers, additional insulation is required to reduce shell losses.

2.7.4.4 Gas-Holder Cover

Like floating covers, gas-holder covers are not fixed in place. The cover has a skirt that extends into the liquid, allowing the cover to float on a bubble of gas stored for off-peak use (Figure 25.48). As with floating covers, hydraulic overpressurization of the tank is difficult because solids will escape from the annular space.

While gas-holder covers provide extra gas storage, they have some distinct disadvantages. They have no surface withdrawal above the corbels, so foam and scum become entrapped and accumulate in the digester. The cover is effectively a large gas–liquid interface, which can lead to corrosion. Furthermore, mixing alternatives are limited because the cover moves and must be ballasted to ensure that it is balanced. Also, because the cover is riding on a gas bubble, the total volume of the reactor is not used for solids digestion (i.e., overall maximum hydraulic capacity is reduced).

2.7.4.5 Membrane Gas-Holder Cover

Membrane digester covers provide maximum gas-storage capacity for an anaerobic digester without a separate structure (Figure 25.49). While effective, these covers have

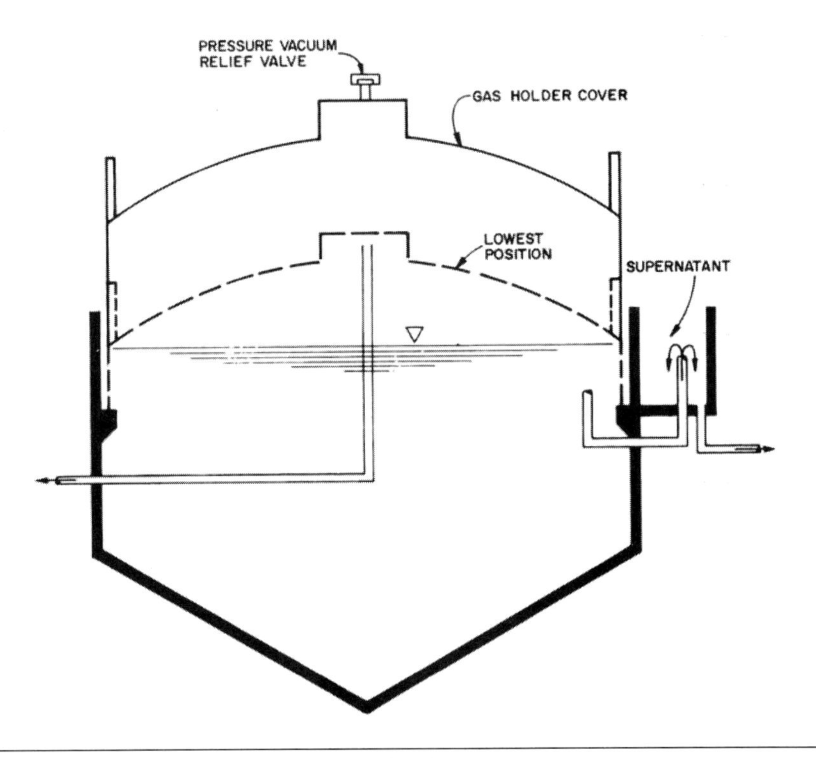

FIGURE 25.48 Schematic of a typical gas-holder cover.

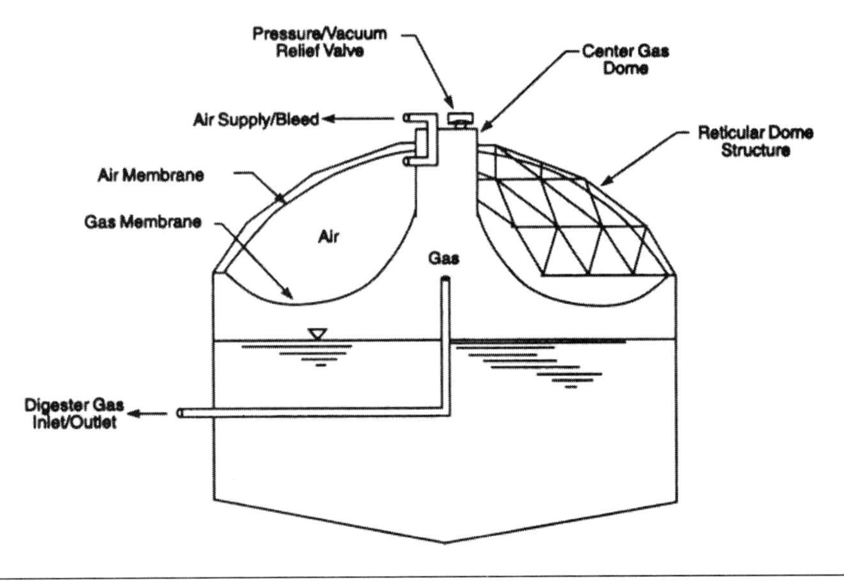

FIGURE 25.49 Schematic of a membrane gas-holder cover.

some design limitations (e.g., cover-mounted mixing equipment). Mixing is limited to either pump mixing, floor-mounted gas mixing, or external draft tubes, which can be difficult to retrofit. Also, there is no scum or foam control. The life expectancy of the material varies—typically 20 years for the outer membrane, and 10 years for the inner, similar to gas-holder covers.

2.7.4.6 Submerged Fixed Cover

A submerged fixed cover converts the top of a cylindrical or silo digester into a system resembling the top of an egg-shaped digester (see Figure 25.50). The sloping top directs gas, foam, and scum to the small gas dome (similar to the Downe's cover). Because the cover is fixed, surface overflow can occur in the gas dome, which serves as a key solids-wastage point. Typically, solids can be circulated from the digester to the dome for foam suppression, and gas is withdrawn from the very top of the gas dome.

As with standard fixed covers, there is the danger of hydraulic overpressurization by wasting at a slower rate than the feed rate to the digester. To help alleviate this concern (especially because the small gas dome has limited volume), an emergency overflow typically is added to the gas dome. A dual U-trap in the line keeps gas from exiting via the emergency overflow (see Figure 25.51).

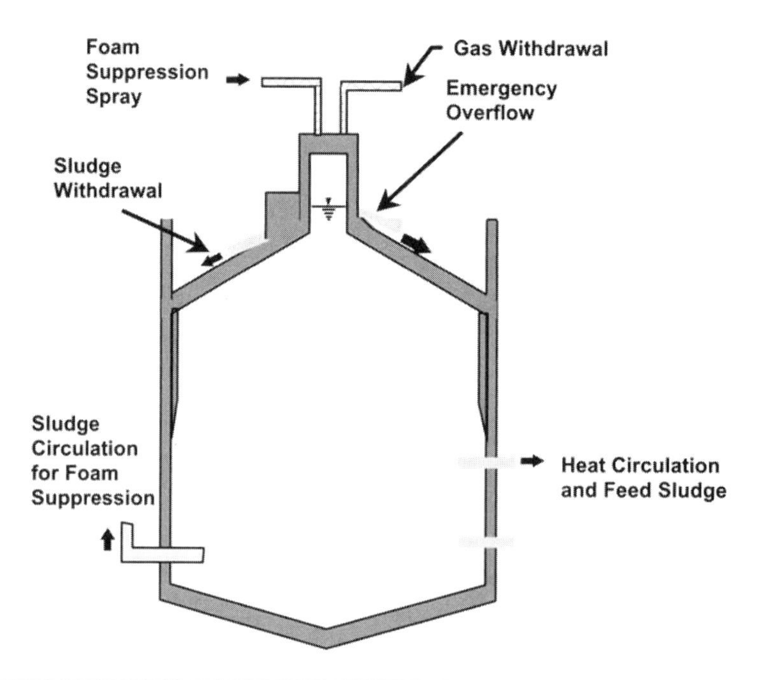

FIGURE 25.50 Schematic of a digester with a submerged fixed cover (courtesy of Brown and Caldwell, Seattle).

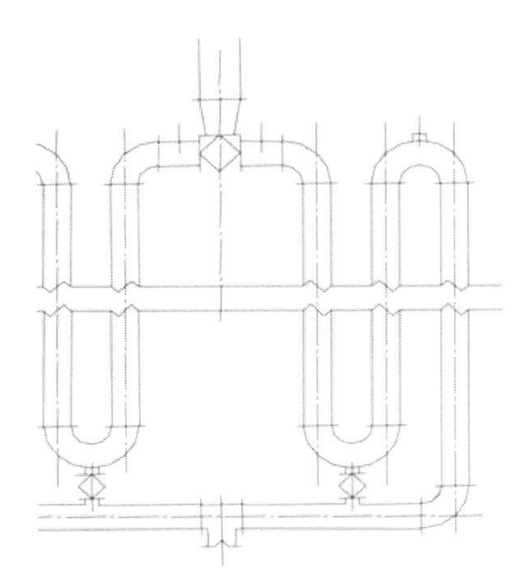

FIGURE 25.51 Emergency overflow apparatus for submerged fixed covers (courtesy of Brown and Caldwell).

While submerged fixed covers have some obvious benefits, they also have some limitations. The small gas dome provides limited gas storage, and any change in liquid volume is quickly realized at the overflow points.

2.7.5 *Digester Feeding Systems*

Raw solids can be introduced to digesters at several locations, although the key criterion is to avoid short-circuiting feed to any exit (withdrawal) point. Solids can be added at the top, bottom, or in a recycle loop; however, if solids will be withdrawn from the top and bottom, feed through the digester sidewalls is typically best. If possible, feed location also should maximize solids dispersion. This is particularly easy with pumped systems if there is more than one return location. Sequencing control valves will allow for a good distribution of solids.

Feeding into a recycle loop should be done with care, because while it may help preheat solids further, raw solids can plug spiral heat exchangers if they are too thick or laden with debris.

2.7.6 *Digester Mixing Systems*

Auxiliary mixing of digester contents is beneficial for the following processes:

- Reducing thermal stratification,
- Dispersing substrate for better contact with the active biomass,

- Reducing scum buildup,
- Diluting any inhibitory substances or adverse pH and temperature feed characteristics,
- Increasing the reactor's effective volume,
- Allowing reaction product gases to separate more easily, and
- Keeping in suspension inorganic material that has a tendency to settle.

Three types of mixing methods typically have been used: mechanical, pumped, and gas recirculation. Mechanical or internal mixers use impellers, propellers, and turbine wheels to mix materials. Problems arise because of the exposed surfaces of the mixers. Shafts and impellers are subject to vibration (due to collected materials) and wear (due to grit and debris).

Mixing via pumped recirculation involves using an external pump to recycle digester contents. The efficiency of this system depends on digester size, net energy input, viscosity, and turnover rate.

Gas-recirculation systems may use tubes, sequentially operated lances, diffusers on the tank bottom, or a 0.3-m-diam (l-ft-diam) tube that releases unconfined bubbles. In each case, the gas is produced, compressed, and circulated through the tank to promote mixing. There are basically two types of gas-recirculation systems: unconfined or confined. Unconfined systems include top-mounted lances and diffusers on the tank bottom. Confined systems discharge gas through draft tubes. Each system has advantages and disadvantages, and the degree of mixing typically depends on the energy input.

2.7.6.1 Mixing Requirements

Most manufacturers of digester mixing equipment can suggest the appropriate type, size, and power level, which depend the digester volume and geometry. These suggestions typically are based on in-house studies and the successful experiences of similar installations. Anaerobic digesters can be mixed via gas, mechanical, or pumped mixing systems (various mixing systems have different advantages and disadvantages). Selection of a mixing system is based on costs; maintenance requirements; process configuration; and the screenings, grit, and scum content of the feed. Suggested parameters for sizing digester mixing systems include unit power, velocity gradient, unit gas flow, and digester volume turnover time. These four parameters are related and can be used to equate manufacturers' recommendations.

Unit power is defined as delivered motor watts per cubic meter (horsepower per 1 000 cu ft) of digester volume. Actual energy applied, viscosity, and digester configuration are not accounted for. Several values have been suggested for unit power selection, ranging from 5.2 to 40 W/m^3 (0.2 to 1.5 hp/1 000 cu ft) of reactor volume. Using laboratory data, Speece (1972) predicted a level of 40 W/m^3 to be sufficient for a complete mix reactor.

The velocity gradient parameter as a measure of mixing intensity was presented by Camp and Stein (1943). It is expressed as the following equation:

$$G = (w/\mu)^{1/2} \tag{25.11}$$

Where

G = root-mean-square velocity gradient (s^{-1});
W = power dissipated per unit volume [W/m^3 or $N/m^2 \cdot s$ (lb-sec/sq ft)]; and
μ = dynamic viscosity [$N\text{-}s/m^2$ or Pa-s (lb-sec/sq ft)] [for water, 7.2×10^{-4} $N\text{-}s/m^2$ or Pa-s at 35°C ($1.5 \cdot 10^{-5}$ lb-sec/sq ft at 95°F)].

and

$$W = E/V \tag{25.12}$$

Where

E = power dissipated [Watts (ft-lb/s)], and
V = tank volume [m^3 (cu ft)].

The power for gas injection can be determined from the following equation:

$$E = P_1\,(Q)\,(\ln P_2/P_1) \text{ (metric units)} \tag{25.13}$$

or

$$E = 2.40\,P_1\,(Q)\,(\ln P_2/P_1) \text{ (English units)}$$

Where

Q = gas flow [m^3/s (cu ft/min)];
P_1 = absolute pressure at surface of liquid [Pa (psi)];
P_2 = absolute pressure at depth of gas injection [Pa (psi)]; and
2.40 = lumped conversion factor for English units.

These equations can be used to determine the necessary power and gas flow of compressors and motors for a gas-injection system. Viscosity is a function of temperature, total solids concentration, and volatile solids concentration. As temperature increases, viscosity decreases; as solids concentration increases, viscosity increases. In addition, as volatile solids increase to more than 3.0%, viscosity increases. Appropriate values of the root-mean-square velocity gradient are 50 to 80 s^{-1}. The lower values can be used for a system using one gas port, or where grease, oil, and scum are suspected problems.

By rearranging the preceding equations, the unit gas flow relationship to the root-mean-square velocity gradient can be solved by the following equation:

$$\frac{Q}{V} = \frac{G^2}{P_1\left(\ln \frac{P_2}{P_1}\right)} \tag{25.14}$$

Suggested values of gas flow/tank volume (Q/V) for a free-lift system range from 76 to 83 mL/m^3·s (4.5 to 5.0 cu ft/min/1 000 cu ft). For a draft tube system, the suggested values range 80 to 120 mL/m^3·s (5 to 7 cu ft/min/1 000 cu ft).

Turnover time is defined as digester volume divided by the flowrate through the draft tube. This concept typically is used only with draft tube gas and mechanically pumped recirculation systems, where such a flowrate actually can be determined. Typical digester turnover times range from 20 to 30 minutes.

2.7.6.2 System Performance

A specific definition of "adequate digester mixing" has not yet been formulated. Various methods (e.g., solids concentration profiles, temperature profiles, and tracer studies) have been used to evaluate mixing system performance.

Solids concentration profiles are used to determine the effectiveness of digester mixing. To use this method, samples are collected at specified depth intervals in the tank [typically 1.0 to 1.5 m (3 to 5 ft)] and analyzed for total solids concentration. Mixing is considered adequate if the solids concentration does not deviate from the average concentration in the digester by more than a specified amount (often 5 to 10%) over the entire digester depth. Allowances are sometimes made for greater deviations in the scum and bottom solids layers. A drawback of the solids concentration profile method, particularly for systems digesting secondary or combined primary and secondary solids, is that these solids often do not stratify significantly, even without mixing, so inefficient mixing cannot be shown by solids concentration profiles alone.

Temperature profiles also have been used to assess mixing effectiveness. The temperature profile method is similar to solids profile method. Temperature readings are taken at specified depth intervals in the digester. Mixing is considered adequate if the temperature at any point does not deviate from the average by more than a specified amount [often 0.5 to 1.0°C (1.0 to 2.0°F)]. A drawback of this method is that the digester may have enough heat dispersion without effective mixing to maintain a relatively uniform temperature profile, particularly in digesters with a long SRT.

The most reliable method currently available for evaluating mixing effectiveness is the tracer test method. In this method, a carefully measured amount of a conservative tracer material (e.g., lithium) is injected as a slug to the digester. (Continuous feed methods also can be used but are typically impractical because of the large amounts of tracer required and the long time required to perform the test.) Samples of digested solids are collected and analyzed for tracer content. For an "ideal" (i.e., completely mixed) digester, the tracer concentration in digested solids leaving the digester at any time is calculated as follows:

$$C = C_0 e^{-\frac{t}{HRT}} \tag{25.15}$$

Where

C = tracer concentration at time t (mg/L);
C_0 = theoretical initial tracer concentration at time $t = 0$ (total mass of tracer injected/total digester volume) (mg/L);
t = elapsed time since injection of tracer (hours); and
HRT = digester hydraulic retention time (hours).

Substituting and taking natural logs, this equation becomes the following:

$$\ln C = \ln C_0 - \frac{v}{V_0} \tag{25.16}$$

Where

v = total volume of solids fed in time t [$F \times t$, where F = average solids feed rate (m^3/h)] (m^3);
V_0 = total digester volume (m^3).

Plotting ln C (y axis) versus v/V_0 (x axis) gives the "tracer washout curve". The slope of this line gives an estimate of the effective digester volume, as follows:

$$V_e = \frac{1}{\text{slope}} \tag{25.17}$$

where V_e = estimated effective digester volume (m^3).

The percentage active volume then is calculated by the following equation:

$$V_{\text{act}} = \frac{V_e}{V_0} \times 100 \tag{25.18}$$

where V_{act} = estimated percent active volume.

This method of estimating mixing effectiveness is the most accurate of the methods discussed (Chapman, 1989). However, because it requires careful monitoring of digester feed and withdrawal rates and a large number of tracer concentration analyses in digested solids, this method is considerably more expensive than any of the other methods discussed.

2.7.7 *Digester Heating Systems*

To be effective, anaerobic digestors need a consistent, reliable heating system.

2.7.7.1 Digester Heating Needs

Anaerobic digesters must be heated to provide suitable environmental conditions for optimal biological activity. Mesophilic digestion need to operate between about 35 and 39°C (95 and 102°F), and thermophilic digestion needs to remain between 50 and 56°C (122 and 133°F). The amount of heat needed varies seasonally, mainly in relation to the raw solids temperature, and in relation to heat losses from the reactor to the environment.

2.7.7.2 Solids Heating

The lion's share of a digester's total heating load is the energy needed to heat raw solids to the temperature needed for anaerobic digestion. This energy is calculated as follows:

$$q = m \times Cp \times T \tag{25.19}$$

Where

q = the heat load [J/h or MJ/h (Btu/hr or mil. Btu/hr)];
m = the mass flowrate of the cake's liquid, treated as water [kg/h (lb/hr)];
Cp = the solids's specific heat or heat capacity [J/kg·°C or MJ/kg·°C (Btu/lb·°F)];
T = the temperature difference between the cold, raw solids and desired heated solids temperature [°C (°F)].

To accurately compute the solids' heating needs, design engineers need to know the actual solids temperature, which typically is rarely recorded. However, the treatment plant's influent and effluent temperatures typically are known and are representative of the raw solids temperature. (A treatment plant does not appreciably change the average temperature of wastewater because of the large mass of water involved.)

2.7.7.3 Digester Heat Losses

In all but the very hottest weather, digesters lose heat to the environment via their roofs, walls, sides, and bottom. However, because digester temperatures typically are near ambient, radiant heat loss is small; virtually all heat is lost via convection. The general formula for heat loss (heat transfer) from these areas is as follows:

$$q = U\,A\,T \tag{25.20}$$

Where

q = the heat load [J/h or MJ/h (Btu/hr or mil. Btu/hr)],
U = the overall heat-transfer coefficient [J/h·m^2·°C (Btu/hr·sq ft·°F)],
A = the surface area [m^2 (sq ft)], and
ΔT = the temperature difference between between the digester and the environment [°C (°F)].

The coefficient U is also the inverse of resistance to heat transfer. For multiple layers, 1/U can be expressed as a series of resistances (Bird et al., 1960):

$$\frac{1}{U} = \frac{1}{h_0} + \frac{x_1}{k_1} + \frac{x_2}{k_2} \ldots + \frac{1}{h_3} \tag{25.21}$$

Where

h_0 and h_3 = film coefficients (J/h·m^2·°C) or (Btu/hr·sq ft·°F),
x = thickness of material (consistent units), and
k = thermal conductivity of material (J/hr·m·°C [Btu/hr·ft·°F]).

Coefficient values and the application of these equations to heat loss from digesters can be found in American Society of Civil Engineers (1959), American Society of Heating, Refrigerating, and Air Conditioning Engineers (2005), Avallone and Baumeiter (1996), and Perry and Green (1997).

2.7.7.4 Heat Sources

The following types of heat sources are available:

- Fired Boilers. Boilers (e.g., steam boilers and hot water boilers) typically are used at treatment plants to produce heat from fuel. Fired boilers burn a fuel to supply heat.
- Cogeneration. *Cogeneration* is the production of both usable heat and electric power from one fuel. So, cogeneration systems (also called *combined heat and power systems*) can deliver heat for solids digestion.
- Water-Source Heat Pumps. A *heat pump* is a mechanical device that extracts heat from one source, and elevates its temperature to make it usable for other applications. Water-source heat pumps can heat water to between 68 and 76°C (155 and 170°F).

- Solar Radiation. Solar energy is a form of clean, renewable energy that is becoming increasingly popular. Because it is not available at night or during inclement weather, solar power typically is not adequate as a sole source of heat. However, it is a carbon free, renewable energy source that can provide a portion of solids heating needs.

2.7.8 Heat Exchangers

2.7.8.1 Heat Exchanger Types

The following types of heat exchangers have been used in anaerobic digestion systems:

- Concentric Tube. One of the oldest, most widely used heat exchangers is the concentric-tube (often called *the tube-in-tube* or *concentric-pipe*) heat exchanger (see Figure 25.52).
- Spiral Plate. Spiral (spiral-plate) heat exchangers also are common (see Figure 25.53). Water temperatures are typically kept below 68°C (154°F) to prevent caking.

FIGURE 25.52 Tube-in-tube heat exchanger at the Littleton-Englewood Wastewater Treatment Plant in California. (courtesy of Brown and Caldwell).

FIGURE 25.53 A spiral heat exchanger (courtesy of Brown and Caldwell and City of Takoma, Washington).

- Multiple Tubes in a Box. A variation of the concentric-tube heat exchanger consists of multiple tubes in a box (see Figure 25.54). The small-diameter tubes have a common inlet and outlet.
- Interior Submerged Coils. Older digesters may have internal heating coils attached to the walls. The coils circulate hot water. They are subject to fouling, which can decrease the heat transfer. Maintaining the coils requires operators to remove the digester from service and empty it.
- Jacketed External Solids Mixers. A heat-jacketed draft-tube mixer also can be used to internally heat solids (see Figure 25.55). However, internal heating systems are seldom used because of the difficulty associated with providing maintenance inside the reactor.

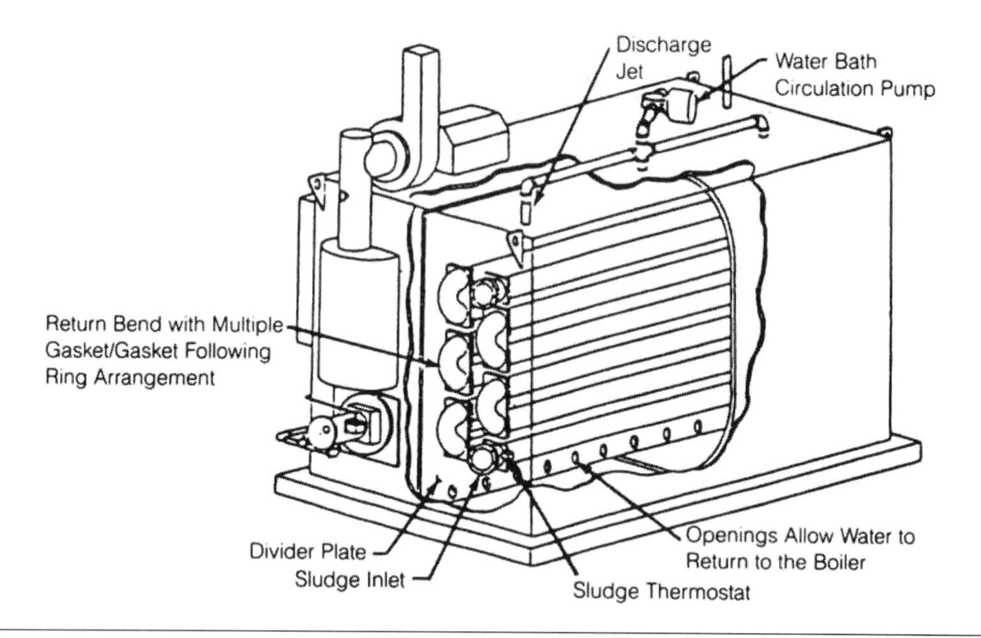

FIGURE 25.54 Cutaway view of a tube-in-box heat exchanger (WPCF, 1987).

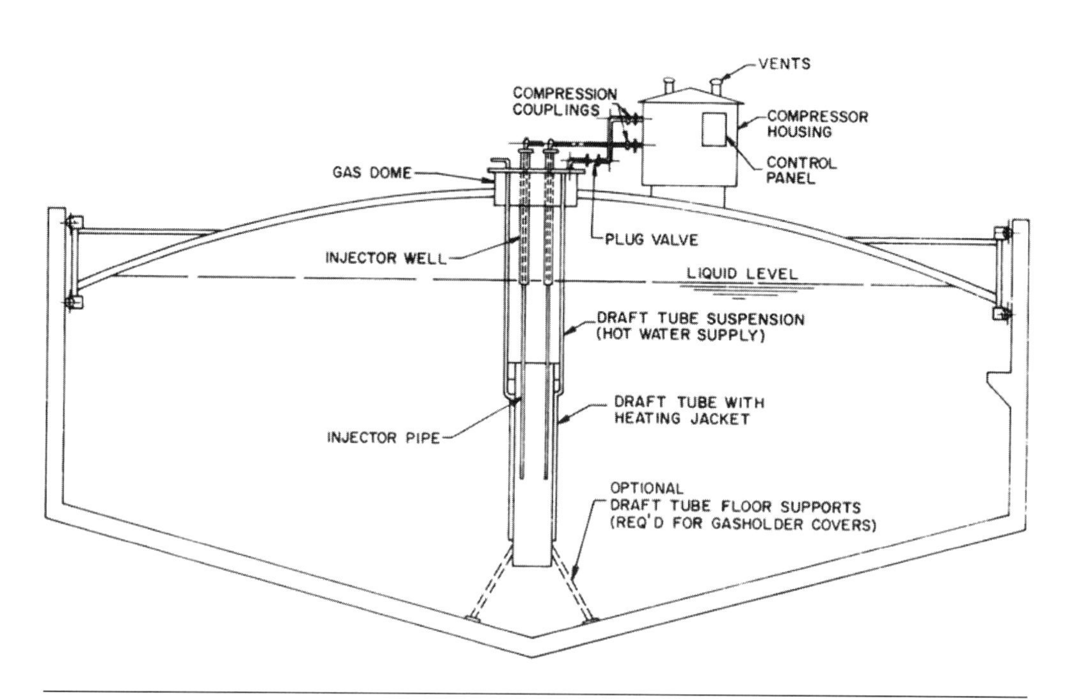

FIGURE 25.55 Single internal draft-tube mixer with heating jacket.

2.7.8.2 *Heat Exchanger Characteristics*

Heat-transfer coefficients for external heat exchangers range from 0.9 to 1.6 $kJ/m^2{\cdot}°C$. Transfer coefficients for internal heating coils range from 85 to 450 $kJ/m^2{\cdot}°C$ (15 to 80 Btu/hr/sq ft/°F) depending on the biomass' solids content.

2.7.9 Steam Heating

The following two types of steam heating systems have been used in anaerobic digestion systems:

- Submerged pipes. The submerged steam pipe (also called a *steam lance*) is a vertical, open-ended, small-diameter pipe that discharges at least 3 m (10 ft) below the liquid surface of the digester. Some treatment plants [e.g., the Hyperion Wastewater Treatment Plant (City of Los Angeles), the JWPCP (of the Los Angeles County Sanitation Districts), and the Rancho Las Virgenes digester complex in Calabasas, California] heat via submerged steam.
- Steam injection. Steam injectors are precision devices that blast a small jet of steam into a stream of solids. The device is outside the digester, and the steam flow is precisely controlled to produce a specific discharge temperature. The amount of steam is adjusted by throttling the plug in the steam injector throat. A few plants [e.g., the Back River Wastewater Treatment Plant in Baltimore, Maryland, the Crystal Lake (Illinois) Wastewater Treatment Plant, and the Spokane (Washington) Wastewater Treatment Plant] use external steam jet injectors to warm their solids.

2.7.10 Heat Recovery

The following three types of heat-recovery systems have been used in anaerobic digestion systems:

- Cogeneration Equipment. Cogeneration equipment typically withdraws heat from hot exhaust gases or cooling equipment. Heat recovery percentages for specific cogeneration equipment are discussed in later sections of this chapter.
- Solids Heat Recovery. At press time, recent increases in fuel costs had increased the value of heat, so recovering heat from digested solids and using it to warm raw solids is becoming more practical and common.
- Gas Compression and Equipment Cooling. At least one treatment plant is capturing the relatively low-temperature heat (40 to 50°C) from other large processes and using it, along with some water-source heat pumps, to warm solids.

2.7.11 *Additional Equipment Options*

2.7.11.1 *Debris Buildup and Foam Control*

Given the heterogeneous nature of raw solids, the accumulation of debris and foam are part of normal digester operations. However, the rate and extent of accumulation can be controlled via proper digester design, best management practices, and wastewater treatment controls.

Debris can affect digesters in two ways. Floating debris can accumulate on the surface of digesters at the gas–liquid interface, forming a thick blanket that can affect mixing and reduce digester capacity. Heavy debris (e.g., grit) can accumulate on the bottom of the digester, consuming process capacity and possibly causing short-circuiting (if the draw-off points are at the bottom of the tank).

Debris can be reduced via pre-screening or grinding solids, or via better screening at the headworks. Most debris and grit enter the solids-handling system via primary solids and scum-removal systems (floating materials). If debris removal can only occur in the digester, several solutions are available (e.g., better mixing to keep material suspended, spray bars to entrain material back into solution, surface withdrawal for floating materials, or grinding and screening). In most cases, a combination of these solutions would be used to mitigate the effects of debris.

Digester foaming can be caused by several factors (e.g., surfactants in influent or hauled wastes, filaments from the secondary treatment system, or digester perturbation). Surfactants can be removed via source identification testing. Filaments can be controlled in the activated sludge system via selectors or chlorine dosing. Foaming typically is associated with erratic feeding after digester perturbation, which can be mitigated via changes in operating practices. Foaming is more difficult to control when associated with a transient event (e.g., a toxic compound), which is difficult or impossible to predict. (In the case of toxic compounds, efforts should be made to identify chronic sources.)

Several design strategies can be used to limit foaming, many of which are limited to the type of cover associated with the digester. The most common approach is to add spray bars, which entrain foam back into the bulk solution. If the cover or tank design allows it, surface withdrawal is an effective means of removing foam (similar to a selector in secondary systems). Typically, surface withdrawal systems are augmented with spray bars to direct flow. Another alternative is to change the mixing system. Gas systems tend to exacerbate foaming because the material interacts with the gas bubbles carrying it out of solution.

Foaming can greatly increase O&M costs for a utility. As foam can escape digesters through the annular space on floating covers or enter the gas system requiring it to be taken off line for cleaning.

Both foam and debris reduce digester capacity either by directly displacing volume or by causing operating levels to be reduced to contain them. Anerobic digestion systems should be designed to minimize foam accumulation.

2.7.11.2 Scaling (Struvite)

Struvite [magnesium ammonium phosphate ($MgNH_4PO_4 \cdot 6H_2O$)] is a white crystalline solid or scale that typically forms in anaerobic digesters, digested solids and centrate piping, solids lagoons and lagoon piping, and dewatering equipment. It forms a hard, tenacious precipitate that adheres to pipes and equipment, reducing pipe flow capacity and overloading motors serving brush aerators (see Figure 25.56).

Struvite typically forms when the concentrations of magnesium, ammonium, and phosphate exceed the solubility limit of struvite. Its formation also is influenced by various conditions in the digester (e.g., pH, temperature, and other chemicals that can compete primarily for phosphate). Anaerobic digestion can promote struvite formation via the release of phosphate and ammonia during stabilization. The amount of phosphorus and ammonium released depends on the digestion process and the wastewater treatment process(es) that generated the raw solids. For example, the phosphorus taken up during the Bio-P process is much more likely to be released in an anerobic digester than in a conventional activated solids system. The amount of magnesium in solids also will affect struvite formation. Because struvite formation (as the chemical formulae suggests)

FIGURE 25.56 Struvite precipitate in piping (courtesy of Brown and Caldwell and City of Sacramento).

is based on equimolar stoichiometric concentrations of magnesium, ammonium, and phosphate, the limiting concentration of each component typically determines the amount of struvite that eventually forms. For example, removing or reducing the concentration of any of these constituents can reduce struvite scaling (buildup).

Systems should be designed to give operators access to pipes and equipment so they can remove struvite. In particular, if a facility being upgraded has a history of struvite, design engineers should ensure that the changes provide access and minimize deposition points.

Struvite control can be both a design and process consideration. However, most options for addressing struvite scaling (e.g., smooth-lined piping and more maintenance) are preventive.

Innovative research is needed to better control and reduce struvite scaling in digestion systems. One commonly used method is adding an iron compound to precipitate phosphorus (one of the three constituents for struvite formation). Staff at a California digestion system added ferrous chloride to control hydrogen sulfide generation and found that it also precipitated phosphate, thereby reducing struvite precipitation. However, phosphorus is an essential nutrient for bacteria, so design engineers should ensure that enough phosphorus remains to meet nutritional needs.

Other control methods that may mitigate struvite precipitation include dilution (to reduce ion concentrations) and operating at a lower pH. Neither is a desirable operating procedure.

Once struvite deposits have formed, they are difficult to remove. Acid washing removes struvite effectively but can be costly and a safety hazard (Barker, 1996). During early stages of formation, struvite can be controlled by frequent cleaning (pigging) of pipelines. Smooth-lined pipes made of PVC or glass-lined materials, and polyethylene- or polytetrafluoroethylene-coated plug valves will resist struvite accumulation better than other materials.

Several plants have found vivianite [$Fe_3(PO_4)2 \cdot 8(H_2O)$, also called *hydrated iron phosphate*] in their systems, especially when the digesters contain high levels of phosphate and also contain iron. Vivianite loses solubility when temperatures rise; it forms quickly when solids containing iron and phosphate are heated. Vivianite is blue, green, or gray-black; turns opaque or dark when exposed to light; and is soluble in hydrochloric acid or nitric acid (HNO_3).

2.7.11.3 *Piping and Cleaning Maintenance*

Piping configurations should be designed to promote maximum flexibility for feeding, recirculating, and discharging solids. Piping should be arranged to provide several points for solids feed, solids withdrawal, and supernatant withdrawal. Because solids pumping is characterized by low velocities and possible solids accumulation in pipelines, design engineers should make provisions for cleaning out and backflushing lines (using treated effluent when available). They also should consider the choice of valves and where they should be placed for most utility (e.g., easy access and manual operations). Design engineers also should include provisions to allow all tanks and pumps to be isolated for maintenance and safety purposes.

Piping should be arranged to accommodate the following operating modes for two-stage digestion: transferring biomass via gravity from the first stage to the second, pumping the biomass from one digester to another, withdrawing supernatant via multiple ports, recirculating solids via several suction and discharge ports, and providing redundancy/backup.

2.7.11.4 Corrosion

An anaerobic digester is a highly corrosive environment, especially because of the release of hydrogen sulfide. Tank interiors, equipment, piping, and any other elements that may contact biogas should be designed to be corrosion-resistant, and all seals and gaskets should be compatible with the material they contact. A system's life expectancy will be significantly reduced if its equipment and materials lack corrosion protection. For a detailed discussion of corrosion, see Chapter 10.

2.7.11.5 Pumping

Pumping the primary means of conveying solids to and from digesters. For a detailed discussion of solids pumping, see Chapter 21.

2.7.11.6 Sampling and Process Monitoring

Anaerobic digestion is sensitive to changes in operating conditions. If uncontrolled, such changes can result in digester upsets and failure. Proper monitoring techniques promote successful operation and ensure process stability and methane production. All process streams should be available for sampling and analysis. Feed, digested solids, supernatant, digester gas, and the heating fluid (hot water) should be analyzed for various constituents and physical conditions. Sampling ports should be incorporated into the design to ensure that operators have adequate access for sampling. Feed typically is analyzed for the following: total solids, volatile solids, pH, alkalinity, and temperature. Digester content and effluent solids should be analyzed for the same parameters and for volatile acids. Digester gas should be analyzed for volume and percentage of methane, carbon dioxide, and hydrogen sulfide. Supernatant should be analyzed for pH, biochemical oxygen demand (BOD), COD, total solids, total nitrogen and ammonia-nitrogen, and phosphorus; and heating fluid for total dissolved solids and pH.

The flowrates of all streams should be monitored by accurate meters. If digesters produce supernatant, this stream should be quantified for use in computing the system's total solids content and, subsequently, the SRT. Additional monitoring requirements (e.g., those for toxics) should be determined on a case-by-case basis.

2.7.11.7 Alkalinity and pH Control

The methanogens in anaerobic digesters are affected by small pH changes, while the acid producers can function satisfactorily in a wide range of pH values. The effective pH range for methane producers is about 6.5 to 7.5, with an optimum range of 6.8 to 7.2. Maintaining this optimum range is important to ensure effective gas production and eliminate digester upsets. Digestion stability depends on the buffering capacity of the digester's contents (i.e., the digester contents' ability to resist pH changes). Alkalinity is important in anaerobic digestion; higher alkalinity values indicate more capacity for resisting pH changes. It is measured as bicarbonate alkalinity and ranges from 1500 to 5000 mg/L as calcium carbonate in anaerobic digesters. The volatile acids produced by the acid producers tend to depress pH. Under stable conditions, volatile acid concentrations range from 50 to 100 mg/L. By maintaining a constant ratio of volatile acids to alkalinity that is less than 0.3, the system's buffering capacity can be maintained.

The bicarbonate alkalinity concentration can be calculated from the total alkalinity [which also includes the alkalinity of volatile acids (e.g., acetate) and ammonium] as follows:

$$\begin{aligned}\text{Bicarbonate alkalinity (mg/L as CaCO}_3\text{)} &= \\ \text{total alkalinity (mg/L as CaCO}_3\text{)} &- \\ [0.71 \times \text{volatile acids (mg/L as acetic acid)}]\end{aligned} \tag{25.22}$$

where 0.71 is a conversion factor to mg/L as $CaCO_3$.

Barber and Dale (1978) developed the following equation to predict how much bicarbonate alkalinity is needed to raise the total alkalinity:

$$D_d = D_{max}\, 1 - \frac{1}{\theta} \tag{25.23}$$

where

D_d = amount added daily to reach set level (mg/L as $CaCO_3$),
D_{max} = required increase (mg/L as $CaCO_3$), and
$1/\theta$ = reciprocal of average detention time or SRT ($days^{-1}$).

Sodium bicarbonate, lime, sodium carbonate, and ammonium hydroxide all have been used successfully to increase the alkalinity of digester contents. However, most well-designed and well-operated digester facilities do not require alkalinity addition as long as the wastewater has sufficient buffering capacity. Design engineers should evaluate the wastewater's alkalinity before determining whether an alkalinity feed system should be constructed.

2.7.12 *Design Example—Thermophilic Digestion*

This design example is for a single-stage thermophilic anaerobic digestion process. The owner has selected thermophilic digestion because the reaction rates are higher than those of the previous mesophilic system, and it can provide better digestion performance in the limited available tankage. After thermophilic digestion, biosolids will be dewatered and sent to a Class A composting facility, so the owner is not concerned about meeting specific pathogen requirements (either Class A or Class B) in the digestion system.

2.7.12.1 Digestion System and Suitability for Thermophilic Operation

Two existing digesters and support systems exist. The tanks and systems are sized and configured as follows:

- Inside diameter = 21.2 m (70 ft);
- Sidewater depth = 7.6 m (25 ft) (at maximum depth);
- Construction method of tanks: reinforced concrete, cast in place;
- Bottom tank configuration = cone-shaped bottom;
- Covers = floating covers, somewhat damaged and corroded;
- Volume of each tank = 2680 m^3 (719 000 gal);
- Mixing system = floor-mounted gas mixing system;
- Heating system = spiral heat exchanger at each digester; and
- Biogas management = hot water boilers and system to heat mesophilic digesters, and supplementary building heating, plus emergency flares.

A structural evaluation confirmed that the floating covers were not in good condition for reuse as either floating or fixed covers. However, the two tanks are in good condition and can perform thermophilic digestion service up to 60°C. The mixing system is at the end of its useful live.

2.7.12.2 Digestion Loading and Operating Conditions

The feed to thermophilic digesters is a mixture of primary solids and thickened WAS that is thickened to 5.5% solids. Trucked FOG waste is another feedstock. Predicted digester feedstock is:

Feed material (% solids and % volatile solids)	Average annual (lb/d*)	Average annual Peak week (lb/d)
Primary solids (5.5% and 78%)	20 600	27 500
Thickened WAS (5.5% and 77%)	18 800	24 800
FOG (8% and 95%)	5 300	13 300
Totals	44 700	65 600

*lb/d × 0.453 6 = kg/d.

Flowrates	Average annual (gpd*)	Peak week (gpd)
Primary solids	45 000	60 000
Thickened WAS	41 000	54 000
FOG	8 000	20 000
Totals	84 000	134 000

*gpd × 0.003 485 = m^3/d.

These loads result in the following volatile solids loading and SRT conditions:

Volatile solids loading condition	Average annual (lb/d/cu ft*)	Peak week (lb/d/cu ft)
Both tanks in service	0.185	0.28
One tank out of service	0.37	0.55

SRT	Average annual (days)	Peak week (days)
Both tanks in service	15.3	10.7
One tank out of service	7.6	5.4

*lb/d/cu ft × 16.02 = kg/m^3·d.

The loading conditions with both tanks in service are easily workable in both average annual and peak week loading conditions. The 15-day SRT is not a limit if Class B digestion is not mandated. For the loading condition with one tank out of service, the volatile solids loading rate and SRT are adequate for average annual condition, but the peak week conditions are higher than typically used for thermophilic digestion. So, one option would be to eliminate some or all of the trucked FOG loading when peak solids loading is high and prolonged.

2.8 Physical Facilities

The choice of anaerobic digestion tanks and equipment, and sometimes the configuration itself, is often affected by the physical space available.

2.8.1 *Tanks and Materials*

The tanks, tank configuration, and system geometry involved depend on the situation. Pancake digesters, which have large diameter-to-height ratios, require the most land for a given volume. These units historically have been the most common in the United States. If land is limited or expensive, then silo or egg-shaped digesters may be an economical choice. However, such digesters are more complex to design and build, and their height may be an aesthetic issue for the neighbors.

Digester tanks typically are made of steel-reinforced concrete that is either cast in place or post tensioned. They often are designed to provide 40- to 50-year service lives, or even longer. Some tanks are made of steel. In the United States, egg-shaped digesters typically have been made of steel, while in Europe, they often are made of reinforced concrete.

Tank construction has become more varied and ingenious in recent years—partly because of increases in construction material and labor costs, but also because wastewater treatment professionals recognized that tank shape and other details can influence digestion performance. For example, the egg shape has superior mixing characteristics, but similar shapes (e.g., silos) may provide nearly equal mixing performance with less complex construction requirements. The tank bottom, top slope, and tank configuration can be critical for good mixing and treatment performance.

Design engineers should select the tank type, shape, bottom and top configuration, and construction materials and methods based on each system's criteria and needs. Such criteria include costs, available area, future expansion needs, desired life expectancy, specific digestion process and temperature regime, specific foundation and structural needs (i.e., seismic requirements), contractor and specialty firm availability, and schedule constraints.

2.8.2 *Pumps and Piping*

Pumps and piping systems should have enough clearance for staff to maintain the equipment, move equipment in and out, and allow for easy cleaning. Piping systems should have cleanouts at periodic intervals, with drains and associated flushing systems nearby. Hot-water flushing is particularly effective for adhesive materials (e.g., solids).

2.8.3 *Mixing Equipment*

The physical considerations for mixers are process-dependent. Pump-based systems should have enough space for pump maintenance and piping cleanout. When designing gas-based mixers, engineers need to consider pipe materials and routes, especially if confined-space issues are possible. When designing draft-tube mixers, engineers must ensure that there is enough space between and around digesters to allow a crane to remove and replace equipment without affecting other facets of digester operations.

2.8.4 *Heating and Heat-Transfer Equipment*

The primary issue for all heat exchangers is cleaning and maintenance. The frequency depends on the type of heat exchanger used, the solids being conveyed, and the specific operating conditions. An effective design will include convenient wash stations, drains, and ample space to clean the heat exchanger. This is particularly critical for tube-in-tube heat exchangers, which often require clearance at both ends of the unit to clean the elbows.

2.8.5 *Cleaning and Safety*

The decision to clean a digester tank is based on several factors (i.e., the degree to which grit and scum accumulation has reduced the digester's effective volume, the condition of internal heating and mixing equipment, the availability of alternate solids-handling equipment, and tank structure). The tanks, mixers, and heaters should be designed for easy access during cleanout operations. At a minimum, there should be access manholes on the top and sides of the digester. The manholes should be at least 0.9 m (36 in.) in diameter, or large enough to enable an operator to use grit- and scum-removal equipment.

Heating and mixing equipment must be maintained throughout the life of the digester, so ideally, most of the cricital equipment should be outside the tanks. However, interior equipment that can be removed during digester operations is often satisfactory. Digesters can be cleaned by in-house staff or by a contractor that specializes in such services. They typically are cleaned every 5 years, but this is only a general guideline—the timing should be based on the plant's specific situation.

Safety is of primary importance during digester cleaning. Anaerobic digesters are confined spaces, so all systems must be designed to ensure that O&M personnel are safe. Before entering a digester, for example, O&M personnel must determine whether the air inside is oxygen-deficient or contains life-threatening gases. Design engineers also need to specify the appropriate personal safety equipment needed when entering the tank for inspection and cleanout. Several pieces of equipment are available to perform safe cleaning operations (see Table 25.15); the items needed depend on the size of the operation.

Other safety equipment should be included to prevent falls, infection, and injuries during system operations.

The gas-collection and -piping system design must include vacuum- and pressure-relief valves, flame traps, and automatic thermal-shutoff valves. Biogas safety also includes protecting against suffocation, asphyxiation, and explosions, so proper enclosures and ventilation must be provided.

Table 25.15 Digester cleaning and safety equipment.

Sludge line valves
Sludge line (permanent)
Sludge line (temporary)
Digester access
Explosionproof vent fan
Explosive level meter
Safe ladder
Self-contained breathing apparatus
Safety harness
Nonskid boots
Explosionproof lights
Water source
Washdown hose
Nozzle with shutoff
Wash water pump
Fixed sludge pump
Portable sludge pump
Turret nozzle
Tripod or hoist
Tank truck
Crane

For more information on safety features, see the *National Electric Code* (NFPA, 1993), *Standard for Fire Protection in Wastewater Treatment and Collection Facilities* (NFPA, 1995), *Recommended Standards for Wastewater Facilities* (Great Lakes, 1997), and *Safety and Health in Wastewater Systems* (WEF, 1994b).

2.9 Digester Gas Handling

This section covers a wide range of issues with respect to digester gas (biogas) characteristics, gas-processing equipment, gas-handling equipment, and gas beneficial use.

2.9.1 Characteristics and Contaminants

Anaerobic digesters continuously produce a valuable, methane-rich gas called *digester gas (biogas)*. It is an important source of renewable energy (see Table 25.16). A fuel's lower heating value (LHV) does not include the heat of vaporized water.

2.9.2 Gas Collection

2.9.2.1 Piping Systems-Piping Material

There are two types of gas piping: aboveground and belowground.

Table 25.16 Typical digester gas compared to typical natural gas.*

	Digester gas		
Item or parameter	**Range**	**Common value**	**Natural gas**
Methane (%) (dry basis)	50–73	60	80–98
Carbon dioxide (%) (dry basis)	30–48	39	0–2
Nitrogen (%) (dry basis)	0.2–2.5	0.5	0.2–10
Hydrogen (%) (dry basis)	0–0.5	0.2	~0
Hydrogen sulfide (ppm_v) (dry basis)	200–3 500	500	<16
Ethane (%) (dry basis)	0	0	0.3–5
Propane (%) (dry basis)	0	0	0.6–5
Butane (%) (dry basis)	0	0	0.5–3
Specific gravity (based on air = 1.0)	0.8–1.0	0.91	0.58
Ignition velocity, maximum (ft/sec)	0.75–0.90	0.82	1.28
Wobbe number			
Higher heating value (HHV) (Btu/cu ft)	600–650	620	1 030–1 050
Lower heating value (LHV) (Btu/cu ft)	520–580	560	930–950

* All percentages listed above are percentages by volume; ppmv = parts per million, by volume; the higher heating value includes the heat of the water of vaporization; the fuel's lower heating value does not include the heat of the water of vaporization; ft/sec × 0.304 8 = m/s; and Btu/cu ft × 37.26 = kJ/m^3

Many wastewater treatment plants built in the 1970s or earlier originally had carbon steel digester-gas piping. Typically, water and hydrogen sulfide caused corrosion or deposits in this piping that forced personnel to replace it with corrosion-resistant gas piping. Gas piping must be sloped to low points for water removal, and ideally, the gas is dried to remove moisture.

Design practice recently standardized on stainless steel for aboveground piping, which is becoming popular. Aboveground gas piping can help reduce corrosion and water-removal concerns.

2.9.2.2 Pressure Loss Considerations

Virtually all combustion equipment used in digester gas-handing systems originally was designed for natural gas, and the differences between natural gas and digester gas are not always appreciated. So, sometimes digester gas piping is undersized.

Typical digester gas has a higher heating value (HHV) of about 23 kJ/m^3 (620 Btu/cu ft), while natural gas has a HHV of 39 kJ/m^3 (1050 Btu/cu ft). So, the pipes must transport about 69% more digester gas [39/23 (1050/620)] to convey the same amount of fuel energy as natural gas. Also, gas-pressure losses are based on the square of the gas flowrate, so digester gas produces about about 2.9 times the gas pressure drop of natural gas [$(39/23)^2$ $(1050/620)^2$], on an equivalent caloric (Btu) or energy-delivery basis. If specific gravity differences are included, typical digester gas [on a caloric (Btu) basis] actually causes about 3.5 to 4 times the gas pressure drop of natural gas per equivalent energy transfer. This may be one reason why digester gas piping is frequently undersized, especially in the smaller diameters.

2.9.3 *Digester Gas Storage*

Some treatment plants use one or more forms of digester gas-storage systems (e.g., low-pressure gas holder or higher-pressure compressed-gas storage) to help them use their gas more effectively.

2.9.3.1 *Low-Pressure Digester Gas Storage*

In an operating digester gas system, biogas constantly is being evolved and used. This is a dynamic system in an essentially constant-volume arrangement of piping and vessels. Gas is produced in the digesters at a variable rate that depends heavily on how recently each digester was fed with raw solids. Meanwhile, the devices using digester gas (e.g., engines and boilers) may have variable or relatively constant gas-consumption rates.

Low-pressure gas-storage systems include flexible-membrane dome covers; dry-seal cylindrical steel gas-holder tanks; and floating, deep-skirted, digester gas-storage covers.

2.9.3.2 *Flexible Membrane Covers*

One newer fabric gas-storage option is the flexible-membrane gas-holder cover. While this is relatively new development, some of these covers have been used successfully for 20 years. Flexible membrane covers provide short-term gas storage to equalize gas pressure, an important consideration when cogeneration systems are involved. Flexible-membrane digester-gas storage systems are available in sizes up to 34 m (110 ft) and for gas pressures up to 4 kPa(16 in. H_2O). They often are less expensive than traditional steel digester covers, and they typically leak less gas than floating digester covers do.

2.9.3.3 *Flexible Membrane Cover Comparison*

A summary of the advantages and disadvantages of flexible membrane covers is noted here. Table 25.17 provides a summary of membrane cover design considerations.

2.9.3.4 *Dry Seal Type Cylindrical Steel Gas Holder Vessels*

The dry-seal (piston) gas holder is a vertical steel tank-within-a-tank gas-pressurization device designed to use its ample weight to keep digester gas pressure virtually constant while gas production or use varies. It is a non-powered technique for supplementing the limited gas-storage volume between the liquid surface and the digester cover. The weighted, movable piston helps maintain a constant digester-gas pressure in the low-pressure gas piping.

Table 25.17 Advantages and disadvantages of flexible membrane covers.

Advantages	Disadvantages
• Could function both as a digester cover and as a gas holder. • Might be available quicker.	• Has no proven method of reporting status of percent storage used. • No clear method for operating the secondary digesters with balanced storage covers in parallel. • Is much more mechanically complex. • Has a long-term durability concern. • Some safety concerns. May not be completely impervious to methane gas leakage. • Requires operation and maintenance on the pressuring air blowers, accessories, and controls.

2.9.3.5 Dry Seal Type Gas Holder

Dry-seal (piston) welded-steel gas holders have been used at many U.S. locations for more than 150 years. Table 25.18 lists some advantages and disadvantages of replacing secondary digester covers with similar welded-steel covers and then adding dry-seal digester gas-holder vessels.

2.9.3.6 Floating Deep Skirted Digester Gas Holder or Gas Storage Covers

Floating digester covers were common years ago, but are less common today because of concerns about gas leakage from the annular space between the outer edge of the cover and the inner digester wall. Another concern is the effect of such gas leakage on air quality. Most digester covers in California are now sealed or fixed covers.

Table 25.18 Advantages and disadvantages of dry-seal gas holders.

Advantages	Disadvantages
• Is mechanically simple; easy to operate and to understand. • The dry-seal vessel height provides a complete and accurate indication of digester gas-storage status. digester gas-storage status. • Emergency repairs, if required, could be performed by local staff.	• When the costs of digester covers and dry seal gas holders are added together, it almost certainly will total more than a gas-membrane cover solution alone. • Will probably not be available as quickly as a gas-membrane cover could be available.

2.9.3.7 *High-Pressure Compressed Digester Gas Storage*

Some treatment plants effectively use a higher percentage of their digester gas via a system of medium- or high-pressure gas compressors and gas storage spheres or horizontal storage tanks (pressure vessels).

When biogas is compressed via high pressure, smaller storage vessels are needed but more electricity and a more expensive compressor are required. Several plants use high-pressure digester gas-storage systems; they frequently operate at pressures of about 7 to 22 kPa (50 to 150 psi).

A few treatment plants have medium-pressure digester gas-storage systems, which operate at a pressure of about 3 kPa (20 psi). Medium-pressure systems sometimes are used only to compress the excess gas produced at night. This gas is used in cogeneration engines the following day during on-peak hours, when electric rates are higher.

2.9.4 *Gas Processing and Equipment*

2.9.4.1 *Sediment and Condensate Traps*

Sediment traps (accumulators) should be properly sized and strategically located in the gas-piping system to collect moisture and remove pipe scale and particulate. Appropriate locations for these devices are downstream of the digester, at the end of long pipe runs, and anywhere the gas may be cooled or compressed, resulting in condensation. If sediment traps are made of carbon steel, design engineers should consider a protective coating. Steel sediment traps may be galvanized by the manufacturer or painted with a corrosion-resistant coating. Stainless steel sediment traps also are available but are significantly more costly.

One of the most common flaws in gas-collection-system designs is an insufficient number of drip traps to remove condensate from piping. Design engineers should install drip traps at all low spots in the gas system and on each sediment trap. Low-pressure drip traps typically are made of low-copper aluminum castings. High-pressure devices should be made of steel or stainless steel. (Corrosion and freezing can be serious problems in gas-collection systems.)

Manually operated drip traps should be used in indoor installations. Float-controlled, automatic drip traps require frequent maintenance to keep the valve from sticking open. They only should be used in outdoor installations (when local codes and safety considerations permit).

2.9.4.2 *Moisture Removal*

When produced, digester gas is saturated with water. However, an increasing number of digester gas treatment technologies and use equipment require dry gas. Digester gas can be dried via several techniques (e.g., refrigerant dryers, desiccant dryers, coalescent filers, and glycol systems). All moisture-removal equipment should be made of corrosion-resistant materials and preceded by sediment and condensate trips.

- Refrigerated dryers are the most common and often the most successful technique for drying digester gas. They use gas heat exchangers and mechanical chillers to cool the gas and condense water so it can be easily removed via physical separation.
- Desiccant dyers are sometimes used in packaged compressed-air dryers or natural gas applications. Because digester gas typically is saturated and often contains other contaminants, desiccant gas dryers must be specifically designated for these conditions to be effective.
- Coalescent filters can be used to remove water from digester gas, but not molecular water vapor. So, these filters are used with other water-removal equipment.
- Glycol is a *hygroscopic substance* (i.e., it naturally attracts water molecules from its surroundings) that often is used to remove moisture from raw natural gas. Glycol systems typically add glycol solution is to the wet gas, provide time for reaction, separate hydrated glycol from the gas, dry the glycol solution, and then recycle it back to the gas. They sometimes are used to process wet landfill gas.

2.9.4.3 Gas-Pressure Boosters

Anaerobic digesters typically produce gas at a pressure of 1 to 2.5 kPa (4 to 10 in. H_2O, which typically is insufficient for the gas-use equipment. However, many boilers, flares, and some gas-use equipment only require an inlet gas pressure of 0.3 to 2 kPa (2 to 14 psi). A gas booster often can make up the difference. Centrifugal gas-booster blowers can increase gas pressure by about 0.6 to 1.5 kPa (4 to 10 psi) (depending on size) and this increase in gas pressure is typically adequate for many applications.

Centrifugal gas-booster blowers are available both as open machines and as hermetically sealed gas blowers. They are available in stainless steel and cast iron. When using centrifugal gas blowers in indoor applications, design engineers should make sure to provide gas-tight seals.

2.9.4.4 Corrosion

Corrosion resistance is important when selecting digester gas piping and gas-handling equipment. Pipes primarily are corroded by acidic solutions that attack the pipe material, reducing its thickness. Corrosion also can cause scale and deposits, which constrict gas flow and plug equipment.

Acidic solutions form when gaseous carbon dioxide or hydrogen sulfide dissolve in water and then condense on the internal pipe surface. They also may be produced by anaerobic bacteria in the piping. Both carbonic and sulfuric acids can be detrimental to digester gas piping.

Both carbon steel and stainless steel are susceptible to corrosion, although stainless steel is less susceptible because of its higher chromium and manganese content. The corrosion rate of carbon steel is 0.05 mm/a (2 mil/yr) or more in wastewater treatment plant environments, while the corrosion rate of Type 316 stainless steel is less than 0.002 5 mm/a (0.1 mil/yr).

The apparent risk, presence, location, rate, and type of internal corrosion damage depends on biogas constituents, system pressure and temperature, system configuration, flow characteristics, solids deposition, piping material, and corrosion mechanism(s) (e.g., general corrosion, pitting corrosion, and under deposit corrosion).

2.9.4.4.1 General Corrosion

General corrosion is a low-level attack against the entire metal surface, with little or no localized penetration. The least damaging form of corrosion, it typically is present in all gas pipes to some degree. General corrosion results in the most uniform circumferential reduction of pipe thickness.

2.9.4.4.2 Pitting Corrosion

Pitting is another common form of corrosion; it results in localized, deep penetration of the metal surface with little general corrosion in the surrounding area. Such concentrated corrosion activity may be attributed to many factors (e.g., surface deposits, electrical imbalance, trapped condensate, and material characteristics).

2.9.4.4.3 Under-Deposit Corrosion

Under-deposit corrosion is a common form of corrosion that results from a buildup of deposits and stagnant acidic condensate or metabolic byproducts. Such deposits include black crystalline iron sulfide, which is pyrophoric (i.e., will spontaneously ignite when exposed to air). The concentration of deposits and acids encourages corrosion cells to form because of differences in the environment surrounding the pipe. The location of under-deposit corrosion can be random, and it can result in severe pitting.

2.9.4.5 High-Pressure Gas Compressors

Rotary screw compressors, sliding vane compressors, and reciprocating-piston gas compressors all have been used to successfully compress digester gas up to 50 kPa (350 psi). All have limitations, and all must be designed for continuous duty with a typically contaminant-laden wet gas. Light duty, intermittent-use air-compression machinery should not be used in digester gas applications. Important design issues with high-pressure digester-gas compressors include:

- Robust structural foundations for the compressors,
- Gas-flow pulsation attenuation,
- Lubricating oil carryover and removal from the gas,
- Moisture condensation in the gas during shutdown, and
- Compressor cooling.

2.9.4.6 Gas Metering and Gas Pressure Monitoring

Treatment plant personnel need accurate, reliable measurement of gas flow. Gas production is a measurement of digester performance. A reliable metering system enables the plant to optimize both the digester and the gas-use system. It promptly alerts operators to gas-system leaks and process fluctuations. It also allows operators to store excess gas properly and helps them plan process schedules. In addition, flow data are needed to calculate digester efficiency and the fuel savings obtained by using digester gas.

Digester gas, which is moist, dirty, and corrosive, is produced at fluctuating rates. The piping and appurtenances are designed to convey gas at low velocities and pressures. These characteristics can cause numerous maintenance problems for metering devices that engineers need to address when designing the gas-collection system and selecting gas meters.

Flow meters that have monitored gas successfully include positive-displacement bellows, shunt flow, turbine, differential-pressure Venturi tube, orifice plate, and flow tube. More recently, vortex shedding devices also have been used. In addition, thermal-dispersion mass flow meters, which use no moving parts, have proven reliable.

The meters should measure each digester's gas production, total gas production (after recirculation), gas sent to each engine or boiler, and gas wasted to the flares. They should resist corrosion and be easily serviced. Provisions removing condensate and lubricating meters typically also are required.

When selecting the device, design engineers also should consider both startup and design flow conditions. Startup (low-flow) conditions may be below the operating range of a meter sized for design flows.

2.9.4.6.1 Gas-Pressure Gauges

Gas-pressure gauges are available in both dial and manometer designs; the manometer typically is used because of the low pressures involved. Larger plants may use pressure transmitters with remote indication in the control room. Gauges indicate the pressure available in the system; they help O&M personnel locate line blockages. Figure 25.57 is a gas-piping schematic that shows typical locations for gauges.

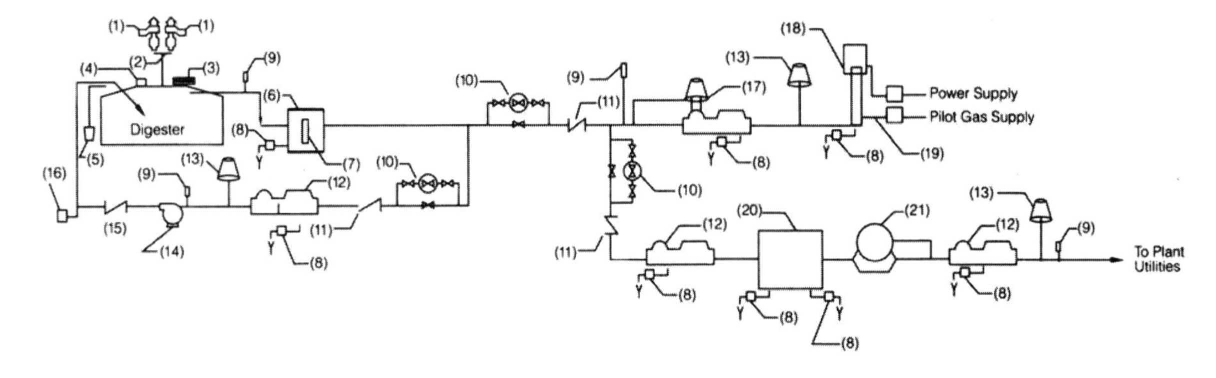

Legend			
Item	Description	Item	Description
1	Press/VAC Relief Valve with Flame Arrester	12	Flame Trap Assembly
2	Three-way Valve	13	Pressure (Explosion) Relief Valve
3	Manhole Cover	14	Blower/Compressor
4	Sampling Hatch Cover	15	High-Pressure Check Valve
5	Cover Position Indicator (for Floating Roof)	16	High-Pressure Drip Trap
6	Condensate and Sediment Trap	17	Pressure Relief and Flame Trap Assembly
7	Sight Glass	18	Waste Gas Burner and Ignition System
8	Low-Pressure Drip Trap	19	Flame Check
9	Manometer	20	Gas Purifier
10	Flowmeter	21	Double Port Regulator
11	Check Valve		

FIGURE 25.57 Diagram of a gas-control system (single-digester gas-handling system).

2.9.4.7 Isolation Valves

Several types of isolation valves (e.g., butterfly valves, plug valves and knife gate valves) have been used successfully on digester gas piping. For information on valve location requirements, see NFPA 320 and NFPA 54.

2.9.4.8 Gas Analysis

The sampling and analytical methods required for digester gas depend on the type and concentration of compound(s) of interest. Many relevant methods and techniques are available, but only should be used under the correct conditions.

Digester gas often is sampled via Tedlar bags or by collecting a gas sample in a methanol impinger.

There are several methods for analyzing digester gas (e.g., gas chromatography and mass spectrometry) (see Table 25.19). If the gas will be tested for siloxanes, or similar organic compounds typically found in small concentrations, then more analytical work may be required.

TABLE 25.19 Methods typically used to analyze biogas.

Method number	Description
ASTM D-3588	Standard Practice for Calculating Heat Value, Compressibility Factor, and Relative Density of Gaseous Fuels
ASTM D-1945	Standard Test Method for Analysis of Natural Gas by Gas Chromatography
EPA TO-14	Determination of Volatile Organic Compounds (VOCs) in Ambient Air Using Specially Prepared Canisters With Subsequent Analysis by Gas Chromatography

Complex sample collection and analysis (typically due to reactivity and low detection-limit requirements) require the use of certified and accredited labs staffed by analysts with the appropriate skills.

2.9.4.9 Gas Safety Equipment

Gas-safety equipment is an important part of the gas-handling system (see Figure 25.57). It protects both personnel and property from explosions and toxic hazards related to digester gas. Systems designed with appropriately sized equipment minimize pressure drop. Equipment selected for easy maintenance ensures a safe operating environment.

2.9.4.10 Waste Gas Combustion

Waste-gas burners (flares) are safety devices used to combust any digester gas not used by the cogeneration system or boilers. Anaerobic digesters generate biogas continuously; waste-gas burners reduce the possibility of odors or gas explosions caused by excess digester gas directly vented to the atmosphere by pressure-relief valves. Whenever a digester gas use system is modified significantly, design engineers also should review the waste-gas burners and their accessories.

Wastewater treatment plants currently use two types of waste-gas burners: conventional and enclosed combustion. Conventional waste gas burners include gas combustion and safety equipment. Conventional waste gas burners consist of a gas burner head, gas supply piping and gas piping safety accessories, and often an ignition system, mounted on a support pedestal within a cylindrical flame shield. Combsion controls are usually limited to a simple system such as an aspirated type gas/air mixing system.

Enclosed combustion flares contain the same digester gas piping and gas safety accessores as do conventional buners, along with much more sophisticated combustion controls to precisely meter the burner air/fuel ratio to achieve more complete gas combustion. The burner head is often mounted much closer to the ground within a much larger enclosed housing to help control the exact airflow to the burner and to provide the required flame residence time for more through combustion. They include controls to carefully optimize gas and air proportions for more complete combustion.

2.9.4.11 Hydrogen Sulfide Removal

Unless controlled, the concentration of hydrogen sulfide in digester gas can range from 150 to 3 000 ppm or more, depending on both the influent's composition and the digester feedstock's characteristics. Major sources of sulfur compounds in influent are the potable water supply and industrial discharges. Sulfates occur naturally in water when urine and protein decompose; they also result from alum treatment in the water supply system. Industries can discharge various sulfur materials to the collection system. In addition, trucked wastes, which are fed directly to anaerobic digesters, often contain sulfur material.

The hydrogen sulfide in digester gas is formed when anaerobic bacteria reduce sulfates and other sulfur material. It may need to be removed from digester gas to reduce corrosion in boilers and engine parts. It also may need to be removed to satisfy local air emissions standards. Hydrogen sulfide is a toxic air pollutant that can create both odor and safety issues, even in minute concentrations. Biogas with a high hydrogen sulfide content can contribute to air pollution. Flaring eliminates the odor problem but produces sulfur dioxide, which is a major cause of acid rain.

Regulators have long studied the effect on emissions of burning gas containing small amounts of hydrogen sulfide. In the refinery industry, U.S. EPA source performance standards for newly constructed burners and combustion units limit hydrogen sulfide levels in fuel gas to 160 ppm or less (Leicht et al., 1986).

One method for removing hydrogen sulfide from biogas is to use a scrubber. Some scrubbing technologies also reduce carbon dioxide concentrations, producing a higher quality biogas. Various chemistries for hydrogen sulfide removal (e.g., Sulfa-Sweet, Lo-Cat, Sulfa-Scrub, Chem-Sweet, and Stretford Process) are available for consideration.

A common dry scrubber (called the *iron sponge*) uses iron oxide-impregnated wood chips (see Figure 25.58). Hydrogen sulfide reacts with iron oxide to form elemental iron, elemental sulfur, and water. The iron sponge is periodically regenerated by removing the sulfur and oxidizing the iron to form iron oxide. Such regeneration could be hazardous because spontaneous combustion is possible if the iron is oxidized too rapidly. The iron sponge method typically is best suited to relatively small gas flows.

A conventional wet scrubber uses a liquid that is maintained at a high pH (via caustic) to enhance hydrogen sulfide absorption. It also may contain an oxidant (e.g., sodium hypochlorite or potassium permanganate) to reduce adsorbent disposal problems and increase its useful life. Wet scrubbers use nozzles or diffuser plates, which

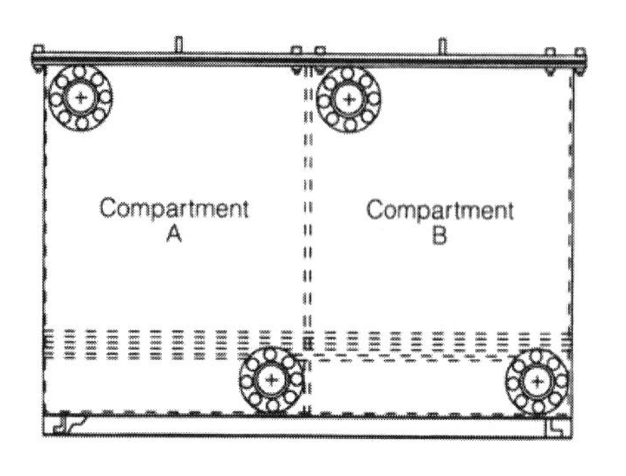

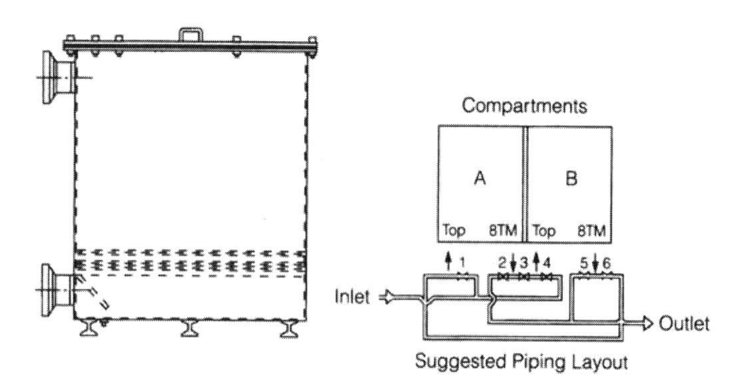

FIGURE 25.58 Schematic of a typical iron sponge used to remove hydrogen sulfide from biogas.

periodically require cleaning. The gas leaving the wet scrubber is saturated with moisture, which must be condensed and removed downstream. Also, because the headloss through wet scrubbers typically is too high for the low-pressure digester gas system, the gas must be compressed before being scrubbed.

Adsorption systems use activated carbon treated with an alkaline material to adsorb hydrogen sulfide from biogas. They typically are only suitable for treating low flows of digester gas (e.g., those from pressure-relief valves). Also, they can spontaneously combust under certain conditions of rapid exposure to fresh air.

Catalytic scrubbers use an aqueous chelated iron catalyst to treat biogas, producing elemental sulfur (see Figure 25.59). The catalyst is reactivated via air in oxidizer vessels.

Instead of gas scrubbing, some plants have added iron salts directly to the digester or plant influent. Iron reacts with sulfide to form insoluble iron sulfide. However, iron salts should not be added to heated solids lines because this results in a rapid buildup of vivianite (ferrous phosphate) scale. Iron salts also can reduce digester alkalinity, so design engineers must make provision to monitor and control the solution strength and dosing rate to avoid lowering the digester's pH.

This method requires a bulk-storage tank, chemical feed pumps, piping, and monitoring equipment. Its chief cost is for chemicals. Although this method's O&M costs are low, it requires more operator skill than the iron sponge method.

2.9.4.12 Siloxane Removal Systems

Siloxanes are a family of anthropogenic organic compounds containing silicon that are becoming increasingly common in many household products (e.g., deodorants, cosmetics, shampoos, dyes, lubricants, dry cleaning fluids, and waterproofing compounds). As a result, volatile siloxanes can be found in landfill gas and digester gas, often at concentrations of a few parts per million or less (see Table 25.20).

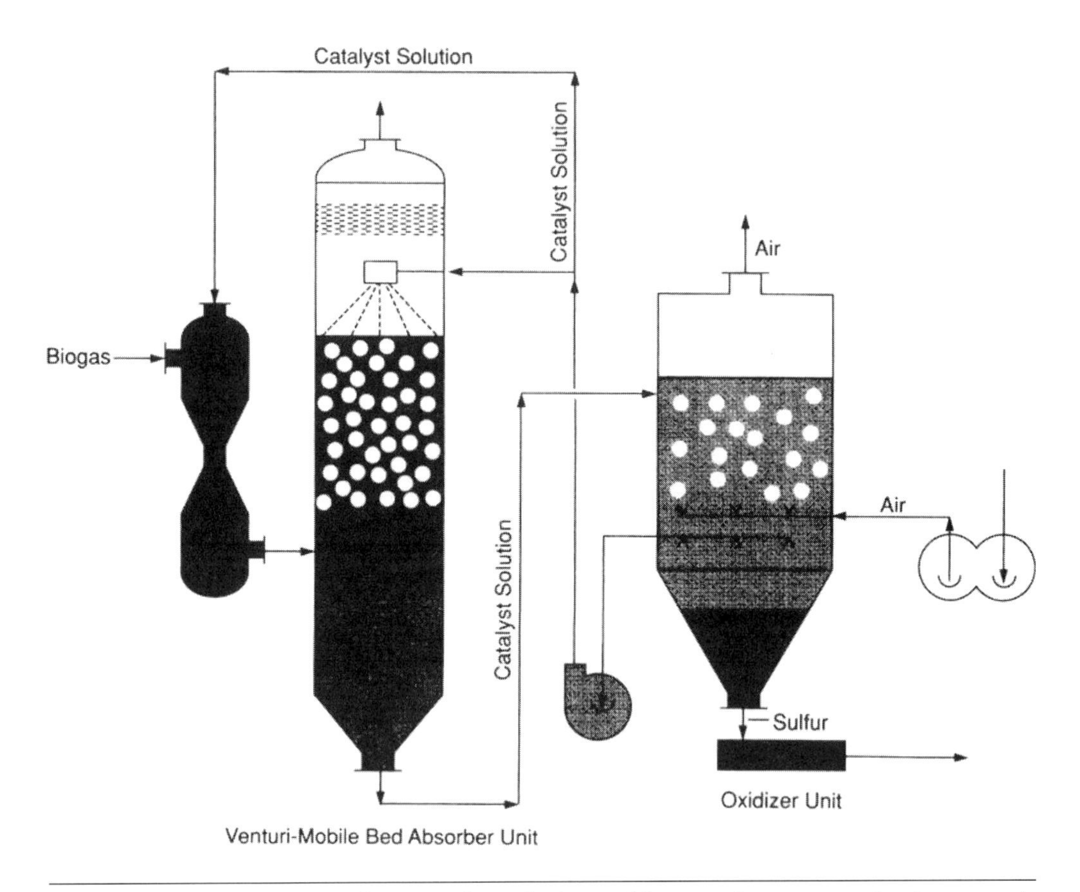

FIGURE 25.59 Schematic of a typical catalytic scrubber used to remove hydrogen sulfide from biogas.

TABLE 25.20 Typical volatile organic siloxanes found in digester gas.

Siloxane species	Formula	Common abbreviation	Molecular weight	Vapor pressure (mm Hg at 77°F)*	Boiling point (°F)	Water solubility (mg/L at 25°C)
Hexamethyldisiloxane	$C_6H_{18}Si_2O$	MM	162	31	224	0.93
Octamethyltrisiloxane	$C_8H_{24}Si_3O_2$	MCM	236	3.9	N/A	0.035
Hexamethylcyclotrisiloxane	$C_{12}H_{18}O_3Si_3$	D3	222	10	275	1.56
Octamethylcyclotetrasiloxane	$C_8H_{24}O_4Si_4$	D4	297	1.3	348	0.056
Decamethylcyclopentasiloxane	$C_{10}H_{30}O_5Si_5$	D5	371	0.02	412	0.017

* 0.555 6(°F −32) = °C.

Siloxanes are difficult to detect and control. Two common siloxanes, hexamethyldisiloxane (MM) and octamethyldisiloxane (MCM), are relatively large linear chain molecules while others are cyclical (similar to benzene rings). Only hexamethylcyclotrisiloxane (D3) and MM are significantly soluble in water at ambient temperatures.

When combusted, siloxanes form tough, often abrasive silicon dioxide deposits. (*Silicon dioxide* is the chemical name for ordinary beach sand.) The combusted siloxanes also promote the formation of other chemical deposits (e.g., calcium, sulfur, iron, and zinc compounds) on them. These deposits often clog engine heads, foul exhaust and intake valves, and coat combustors and fuel injectors. They also cover exhaust catalysts, boiler surfaces, and exhaust heat-recovery equipment tubes.

Several treatment plants have had success removing siloxanes from digester gas. The best siloxane-control systems typically include moisture removal upstream of activated carbon (see Table 25.21). To maximize media life, both water and hydrogen sulfide should be removed from biogas before activated carbon treatment.

2.9.4.13 Carbon Dioxide Removal

As global concern about climate change increases, some treatment plants have begun to monitoring carbon dioxide. Dry digester gas typically contains about 60% carbon dioxide by volume, or about 55 to 60% carbon dioxide on a mass basis. Techniques for

TABLE 25.21 Advantages and disadvantages of various siloxane-removal systems.

General type	Advantages	Disadvantages
Water scrubber	Continuous process Can be used upstream of others technologies	Only removes water soluble siloxanes
Refrigeration (to 40°F*)	Continuous process Can be used upstream of others technologies	Only removes a fraction of the siloxanes
Refrigeration (to below 0°F)	Continuous process Also removes water	All units to date have had significant freezing problems Consumes more electricity
Activated carbon adsorption	Proven technology that works very effectively Also removes H_2S and other trace organics	Batch type process Media must be replaced or regenerated
Silica gel systems	An alternative to carbon Can be used upstream of others technologies	Very limited operation experience

* 0.555 6 (°F = 32) = °C.

removing carbon dioxide from digester gas include pressure swing adsorption, temperature swing adsorption, cryogenic refrigeration, and amine treatment.

2.9.4.13.1 Pressure Swing Adsorption

In pressure swing adsorption, gas constituents adsorb to the surface of a media at one pressure (typically high), and are released at another pressure (typically much lower).

2.9.4.13.2 Temperature Swing Adsorption

Temperature swing adsorption might be the most common technique for removing carbon dioxide. In this process, carbon dioxide adsorbs to a media at a low temperature {typically at or near ambient temperatures [10 to 32°C (50 to 90°F)]}. Once the adsorption media is saturated, carbon dioxide is expelled from it by heating the media to typically 150 to 200°C (300 to 400°F).

2.9.4.13.3 Cryogenic Refrigeration

Cryogenic refrigeration takes advantage of the fact that carbon dioxide freezes at a warmer temperature [−79°C (−110°F) than methane does [−182°C (−297°F)]. Cryogenic systems can work well but require a significant amount of mechanical energy to refrigerate the gas.

2.9.4.14 Amines

Amines [e.g., monothanolamine (MEA) and diethanolamine (DEA)] are a class of substances derived from ammonia. They frequently are used to remove both hydrogen sulfide and carbon dioxide from raw or sour natural gas.

2.9.5 Gas Use—Boilers

Boilers extract usable heat energy from a fuel, typically via combustion. Historically, this is the most common technique for capturing a digester gas' energy. Even treatment plants that use engine generators or gas turbines need boilers as standby or supplemental heating equipment. Boiler sizes range from about 100 000 to more than 1 billion kJ/h (Btu/hr).

Emissions controls are becoming increasingly important features for boilers throughout the United States. Engineers should address air quality regulations as part of all boiler designs. If designed appropriately, specially modified boilers can meet nearly all air quality regulations.

2.9.5.1 Fire-Tube Boiler

Packaged fire-tube boilers are the most common type of boilers used in treatment plants. They are available in sizes from about 2 to 30 mil. kJ/h (2 to 30 mil. Btu/hr).

2.9.5.2 Fire-Box Boilers

Fire-box boilers are a special type of fire-tube boiler with an oversized combustion chamber. This chamber may help properly combust the relatively low caloric (Btu) content of digester gas. Fire-box boilers range in size from about 2 to 10 mil. kJ/h (2 to 10 mil. Btu/hr).

2.9.5.3 Water-Tube Boilers

Water-tube boilers are similar to fire-box boilers, except that the combustion chamber is a horizontal insulated gas-tight compartment containing multiple water filled steel tubes that are heated by the combustion gases. They contain less water internally, so that water tube boilers often can warm up quicker and start up faster than fire-tube boilers. Those with flexible tubes are particularly resistant to thermal shock and less vulnerable to siloxane-caused silicon deposits on the tubes. Water-tube boilers are available in sizes ranging from less than 8 to more than 20 mil. kJ/h (8 to 20 mil Btu/hr).

2.9.5.4 Cast-Iron Boilers

Cast-iron sectional boilers are sometimes used in retrofit applications because they can fit through small doorways. These boilers are smaller, typically available in sizes from 300 000 to 10 mil. kJ/h (300 000 to 10 mil. Btu/hr).

2.9.6 Gas Use—Combined Heat and Power (Cogeneration)

Treatment plants larger than about 20 to 40 ML/d (5 to10 mgd) are possible candidates for digester gas cogeneration (combined heat and power systems). Typical cogeneration systems are based on internal-combustion engines, microturbines, or gas turbines.

2.9.6.1 Reciprocating Internal-Combustion Gas Engines

Most wastewater treatment plants that have digester gas cogeneration systems use reciprocating internal-combustion engines. Major engine manufacturers have recently developed many advanced internal-combustion engines to improve fuel economy, reduce maintenance, and lower exhaust emissions.

2.9.6.1.1 Reciprocating engines

Reciprocating engines are the most widely used technology in digester-gas cogeneration applications.

2.9.6.1.2 Advanced reciprocating engine systems

With the goals of significantly better fuel economy and lower exhaust emissions, several manufacturers developed models that they used to create technically modern, progressive spark-ignition, lean-burn engines. These engines are called advanced reciprocating engine systems (ARES). With their higher fuel efficiency, these technologies could enable wastewater treatment plants to produce substantially more electric power to offset energy costs using existing digester-gas production.

Advanced reciprocating engine systems have been in service since 2005 or 2006; they have an electrical output of about 1 000 to 3 000 kW. The more fuel-efficient engines produce more power using less digester gas than the older engines now in service at some treatment plants. They also operate at a gas pressure less than 0.6 kPa (4 psi) and so often can be used with many existing digester gas systems.

2.9.6.1.3 Dual-Fuel Engine-Generator

Dual-fuel (gas–diesel) engines are compression-ignition, not spark-ignition, engines. To ignite, they simultaneously burn gas and a small amount of diesel fuel pilot oil. These engines must use some diesel fuel as pilot fuel, but their controls also allow automatic switchover to 100% diesel fuel operation without changing load if the gaseous fuel supply is interrupted. This capability is a beneficial feature for standby units because they can start and operate even during power failures.

Dual-fuel engines typically use 1 to 5% diesel fuel oil, but many can, if necessary, operate on 1 to 100% diesel fuel. Such fuel flexibility is an excellent advantage, especially if the gaseous fuel supply is disrupted. This option includes storage and handling equipment for diesel fuel, along with 11-kPa (75-psi) gas compressors to supply gaseous fuel to these engines.

2.9.6.2 Combustion Gas Turbine Generators

Combustion gas turbines are available in sizes ranging from 250 to 250 000 kW. They typically are used at wastewater treatment plants with influent flows of 300 ML/d (80 mgd) or more. They also are widely used in new large commercial electric-power plants. Gas turbines are an attractive option for generating electricity because they have several important characteristics (see Table 25.22).

A few U.S. treatment plants have been successful in using gas turbines fueled by low-Btu digester gas. Most probably will require some form of exhaust emissions control. There are three types of emission-control systems available for turbines: wet technologies; catalytic converters; and dry, low-nitrogen oxides (NO_X) combustors. Wet technologies (e.g., water or steam injection directly into the turbine's combustion zone) can substantially reduce exhaust emissions, but they require a continuous flow of ultraclean water. This is both expensive and time-consuming. Catalytic converters, which often follow water or steam injection, are expensive and simply not appropriate for digester gas fuel without extensive, reliable fuel treatment. Dry, low-NO_X combustors are the newest and most attractive technology, and may be the only one appropriate for digester gas operations. Not all gas turbine manufacturers offer this technology, and

Table 25.22 Advantages and disadvantages of gas turbines.

Advantages	Disadvantages
Suitably equipped gas turbines can burn several fuels, including digester gas.	Gas turbines are less-efficient electricity generators than engines.
Gas turbines are available from several experienced manufacturers in sizes from about 250 to 250 000 kW.	Gas turbines lose power and fuel efficiency in ambient air temperatures above 60°F*.
Gas turbines have few moving parts and typically require less maintenance than internal-combustion engines.	Gas turbines lose power and fuel efficiency at high elevations.
High-pressure steam can be produced from the hot turbine exhaust gasses sufficient in large gas turbines to provide another 50% electric generation capacity.	Gas turbines require high-pressure, clean fuel.

* 0.555 6(°F −32) = °C.

many have dry-NO_X units that are, at best, experimental. Most of the newer, more advanced gas turbines are available with low-NO_X combustors.

Gas turbines require slightly less maintenance than reciprocating engines, but service is highly specialized and expensive. So, it is important that the treatment plant have local service support.

The fuel should be free of condensation and particles larger than 5 μm. The inlet pressure should be 17 to 36 kPa (120 to 250 psi). A high-pressure gas compressor and a moisture separator or filter probably would be needed to meet these requirements.

Gas turbines require a fuel gas-booster compressor to supply the required 29 kPa (200 psi) to the combustion chamber. Some turbine models are available in a "dry low emissions" version. Other gas turbines can meet NO_X emission standards via a selective catalytic reduction system.

Catalysts are not suitable for use with digester gas unless the gas has been thoroughly and reliably cleaned of impurities. It has been tried unsuccessfully at the Los Angeles County Sanitation Districts' Carson plant, and at the Sacramento Regional Wastewater Treatment Plant's cogeneration facility. Various contaminants in digester gas quickly poison the noble metals in the catalyst.

2.9.6.3 Microturbines

A highly publicized new technology, microturbines are small, high-speed gas turbines ranging from 30 to 250 kW (see Table 25.23). Many were originally developed from large engine turbochargers and use new technologies (e.g., extended-surface recupera-

TABLE 25.23 Advantages and disadvantages of microturbines.

Advantages	Disadvantages
Are available is smaller sizes, down to 30 kW for small-capacity plants.	Are inefficient electricity producers. Their electrical-generation efficiency typically is only 24 to 30%.
Produce fewer exhaust emissions than some other types of digester gas cogeneration equipment.	Require significant gas cleanup, including moisture and siloxane removal.
Are available as modular, fully packaged equipment.	Even with appropriate heat-recovery equipment, they supply relatively little heat for the quantity of digester gas consumed.
Are quiet and can be used outdoors.	The digester gas must be compressed to between 75 and 100 psi*.

* psi × 6.895 = kPa.

tors, air bearings, and ultra-fast operating speeds). Recently, interest has grown in using microturbines for distributed generation and cogeneration.

All gas turbines—including microturbines—generate less power when installed at high elevations and when ambient temperatures exceed 15°C (59°F). If installed at a site with an elevation of 1 295 m (4 250 ft), for example, the gas turbine's performance would be about 20 to 25% less than that of one installed at sea level, depending on the inlet combustion air temperature.

Because of their comparatively small size and output, microturbines have been attractive to treatment plants with smaller flows [average flows as low as 15 ML/d (4 mgd)] than typically suitable for digester gas-fueled cogeneration systems. For example, a 57-ML/d (15-mgd) treatment plant in southern California installed a 250-kW microturbine to process its digester gas. Additionally, between 2000 and about 2004, several California treatment plants installed microturbines without sufficient digester gas treatment and the plant staffs had difficulty operating and maintaining them. Several of the micrturbines have been shut down.

2.9.6.4 Steam Turbines and Steam Boilers

A few U.S. wastewater treatment plants are large enough [more than 400 ML/d (100 mgd)] to produce and burn digester gas in large steam boilers and then generate high-pressure steam and electricity via a steam-driven rotating steam turbine generator.

For smaller treatment plants, this steam boiler–turbine technology is an inefficient power generator. Superheated, very high pressure steam, typically above 130 kPa (900 psia) is required for efficient steam turbine generator performance and steam turbines less than about 10 MW in size are not physicaslly large enough to be built with the relatively close maching tolerences and thus higher mechanical efficiencies of much larger steam tubines. Also, the high-pressure steam boiler must be continuously staffed (around the clock) by a licensed steam-boiler operator.

2.9.7 *Gas Cleanup and Sale*

The cost of natural gas in the United States has risen dramatically since about 1995. This has made it more economical to clean up digester gas.

Table 25.24 characterizes pipeline-quality digester gas. Once carbon dioxide, hydrogen sulfide, and water are removed, digester gas also can be used for direct pipeline injection. The applicability of scrubbing and selling digester gas to the local natural gas utility depends entirely on the gas utility's interest and willingness to offer a competitive price for the methane cleansed from digester gas.

2.9.8 *Solids Drying*

Any digester gas not used for process heating can be used for other purposes (e.g., solids drying). Fuel is one of the largest costs associated with solids drying, so using biogas can offset some of the long-term O&M costs.

TABLE 25.24 Characteristics of typical digester gas.*

	Digester gas	
Item or parameter	**Range**	**Common value**
Methane (%) (dry basis)	50–70	60
Carbon dioxide (%) (dry basis)	30–45	39
Nitrogen (%) (dry basis)	0.2–2.5	0.5
Hydrogen (%) (dry basis)	0–0.5	0.2
Water vapor (%)	5.9–15.3	6
Hydrogen sulfide (ppm_v) (dry basis)	200–3 500	500
Siloxanes, total (ppb_v)	100–4 000	800
Ammonia (ppb_v)	100–2 000	1 000
Carbon disulfide (ppb)	200–900	500
Specific gravity (based on air = 1.0)	0.8–1.0	0.91
Higher heating value (HHV) (Btu/cu ft)	600–650	620
Lower heating value (LHV) (Btu/cu ft)	520–580	560

* All percentages listed above are percentages by volume
ppmv = parts per million, by volume.
ppbv = parts per billion, by volume.
As produced, mesophilic digester gas at 37 °C (98 °F) is saturated with water and contains 5.9 to 6% water vapor. Thermophilic digester gas at 55°C (131°F) contains about 15% water vapor.
A fuel's higher heating value includes the heat of the water of vaporization.

2.9.9 *Emerging Technologies—Fuel Cells*

A *fuel cell* is an electrochemical device that combines hydrogen with oxygen to continuously produce electricity. The hydrogen is extracted from the fuel delivered to the unit, while the oxygen is simply obtained from the air. The popularity of fuel cells is due to their high power-generation efficiency, vibration-free operation, clean exhaust emissions, and technical novelty. Fuel cells are quiet; their accessories generate what little noise they produce. Stationary fuel cells are available as fully modular units in sizes of 200-kW and larger. They are readily installed outdoors.

Fuel cells are used today at a growing number of municipal and industrial wastewater treatment plants. They are becoming an increasingly proven technology. The current economic viability of fuel cells, however, depends largely on funding assistance, often via available grants.

2.9.9.1 *Representative Digester Gas Fuel Cell Plants*

An increasing number of wastewater treatment plants are using digester-gas fuel cells.

2.9.9.2 *Types of Fuel Cells*

Four types of fuel cells are in development or commercially available: phosphoric acid fuel cells, carbonate fuel cells, solid oxide fuel cells, and proton exchange membrane fuel cells.

2.9.9.2.1 Phosphoric Acid Fuel Cells

Phosphoric acid fuel cells were the first commercial fuel cells used at treatment plants. The phosphoric acid system is the most mature technology. At least 10 municipal wastewater treatment plants have installed 200-kW digester-gas phosphoric acid fuel cells, and some have more than 7 years of operating experience with this technology. So, phosphoric acid fuel cells should be considered a proven technology, not a developing or experimental one.

Several of the initial digester-gas phosphoric acid fuel cell installations at treatment plants are best characterized as developmental or experimental applications. These early units might not accurately characterize current fuel cell offerings.

2.9.9.2.2 Carbonate Fuel Cells

Many newer fuel-cell installations use the carbonate fuel cell (sometimes called the *molten carbonate fuel cell* or *direct carbonate fuel cell*) (see Table 25.25). Portions of the molten carbonate fuel cell (e.g., the reformer and the inverter) are similar to those in phosphoric acid fuel cells. One important difference is the lithium and potassium carbonate electrolyte solution, which allows electrons to transfer within the unit.

TABLE 25.25 How the carbonate fuel cell works.

Process or component	Chemical reaction
Within the internal steam reformer, the pure methane is converted to hydrogen gas via a reaction with steam.	$CH_4 + 2H_2O_{(steam)} \rightarrow 4H_2 + CO_2$
After the reformer, hydrogen combines with carbonate at the anode. The reaction produces water and carbon dioxide.	$H_2 + CO_3^{=} \rightarrow H_2O + CO_2 + 2e-$
The carbonate is the electrolyte media allowing electrons to flow from anode to cathode.	No chemical reaction
At the cathode, oxygen from the air completes the carbonate balance, and the flow of electrons is finished.	$2CO_2 + O_2 + 2e- \rightarrow 2CO_3^{=}$
At the electric power inverter, direct current is converted to alternating current at 480 V.	No chemical reaction

Like phosphoric acid fuel cells, a carbonate fuel cell is a mature technology with a proven track record. Design engineers should consider both technologies when evaluating digester gas applications (see Table 25.26).

2.9.9.3 Fuel Cell Components

A fuel cell consists of several main process modules: the gas-cleanup unit, reformer, cell stack, and inverter.

2.9.9.3.1 Gas-cleanup unit

This module purifies digester gas or natural gas, removing all potential contaminants. Fuel cell stacks are exceptionally sensitive to certain impurities, so only exceptionally pure, clean, and pressurized methane gas leaves this module for the reformer.

2.9.9.3.2 Reformer

This device combusts a tiny amount of fuel to produce steam. The reformer mixes this pressurized, high-temperature steam with pure methane from the gas-cleanup module to produce the hydrogen gas essential to fuel cell operations.

2.9.9.3.3 Cell stack

The cell stack uses hydrogen gas to produce electricity. Hydrogen gas and oxygen ions in the carbonate, or similar, electrolyte react to produce the constant flow of electrons needed to produce electricity.

TABLE 25.26 Comparison of two types of digester-gas fuel cells.*

Item	Phosphoric aid	Molten carbonate	Remarks
Representative fuel cell manufacturer	United Technology Corporation (UTC)	Fuel Cell Energy	UTC purchased ONSI fuel cell business
Modular sizes, electrical output	200 kW	300-kW and 1 200-kW units	FCE recently increased their unit capacity
Electrical efficiency, percent	36–40	45–47	New equipment performance, typical
Operating electrolyte temperature (°F)	375	1 200	Operating temperature affects start-up time
Recoverable heat output at 180 to 200°F (Btu/hr)	2.0 million (for units totaling 1 MW)	1.49 million (for 1.2 MW unit)	Per manufacturer's performance claims
Expected performance degradation (%)	About 2% yearly capacity decrease	2 to 3% yearly capacity decrease	Capacity drops off as the cell stack fouls
Water consumption, Continuous (gpm)	1.6 gpm for 1-MW unit	2 gpm for 1.2-MW unit	Water is used to make steam in the reformer
Required digester-gas fuel pressure (psi)	20–30	About 25	Gas compressors are required
Supplemental natural gas to the digester gas	Not required	Recommends 10% natural gas	Strongly recommended by fuel cell supplier

* 0.555 6(°F −32) = °C; Btu/hr × 0.293 1 = W; gpm × 5.451 = m^3/d; psi × 6.895 = kPa.

2.9.9.3.4 Inverter

The inverter consists of electrical devices that convert the direct current (DC) electric power created by the fuel cell into alternating current (AC) and transforms this AC power into the required system voltage.

2.9.9.4 Emerging Technologies—Solid Oxide and Proton Exchange Membrane Fuel Cells

At press time, solid oxide fuel cells and proton exchange membrane (PEM) fuel cells were not yet ready for long-term digester-gas use (see Figure 25.60). Neither technology was in full-scale service with digester gas fuel at a wastewater treatment plant at the time of this document.

2.9.10 Emerging Technologies—Stirling Cycle Engines

A stirling cycle engine is a possible cogeneration technology that uses an external combustion process to convert heat into mechanical power. The manufacturer(s) claim that

FIGURE 25.60 A fuel cell installed at a King County, Washington, facility (courtesy of King County, Washington).

the engines require only limited fuel treatment and can operate on a low-pressure fuel source. They also have fewer emissions than reciprocating engines. A 55-kW stirling cycle engine recently was installed for demonstration testing in Oregon. The availability of stirling cycle engine manufacturing, however, is very limited; interested organizations should check on the status of available vendors.

2.9.11 *Digester Gas Use Technology and Heat Recovery*

Anaerobic digesters require a constant, reliable supply of heat to ensure optimal biological activity in the reactors. So, the first requirement of any digester-gas use technology is to reliably satisfy the treatment plant's heating needs. This means providing

- An adequate quantity of digester heat,
- A reliable heating source or heat-recovery system, and
- A consistent heating supply when used with the plant's heating water loop [typically a heating loop at 60 to 80°C (140 to 180°F)].

The type of heat and relative amount of recoverable heat is shown in Table 25.27.

2.9.11.1 Internal Combustion Engine Heat Recovery

Large stationary internal-combustion engines sometimes are used to drive big air blowers, electric generators, and large pumps at wastewater treatment plants. Recovering heat from gaseous fueled reciprocating engines is an established practice used at many treatment plants. Traditionally, engine jacket water at 80 to 110°C (180 to 230°F) and heat from hot-engine exhaust gases [340 to 540°C (650 to 1 000°F)] are common sources of recovered engine heat. Lower-temperature engine lubricating oil heat and turbocharger aftercooler heat at only about 50 to 60°C (120 to 140°F) typically is wasted to an air-cooled radiator or a water-cooled waste-heat exchanger.

The newer lean-burn reciprocating engines (e.g., ARES) incorporate two-stage aftercoolers or intercoolers. In these advanced gas-fueled engines, heat from the first turbocharger aftercooler is combined with the engine-jacket cooling-water system for better use in heat-recovery applications. This arrangement improves engine turbocharger performance and makes more of the engine's total heat available at a higher, more economically usable temperature.

2.9.11.2 Fuel Cell Heat Recovery

The chemical reactions in the current generation of fuel cells are exothermic, and they generate enough heat to vaporize the water chemically produced during the reactions. Excess fuel-cell heat is often captured and used productively.

Table 25.27 Heat recovery from cogeneration systems.

	Combined heat and power technology				
Parameter	**Microturbine**	**Fuel cells**	**Combustion gas turbines**	**Advanced internal-combustion engines**	**Steam boilers + steam turbines**
Forms of heat energy available from this technology	Hot water Hot exhaust gases	Hot water Hot exhaust gases	Low-pressure steam High-pressure steam and/or hot exhaust gases	Hot water Some low-pressure steam	Low-pressure steam High-pressure steam and/or hot exhaust gases
Portion of the biogas fuel energy available as heat energy		20–25%	40–50%	35–45%	45–55%

Based on vendor performance data, a 1 400-kW (1.4-MW) fuel cell assembly produces about 8 300 kg/h (18 300 lb/hr) of 370°C (700°F) exhaust consisting of steam and clean hot gases. When passed through a heat-recovery heat exchanger, this exhaust can produce about 2.1 mil. kJ/h (2.2 mil. Btu/hr) of fuel cell heat while cooling to about 120°C (250°F). Although slightly more fuel-cell exhaust heat could be captured, it would only be about 50 to 60°C (120 to 140°F), which would be too cool for the treatment plant's heating water loop and its digester heat exchangers. The 2.1 mil. kJ/h (2.2 mil. Btu/hr) of fuel cell heat is enough to meet summer heating needs, but the heating boiler must be used for the rest of the year. The boiler is digester-gas fueled, so for most of the year, a portion of the available digester gas must be diverted to operate the boiler, which supplements the fuel cell's heat-recovery process.

2.9.12 *Air Emissions; Limits and Control Options, Greenhouse Gases*

2.9.12.1 Criteria Pollutants

The traditional air pollutants of concern (criteria pollutants) from gas-combustion equipment are NO_X, carbon monoxide, SO_X, non-methane hydrocarbons (NMHC), and PM_{10}:

- Nitrogen oxides [e.g., nitrogen dioxide (NO_2) and nitric oxide (NO)] traditionally have been the most significant criteria pollutants. They are formed via combustion from nitrogen in the air. This group typically does not include nitrous oxide (N_2O).
- Carbon monoxide is formed via the partially complete combustion of methane (CH_4). Its emissions are controlled by combustion modifications.
- Sulfur oxides (e.g., sulfur dioxide) typically form when the hydrogen sulfide in digester gas combusts. It is controlled by eliminating hydrogen sulfide from the gas.
- Non-methane hydrocarbons typically are not found in significant quantities in digester gas.
- Particulate matter can include both PM_{10} and $PM_{2.5}$. PM_{10} are particles larger than 10 μm, while $PM_{2.5}$ are particles larger than 2.5 μm.

2.9.12.2 Greenhouse Gases

One project consideration that has become a significant public concern is the reduction of greenhouse gases (climate change emissions). There are three greenhouse gas of concern in digester gas use evaluations: carbon dioxide (CO_2), methane (CH_4), and nitrous oxide (N_2O).

2.9.12.2.1 Carbon dioxide

Probably the best known greenhouse gas, carbon dioxide is a relatively heavy gas. Digester gas can contain as much as 40% carbon dioxide by volume or about 60% carbon dioxide on a per weight basis. For most of the digester gas use processes under consideration, the carbon dioxide initially in digester gas passes through unreacted and unchanged.

Carbon dioxide is formed from the complete combustion of any fuel that contains carbon (e.g., methane). Any boiler, flare, incinerator, or power-generation technology that combusts methane will produce a corresponding predictable amount of carbon dioxide. The basic reaction for this chemical reaction is as follows:

$$CH_4 + 2O_2 \rightarrow CO_2 + 2H_2O \tag{25.24}$$

In other words, when 1 mole of methane is combusted completely, it will form exactly 1 mole of carbon dioxide. This conversion is essentially the same in a boiler, engine, gas turbine, or flare. Because methane has a molecular weight of 16 and carbon dioxide has a molecular weight of 44, each completely combusted kilogram of methane will produce 44/16 = 2.75 kg of carbon dioxide.

Biogenic carbon dioxide is carbon dioxide produced by life processes. It is not included in greenhouse gas inventories.

2.9.12.2.2 Methane

Methane is the principal component of both digester gas and natural gas. It is a light gas with a specific gravity of less than 1.0. Methane is an exceptionally important greenhouse gas; its global warming potential is 21 to 23 times that of carbon dioxide. From a greenhouse gas perspective, completely combusting all methane without atmospheric release is vital.

In fuel cells, methane gas first reacts with steam to produce hydrogen gas, as follows:

$$CH_4 + H_2O_{(steam)} \rightarrow 3H_2 + CO \tag{25.25}$$

Then, the carbon monoxide (CO) combines with atmospheric oxygen (O_2) to produce carbon dioxide:

$$CO + ½O_2 \rightarrow CO_2 \tag{25.26}$$

When totally combusted, 1 mole of methane produces 1 mole of carbon dioxide.

Low-NO_X boilers, lean-burn engines, and combustion gas turbines operate with an abundant amount of excess air (up to 70% or more) in their carefully controlled combustion chambers to ensure virtually complete oxidation of all methane in their fuel.

However, traditional waste-gas burners are not precisely controlled combustion devices, and a substantial portion of the methane in digester gas passes unburned through the flare. The amount flared is difficult to measure because the conditions in one part of the flame differ greatly from another, according to the wind direction. Getting an accurate and true sample is next to impossible.

Unburned methane is an exceptionally powerful greenhouse gas, so alternatives that must flare some of the digester gas via a conventional waste-gas burner contribute much more greenhouse gas than alternatives that fully combust all the digester gas to generate electricity or that consume all the gas via other means.

Methane also is released to the atmosphere whenever digester gas is discharged from pressure-relief valves. Again, such discharges are difficult to measure; but fortunately, they are unusual.

2.9.12.2.3 Nitrous Oxide Emissions

Nitrous oxide indirectly serves as a greenhouse gas because it produces tropospheric ozone when its molecules break down. With a global warming potential of 310 (based on carbon dioxide = 1), nitrous oxide is particularly a concern, even in small quantities. It can be formed during methane combustion as an intermediate combustion byproduct. Nitrous oxide emissions are a function of many complex combustion dynamics and combustion equipment. For example, higher combustion-zone temperatures destroy nitrous oxide.

Emissions factors for nitrous oxide are varied. Most gas equipment has little or no information about nitrous oxide emissions, partially because it typically is produced in tiny amounts during combustion. One common guideline (AP-42) lists emissions factors for nitrous oxide from natural gas combustion as 0.01 to 0.034 g/m^3 (0.64 to 2.2 lb/mil. cu ft). A nitrous oxide emissions factor of 0.1 kg nitrous oxide per tetrajoule (TJ) also is used for natural gas. Few actual, reliable nitrous oxide emissions factor data are available for digester gas, but they typically should be similar to natural gas emissions. Many combustion authorities do not consider nitrous oxide to be a component of traditional nitrogen oxides emissions.

2.9.12.3 Greenhouse Gases and Power-Generation Efficiency

Another metric is based on the amount of electricity produced per pound of carbon dioxide released. As expected, more energy-efficient power-generation technologies (e.g., fuel cells and ARES) are the leaders in this area among fuel-burning power generators. Also, the net electrical output—after subtracting auxiliary electrical loads and the digester gas used to fire a supplemental heating boiler—is much more important than the gross power-generation efficiency.

2.9.12.4 Digester Gas Use Greenhouse Gas Concerns

When selecting a digester gas use application, design engineers should address the following greenhouse gas emission concerns:

- The digester gas use application should be selected to avoid venting or indirectly releasing any gas because of methane's high greenhouse gas potential.
- Digester gas use applications should flare as little of the gas as possible due to the incomplete combustion characteristics of virtually all traditional waste-gas burners.
- The digester gas use application should be designed to produce—directly or indirectly—as little nitrous oxide as possible.
- In cogeneration applications, the gas use technology should produce as much net usable electricity as possible to reduce the electric utility's carbon dioxide emissions.

3.0 AEROBIC DIGESTION

Stabilization during aerobic digestion occurs from the destruction of degradable organic components and the reduction of pathogens by aerobic, biological mechanisms. Aerobic digestion is a suspended-growth biological treatment process based on biological theories similar to those of the extended aeration modification of the activated sludge process. The objectives of aerobic digestion, which can be compared to those of anaerobic digestion, include producing a stable biosolids via oxidizing organisms and other biodegradable organics, reducing mass and volume, reducing pathogens, and conditioning solids for further processing.

Advantages of the aerobic process compared to anaerobic digestion are the production of an inoffensive, biologically stable product, lower capital costs, simpler operational control with reductions in volatile solids concentrations slightly less than those achieved in anaerobic digestion, safer operation with no potential for gas explosion and less potential for odor problems, and discharge of a supernatant with a 5-day BOD (BOD_5) concentration typically less than that found in the anaerobic process. In addition, it is less prone to upsets and less susceptible to toxicity.

The primary disadvantage typically attributed to aerobic digestion is the higher power cost associated with oxygen transfer. Recent developments in aerobic digestion (e.g., highly efficient oxygen-transfer equipment and research into operation at elevated temperatures) may reduce this concern. Other disadvantages cited include the process' reduced efficiency during cold weather, its inability to produce a useful byproduct (e.g., methane gas from anaerobic digestion), and the mixed results achieved during mechanical dewatering of aerobically digested solids.

Conventional aerobic digestion has been used for more than 35 years. The autothermal thermophilic process modification has been used at many plants and will be discussed further in this section. This section will also discuss the development of mesophilic aerobic digestion, based on research initiated by Elena Bailey and Glen Daigger (2000) in the early 1990s in response to the new performance requirements of the rules and regulations for beneficial reuse (U.S. EPA, 1992). Results from their research were presented during a series of aerobic digestion workshops from 1997 to 2001 at Water Environment Federation conferences. Each year, the work presented during the workshops was compiled in books, resulting in a five-volume series of books or compact discs that were available to the public. *Aerobic Digestion Workshops* Volumes I through V include comprehensive design guidelines, operational data, and extensive research focused entirely on aerobic digestion (Daigger et al., 1997, 1998, 1999, 2000, 2001). Thay have been used by both engineers and operators as reference manuals.

3.1 Process Applications

Aerobic digestion typically is used in plants with design capacities of less than 19 000 m^3/d (5 mgd), although small plants use other digestion processes. It has been used successfully in extended aeration activated sludge facilities, both with and without primary settling, and in many package treatment facilities. In many extended-aeration facilities, a sufficient SRT is maintained to provide aerobic digestion in the aeration system. Although this approach is somewhat inefficient in terms of oxygen use, the initial cost reduction and overall system simplicity can benefit small systems. Design engineers should note, however, that this approach may not meet state and federal regulations for stabilization.

Aerobic digestion has been used successfully in facilities with capacities up to 1.89×10^5 m^3/d (50 mgd). In these facilities, mixed primary and biological solids are most often handled, and their oxygen requirements are greater than those of waste biological solids alone. Because of the high energy cost for aeration in such facilities, it may be more economical to anaerobically digest primary solids separately while aerobically digesting biological solids.

In cases where other disposal methods are not readily available, screenings, grease, and skimmings are treated in aerobic digesters. These streams should have a low inorganic content and pass through grinders before they are added to the aerobic digestion system. Even with thorough grinding, recombination of stringy material, resulting in clogging or fouling of aeration equipment, is a potential problem. Grease and skimmings are likely to accumulate as digester scum unless special provisions are made to keep this material in suspension (e.g., relatively intense surface mixing).

3.2 Process Theory

Aerobic digestion is based on the biological principle of endogenous respiration. *Endogenous respiration* occurs when the supply of available substrate (food) is depleted and microorganisms begin to consume their own protoplasm to obtain energy for cell-maintenance reactions.

During digestion, cell tissue is oxidized aerobically to carbon dioxide, water, and ammonia or nitrates. Because aerobic oxidation is exothermic, heat is released during the process. Although digestion should theoretically go to completion given an infinite SRT, in actuality only 75 to 80% of cell tissue is oxidized. The remaining 20 to 25% is composed of inert components and organic compounds that are not biodegradable. The material that remains after digestion is complete exists at such a low energy state that it is essentially biologically stable. So, it is suitable for a variety of disposal options.

Aerobic digestion actually involves two steps: direct oxidation of biodegradable matter and subsequent oxidation of microbial cellular material by organisms. These processes are illustrated by the following formulas (U.S. EPA, 1979):

$$\text{Organic matter} + NH_4 + O_2 \xrightarrow{\text{bacteria}} \text{Cellular material} + CO_2 + H_2O \quad (25.27)$$

$$\text{Cellular material} + O_2 \xrightarrow{\text{bacteria}} \text{Digested biosolids} + CO_2 + H_2O + NO_3 \quad (25.28)$$

Equation 25.27 describes the oxidation of organic matter to cellular material, which then is oxidized to digested biosolids. The process represented by Equation 25.28 is typical of endogenous respiration and is the predominant reaction in aerobic digesters.

Because of the need to maintain the process in the endogenous respiration phase, aerobic digestion typically is used to stabilize WAS. Because primary solids contain little cellular material, most of the organic and particulate material in primary solids is an external food source for the active biomass in biological solids. So, longer retention times are required to accommodate the metabolism and cellular growth that must occur before endogenous respiration conditions are achieved.

Using the formula $C_5H_7NO_2$ as representative of a microorganism's cell mass, the stoichiometry of aerobic digestion can be represented by either of the following equations:

$$C_5H_7NO_2 + 5O_2 \rightarrow 5CO_2 + 2H_2O + NH_3 + \text{Energy} \quad (25.29)$$

$$C_5H_7NO_2 + 7O_2 \rightarrow 5CO_2 + 3H_2O + NO_3 + H^+ + \text{Energy} \quad (25.30)$$

Equation 25.29 represents a system designed to inhibit nitrification (because it is oxygen limited); nitrogen appears in the form of ammonia. The stoichiometry of a system in which nitrification occurs is represented by Equation 25.30, where nitrogen appears in the form of nitrates.

Theoretically, about 50% of the alkalinity consumed by nitrification can be recovered by denitrification. If excessive pH depression is a problem (as a result of alkalinity consumption by nitrification), it may be possible to control this problem by periodic denitrification or the addition of lime. Denitrification can be accomplished by periodically turning off the aerators while continuing to mix the digester (if the facility is designed with a draft tube aerator containing an air sparger).

As indicated by Equation 25.30, nitrification during aerobic digestion increases the concentration of hydrogen ions and subsequently decreases pH if the solids have insufficient buffering capacity. As in the activated sludge process, about 7 kg of alkalinity is destroyed per kilogram of ammonia oxidized (7 lb/lb). The pH may drop as low as 5.5 during long aeration times, but aerobic digestion does not seem to be adversely affected.

Equations 25.29 and 25.30 indicate that, theoretically, 1.5 kg of oxygen is required per kilogram of active cell mass (1.5 lb O_2/lb) in the non-nitrifying system, while 2 kg of oxygen per kilogram of active cell mass (2 lb O_2/lb) is required when nitrification occurs. Actual oxygen requirements for aerobic digestion depend on such factors as operating temperature, inclusion of primary solids, and the SRT in the activated sludge system.

3.3 Process Design

3.3.1 *General*

This discussion on the design of aerobic digestion systems is directed toward conventional aerobic systems (i.e., systems operating at temperatures between 20 and 30°C that use air as the oxygen source for biological activity).

Factors that govern the design of aerobic digestion systems include desired reduction in volatile solids, influent quantities and characteristics, process operating temperature, oxygen-transfer and mixing requirements, tank volume/detention time, and method of system operation. However, since U.S. EPA's 40 CFR 503 regulations went into effect, the overriding factor in the design of these systems has been meeting the requirements for vector-attraction and pathogen reduction.

3.3.2 *Reduction in Volatile Solids*

The primary purpose of aerobic digestion is to produce biosolids that are stabilized and amenable to various disposal options. Here, *stabilized* implies that the biological organisms, particularly pathogens, have been reduced to a level at which the use or disposal of the biosolids will not result in a significant adverse environmental impact.

Aerobic digestion can reduce volatile solids by 35 to 50%. Part 503 regulations require that a 38% volatile solids reduction be met to attain the vector-attraction reduction requirements. The regulations governing land application of biosolids classify aerobic digestion as a process to significantly reduce pathogens (PSRP) that can produce biosolids suitable for land application if the specified time/temperature requirements are met.

Researchers have, however, suggested that volatile solids reduction may not be a valid indication of stabilization (Hartman et al., 1979; Matsch and Drnevich, 1977). Other parameters [e.g., the residual rate of oxygen demand, pathogen levels, odor-producing potential, specific oxygen uptake rate (SOUR), or oxidation/reduction potential] may be more indicative of stabilized, aerobically digested biosolids.

3.3.3 *Feed Quantities/Characteristics*

Aerobic digestion typically is used to stabilize biological solids (e.g., WAS). The process has been used to stabilize primary and biological solids mixtures; however, the associated retention times and oxygenation requirements are substantially increased to obtain a stabilization level equal to that achieved with biological solids. Because the mechanism of aerobic digestion is similar to that of the activated sludge process, the same concerns regarding variations in influent characteristics and levels of biologically toxic materials apply, although a dampening effect will occur as a result of upstream treatment processes.

The heavy metals accumulated in activated solids via precipitation and adsorption (which can occur when pH is greater than 7) can from resolubilize under low-pH conditions in the digester, resulting in toxicity.

The influent concentration is important in the design and operation of an aerobic digestion process. While thickening solids will increase the digester's oxygen requirements, thickening will result in longer SRTs, smaller digester volume requirements, easier process control (less decanting), and therefore more volatile solids destruction.

3.3.4 Operating Temperature

The aerobic digester's operating temperature is a critical parameter. A frequently cited disadvantage of the aerobic process is the variation in process efficiency that results from changes in operating temperature. Changes in operating temperatures are closely related to ambient temperatures because most aerobic digestion systems use open tanks.

Aerobic digestion systems typically are operated in the mesophilic zone of bacterial action (between 10 and 40°C). There has been more research into the operation of aerobic systems in other temperature zones: the cryophilic zone (less than 10°C) and the thermophilic zone (more than 40°C).

Because aerobic digestion is a biological process, the effects of temperature can be estimated by the following equation:

$$(K_d)T = (K_d)_{20^\circ C}\, q^{T-20} \qquad (25.31)$$

Where

K_d = reaction rate constant (time);
q = temperature coefficient; and
T = temperature (°C).

The reaction rate constant indicates the destruction rate of volatile solids during digestion. The reaction rate constant typically increases when the system's temperature increases, implying an increase in the digestion rate.

Temperature coefficients ranging from 1.02 to 1.10 have been reported. The average temperature coefficient is 1.05.

The rate of biological processes in the mesophilic temperature range typically increases with temperature (see Figure 25.61). Above a critical temperature, the process

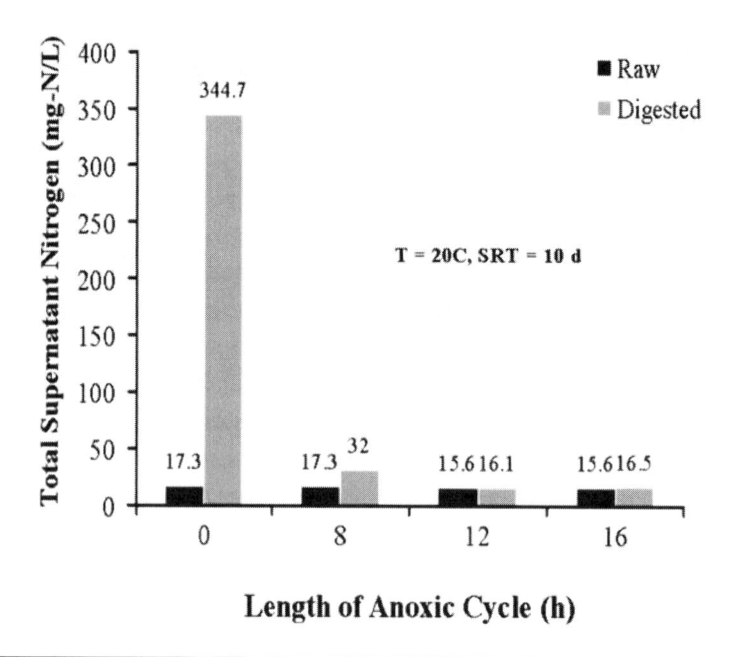

FIGURE 25.61 Effect of anoxic-cycle duration on total nitrogen concentration in filtrate (Al-Ghusain et al., 2004).

will be inhibited. One study showed a maximum volatile solids destruction rate at 30°C, with a rate reduction at higher temperatures (Hartman et al., 1979). This is in contrast with the data in Figure 25.62 and indicates the importance of obtaining rate data applicable to the system being designed.

3.3.5 Oxygen-Transfer and Mixing Requirements

The biological reaction that occurs during aerobic digestion requires oxygen for the respiration of cellular material in activated sludge and, in the case of mixtures with primary solids, the oxygen needed to convert organic matter to cellular material. In addition, proper system operations require adequate mixing of the contents to ensure proper contact of oxygen, cellular material, and organic matter (food source). Because introducing oxygen to maintain the biological process typically mixes the contents in the process, these parameters are interrelated.

In aerobic digestion systems that strictly treat biological solids in the range of 1 to 2%, the need for adequate mixing typically will govern the capacity of the oxygenation

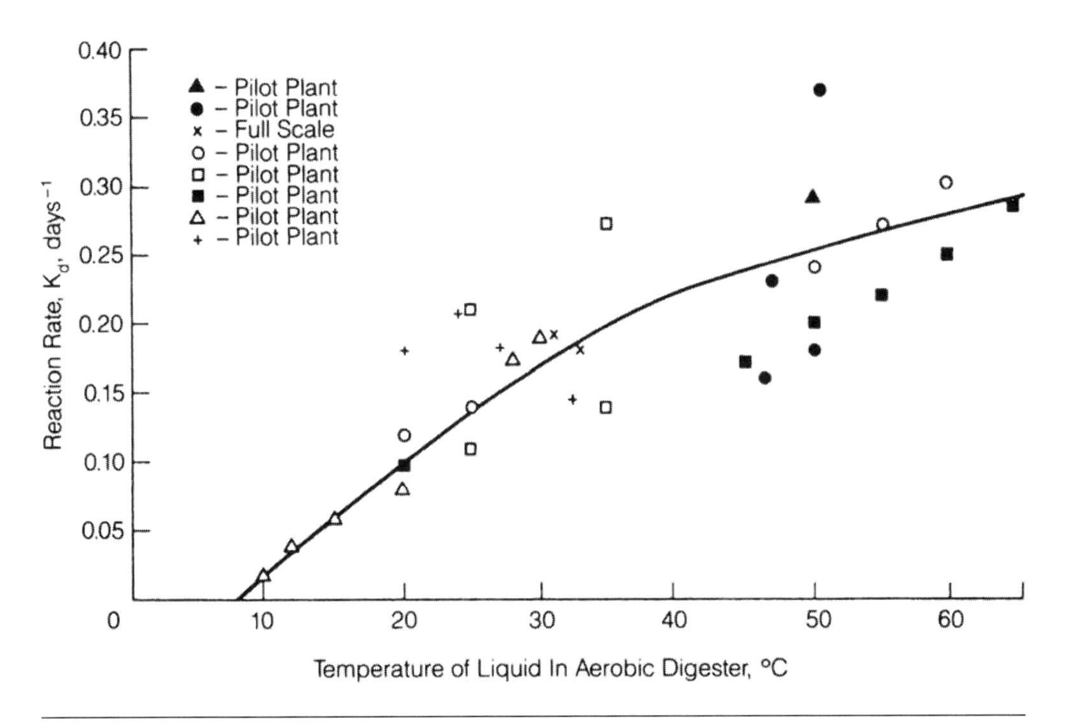

FIGURE 25.62 Experimentally determined reaction rate (K_d) versus aerobic digester liquid temperature. The value of K_d depends on solids characteristics and digester operating conditions (e.g., pH, TSS, and oxygen level) (U.S. EPA, 1978).

equipment. However, the use of designs with thicker feed concentration (3 to 6%), process air requirements will eventually govern design parameters.

Systems treating primary and biological mixtures require more oxygen for the biological oxidation process, and in most cases, this requirement will govern mixing equipment size.

Actual mixing requirements typically range from 10 to 100 W/m^3 (0.5 to 4.0 hp/1 000 cu ft) of digester volume; however, this value will vary depending on tank geometry and type of mixing device. Design engineers should consult an experienced equipment manufacturer to determine actual mixing requirements.

Equations 25.29 and 25.30 indicate that aerobic digestion theoretically requires 1.5 to 2.0 parts of oxygen per part of applied organic cell mass, depending on whether nitrification is inhibited or allowed to proceed. Design experience has shown that 2.0 parts of oxygen per part of organic cell mass destroyed is a standard minimum value for biological stabilization. Including primary solids in the digestion process requires another 1.6 to 1.9 parts of oxygen per part of volatile solids destroyed to convert the organic matter to cell tissue and satisfy the endogenous demand of the resulting cell mass. Some U.S. state regulators (e.g., Wisconsin) stipulate in their design standards for wastewater treatment plants that the aeration system should account for another 0.91 kg (2 lb) of oxygen per pound of BOD_5 applied by primary solids in the aerobic digestion design.

Aerobic digestion is a continuation of the suspended-growth or fixed-film biological treatment process; proper sizing of oxygen-transfer equipment for aerobic digestion systems must reflect the reduction in oxygen demand that occurs in the biological process. In addition, oxygen requirements for aerobic digestion are increased if the secondary system is nitrifying. Estimations of the oxygen required because of nitrification must be reflected not only in the raw wastewater's ammonium concentration, but in the conversion of organic nitrogen to ammonium in the secondary process and in primary solids.

The oxygen requirements for an aerobic digester treating biological solids can be approximated from the following equation:

$$\mathrm{Ri} = K\,(1.67\,\mathrm{Si} - O_a) \qquad (25.32)$$

Where

Ri = actual aerobic digester oxygen requirements [kg/d (lb/d)];

Si = raw wastewater BOD5 load equivalent to the waste solids under oxidation [kg/d (lb/d)];

O_a = oxygen consumed in main aeration basin [kg/d (lb/d)]; and

K = constant equal to 1.0 when main aeration does not nitrify and 1.24 when main aeration does nitrify.

Oxygen requirements for aerobic digestion systems typically represent airflow rates of 0.25 to 0.33 $L/m^3 \cdot s$ (15 to 20 cu ft/min/1 000 cu ft) for WAS. Airflow rates increase to a range of 0.40 to 0.50 $L/m^3 \cdot s$ (25 to 30 cu ft/min/1 000 cu ft) for a mixture of primary solids and WAS (Benefield and Randall, 1980). The typical solids content in these parameters will be between 1% and 2% and will possibly require more oxygen for increased solids content. Dissolved oxygen levels in the aerobic digester typically are maintained at about 2 mg/L; however, this level may be reduced if the oxygen uptake rate is less than 20 mg/L·h.

After the requirements for adequate mixing and oxygen transfer have been separately computed, the larger of the two requirements will govern overall system design. If the mixing requirement exceeds the oxygen-transfer requirement, design engineers should consider providing supplemental mechanical mixing rather than overdesigning the oxygen-transfer system. The increased capital cost of supplemental mechanical mixers must be balanced against the power and maintenance costs of more aeration to determine the optimum configuration.

3.3.6 Tank Volume and Detention Time Requirements

The required volume of an aerobic digester typically is governed by the detention time needed to achieve a desired reduction in volatile solids. In the past, the detention time required to reduce volatile solids by 40 to 45% typically ranged from 10 to 12 days at an operating temperature of about 20°C (Metcalf and Eddy Inc., 1991). Part 503 regulations discuss a 38% volatile solids reduction that stabilization processes must achieve, but more importantly, the new designs must focus on pathogen reduction as the controlling factor. Although volatile solids destruction will continue as detention times increase, the oxidation rate significantly decreases, and continuing digestion past the typical detention time is not economical. Full-scale aerobic digestion studies have shown that a total aeration time (including time in the extended aeration process) of 35 to 50 days was required to consistently meet Part 503's vector-attraction reduction requirement (503.33) for the SOUR requirement of less than 1.5 mg oxygen/g·h volatile solids at wastewater treatment plants located at higher altitudes with colder climates (Maxwell et al., 1992). The total required aeration time depended significantly on operating temperature and WAS biodegradability. It has been reported that aerobic digesters with untypically long detention times produce biosolids that are significantly more difficult to dewater (U.S. EPA, 1979). However, before the Part 503 regulations, many state agencies adopted U.S. EPA's standards for aerobic digestion, which required a residence time of 60 days at 15°C and 40 days at 20°C and focused on pathogen reduction rather than on volatile solids reduction.

The reduction in biodegradable solids during digestion typically is described by a first-order biochemical reaction at constant volume conditions, similar to the following:

$$dM/dt = K_d M \qquad (25.33)$$

Where

dM/dt = rate of change of biodegradable volatile solids per unit of time (mass/time);

K_d = reaction rate constant (time^{-1}); and

M = concentration of biodegradable volatile solids remaining at time t (mass/volume).

The time factor in Equation 25.33 represents the SRT in the aerobic digester. Such factors as the method of digester operation, operating temperature, and the SRT of the activated sludge system may make the time factor equal to or greater than the system's theoretical HRT. Using the biodegradable portion of volatile solids in the equation recognizes that about 20 to 35% of WAS from wastewater treatment plants with primary treatment systems is nonbiodegradable. The percentage of nonbiodegradable volatile solids for WAS from contact stabilization processes (no primary tanks) ranges from 25 to 35%.

The reaction rate constant is a function of the type of residuals being digested, operating temperature, SRT of the system, and solids concentration in the digestion system. Figure 25.61 depicts a graph of the change in reaction rate constant versus increasing operating temperature. The results of one study with WAS at a temperature of 20°C indicated that the reaction rate constant declined as the digester's suspended solids level increased (Reynolds, 1973).

The product of temperature and SRT appears to correlate with the percentage of volatile solids destruction that can be achieved during digestion. The selection of a desired percentage of reduction in volatile solids, coupled with an assumed operating temperature for the system, can be used to estimate the required digester SRT (see Figure 25.63).

As the temperature and residence time in the digester increase, the fractional increase in volatile solids reduction diminishes. In the United States, the final regulations (40 CFR 257 and 503) require that a aerobic digestion reduce volatile solids by 38% to meet vector-attraction reduction objectives. This reduction may be difficult to obtain in an aerobic digester if significant biodegradable solids reduction has already occurred in a secondary treatment system with a long SRT. During the 1980s, design engineers sometimes were allowed variances from state regulators, in which credit was given for

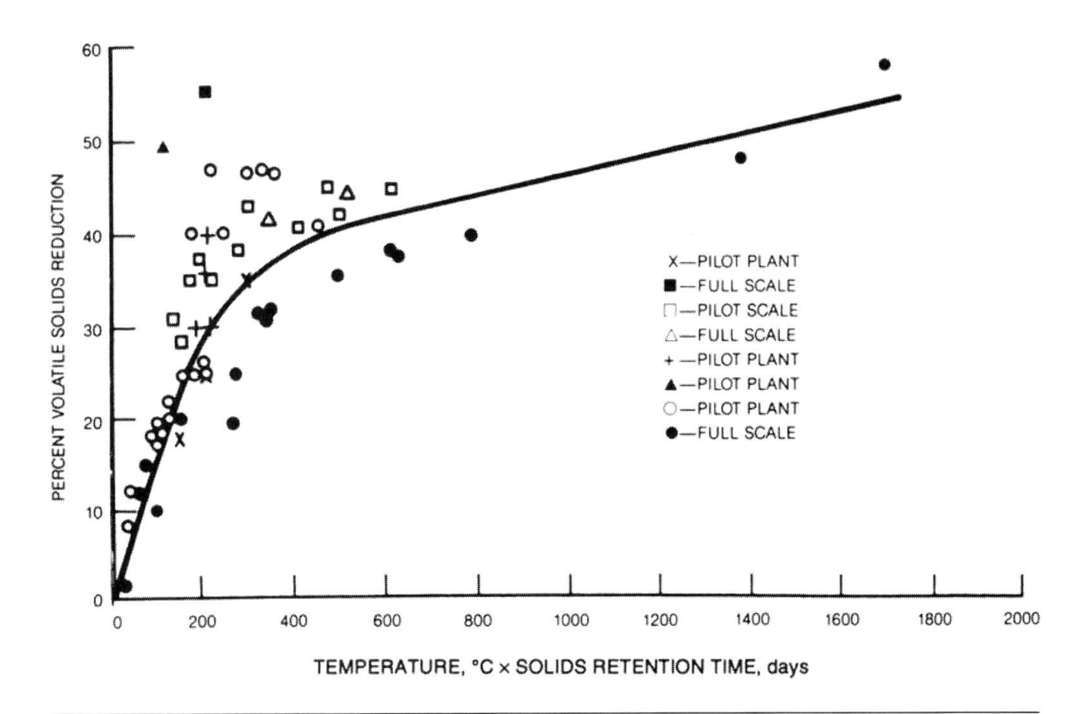

FIGURE 25.63 Volatile solids reduction as a function of digester liquid temperature and solids retention time.

the volatile solids reduction that occurred in extended aeration facilities. Therefore, the overall volatile solids reduction took into account the waste flow through both the activated sludge and solids-handling processes. This variance allowed municipalities to meet 38% volatile solids reduction through their facilities.

With the promulgation of Part 503, however, state regulators no longer allow design engineers to count the reduction of volatile solids through the secondary treatment process as part of the 38% reduction that must occur during aerobic digestion or other stabilization processes. If the aerobic digester does not achieve 38% reduction, the regulations have stated that vector-attraction reduction can be demonstrated by aerobically digesting a portion of the previously digested material that has a solids concentration of 2% or less in the laboratory in a bench-scale unit for 30 more days at 20°C. If the sample's volatile solids concentration is reduced by less than 15%, vector-attraction reduction is achieved. In addition, based on Part 503 regulations, design engineers and state regulators are using the SOUR criteria of less than 1.5 mg of oxygen per g/h total solids at a temperature of 20°C for digested solids instead of the 38% volatile solids reduction criterion.

One method of determining the volume of a continuously operating aerobic digester is to apply the following formula:

$$V = Q_i(X_i + YS_i)/X\,(K_d P_v + 1/\mathrm{SRT}) \tag{25.34}$$

Where

V = volume of the aerobic digester [L (cu ft)];
Q_i = influent average flowrate [L/d (cu ft/d)];
X_i = influent suspended solids (mg/L);
Y = portion of the influent BOD consisting of raw primary solids (%);
S_i = influent digester BOD_5 (mg/L);
X = digester suspended solids (mg/L);
K_d = reaction rate constant (d&1);
P_v = volatile fraction of digester suspended solids (%); and
SRT = solids retention time (days).

The term YS_i in Equation 25.34 can be disregarded if no primary solids are included in the load to the aerobic digester. Equation 25.34 should not be used to compute digester volumes in systems where significant nitrification will occur.

Research by Benefield and Randall (1980) has resulted in the development of equations for determining required digester detention times. These proposed equations result from an analysis of the kinetics associated with the digestion process, and the understanding that a portion of the volatile solids in the process are nonbiodegradable and a portion of the nonvolatile solids are solubilized from an analysis of the microbial cells contained within the solids. The basic equation is as follows:

$$t_d = (X_i - X_e)/(K_d)(D)(X_{\mathrm{oad}})(X_i) \tag{25.35}$$

Where

t_d = digester detention time (days);
X_i = TSS concentration in influent (mg/L);
X_e = TSS concentration in effluent (mg/L);
K_d = reaction rate constant for the biodegradable portion of the active biomass (d^{-1});
D = biodegradable active biomass in the influent that appears in the effluent (%); and
X_{oad} = percentage of active biomass that is biodegradable in the influent.

In instances in which there is a mixture of primary solids and WAS the inclusion of the factors describing the primary solids component (refer to Benefield and Randall, 1980, or Grady et al., 1999, for more detail on this less common aerobic digestion design).

Equation 25.35 supports the assumption that for equivalent solids reduction and constant solids loading, SRT must be increased as the active fraction of the influent biomass decreases. Actual operating experience indicates that for systems with a low fraction of active biomass in the feed, which is typical of extended aeration systems, this trend does not hold. For these systems, the detention times computed by Equation 25.35 may be reduced in proportion to the decrease in the active fraction of the biomass.

Another rendition of the relationships shown in Equation 25.35 is described and exemplified by Grady et al. (1999) while expressing it in the Activated Sludge Model 1 (ASM1) terminology.

3.3.7 Summary of Design Parameters

The promulgation of 40 CFR 503 has substantially changed the typical design parameters used for the standard aerobic digestion process. For example, on early aerobic digester designs that were receiving only WAS, design engineers would use a residence time of 10 to 15 days. Then, U.S. EPA focused on pathogen reduction and set forth regulations in which the residence time had to be 40 days at a temperature of 20°C and 60 days if the wastewater temperature was 15°C. Part 503 require retention times only if design engineers want to meet Class B pathogen-reduction criteria without the need to periodically monitor for fecal coliform. If design engineers use aerobic digestion for stabilization and does not meet the regulations that stipulate 40 days at 20°C or 60 days at 15°C, then they must monitor fecal coliform. In both cases, design engineers must demonstrate that vector-attraction reduction has been achieved by a 38% reduction in volatile solids, an oxygen uptake rate of less than 1.5 mg of oxygen per gram of total solids per hour, or a volatile solids reduction of less than 15% on further testing after 30 days of more digestion. If design engineers need the 40 days at 20°C or 60 days at 15°C (or a linear interpolation between them), then they do not have to monitor fecal coliform, but must show vector-attraction reduction.

An important deviation from this rule is obtained in the design of two-stage or batch operation, in which we have a 30% reduction in the time required to obtained the pathogen and vector attraction reduction specified by the U.S. EPA regulations (U.S. EPA, 2003). This credit results in that the time required is now reduced from 40 days at 20°C (68°F) to 28 days at 20°C (68°F), and from 60 days at 15°C (59°F) to 42 days at 15°C (59°F). These reduced times are also more than sufficient to achieve adequate vector attraction reduction.

3.3.7.1 Aeration and Mixing Equipment

Several devices (e.g., diffused air, mechanical surface aeration, mechanical submerged turbines, jet aeration, and combined systems) have been used successfully to accommodate an aerobic digester's oxygenation and mixing requirements. The design of diffused-air systems for aerobic digesters is similar to the design of those used in standard activated sludge systems. Diffusers typically are located near the tank bottom. They also can be placed along one side of the tank to produce a spiral or cross-roll pattern, or they may be installed as a floor-mounted grid system. Airflow rates of 0.33 to 0.67 $L/m^3{\cdot}s$ (20 to 40 cu ft/min/1 000 cu ft) typically are required to ensure that mixing is adequate. The airflow rates needed to meet oxygen transfer requirements depend on digester loading.

Both fine-bubble and coarse-bubble diffusers have been used in aerobic digesters. Diffuser plugging is a potential problem in aerobic digesters, especially in those whose operation includes periodic settling and supernatant removal. While the air is turned off, solids can enter the air piping and adhere to the inner walls of piping or diffusers. Nonclog and porous media devices are more resistant to this type of plugging than large-bubble, orifice diffusers. However, surface fouling of porous diffusers can occur.

Diffused-air systems provide the following advantages: oxygen transfer is controlled by varying the air-supply rate; the introduction of compressed air to the digester typically adds heat to the system, which minimizes temperature loss during cold weather; and overall heat loss from the system is minimized because of the relatively small degree of surface turbulence. Advantages of diffused-air systems may be outweighed by clogging problems that can occur in aerobic digesters. If a diffused-air system is to be used, it is imperative that provisions be included for easy removal of the diffuser device and air drop pipes for cleaning.

An alternative to the floor-covering diffuser systems is a full range of high shear, nonclog aeration equipment designed specifically for high solids concentrations (4 to 8% solids). This aeration system is combined with an adjustable, above-water orifice to allow for varying the air provided to meet demand, and nonclog diffusers ensure that plugging will not occur during anoxic operation. The shear tube and draft tubes provide mixing and shearing to transfer oxygen and achieve volatile reduction with high solids (see Figures 25.64 and 25.65). The limitation of this system is that it works better when the liquid is more than 6.1 m (20 ft) deep (Daigger et al., 1997).

In summary, single-drop aeration with shear tubes or draft tubes systems

- Are specifically designed for higher solids concentrations (4 to 8% suspended solids),
- Tend to add heat to the digester,

FIGURE 25.64 A typical draft tube system (in this case used to treat a mixture of primary and secondary waste at Paris, Illinois). The picture was taken after conversion from anaerobic to a prethickened, two-stage-in-series, aerobic digestion (Daigger et al., 1997).

- Do not require maintenance because they are nonclog systems, and
- Require tanks with more than 6.1 m (20 ft) of liquid depth.

Mechanical surface aerators typically are floating, pontoon-mounted devices of either low- or high-speed design. Low-speed aerators are more often used in aerobic digesters. Compared to diffused-air systems, mechanical surface-aeration systems typically are simpler and easier to maintain, and less prone to fouling. Disadvantages typically attributed to surface aeration include the lack of control of the oxygenation rate, performance deterioration if excessive foam is present, more potential for foaming because of high surface turbulence, increased heat loss from the system, and the potential for ice accumulation during winter in cold climates as a result of the device's splashing.

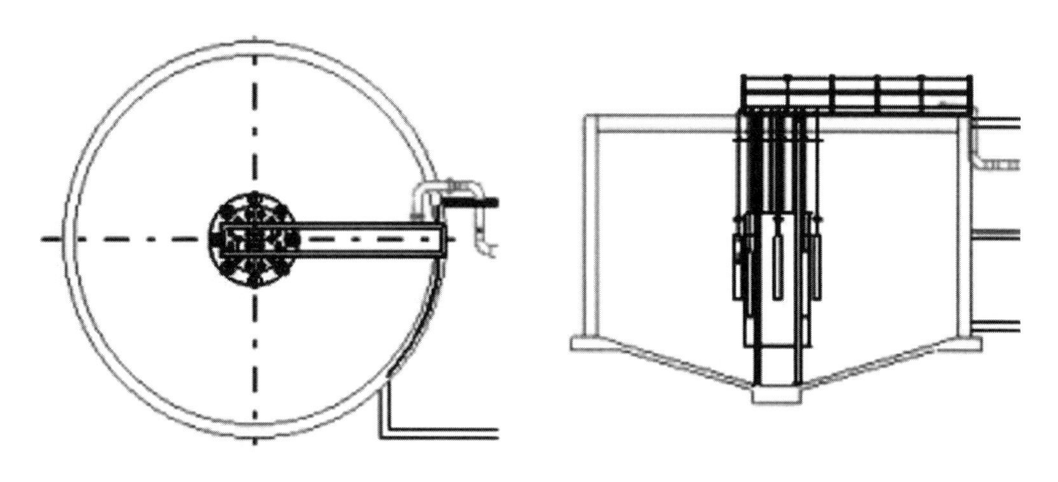

FIGURE 25.65 Plan and section views of a digester basin at Paris, Illinois. Basin is 14 m diam × 9.4 m deep (45 ft diam × 31 ft deep) (Daigger et al., 1997).

Mechanical submergence turbine aerators (and other combined mechanical mixing and diffused-air systems) provide several advantages and eliminate some disadvantages of the diffused-air and surface aerator devices. Oxygenation rates can be controlled by varying the airflow rate to the submerged impeller. Because impellers are submerged, they are not as sensitive to foaming conditions as surface aerators and avoid the ice and heat dissipation problems associated with surface aerators.

Additionally, the submerged unit can be operated as a mixer only, thereby promoting denitrification.

Jet aeration devices provide many of the advantages of submerged turbines. These devices typically are more easily installed and have a somewhat higher overall oxygen-transfer efficiency than submerged turbines. Problems with device plugging have occurred in the past when liquid flow paths were not large enough to pass the stringy solids typically found in aerobic digesters. Their use may promote floc shear and subsequent dewatering difficulties. (For more information on the relative merits and design features of aeration systems from the perspective of their more common use for oxygenating activated sludge systems, see Chapter 14.)

3.3.7.2 Piping Arrangements

Specific piping requirements for aerobic digesters include provisions for feeding solids, decanting supernatant, withdrawing digested solids, and supplying air for aeration, when applicable. In aerobic digestion systems designed with settling basins, digested solids and supernatant are removed in the settling basin.

Solids also are returned to the aerobic digester from the settling basin to maintain the required SRT. Consideration should be given to a separate air supply to the aerobic digester, especially if the liquid level varies because of supernatant decanting.

When the liquid level in the digester is lower than that in the aeration tank, the digester will "rob" the aeration tank of air unless the air supply is separate or there is pressure compensation.

Batch-operated aerobic digesters can be designed to remove solids and supernatant via pumping or gravity. If a fixed supernatant-removal system is used, enough flexibility should be provided to allow supernatant to be removed over a relatively wide range of depths. At a minimum, two supernatant withdrawal lines located at different depths are advisable. Alternatively, floating decanter devices can be used for effective supernatant removal.

Digested solids typically are withdrawn from the low point of each tank. Only one solids feed inlet is necessary if the digester is designed with adequate mixing.

The digester should be fed often enough to avoid localized shock loading. An emergency digester overflow should be provided if the potential for overfilling exists.

Strategically located hose–gate connections flush out solids lines with plant effluent. Water (effluent) sprays provided for foam control are seldom used because of the quantities of fluid that would be added to the system. Drains or sumps for dewatering and cleaning basins should be provided.

3.3.7.3 Instrumentation and Controls

Aerobic digestion typically is controlled manually. The operating variables that currently lend themselves to automatic control are dissolved oxygen and tank level. The dissolved oxygen signal can be used to control the aeration system so it maintains an optimum dissolved oxygen level (typically between 1 and 2 mg/L). This can conserve energy. Low- (and occasionally high-) dissolved oxygen conditions can trigger an alarm to allow operators to take corrective action. However, except for inadvertent digester overloads, dissolved oxygen changes in aerobic digesters typically are minimal, and maintaining dissolved oxygen monitors may be time-consuming.

A tank-level signal, useful in preventing overfilling, can be used for on–off control of digester feed pumps. Intermittent feeding of primary solids to digesters via timer-controlled pumps has been used. This controlled feeding technique also has been used for waste biological solids feeding. Automatically controlled feeding is successful when care is taken to establish the proper time as plant-operating conditions change. With manually controlled feed systems, a tank level signal can be used to warn of a high-level condition.

Withdrawal of solids from batch-operated aerobic digesters typically is manually controlled and done intermittently. Manual withdrawal allows rational reaction to variable solids production, solids concentration, digestion rates, and capacity of subsequent processing.

3.3.7.4 Considerations of Equipment Selection

Flexibility and maintainability are key criteria when selecting aerobic digester equipment. The major equipment items of concern are piping and aeration and mixing equipment.

Piping systems require valves that resist clogging (e.g., eccentric plug valves). All feed and withdrawal piping (e.g., feed solids, digested solids, and supernatant lines) should have provisions (e.g., cleanable sight glasses or flow meters) for confirming that liquid or solids are flowing through the line during operations. Flow meters on some positive-displacement pumps will not be effective for this purpose, however, because pulsations can give the impression of positive flow when no net solids movement is occurring.

The aeration and mixing system(s) should be designed to facilitate maintenance. Access to surface aerators or mixers also should be provided. Swing-arm (knee-joint) or lift-out diffuser assemblies simplify the maintenance and cleaning of diffusers.

Consideration also should be given to multiple tanks, so one tank can be completely drained for maintenance without interrupting the process.

3.3.7.5 Design for Safety

Although subject to the safety hazards typically associated with mechanical and electrical equipment, aerobic digestion does not involve the explosive and toxic gases generated in anaerobic digestion. Safety considerations for aerobic digesters are similar to those for activated sludge basins. For example, placing life preservers with safety lines at intervals around the digesters can prevent drowning accidents. Adequate lighting around the tanks allow for safe nighttime O&M. Nonslip, corrosion-resistant grating should be used for access walkways. (For more information on safety issues in wastewater treatment plants, see Chapter 6.).

3.3.7.6 Design for Operability

All systems, including aerobic digestion, should be designed with operability in mind. Design engineers should consider the following operability issues:

- Aeration system selection for ease of maintenance (periodic diffuser cleaning);
- Location, number, and type of monitoring instruments to enhance control capability;
- Location, number, and type of supernatant-withdrawal devices;
- Aboveground or belowground installation of digestion reactor (ease of temperature control versus accessibility);
- Ability to mix and aerate the system independently; and
- Access to the tank and other equipment for maintenance.

3.4 Process Description

3.4.1 *Conventional (Mesophilic) Aerobic Digestion*

Aerobic digestion may be used to treat WAS, mixtures of WAS or trickling filter solids and primary solids, waste solids from extended aeration plants, or waste solids from membrane bioreactors (MBRs). Aerobic digestion treats solids that are mostly a result of the growth of biological mass during wastewater treatment. Aerobically digested biosolids are less likely to generate odors and have fewer bacteriological hazards than unstabilized solids.

3.4.1.1 Process Design

The design of conventional aerobic digestion facilities is based on the principles described in Section 3.3.

3.4.1.2 Process Performance and Operation

Class B biosolids are biosolids in which the pathogens levels are unlikely to pose a threat to public health and the environment under specific use conditions (U.S. EPA, 2003). Class B biosolids cannot be sold or given away in bags or other containers or applied on lawns or home gardens. They typically are land-applied or landfilled.

Conventional aerobic digestion typically produces Class B biosolids. The Class B biosolids criteria that a conventional mesophilic aerobic digestion system typically is designed to meet are as follows.

A system with one digester could be designed to meet the following criteria.

1. Meet one of the following pathogen-reduction requirements:
   - 60-day SRT at 15°C or 40-day SRT at 20°C, or
   - Fecal coliform density of less than 2 mil. most probable number (MPN)/g total dry solids.
2. Meet one of the following vector-attraction reduction requirements:
   - At least 38% VSR during biosolids treatment, or
   - A SOUR of less than 1.5 mg/g·h of total solids at 20°C (68°F).
3. Solids also could meet vector attraction reduction requirements if they had less than 15% more VSR after 30 days of further batch digestion at 20°C (68°F).

A system with multiple digesters could be designed to meet the following criteria (U.S. EPA, 1999a):

1. Meet both pathogen-reduction requirements:
   - Fecal coliform density of less than 2 mil. MPN/g total dry solids, and
   - 42-day SRT at 15°C or 28-day SRT at 20°C. In this case, because regulators must approve the process as a PSRP-equivalent alternative, plant operators should

demonstrate experimentally that the resulting biosolids both contain low enough levels of microbes and meet one of the vector attraction reduction requirements listed above.

2. Meet one of the following vector-attraction reduction requirements:
   - At least 38% VSR during biosolids treatment, or
   - A SOUR of less than 1.5 mg/g·h of total solids at 20°C (68°F).

The important factors when controlling aerobic digestion operations are similar to those for other aerobic biological processes (see Table 25.28) (Stege and Bailey, 2003). Operators should monitor the primary process indicators (e.g., temperature, pH, dissolved oxygen, odor, and settling characteristics, if applicable) daily. Monitoring helps control process performance and serves as a basis for future improvements. The secondary indicators (e.g., ammonia, nitrate, nitrite, phosphorus, alkalinity, SRT, and SOUR) are useful in monitoring long-term performance and for troubleshooting problems associated with the primary indicators. While monitoring and controlling these parameters is important, the degree of control that can be exercised on each parameter varies.

Analysis frequency should be increased during startup and whenever large changes are made to operating conditions (e.g., solids flowrate, solids source, change in polymer, large increase or decrease in feedstock's solids concentration, or large increase or decrease in feedstock's temperature).

TABLE 25.28 Monitoring parameters for aerobic digestion performance (adapted from WEF, 2007; primary source is Stege and Bailey, 2003).

		Operating range		
Monitoring parameter*	**Frequency**	**Minimum**	**Nominal**	**Maximum**
Temperature (°C)	Daily	15	20	37
pH	Daily	6.0	7.0	7.6
Dissolved oxygen (mg/L)	Daily	0.1	0.4 to 0.8	2.0
Alkalinity (mg/L as $CaCO_3$)	Weekly	100	>500	—
Ammonia-nitrogen (mg/L)	Weekly	—	<20	40
Nitrate (mg/L)	Weekly	—	<20	—
Nitrite (mg/L)	As required	—	<10	—
SOUR (mg oxygen/h/g of total solids)	As required	—	<1.5	—
Phosphorus (mg/L)	As required		<5	

* $CaCO_3$ = calcium carbonate and SOUR = specific oxygen uptake rate.

3.4.2 *Autothermal Thermophilic Aerobic Digestion*

Autothermal thermophilic aerobic digestion (ATAD) uses mixing energy to achieve operating temperatures of 40 to 80°C (see Figure 25.66). It relies on sufficient levels of oxygen, volatile solids, and mixing to allow aerobic microorganisms to degrade organics into carbon dioxide, water, and nitrogen in exothermic reactions. If sufficient insulation, SRT, and adequate solids concentrations are provided, the process can be controlled at thermophilic temperatures to achieve greater than 38% VSR and meet Part 503's Class A pathogen requirements.

Autothermal thermophilic aerobic digestion has been studied since the 1960s. Much of the developmental work was done by Popel (1971a, 1971b), who, along with

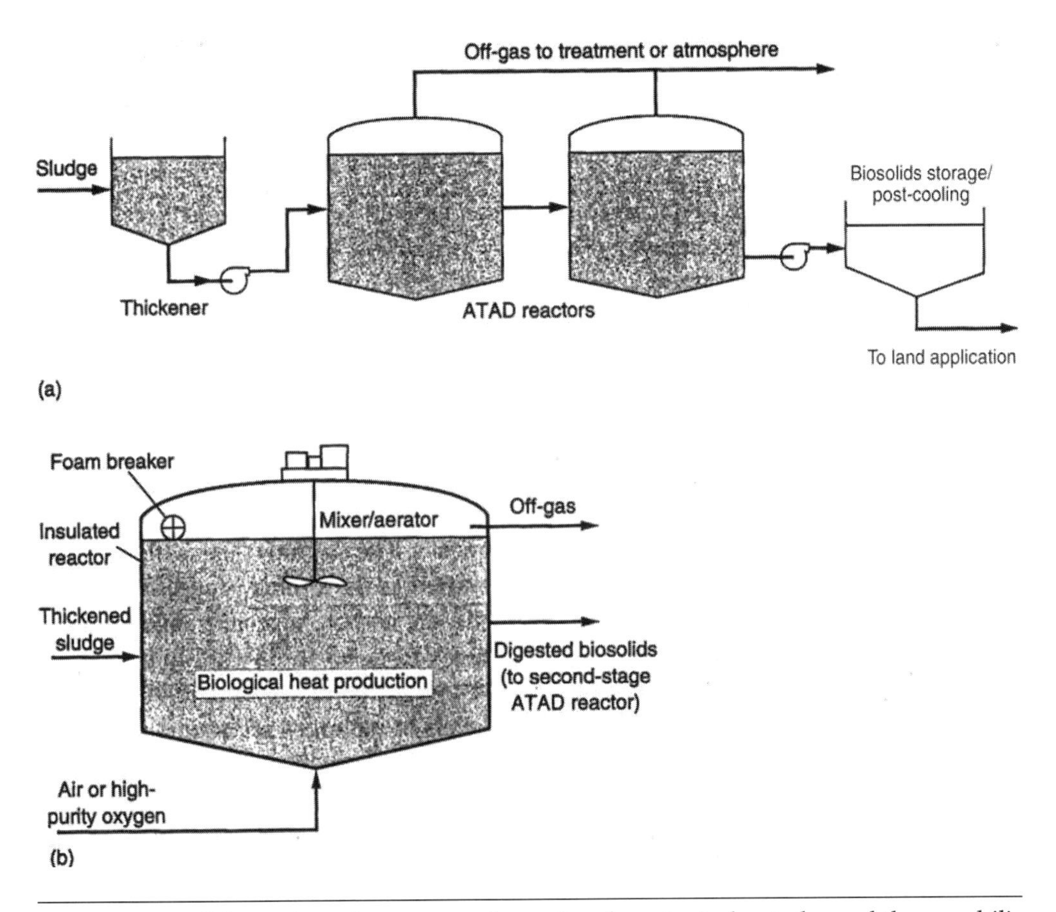

FIGURE 25.66 Schematic and reactor configuration for a typical autothermal thermophilic aerobic digestion system (from Metcalf & Eddy, *Wastewater Engineering: Treatment and Reuse*, 4th ed. Copyright © 2003, The McGraw-Hill Companies, New York, N.Y., with permission).

his coworkers, studied animal manure and wastewater residuals in Germany. They developed an aspirating aeration device that was key to the process success. Research in the United States was done by Matsch and Drnevich (1977) using pure oxygen and by Jewell and Kabrick (1980) using air with submersible aeration devices.

[For more information on ATAD, see *Assessment of Innovative Technologies for Wastewater Treatment: Autothermal Aerobic Digestion (ATAD)* by Stensel and Coleman (2000). This report provides detailed information on the history, design, O&M, and performance of ATAD systems. It is an updated version of U.S. EPA's *Autothermal Thermophilic Aerobic Digestion of Municipal Wastewater Solids* (U.S. EPA, 1990).]

3.4.2.1 Advantages and Disadvantages

The major advantages of ATAD are as follows:

- Shorter retention times (smaller volume required to achieve a given suspended solids reduction) to about 5 to 6 days to achieve 30 to 50% VSR, similar to conventional aerobic digestion;
- Greater reduction of bacteria and viruses compared with mesophilic anaerobic digestion (Metcalf and Eddy, 2003); and
- When the reactors are well mixed and maintained at 55°C and above, pathogenic viruses, bacteria, viable helmith ova, and other parasites can be reduced to below detectable levels, thus meeting the pathogen reduction requirements for Class A biosolids.

The major disadvantages of ATAD are as follows:

- Poor dewatering characteristics of ATAD biosolids (Daigger et al., 1998);
- Objectionable odors;
- Lack of nitrification and/or denitrification (Daigger et al., 1998);
- High capital cost; and
- Foam control is required to ensure effective oxygen transfer (Metcalf and Eddy, 2003).

This process is relatively stable, self-regulates its temperature (because of the heat released during oxidation of the biomass), recovers quickly from minor process upsets, and is not greatly affected by ambient air temperatures. Autothermal thermophilic aerobic digestion is gaining popularity because it can produce Class A biosolids. In 2003, there were 35 ATAD systems operating in North America. More than 40 plants are operating in Europe (Stensel and Coleman, 2000).

3.4.2.2 Process Design

The following design parameters were adapted, in part, from Stensel and Coleman (2000).

3.4.2.2.1 Nitrification is Inhibited

Because of the high operating temperatures involved, ATAD inhibits nitrification and so the system's pH is typically between 8 and 9. Aerobic destruction of volatile solids occurs as described by Equation 25.27, without the subsequent reactions described in Equation 25.28. Also, most ATAD systems may be operating under microaerobic conditions, in which oxygen demand exceeds oxygen supply (Stensel and Coleman, 2000) Ammonia is released as a result of digestion, and the ammonia-nitrogen produced will be present in both gas and solution at concentrations of several hundred milligrams per liter.

3.4.2.2.2 Effect of Liquid Sidestreams that Contain Ammonia-nitrogen

Most of the ammonia-nitrogen will be recycled to the wastewater treatment train via sidestreams from the odor-control and dewatering systems. If effluent nitrogen and phosphorus limits are low, then these recycle streams could hurt plant performance. If the treatment plant will include both ATAD and a biological nutrient removal process, then the liquid sidestreams from both the odor-control and dewatering systems need to be accounted for or else treated separately. (For more information on sidestream treatment, see Chapter 17.)

3.4.2.2.3 Foam

Autothermal thermophilic aerobic digestion generates a substantial amount of foam as cellular proteins, lipids, and FOG are broken down and released into solution. The foam contains high concentrations of biologically active solids, which provide insulation. It is important to manage foam effectively (via foam cutters or spray systems) to ensure effective oxygen transfer and enhanced biological activity. A freeboard of 0.5 to 1 m (1.65 to 3.3 ft) is recommended (Stensel and Coleman, 2000).

3.4.2.2.4 Equipment Design

Table 25.29 shows recommended design parameters for ATAD systems (Stensel and Coleman, 2000).

3.4.2.2.5 Prethickening

Thickening or blending facilities may be required to ensure that ATAD influent contains more than 4% solids.

3.4.2.2.6 Basin Configuration

The system should include two or more enclosed, insulated reactors in series. Both reactors need mixing, aeration, and foam-control equipment.

Both continuous and batch processing are acceptable. To comply with Class A biosoids requirements, a batch process should be used. In this case, pumps should be

TABLE 25.29 Recommended design parameters for ATAD digester systems (Stensel and Coleman, 2000).

Parameters	Range	Typical
Number of reactors	2–3	2
Prethickened solids	4–6%	4%
Reactors in series		Yes
Total HRT* in reactors	4–30 days	6–8 days
Temperature—stage 1	35–60°C	40°C
Temperature—stage 2	50–70°C	55°C

* HRT = hydraulic retention time.

designed to withdraw and feed the daily allotment of solids in 1 hour or less. The reactor then should be isolated for the remaining 23 hours each day, at a minimum temperature of 55°C.

This strategy maximizes the system's pathogen-destruction potential and reduces the chance for contamination. To prevent contamination with raw incoming feed, a specific volume of solids is removed on a daily basis from the second-stage reactor (which is operating in a range of 55 to 65°C). After the solids are removed, biomass from the first-stage reactor is transferred into it as a batch. The second stage is then not disturbed until the next batch is loaded 24 hours later. With this operating method, biomass that has been transferred from the first-stage reactor is maintained at a thermophilic temperature for a minimum period of 24 hours. Raw feed is then introduced to the first stage to make up the volume removed. This feeding approach isolates the reactors from each other and reduces the potential for contamination of the product.

3.4.2.2.7 Post-Process Storage and Dewatering

Post-process cooling is necessary to consolidate solids and enhance dewaterability. Typically, 14 to 20 days of SRT may be necessary, unless heat exchangers are used to cool the biosolids.

3.4.2.3 Process Performance and Operation

3.4.2.3.1 Volatile Solids Reduction

The volatile solids reduction achieved by the process depends on the feedstock(s), SRT, operating temperature, and reactor loading. In Germany, systems with detention times longer than 4 days reportedly had VSRs of 40% or more. The Haltwhistle, U.K., plant reported 30 to 40% VSR during a 2-year period. The Whistler ATAD system has achieved 70% VSR (Kelly et al., 1993). A comparison of several ATAD plants found that

VSR ranged from 28.5 to 53.8% (Schwinning et al., 1997). The Bowling Green, Ohoi, plant reported a VSR of 75% (Scisson, 2006).

3.4.2.3.2 Pathogen Reduction

German regulations require ATAD systems to produce biosolids containing no more than 1000 enterobacteria/mL. In fact, the German government considers ATAD to be a process capable of producing a "pasteurized (hygienic) solids"—a status similar to the PFRP designation in Part 503. (The U.S. Environmental Protection Agency has indicated that the ATAD system should qualify as a PFRP if the time and temperature requirements for Class A pathogen reduction are met.) Twice a year, Germany's regional health districts sample biosolids from treatment plants that use ATAD and analyzes them for a number of parameters. If the biosolids meet the coliform limit of 1 000 CFU/mL, along with other organic and inorganic criteria, they are deemed "acceptable" for agricultural use.

The Haltwhistle, U.K., facility reported a more than 4-log pathogen reduction via its ATAD system (Murray et al., 1990). Canadian facilities using ATAD had less than 100 MPN/wet gram of fecal coliform and fecal streptococci in 7 of 12 samples of their biosolids (Kelly, 1991), whereas *Salmonella* was not detected in any of the samples. Tests by Jewell and Kabrick (1980) in Binghamton, New York, showed that an ATAD system with a 24-hour SRT at 45°C reduced *Salmonella* and virus concentrations below detection limits.

3.4.2.3.3 Odor Control

Deeney et al. (1991) toured six facilities in Germany and found that none emitted offensive odors; rather, they discharged "musty" odors similar to those emitted from conventional aerobic digesters. Three facilities reported occasional odors: one when reactor temperatures exceeded 65°C, and the other two during system feeding. Two of the ATAD reactors near residential areas are equipped with exhaust gas scrubbers—an application considered to be somewhat experimental. Scrubber performance is reported to be good; however, occasional odors still have been noted.

The Banff facility uses a water scrubber on ATAD exhaust. Its dewatered biosolids exhibited no odors and seemed well stabilized. The Haltwhistle plant had no odor complaints. The Salmon Arm facility sends exhaust gases to a trickling filter; no odor problems have been reported. The Ladysmith and Gibsons facilities discharge ATAD exhaust to biological filters. The Bowling Green, Ohio, plant reports no odors after a year of operation (Scission, 2006).

The Glenbard, British Columbia, has a pure-oxygen ATAD system that emits "rotten broccoli" odors during operations. When analysts tested the offgas, they found dimethylsulfide, which is an indicator of anaerobic conditions. At the Salmon Arm and Whistler facilities, testing showed that the offgas contained hydrogen sulfide, methyl disulfide, dimethylsulfide, ammonia, and unidentified organic compounds (Kelly et al., 1993). Reports on plants in Colorado and Pennsylvania cite odor issues and the need for odor control (Bowker and Trueblood, 2002; Hepner et al., 2002). Meanwhile, several facilities in North America and Europe emitted odors and needed to implement odor controls (Layden et al., 2007).

Typically, odors can be minimized if the ATAD system maintains proper operating temperatures and is adequately mixed and aerated. Further odor-control measures (e.g., water scrubbers, biofilters, compost/soil filters, and offgas diversion to other trickling filters or activated sludge reactors) depend on the ATAD system's proximity to residences and the public.

3.4.2.3.4 Dewaterability

Autothermal thermophilic aerobic digestion produces biosolids with small flocs and, therefore, a large surface area that requires more polymer during dewatering operations (Kelly et al., 2003). In fact, conditioning chemical costs could offset the benefits of ATAD (Agarwal et al., 2005) if the goal is a dewatered biosolids containing 20 to 30% solids, because it can cost five to ten times more to chemically condition ATAD solids than undigested solids (Murthy et al., 2000a), and about two to three times more to chemically condition ATAD solids than anaerobically digested solids (high-rate mesophilic) (Spinosa and Vesilind, 2001).

The system's high temperature contributes to dewatering challenges because it promotes cell lysis and the release of proteins to liquid. These proteins, along with extracellular polymeric substances, alter the biosolids' conditioning polymer requirements. However, if operating temperatures exceed 70°C, the dewatering properties of biosolids actually improve because the production of extracellular substances decreases (Zhou et al., 2002).

Investigators have tried several methods for improving the dewaterability of ATAD solids:

- Sequential polymer dosing using iron and anionic polymer, or cationic and anionic polymers (Murthy et al., 2000a; Agarwal et al., 2005);
- Post-ATAD mesophilic aeration of biosolids (Murthy et al., 2000b); and
- Electrical arc treatment (Abu-Orf et al., 2001).

These methods expanded our knowledge of the rheology, characteristics, and other dewaterability factors of ATAD solids, but it is evident that more investigations are needed.

3.5 Process Variations

Investigators have tested several variations on standard mesophilic aerobic digestion in recent years. The more notable variations are high-purity oxygen aeration and dual digestion.

3.5.1 *High-Purity-Oxygen Aeration*

This aerobic digestion system uses high-purity oxygen rather than air. Recycle flows and the resultant biosolids are similar to those obtained via conventional aerobic digestion. Typical influent solids concentrations may vary from 2 to 4%. High-purity-oxygen aerobic digestion works well in cold weather climates because of its relative insensitivity to changes in ambient air temperatures because of the increased rate of biological activity and the exothermal nature of the process.

High-purity-oxygen aerobic digestion is conducted in either open or closed tanks. Because the digestion process is exothermic in nature, the use of closed tanks will result in a higher operating temperature and a significant increase in the VSR rate. The high-purity-oxygen atmosphere in closed tanks is maintained above the liquid surface, and the oxygen is transferred to the solids via mechanical aerators. In open tanks, oxygen is introduced to solids by a special diffuser that produces minute oxygen bubbles. The bubbles dissolve before reaching the air–liquid interface (Metcalf and Eddy, 2002). High operating costs are associated with high-purity-oxygen aerobic digestion because of the oxygen generation requirement. As a result, high-purity-oxygen aerobic digestion is cost-effective typically only when used with a high-purity-oxygen activated sludge system. Neutralization may be required to offset the system's reduced buffering capacity (Metcalf and Eddy, 2002).

3.5.2 *Combined Stabilization Processes*

3.5.2.1 *Combined Aerobic and Anaerobic Digestion*

The fusion of two stabilization process in ATAD and conventional mesophilic anaerobic digestion are covered in the dual digestion sub-section of the anaerobic digestion section. Currently the research has gone to using aerobic digestion as a post treatment for conventional mesophilic anaerobic digestion and has been demonstrated by Kumar et al. (2006a, 2006b) and Parravicini, et al. (2008). The advantages of these systems include improved volatile solids destruction, nitrogen removal from return streams, and improved dewaterability.

3.5.2.2 *Aerobic Digestion + Drying*

Aerobic digestion has been used as a conditioning step for solids that will undergo drying to stabilize them to a level that will minimize the risks for odor production during the drying process.

3.6 Design Techniques to Optimize Aerobic Digestion

3.6.1 Prethickening

3.6.1.1 Advantages of Prethickening

The main advantages of this technique include

- Increased SRT and VSR;
- Accelerated digestion and pathogen destruction rate as a result of the oxidation of biodegradable organic matter, which elevate digester temperatures {via its heat of combustion [about 3.6 Kcal/g (6 500 BTU/lb) of VSS destroyed]} and accelerate digestion and pathogen destruction rates (Grady et al., 1999); and
- The resulting temperature increase can be significant when feeding thicker solids.

If the heat generated by volatile solids can be captured, it can be used to control the reactor's temperature. This could be beneficial in cold climates. Because conventional aerobic digestion typically operates in the range of 15 to 35°C, it is classified as a mesophilic process. (It is also important to avoid excessive temperatures, especially in summer months.)

3.6.1.2 Disdvantages of Prethickening

Small plants need another unit process, which may increase in labor and O&M. Solids can only be prethickened to a maximum solids concentration wherein oxygen can be successfully transferred in the solids using available aeration equipment, and to the maximum extent that solids rheology does not significantly affect mixing characteristics in the aerobic digester.

3.6.1.3 Categories of Prethickening

Prethickened aerobic digestion is divided into five major categories (as described below) based on the thickening treatment processes used to increase the feed cake's solids concentration:

3.6.1.3.1 Batch Operation or Decanting of Aerobic Digester

Batch operation involves the practice of manually decanting digested solids. Originally, aerobic digestion was operated as a draw-and-fill process, a concept still used at many facilities. Solids are pumped directly from the clarifiers or SBRs to the aerobic digester. The time required to fill the digester depends on the tank volume available and the volume of solids. When a diffused-air aeration system is used, the solids being digested are aerated continually during the filling operation. When the solids are removed from the digester, aeration is discontinued, and the biosolids are allowed to settle. The clarified supernatant is then decanted and returned to the treatment process. The removed biosolids contain between 1.25 and 1.75% solids.

Disadvantages of this process include the following:

- Basins are sized based on low solids concentration and high water content (i.e., large volumes are required);
- High capital cost;
- High O&M cost;
- No control of alkalinity, temperature, ammonia, nitrates, and phosphorus; and
- Difficult to meet stringent limits on supernatant.

3.6.1.3.2 Continuous-Feed Operation with Post-Sedimentation

This mode of thickening treatment process consists of a continuous-feed operation using sedimentation (e.g., a gravity thickener) after digestion. This is typically a continuous aerobic digestion process that closely resembles the activated sludge process. Solids are pumped directly from the clarifiers, SBR, or MBR into the aerobic digester. The digester operates at a fixed level, with the overflow going to a solid–liquid separator. Thickened and stabilized solids are removed for further processing. Continuous operation typically produces biosolids with lower solids concentrations. The Category B process can produce marginally better effluent than Category A, because the aerobic digestion basin is operated at a fixed level and the aeration-transfer efficiency is optimized.

For continuous-feed digesters, the process can be improved by:

- Adjusting the rate of settled return solids to obtain the best balance between return solids concentration and supernatant quality,
- Adjusting the settling chamber's inlet and outlet flow characteristics to reduce shortcircuiting and unwanted turbulence (which hinders solids concentration), and
- Modifying the weir and piping arrangements.

Disadvantages of this process include the following:

- Basins are sized based on low solids concentration and high water content (i.e., large volumes are required);
- High capital cost;
- High O&M cost;
- No control of alkalinity, temperature, ammonia, nitrates, and phosphorus;
- Difficult to meet stringent limits on supernatant effluent; and

- If nitrification and denitrification are not controlled between the digester and thickener, it leads to anaerobic conditions and undesirable odors.

3.6.1.3.3 Gravity Thickener in Loop with Aerobic Digestion

This process typically consists of two main phases (in-loop and isolation) and four main basins (two digesters, one premix basin, and a gravity thickener). For feeds from SBRs and MBRs, more basins are incorporated into the design to optimize flexibility; however, the four basins are still the main components of the process. During the in-loop phase, a digester, a premix, and a thickener operate in a loop, which reduces volatile solids, reduces ammonia, and increases solids concentration. The in-loop thickener has two main functions: it thickens and denitrifies. The in-loop digester acts as a volatizer and reduces most volatile solids. During the isolation phase, no contamination occurs in the digester, which completes the additional pathogen reduction needed to meet solids requirements. The digesters are fed in batches either 8, 16, or 24 times per day; however, the process is considered a "modified batch process" because one digester is fed in short-batch intervals for an extended period—typically 10 to 20 days (equal to the length of the in-loop phase)—before it enters the isolation phase (10 to 20 days). This process produces biosolids containing 2.5 to 3% solids.

Advantages of this process include the following:

- Process provides all the benefits of aerobic–anoxic operation;
- Process provides all the benefits of staged operation;
- Good control of alkalinity without needing to turn air on and off;
- Can easily meet stringent limits on supernatant effluent for ammonia, nitrates, phosphorus, and total suspended solids (TSS);
- Process provides better SOUR and pathogen reduction compared to the processes in Sections 3.6.1.3.1 and 3.6.1.3.2 as a result of true isolation;
- Moderate capital cost;
- Low O&M cost; and
- No polymer required, achieving 3% solids.

3.6.1.3.4 Membranes for In-Loop Thickening with Aerobic Digestion

Membrane technology is fairly new in the United States and only has been used in Europe and Japan for 16 years. Its thickening applications are even more limited; the oldest installations that operate successfully with no odor issues, have typical cleaning frequencies, and have no membrane replacements date back to 1998. Applications range from 3 to 5% solids concentrations, operating in continuous or batch mode, in isolation and in series. The process incorporates a wastewater membrane suitable for high solids (e.g., flat plate, or hollow fiber). This process can be used in lieu of any process listed in Sections 3.6.1.3.1 through 3.6.1.3.3; however, it is not recommended that design solids exceed 3.5% for single-stage systems.

Membranes can be fitted into existing basins to provide both thickening and digestion at the same time, because airflow is required to clean the membranes.

Designs include two-, three-, four-, or five-basin configurations, operating in batch or in series (see Figures 25.67 and 25.68) (Daigger et. al., 2001).

Advantages of this process include the following:

- Provides the best control of supernatant effluent, because there is no danger of solid overflow resulting in poor supernatant (separate scum removal is not required);
- Requires small footprint, which is ideal for high-rate digestion;
- Provides good temperature control;
- Provides all the benefits of staged operation;
- Provides all the benefits of aerobic–anoxic operation;
- Provides better SOUR and pathogen reduction compared with the processes in Sections 3.6.1.3.1 and 3.6.1.3.2, as a result of true isolation;
- Moderate capital cost (smaller footprint); and

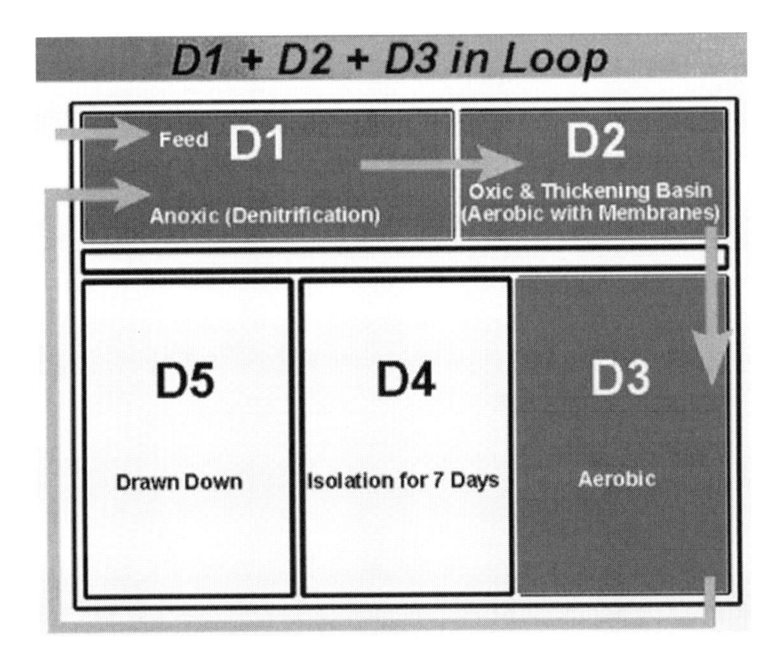

FIGURE 25.67 A five-stage batch operation setup using membranes for in-loop thickening as part of an aerobic digestion system (Daigger et al., 2001).

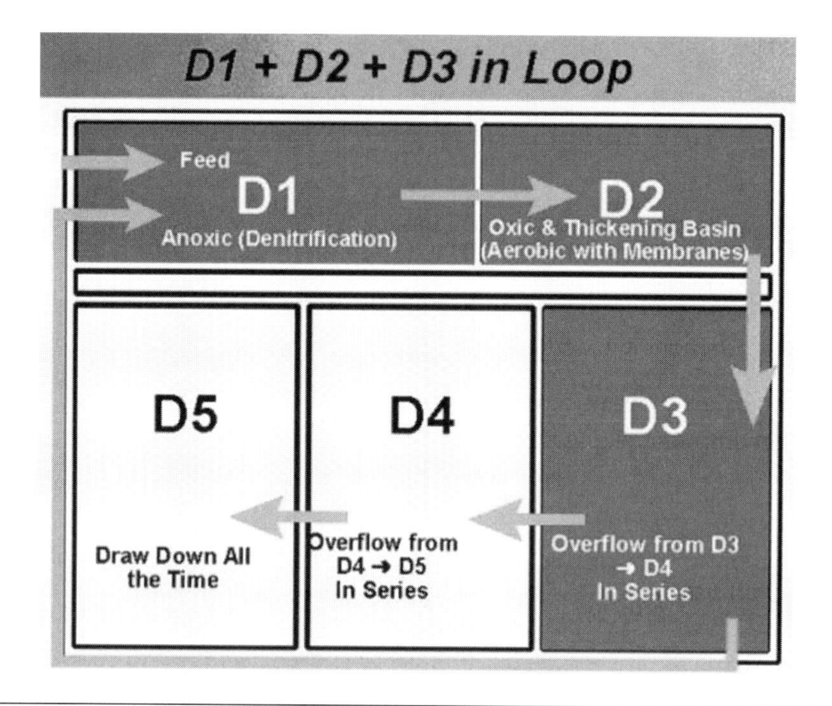

FIGURE 25.68 A five-stage, in-series operation setup using membranes for in-loop thickening as part of an aerobic digestion process (Daigger et al., 2001).

- Low O&M cost (less air is required because the viscosity factor is lower, so the airflow requirement is lower than that for thickened systems using polymer).

Following nitrification and denitrification, phosphorus release is reduced. If a bio-phosphorus permit applies on the liquid sidestream, the permeate can be treated with alum or ferric chloride to fix phosphorus so it can be removed with the solids (for more details on phosphorus, see Section 3.6.3).

Additional phosphorus-removal advantages that membranes offer include the following:

- Membrane plants produce low TSS;.
- The permeate from a membrane digester is collected in the aerobic phase (rather than the supernatant or decant typically collected from the anoxic phase in the other four systems); and
- Most membrane systems in thickening applications operate at about 0.5 to 1 mg/L dissolved oxygen, which is ideal for this application.

Another factor that affects phosphorus, ammonia, and nitrogen is pH. Membrane systems include an anoxic zone to balance alkalinity, so pH balancing is always integral to the process.

3.6.1.3.5 Using Any Mechanical Thickener Before Aerobic Digestion

In this process the thickening treatment process uses any mechanical thickener (e.g., a gravity belt thickener, DAF mechanism, centrifuge, or drum thickener) before aerobic digestion. This process uses mechanical prethickening devices that use polymers as conditioning agents to maximize thickening. The designer can choose the ideal mechanical device and desired operating solids concentration (e.g., 4, 5, or 6%) to minimize the aerobic digestion basins. This process gives ultimate process control to operators to meet performance requirements in summer and winter by modifying the mechanical device's operating schedule as desired. For this process, two digesters in series are recommended, as a minimum (in-series operation will be addressed later in this chapter). Because of the process' flexibility and reliability, and the capital and O&M cost savings, this process is preferred in treatment plant designs that can handle more than 7 600 m^3/d (2.0 mgd) of influent.

Advantages of this process include the following:

- Provides the ultimate process control;
- Provides capabilities to maximize reduction in volume (by choosing the ideal machine and ideal solids concentration for the application);
- Minimizes footprint, which is ideal for high-rate digestion (a deep tank is necessary);
- Provides the best temperature control when flexibility is included in the design (cold weather not an issue with these systems, but provisions may needed to prevent thermophilic conditions in summer);
- Provides all the benefits of staged operation;
- Provides all the benefits of aerobic–anoxic operation;
- After nitrification and denitrification, phosphorus release is reduced so phosphorus can be fixed and removed with solids (if a biophosphorus permit applies on the liquid stream, treat the permeate with alum or ferric chloride);.
- Provides excellent SOUR and pathogen reduction;
- Moderate capital cost;
- Low O&M cost. (The alpha values and transfer efficiency are lower in digesters operating at 4 to 6% than in those operating at 2 to 3%. However, because of the reduced basin volume required for these systems, the airflow required for both process and mixing is comparable. Mixing requirements typically are higher

than process air requirements for systems with lower solids concentration, so the overall operating horsepower for higher solids is less.)

- Supernatant is not produced in prethickened mode because the liquid in the digesters is thixotropic.

3.6.2 *Basin Configuration—Staged or Batch Operation (Multiple Basins)*

Traditionally, aerobic digesters have been designed with one basin. If multiple basins were supplied, they typically were operated in parallel. Multiple tanks in series or in isolation operated in a batch operation have proven to improve both pathogen destruction and SOUR.

According to U.S. EPA, solids can be aerobically digested using a variety of process configurations, including continuously fed or fed in a batch mode, in a single- or multistage (U.S. EPA, 1999b). Single-stage completely mixed reactors with continuous feed and withdrawal are the least effective option for bacterial and viral destruction, mainly because organisms that have been exposed to the digester's adverse conditions for a short time can leak through to the biosolids.

Probably the most practical alternative is staged operation (e.g., using two or more completely mixed digesters in series); it greatly reduces the amount of processed solids passing from inlet to outlet. If the kinetics of the pathogen density reduction is known, design engineers can estimate how much improvement can be made by staged operation.

Farrah et al. (1986) have shown that the decline in densities of enteric bacteria and viruses follow first-order kinetics. If first-order kinetics are assumed to be correct, it can be shown that a 1-log reduction of organisms is achieved in a two-stage reactor (with equal volume in each stage) compared to a one-stage reactor. Direct experimental verification of this prediction has not been done, but Lee et al. (1989) have qualitatively verified the effect.

It is reasonable to give credit to an improved operating mode; however, because not all factors involved in the decay of microorganism densities are known, some factor of safety should be introduced. For staged operation (using two stages with about equal volume), it is recommended that the required time be reduced to 70% of that needed for single-stage aerobic digestion in a continuously mixed reactor. This allows a 30% reduction in time instead of the 50% estimated from theoretical considerations. The same reduction is recommended for batch operation or for more than two stages in series. Thus, the time required would be reduced from 40 to 28 days at 20°C (68°F) and from 60 to 42 days at 15°C (59°F). These reduced times are also more than sufficient to achieve adequate vector attraction reduction. (For more information on this topic, see Section 3.7.4).

The benefits of a two-stage reactor system (in series or in isolation) include:

- Improvement in pathogen destruction;
- Improvement in SOUR;
- Capital cost reduction of aeration equipment;
- Capital cost reduction of tankage;
- Airflow reduction as a result of process requirements in the first digester; and
- Airflow reduction as a result of mixing requirements in the second digester.

3.6.3 Aerobic–Anoxic Operation

As described in Section 3.2 and Chapter 14, the oxidation of biomass produces carbon dioxide, water, and ammonia (see Equation 25.29). Ammonia then combines with some of the carbon dioxide to produce a form of ammonium bicarbonate. Often, partial nitrification occurs, and a portion of the nitrogen remains as ammonia. The system will nitrify until the pH drops enough to begin inhibiting nitrifying bacteria. This only occurs when the feed solids contain an inconsequential amount of alkalinity. Both ammonia and nitrate could be produced simultaneously during the "decant" ("supernating") phase of digestion, when the accumulated ammonia in the solids cannot be removed because there is not enough alkalinity to drive the reaction. This results in odors.

If the oxygen in nitrate can be used to stabilize biomass, then both nitrification and denitrification can occur in the reactor. Oxidizing biomass with nitrates both releases ammonia and produces nitrogen gas plus bicarbonate (a form of alkalinity).

Equation 25.36 is a balanced stoichiometric equation of nitrification and denitrification. It illustrates how oxidized biomass is converted to carbon dioxide, nitrogen gas, and water.

$$C_5H_7NO_2 + 5.7O_2 \rightarrow 5CO_2 + 3.5H_2O + 0.5N \qquad (25.36)$$

In Kuwait, investigators studied aerobic digestion in a controlled environment at 20°C and a 10-day SRT and optimized the duration of the anoxic stage at 8 to 16 h/d (Al-Ghusain et al., 2004). Figure 25.61 shows that total nitrogen removal in filtrate is optimized at 8 hours of anoxic cycle (Al-Ghusain et al., 2004).

3.7 Process Considerations for Designers

3.7.1 Specific Oxygen Uptake Rate

The microorganisms' rate of oxygen use depends on the biological oxidation rate. The U.S. Environmental Protection Agency selected a SOUR of 1.5 mg/g·h of oxygen of total solids at 20°C (68°F) to indicate that an aerobically digested solids has been adequately reduced in vector attraction.

The oxygen uptake rate is used to determine the level of biological activity and resulting solids destruction in the digester. The specific oxygen uptake rate is becoming a more common testing procedure among operators than the traditional VSR.

The specific oxygen uptake rate is a quick test and is independent of the initial value in the system or reduction of SOUR in the upstream process. On the other hand, volatile solids reduction is a percentage of the incoming volatile level to the digester system. During active aerobic digestion for staged operation, the typical oxygen uptake rate is between 3 and 10 mg/g·h of total solids in the first-stage digester, compared with a range of 10 to 20 mg/g·h in the active phase of the activated sludge process.

If primary solids are added to the first digester, its uptake rate may range from 10 to 30 mg/g·h of total solids.

The oxygen uptake rate for aerobically digested biosolids ranges from 0.1 to 1.0 mg/g·h of total solids, well below the required 1.5 mg/g·h by U.S. EPA standards.

3.7.2 Pathogen Reduction

Like solids reduction, little pathogen reduction can be expected at temperatures less than 10°C (50°F). On the other hand, significant reduction may be achieved at temperatures higher than 20°C (68°F). Although U.S. EPA standards allow for operations at 15°C (59°F), for economic reasons and reliable performance, the plant should be designed and operated at a minimum of 20°C (68°F).

3.7.3 Volatile Solids Reduction and Solids Reduction

Aerobic digestion destroys VSS. If a membrane thickening digestion option is selected, fixed suspended solids (FSS) also can be reduced in the permeate sent to the headworks or to the effluent channel when blending because of the higher liquid–solid separation mechanisms of the membrane. This occurs because both the organic and inorganic material in biodegradable suspended solids are solubilized and digested. However, the components of VSS and FSS are not equal, so they typically will not be destroyed in the same proportion.

Primary solids and WAS from a system with a short SRT will contain relatively high fractions of biodegradable material, as opposed to WAS from a system with a long SRT, which will contain a low fraction of biodegradable material and a high fraction of biomass debris (Grady et al., 1999).

In a study performed to determine the minimum SRT required to meet Class B requirements, by meeting both pathogens and VSR operating at minimum temperatures as concluded by Lu Kwang Ju (Daigger et al., 1999), two systems were evaluated.

One system included two basins in series and the other three basins in series. Table 25.30 shows VSRs at different temperatures and SRTs (Daigger et al., 1999).

The solids used in this study had a very low digestible fraction. It was concluded that, even though all systems met the pathogen-destruction requirement, an SRT of 29 days was required to exceed the required 38% VSR. This confirms that VSR depends on the source of the solids, and, if it has low fraction of digestible organic content, it is difficult to meet U.S. EPA's minimum requirements.

3.7.4 Solids Retention Time × Temperature Product

A significant factor in the effective operation of aerobic digesters, SRT is the total mass of biological solids in the reactor divided by the average mass of solids removed from the process each day. Typically, an increase in SRT increases VSR.

Based on the discussion above, with respect to temperature, degradable solids, nondegradable solids and the effect on VSR, the SRT × temperature product (days °C) curve can be used to design digester systems, taking into consideration not only the total days °C that the digester system will have to meet, but also the quality of the source. The original U.S. EPA SRT × temperature curve, developed in the late 1970s (U.S. EPA, 1978, 1979) was updated by incorporating data from two extensive pilot studies conducted by Lu Kwang Ju over 3 years, as well as data from three full-scale installations (Daigger et al., 1999). Figure 25.69 shows a process design based on 600 days °C, assuming the feed has relatively high degradable solids content (Daigger et al., 1999). Figure 25.70 shows another operation if the feed contains low degradable solids (Daigger et al., 1999). Both systems can coexist, from a design standpoint, if prethickening is incorporated into the design to allow an increase or decrease of SRT, as necessary.

3.7.5 Nitrogen Removal in Biosolids

The conventional aerobic digestion process, operated in the aerobic–anoxic mode, provides full nitrification and denitrification and a reduction in total nitrogen. The annual

Table 25.30 Volatile solids reduction at minimum operating temperatures and minimum solids retention time (Daigger et al., 1999).

Volatile solids reduction data from both studies in all basins						
°C	8–10°C	12°C	21°C	21°C	23°C	31°C
Two basins in series (no. of days)	19.25	13.75	13.75	13.75	19.25	19.25
Volatile solids removal	27%	31%	31%	31%	28%	31%
Three basins in series (no. of days)	29.25				29.25	29.25
Volatile solids removal	28%				32%	40%

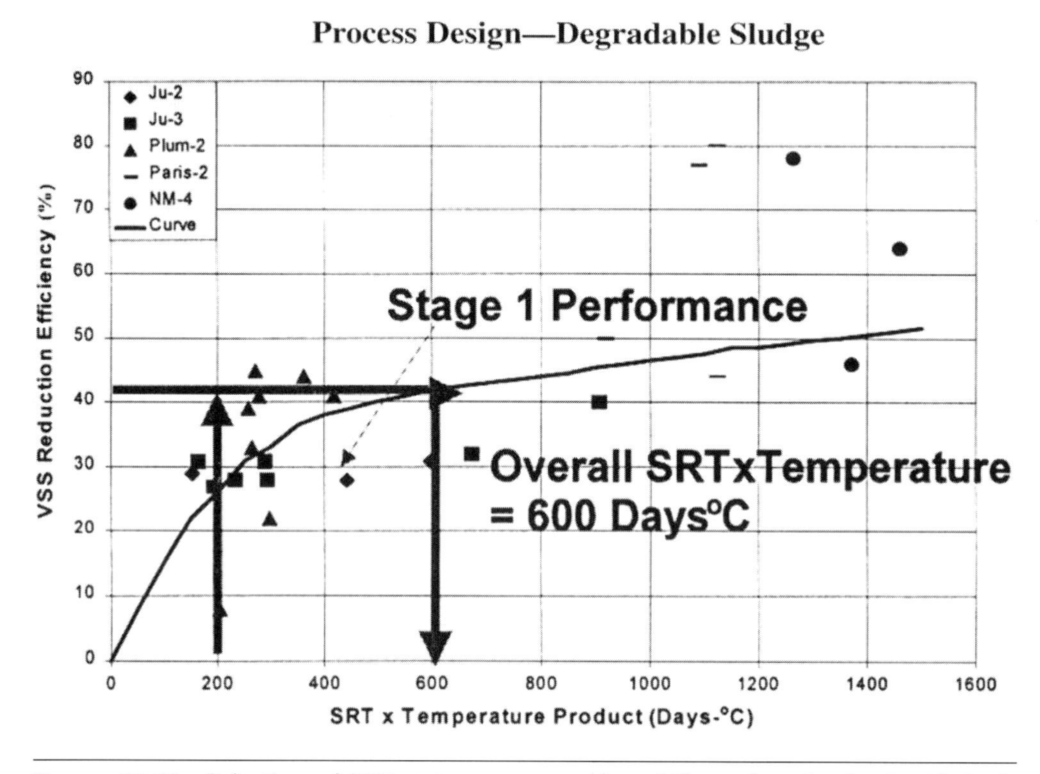

FIGURE 25.69 Selection of SRT × temperature (days °C) product for feed with high degradable solids content (Daigger et al., 1999).

design loading rates for land application of biosolids typically are limited by the nitrogen loading rate. Because nitrification is inhibited in the ATAD process, the most ideal process to meet these requirements is the conventional process operated in the aerobic–anoxic mode.

3.7.6 *Phosphorus Reduction in Biosolids and Biophosphorus*

Both anaerobic and aerobic digestion release phosphorus; however, if the conventional aerobic digestion process is operated in an aerobic–anoxic mode or under low dissolved oxygen conditions, phosphorus release is minimized.

From an overall mass-balance prospective, the phosphorus entering the digester ends up either in waste solids or in effluent. If it cannot move forward with the solids, then phosphorus will be recycled back to the head of the plant. Phosphorus will be released from both anaerobic and aerobic digestion; however, the release is lower from aerobic processes than from anaerobic processes.

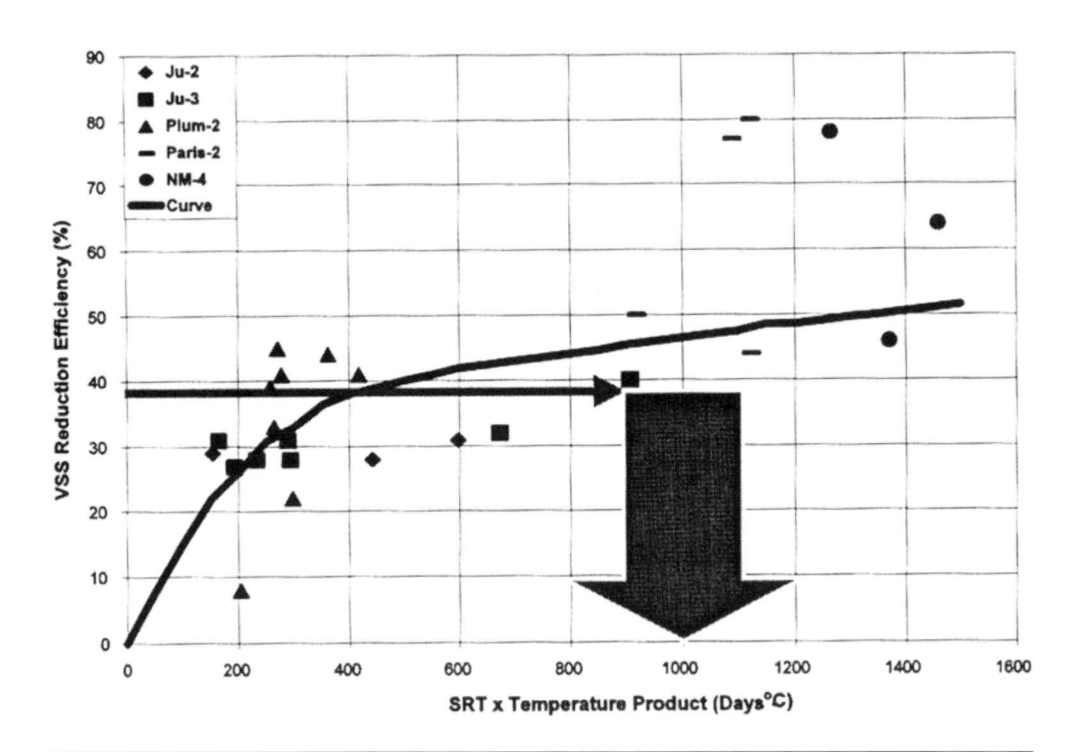

FIGURE 25.70 Selection of SRT × temperature (days °C) product for feed with low degradable solids content (Daigger et al., 1999).

It has been proven, through the experimental work of Lu Kwang Ju and in full-scale installations, that cyclic operation of air on and off or under low dissolved oxygen conditions (which provides simultaneous nitrification and denitrification) minimize phosphorus release as measured in the permeate, filtrate, or supernantant associated with the aerobic digester (Daigger et al., 2001). As shown in Figure 25.71, when solids from a biophosphorus facility (Ozark wastewater treatment plant, Kansas) were digested under fully aerobic conditions, the phosphorus release was in the range 120 to 150 mg/L (Daigger et al., 2001). If they were digested under cyclic operations, phosphorus releases range from 70 to 90 mg/L after 500 hours of operation.

Another factor that affects phosphorus, ammonia, and nitrogen is pH. Conclusions from the same study showed that a pH less than 6.0 should be avoided, because it encourages inorganic metal phosphates to dissolve (Daigger et al., 2001).

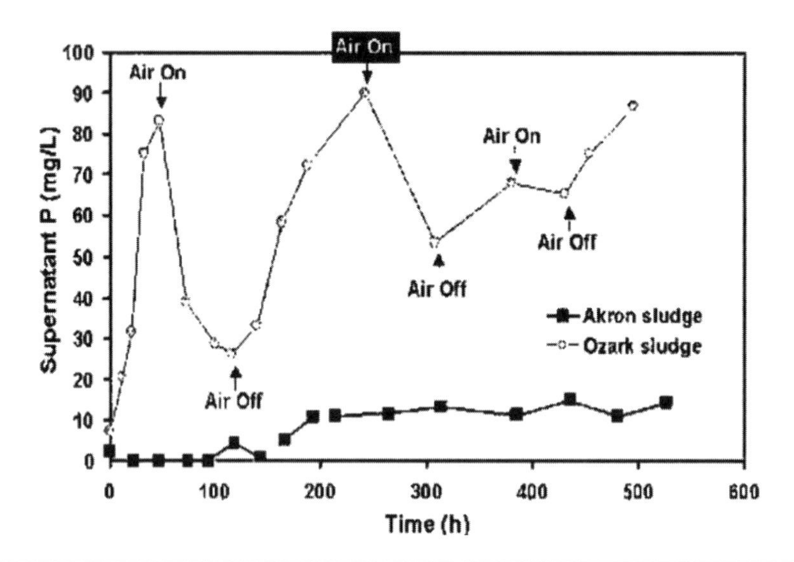

FIGURE 25.71 Polyphosphorus release and uptake of phosphorus during solids digestion (Daigger et al., 2001).

3.7.6.1 Option I: Liquid Disposal—No Restriction of Phosphorus on Land Application

Assuming that a system is designed as a two-stage-in-series, prethickened aerobic digester using a mechanical thickener that can thicken to 6% solids, the expected solids leaving the system would be about 3.6%. Figure 25.72 shows a typical prethickened application for liquid disposal with no phosphorus limit restriction on land application per Jim Porteous (Daigger et al., 2000).

The necessary steps for Option I are as follows:

- The main design consideration, in this case, is that the solids being fed to the prethickening device remain aerobic.
- To prevent anaerobic conditions, solids should be wasted directly from the liquid sidestream to the prethickening device; or
- If storage is required before prethickening, the detention time should be minimized.

3.7.6.2 Option II: Dewatering, Post-Thickening, and Supernating, with Limit Restriction of Phosphorus on Land Application

Option II assumes that a system is designed as a three-stage-in-series, prethickened aerobic digester using a mechanical thickener that can thicken to 6% solids; however, it is

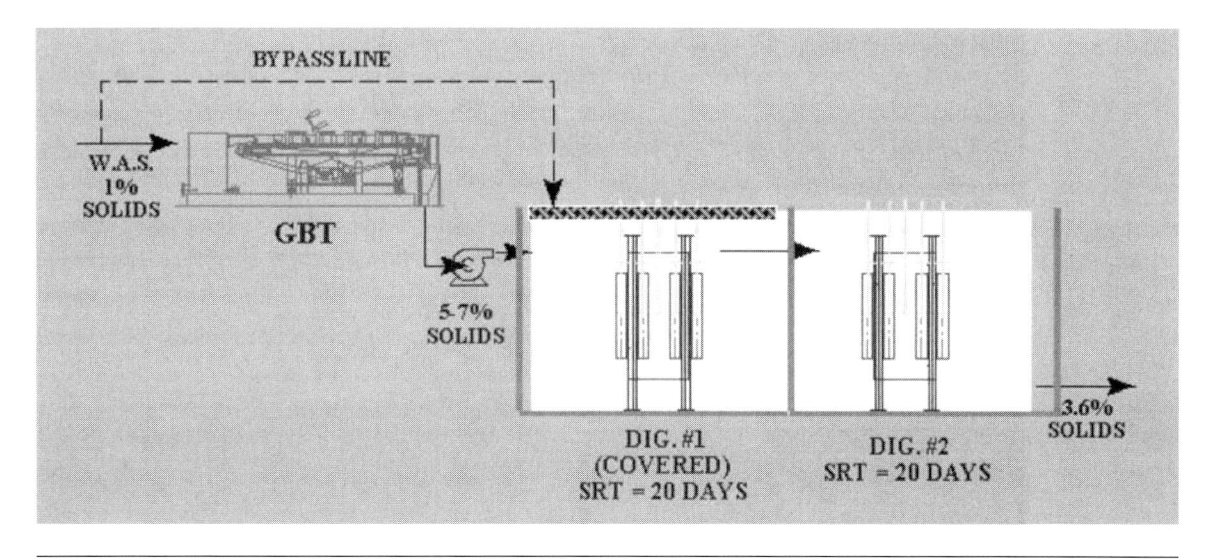

FIGURE 25.72 Option I: prethickened liquid disposal with no phosphorus limit restriction on land application. (GBT = gravity belt thickener) (Daigger et al., 2000).

also designed to post-thicken. In this case, it is assumed that there is a restriction on land and that the prethickening device will be bypassed during the summer. Figure 25.73 shows such a dewatering option, with a phosphorus restriction on land application (Daigger et al., 2000). The necessary steps for Option II are as follows:

- Phosphorus will be released during digestion and returned to the headworks.
- Nitrification can be controlled in first digester to encourage struvite production, which will be separated.
- Add alum or ferric chloride to the supernatant and filtrate. Chemicals are only required for phosphorus removal. There is little else in the supernatant, so chemicals for phosphorus removal are minimized.
- Phosphorus is now fixed and can be removed with the solids.

3.7.7 Supernatant Quality of Recycled Sidestreams

Careful monitoring of solids–liquid separation in continuous and batch-feed digesters increases aerobic digester performance. The supernatant liquid (filtrate) should contain low levels of soluble BOD, TSS, and nitrogen in both batch and continuous-flow operations. Table 25.31 lists characteristics of "acceptable" supernatant values from aerobic digestion processes (Metcalf and Eddy, 2003).

Although the values are considered "acceptable", using the techniques described in this chapter, the process can be controlled and optimized to provide better supernatant

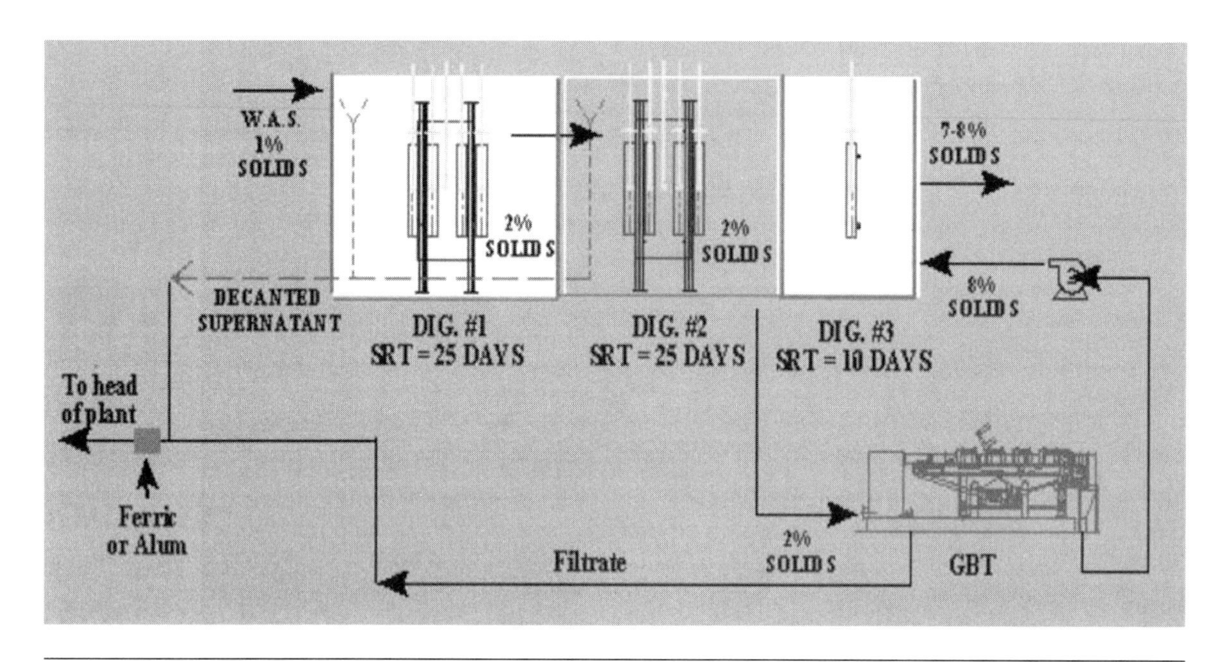

FIGURE 25.73 Option II: dewatering post-thickening, with phosphorus limit restriction on land application. (GBT = gravity belt thickener) (Daigger et al., 2000).

than what is described in Table 25.31. To improve these parameters, it is necessary to operate in aerobic–anoxic operation. Controlling the availability of necessary quantities of carbon source required for denitrification to allow for full nitrification and denitrification will enhance supernatant quality. Maintaining a neutral pH of 7.0 also will enhance nitrification and denitrification. For more information, refer to Figures 25.61 and 25.74 on the effect of both anoxic-cycle length and temperature on total nitrogen levels in filtrate.

TABLE 25.31 Acceptable characteristics for supernatant from aerobic digestion systems (from Metcalf & Eddy, *Wastewater Engineering: Treatment and Reuse*, 4th ed. Copyright © 2003), The McGraw-Hill Companies, New York, N.Y. with permission).

Parameter	Acceptable range	Acceptable values
pH	5.9–7.7	7.0
5-day BOD (mg/L)	9–1700	500
Filtered 5-day BOD (mg/L)	4–173	50
Suspended solids (mg/L)	46–2 000	1 000
Kjeldahl nitrogen (mg/L)	10–400	170
Nitrate-nitrogen (mg/L)	0–30	10
Total phosphorus (mg/L)	19–241	100
Soluble phosphorus (mg/L)	2.5–64	25

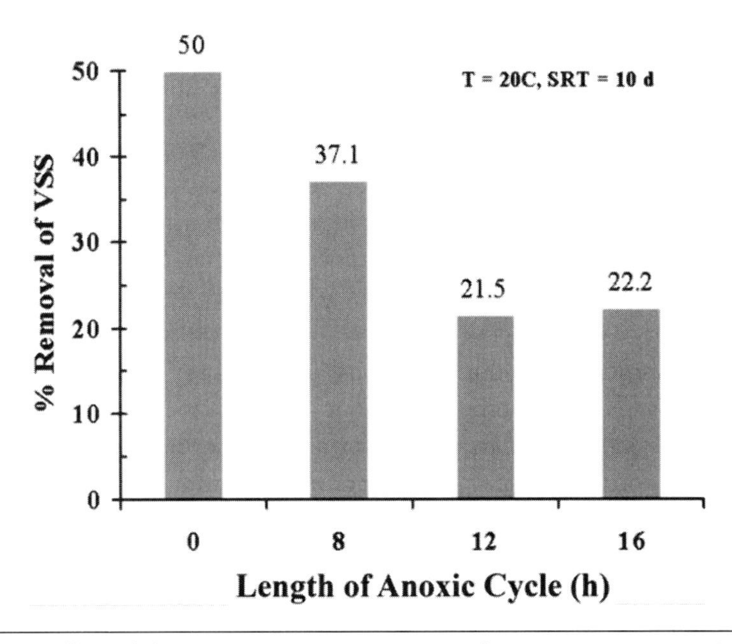

FIGURE 25.74 Effect of digestion temperature on total nitrogen concentration in filtrate (Al-Ghusain et al., 2004).

Table 25.32 shows data from an installation in Stockbridge, Georgia, that uses a gravity thickener in loop with an aerobic digester to allow nitrification to occur in the digester and denitrification and recovery to occur in the thickener (Stege & Bailey, 2003). Thickener blanket data indicate good settling. Average data for TSS, ammonium, and phosphorus indicate both nitrification and denitrification, and phosphorus removal with TSS removal. The solids blanket during the collection of these data points varied between 2.4 and 4.1 m (8.0 and 13.5 ft). In comparison, data presented by Stege

TABLE 25.32 Data from a gravity thickener–aerobic digester in-loop process at the Stockbridge, Georgia, Wastewater Treatment Plant (courtesy of Stantec Consulting).

Parameter	Actual data from a gravity thickener–aerobic digester in-loop system	Compared with acceptable typical values from Table 25.31
pH	6.5–7.1	7.0
Suspended solids (mg/L)	10–50	1 000
Total Kjeldahl nitrogen (mg/L)	2.5–4	170
Nitrate-nitrogen (mg/L)	—	30
Total phosphorus (mg/L)	0.3	100
Thickener blanket	8–13.5	10

and Bailey (2003) were used to produce Table 25.32 to demonstrate the capabilities of optimization of the techniques and the effect they have on supernatant quality.

3.7.8 *Aerobic Digester Design Example*

3.7.8.1 Standard Design: Single Tank

Design a mesophilic aerobic digester processing solids from a non-primary solids secondary biological treatment process. The following conditions are applicable to the design:

Secondary solids concentration	0.8%
Total solids	1 144 kg/d (2 522 lb/d)
Volatile solids	894 kg/d (1971 lb/d) or 78.15%
Decanted solids concentration	1.5%
Minimum liquid temperature (winter)	15°C (59°F)
Maximum liquid temperature (summer)	30°C (86°F)

The biosolids must meet Class B conditions in a single tank configuration.

3.7.8.1.1 Determine the Digester Volume

3.7.8.1.1.1 Determine the SRT Required to Meet Class B Requirements

Using the worst-case condition [the winter temperature (15°C)], the SRT required to meet Class B requirements at this temperature is 60 days.

3.7.8.1.1.2 Determine the Decanted Solids Volume

In SI units: 1 144 kg/d/1 000 kg/m^3/1.75% = 65.37 m^3/d

In U.S. customary units: 2 522 lb/d/8.34 lb/gal/1.75% = 17 268.91 gpd

3.7.8.1.1.3 Determine the Digester Volume

65.37 m^3/d (17 268.91 lb/d) × 60 days = 3 922.2 m^3 (1.04 mil. gal)

3.7.8.1.2 Determine the Oxygen Requirements

3.7.8.1.2.1 Mixing Air Requirements

From Section 3.7.1, we use the average value of 0.5 L/m^3·s (30 cu ft/min/1000 cu ft):

In SI units: 3 922.2 m^3 × 0.5 L/m^3·s = 1 961.1 L/s

In U.S. customary units: 1.04 mil. gal/7.48 gal/cu ft/1 000 × 30 cu ft/min/1 000 cu ft = 4 155.3 cu ft/min

3.7.8.1.2.2 Process Air Requirements

From Figure 25.69, determine the amount of VSR expected:

Winter: (Solids temperature × SRT required per regulation at temperature)
15°C × 60 days = 900°C·days, which is ~ 45% VSR

Summer: (Solids temperature × SRT to meet Class B in winter)
30°C × 60 days = 1800°C·days, which is ~ 55% VSR

Calculate the VSR:

Winter: 894 kg/d (1971 lb/d) × 45% = 402.3 kg/d (887 lb/d)

Summer: 894 kg/d (1971 lb/d) × 55% = 491.7 kg/d (1084 lb/d)

Calculate the oxygen demand at 2 lb O_2/lb VSR:

Winter: 402.3 kg/d (887 lb/d) × 2 lb O_2/lb VSR = 804.6 kg O_2/d (1774 lb O_2/d)

Summer: 491.7 kg/d (1 084 lb/d) × 2 lb O_2/lb VSR = 983.4 kg O_2/d (2168 lb O_2/d)

Calculate the Process Air requirement, assuming 0.56 AOR/SOR and 14% OTE for coarse bubble diffusers:

$$\text{Winter: } \frac{1774 \text{ lb } O_2 / \text{d}/0.56 \text{ AOR/SOR}/1440 \text{ min/d}/14\% \text{ OTE}}{0.2315 \text{ lb } O_2 / \text{lb air} \times 0.075 \text{ lb air/cu ft}}$$

$$= 905 \text{ cu ft/min } (427 \text{ L/s})$$

$$\text{Summer: } \frac{2168 \text{ lb } O_2 / \text{d}/0.56 \text{ AOR/SOR}/1440 \text{ min/d}/14\% \text{ OTE}}{0.2315 \text{ lb } O_2 / \text{lb air} \times 0.075 \text{ lb air/cu ft}}$$

$$= 1106 \text{ cu ft/min } (522 \text{ L/s})$$

The air requirements that govern the design are based on mixing, so the blower is sized based on 1961.1 L/s (4155.3 cu ft/min).

3.7.8.1.3 Determine the Blower Power

Assuming an approximation of 20 cu ft/min/blower hp (12.657 L/s/kW) we get:

4155 cu ft/min/20 cu ft/min/hp = 207.76 hp

or

1961.1 L/s/56.2 L/s/kW = 154.93 kW

3.7.8.2 Optimizing the Single Tank Conventional Design by Thickening

As discussed in Sections 3.3.7 and 3.4.1.2, aerobic digester design can be optimized by setting a batch or a staged design comprising at least two tanks. The design example for a two-tank design is as follows:

Secondary solids concentration	0.8%
Total solids	1144 kg/d (2522 lb/d)
Volatile solids	894 kg/d (1971 lb/d) or 78.15%
Thickened solids concentration	3.0%
Minimum liquid temperature (winter)	15°C (59°F)
Maximum liquid temperature (summer)	30°C (86°F)

The biosolids must meet Class B requirements in a two-tank-in-series configuration.

3.7.8.2.1 Determine the Digester Volume

3.7.8.2.1.1 DETERMINE SRT REQUIRED FOR CLASS B SOLIDS REGULATIONS

Using the worst case condition which is the winter temperature at 15°C determine the solids retention time required to meet Class B (as expressed on section 3.7) at this temperature and applying the 30% credit for staged operation we can use 42 days.

3.7.8.2.1.2 DETERMINE THE THICKENED SOLIDS VOLUME

In SI units: $1\,144\ \text{kg/d}/1\,000\ \text{kg/m}^3/3.0\% = 38.13\ \text{m}^3/\text{d}$

In U.S. customary units: $2\,522\ \text{lb/d}/8.34\ \text{lb/gal}/3.0\% = 10\,073.53\ \text{gpd}$

3.7.8.2.1.3 DETERMINE THE DIGESTER VOLUME

$$38.13\ \text{m}^3/\text{d}\ (10\,073.53\ \text{gpd}) \times 42\ \text{days} = 1601.46\ \text{m}^3\ (423\,088.44\ \text{gal})$$

Each digester volume is $1601.46\ \text{m}^3$ (423 088.44 gal)/2 = $800.73\ \text{m}^3$ (211 544.22 gal)

3.7.8.2.2 Determine the Oxygen Requirements

3.7.8.2.2.1 MIXING AIR REQUIREMENTS PER DIGESTER, ASSUMING EQUAL VSR IN BOTH DIGESTERS

From Section 3.7.1, we use the average value of $0.5\ \text{L/m}^3\cdot\text{s}$ (30 cu ft/min/1000 cu ft):

In SI units: $800.73\ \text{m}^3 \times 0.5\ \text{L/m}^3\cdot\text{s} = 400.27\ \text{L/s}$ (848.44 cu ft/min)

In U.S. customary units: $211\,544.22\ \text{gal}/7.48\ \text{gal/cu ft}/1000 \times 30\ \text{cu ft/min}/1\,000\ \text{cu ft} = 848.44\ \text{cu ft/min}$

Total air requirements for both digesters is 800.73 (1696.87 cu ft/min).

3.7.8.2.2.2 PROCESS AIR REQUIREMENTS

From Figure 25.69, determine the amount of VSR expected:

Winter: (Solids temperature × SRT required per regulation at temperature)
15°C × 42 days = 630°C·days, which is ~ 42% VSR

Summer: (Solids temperature × SRT to meet Class B in winter)
30°C × 42 days = 1260°C·days, which is ~ 49% VSR

Calculate the VSR (assuming equal VSR in both digesters):

Winter: 894 kg/d (1971 lb/d) × 42% = 375.5 kg/d (827.8 lb/d)

Summer: 894 kg/d (1971 lb/d) × 49% = 438 kg/d (965.8 lb/d)

Calculate the oxygen demand at 2 lb O_2/lb VSR:

Winter: 375.5 kg/d (827.8 lb/d) × 2 lb O_2/lb VSR = 751 kg O_2/d (1655.6 lb O_2/d)

Summer: 438 kg/d (965.8 lb/d) × 2 lb O_2/lb VSR = 876 kg O_2/d (1931.6 lb O_2/d)

Calculate the process air requirement. Assume 0.56 AOR/SOR and 14% OTE for coarse-bubble diffusers:

$$\text{Winter: } \frac{1655.6 \text{ lb } O_2 \text{ / d/0.56 AOR/SOR/1440 min/d/14\% OTE}}{0.2315 \text{ lb } O_2 \text{ / lb air} \times 0.075 \text{ lb air/cu ft}}$$

$$= 844.62 \text{ cu ft/min (398.6 L/s)}$$

$$\text{Summer: } \frac{1931.6 \text{ lb } O_2 \text{ / d/0.56 AOR/SOR/1440 min/d/14\% OTE}}{0.2315 \text{ lb } O_2 \text{ / lb air} \times 0.075 \text{ lb air/cu ft}}$$

$$= 985.43 \text{ cu ft/min (465 L/s)}$$

The air requirements that govern the design are based on mixing again, but with much lower air demand than the single-tank design with thinner solids.

3.7.8.2.3 Determine the Blower Power

Assuming an approximation of 20 cu ft/min/blower hp (12.657 L/s/kW) we get:

For both digesters 1696.87 cu ft/min/20 cu ft/min/hp = 84.84 hp

For both digesters 800.73 L/s/12.657 L/s/kW = 63.26 kW

The optimization results via thickening are clearly evident in the smaller digester volume, air requirements, and blower power. The only caution on using this technique is the limited selection of diffusers that can handle 3% solids in continuous service.

Also, this savings could be increased by reducing air requirements in the second stage, because the VSR in the first digester is significant.

4.0 COMPOSTING

Composting is a biological process in which organic matter is decomposed under controlled, aerobic conditions to produce humus. Any organic material can be composted under almost any conditions. However, operators can accelerate the process by using the proper blend of materials and controlling the temperature, moisture content, and oxygen supply. The resulting compost is stable and can be safely used in many landscaping, horticulture or agriculture applications.

Composting can be readily used to treat both unstabilized solids and partially stabilized biosolids. In both solids and biosolids composting, operators control several essential process variables to optimize the material's decomposition/stabilization rate:

- Solids content;
- Carbon-to-nitrogen (C:N) ratio;
- Aerobic conditions; and
- Temperature.

Via process-control methods, operators typically can cause the composting mass to achieve thermophilic temperatures, which destroy pathogens. Well-stabilized compost can be stored indefinitely and has minimal odor, even if rewetted. It is suitable for a variety of uses (e.g., landscaping, topsoil blending, potting, and growth media) and can be distributed to the public for gardening. It also can be used in agriculture to control erosion, improve the soil's physical properties, and revegetate disturbed lands. Local markets may be developed in urban and nonagricultural areas, as well as in agriculture and mine revegetation.

4.1 Process Variables

Although a wide variety of composting technologies are available, they all are designed to control the essential variables mentioned above.

4.1.1 Solids Content

The initial solids content depends on how much amendment or bulking agent is mixed with dewatered cake. For good process performance, the dewatered cake should contain between 14 and 30% solids. It then is blended with drier materials (e.g., wood chips, sawdust, shredded yard waste, and ground pallets) to achieve a solids content of about 38 to 45%.

The target solids content depends on the composting technology used. Solids content is controlled throughout the process via aeration, material agitation, or both.

4.1.2 Carbon-to-Nitrogen (C:N) Ratio

The amount of carbon and nitrogen used by microorganisms depends on the composition of the microbial biomass. Ideally, the ratio of available carbon to nitrogen is between 25:1 and 35:1. If the ratio is less than 25:1, excess nitrogen will be released as ammonia, reducing the compost's nutrient value and emitting odor. If the ratio exceeds 35:1, organic material will break down more slowly, remaining active well into the curing stage (Poincelot, 1975). Wastewater residuals typically have a carbon-to-nitrogen ratio between 5:1 and 20:1. Adding an amendment or bulking agent increases the carbon content, improving both the energy balance and the mixture's carbon-to-nitrogen ratio.

Calculating the carbon-to-nitrogen ratio is complicated, because some of the carbon becomes available more slowly than the nitrogen (Kayhanian and Tchobanoglous, 1992). If wood chips are the bulking agent, for example, only a thin surface layer of the wood provides available carbon. The carbon in sawdust, on the other hand, is more readily available to degradation.

4.1.3 Maintaining Aerobic Conditions

Microbial oxygen demand during composting can reduce the available oxygen in air to as low as 3 to 5% in as little as 15 minutes. Aerobic conditions are maintained via forced or convective aeration, material agitation, or both, depending on the composting technology used.

4.1.4 Maintaining Proper Temperatures

At first, the challenge is to heat the material up to the thermophilic range as quickly as possible. Then, the challenge is removing excess heat to maintain the process in the thermophilic range. Near the end of composting, the goal is to dry the material without removing too much heat. All of these are achieved using aeration, agitation, or both.

4.1.5 Microbiology

Three major categories of microorganisms involved in composting are bacteria, actinomycetes, and fungi. Bacteria are responsible for decomposing a major portion of organic matter. At mesophilic temperatures [lower than 40°C (104°F)], bacteria metabolize carbohydrates, sugars, and proteins. At thermophilic temperatures (higher than 40°C), they decompose protein, lipids, and the hemicellulose fractions. Bacteria also are responsible for much of the heat produced.

Actinomycetes are microorganisms common to soil environments. They metabolize a wide variety of organic compounds (e.g., sugars, starches, lignin, proteins, organic acids, and polypeptides). Their role in composting is unclear. Waksman and Cordon (1939) indicated that this group attacks hemicellulose but not cellulose. Stutzenberger (1971) isolated a thermophilic actinomycete that may be important in cellulose degradation.

Fungi are present at both mesophilic and thermophilic temperatures. Chang (1967) indicated that mesophilic fungi metabolize cellulose and other complex carbon sources. Their activity is similar to that of actinomycetes; both typically are found in the exterior portions of compost piles. Golueke (1977) suggested that this phenomenon is related to the organisms' aerobic nature, because most fungi and actinomycetes are obligate aerobes.

Microbial activity during composting occurs in three basic stages: mesophilic, when temperatures in the pile range from ambient to 40°C (104°F); thermophilic, when temperatures range from 40 to 70°C (104 to 158°F); and a cooling period associated with a reduction in microbial activity and the completion of composting. The optimum temperature in the thermophilic range seems to be between 55 and 60°C (131 and 140°F) where the maximum rate of volatile solids destruction occurs.

Biological solids, newly harvested wood wastes, and yard wastes provide a diverse population of microflora that can respond to changes in temperature and substrate. Under most circumstances, an inoculum of pure cultures does not significantly enhance composting. Sawdust decomposition, however, can be accelerated by inoculating a cellulose-decomposing fungus and adding nutrients.

4.1.6 Energy Balance

Heat is generated when organic carbon converts to carbon dioxide and water vapor. The fuel is provided by rapidly degraded volatile solids. Heat primarily is removed by the evaporative cooling promoted by aeration and agitation. Some heat also is lost at the pile surface. The process temperature will not rise if heat is lost faster than it is generated.

Haug (1980) provides a detailed discussion of the energy balance, concluding with the following relationship:

$$W = \frac{\text{Weight of water evaporated}}{\text{Weight loss of volatile solids}} \tag{25.37}$$

If W is below 8 to 10, enough energy should be available for heating and evaporation. If W exceeds 10, the mix will remain cool and wet. This generalization is based on heat of vaporization and does not consider the effect of ambient conditions on evaporation and surface cooling.

4.2 Process Objectives

The primary objective of composting is to produce a nutrient-rich soil amendment that complies with federal, state, and local requirements for beneficial use of biosolids. The compost must meet both environmental and public health requirements, and be attractive for use. This primary objective is met via the following process objectives: pathogen reduction, maturation, and drying.

4.2.1 *Pathogen Reduction*

There are five type of pathogens in wastewater residuals: bacteria, viruses, protozoa cysts, helminthic (parasitic worm) ova, and fungi. The first four groups often are called *primary pathogens* because they can invade typically healthy persons and cause diseases. Fungi are called *secondary pathogens* because they typically only infect persons with weakened respiratory or immune systems.

Heat is one of the most effective methods for destroying pathogens. Table 25.33 summarizes time-and-temperature relationships for inactivating pathogens in actual composting operations. Note that temperatures measured in a composting pile or vessel may not be uniform because of variations in heat loss, solids-mixture characteristics, and airflow.

TABLE 25.33 Temperature exposure required for pathogen destruction in compost (Knoll, 1964; Morgan and MacDonald, 1969; Shell and Boyd, 1969; Wiley and Westerberg, 1969).

	Exposure time for destruction at various temperatures hours		
Microorganisms	**45–55°C**	**60°C**	**65°C**
Salmonella Newport		25	
Salmonella	168	116	
Poliovirus type 1		1.0	
Candida albicans		72	
Ascaris lumbricoides		4.0	1.0
Mycobacterium tuberculosis			336

Composting in the thermophilic range should eliminate practically all viral, bacterial, and parasitic pathogens (WEF, 2007). However, some fungi (e.g., *Aspergillus fumigatus*) are thermotolerant and, therefore, survive.

Los Angeles' data on windrow composting showed that bacterial concentrations were markedly reduced within 15 days (Iacoboni et al., 1980). At 20 days, no *Salmonella* was detected. Fecal and total coliforms survived windrow composting in the cool, humid climate, but *Salmonella* was eliminated after 14 days. Studies using an F_2 bacteriophage virus (as an indicator of virus destruction) showed it could survive for as long as 45 days in digested solids and more than 55 days in undigested solids.

Static-pile composting data show that total coliforms, fecal coliforms, and *Salmonella* were not detected after 10 days of composting when temperatures exceeded 55°C (131°F) for several days. Further studies using an F_2 bacteriophage virus revealed that static-pile composting destroyed the indicator in 14 days.

Salmonella can regrow in finished compost. However, parasite ova and virus cannot. Regrowth can be reduced by not using the same equipment to handle both raw feed and finished compost or by cleaning the equipment before handling finished compost.

Many microorganisms can function as secondary pathogens, although composting conditions favor the growth of some more than others. Millner et al. (1977) report that the fungus *A. fumigatus* Fres has been isolated at relatively high concentrations from finished compost and from compost-pile zones at less than 60°C (140°F). Other secondary fungi occasionally isolated from compost are *M. pusillus* and *M. miebei*. Common to composting operations, these fungi typically are found in backyards, decayed leaves, grass, commonly available organic soil amendments, and ventilation ducts.

During certain composting operations, more *A. fumigatus* spores are released to the atmosphere. In windrow and reactor studies at the Los Angeles County Sanitation District in California, LeBrun (1979) found that compost feedstock contained 1000 to 10 000 colony-forming units (CFU)/g; after composting, biosolids contained 10 CFU/g. Exposure to airborne spores can be minimized by controlling dust. So, compost should not be allowed to become too dry, and workers should be provided with dust masks when working in dusty areas.

4.2.2 *Maturation*

Maturation refers to the conversion of a solids–amendment mixture's rapidly biodegradable components into substances similar to soil humus, which decomposes slowly. Insufficiently mature compost will reheat and generate odors when stored and rewetted. It also may inhibit seed germination (by generating organic acids) and plant growth (by removing nitrogen as it decomposes in soil). *Stability* refers to the reduction in microbial-degradation rate of the mixture's biodegradable components. Stabilization is achieved by maintaining optimal conditions for a sufficient period of time. Cellulose materials (e.g., wood and yard wastes) take longer to decompose than wastewater residuals, so screening out the bulking agent may improve stability.

There are a number of testing methods and standards for measuring compost stability or maturity, but none is universally accepted (Jimenez and Garcia, 1989). The standards associated with each test are still tentative, and much work needs to be done to correlate test results with odor generation and plant growth. A complete assessment of maturity may require multiple tests.

Volatile solids (as a percentage of total solids) is not a good measure of stability because it fails to account for the biodegradation rate.

Respiration tests, which measure carbon dioxide production or oxygen demand, better represent stability but are sensitive to test conditions. Carbon dioxide production typically is measured directly on the mixture in an incubator. Incubators are useful compost simulators that can effectively measure carbon dioxide productions in both highly unstable samples (from early in the process) to highly stable samples such as finished compost. Oxygen uptake rates can be measured on the mixture, or in an aqueous extract via the specific oxygen uptake rate (SOUR) test. Mature compost should have a carbon-to-nitrogen ratio that is less than 20:1. Available carbon in compost can deplete the nitrogen in soil that microorganisms typically use.

Seed-germination and root-elongation tests measure phytotoxicity caused by organic acids in compost. They are performed by germinating seeds (e.g, cress) in a filtered extract of compost and comparing them with a control using distilled water.

4.2.3 *Drying*

To dry compost, operators provide enough aeration or agitation to facilitate the removal of water vapor. This increases the solids content from about 40 to 55% or more. Drying is critical in processes that include screening, because screens do not perform well if the compost contains less than 50 to 55% solids.

4.3 Description of Composting Methods

Although composting is a naturally occurring biological process, the degree of control imposed on a system can range from periodically turning a pile or windrow to the more involved enclosed or in-vessel system with mechanical agitation and forced aeration.

In an attempt to respond to local and regional needs, a number of composting methods have evolved. These methods offer the following benefits: accelerating a naturally occurring biological process; providing for process control over variables such as moisture, carbon, nitrogen, and oxygen; containing odors and particulates; reducing land area requirements; reliably producing consistent product quality; and integrating aesthetically pleasing facilities into local and regional sites.

4.3.1 Aerated Static-Pile Composting

Aerated static-pile composting is also called the *Beltsville Method* because it was developed in Beltsville, Maryland, in the 1970s by the U.S. Department of Agriculture. As the name suggests, it involves aerating piled feedstock (see Figure 25.75). This flexible method is popular in the United States.

In this method, the solids–amendment mixture is constructed into a 2- to 4-m-deep (6- to 12-ft-deep) pile over an aeration floor (plenum) and then covered with a 150- to 300-mm-deep (6- to 12-in.-deep) insulating blanket of wood chips or unscreened finished compost to ensure that all of the mixture will meet the temperature standards for pathogen and vector-attraction reduction. Small operations may construct individual piles, while large ones may divide a continuous pile into sections representing each day's contribution. The mixture typically remains in the pile for 21 to 28 days while the plenum forces air through the material to provide an aerobic composting environment. Then the piles are broken down, and the material is either moved directly to a curing

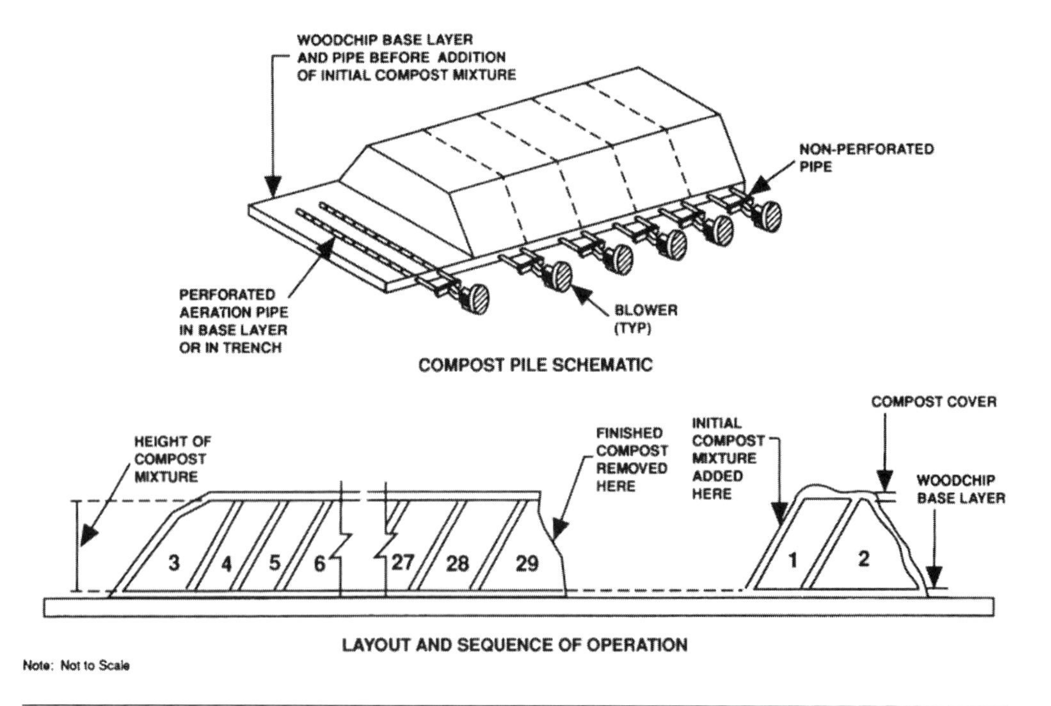

FIGURE 25.75 Schematic of an aerated static-pile composting system.

area, or screened and then moved to the curing area. Compost must contain at least 50 to 55% solids before screening. In some facilities, an intensive drying step (with a higher aeration rate than active composting) precedes screening. Compost typically is cured for at least 30 days to further stabilize the material. Some facilities screen the compost after curing (rather than before curing).

Aerated static-pile composting originally was developed for outdoor sites, but many systems are either partially or fully enclosed to control odors or facilitate operations during unfavorable environmental conditions (e.g., temperature or rainfall extremes).

4.3.2 *Windrow Composting*

In windrow composting, the solids–amendment mixture is formed into long parallel windrows whose cross-sections are either trapezoidal or triangular (see Figure 25.76). The material then is turned periodically by a front-end loader or a dedicated windrow-turning machine to release moisture, expose more particles to the air, and loosen (fluff) the material to facilitate air movement through the windrow.

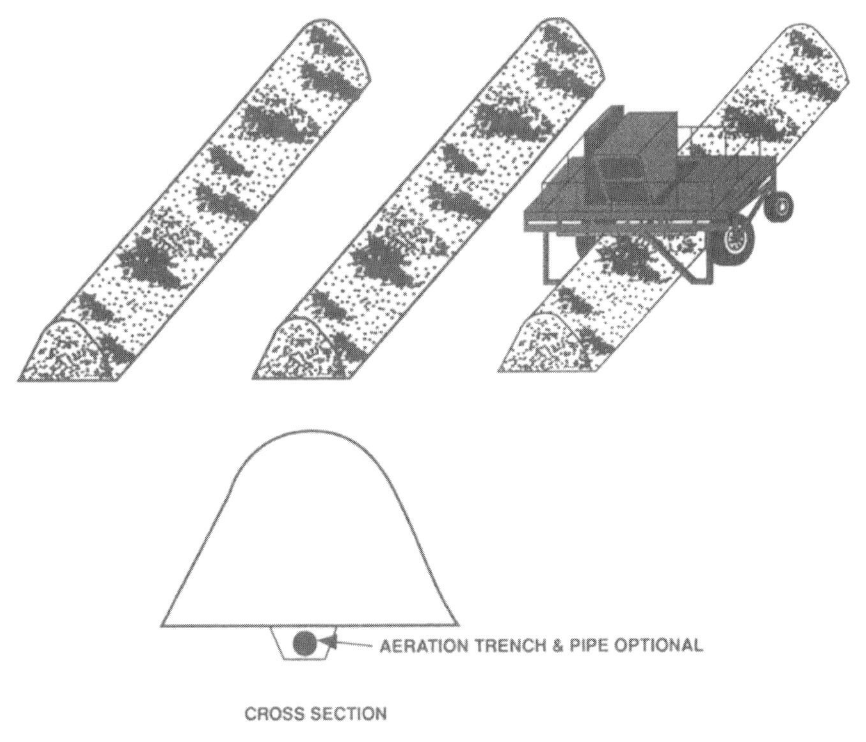

FIGURE 25.76 Schematic of a windrow composting system.

In the aerated windrow method, windrows are constructed over air channels to protect aeration piping from the turning equipment. Air can either be forced up through the windrow or pulled down through the windrow into the channel. The windrows are turned periodically to expose more particles to air. Aeration and turning optimize the composting rate and release of moisture.

Windrow composting occurs at open outdoor sites or covered sites. This system needs more space than other composting technologies because of pile geometry and the room needed to maneuver a windrow-turning machine.

4.3.3 *In-Vessel Composting*

In-vessel systems typically combine aeration with some type of automated material movement in a reactor. A wide variety of such systems has been developed over the years, but only a few have been installed in more than one or two sites.

The solids retention time ranges from about 10 to 21 days, depending on system-supplier recommendations, regulatory requirements, and costs. It also should be based on desired product characteristics—especially stability—and take into account the overall solids residence time in the entire composting operation (all process phases). Once discharged from the reactor, the composted biosolids typically must be further stabilized for 30 to 60 days to achieve the desired product stability.

There are basically three types of in-vessel composting systems: vertical plug-flow reactors, horizontal plug-flow reactors, and agitated bay systems. Vertical plug-flow reactors are made of steel, concrete, and/or reinforced fiber-glass panels (see Figure 25.77). A mix of dewatered cake, amendment, and recycled solids is loaded in the top of the reactor, where it is aerated but not agitated (mixed). It moves as a plug to the bottom of the reactor, where it is removed via a traveling auger.

Horizontal plug-flow reactors are similar to vertical ones, except that the solids–amendment mixture is moved laterally through the reactor by a hydraulic ram (see Figure 25.78).

Agitated-bay reactors are open-topped bays with with blowers and piping systems that supply air from the bottom (see Figure 25.79). Unlike plug-flow reactors, they also have mechanical devices that periodically agitate the mixture during its stay in the reactor. These systems are designed to function much like aerated windrows. A variety of methods are used to transfer compost from the reactors.

The most commonly used in-vessel system is the horizontal agitated-bed reactor. These reactors are rectangular, aerated from the bottom with independently programmable aeration zones, and enclosed in a building. A loader places the solids–amendment mixture into the front end. The agitation device is completely automatic, operates only in agitation mode, and typically makes one pass through the reactor each day. The

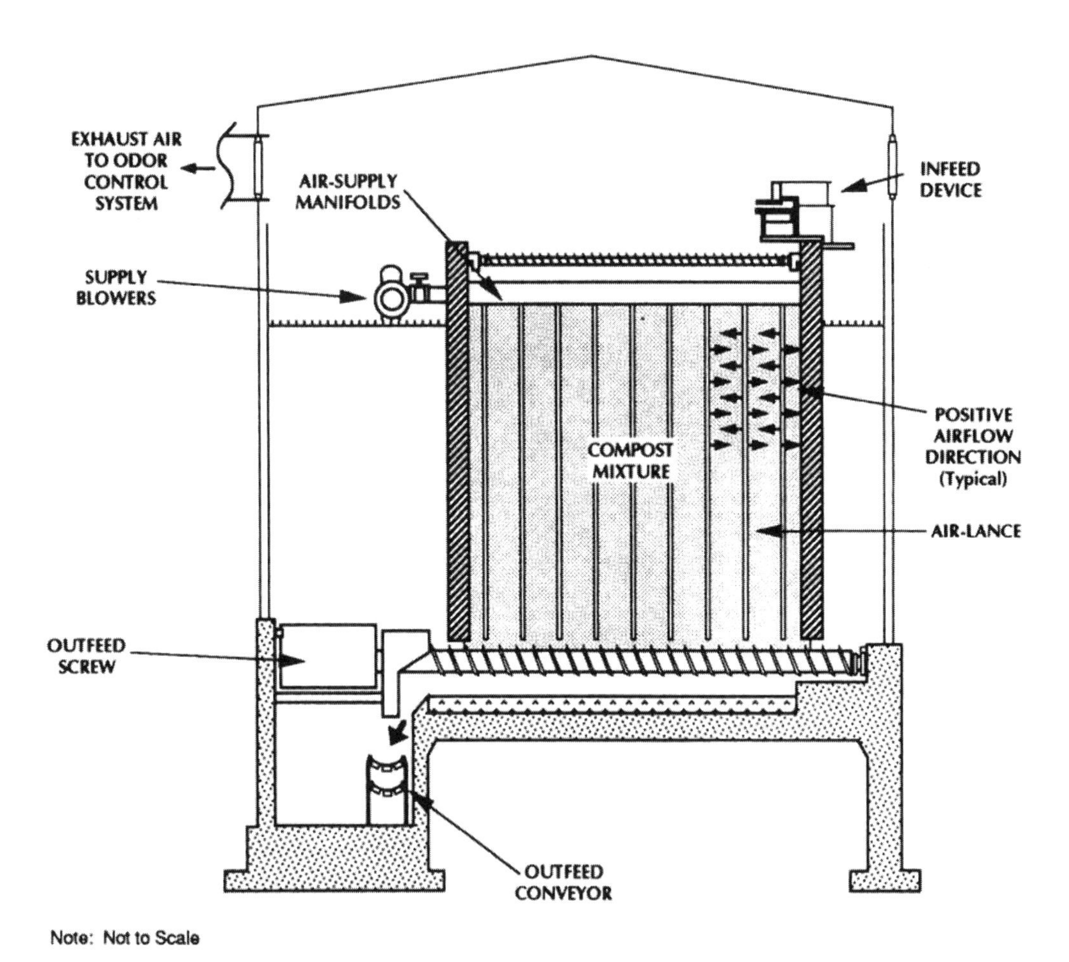

FIGURE 25.77 Cross-section of a vertical plug-flow reactor (rectangular design, made of steel).

composting material is dug out and redeposited about 4 m (11 ft) behind the machine until it has moved through the entire length of the reactor.

4.3.4 Comparison of Composting Methods

Of the three technologies discussed above, aerated static-pile is the most commonly used (see Table 25.34) (Biocycle, 1993; NEBRA, 2007) .

Table 25.35 lists the advantages and disadvantages of five composting technologies based on physical facilities, processing aspects, and O&M. None is appropriate for every situation. The choice depends on many factors (e.g., climate, siting considerations,

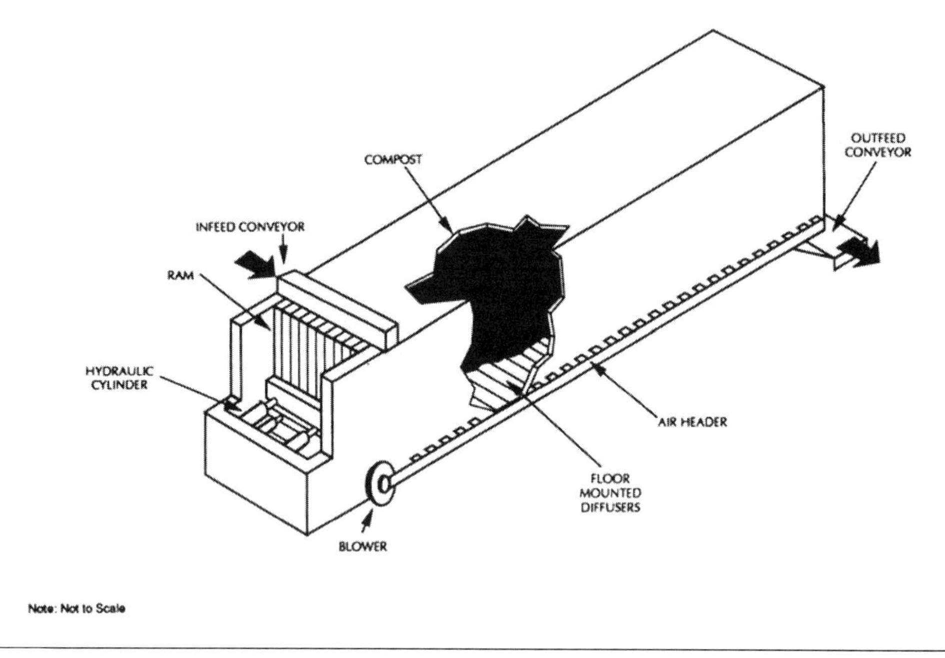

FIGURE 25.78 Schematic of a horizontal plug-flow reactor.

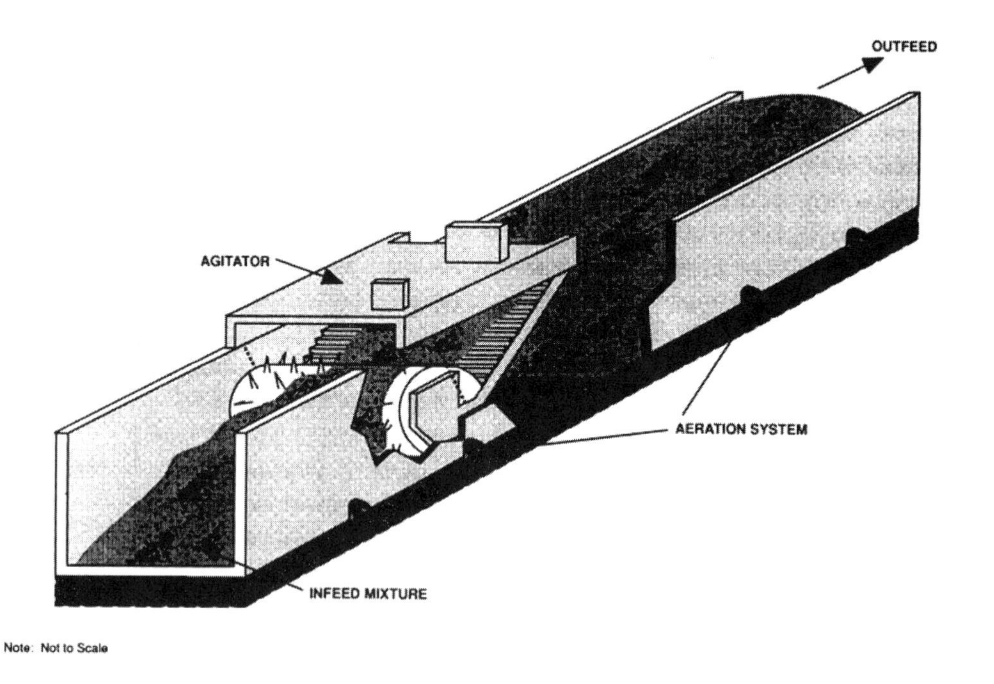

FIGURE 25.79 Schematic of a horizontal agitated-bed reactor.

TABLE 25.34 Representative list of composting facilities.

Facility	Type	Capacity (dry ton/d)*
Inland Empire Regional Composting Facility, California	Enclosed ASP	100
Davenport, Iowa	Enclosed ASP	25
Columbus, Ohio	Outdoor ASP	28
Rockland County, New York	Agitated bay	25
Hamilton, Ohio	Horizontal plug flow	15
Schenectady, New York	Vertical plug flow	50
Hawk Ridge, Unity, Maine	Tunnel (ASP)	15

* ton/d × 0.907 2 = Mg.

TABLE 25.35 Key advantages and disadvantages of composting systems.

Composting technology	Advantages	Disadvantages
Aerated static pile	• Adaptability to various bulking agents • Flexibility to handle changing feed conditions and peak loads (volume not fixed) • Relatively simple mechanical equipment	• Relatively labor intensive • Relatively large area required • Operators exposed to composting piles • Potentially dusty working environment
Windrow	• Adaptability to various bulking agents. • Flexibility to handle changing feed conditions and peak loads (volume not fixed) • Relatively simple mechanical equipment • Requires no fixed mechanical equipment	• Very large area required • Relatively labor intensive • Operators exposed to composting piles • Dusty working conditions
Vertical plug flow	• Completely enclosed reactors in some systems improve ability to control odors • Relatively small area required • Operators not exposed to composting material	• Single outfeed device per reactor (large reactors), potential bottleneck • Potential inability to maintain uniform aerobic conditions throughout reactor • Relatively maintenance intensive. • Limited flexibility to handle changing conditions • Materials-handling system may limit choice of bulking agents
Horizontal plug flow (tunnel)	• Completely enclosed reactors improve ability to control odors • Relatively smaller area required (composting mix compacted) • Operators not exposed to composting material	• Fixed-volume reactors (no flexibility) • Limited ability to handle changing conditions. • Relatively maintenance intensive • Materials-handling system may limit choice of bulking agents
Agitated bin	• Mixing enhances aeration and uniformity of compost mixtures • Ability to mix compost (advantage in handling some bulking agents • Adaptability to various bulking agents	• Fixed-volume reactors (no flexibility) • Relatively large area required • Potentially dusty working environment • Operators exposed to composting piles • Relatively maintenance intensive

operational concerns, and sensitivity to odors). Design engineers should consider the following factors when slecting a composting technology:

- Physical facilities (availability of space, materials-handling system complexity, aeration equipment, and degree of enclosure);
- Process considerations (e.g., uniform aeration, aeration type, availability of different bulking agents, adaptability to changes in volume of feed solids, and odor emissions/odor control); and
- O&M issues (e.g., labor requirements, energy requirements, operator exposure, dust generation, and degree of maintenance).

4.4 Process Considerations for Designers

This section provides ranges of design parameters for each stage of the composting process and identifies the design criteria essential to successful operation. Consideration is made for each type of composting technology.

4.4.1 *Bulking Agents and Amendments*

All composting technologies require mixing sufficient quantities of bulking agent with dewatered solids to adjust the initial solids content and provide porosity. Bulking agents also provide supplemental carbon to adjust the carbon-to-nitrogen ratio and energy balance. Table 25.36 lists some typically used bulking agents and their characteristics.

Although yard debris can be used as a bulking agent, grass clippings and substantially green yard waste are unsuitable because of their high water and nitrogen content, and lack of porosity. If grass clippings and substantially green yard waste are composted, they also will require supplemental bulking agent.

4.4.2 *Characteristics of the Solids–Amendment Mixture*

The ratio of bulking agent to biosolids depends on the available agent's characteristics and the desired solids content. For example, if dewatered cake contains 18 to 24% solids and the agent (a blend of woody yard debris) contains 55 to 65% solids, then the bulking agent-to-biosolids ratio must be 3:1 or 4:1 (by volume) to produce a mixture containing 40% solids. To produce a mixture containing 45% solids, the ratio should be 5:1 or 6:1. To produce a mixture containing 38% solids, the may be as low as 2.5:1. (Below 2.5:1, the mixture probably will not be porous enough to promote decomposition.)

The initial solids content needed depends on the composting technology used (specifically, the amount of agitation and aeration involved):

- Aerated static-pile systems need a mixture containing 40 to 45% solids. Wetter mixtures will lose heat energy to evaporation, thereby slowing the process.

TABLE 25.36 Types and characteristics of bulking agents.

Bulking agent	Characteristics
Wood chips (1 to 2 in.*)	• Must typically be purchased • High recovery rate in screening (60 to 80%) • Good source of supplemental carbon
Chipped yard or land-clearing debris	• May be available as waste material • Low recovery rate in screening (40 to 60%) because of higher percentage of fines • Green waste fraction adds nitrogen; more may be needed for C:N ratio • Good source of supplemental carbon
Ground waste lumber	• May be available as waste material • May be poor source of supplemental carbon if old and extremely dry because more volatile forms of carbon will be missing
Leaves	• Insufficient porosity to be used alone • Rapidly available source of supplemental carbon • Available as waste material • Not recovered by screening and adds to compost volume
Sawdust	• Insufficient porosity to be used alone • Rapidly available source of supplemental carbon • Must be purchased and typically is expensive • Not recovered by screening and adds to compost volume
Shredded paper	• Insufficient porosity to be used alone • Rapidly available source of supplemental carbon • Available as waste material • Not recovered by screening and adds to compost volume

* in. × 25.4 = mm.

Drier mixtures may not provide enough moisture to complete the biological process.

- Turned windrow systems need a mixture containing about 45% solids. In wet climates, however, the mixture should be slightly drier to compensate. Wetter mixtures will not be porous enough to allow for convective airflow.
- Automated loading tunnel and vertical plug-flow systems need a mixture containing 40 to 45% solids. Agitated bay systems, however, need a mixture containing 38 to 40% solids because the frequent agitation and forced aeration will dry the material much faster than other systems. Experience has shown that an agitated bay system can lose as much as 2% moisture during one agitation period.

It is critical that the mixture has uniform porosity and that all particles of cake be in close contact with the bulking agent. Dewatered cake with 18 to 25% solids should be mixed with a bulking agent so each wood chip or other bulking agent particle is coated with a thin layer of solids. Dewatered cake with 30 to 35% solids will break into clumps that must be uniformly small and mixed with the bulking agent. (Large clumps and balls will become anaerobic, leading to excessive odors.) If mixing is not uniform, zones with a disproportionate amount of bulking agent will divert the flow of air, allowing other zones to become anaerobic.

4.4.3 *Calculation of Materials Balance*

Materials-balance calculations track the weight and volume of each material through each stage of the composting process. Table 25.37 shows a typical materials balance for 1 dry ton of biosolids (20% solids) in an aerated static-pile process (see Figure 25.75).

In this process, solids were mixed with yard waste, stacked over a layer of yard waste to provide air distribution, and covered with a layer of unscreened compost. The entire pile (except for the volume reserved for the cover layer) was screened after composting, and the oversized particles were recycled as a bulking agent. Screening typically recovers between 50 and 80% of the bulking agent (by volume), so it must be supplemented with makeup bulking agent. The recovery rate depends on the compost's moisture content (stickiness), the bulking agent's particle size, and the screen's loading rate. Because some of the bulking agent is recycled, it is important to account for all of this material and balance recycled and new bulking agent so all recycled agent is used.

The required input assumptions are the density of each material, the volatile solids reduction of each input, and the screen's recovery efficiency.

4.4.4 *Temperature Control and Aeration*

In the United States, each state is responsible for regulating biosolids use within its borders. However, the federal government has issued minimum guidelines that all states must meet: 40 *CFR* 503 (typically called *Part 503* or *the 503 regulations*). These regulations require that solids treatment processes meet certain requirements to produce a biosolids that will not endanger the environment or public health. The specific requirements for composting depend on the technology used.

In addition to meeting regulatory requirements, composting systems also need to control temperatures to optimize decomposition. The optimum temperature range for volatile solids destruction is about 55 to 60°C (131 to 140°F). Part 503 regulations require pathogen kill temperatures of 55°C for aerated static pile and in-vessel systems, for 14 days for windrow systems with 5 turnings during the 14-day period. Fourteen days with an average temperature of 45°C with a minimum of 40°C are required for vector

TABLE 25.37 Materials balance for 1 dry ton of biosolids in aerated static-pile composting.*

Material	Volume (cu yd)	Total weight (ton)	Dry weight (ton)	Volatile solids (ton)	Bulk density (lb/cu yd)	Solids content (%)	Volatile solids (%)
Biosolids	6.3	5.0	1.0	0.5	1 600	20.0%	53.0%
Yard waste (processed)	10.9	3.3	1.8	1.3	600	55.0%	70.0%
Wood waste	0.0	0.0	0.0	0.0	500	60.0%	95.0%
Screened recyled bulking agent	9.9	3.5	1.9	1.8	695	55.0%	93.0%
Unscreened recycle	0.0	0.0	0.0	0.0	780	55.0%	88.6%
Mixture	25.7	11.7	4.7	3.6	911	40.1%	75.7%
Base (recycled bulking agent)	1.4	0.5	0.3	0.3	695	55.0%	93.0%
Cover (unscreened)	2.9	1.1	0.6	0.5	780	55.0%	88.6%
Composting losses		9.4	0.4				
Cover (unscreened)	2.9	1.1	0.6	0.5	780	55.0%	88.6%
Screen feed	21.5	8.4	4.6	3.5	780	55.0%	74.8%
Recycled bulking agent	11.4	4.0	2.2	2.0	695	55.0%	93.0%
Curing	9.9	4.4	2.4	1.4	900	55.0%	58.6%
Curing losses		0.2	0.1	0.1			
Compost to storage	9.5	4.3	2.3	1.3	900	55.0%	56.9%

Assumptions:

Recovery by screening

- Yard waste 50% by volume
- Wood waste 70% by volume
- Recycled bulking agent 50% by volume
- Pile base 95% by volume

Processing losses

- Losses during composting 10% of volatile solids
- Losses during curing 5% of volatile solids

* cu yd × 0.764 6 = m^3; lb/cu yd × 0.593 3 = kg/m^3; ton × 0.907 2 = Mg.

attraction reduction. In addition to maintaining certain regulatory dictated temperatures it is also desirable to prevent material temperatures from climbing too high. Pile temperatures in excess of 70°C inhibit the biological decomposition process. Also, if high temperatures persist for periods longer than several weeks, the potential of spontaneous combustion can occur in very dry material (>75% solids).

In turned windrow operations the temperature and oxygen content are controlled by the porosity of the windrows and the frequency of turning. Initial porosity is controlled by thorough blending of the feedstock and having the proper bulking agent to biosolids mix ratio. Once the windrows are in place both temperature and oxygen content are controlled by turning of the windrow. Turning incorporates oxygen, and releases heat and moisture. Although turning releases heat, the pile temperature will spike upwards shortly after turning. This is the result of the redistribution of feedstock and the infusion of oxygen. These spikes are typically short lived (a few hours).

In aerated static-pile and in-vessel composting systems forced aeration is used to supply oxygen and maintain aerobic conditions within the material, control temperatures, and remove moisture. In the first one to two days of composting increasing airflow typically kick-starts the process and causes pile temperatures to rise quickly. However throughout the rest of the process as the rate of airflow is increased in a forced aeration system, the pile temperature decreases and the rate of water vapor removal increases. As with a turned windrow system, agitation releases heat and water vapor.

Higgins et al. (1982) reported that an aeration rate of 34 m^3/Mg·h (1 100 cu ft/hr/dry ton) provided adequate drying and high enough temperatures for pathogen destruction. Early in the composting process, higher aeration rates may be needed to prevent excessive pile temperatures.

To maintain temperatures less than 60°C during peak activity, aeration rates may need to approach 300 m^3/Mg·h (10 000 cu ft/hr/dry ton). Such aeration capacity may be impractical in large systems. Practical aeration capacities are in the range of 90 to 160 m^3/Mg·h (3 000 to 5 000 cu ft/hr/dry ton) of wastewater solids. Aeration in this range will control temperatures throughout most of the composting period and provide adequate moisture removal capacity. Higher aeration rates are possible, but require more energy, larger and more closely spaced piping or trenches, and larger odor-collection and -treatment systems (if provided). If highly reactive bulking materials are used (e.g., ground-up leaves) the mass of bulking agent and dewatered cake may enter into the sizing of aeration capacity.

4.4.5 Detention Time

The time required to stabilize organic material typically is divided between an active composting stage and a curing stage (see Figure 25.80). When the U.S. Department of Agriculture (USDA) developed the aerated static-pile process, researchers found that 21 days of aerated composting followed by 30 days of unaerated curing would adequately stabilize a raw feed with wood chips as the bulking agent. However, to create fully stabilized compost suitable for any use, another 20 days or more of detention time is recommended. This detention time criterion has been codified in a number of state regulations and incorporated into some design standards. Most horizontal agitated-bed systems are designed for 21 days of aerated composting followed by curing. However, other in-vessel systems use shorter active composting times (often 14 days) to minimize

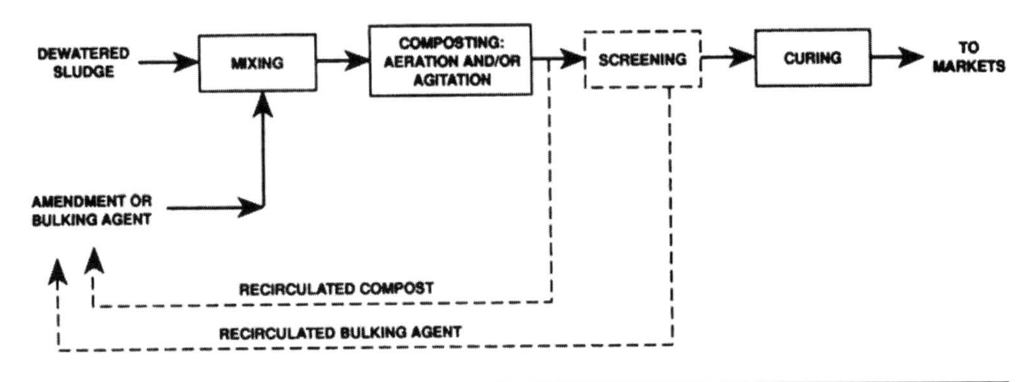

FIGURE 25.80 Generalized composting flow chart (dashed lines indicate optional steps; screening may follow curing; drying step may precede screening).

the system's capital costs. Additional detention time outside the vessel (in the form of windrow or static pile systems) are typically added to these systems.

Detention time is affected by the bulking agent or amendment, carbon-to-nitrogen ratio, and pH. An amendment that is not screened out may continue to decompose, prolonging the curing period. An excessively high carbon-to-nitrogen ratio may have the same effect. The composting process has no fixed endpoint because the organic materials continue to decompose after the compost is considered stable. One test for stability is based on a respiration rate measured as a rate of carbon dioxide evolution. A respiration rate of 3 mg CO_2/g organic carbon per day typically indicates that compost will be free of fecal odor and phytotoxic effects.

Another test measures oxygen consumption. Jimenez and Garcia (1989) report that a compost taking up 0.96 mg O_2/g of organic carbon per day is considered stable. This is equivalent to 1.4 mg CO_2/g of carbon per day.

4.5 General Design Considerations

The essential elements of composting facility designs involve handling large volumes of material and air. The relative importance of each depends on the composting technology used. For windrow operations, air handling is negligible or nonexistent; for enclosed operations, proper air handling is vital.

4.5.1 Site Layout

As with any facility, layout is dictated by the available site; however, there are a few items to keep in mind. Because of the large volumes of material handled, all composting operations involve the use of heavy equipment (e.g., front end loaders and trucks). Concrete or high-durability asphalt-paved pads serve best for bulking agent storage, mixing, pile construction, screening, curing, and finished compost storage areas. Runoff from any areas exposed to raw feedstock must be collected and treated. Typically, plants are designed with covered areas for bulking agent storage and composting aerated static-pile systems. Covered facilities can be operated under adverse weather conditions and generate minimal runoff.

4.5.2 *Material-Handling Systems*

Material typically is moved around a composting facility by either a front-end loader or conveyor. In-vessel technologies have special equipment for moving material in the vessel for portions of processing, but they still rely on the loaders and conveyors for most of the material movement.

Bulking agents, biosolids, and finished compost have relatively low densities, so light-material, large-volume buckets can be used on front-end loaders. Rollout or pushout buckets are also advantageous because they allow for more vertical and horizontal reach.

The two most commonly used conveyors are belt and screw conveyors. Belt conveyors must have shallow incline angles, or material will tumble or flow backwards. The incline angle depends on the material being conveyed, but typically the maximum is 15 to 20 deg. Screw conveyors allow steeper inclines but will require more maintenance. Biosolids typically have some grit and can abrade screws.

4.5.3 *Bulking Agent Storage and Handling*

Ideally, storage is provided for a 15- to 30-day supply of bulking agent. A paved, covered storage area minimizes excessive moisture accumulation in both bulking agent and finished compost.

Enclosing the unloading and conveying facilities minimizes the spread of dust and particulate, and protects the equipment from the adverse effects of wet and cold weather. Because the dust could explode, the design of any enclosure should adhere to applicable explosion hazard standards.

4.5.4 *Mixing*

As previously indicated, dewatered solids must be well mixed with a bulking agent to ensure uniformity and good airflow characteristics during composting. Therefore, a mechanical mixing system typically is included in the design. A good mixture consists of bulking agent particles uniformly coated with solids containing no balls of dewatered solids that are more than 126 mm (5 in.) in diameter. Immediately mixing dewatered solids with a bulking agent minimizes storage-facility size and the potential for odor generation. A solids and bulking agent mixture can be stacked and conveyed more easily than dewatered cake alone.

Several types of mechanical mixing systems are available:

- Front-end loaders portion feedstock in discrete piles and "toss" the material several times until it is blended (much like how a salad is tossed). Mixing is time consuming, and not particularly effective. Loaders are best suited for small facilities and as backup for another mixing system.
- Batch mixers are stationary, truck-mounted, or trailer-mounted hoppers equipped with internal paddles or augers that mix the material. The blended batch is discharged via a short, side-mounted conveyor with a slide gate. Batch mixers also have internal scales and a weight display to help operators portion the material. They typical are loaded by front-end loaders but also can be loaded via a conveyor from a live-bottom hopper. Batch mixers are well suited for small and medium facilities.
- Continuous mixers (e.g., pug mills and plow mixers) are the most automated, complex mixing systems. In these systems, feedstocks first are loaded into separate live-bottom hoppers, which have variable-speed augers to meter the correct portions of each feedstock. (Feedstocks are weighed by elements in the hoppers' discharge conveyors.) The material is conveyed to the mixers for blending, and afterward, conveyors discharge the blended material into or near the composting piles or vessels. Continuous mixers are found only at medium and large facilities, where their capital costs are offset by labor cost savings.
- Windrow turners are mobile machines designed to mix materials that have been layeared on a concrete pad. The machines vary in size and complexity. Small machines towed by a tractor can stack material 0.6 to 0.9 m (2 to 3 ft) high. Large self-propelled machines can form piles about 2.4 m (8 ft) high. This technology works well in windrow composting operations.
- A horizontal agitated-bed reactor also provides mixing. However, the material should be premixed before being loaded into such reactors to optimize the reactors' SRT.

Mixing and storage areas are odorous and, if enclosed, typically need at least six air changes per hour for effective odor control and personnel safety. Design engineers should consider treating exhaust airstreams from these process areas before discharging to the atmosphere.

4.5.5 *Leachate*

All composting processes produce leachate that must be treated. Some common sources of leachate sources in composting operations include:

- Aeration pipes and ducts;
- Building ventilation ductwork;
- Composting piles;
- Washdown water for all mobile and stationary equipment;
- Biosolids and recycled bulking agent storage areas; and
- Site drainge from areas exposed to unfinished compost, recycled bulking agent, and biosolids.

Aeration fans and ductwork—especially in negative-mode aeration—must be equipped with drains and cleanouts in all low areas. Even in positive aeration, ventilation fans and ducts will collect condensation, so they must have adequate drains and cleanout access. Drains for compost piles are often part of the aeration floor and must be equipped with traps to prevent air from short-circuiting the process.

All stationary and mobile equipment must be washed down periodically to keep it in good working order. For mobile equipment, a designated washdown area is often part of the facility. For stationary equipment, drains must be provided. All conveyor and other equipment pits also should have drains both for washdown and condensation, which occurs in enclosed facilities.

All of the water from the above sources, as well as any water that contacts unfinished compost or raw materials, must be collected and treated. Water is a byproduct of decomposition, so leachate can contain soluble organics, nutrients and other material that cannot be released to the environment. Leachate should be discharged to a sanitary sewer, recycled to the treatment plant's headworks, or treated onsite.

4.5.6 *Aeration and Exhaust Systems*

Compost typically can aerated either by forcing air up through the material (positive aeration) or pulling it down through the material (negative aeration). Positive aeration typically requires less energy than negative aeration to move the same volume of air. In positive aeration, the air is cooler, drier, and therefore, has less volume.

Negative aeration is better in enclosed and worker-occupied operations because it captures most of the material's odors and moisture, preventing them from entering the air above the pile (where greater airflows are required to capture and treat such emissions). However, condensation accumulates in the ductwork and blowers, so ample drainage must be provided.

Most in-vessel systems use positive aeration for system-specific reasons. For example, in-tunnel systems have little headspace above the piles and are not occupied by workers during active composting, so there is no advantage to negative aeration. Negative aeration is popular in aerated static-pile operations because it directly captures odors and moisture. Many aerated static-pile operations are configured to allow for both negative and positive aeration. During decomposition, negative aeration captures odors and moisture. Afterward, positive aeration can provide more air to accelerate drying before the compost is screened.

Figures 25.81 through 25.84 show several air-floor configurations for aerated static-pile systems. For example, agitated bay systems use perforated pipe embedded in a gravel plenum. No matter which configuration is used, it is vital that it be designed to deliver air evenly the entire length of the pile. Three methods are used to accomplish this:

- Provide progressively more air outlets along the pipe or trench so friction headloss is offset by reduced velocity loss through the outlets;
- Change the cross-section of the pipe or trench to provide a constant air velocity along the entire length; or
- Use a combination of the two.

Pipes or trenches typically are spaced 1 to 2 m (3 to 6 ft) apart on a layer of wood chips, which help distribute airflow. The spacing depends on the size of the pipe or trench; larger elements require more space. If the pipes or trenchew are too far apart, however, anaerobic zones develop in the bottom of the piles because air always seeks the path of least resistance.

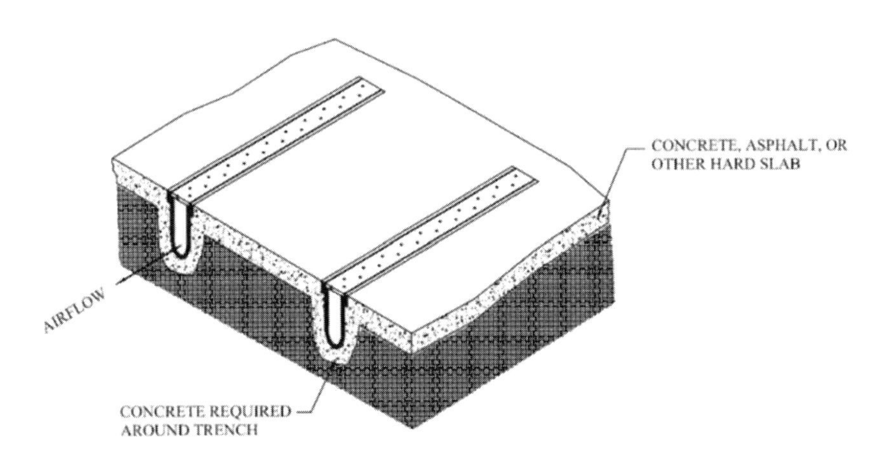

FIGURE 25.81 Composting floor with aeration trenches.

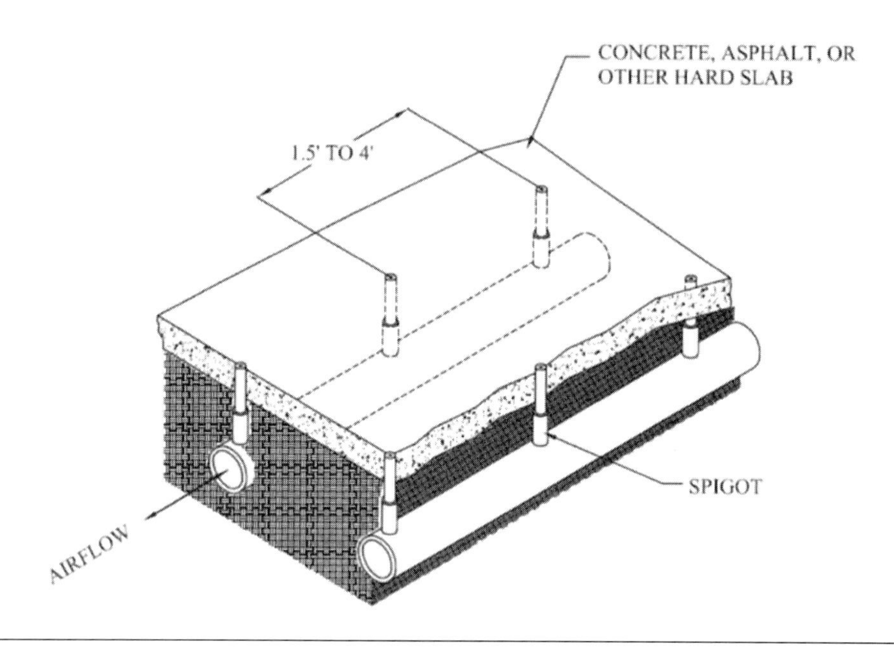

FIGURE 25.82 Composting floor with embedded pipe and spigot aeration system.

In-vessel systems may use continuous plenums or gravel floors with permanent piping. All of them will require regular cleaning. Many gravel plenums develop a hard pan on the surface that will block and redirect airflow if not routinely removed.

In aerated static-pile systems, the air outlets in pipes or trenches can become blocked with material, particularly when equipment moves over the outlets to add and

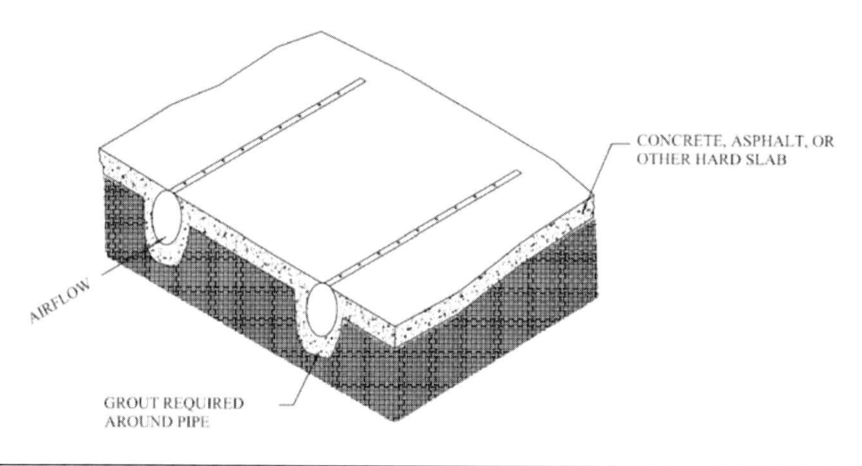

FIGURE 25.83 Composting floor with embedded pipe aeration system.

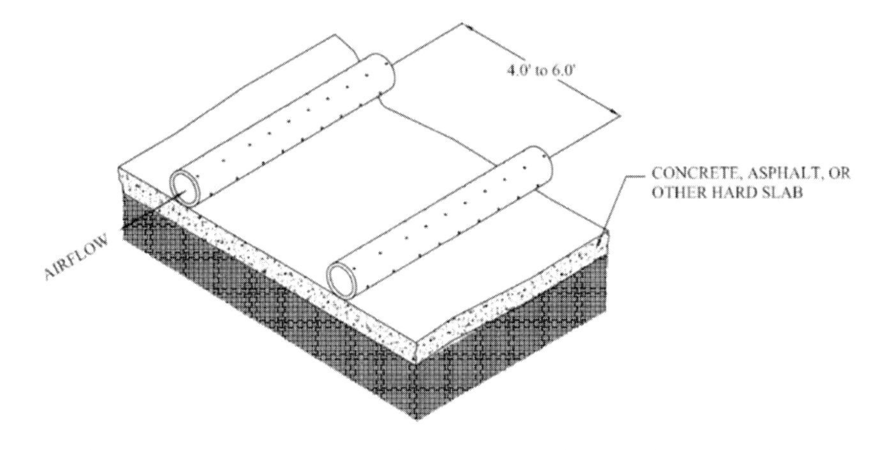

FIGURE 25.84 Composting floor with aeration pipes on slab.

remove material. The outlets must be cleaned after every one or two uses, using either water or compressed air.

In negative aeration, the air initially is hot and virtually saturated, but cools slightly as it moves through the duct. So, condensation forms and must be removed via frequent drains and cleanout. Drains traps also will be needed to prevent airflow from short-circuiting. Because of the heat and moisture, negative aeration systems need corrosion-resistant ductwork. Fiberglass, PVC, polyethylene, and stainless steel have all been used.

When operating composting systems with forced aeration, O&M personnel need to be able to:

- Monitor and record pile temperatures and
- Control aeration quantities based on oxygen demand, temperature, and moisture removal requirements.

The simplest control system involves measuring and recording pile temperatures manually and controlling aeration blowers via a manually adjusted cycle timer. The most complex system includes a temperature feedback control where outputs from temperature probes in the piles are connected to a computer, which adjusts the aeration rates based on temperature readings using a preset control strategy.

All control systems used there are certain rules that should be observed in controlling the aeration:

- During active composting (when the material is heating up), the blowers typically should not be off for more than 15 minutes in a cycle. Murray and Thompson

(1986) reported significant oxygen depletion after 12 to 15 minutes without aeration (see Figure 25.85).

- The temperature of the composting material should be measured directly, whenever possible. Although this seems obvious, some systems measure temperatures via sensors in the ductwork or in the walls contacting the piles (to avoid damage from agitation). Such sensors consistently provide measurements that are lower than the actual pile temperature, causing the material to be underaerated.

Because the material's aeration demand constantly changes during composting, blowers must be able to provide a varying amount of airflow to the material. Design engineers can either provide single- or two-speed blowers, which will operate intermittently, or provide variable-frequency drives so the blowers can run continuously but the airflow is varied based on process needs. Another option is equipping single- or two-speed blowers with motorized dampers to regulate the amount of airflow supplied to each pile based on process needs.

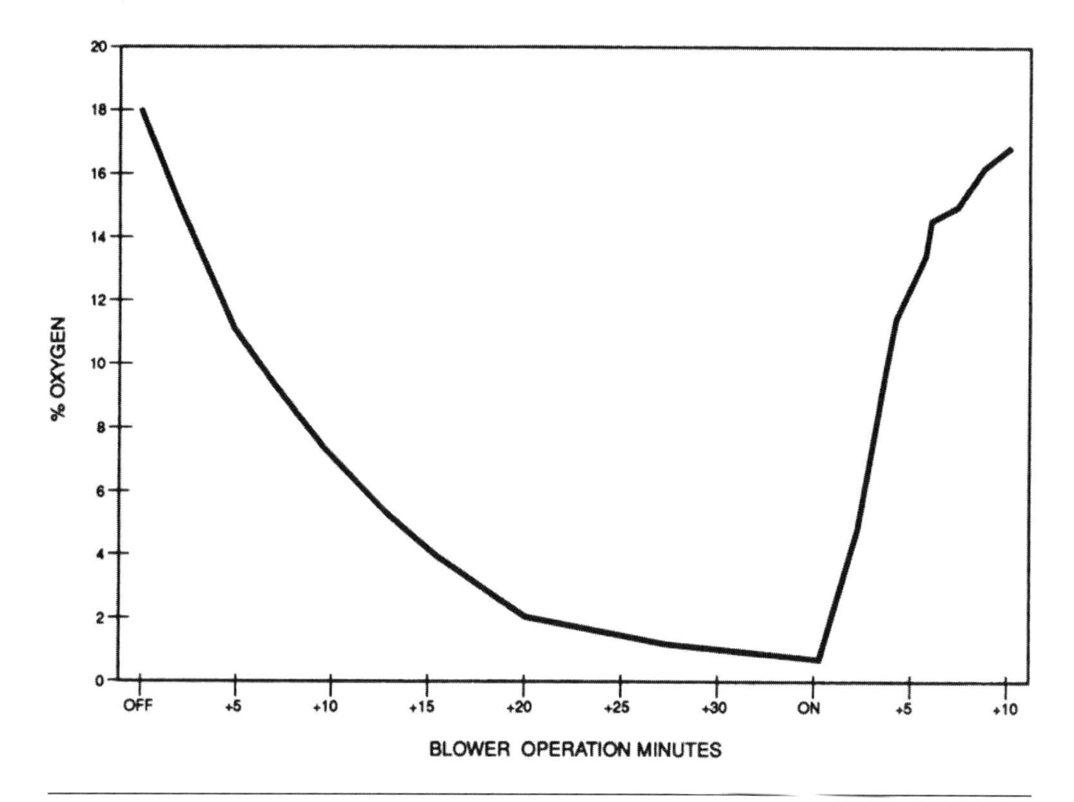

FIGURE 25.85 Oxygen depletion and regeneration in an active compost pile (Murray and Thompson, 1986, with permission from *BioCycle*).

4.5.7 *Ventilation*

The National Fire Protection Association issues ventilation rules related to fire prevention; however, in enclosed composting facilities, ventilation rates typically must be larger to control odors, prevent fogging, and ensure that workers are safe. In cold climates, heavy building insulation is needed to prevent fogging in winter and avoid worker heat stress in summer.

Because the material often is moved around a facility by a front-end loader, design engineers must consider what happens when doors are left open for extended periods. How will that affect ventilation rates and ductwork design? They also must locate the air-collection points in the building to avoid dead air spaces, where ammonia and other compounds can accumulate.

4.5.8 *Screening*

Except for leaves and sawdust, bulking agents can be screened out fo the finished compost and reused. This reduces bulking agent costs by 50 to 80%. Screening also produces more uniform, aesthetically pleasing compost, thereby improving its marketability. Vibrating deck screens and rotating screens typically are used. All screen must have a self-cleaning feature (e.g., rotating brushes in rotating trammel screens, or a layer of balls between the decks of a vibrating deck screen).

Vibrating deck screens and rotary trammel screens can separate material into multiple sizes, which can be useful if some markets (e.g., turf top dressing) demand a product with fine particles.

4.5.9 *Product Curing and Storage*

Composting basically has two phases: a rapid decomposition period (14 to 21 days) followed by a longer, slower one with significantly lower oxygen and moisture-removal demands. This second phase (called *curing*) is typically about 30 days long and is needed to produce a stable, usable product. Sometimes curing consists of merely stockpiling the material, but this can prolong curing time and increase the danger of fires. Low-rate aeration better controls curing time and product stability. Also, the curing material should be covered to control its moisture content and thereby prevent the material from compacting and going anaerobic. This is especially important if the material was screened first.

4.5.10 *Odor Control*

Odor control may be the composting industry's greatest challenge. Most conflicts over and suspensions of composting operations have been caused by odors or concerns about potential odors. Composting is inherently odorous as a result of the production and removal of volatile products of decomposition. Current design practices include more emphasis on enclosing operations, capturing and treating exhaust air, and improving process control to reduce odors at the source.

4.5.10.1 Odor Sources in Composting

Every stage in the composting process is a potential source of odors (see Tables 25.38 and 25.39). Odor sources can be divided into the following three categories:

- *Active sources* are those that exist when material is being actively handled (e.g., during mixing, screening, and dewatering). Odors from these sources occur during working hours.

TABLE 25.38 Typical odor sources in composting operations.

Odor source	Category*
Dewatered sludge transport and storage	
Open trucks en route	A
Open trucks parked on site	A
Dumping operations	C
Untreated ventilation from storage facilities	C
Open conveyors	A
Spillage from trucks	H
Spillage around storage facilities	H
Residue on empty trucks	H
Puddles from truck washing	H
Tire trucking of spillage sludge	H
Mixing	
Surface emissions from mixing by front-end loader or batch mixer	A
Untreated ventilation from pug mills	A
Mix left on paved surface after day's activities	H
Residue on equipment	H
Pile building	
Surface emission from materials-handling activities	A
Spillage of mix left on paved surface after day's activities	H
Residue on equipment	H
Sludge balls from poor mixing	H
Surface emissions from pile before placement of blanket layer	A
Composting	
Surface emission from active piles	C
Leachate puddles at base of piles	H
Aeration	
Blowers exhaust	C
Leakage of condensate from aeration piping	H
Leakage of exhaust from piping and blower housing	H

* A = active source; C = continuous sources; and H = housekeeping sources.

TABLE 25.39 Odor compounds and sources (Verscheueren, 1983; WEF, 1995).

Class compounds	Odor threshold, ppm	Likely source at treatment plant	Pathway of formation/release
Inorganic sulfur			
Hydrogen sulfide	0.000 47	Septic wastewater or sludge	Anaerobic reduction of sulfate to sulfide or anaerobic breakdown of amino acids.
Organic sulfur			
Mercaptans			
Ethyl mercaptan	0.000 19	Sludge or wastewater subjected to anaerobic conditions	Anaerobic and aerobic breakdown of amino acids
tert-Butyl mercaptan	0.000 08		
Allyl mercaptan	0.000 05		
Organic sulfides			
Dimethyl sulfide	0.001	Composting	Aerobic oxidation of mercaptans.
Dimethyl disulfide	0.002		
Inorganic nitrogen			
Ammonia	0.037	Composting; processing of anaerobically digested biosolids	Anerobic decomposition of organic nitrogen volatilization at high pH, temperature
Organic nitrogen			
Methylamine	0.021	Solids processing	Anaerobic decomposition of acids
Ethylamine	0.83		
Dimethylamine	0.047		
Fatty acids			
Acetic acid	0.001–1.0	Sludge subjected to anaerobic conditions	Anaerobic decomposition
Propionic acid	0.005–0.05		
Gutyric acid	0.000 01–0.01		
Aromatics			
Acetone	0.05–400	Preliminary and primary wastewater treatment processes, solid processing, and composting	Present in wastewater contribution from industry; breakdown of lignins
Methylethyl ketone	1–12	Composting,wood-based bulking agents	
Terpenes	Varies		Present in wood products, such as wood chips, sawdust

- *Continuous sources* are those that originate in the aeration and storage areas. These may be point sources (e.g., blower exhaust) or area sources (e.g, pile and windrow surface emissions). Odors from these sources may occur 24 hours a day.
- *Housekeeping sources* are those related to material spills, unclean equipment, and condensate on ground surfaces. Such odors can persist after daily activity has stopped, so they are continuous sources.

4.5.10.2 *Odor Measurement*

The concentrations of individual compounds can be measured via standard analytical methods. For example, a simple apparatus consisting of a manual pump and a colorimetric adsorption tube can be used in the field. (Tubes are available for a number of the compounds listed in Table 25.39.) For more accurate and complete results, samples should be collected (in bags, stainless steel vacuum canisters, or tubes filled with adsorbent) and analyzed via gas chromatography in a laboratory.

However, the odor of composting typically is a mixture of compounds that cannot be quantified as a sum of individual constituents. Such odors can only be directly measured by the human nose (*sensory analysis*). Odor samples can be captured in Tedlar bags for sensory analysis at another location. Several methods have been developed to quantify odor concentrations using a panel of human subjects; these are described in detail in Chapter 7.

4.5.10.3 *Containment and Treatment*

The level of odor containment and control is dictated by the proximity of neighbors and local regulations. Design engineers must take care to provide for adequate capture of emissions under all operating conditions. For example, failing to account for material-movement operations that require open doors will lead to fugitive emissions.

Once contained and captured, odors can be treated or exhausted. Treatment typically is required. A wide variety of treatment technologies are available (see Chapter 7). Organic media biofilters have been used extensively at composting facilities for several reasons:

- They have proven effective in treating compost odors;
- They are inexpensive and easy to operate;
- Composting facilities typically are large enough to have space for the biofilter;
- The materials used for biofilter media are readily available at composting facilities; and
- The equipment used to replace biofilter media (e.g., front-end loaders) is available at any composting facility.

For more details on odor removal, see Chapter 7.

4.5.11 Design Example

When designing any in-vessel system, engineers need details from vendors; in fact, a vendor often is selected before the detailed design proceeds. When designing an aerated static-pile system, on the other hand, the details are not vendor-dependent. Below is an example of an aerated static-pile system design. The following design criteria apply to this example:

- 20 dry ton/d of cake containing 20% solids;
- Operations occur 7 days per week;
- The bulking agents are yard waste supplemented by ground wood waste;
- All storage, mixing, active composting, and screening operations are fully enclosed;
- Enough covered storage space for 30 days' worth of new bulking agent;
- Enough storage space for 1 day's worth of feedstock biosolids;
- Enough covered storage space for 7 days' worth of recycled bulking agent;
- 21-day minimum SRT in active composting area;
- 28-day minimum SRT in aerated curing area; and
- Enough outdoor storage for 90 days' worth of finished product.

In this exercise, the various areas of the aerated static-pile system will be sized. Each area's size depends on the types of vehicles expected and the site topography.

First, design engineers must develop a materials balance for the facility (see Table 25.40). The total amount of bulking agent that needs to be recycled is

$$\begin{aligned}\text{Total recycled bulking agent} &= \text{Screened recycled input} + \\ &\quad\ \text{Base (recycled bulking agent)} \\ &= 204 \text{ cu yd} + 28 \text{ cu yd} \qquad (25.38) \\ &= 232 \text{ cu yd } (177 \text{ m}^3)\end{aligned}$$

Comparing the materials balance with Table 25.37 and multiplying those values by 20, design engineers determine that adding drier ground-wood waste reduced the overall amount of active-composting feedstock from 393 to 389 m^3 (514 to 509 cu yd). It also reduced the amount of product produced from 145 to 138 m^3 (190 to 180 cu yd). The effect on product production is larger than that on feedstock volume and, therefore, facility size.

The following areas are constructed with concrete walls on three sides: active composting, curing, and storage areas for biosolids and all bulking agents (see Figure 25.86).

TABLE 25.40 Materials balance for design example.*

Material	Volume (cu yd)	Total weight (ton)	Dry weight (ton)	Volatile solids (ton)	Bulk density (lb/cu yd)	Solids content (%)	Volatile solids (%)
Biosolids	125.0	100.0	20.0	10.6	1 600	20.0%	53.0%
Yard waste (processed)	180.0	54.0	29.7	20.8	600	55.0%	70.0%
Wood waste	26.7	6.7	4.0	3.8	500	60.0%	95.0%
Screened recycled bulking agent	204.1	70.9	39.0	36.3	695	55.0%	93.0%
Unscreened recycle	0.0	0.0	0.0	0.0	780	55.0%	88.6%
Mixture	508.9	231.6	92.7	71.5	910	40.0%	77.1%
Base (recycled bulking agent)	28.3	9.8	5.4	5.0	695	55.0%	93.0%
Cover (unscreened)	56.5	22.1	12.1	10.7	780	55.0%	88.6%
Composting losses		182.8	7.1	7.1			
Cover (unscreened)	56.5	22.1	12.1	10.7	780	55.0%	88.6%
Screen feed	424.0	165.4	91.0	69.3	780	55.0%	76.2%
Recycled bulking agent	232.2	80.7	44.4	41.3	695	55.0%	93.0%
Curing	188.2	84.7	46.6	28.1	900	55.0%	60.3%
Curing losses		3.8	2.1	2.1			
Compost to storage	179.8	80.9	44.5	26.0	900	55.0%	58.4%

Assumptions:

Recovery by screening

- Yard waste — 50% by volume
- Wood waste — 70% by volume
- Recycled bulking agent — 50% by volume
- Pile base — 95% by volume

Processing losses

- Losses during composting — 10% of volatile solids
- Losses during curing — 5% of volatile solids

* cu yd × 0.764 6 = m^3; lb/cu yd × 0.593 3 = kg/m^3; ton × 0.907 2 = Mg.

For most front-end loaders, the maximum height will be 3 to 3.6 m (10 to 12 ft). In this example, the maximum height (H) is 3 m (10 ft). The following equation represents the total volume for a given area; it is manipulated to determine the desired value. For example, a narrow site may limit the allowable length (L). As a general rule, the width (W) should be at least 4.6 m (15 ft) for each days' worth of material. This is wide enough for a front-end loader to dig out the material while leaving the piles around it intact.

$$\text{Pile volume/d} \times \text{Number of days} \times 27 \text{ cu ft/cu yd} = H \times (L - H/2) \times W \qquad (25.39)$$

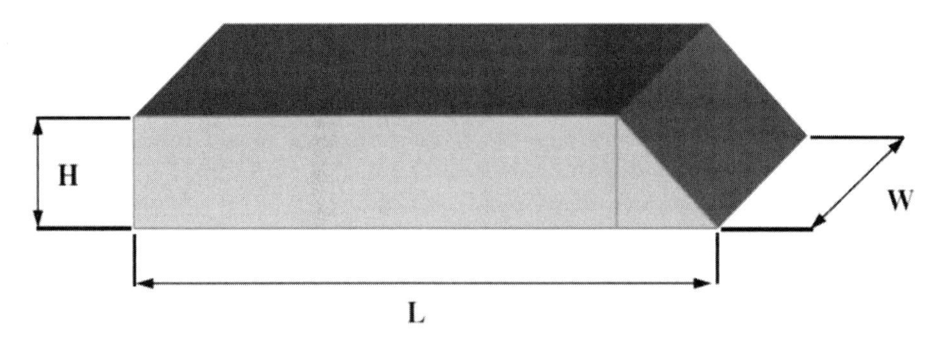

FIGURE 25.86 Sizing of aerated static pile (for design example).

For the bulking agent storage area (for 30 days' worth of ground yard waste) with an assumed length of 30 m (100 ft) and height of 3 m (10 ft),

$$W = \frac{204 \times 27 \times 30}{10 \times (100 - 10/2)} = 157 \text{ ft} \tag{25.40}$$

For a biosolids storage area with an assumed width of 9 m (30 ft) and height of 0.9 m (3 ft),

$$L/\mathrm{d} = \frac{127 \times 27 \times 1 + 3/2}{3 \times 30} = 38 \text{ ft (11 m)} \tag{25.41}$$

Biosolids typically are dense and gelatinous, so they do not stack well. Biosolids containing 18 to 24% solids will only stack about 0.9 or 1.2 m (3 or 4 ft) high. Wetter biosolids will not stack more than 0.3 m (1 ft) high.

When sizing the active compost area, design engineers should keep in mind that most composting facilities will put 1 days' worth of material in each bay. (Small facilities may put 2 or 3 days' worth of material in a bay.) The number of bays is depends on the desired SRT (21 days is the usual minimum). Two extra bays should be provided to allow for one bay to be torn down and another to be constructed without reducing SRT. Aerated static-pile facilities typically have active composting areas that are constructed with multiple bays on either side of a center aisle. A bay on one side typically serves as a mixing surge area (depending on the mixing method selected). There is no physical obstruction between bays (see Figure 25.75).

Below is the length of the active compost hall, based on an assumed width of 6 m (20 ft) and a mixture depth of 2.4 m (8 ft). Although the overall pile depth will be 3 m (10 ft), design engineers need to allow for 0.3 m (1 ft) of plenum layer and 0.3 m (1 ft) of cover layer. The minimum bay width should be 4.6 m (15 ft) so front-end loaders have enough space to build and tear down 1 days' worth of material.

$$L / \text{bay} = \frac{509 \times 27 \times 1 + 8 / 2}{8 \times 20} = 86 \text{ ft } (26 \text{ m}) \tag{25.42}$$

When calculating the composting hall's overall width, design engineers need to include allowances for the piles, the center aisle, and the aeration blowers (which typically are housed behind the piles). The minimum allowance for the blower gallery depends on the size of the blowers and ductwork, and the access to the area. If the only access to the blower gallery is from the ends of the compost building, the gallery must be wide enough to move blowers without dismantling them. If access doors can be put closer to the blowers, the hall can be narrower. In this example, a 4.6-m-wide (15-ft-wide) gallery is assumed.

The center aisle must be at least 9 m (30 ft) wide so front-end loaders have enough maneuvering space to construct and tear down piles. If the material from the active compost pile will be loaded directly onto trucks, which will deliver it to another location for curing, then the center aisle should be at least 13.7 to 15.2 m (45 to 50 ft) wide.

$$\begin{gathered}\text{Hall width} = 2 \times (86 + 15) + 30 + 4 \times 1 \\ \text{(allowance for concrete walls)} = 236 \text{ ft } (72 \text{ m})\end{gathered} \tag{25.43}$$

$$\begin{gathered}\text{Compost hall length} = 20 \times 12 + 2 \\ \text{(allowance for concrete walls)} = 242 \text{ ft } (74 \text{ m})\end{gathered} \tag{25.44}$$

In most aerated static-pile facilities, each bay (1 day's worth of material) is aerated separately (typically one blower per bay). This configuration provides the most flexibility and least interruption in operations if a blower goes out of service. At small facilities, however, one continuously running blower can serve several bays. In this example, one blower will be used for each bay.

$$\begin{aligned}\text{Required aeration} &= \frac{5\,000 \text{ cu ft/h/dry ton} \times 20 \text{ dry ton biosolids}}{60 \text{ min/hr}} \\ &= 1\,667 \text{ cfm } (787 \text{ L/s})\end{aligned} \tag{25.45}$$

4.6 Health and Safety Considerations

Potential dangers associated with composting systems include poorly ventilated areas, areas where exhaust gas is discharged, conveyors, and heavy equipment traffic. The primary concerns include:

- Fog generation in cold weather: Dense fog in a building with heavy equipment is an obvious hazard; it also may prevent others from seeing an injured worker.

- Worker heat stress: Composting generates significant amounts of heat. During warm weather, enclosed composting facilities can easily exceed 100°F for prolonged periods.
- Unsafe chemical concentrations: If not properly ventilated pockets of dead air can develop unhealthy concentrations of compounds (e.g., ammonia).
- Dust—Near screening operations and in high traffic areas, dust levels can exceed Occupational Safety and Health Administration (OSHA) limits if not properly contained and captured. Design engineers should provide screens with hoods connected to dust collectors. High-traffic areas should be cleaned regularly to prevent dust buildup.

Material-handling equipment (e.g., conveyors and screens) has exposed moving parts and poses a worker hazard. The main safety concerns are at the points of material transfer and locations of exposed belts. To minimize the possibility of material spilling or accumulating at transfer points, design engineers should provide emergency pull-cords along the full length of conveyors, as well as interlocks to shut down all material-handling operations in the event of an emergency.

Wood chips and compost piles may contain high concentrations of the airborne fungus *A. fumigatus*, which naturally occurs in grass and leaves. Although typically not harmful, *A. fumigatus* may cause aspergillosis in individuals with extreme susceptibility. Personnel with respiratory problems, that exhibit adverse physical reactions, or who have histories of suppressed immune response should not work in a composting or wastewater treatment facility.

5.0 ALKALINE STABILIZATION

Adding alkaline chemicals to solids is a reliable stabilization method that wastewater treatment plants have practiced since the 1890s. The chemicals traditionally used are quicklime and hydrated lime.

In recent years, a number of advanced alkaline-stabilization technologies have emerged. These technologies, which use new chemical additives, special equipment, or special processing steps, all claim advantages over traditional lime stabilization (e.g., enhanced pathogen control and a more publicly acceptable product). They also produce a biosolids that sometimes is called *artificial soil* because it has been successfully used as a soil substitute.

Lime is the most widely used and one of the least expensive alkaline materials available in the wastewater industry. It has been used to reduce odors in privies, increase pH in stressed digesters, remove phosphorus in advanced wastewater treatment processes, treat septage, and condition solids before and after mechanical dewatering. It is also the principal stabilizing chemical at municipal wastewater treatment plants with capacities ranging from 379 m^3/d to about 1.13 mil. m^3/d (0.1 to 300 mgd) (U.S. EPA, 1979). Larger plants that have used the process include those in Pittsburgh, Pennsylvania; Memphis, Tennessee; and Toledo, Ohio; as well as the Blue Plains Wastewater Treatment Plant in Washington, D.C. According to U.S. EPA's *1988 Needs Survey of Municipal Wastewater Treatment Facilities*, more than 250 plants use lime stabilization (U.S. EPA, 1989). According to a 2007 Northeast Biosolids Management Association survey, 900 of the 4 800 facilities surveyed—18% of facilities surveyed and 12% of the total volume of biosolids produced—used some form of alkaline stabilization (NEBRA, 2007). These results emphasize that alkaline stabilization primarily is used by smaller treatment facilities.

Alkaline-stabilized biosolids can be beneficially used in many ways, depending on the particular quality requirements and associated standards. Traditional lime stabilization is classified in U.S. EPA's *Standards for the Use or Disposal of Sewage Solids* as a Class B process (PSRP) (U.S. EPA, 1999b). Many of the advanced alkaline stabilization technologies meet U.S. EPA's definition of a Class A process (PFRP).

Many of the beneficial use and disposal options for alkaline stabilized biosolids are further discussed in Oerke (1999).

5.1 Stabilization Objectives

The purposes of alkaline stabilization may include

- To substantially reduce the number and prevent the regrowth of pathogenic and odor-producing organisms, thereby preventing biosolids-related health hazards;
- To create a stable product that can be stored; and
- To reduce the short-term leaching of metals from biosolids not incorporated with natural soil.

Several studies have demonstrated that both liquid and dry lime stabilization achieve significant pathogen reduction, provided that a sufficiently high pH or temperature is maintained for an adequate period of time (Bitton et al., 1980; Christensen, 1982). Table 25.41 lists bacteria levels measured during full-scale studies at the Lebanon, Ohio, wastewater treatment plant; it shows that liquid lime stabilization at pH 12.5 and a 25% dose (dry-weight) reduced total coliform, fecal coliform, and fecal streptococci concentrations by more than 99.9%. Also, the numbers of *Salmonella* and *Pseudomonas aeruginosa* were reduced below the level of detection. In addition, Table 25.41 shows that pathogen concentrations in liquid lime-stabilized biosolids ranged from 10 to 1 000 times less than those in anaerobically digested biosolids from the same wastewater treatment plant.

TABLE 25.41 Bacteria reduction via liquid lime stabilization at Lebanon, Ohio (U.S. EPA, 1979).

Type of solids	Bacterial density, number/100 mL				
	Total coliforms[a]	Fecal coliforms[a]	Fecal streptococci	*Salmonella*[b]	*Ps. aeruginosa*
Raw sludge					
Primary	2.9×10^9	8.2×10^8	3.9×10^7	62	195
Waste activated	8.3×10^8	2.7×10^7	2.7×10^7	6	5.5×10^3
Anaerobically digested biosolids					
Mixed primary and waste activated	2.8×10^7	1.5×10^5	2.7×10^5	6	42
Lime stabilized					
Biosolids[c]					
Primary	1.2×10^5	5.9×10^3	1.6×10^4	<3	<3
Waste activated	5.2×10^5	1.6×10^4	6.8×10^3	<3	13
Anaerobically digested	18	18	8.6×10^3	<3	<3

[a] Millipore filter technique used for waste activated sludge. Most probable number technique used for other sludges.
[b] Detention limit = 3.
[c] To pH equal to or greater than 12.0.

Christensen (1987) researched the pathogen-reduction performance of dry lime stabilization using dry quicklime doses of 13 and 40% (as calcium hydroxide; dry-weight basis). His results indicated that dry lime stabilization can reduce fecal coliform and streptococcus pathogens by at least two orders of magnitude. This was as good as, and in some cases better than, the results of standard liquid lime stabilization and liquid lime conditioning followed by vacuum filtration (see Figures 25.87 and 25.88). No growth of either fecal coliform or fecal streptococci occurred by the seventh day (Westphal and Christensen, 1983). Westphal and Christensen (1983) also reported that alkaline-stabilization processes used to reduce the densities of fecal coliform and fecal streptoccus performed as well as or better than mesophilic aerobic digestion, anaerobic digestion, and mesophilic composting (see Table 25.42). Additional discussions of lime treatment and the control of bacterial, viral, and parasitic pathogens are reviewed in reports by Christensen (1987) and Reimers et al. (1981).

Another study of dry lime stabilization showed a 4- to 6-log reduction of fecal streptococcus at about pH 12 (Otoski, 1981). Such a treatment scheme can yield Class A biosolids by using a 2:1 lime dose at 20% solids (dry-weight basis) to raise the solids temperature to more than 70°C for 30 minutes and meet pathogen-reduction requirements via the heat of lime hydration.

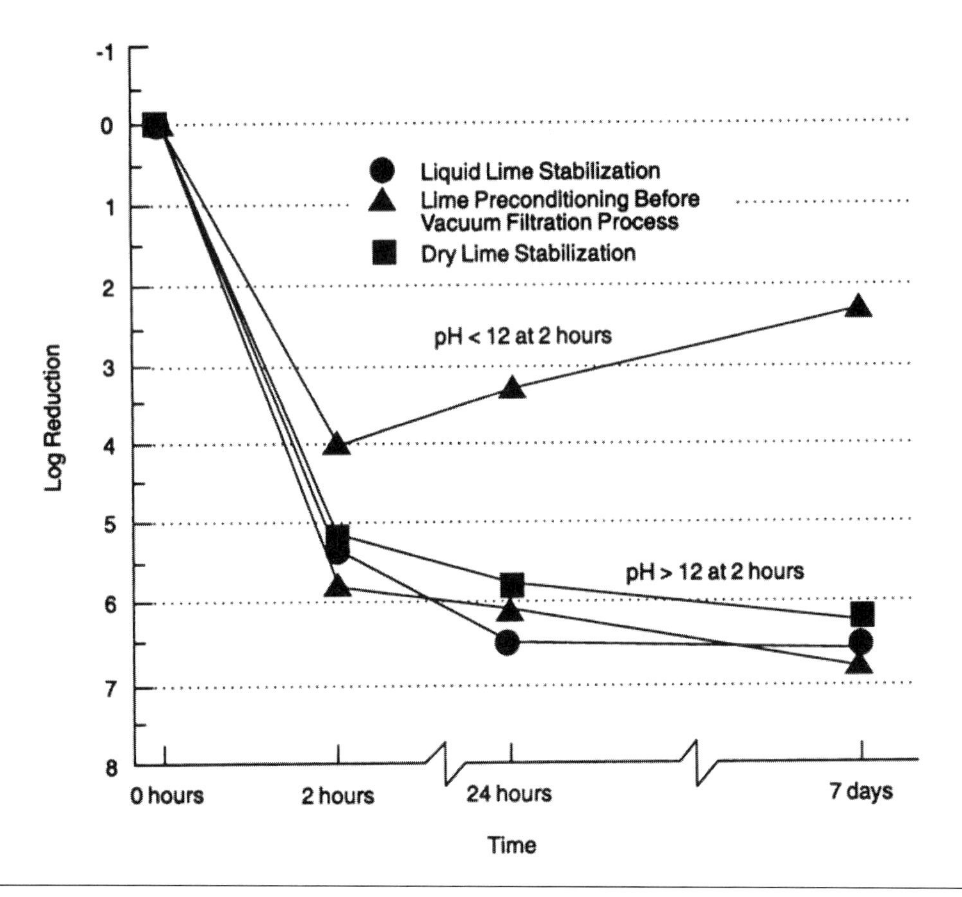

FIGURE 25.87 Average fecal coliform inactivation via two liquid lime stabilization processes and one dry lime stabilization process (Westphal and Christensen, 1983).

Treating dewatered cake with cement-plant kiln dust (alone or with a small amount of quicklime) reduces pathogenic microbe populations below U.S. EPA's Class A standard (Burnham et al., 1992). Both laboratory- and large-scale field tests have shown that indigenous and seeded populations of *Salmonella*, poliovirus, and *Ascaris* ova can be eliminated within 24 hours if the treated biosolids are contained at pH 12 and 52°C for 12 hours.

Although there is little information quantifying virus reduction during lime stabilization, lime has been identified as an effective viricide. Qualitative analysis has indicated substantial survival of higher organisms (e.g., hookworms and amoebic cysts) after 24 hours at high pH (Farrell et al., 1974). It is unknown whether prolonged contact eventually destroys these organisms. Class A alkaline stabilization processes that main-

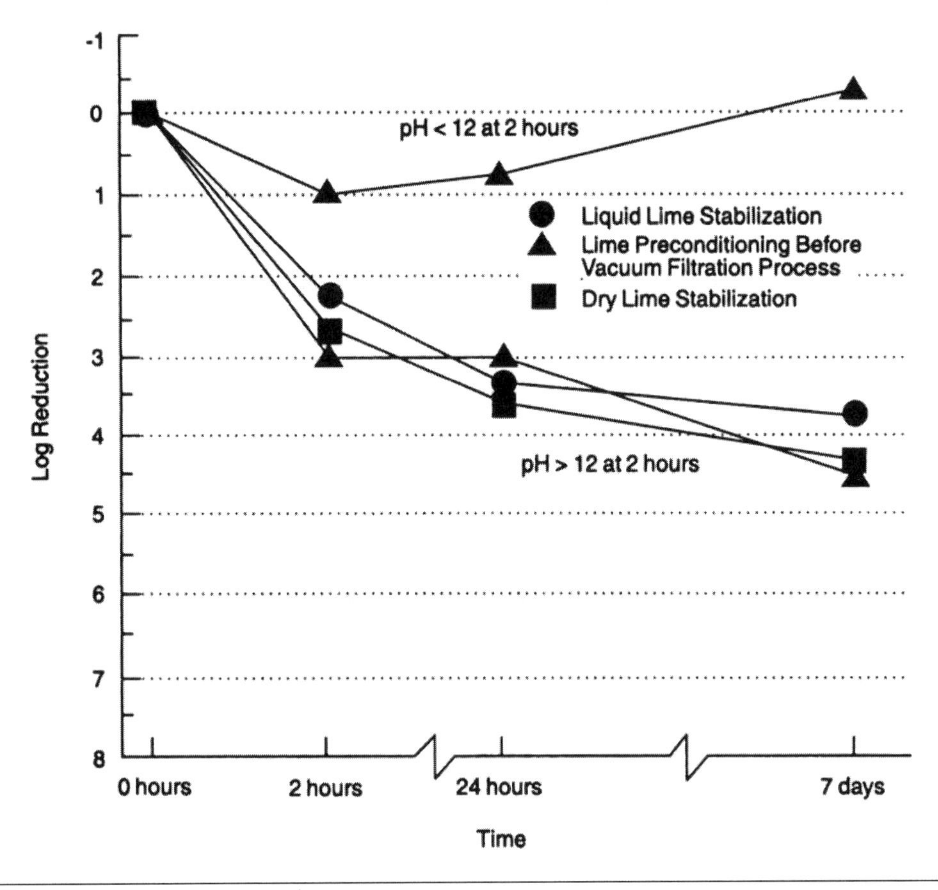

FIGURE 25.88 Average fecal streptococci inactivation via two liquid lime stabilization processes and one dry lime stabilization process (Westphal and Christensen, 1983).

tain 70°C for 30 minutes have been shown to kill *Ascaris* ova. Studies have shown that a high pH has little effect on parasites (e.g., toxocara, mites, and nematodes) (U.S. EPA, 1975). Comparisons of parasite types in lime-stabilized and anaerobically digested solids showed similar parasite types and densities in both solids.

Alkaline stabilization is a simple process. An alkaline chemical is added to feed solids to raise its pH, and adequate contact time is provided. At pH 12 or higher, with sufficient contact time and thorough lime–feed cake mixing, pathogens and microorganisms are either inactivated or destroyed. The chemical and physical characteristics of the resulting biosolids also are altered. The chemistry of the process is not well understood, although it is believed that some complex molecules are split by reactions (e.g., hydrolysis and saponification) (Christensen, 1982). It is also now understood that

TABLE 25.42 Bacteria reduction via various stabilization processes.

Process	Fecal coliform	Fecal streptococci
Anaerobic digestion (35°C)		
Mean	1.84	1.48
Range	1.44–2.33	1.1–1.94
Aerobic digestion		
20°C[a]	1	1
30°C[a]	2	1.64
Composting	≥4	2.9
Liquid lime stabilization		
Raw primary	5.1	2.4
Waste activated	3.2	3.2
Mixed primary and trickling filter humus, 4% solids	2.6	1.8
Storage[b]		
10°C	—	1
20°C	—	1.5
30°C	—	2.0

[a] Laboratory study, 35-day detention time.
[b] Laboratory study, 30-day detention time.

high pH releases gaseous ammonia from biosolids. Gaseous ammonia has been shown to be an effective disinfectant.

To meet Class B stabilization requirements, the pH of the feed cake–chemical mixture must be elevated to more than pH 12.0 for 2 hours and then maintained above pH 11.5 for another 22 hours to meet vector-attraction reduction (VAR) requirements. To meet Class A stabilization requirements, the elevated pH is combined with elevated temperatures (70°C for 30 minutes or other U.S. EPA-approved time and temperature combinations listed in U.S. EPA, 1999b). As long as the pH remains above 10 to 10.5, microbial activity and the associated odorous gases are greatly reduced or eliminated (U.S. EPA, 1979). However, other odorous gases (e.g., ammonia and trimethylamine) may be produced under high-pH and -temperature conditions.

5.1.1 Process Application

Although both small and large treatment plants have used lime stabilization, this process is more common at small facilities. It typically is more cost-effective than other chemical stabilization options. Relatively large plants have typically used lime stabilization as an interim process when their primary stabilization process (e.g., anaerobic or aerobic digestion) was temporarily out of service. Lime stabilization also has been used to supplement the primary stabilization process during peak solids production periods.

Lime-stabilized biosolids may be land-applied, benefiting large agricultural areas with acidic soils. However, because of the inert solids and reactions involved, lime-stabilized biosolids have lower concentrations of available nutrients (e.g., nitrogen and phosphorus) than a comparable mixture of biologically stabilized primary and WAS. (For more information on biosolids use and disposal considerations, see Chapter 27.)

5.1.2 Process Fundamentals

5.1.2.1 pH Elevation

Effective lime stabilization depends on raising the pH high enough and maintaining it at that level long enough to halt or substantially retard the microbial reactions that otherwise could lead to odor production and vector attraction. The process also can inactivate viruses, bacteria, and other microorganisms.

Lime stabilization involves a variety of chemical reactions that alter the chemical composition of solids. The following equations (simplified for illustrative purposes) show the types of reactions that may occur:

Reactions with inorganic constituents:

Water: $$CaO + H_2O = Ca(OH)_2 \tag{25.46}$$

Calcium: $$Ca^{2+} + 2HCO_3^- + CaO \rightarrow 2CaCO_3 + H_2O \tag{25.47}$$

Phosphorus: $$2PO_4^{-3} + 6H^+ + 3CaO \rightarrow Ca_3(PO_4)_2 + 3H_2O \tag{25.48}$$

Carbon dioxide: $$CO_2 + CaO \rightarrow CaCO_3 \tag{25.49}$$

Reactions with organic constituents:

Acids: $$RCOOH + CaO \rightarrow RCOOCaOH \tag{25.50}$$

Fats: $$\text{"Fat"} + CaO \text{ fatty acids} \tag{25.51}$$

Lime initially raises the pH of solids. Then, reactions occur (e.g., those in the equations above) that will lower the pH unless excess lime was added. The amount of excess lime needed depends on the length of time that a high pH must be maintained (e.g., during extended storage).

Biological activity produces compounds (e.g., carbon dioxide and organic acids) that react with lime. If biological activity is not sufficiently inhibited during alkaline stabilization, these compounds will reduce the pH, which could result in incomplete stabilization.

5.1.2.2 Heat Generation

If quicklime (or any compound with high quicklime concentrations) is added to solids, it initially reacts with the water in solids to form hydrated lime. This exothermic reaction releases about 15 300 cal/g·mol (2.75×10^4 Btu/lb/mol) (U.S. EPA, 1982). The reaction between quicklime and carbon dioxide is also exothermic, releasing about 4.33×10^4 cal/g-mol (7.8×10^4 Btu/lb/mol).

Both reactions can raise the temperature substantially, particularly in solids cake with a low moisture content. For example, adding 45 g (0.1 lb) of quicklime per gram of solids to a cake containing 15% total solids can result in a temperature increase of more than 10°C (50°F), as the following formula demonstrates:

$$(0.1 \text{ lb CaO})\,(1 \text{ lb mol}/56 \text{ lb})\,(27\,500 \text{ Btu/lb mol}) = 49 \text{ Btu } (52 \text{ kJ}) \quad (25.52)$$

$$(49 \text{ Btu})\,(1/0.85 \text{ lb } H_2O)\,(1°\text{F/lb } H_2O/\text{Btu}) = 58°\text{F } 14°\text{C} \quad (25.53)$$

In actual practice, temperature increases will be smaller, although they can be substantial. Sometimes they can be sufficient to contribute to pathogen destruction during lime stabilization.

5.1.3 Process Description

Several alkaline-stabilization technologies are available. Each system has advantages and disadvantages, so design engineers should evaluate them and select the appropriate process on a case-by-case basis.

5.1.3.1 Liquid Lime (Pre-lime) Stabilization

In liquid lime (pre-lime) stabilization, a lime slurry is added to feed solids to meet Class B stabilization requirements (see Figure 25.89). The lime typically is added to thickened solids at wastewater treatment plants that land-apply liquid biosolids (e.g., subsurface injection on agricultural land). This practice typically has been limited to smaller treatment plants or those with nearby land-application or use sites. That said, a Washington Suburban Sanitary Commission treatment plant in Piscataway, Maryland, has used pre-lime stabilization followed by belt filter-press dewatering to create a biosolids suitable for hauling longer distances. Because the biosolids were pre-limed, Piscataway operators claim that they have low odor characteristics. However, equipment scaling remains a concern at this facility.

Another liquid lime stabilization method involves conditioning solids or septage with lime before dewatering. The lime typically is combined with other conditioners (e.g., aluminum or iron salts) to improve solids dewatering. This method primarily has been used with vacuum filters and recessed-plate filter presses; in such cases, the lime dose needed to condition solids typically exceeds that required to stabilize them.

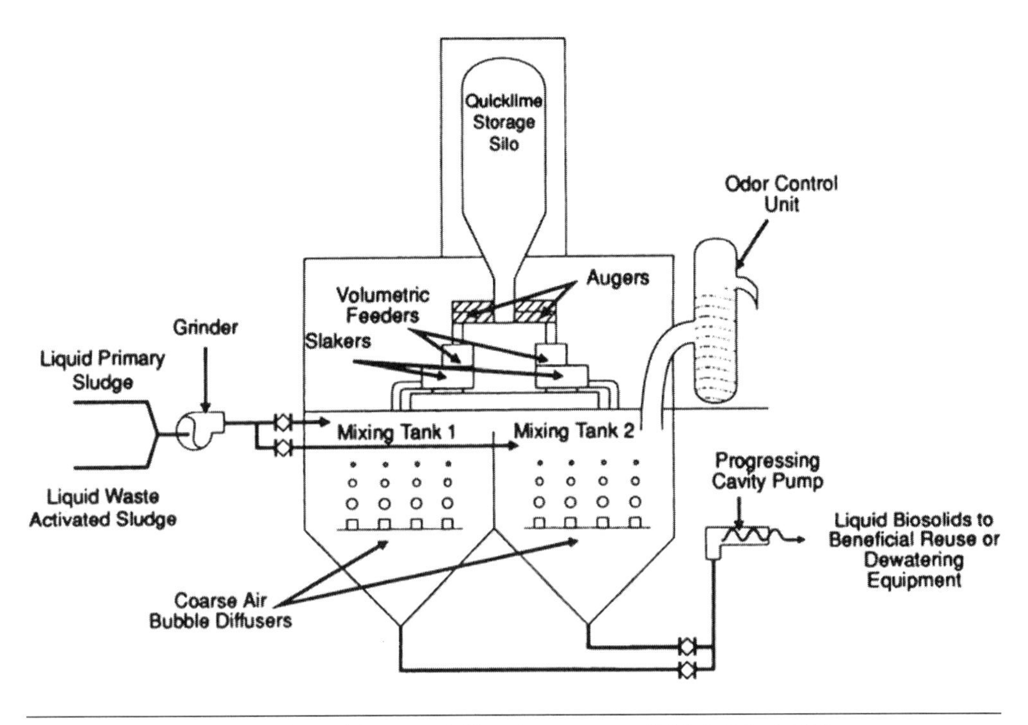

FIGURE 25.89 Typical liquid lime stabilization system (U.S. EPA, 1979).

5.1.3.2 Dry Lime (Post-lime) Stabilization

In dry lime (post-lime) stabilization, dry quicklime or hydrated lime is added to dewatered cake. This process has been practiced at wastewater treatment plants since the 1960s (Stone et al., 1992). The lime typically is mixed with the cake via a pug mill, plow blender, paddle mixer, ribbon blender, screw conveyor, or similar device. Figure 25.90 is a process schematic for a typical dry lime stabilization system with a pneumatic lime-conveyance system.

Quicklime, hydrated lime, or other dry alkaline materials can be used in this process, although the use of hydrated lime typically is limited to smaller installations. Quicklime is less expensive and easier to handle than hydrated lime, and the heat of hydrolysis released when quicklime is added to dewatered cake can enhance pathogen destruction.

If enough dry alkaline material is added to feed solids, the resulting biosolids can meet either Class B or Class A requirements.

5.1.3.3 Advanced Alkaline Stabilization Technologies

Typical advantages and disadvantages of advanced alkaline stabilization are shown in Table 25.43.

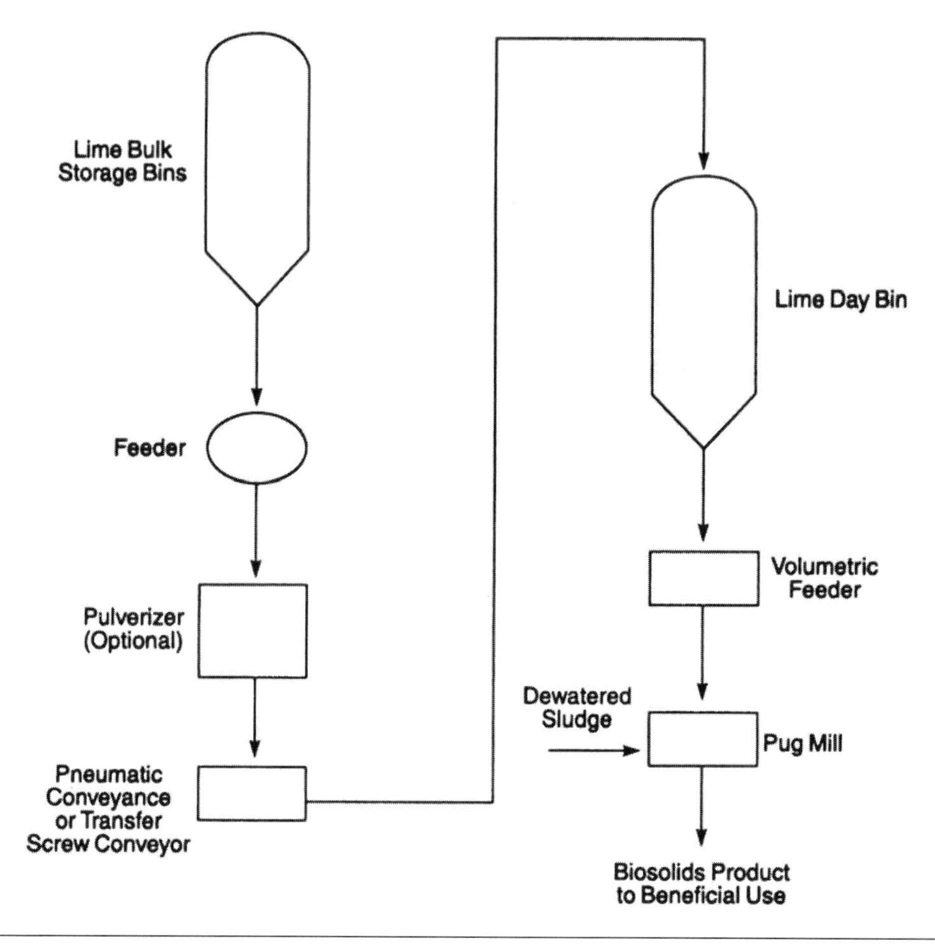

FIGURE 25.90 Process schematic of a typical dry lime stabilization system (Oerke and Rogowski, 1990).

In the last 30 years, alkaline-stabilization methods have been developed that use materials other than lime; these methods are being used by a number of municipalities. Most of those that rely on additives (e.g., cement kiln dust, lime kiln dust, Portland cement, or fly ash) are modifications of traditional dry lime stabilization. The most common modifications include the use of other chemicals, a higher dose (depends on the chemical), and supplemental drying. These processes alter the feed material's characteristics and, depending on the process, increase stability, decrease odor potential, reduce pathogens, and otherwise enhance the resulting biosolids.

Many of the processes are proprietary. The following descriptions illustrate the scope of processes available to municipalities. [For more detailed case-study planning,

TABLE 25.43 Typical advantages and disadvantages of advanced alkaline stabilization processes (WEF, 2007).

Advantages	Disadvantages
Meets Class A stabilization requirements	High annual cost
Multiple product markets	High chemical use
Typically lower capital cost when compared with other Class A stabilization processes	Extensive odor-control systems required to treat ammonia and other offgases
Proven with more than 40 installations in U.S.	Dewatering facilities required
Easy to operate, start up, and shut down	Some proprietary processes; annual patent fee could be required
Metal concentrations in biosolids are diluted	Worker safety concerns with dust from alkaline chemical and ammonia offgas
Product has value as a liming agent	Increase on total solids/chemical mass to transport
Enclosed facilities for better odor control	Product not appropriate for alkaline soils
Properly stabilized product is easy to handle and can be stored in smaller storage facilities	

design, and operational considerations on advanced alkaline-stabilization processes, see *Technology Evaluation Report: Alkaline Stabilization of Sewage Solids* (Engineering–Science Inc. and Black and Veatch, 1991).]

Pasteurization processes use the exothermic reaction of quicklime with water to raise process temperatures above 70°C. They then maintain this temperature for more than 30 minutes, as required by federal regulations for add-on pasteurization to meet Class A criteria. This pasteurization reaction must occur under carefully controlled and monitored mixing and temperature conditions to ensure that all solids particles are uniformly treated and pathogens are inactivated by the heat generated during the reaction.

The process produces a soil-like material that is nonviscous and, therefore, not subject to liquefaction under mechanical stress. Varying the process additives and mixing ratios results in a range of biosolids-derived materials suitable for use as daily, intermediate, and final landfill cover or in land reclamation (Sloan, 1992). Figure 25.91 is a process schematic for a typical pasteurization process. In a variation of this process, pasteurization occurs in a heated and insulated vessel reactor, where temperatures are maintained at 70°C or higher for at least 30 minutes.

A chemical stabilization/fixation process typically involves adding pozzolanic materials to dewatered cake (see Figure 25.92). Such materials cause cementitious reactions and produce, after drying, a soil-like material containing about 35 to 50% solids. To date, this soil-like product has been used only as landfill cover material. In many cases, the treated material is further dried at the landfill for 2 to 3 days in small windrows. Class A or PFRP equivalency has not yet been proven (Oerke and Rogowski, 1990; Reimers et al., 1981).

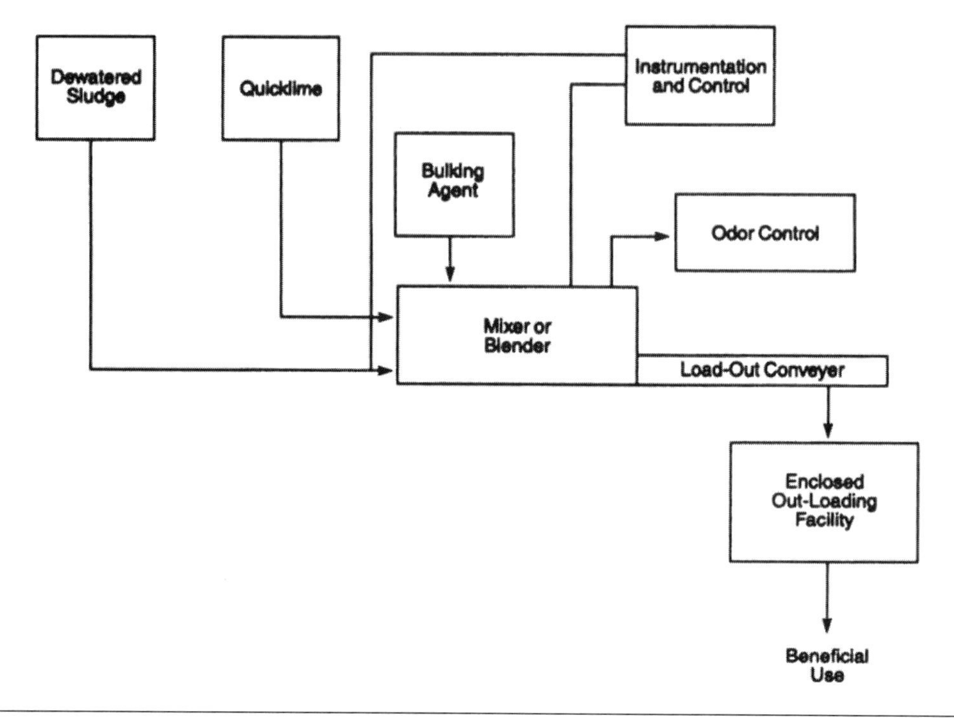

FIGURE 25.91 Process schematic of a typical pasteurization system.

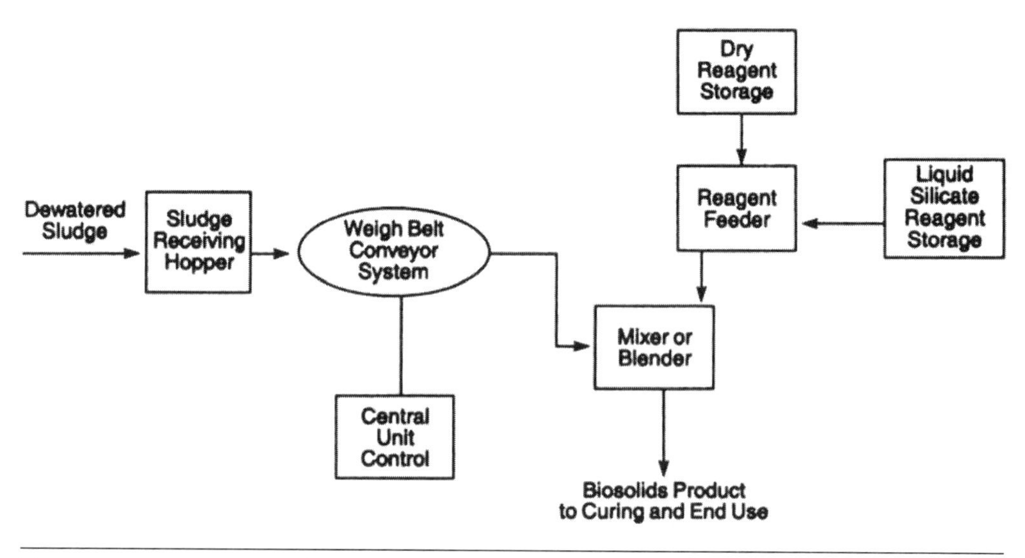

FIGURE 25.92 Process schematic of a typical chemical stabilization system.

One proprietary process (the N-Viro process) combines advanced alkaline stabilization with accelerated drying (AASAD) (see Figure 25.93). The U.S. Environmental Protection Agency has approved two versions of this technology as systems that produce Class A biosolids. Both versions involve adding quicklime, cement-plant kiln dust, lime-plant kiln dust, alkaline fly ash, or other alkaline admixtures and further processing the solids to stress pathogens via pH, temperature, ammonia, salts, and dryness (Burnham et al., 1992). In one version, chemical addition is followed by raising the material's temperature to between 52 and 62°C for at least 12 hours so the heat generated by the chemical reaction can further reduce pathogens. The second version uses chemical addition to raise the solids' pH above 12 and then mechanically dries the material in windrows or a rotary drum dryer to produce biosolids containing 50 to 60% solids. The biosolids predominantly are used as an agricultural liming agent, a soil conditioner, landfill cover, or a component of blended topsoil.

A second proprietary process (RDP, envessel pasteurization) uses electrically generated heat to supplement the heat generated by quicklime, which purportedly reduces lime consumption. An electrically heated screw auger transfers the solids–lime mix to an enclosed reactor, where the material is held for 30 minutes at 70°C to achieve pasteurization.

A third proprietary process (Bioset) uses sulfamic acid to supplement the heat produced by quicklime. (Both lime and water, and lime and sulfamic acid react exothermically.) The process occurs in a pressurized vessel to achieve pasteurization conditions. Bioset has applied to the U.S. EPA Pathogen Equivalency Committee for certification of

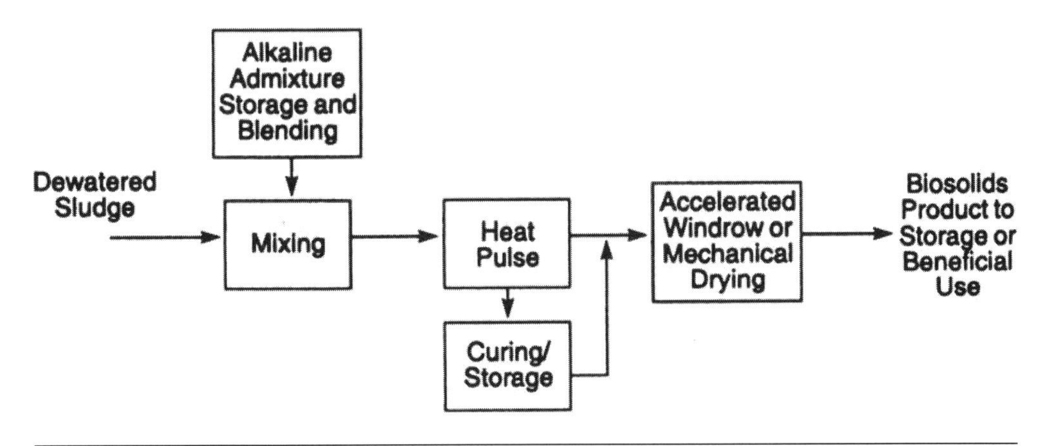

FIGURE 25.93 Process schematic of a typical alkaline stabilization system with a subsequent drying process.

its process as a Class A biosolids technology under the "process to further reduce pathogens (PFRP)" alternative in 40 CFR 503.

5.1.4 *Process Variations*

Several alternative approaches or modifications to the basic alkaline stabilization process have been developed. Some evolved from other treatment processes. For example, lime-treated primary solids have been combined with raw secondary solids to remove phosphorus (Paulsrud and Eikum, 1975). Existing digesters (or other available tanks) have been used to thicken alkaline-stabilized biosolids before dewatering and disposal (Farrell et al., 1974).

Another alternative uses two mixing vessels: the pH is raised above 12 in one, and the other provides adequate contact time and excess lime addition to keep pH within the desired range (Counts and Shuckrow, 1975).

Waukegan, Illinois, mixes fly ash and dewatered cake at ratios between 2.0:1 and 2.5:1 to produce Class B alkaline-stabilized biosolids. Personnel used this structurally stable material to "build" a biosolids-only monofill, rather than buying and importing fill material (Byers and Jensen, 1990).

5.2 Advantages and Disadvantages

Both liquid and dry lime stabilization processes are reliable, compact, relatively inexpensive to install, and easier to operate than many other stabilization processes. Many wastewater utilities that use lime stabilization have indicated that the process greatly reduces odors if the mixing is thorough (Kampelmacher and van Noorle Jansen, 1972; Westphal and Christensen, 1983). However, odor experiences with lime stabilization have been mixed and are typically the result of variations in operating procedures. This process' pathogen reduction has been reported to be as effective as or better than digestion processes (U.S. EPA, 1979).

Nevertheless, there are disadvantages. Compared to digestion, alkaline stabilization does not reduce solids mass. In fact, it increases mass because of the added lime and resulting chemical formations; the amount to be handled is essentially proportional to the chemical dose. The increase in mass may increase transportation costs for bisolids use or disposal, but such costs may be offset by capital and O&M savings (from using alkaline stabilization rather than another process). Also, the weight typically increases more than the volume, which actually may shrink because of lime slaking. Slaking raises the temperature of solids, causing water to evaporate.

Stabilized solids are a source of nitrogen, phosphorus, and beneficial organic matter that can be land-applied on farms. However, alkaline-stabilized biosolids typically contain less soluble nitrogen and phosphorus (on a dry-weight basis) than aerobically or anaerobically digested biosolids. The biosolids also may partially or fully replace liming agents on acid soils because it elevates soil pH and, therefore, restricts plant uptake of metals. However, metal ions only are immobilized as long as the biosolids' pH remains high. Also, alkaline-stabilized biosolids may not be appropriate in areas where the soils are naturally alkaline (e.g., many parts of the western United States).

Another disadvantage is that the system has difficulty consistently providing thorough mixing. Also, alkaline stabilization produces ammonia and possibly other odorous gases that should be treated before being exhausted.

5.3 Applicability

Alkaline stabilization has been used in numerous biosolids-management programs (Oerke, 1999). Below are some typical situations in which alkaline stabilization has been used:

- Traditional dry lime stabilization is a cost-effective technology for land-applied or landfilled biosolids. However, because biosolids are not destroyed, it is more cost-effective when hauling distances are short.
- Traditional liquid lime stabilization is appropriate at small treatment plants, where the small volume of biosolids produced can be readily land-applied. It is also practical at small plants that store biosolids for later transportation to larger facilities for further treatment or disposal.
- Because chemicals are the main operations and maintenance expense in this process and because the process has great flexibility, alkaline stabilization may be a cost-effective option for facilities that only operate seasonally or whose solids production are variable.
- Advanced alkaline stabilization may allow municipalities to operate a biosolids distribution and marketing program at a lower capital cost than other technologies (e.g., in-vessel composting or heat drying).
- Because a well-maintained alkaline-stabilization system can be quickly started (or stopped), it can be used to supplement existing solids treatment capacity or substitute for incineration and drying facilities during fuel shortages. It also can treat the total solids production when existing facilities are out of service for cleaning or repair.
- Alkaline stabilization systems have comparatively low capital costs, so they may be cost-effective for plants with short service lives.
- Alkaline stabilization typically is used to treat septage, reducing odors before the material is land-applied or discharged to wastewater treatment plants. (The U.S. Environmental Protection Agency's *Standards for the Use or Disposal of Sewage*

Solids (1993) require that septage be treated with lime and maintained at pH 12 for 30 minutes before land application.)
- Alkaline stabilization may be added to processes (e.g., overloaded digesters) that have inadequate pathogen reduction. However, strong ammonia odors typically are generated when anaerobically digested solids are treated with alkaline materials.

5.4 Design Considerations

Because product quality and process design are interdependent, the importance of defining both process and product goals cannot be overemphasized. Engineers should evaluate a number of design criteria before implementing an alkaline stabilization process (see Table 25.44). Although they vary from site to site, typical design criteria include:

- Sources and characteristics of feed cake (e.g., quantity, type, quality, and solids content);
- Contact time, pH, and temperature;
- Alkaline chemical types and doses;
- Solids concentration of the feed cake–chemical mixture;
- Energy requirements;
- Storage requirements; and
- Pilot-scale test results.

The desired product is also an important design criterion. For more information on biosolids use considerations, see Section 7.8.

5.4.1 *Feed Characteristics*

The amount, sources, and composition of the feed cake determine the overall size of the alkaline stabilization system. Variable thickening or dewatering performance is an important consideration because poor performance significantly increases the size of the stabilization system. The dewatered cake's solids concentration affects both chemical dose and system size. Equipment capacities must be able to accommodate the volume of feed cake to be processed. The system will need larger equipment and more alkaline chemical to process a "wet" cake (10 to 15% solids) than a drier one (20 to 25% solids).

The feed cake's nutrient content affects the biosolids characteristics. The agronomic benefit of an alkaline-stabilized biosolids depends on the amount of plant nutrients it contains and the need for a liming agent at the application site. Alkaline stabilization may be advantageous for untreated solids with relatively high metal concentrations because alkaline additives dilute metals (on a dry-weight basis) and immobilize some trace metals.

TABLE 25.44 Typical advanced alkaline stabilization design criteria (Fergen, 1991).

Item	Description or equipment	Parameter	Units	Range of value		Selected design value
				Minimum	Maximum	
Materials	Sludge	Solids	Percent	20	30	25
		Density	lb/cu ft[a]	45	55	50
	Alkaline bulking chemical	Solids	Percent	90	98	95
		Density	lb/cu ft[a]	50	65	65
	Lime	Solids	Percent	90	96	95
		Density	lb/cu ft[a]	55	60	60
	Stabilized product	Solids	Percent	55	65	60
		Density	lb/cu ft[a]	65	75	75
Curing	Technology	Windrow				
	Detention time	Average	Days	3	7	6
		Peak	Days	3	7	4
	Temperatures		°C	—	—	52 to 12 hr
	Pile dimensions	Bottom width	ft[b]	6	14	10
		Mix height	ft[b]	2	3	3
		Top width	ft[b]	4	8	6
		Area/unit length	sq ft/ft[c]	10	33	24
		Pile spacing	ft[b]	—	—	5
	Pile turning		lb/d[d]	—	—	1 (typical)
Odor control	Building air	Number of stages	Number	1	3	1
		Air changes	Number/hr	6	15	12
	Product storage	Number of stages	Number	1	3	1
Storage	Sludge	Days of storage	Days	0	1	1
	Chemicals	Days of storage	Days	5	30	5
	Product	Days of storage	Days	80	180	60

[a] lb/cu ft × 16.02 = kg/m^3.
[b] ft × 0.304 8 = m.
[c] sq ft/ft × 0.304 8 = m^2/m.
[d] lb/d × 0.453 6 = kg/d.

The type of solids also should be considered. For example, anaerobically digested biosolids contain 5 to 8 times more ammonia-nitrogen than other solids. All of this ammonia-nitrogen would volatilize at the elevated pH required for alkaline stabilization, increasing the potential for odors. Anaerobically digested biosolids treated in alkaline stabilization systems also may release odors related to other nitrogen compounds (e.g., amines). Alkaline-unstable polymers also can contribute to the formation of odorous methyl amines. As with all solids-processing systems, odor-control facilities typically are required at alkaline-stabilization systems near residences or sensitive commercial areas.

5.4.2 Contact Time, pH, and Temperature

Contact time and pH are directly related because the pH must be maintained at the required level for enough time to destroy pathogens. The treatment chemical must have enough residual alkalinity to maintain a high pH in the biosolids until they are used or disposed. The high pH prevents odorants and pathogenic organisms from growing or reactivating.

A drop in pH (*pH decay*) occurs when biosolids absorb atmospheric carbon dioxide or acid rain (which forms a weak acid when dissolved in water), which gradually consume the residual alkalinity. The pH gradually decreases, eventually dropping below 11.0. Bacterial action then resumes, and the renewed production of organic acids causes the pH to continue decaying (similar to the reactions in anaerobic digestion).

The pH typically drops during stabilization, so it should be raised to and maintained at more than pH 12. Biosolids do not have to be inside a contact vessel as long as the pH can be monitored to ensure that it remains at the desired value for the desired time.

5.4.3 Alkaline Chemical Types and Doses

The types and doses of alkaline chemicals are important design criteria. The quality of the chemicals (e.g., lime, cement-plant kiln dust, Portland cement, and lime-plant kiln dust) should be consistent. Different types or sources of additives produce different biosolids textures and granularities. Lime is available from numerous sources, ranging from a high calcium lime in oyster or clam shells to a relatively low-calcium dolomitic lime. Major considerations when selecting a chemical include economics, availability, desired mixing, and desired product characteristics.

Some alkaline reagents (e.g., cement kiln dust, lime kiln dust, and fly ash) are considered industrial byproducts, and design engineers must ensure that this material does not introduce contaminants or additional pollutants that jeopardize biosolids quality. Cement kiln dust from hazardous waste kilns, for example, should be avoided. Also, the characteristics of a byproduct can vary from one location to the next, so consistent vendor quality-control procedures are essential. The material from one kiln or furnace will remain fairly consistent, provided that operating conditions do not drastically change.

Treatment plant personnel should develop a quality assurance/quality control program that includes frequent sampling and analysis to ensure that biosolids quality is consistent. Because the quality of alkaline additives may directly affect biosolids quality, adequate monitoring and proper management are important. More importantly, pilot- or bench-scale testing should be performed to determine how variations in alkaline additives will affect product quality and how the process and chemical doses should be adjusted to compensate for such variations.

The two predominant types of lime are quicklime (calcium oxide) and calcium hydroxide. Slaked lime in a liquid slurry (carbide lime) is also available. Carbide lime is a byproduct of manufacturing welding-grade acetylene from calcium carbide. Its application principles are the same as those for calcium hydroxide or quicklime in slurry form, so carbide lime is not specifically discussed here.

Design engineers should slect the type of lime based on economics and materials-handling characteristics (e.g., alkaline-material particle size). Calcium hydroxide costs about 30% more to produce and transport than quicklime, but it requires less equipment onsite because it already has been hydrated (slaked). Calcium hydroxide typically is economical for use at small facilities, but if more than 9000 to 13 000 m^3/d (3 to 4 ton/d) is needed, quicklime should be considered.

Quicklime typically requires slaking equipment on site. Dry lime stabilization (i.e., adding quicklime directly to dewatered cake) does not require that the chemical be slaked first, but additional handling precautions must be addressed because of the exothermic reaction of quicklime and water. Dry lime stabilization also eliminates lime sidestreams and the related abrasion and scaling of piping and mechanical equipment.

The required doses of specific chemicals will depend on the type of feed solids (e.g., primary, WAS, trickling filter, or septage), its quality and chemical composition (including organic content), its solids concentration, the desired final product characteristics, and the type and quality of the alkaline material.

Table 25.45 shows the range of liquid lime doses required to maintain pH 12 for 30 minutes (U.S. EPA, 1979). Numerous researchers have confirmed these doses (Ramirez and Malina, 1980).

The chemical dose is affect by the feed cake's chemical composition, which depends on the type of solids and the treatment process used (e.g., chemical coagulation). Another factor that affects chemical dose is solids concentration (see Figure 25.94) (U.S. EPA, 1975). Table 25.46 shows a wide range of lime doses (from 10 to 60% on a

TABLE 25.45 Lime dose required for liquid lime stabilization at Lebanon, Ohio[a] (U.S. EPA, 1979).

Type of sludge	Average solids concentration, %	Average lime dosage, lb calcium hydroxide/lb[b] dry solids	Average pH	
			Initial	Final
Primary sludge[c]	4.3	0.12	6.7	12.7
Waste activated sludge	1.3	0.30	7.1	12.6
Anaerobically digested combined	5.5	0.19	7.2	12.4

[a] Dose required to maintain pH 12 for 30 minutes.
[b] lb/lb × 1 000 = g/kg.
[c] Includes waste activated sludge.

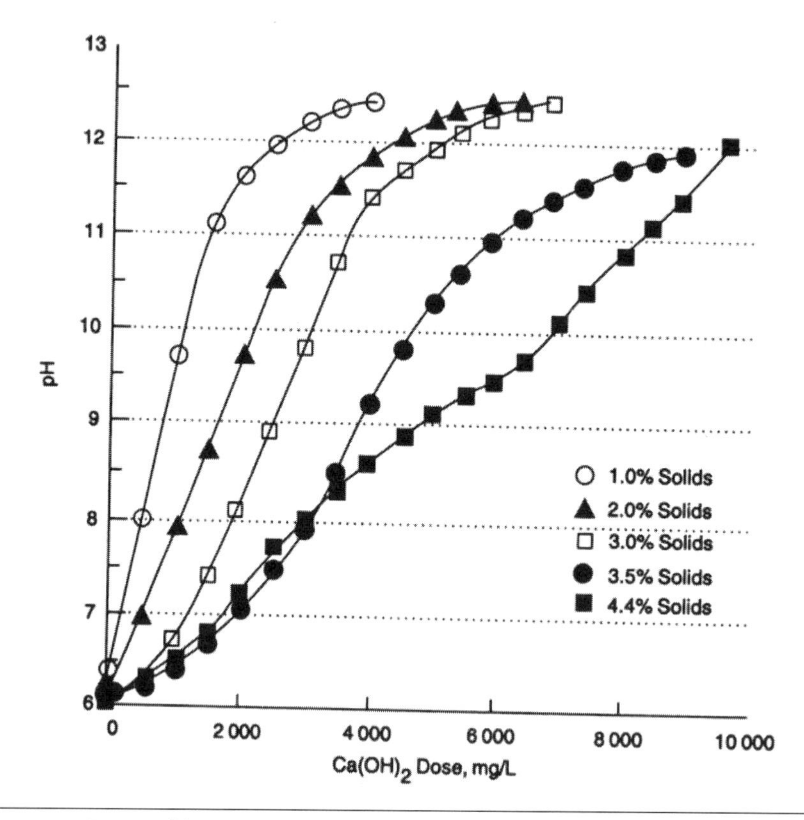

FIGURE 25.94 Dose of liquid lime required to raise the pH in a stabilization system feedstock (primary solids and trickling filter humus) with various solids concentrations (U.S. EPA, 1975).

TABLE 25.46 Liquid lime stabilization doses required to keep pH above 11.0 for at least 14 days (Farrell et al., 1974).

Type of raw sludge	Lime dose, lb calcium hydroxide/lb suspended solids[a]
Primary sludge	0.10–0.15
Activated sludge	0.30–0.50
Septage	0.10–0.30
Alum sludge[b]	0.40–0.60
Alum sludge[b] plus primary sludge[c]	0.25–0.40
Iron sludge[b]	0.35–0.60

[a] lb/lb × 1 000 = g/kg.
[b] Precipitation of primary treated effluent.
[c] Dry-weight basis.

dry-weight basis). As the solids concentration increases, the required dose typically increases. The required dose per unit mass of solids tends to be somewhat higher for dilute feeds (less than 2.0% solids) because more lime is required to raise the pH of water. However, liquid lime requirements are more closely related to the feed cake's total mass than to its volume when its solids concentration ranges from 0.5 to 4.5% (U.S. EPA, 1979). Thickening solids to reduce the volume may have little or no effect on lime requirements because the mass is not significantly changed.

Minimum lime doses of 25 to 40% (on a dry-weight basis as calcium hydroxide) typically are required for liquid lime Class B stabilization before vacuum filtration (see Figure 25.95). The curves in Figures 25.95, 25.96, and 25.97 show the characteristic pH drop that occurs when not enough liquid lime is added. When the dose is too low, the pH of the feed cake–lime mixture initially may reach 12 but then rapidly decay.

Minimum doses of 15 to 30% (on a dry-weight basis as calcium hydroxide) typically are required for effective dry lime stabilization (see Figure 25.98). Figure 25.99 shows the theoretical dry lime stabilization dose for both Class B and Class A stabilization. The lower line shows maximum pH requirements, and the upper line shows Class A temperature requirements. Figure 25.99 is based on a quicklime dose requirement of 25% (dry-weight basis). Design engineers should note that while the quicklime requirement for Class B stabilization theoretically increases as the solids concentration increases, the quicklime requirement for Class A stabilization decreases as the solids concentration increases because lime is used to heat the cake to achieve Class A disinfection (a lower solids concentration will mean that more mass of water needs to be heated using quicklime to the required temperature), whereas lime is used to raise pH for Class B (Lue-Hing et al., 1992).

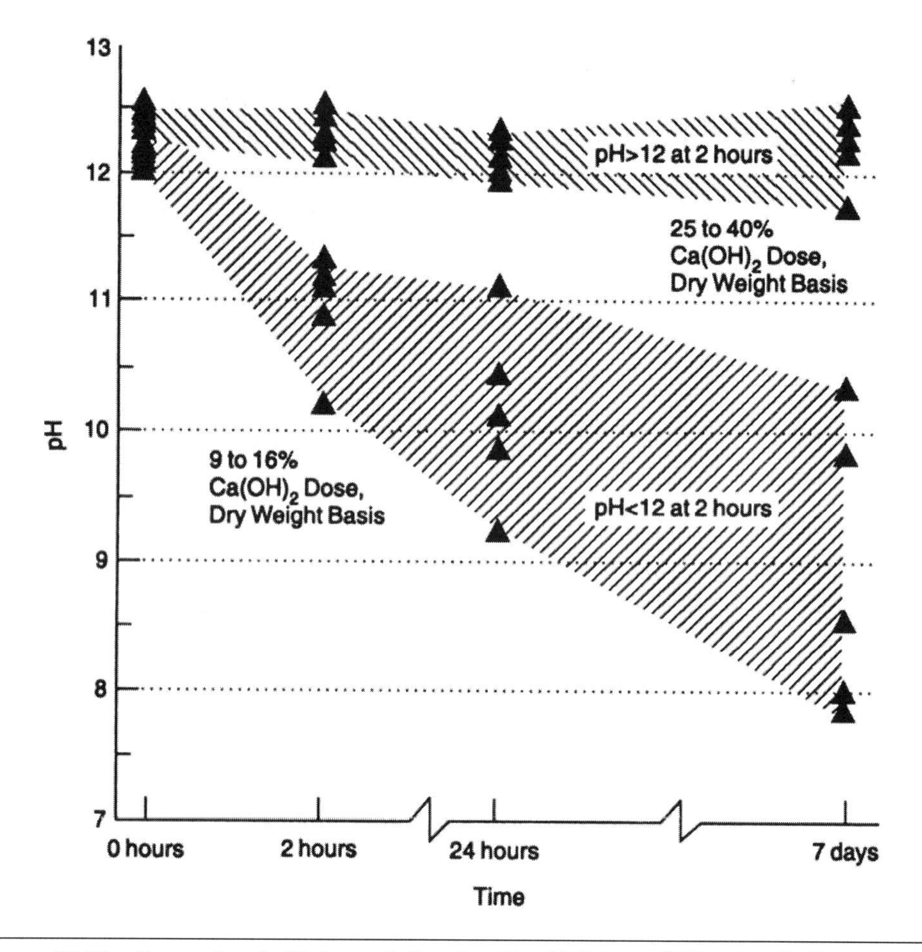

FIGURE 25.95 Example of pH decay following liquid lime stabilization before vacuum filtration (Westphal and Christensen, 1983).

The following assumptions were used for the Class A temperature requirements in Figure 25.99:

- The feed cake's temperature was 20°C (68°F);
- All of the quicklime reacted with water in the feed cake to produce heat [1 140 kJ/kg (490 Btu/lb) of quicklime];
- Quicklime is 100% calcium oxide (this value typically is 90%);
- The feed solids' specific heat is 0.25; and
- There was no heat loss from the feed to the air or the equipment.

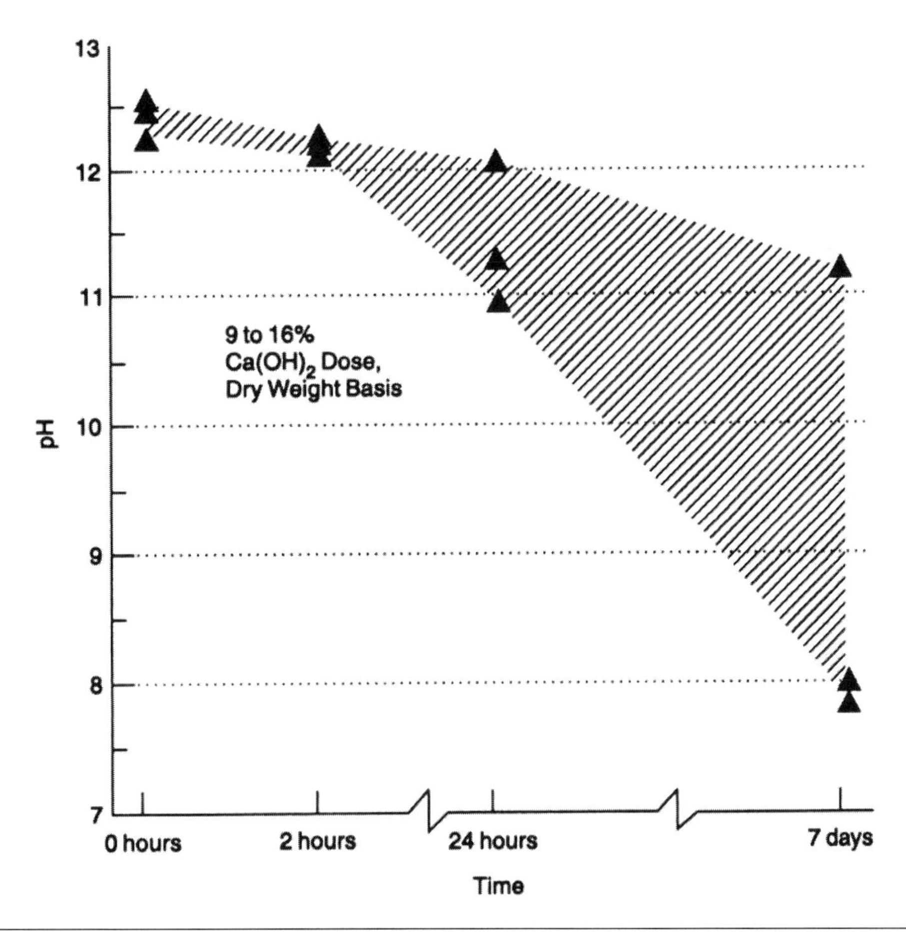

FIGURE 25.96 Example of pH decay following liquid lime stabilization (Westphal and Christensen, 1983).

Such conditions rarely exist in practice, so the amount of quicklime actually needed to meet Class A requirements can be up to 50% more than that indicated in Figure 25.99. To produce a drier, more easily crumbled biosolids, design engineers should increasing quicklime dose by as much as twice the value shown in the table.

Chemical doses for advanced alkaline-stabilization technologies depend on the process, chemical, and biosolids requirements. Material balances should be used to size alkaline-stabilization facilities and determine initial and final solids characteristics. Table 25.47 shows a typical material balance for an advanced alkaline stabilization facility, assuming a 65% chemical dose (wet-weight basis). Design engineers should note that a lime dose expressed on a wet-weight basis is four times greater than a dose

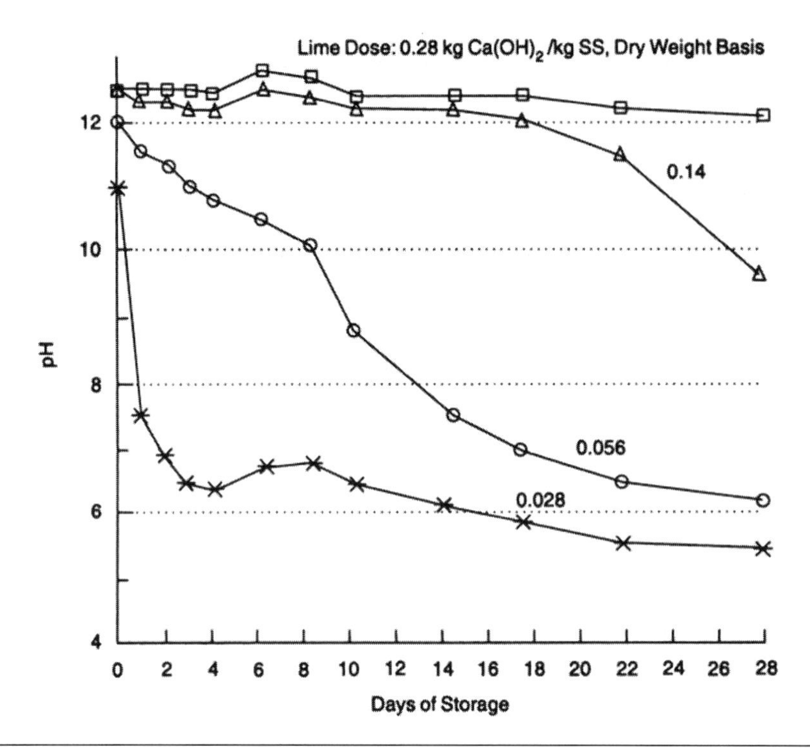

FIGURE 25.97 Change in pH during storage of raw primary solids that had been stabilized using various liquid lime doses (Farrell et al., 1974).

expressed on a dry-weight basis for a dewatered cake with a 25% solids concentration. For example, a chemical dose of 65% (wet-weight basis) is equal to about 245% on a dry-weight basis.

Design engineers can use the data in Tables 25.44, 25.45, 25.46, and 25.47 for preliminary design of liquid and dry lime stabilization facilities; however, the required dose should be determined on a case-by-case basis because of the many factors involved (Farrell et al., 1974). To prevent pH decay and the associated regrowth of organisms, the lime dose may have to be higher than that necessary for stabilization (Ramirez and Malina, 1980). The exact dose for any particular feed cake can be estimated via laboratory testing.

5.4.4 Solids Concentration of Feed/Chemical Mixture

The solids concentration of the feed cake–chemical mixture is an important design consideration for material-handling purposes. Regulations may require a minimum solids concentration (e.g., for landfilling or extended storage). The final product solids

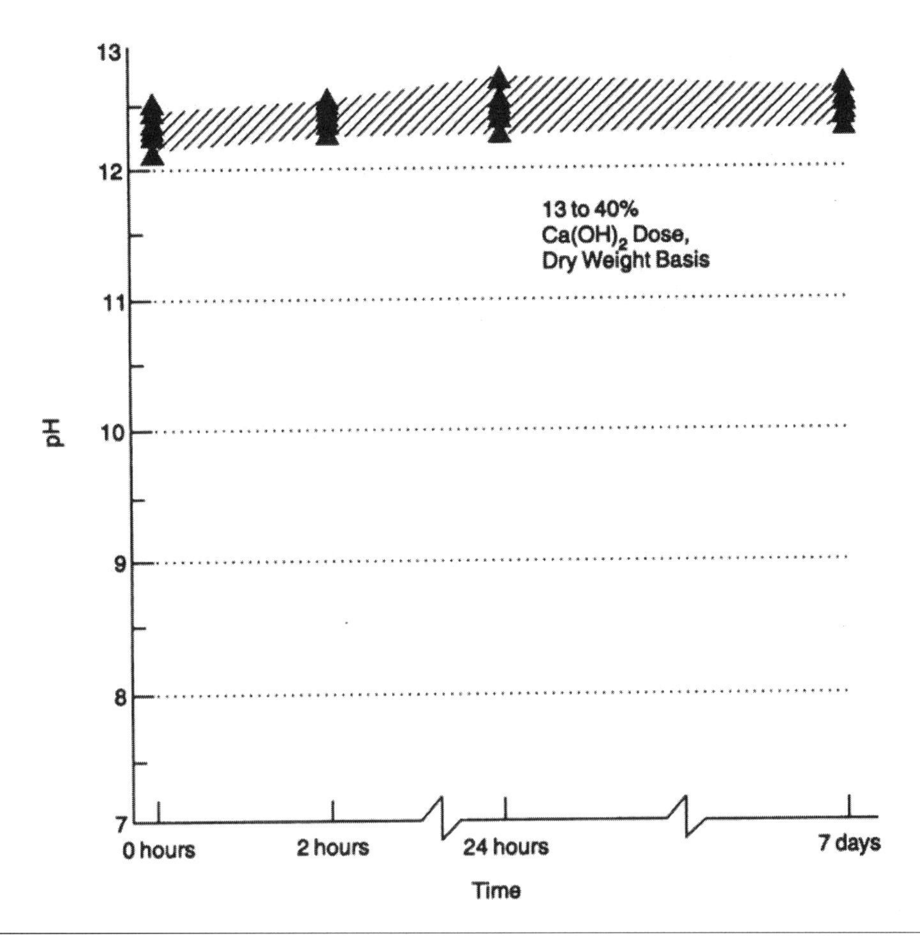

FIGURE 25.98 Example of pH decay after dewatered cake (a mixture of raw primary solids and waste activated sludge) was stabilized with dry lime (Westphal and Christensen, 1983).

concentration (dryness) and granularity also affect the type of biosolids trucks and application/disposal equipment needed.

The solids concentration of the initial feed cake–chemical mixture also affects any supplemental drying step in advanced alkaline-stabilization processes. The alkaline additive causes chemical reactions to occur that increase the mix's apparent solids content. This increase in solids is caused by the addition of solids (treatment chemical), chemical binding, and evaporation of water from the feed cake. The alkaline material—particularly quicklime—produces a fast reaction that increases temperature in a matter of minutes. Thorough mixing of feed cake and alkaline material is important to achieve

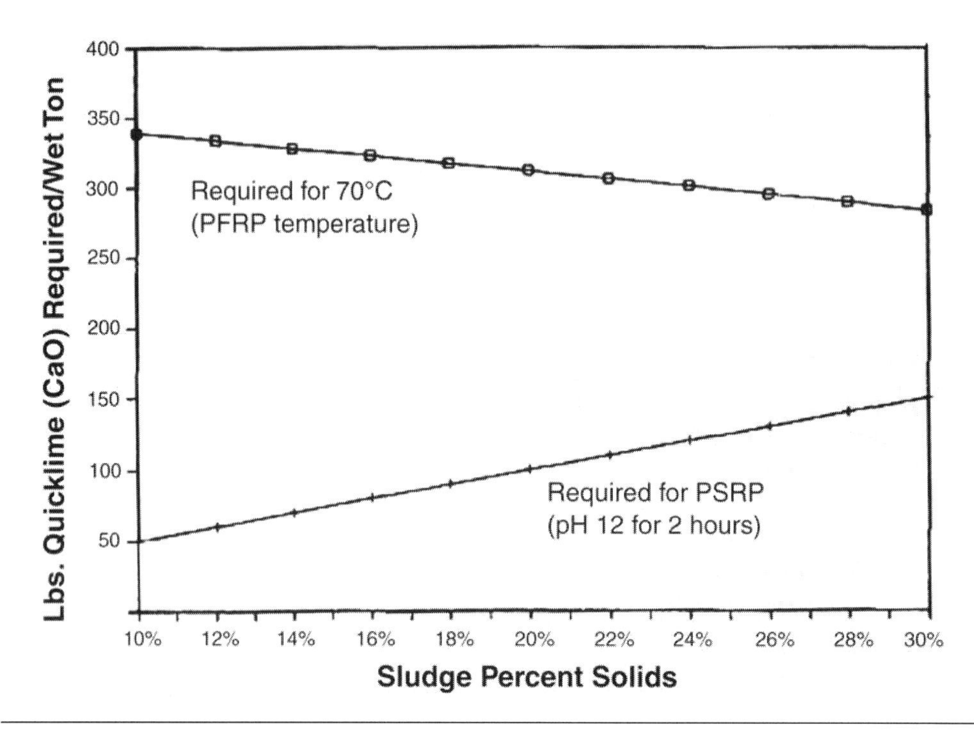

FIGURE 25.99 Theoretical dry lime requirements to stabilize cake with various solids concentrations so it meets Class B or Class A standards (Lue-Hing et al., Eds., 1992).

the target solids content and pathogen destruction, and to reduce residual odors (e.g., ammonia) in biosolids.

A high chemical dose can produce the desired solids concentration, thereby reducing or eliminating the need for supplemental drying, but this practice may be prohibitively expensive. Adding other bulking materials (e.g., fly ash, wood ash, sawdust, sand, and soil) can increase biosolids dryness and improve handling characteristics without increasing the chemical dose. Mechanical mixing in a windrow operation enhances drying, blends the material, and releases trapped ammonia and other volatile gases created during dewatering, resulting in a more homogeneous product. The final design should reflect the best balance between the chemical dose and the amount of subsequent drying required.

5.4.5 *Energy Requirements*

In liquid lime-stabilization processes, energy principally is needed to mix solids with the lime slurry. In dry alkaline stabilization, mixing energy requirements are minimal; they depend on solids throughput, chemical dose, and mixer type.

TABLE 25.47 Typical materials balance for advanced stabilization facilities (Fergen, 1991).[a]

		Solids balance				
Process	Item	Solids content, %	Volume, cu ft[b]	Total weight, ton	Dry weight, ton	Bulk density, lb/cu ft[c]
Mixing	Sludge cake	25.0	6 400	160.0	40.0	50.0
	Chemicals	95.0	3 186	103.5	98.4	65.0
	Initial mix	52.5	9 586	263.5	138.4	55.0
Windrow	Initial mix	52.5	9 586	263.5	138.4	55.0
	Evaporation loss	—	3 437	32.9	—	—
	Product	60.0	6 149	230.6	138.4	75.0

Chemical dose:
Wet-weight basis 65%
Dry-weight basis 245%
[a] For peak conditions multiply all the values by peaking factor (except density and percent solids).
[b] cu ft × 28.32 = L.
[c] lb/cu ft × 16.02 = kg/m^3.

Energy also may be needed for transport vehicles (e.g., feed cake, chemicals, and biosolids), and air ventilation and scrubbing equipment (for ammonia and odor control).

5.4.6 Storage Requirements

The system's storage facilities should be tailored to the facility's actual needs. Both intermediate and final storage should be provided.

5.4.6.1 Intermediate Storage

Some advanced alkaline-stabilization processes (e.g., N-Viro) require intermediate storage for the heating step to achieve Class A stabilization requirements. The objective of this step is to contain the heat produced during the exothermic reaction, so less chemical is needed. Intermediate storage units can include insulated steel, live-bottom hoppers; concrete bunkers; or an uninsulated stockpile in an open concrete pad.

5.4.6.2 Product Storage

Biooslids storage is another important design consideration. Facilities need adequate storage capacity if its biosolids markets are seasonal or have not been established. The amount of storage needed depends on both the type of biosolids and the distribution and marketing methods involved. At least 30 to 90 days' worth of storage should be provided if biosolids curing is required, it also is needed to accommodate road and weather conditions, as well as fluctuations in the biosolids-marketing and -distribution schedule (WEF, 1994a). Facility personnel try to develop markets for biosolids as an agricultural fertilizer, liming agent, or soil amendment. Until such markets have been established, however, the biosolids must be stockpiled or discarded. Also, the demand for agricultural products is seasonal, so the facility must have provisions for stockpiling during low-demand periods.

On the other hand, if the material will be used as landfill cover, storage requirements probably will be minimal (e.g., weekend storage may be needed if the landfill only operates 5 days per week).

The solids concentration and long-term stability of biosolids are important considerations when designing storage facilities. Biosolids storage facilities should be sized to meet each facility's actual needs, including storage during scheduled and unscheduled equipment maintenance, if maintenance includes downtime. The storage facility should be designed to prevent deterioration of product quality during inclement weather. In many climates, covered storage may be desirable. If uncovered storage is used, provisions should be made for leachate and runoff collection to avoid ponding and, in some instances, treatment. Runoff from stockpiles of alkaline-stabilized biosolids can result in stagnation and septic odors (Engineering–Science Inc. and Black and Veatch, 1991).

5.4.7 Pilot-Scale Testing

Because the quality and consistency of feed cake are site-specific, engineers must perform qualitative and quantitative analyses to determine the appropriate chemical doses and process design parameters. Pilot-scale testing should be used to determine optimum chemical doses and mixer performance. It also allows a municipality to evaluate various operating procedures and end-use products.

Engineers should conduct bench- and pilot-scale tests before implementing an alkaline stabilization process. Four primary areas to be evaluated include

- Process requirements (e.g., alkaline material types and doses);
- Equipment (e.g., energy requirements);
- Biosolids quality (e.g., desired solids concentration and granularity); and
- Odor generation and control.

Process concerns include the chemical types and doses; solids concentration in feed cake and biosolids; and other process steps (e.g., supplemental heating and drying), as required. Engineers must determine the chemical dose that will meet pH, solids content, heat rise, and biosolids requirements. It can be estimated in bench-scale tests using carefully measured volumes of feed cake mixed with various doses of chemicals. All pilot-scale testing should include generating mass-balance calculations to ensure consistency between chemical doses and solids concentration in biosolids. Where possible, full-scale pilot equipment should be used to assess actual chemical doses and mixing performance.

It is extremely important that testing conditions be controlled to simulate field conditions to the greatest extent possible. During winter, for example, the system may need a different dose or a modified formula of the chemical to achieve the desired biosolids. In large-scale windrow drying operations, carbon dioxide mixed into the product can lower the pH, so the alkaline-material dose may have to be increased to compensate. Samples should be cured in the same type of closed or open containers that will be used in the full-scale system.

Engineers should test the initial solids concentration of the feed–chemical mixture for compatibility with the proposed drying technique. It also may be useful to investigate various chemical doses in different drying/curing configurations. The chemical dose can significantly affect the drying rate and corresponding drying-area requirements.

Engineers also can use pilot-scale tests to evaluate equipment requirements. The goal of such testing is to determine the equipment, energy, and chemical needed to produce biosolids compatible with the next processing step or desired use. For example, an inappropriate paddle configuration or operating speed on a pug mill at a dry lime stabilization system resulted in an undesirable material. Proper mixing is necessary not only to achieve the desired biosolids characteristics, but to ensure that the alkaline additive has been thoroughly blended. Excessive mixing energy can result in a nongranular mass that is difficult to handle.

Other process parameters that should be considered during pilot- or bench-scale testing include odor emissions; concentrations of plant nutrients, metals, and organic chemicals; and compatibility of the alkaline material dose with the dewatering polymer. Some polymers may deteriorate in high alkaline conditions, exhibiting strong trimethylamine ("dead fish") odors (Jacobs and Silver, 1990). Engineers should test various doses of alkaline material with different polymers to determine their effects on biosolids odors and physical characteristics (e.g., compaction and granularity).

The final item evaluated in pilot-scale tests is the product. Pilot- and bench-scale testing provide excellent opportunities to investigate biosolids quality and marketability before beginning full-scale production. It is helpful to invite prospective users to observe pilot-scale tests or implement small-scale demonstration programs to encourage interest in the product. Physical characteristics (e.g., solids content, pH decay, leachability, permeability, or unconfined compressive strength) should be evaluated if the product will be landfilled or to stabilize slopes. Biosolids quality also should be tested to provide the data and documentation required for regulatory approval (WEF, 1995).

5.5 Description of Physical Facilities

5.5.1 *Solids Handling and Feed Equipment*

Cake-handling equipment chiefly consists of belt and screw conveyors and pumps. Belt conveyors typically are used to move solids horizontally or at gentle slopes. Belt conveyor problems typically include minor spills, slips, and frequent bearing maintenance. Screw conveyors also are used to transfer dewatered cake to the alkaline-stabilization mixer or storage hopper. Screw conveyors and high-pressure cake pumps can physically "condition" dewatered cake, making it difficult (sometimes impossible) to homogeneously mix with a dry alkaline chemical. Some screw conveyors tend to roll the cake–chemical mixture into "balls". Pumps can compact dewatered cake into a long tube that must be broken up during mixing. The rolled balls and compacted cake, which may be desirable or undesirable depending on the final objective, can be especially critical for the resulting biosolids (Oerke and Stone, 1991).

Although alkaline-stabilization processes are relatively simple, a regular inspection and maintenance program is essential. The conveyance system and other moving parts must be closely monitored for wear. If only one conveyor feeds the alkaline-stabilization process, it must be routinely inspected, maintained, and calibrated because conveyance-system downtime can delay or halt stabilization. If multiple process trains are used, bypasses and crossovers should be provided to avoid excessive downtime. Also, engineers should design the alkaline-stabilization system to be as close as possible to both the dewatering equipment and the storage system.

Design engineers should seriously consider using redundant process and storage trains to allow for routine maintenance and calibration, as well as operational flexibility, without downtime. Another option (although less desirable) is using temporary portable units, which can be placed in operation in a matter of hours or days, if necessary. Storage hoppers or bunkers may be placed between the dewatering and alkaline-stabilization systems to dampen variations in dewatering system output and to allow each process to operate independently.

5.5.2 *Alkaline Material Storage and Feeding*

Alkaline stabilization requires special chemical-storage and -feeding equipment. Traditionally, an alkaline chemical storage system should be able to meet at least 7 days' worth of demand (although a 2- to 3-week supply is preferred). The absolute minimum storage capacity recommended is 200% of the volume of the bulk chemical shipment, depending on the distance between the chemical supplier and user (Oerke, 1991). Calcium hydroxide can be stored up to 1 year. Quicklime deteriorates more rapidly; it should not be stored longer than 3 to 6 months.

Because some advanced alkaline stabilization processes have high chemical demand, traditional design criteria can result in an excessively large storage capacity; however, design engineers can consider a smaller capacity (2 to 3 days of chemical use) so long as chemical-delivery arrangements are reliable. The costs associated with daily chemical delivery should be compared to those of extra storage capacity.

Quicklime can be stored in lump or pebble form and ground onsite to reduce the potential for reaction with moisture during storage, especially if the alkaline material will be stored for up to 6 months.

Alkaline material is stored in steel silos with hoppers that have a side slope of at least 60 degrees. Bulk-storage silos and day chemical bins, if used, should be equipped with dust collectors and live-bottom bins, hopper agitation, or air pads to facilitate unloading and reduce clogging or bridging.

There are potential problems with any chemical, however. During storage, lime can react with carbon dioxide in the air to form a calcium carbonate coating on lime particles, making them less reactive. Quicklime and other alkaline materials readily react with moisture from the air (slake), leading to caking that can interfere with feeding and slaking. Therefore, lime should be stored in dry facilities and protected against moisture to prevent accidental slaking. Also, because slaking generates heat, quicklime should not be stored near combustible materials.

Dry alkaline materials can be conveyed mechanically via a screw conveyor if the distance from the bulk-storage silo to the chemical-addition point is short. Dry alkaline materials also can be pneumatically conveyed under either pressure or vacuum. Each type has its benefits. Vacuum systems have fewer dust problems because any leaks are into the system, not out of it. Pressure systems can move more material. Pneumatically conveyed air should be predried to reduce hydration and other moisture-related problems. Pneumatic conveyance systems may have problems maintaining homogeneous chemical bulk densities, however, if a variety of alkaline materials is used (Rubin, 1991).

A wide variety of chemical feed equipment (e.g., volumetric screw feeders, rotary airlock feeders, and gravimetric feeders) is available. A volumetric feeder delivers a constant volume of alkaline material, regardless of its density. A gravimetric feeder delivers a constant mass of alkaline material and provides more accurate control. However, it costs about twice as much as the volumetric type. Design engineers should evaluate feeders to determine which is appropriate for a given application (Rubin, 1991). The feed equipment should be isolated from the storage silo via a slide gate or similar device so the metering equipment can be removed easily if it becomes jammed.

Most chemical feed systems have dust problems. Poorly fitting slide gates and leaking feeders are obvious sources of dust. Also, the vertical drop between the feeding equipment and the process mixers should be reduced or enclosed to reduce dust problems.

Moisture can be generated during mixing that may rise into the chemical feed and storage equipment. For example, lime backups in pipes primarily are caused by moisture generated during the mixing process in the pug mill. Powdery lime is hygroscopic and tends to pack in the corners of the storage hopper. Venting the mixer away from chemical feed and storage equipment can reduce such problems.

5.5.3 *Liquid Lime Chemical Handling and Mixing Requirements*

Lime typically is fed to liquid solids in slurry ("milk of lime") form. Dry lime cannot be added to liquid solids effectively because caking will occur.

After being mixed into a slurry, both calcium hydroxide and slaked quicklime are chemically the same, and the same feeding processes can be used for both. Lime slurry can be prepared via either the batch or the continuous method. The batch method consists of dumping bagged lime into a mixing tank. The contents of slurry tanks are agitated by compressed air, water jets, or mechanical mixers. To ensure initial wetting and dispersion, a mechanical mixer needs about about 200 kW/m^3 to handle a calcium hydroxide slurry at a concentration of 120 kg/m^3 (Beals, 1976).

The slurry then is metered into the mixing tank. This may be the most troublesome step in the process. The slurry can react with bicarbonate alkalinity in the makeup water and with atmospheric carbon dioxide to form calcium carbonate scale that can plug lines. The magnitude of this problem increases as transfer distances increase and more bicarbonate or carbon dioxide contacts the slurry. So, slurry tanks should be as close to the mixing basin as possible, and design engineers should avoid using cascading weirs or other equipment that causes turbulence.

The basic difference between using quicklime and calcium hydroxide to stabilize dewatered cake is quicklime requires slaking equipment. Slaking can be done on either a batch or continuous basis. The batch method is more appropriate for small-scale facilities; however, the use of quicklime typically is less advantageous for such facilities. *Slaking* consists of mixing quicklime and water to create either a lime paste (water-to-lime ratio of 2:1) or slurry (water-to-lime ratio of 4:1). The paste should be held for about 5 minutes to allow complete hydration in the slaking chamber; the slurry should be held for 30 minutes. The hydration reaction is exothermic (i.e., releases heat). Proper slaking requires heat, but localized boiling and spattering could make conditions hazardous. After slaking, the paste enters a chamber where grit is removed and the paste is diluted to the desired concentration.

The appropriate automation equipment for continuous slaking largely depends on the proportion of lime to water, which in turn depends on the type of lime and mixing equipment used.

A stabilization tank is recommended downstream of the slaker to ensure that all chemical reactions between calcium hydroxide and dissolved solids in the water have been completed. This reduces scaling in downstream portions of the system. Slakers should discharge lime slurry directly to the stabilization tank, if possible, and it should be detained in the tank for at least 15 minutes. Adequate mixing is required to keep particles in suspension and prevent short-circuiting.

If baffles are required to prevent vortex formation, they should be designed to prevent solids from building up in the corners (depending on tank geometry).

A cleaning system should be provided that uses dilute hydrochloric acid to remove calcium carbonate scale from pumps and piping. So, the pumps' and pipes' materials of construction must be compatible with both acidic and caustic environments.

To facilitate scale removal, design engineers should use flexible piping or open troughs to convey lime slurry whenever possible. Lime slurry may be abrasive, particularly if low-grade pebble lime is used, so equipment and materials should be selected accordingly.

The mixing tank's primary purpose is to provide adequate mixing and contact time for the dewatered solids and lime slurry. The recommended contact time is about 30 minutes after the pH reaches pH 12.5. Mixing time is site-specific, so engineers should conduct bench- or pilot-scale tests whenever possible.

The tank can be constructed of mild steel. Its size depends on whether mixing will be done on a batch or a continuous basis.

Batch mixing tanks typically are used at smaller facilities. Such tanks should be sized to treat a day's worth of solids in one batch because many small plants only have one staffed shift. With adequate capacity, these tanks also can thicken the solids via gravity after stabilization. If a tank is used for both stabilization and thickening, then special equipment must be used to withdraw the thickened biosolids.

In continuous mixing systems, the pH and volume are held constant, and automated lime-feeding equipment is requried. The primary advantage of continuous mixing facilities is that a smaller tank may be used than is required in batch mixing. Because pH is important, tank contents should be closely monitored and maintained at a pH above 12 for at least 2 hours after mixing.

Both systems must provide enough mixing to keep solids in suspension and distribute lime efficiently. The two most common mixing systems are diffused-air and mechanical. Although both have been successful, diffused air is more widely used.

Diffused-air mixers have at least two important advantages over mechanical mixers. The first is more aeration, which in batch operations, helps keep dewatered solids fresh before the lime is added. The second is less potential for debris to foul the equipment (however, "nonclog" mechanical mixers are available).

Diffused-air systems also have several disadvantages. One is that ammonia stripping creates odors and reduces the biosolids' fertilizer value. Ammonia release also can be hazardous, so adequate ventilation must be provided. Another disadvantage is that the mixture absorbs carbon dioxide from air, so more lime is needed (because some of it reacts with the carbon dioxide). Finally, because gases (e.g., ammonia) are stripped, the facilities must be enclosed and the offgas may require treatment.

The design criteria for mixing facilities are similar to those for aerobic digestion systems. If design engineers select a diffused-air mixer, coarse-bubble diffusers should be used. Diffusers typically are mounted along one wall of the tank to induce a spiral-roll mixing pattern. Airflow rates of 0.3 to 0.5 $L/m^3 \cdot s$ have successfully been used for mixing (Beals, 1976). Airflow requirements may be higher if mixing thickened feeds.

The design criteria for mechanical mixers are based on bulk fluid velocity and impeller Reynolds number. Table 25.48 lists the various sizes of mechanical mixers required for various volumes. The data are based on both maintaining bulk fluid velocity (i.e., turbine agitator pumping capacity divided by cross-sectional area of mixing tank) at more than 0.13 m/s and an impeller Reynolds number at more than 1000. The mixer sizes listed are adequate for mixing feeds with concentrations of up to 10% dry solids and viscosities up to 1 Pa/s (1000 cP).

When feed solids are conditioned in mixing tanks before thickening or dewatering, engineers must carefully consider the mixing design to prevent floc shearing. Typically, lower mechanical mixer speeds and larger turbine diameters are required. Mechanical mixers also should have variable-speed drives to allow for process control.

The American Water Works Association (1983) and the National Lime Association (1988) have published several documents on selecting lime and lime-handling equipment, as well as on designing lime-application systems. These should be consulted for more design information.

5.5.4 Dewatered Cake/Chemical Mixing for Dry Alkaline Stabilization

The most critical component of dry-alkaline stabilization is mixing (blending) dewatered cake and alkaline material. The goal is to provide intimate contact between cake and chemical, so the pH of the entire mixture is adjusted. Inadequate mixing has led to incomplete stabilization, odors, and dust problems at several dry alkaline-stabilization facilities in the United States (Oerke and Stone, 1991; WEF, 1995).

TABLE 25.48 Mechanical mixer specification for liquid lime stabilization (Counts and Shuckrow, 1975).

Tank size		Tank diameter		Motor size		Shaft speed,	Turbine diameter	
m^3	gal	m	ft	kW	hp	rpm	m	ft
19	5 000	2.9	9.5	6	7.5	125	0.8	2.7
				4	5	84	1.0	3.2
				2	3	56	1.1	3.6
57	15 500	4.2	13.7	15	20	100	1.1	3.7
				11	15	68	1.3	4.4
				7	10	45	1.6	5.3
				6	7.5	37	1.7	5.6
114	30 000	5.2	17.2	30	40	84	1.5	4.8
				22	30	68	1.6	5.1
				19	25	56	1.7	5.5
				15	20	37	2.1	6.8
284	75 000	7.1	23.4	75	100	100	1.6	5.2
				56	75	68	1.9	6.2
				45	6	56	2.0	6.6
				37	50	45	2.2	7.3
380	100 000	7.8	25.7	93	125	84	1.8	6.0
				75	100	68	2.0	6.5
				56	75	45	2.4	7.8

[a] Bulk fluid velocity >0.13 m/s (26 ft/min).
Impeller Reynolds > 1 000.
Mix tank configuration:
Liquid depth equals tank diameter.
Baffles with a width of 1/12 the tank diameter placed at 90-deg spacing.

Both batch and continuous mixing systems are available. A mechanical mixer (e.g., a pug mill or plow blender) typically is used (see Figures 25.100 and 25.101). Diffused air is not used for mixing lime with cake. Mixers typically are selected based on experience and trial-and-error testing. Many mixer manufacturers have mobile pilot-scale units available, and engineers should use such equipment whenever possible to evaluate and select the most-effective mixer.

Thorough mixing is an art; many variables affect the mixing process and, therefore, the resulting biosolids characteristics. Dewatered cake and chemicals are added together at the "head" of the mixer, and the proportions are important. The mixing characteristics of a dewatered cake depend on the solids concentration, polymer used to condition solids before dewatering, stabilization chemical and dose, temperature, mixing intensity, SRT, and mixer's surface area per volume of exposed cake. When selecting a mixer, design engineers also should consider minimum and maximum cake

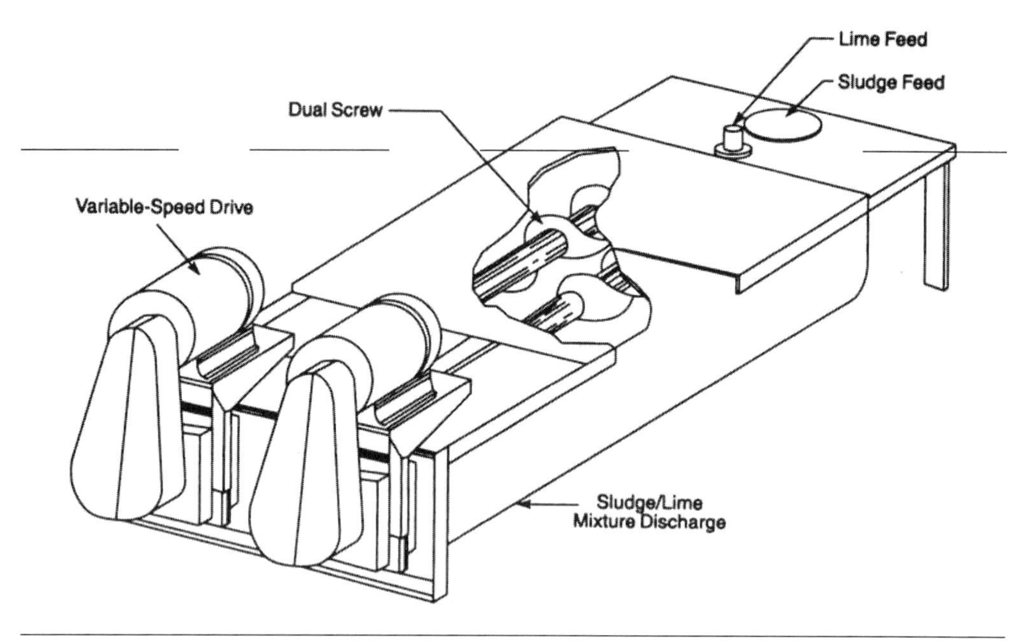

FIGURE 25.100 Typical dual-screw pug mill.

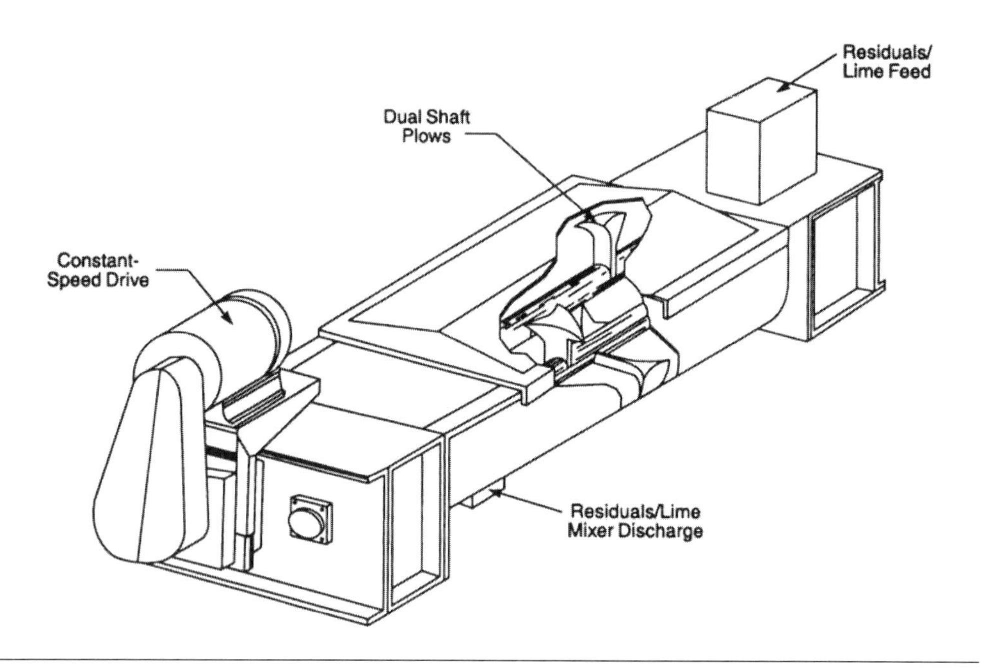

FIGURE 25.101 Typical plow blender.

production, hours of operation, and other operating conditions. To adapt to variations in mixing conditions, mixers can be equipped with variable-speed drives, adjustable paddle configurations, weir plates, and other options that adjust mixing intensity and retention time (Christy, 1992).

The physical characteristics of the resulting biosolids depend on the mixing parameters. Its physical consistency can range from sticky and plastic to granular and dusty. The goal of mixing is to produce a product compatible with the next processing step or intended use. Biosolids characteristics may continue changing up to several days after mixing because of ongoing chemical reactions, temperature, and other parameters.

5.5.5 Space Requirements

Depending on site constraints, the type of process used, and the amount of solids to be processed, site preparation for alkaline stabilization processes typically is minimal. The equipment typically can be arranged to accommodate various site constraints. Because they are relatively simple to operate and do not require extensive, complex equipment, alkaline-stabilization processes can be implemented quickly in a relatively small space. Figure 25.102 shows the layout for a 3.4×10^5 m^3/d (100-ton/d) advanced alkaline-

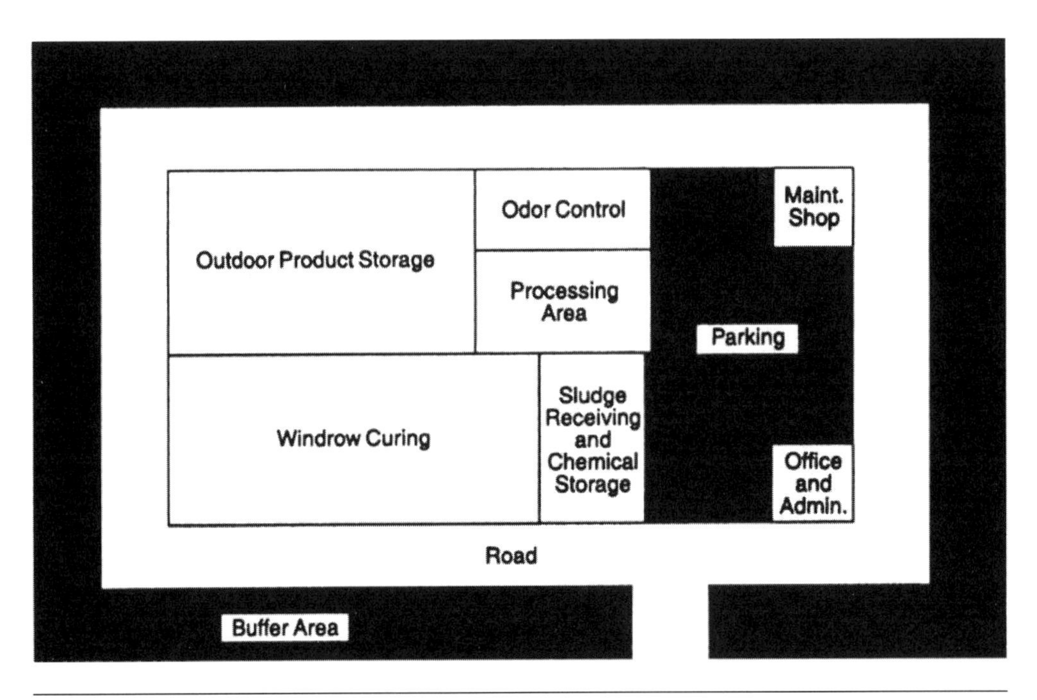

FIGURE 25.102 Layout for a 91-Mg/d (100-ton/d) advanced alkaline stabilization facility (Fergen, 1991).

stabilization system. Space is needed for solids-processing, drying (if necessary), and biosolids storage. Mobile, skid-mounted equipment can be used for backup or in emergencies; it also can be used in demonstration programs to encourage interest in biosolids.

Land requirements depend on the process to be used; solids type, characteristics and volume; and the specific site. Drying/curing area needs are typically 25 to 34 m^2/Mg (300 to 400 sq ft/wet ton) of processed cake, but the area needed also depends on the overall amount of material to be dried or cured, and the drying method used. The size of the drying/curing building can be significantly reduced by increasing the alkaline chemical dose or using a mechanical dryer.

Design engineers should consider storing product offsite if not enough area is available onsite. Landfills typically can provide space to accommodate drying/curing, but the drying and storage areas must be relocated as landfilling progresses. Also, outdoor drying/curing sites at landfills can cause odor complaints. In addition, the drying area must be easily accessible and large enough for trucks to unload biosolids without excessive maneuvering. Land also is required to accommodate additional truck traffic onsite.

If the alkaline stabilization process is located at a wastewater treatment plant, the access roads probably already exist. Sufficient access should be provided for regular delivery of alkaline materials. This truck access should not interfere with the traffic associated with the process or with biosolids distribution.

5.5.6 *Economic Considerations*

Economics is another important factor when selecting a solids-management option. Design engineers should evaluate the costs of an alkaline-stabilization process based on total life-cycle costs via present-worth, equivalent annual cost, or similar approaches. The cost of hauling and land-applying biosolids can be significant and must be included in the cost analysis (Jacobs et al., 1992). In addition, the costs of a privatized option (if the preferred procurement method) should be compared to those of publicly owned and operated options.

Annual O&M costs include labor, chemical costs, fuel, utilities, maintenance costs, and transportation (e.g. chemicals to the facility and biosolids to use or disposal sites). Other annual costs may include public education, public relations, biosolids marketing, soil testing, agronomic testing, and analyses. An owner should exercise caution when examining annual costs at other facilities because they include a number of site-specific factors (e.g., power, labor, and distance of the chemical supplier from the facility). Moreover, minimum biosolids-production amounts specified in the contract also affect total annual costs. When alkaline-stabilization technologies are operated under private contracts, the negotiated contract must accurately reflect actual biosolids production.

Many site-specific factors (e.g., physical layout, solids type and characteristics, biosolids use, local regulations, local biosolids market, and local climate) influence costs and make economic evaluations and comparisons difficult. Also, at some facilities, existing equipment has been retrofitted for use in alkaline stabilization.

Design engineers should consider the flexibility (adaptability) of alkaline stabilization and the use of existing facilities when evaluating solids-management options. Although not always possible, municipalities can save money if existing equipment is used in the process train.

5.6 Other Design Considerations

This section highlights and summarizes some of the O&M issues pertaining to alkaline stabilization. In general, most alkaline stabilization technologies are relatively simple and not equipment intensive; in addition, staff requirements are low compared to those for other stabilization processes. Operating considerations that must be addressed include startup issues, labor requirements, health and safety considerations, feed and product quality monitoring, maintenance, odors, dust, drying, procurement options, and process performance (Oerke, 1991).

5.6.1 Startup Issues

Startup issues associated with alkaline stabilization are installation-specific. The greatest concerns include equipment performance; process verification; physical, chemical, and biological product quality (to verify regulatory compliance and ensure product acceptance); and, if privatized, contractor performance.

While alkaline stabilization processes are not as equipment-intensive as other stabilization processes, some equipment problems and operating difficulties may occur during startup. To make the most of the dry chemical dose and produce the desired product, operators may need to vary the mixer-paddle speed and retention times. All process equipment should be tested at rated capacity during the startup period. The project team also should test a significant amount of representative feed and verify that dose measurement is accurate and mixing is homogeneous. If several types of alkaline materials are to be used, each should be tested with the system storage and feeding equipment to verify acceptable operation.

Regulators should be consulted during the design process to verify the parameters to be monitored for process approval. Monitoring results should be submitted to them as soon as possible to initiate the approval process. Permit delays are not uncommon, and appropriate measures should be taken to avoid them if at all possible. Frequent, continued communication with key regulators can facilitate the approval process. Also, the federal permitting authority (U.S. EPA region or delegated state agency) must be appropriately notified before startup.

Startup operations provide an opportunity to vary process parameters and evaluate the effects of these changes on product quality. Although the effect of various chemical doses and drying times should have been evaluated during pilot- or bench-scale testing, pilot-scale test conditions do not always adequately simulate full-scale operations.

An advantage of alkaline stabilitzation is the ability to start up operations quickly. A mobile, outdoor processing unit can be fully operational in about 10 days or less, depending on the amount of material to be processed.

5.6.2 Health and Safety Considerations

With few exceptions, health and safety considerations for alkaline stabilization are no different than those for typical wastewater treatment plant operations. Standard OSHA requirements (e.g., the use of safety glasses and hard hats) are maintained.

Dust generated from alkaline materials probably is the most significant health concern. Alkaline materials are caustic and cause skin burns and irritation and discomfort to moist surfaces (e.g., eyes, lips, and sweating arms); therefore, readily accessible eyewashes and showers should be provided at various locations throughout a wastewater treatment plant. Operators working in dusty environments or servicing alkaline storage and feeding equipment should be supplied with proper work clothing and safety equipment (e.g., gloves, proper respirators, and eye protection).

Ammonia is another safety concern, especially if anaerobically digested solids are processed, because it is likely that considerable ammonia gas will be released (about 6 to 10 times more likely than from raw solids). Ammonia emissions can be controlled via proper ventilation of mixers, storage hoppers, and loading areas. Strong releases of ammonia may be experienced during mechanical aeration or mixing, and during drying. So, mixing equipment should be enclosed and vented to odor-control facilities if at all possible. In some areas, it may be necessary to provide operators with respirators, depending on the amount of ammonia released to meet OSHA requirements.

Special safety measures may be required for drying areas. The layer of fine, operations-related dust that tends to settle in the drying area can be slippery on concrete or asphalt surfaces. During wet weather, a layer of mud may form outdoors on the drying pad. Mud is slippery and may pose a hazard to pedestrians and vehicle traffic. Special precautions should be taken to improve safety via good housekeeping practices.

5.6.3 Process Monitoring and Control

Feed cake and alkaline materials must be monitored frequently, so operators can adjust the process, as needed, to achieve adequate stabilization and a consistent product. The effects of incomplete stabilization are not readily apparent and may not be seen at a wastewater treatment plant; therefore, proper process control is important. Operators must be aware that acceptable dewatering characteristics and the absence of odors alone are not good indicators of adequate stabilization.

Monitored characteristics include the total solids concentration, pH, and temperature of both feed cake and biosolids. For Class A (PFRP) products, fecal streptococci also must be monitored at the frequency specified in 40 CFR 503.16 (U.S. EPA, 1993). In addition, metals must be monitored if the product will be used for agricultural purposes. If the product will be landfilled, toxic characteristics leaching procedure (TCLP) tests must be performed. Quality control data may be required for regulatory approval; the method and frequency depend on regulatory requirements. In some cases, odor characterization and emissions monitoring also may be required.

Operators can adjust the chemical dose in response to manual measurements of temperature and pH or visual inspections of the feed–chemical mixture. However, some automatic process control can be incorporated if desired. For example, thermocouples can be used to measure the heat pulse in an enclosed vessel. The chemical feed rate can be controlled by pacing it with the incoming-feed flowrate or dewatered cake via a weigh belt or similar means.

Using programmable logic controllers to monitor the chemical feed system helps produce a consistent product. A typical system may include an electronic chemical meter linked to feed-cake belt weigh scales; then, chemical feed is automatically controlled based on the weight of feed cake. Special care should be taken to keep the weigh scales frequently calibrated and correctly operating to ensure that the appropriate dose of alkaline materials is added. Solids weighing and volumetric systems should be calibrated every month. An automated system also decreases the number of personnel needed to operate the process.

Sensors—particularly pH electrodes (both laboratory and automatic process-control units)—must be properly cleaned, calibrated, and maintained. Special pH electrodes are necessary for routine measurements of more than pH 10. The pH must be monitored carefully to ensure that it is kept high enough for long enough to meet regulatory requirements. Portable pH pen probes are acceptable for process monitoring. A qualified laboratory should perform microbiological examinations for indicator organisms (e.g., fecal coliforms and fecal streptococci) regularly.

5.6.4 Odor Generation and Control

Odors and odor control are important issues when evaluating alkaline stabilization as a solids management option. Inadequate control and treatment of odors can be detrimental to a solids management program. Local conditions (e.g, weather, other sources of odors, and the characteristics of the odor-causing compounds) will influence the selection and design of an odor-treatment system. There are many site-specific factors that should be considered when developing a publicly acceptable odor-control program. A successful odor-control effort includes the following elements:

- Initial site selection,
- Proper process performance,
- Reduced biosolids storage time and volume,
- Identification of odor sources and odor-causing compounds,
- Meteorological modeling at different heights,
- Distance to nearest receptors, and
- Appropriate odor-control technology and equipment.

Ammonia is the odor most typically encountered at alkaline-stabilization facilities. Adding alkaline materials raises the pH, which causes the dissolved ammonia in dewatered cake to volatilize. Although the odors tend to dissipate quickly, the ammonia levels in mixing and drying areas can be high if the gas is not collected and treated. Also, if adequate ventilation is not provided, operators may need to wear respirators. So, appropriate odor-control equipment should be provided to ventilate and scrub the air to remove ammonia, thereby reducing odor problems and increasing public acceptance. As pH and temperatures rise, the intensity of ammonia emissions in the processing area may mask other, more prevalent odors that do not readily dissipate (e.g., trimethylamines). So, an odor survey should be performed to identify the sources of odors and characterize the odorants.

In addition to an odor survey, an assessment of meteorological conditions and atmospheric dispersion should be performed. Atmospheric data should be collected on wind speed and direction, temperature, and inversion conditions. This information typically is available from local weather stations and can be used to determine the effect of odor on residents near the alkaline-stabilization facility and the degree of odor control needed to meet community odor standards.

An effective odor-control program involves operational monitoring and may include bench-scale testing (to determine ammonia emissions at various chemical doses) and gas chromatography/mass spectrometry testing. After odors have been characterized, they must be collected and treated. Pilot-scale testing helps check the effectiveness of a proposed treatment option and its chemistry. At many alkaline-stabilization facilities, odor control primarily consists of diluting odors via open-air drying. If drying operations are enclosed, odors can be diluted and dispersed via rooftop ventilation. However, if large quantities of materials are processed in a densely populated area, a combination of dispersion and chemical scrubbing should be seriously investigated. It is important that the odor-control program be responsive to odor complaints. Depending on meteorological conditions and the sources and types of odors, operational or process modifications may be necessary to resolve the problem.

Initially, wastewater treatment professionals thought that alkaline- and advanced alkaline-stabilization facilities did not need odor-control systems. However, numerous odor concerns and complaints have made it clear that odor-control systems should be strongly considered and may be necessary for alkaline-stabilization systems near populated areas. Such systems may consist of enhanced ventilating systems and simple one-stage chemical scrubbers designed to remove ammonia only in the feed–chemical mixing area. They also can be state-of-the-art, three-stage, packed tower–mist scrubber–packed tower systems with air dispersion stacks, designed to treat high volumes of foul air containing particulate, ammonia, amines, dimethyldisulfide, mercaptans, and hydrogen sulfide generated from all areas of the solids treatment train. These sophisticated odor-control systems use sulfuric acid, sodium hypochlorite, and sodium hydroxide to neutralize and oxidize odorants (see Chapter 7 for more information).

5.6.5 *Dust*

Dust is inherent in alkaline stabilization systems. Alkaline materials-handling systems can create significant dust problems, particularly if fine-textured materials (e.g., hydrated lime, cement-plant kiln dust, or lime-plant kiln dust) are used. The alkaline materials-handling system should be designed with provisions for reducing dust production. Excessive alkaline dust also affects odor-control scrubber performance (e.g., acid chemical requirements).

5.6.6 *Sidestream Effects*

Alkaline stabilization processes typically have little effect on wastewater treatment plant operations. Minimal sidestreams result from site drainage, product stockpile leachate, and runoff if the storage area is not covered. However, a potential sidestream plant load is ammonia recycle if ventilation and odor-control acid scrubbers are installed.

5.6.7 Drying

Supplemental drying, if required in the process, also requires special consideration. Drying may make the product easier to handle, and the type of drying system will affect biosolids characteristics. For example, if the material is set out on a pad for solar drying without mechanical turning, it may dry in large clumps that would be incompatible with land application via granular fertilizer spreaders.

The duration of drying or curing depends on environmental conditions, chemical dose, windrow configuration, and initial and final solids concentrations. It also depends on the time required to achieve the process goal (i.e., for the heat of reaction to occur and for the pH to rise enough to destroy pathogens). Both drying and curing modify physical properties to attain the desired solids concentration and biosolids characteristics. The duration of drying depends on windrow size and weather conditions (if the drying facility is not enclosed).

5.6.8 *Process Performance*

Properly designed and operated alkaline-stabilization systems reduce odors, odor-production potential, and pathogen levels.

5.6.8.1 Odor Reduction

With proper mixing, alkaline-stabilization systems substantially reduce odor. One source of odors in solids-processing facilities, hydrogen sulfide, essentially is eliminated after the alkaline chemical is added and the pH rises to 9 or higher, because hydrogen sulfide is converted to nonvolatile ionized forms (see Figure 25.103). When air mixing systems are used, ammonia odors initially increase as a result of ammonia stripping. Once these odors have been emitted and dispersed or treated, odors can be reduced by a factor of 10 (Westphal and Christensen, 1983).

Other odorous gases emitted at high pH and temperature (e.g., trimethylamine) must be considered and dispersed or treated.

5.6.8.2 Settling and Dewatering Characteristics

Lime stabilization improves solids settling and dewatering characteristics. Lime alone has been used in the past as a conditioner before dewatering (although lime conditioning and lime stabilization are different processes). Precipitates associated with excess lime addition [primarily $Ca(CO_3)$ and unreacted $Ca(OH)_2$] act as bulking agents, increasing porosity while resisting compression.

Limited reports of lime-stabilized thickening and dewatering processes show mixed results. One study showed improved thickening (U.S. EPA, 1975). Two studies showed slightly better to slightly poorer dewatering on sand drying beds, compared to solids that were not lime-stabilized (Novak et al., 1977; U.S. EPA, 1975).

Design engineers should use caution when designing mechanical dewatering systems for lime-stabilized solids. If the design does not include proper preventive measures, scaling problems (e.g., deposition of $CaCO_3$ and other precipitates) can occur, resulting in higher O&M costs.

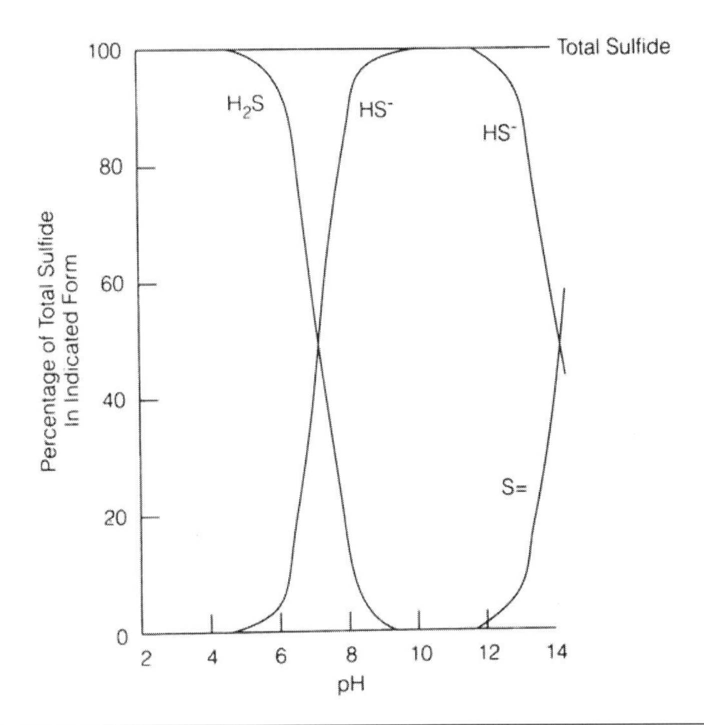

FIGURE 25.103 Effect of pH on speciation of hydrogen sulfide.

5.6.9 Procurement Options

Private firms offer many advanced alkaline-stabilization technologies involving proprietary processes or specialized equipment. Such technologies also involve royalty fees, quality-control fees, or sole-source equipment. Additionally, some firms may offer turnkey design–build facility procurement options or require privatization of various types of solids-processing services.

5.7 Process Considerations for Designers

5.7.1 Dosage Criteria

5.7.1.1 Class B Stabilization

Class B stabilization is achieved by adding enough lime (or its equivalent if using alkaline byproducts) to raise the pH to 12 for 2 hours and then hold it at 11.5 or higher for another 22 hours. The pH must be measured at a temperature of 25°C or corrected to 25°C.

Figure 25.99 shows the theoretical lime dose rates needed to achieve the design pH criteria. However, design engineers always should conduct bench-scale tests with the lime type and grade to be used during full-scale operations. Dose rates depend on the cake's solids content; more lime is needed when solids content is low (13 to 18%) or high (> 25%). Limed biosolids should be tested to ensure that they meet both the pH criteria for Class B disinfection and the Class B coliform limit.

5.7.1.2 *Class A Stabilization*

Lime stabilization meets Class A pathogen requirements by using the exothermic reaction of CaO and water in the biosolids to generate heat. Alkaline-stabilization processes can meet Class A requirements under Alternative 1 (time and temperature) or Alternative 5 (pasteurization); both are based on the assumption that every particle of biosolids will be exposed to 70°C for 30 minutes. This requirement can be met by treating batches of solids with lime in a closed container. Alkaline-stabilization processes that operate in continuous mode may need the specific approval of EPA's Pathogen Equivalency Committee to be accepted as a Class A process.

Several proprietary technologies have been approved by the committee (e.g., N-Viro's AASSAD process) or achieve pasteurization via a combination of lime and other sources of heat (e.g., RDP, Bioset). RDP envessel pasteurization uses an electrically heated screw to provide more heat. The Bioset process uses sulfamic acid to generate extra heat via an exothermic reaction in a pressurized reactor. The alkaline doses for these processes are given in Table 25.49 (EPA, in press).

5.7.1.3 *Class B Odor Control*

Raising pH into the high alkaline range not only stabilizes solids but also provides short-term odor control. However, the lime doses for Class B disinfection only raise pH above 12 temporarily. To control odors days or several weeks, the dose should be above the minimum for Class B stabilization. Although bench-scale testing is the best way to determine the optimum lime dose for odor control, a good general rule is to double the disinfection dose. Odors also can be controlled effectively by adequate mixing to ensure that there are no pockets of biosolids not in contact with lime.

TABLE 25.49 Mass balance for various alkaline-stabilization alternatives.

		dose rate (% wet wt.)		
B	Generic	2–5	None	15–30
A	Generic	10–20	None	23–35
A	N-Viro	12–20	FA, CKD, LKD	50–65
A	RDP	15	Electrical heat	23–35
A	Bioset	15	Sulfamic acid	20–30

Final solids based on cake solids of 25–25%.

5.7.2 *Lime Type and Gradation*

The suitable treatment agents are all lime-based materials. Lime is an alkaline earth material that produces a pH of 12.4 at 25°C when mixed with water. It is found in two forms: calcium oxide (CaO) and calcium hydroxide [$Ca(OH)_2$]. Calcium oxide (also called *quicklime* or *hot lime*) is the result of heating limestone [calcium carbonate ($CaCO_3$)] enough to drive off carbon dioxide (CO_2). When mixed with water, CaO forms a fine white powder [$Ca(OH)_2$, also called *hydrated lime*] and gives off considerable heat (called *heat of hydration*).

Many industrial processes have byproducts that contain usable amounts of lime [e.g., industrial scrubber sludge, fly ash (from incinerators that burn coals containing limestone), cement-plant kiln dust, lime-plant kiln dust, and dry industrial flue-gas scrubbing byproducts]. If used to treat solids, however, these alkaline agents must be carefully evaluated and monitored because their concentrations of free (active) lime content and contaminants vary.

Commercial quicklime grades can vary from several inches in diameter to material passing a #100 sieve. The National Lime Association (1990) lists the following five grades:

- Lump lime [50.8 to 203.2 mm (2 to 8 in.) in diameter];
- Pebble lime [the most common form, ranging from 6.35 to 50.8 mm (0.25 to 2 in.) in diameter];
- Granular lime (100% passes though a #8 sieve, and 100% is retained on a #100 sieve);
- Ground lime (100% passes through a #8 sieve, and 40 to 60% passes through a #100 sieve); and
- Pulverized lime (100% passes through a #20 sieve, and 85 to 95% passes through a #100 sieve).

The following quicklime definitions will help in relieving the confusion of so many terms:

- *Unslaked quicklime fines* (*calcium oxide fines*) are quicklime particles that typically are less than 9.5 mm (3/8 in.) in diameter and have not been mixed with water;
- *Pulverized calcium oxide* is quicklime that has been mechanically ground into particles that typically are less than 60 mesh;
- *Granular calcium oxide fines* is quicklime that has been ground into particles that are larger than pulverized calcium oxide (i.e., there are no dust-sized particles);
- *Unslaked CaO fines* are small quicklime particles that have not been mixed with water; and
- *Unhydrated calcium oxide* is any quicklime that has not been hydrated (*slaked*).

Lime's reactivity with water is measured by the slaking rate (as defined in AWWA specification B202-93, Sec. 5.4). Small-pore limes react need 20 to 30 minutes to fully react with water, forming $Ca(OH)_2$ with a slow heat rise. A moderately reactive lime needs 10 to 20 minutes to react with water, forming $Ca(OH)_2$ and raising the temperature to 40°C in 3 to 6 minutes. A highly reactive lime fully reacts with water within 10 minutes and raises the temperature to 40°C within 3 minutes. Design engineers can use the slaking rate to evaluate the suitability of various industrial byproducts.

Solids should be treated with a moderately or highly reactive lime to ensure that the CaO fully converts to $Ca(OH)_2$. For the reaction to generate a high pH that migrates throughout the solids, there must be a continuous film of water throughout the material. Otherwise, the lime may not be fully hydrated or the hydroxide ions may not migrate throughout the solids. This can and does result in improper pH measurements, improper doses, and therefore, unstabilized solids. If calcium oxide must be pulverized, it should be pulverized at the point of application to prevent air slaking and ensure the desired reactivity.

5.7.3 *Mixing Requirements*

In a survey of 19 wastewater treatment plants in Pennsylvania, the Pennsylvania Departement of Environmental Protection examined process variables (e.g., biological treatment and lime dose) and their effects on odor, as determined by an odor panel (EPA, in press).

Results showed a wide range in lime dose and in solids content before treatment. Centrifuged solids tended to be more odorous than belt-pressed solids, but there was no clear relationship between other process variables and odor.

The agency selected two of the surveyed wastewater treatment plants to study the effect of lime dose and mixing time on pH decay and on odor. Researchers used two parameters to indicate mixing efficiency: total Ca (as measured by EDTA titration) and pH (as measured by a flat-surface pH electrode) (EPA, in press). Higher, relatively stable Ca concentrations throughout solids in the mixing vessel indicated that solids and lime were well mixed. The flat-surface pH electrode measured actual pH in the solids–lime mixture more accurately than the traditional slurry method. (In the slurry method, water is added to the solids–lime mixture before pH measurement; this dissolves any unreacted lime, producing a falsely high pH reading.)

In the first study, researchers added CaO to cake at 4.5 and 11.7% (wet weight) and mixed them for 15 and 45 seconds. Results showed that 15 seconds were inadequate; there was much higher variability in Ca and pH at 15 seconds of mixing. Results also showed that a CaO dose of 4.5% would raise pH above 12, but only at the longer mixing time.

The slurry method indicated that pH dropped below 12 after 15 days at the lower CaO dose and shorter mixing time. The flat-surface electrode, however, showed that the lower CaO dose and shorter mixing time never achieved pH 12.

Increasing CaO dose and mixing time decreased odor generation. They also reduced the generation of NH_3 and amines, an indicator of biological decomposition. Biological decomposition can result in increased odors.

Odor increased in all limed solids up to 15 days, but decreased thereafter for solids with the higher CaO dose and the longer mixing time. The study also showed that NH_3 and amines greatly increased after 15 days in the solids with the lower CaO dose and shorter mixing time.

In the second study, researchers examined the plant-scale effect of optimizing CaO and solids mixing on pH decay and odor generation. To optimize mixing, researchers added CaO to the solids upstream of the mixer to increase contact time. They then compared samples of limed solids from the existing operation with those from the optimized operation.

Results showed that optimizing mixing reduced variations in Ca levels in solids, prevented pH from decaying, and decreased odor, NH_3 and amine generation for up to 20 days.

5.7.3.1 *Measuring Mixing Efficiency*

5.7.3.1.1 Identifying Issues

The District of Columbia Water and Sewer Authority's (DCWASA's) Blue Plains Advanced Wastewater Treatment Facility has used lime stabilization to achieve Class B pathogen standards for many years. While fecal coliform results always met the regulatory limit (<2 mil. CFU/g), they were inconsistent. Odors also were inconsistent, according to empirical evidence gathered in the field. Field inspectors said the odors resembled those of rotten eggs or rotting cabbage (methyl mercaptan, dimethyl sulfide, dimethyl disulfide); rancid meat (volatile fatty acids); and fecal matter (indole, skatole). Some odorants were confirmed via a gas chromatograph (Kim et al., 2003). All of these odors are the products of anaerobic microbial activity, indicating less than optimum microbial inactivation. Personnel suspected these inconsistencies in odor and fecal coliform destruction were related, and that poor mixing might be responsible. So, they implemented changes based on a series of studies, and DCWASA solids now consistently contain less than 1000 CFU fecal coliforms and emit few odors.

Most facilities using lime stabilization rely on the results of pH tests at 2 and 24 hours to indicate if the material is in compliance with EPA Class B standards. However, EPA assumes that complete, efficient mixing occurs and that a pH >12 after 2 hours and >11.5 after 24 hours indicate a stabilized, low-odor material. The standard pH test (the slurry method) involves adding water to and stirring the sample before measurement, so although the test is a good indicator of whether the sample contains enough lime, it does not indicate whether the sample was well mixed before testing. So, pH results may be consistent while the final product has wide swings in quality (fecal coliform levels and odors).

Facilities experiencing odor complaints should determine whether the solids have consistent concentrations of fecal coliforms and odorants. If results indicate that fecal coliform levels are inconsistent or considerably above 1000 CFU, or that odors [measured either by a nose (qualitatively) or by a reduced-sulfur meter or tubes (quantitatively)] are inconsistent or intolerably offensive, then the lime was not thoroughly incorporated into the solids. A set of simple, inexpensive tests can help identify solutions.

Efficient, adequate mixing is affected by at least five factors: lime gradation, cake dryness, residence time in the mixer, mixer type, and conveyance method before mixing. Once operators have a tool to measure mixing efficiency, they can adjust one or more of these factors to achieve the desired product quality.

5.7.3.1.2 Establishing a Benchmark for Good Mixing

If investigators suspect poor mixing, they should start by establishing parameters consistent with sufficient mixing that they can use when comparing results. A simple means of determining mixing efficiency is a calcium test, which requires a 1 g sample. In well-mixed solids, each 1 g sample would contain solids and calcium in the required ratio (i.e., 15% lime on a dry weight basis). In poorly mixed solids, one sample might contain no calcium, another might contain a high percentage of calcium, and others would bear results in between. A large sample set (e.g., 12 to 15 samples) with a high standard deviation would indicate poor mixing, while one with a low standard deviation would indicate well-mixed biosolids. Staff can conduct a bench-scale test in which they mix with solids with lime (as delivered) and determine parameters for well-mixed material. The results then are compared to plant results to grade the performance of full-scale operations.

A bench-scale setup can use a simple bread mixer. Start with unlimed dewatered material, and add lime at the prescribed dose (e.g., 15% on a dry weight basis). Operate the mixer, stopping and sampling after 10, 20, 30, 40, 60, and 90 seconds. Each time the mixer is stopped, take fifteen 1-g samples for calcium analysis. (Fewer samples may be adequate, but calcium tests are inexpensive and more data will provide clearer results). Mixing probably is inadequate at 10 seconds and probably sufficient at 90 seconds. The sample set with the smallest standard deviation is the plant-specific benchmark for a well-mixed product. It is important to conduct this bench test on cake collected just before it enters the mixers because the dewatering and conveyance methods will affect mixing results. The data in Figure 25.104 was generated during the bench-scale testing

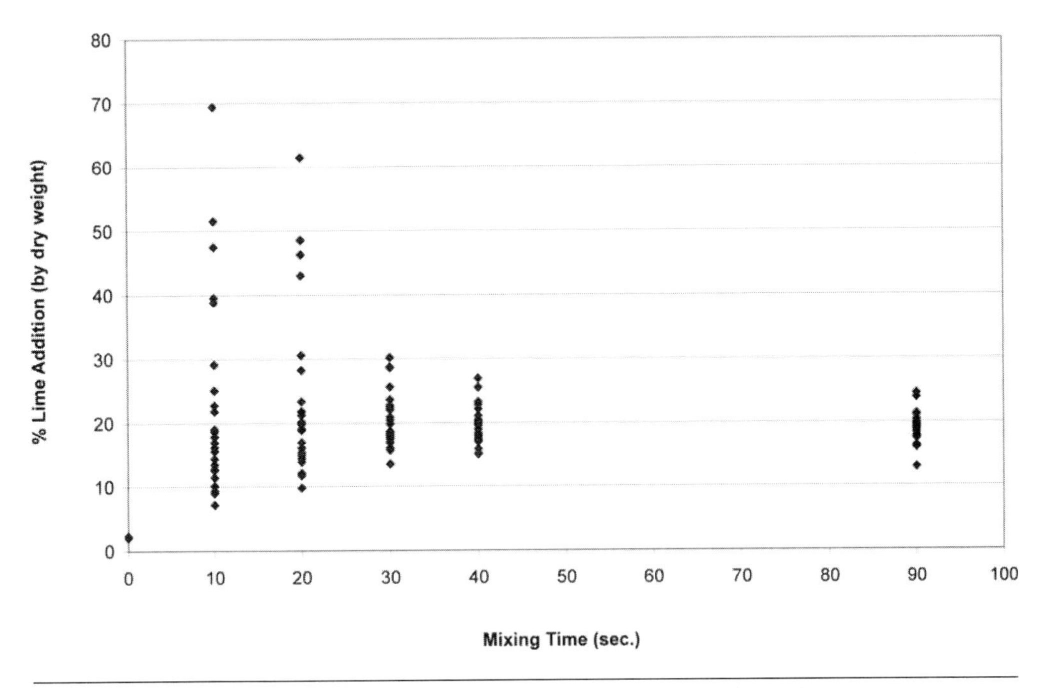

FIGURE 25.104 Results of DCWASA bench-scale mixing test for calcium content.

phase of DCWASA's research (North et al., 2008a); they show that standard deviation decreased as mixing time increased.

5.7.3.1.3 Measuring Performance of Full-Scale Plant Operations

The next step is to take 15 samples from the full-scale operation, analyze them, and calculate their standard deviation. If this standard deviation is higher than the minimum achieved in the lab, then the mixing system can be improved. If the full-scale and minimum bench-scale standard deviations are identical, then better mixing and product quality are unlikely. DCWASA found that when odors were high, the standard deviations of its full-scale sample sets were close to that for the 15- to 20-second samples in bench-scale testing, indicating that the full-scale mixer was far from providing optimum mixing during these periods (North et al., 2008b).

5.7.3.1.4 Mixing Energy and Odor Suppression

At the Blue Plains plant, the minimum standard deviation of the sample sets was about 2.6 (which occurred at about 40 seconds of bench-scale mixing). Results are plant-specific, but this number gave DCWASA operators a tool to measure lime–solids mixing and mixer performance, as well as improve product quality. Figure 25.105 shows the relationship between mixing energy (time, in this case) and reduced sulfur compounds (odors) for the samples in Figure 25.104. Not surprisingly, odors are minimized when good mixing occurs.

5.7.3.1.5 Mixing Energy and Fecal Coliform Destruction

Figure 25.106 shows the relationship between mixing energy (time) and fecal coliform results for the samples in Figure 25.104. Again, fecal coliforms are minimized when good mixing occurs. Surprisingly, minimizing fecal coliforms in this Class B stabilization process yielded results (CFU <1000) consistent with Class A bioolids.

5.7.3.2 Optimization of Mixing—Examining Five Factors Affecting Mixing

5.7.3.2.1 Factor 1: Lime Gradation

To ensure proper mixing, personnel periodically must test lime deliveries via a sieve analysis and compare results to the required lime specifications. If the delivered lime is too coarse, mixing energy may be inadequate. This simple test can help ensure adequate mixing, low odors, and proper stabilization.

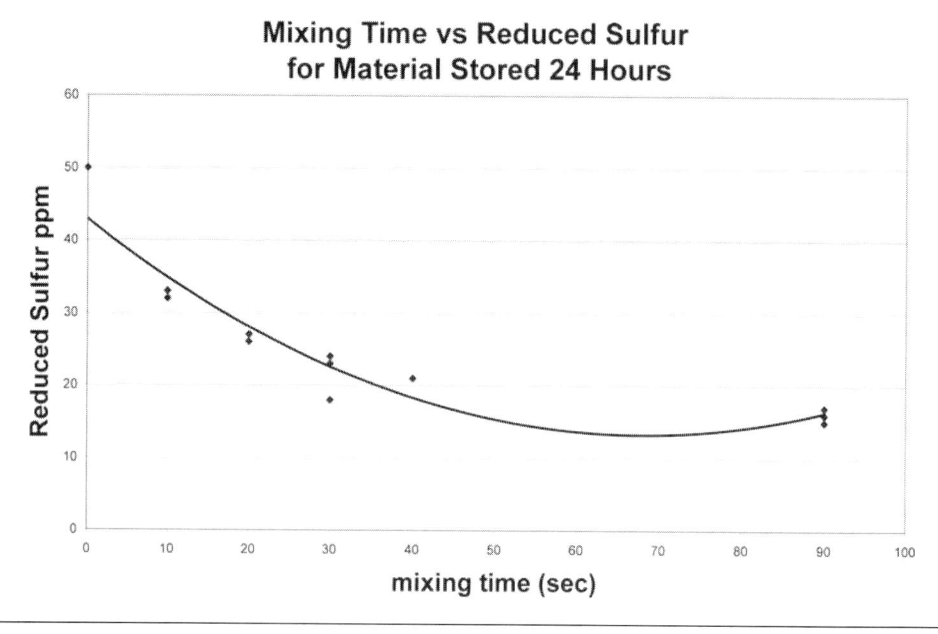

FIGURE 25.105 Relationship between mixing energy and reduced sulfur compounds in DCWASA bench-scale mixing test.

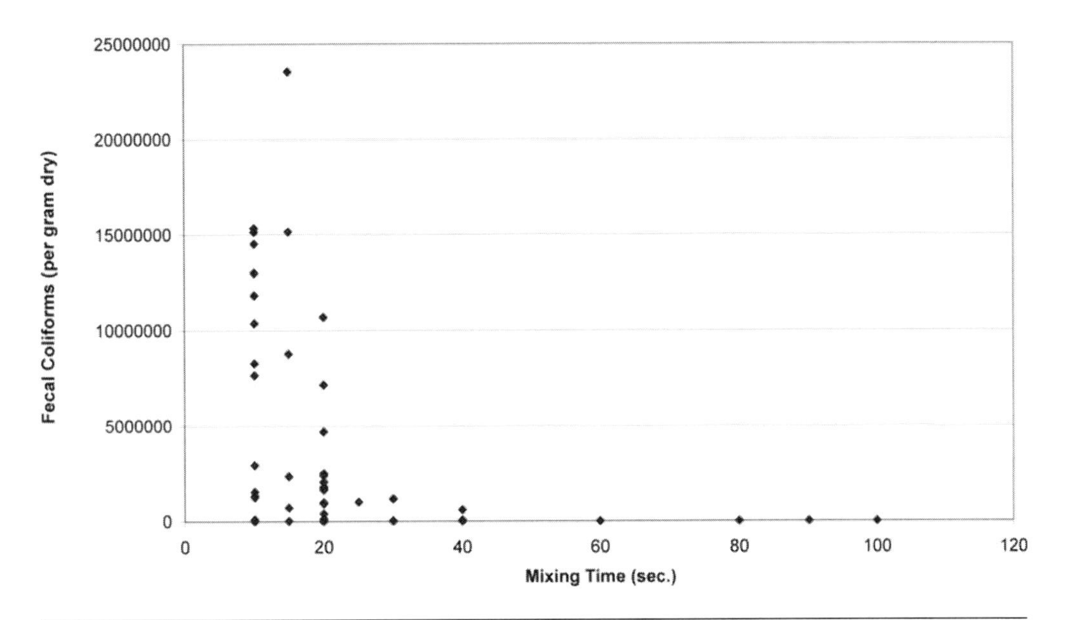

FIGURE 25.106 Effects of mixing on fecal coliform concentrations in DCWASA bench-scale mixing test.

Coarser lime requires more mixing energy for adequate incorporation. Figure 25.107 shows results from the DCWASA bench tests for mixing solids with different grades of lime. The lime used in both DCWASA dewatering trains was subjected to sieve analysis, and results showed that the lime used in the WASA 1 train (operated by WASA employees) was coarser than that used in the WASA 2 train (lime supplied by and equipment operated by contractor). Both limes were mixed with raw solids from one source, and a third sample was mixed with the coarser lime ground to match the sieve analysis of the WASA 2 lime. The results of the coliform analysis show that the samples with finer lime stabilize much sooner than the others. Also, the sample with Lime 1 ground to the fineness of Lime 2 stabilized at a similar rate, showing that the difference between Lime 1 and Lime 2 was primarily due to gradation, not other characteristics.

5.7.3.2.2 Factor 2: Cake Dryness

A dewatered cake's solids content can dramatically affect mixing energy requirements. Drier solids require more energy for adequate lime mixing. Small differences in percent

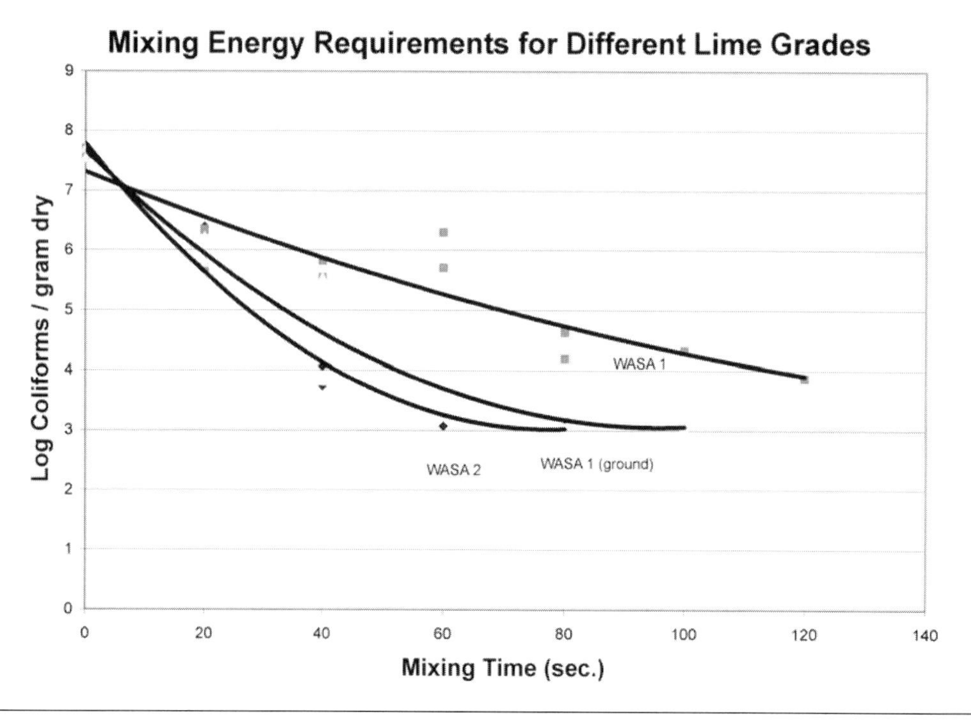

FIGURE 25.107 Mixing energy required for various grades of lime.

solids (3 to 4 %) can double the required mixing energy. Figure 25.108 shows the mixing energy required for proper stabilization of low, medium, and high cake solids. Cakes with higher solids concentrations require much more mixing energy to minimize fecal coliform concentrations. This is an important consideration when assessing mixing problems, because dewatering facilities sometimes can produce inconsistent cake solids. Inconsistencies in odors and fecal destruction might be attributable to inadequate mixing when the cake had high solids concentrations.

Engineers should design lime-mixing facilities to handle the maximum solids content expected in dewatered cake. If hauling costs are not paramount, a treatment plant might consider scaling back the dryness of the cake solids slightly to help ensure adequate mixing and stabilization. If hauling costs are a major portion of the budget, better mixing equipment (or another fix mentioned in this section) might be required.

5.7.3.2.3 Factors 3 and 4: Mixer Residence Time and Mixer Type

A system's mixing efficiency can be affected by the type of mixer used and the equipment configuration. A facility that adequately mixes lime with a specific piece of equipment for years may run into problems if dewatering-system upgrades (e.g., high solids

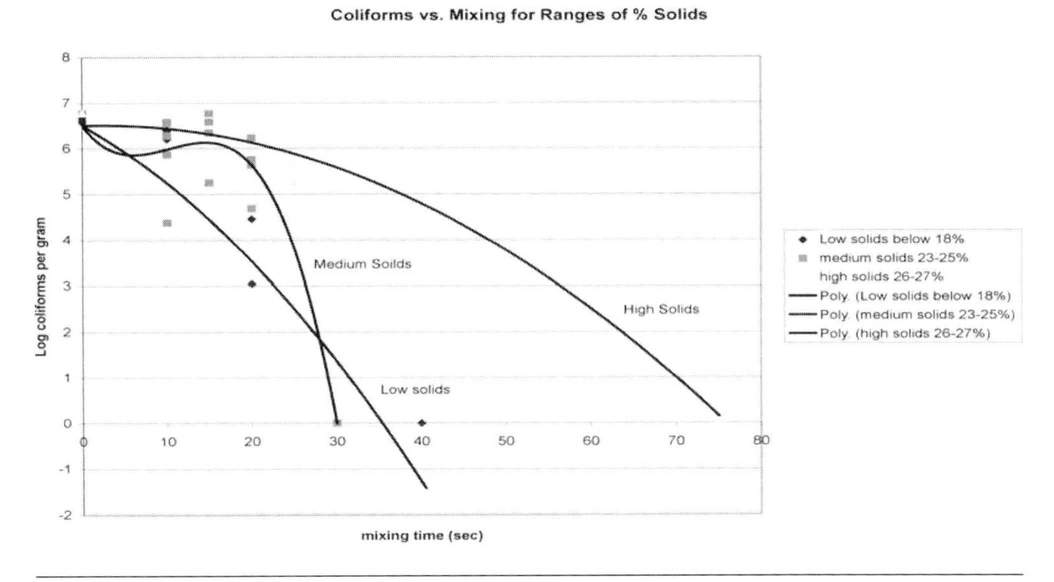

FIGURE 25.108 Mixing energy required for various solids concentrations.

centrifuges or changes in polymers) produce cake containing more solids. To ensure that enough mixing energy can be provided, operators need to examine whether existing mixers should be modified or replaced.

Often, existing mixers can be modified to increase agitation or residence time. For example, plow blenders have removable weirs that are designed to keep the material in place longer. Other blenders come with openings so chopper blades can be easily installed to enhance mixing. If a unit is not achieving optimum mixing, operators should install all optional equipment designed to enhance mixing and residence time. If the unit still cannot achieve optimum mixing after considering Factors 1, 2, and 3, operators must consider replacing it. Before considering larger mixers, however, staff should examine the conveyance system.

5.7.3.2.4 Factor 5: Conveyance Method Before Mixing

Figure 25.109 shows a plan of part of the solid conveyance system at the DCWASA Blue Plains Advanced Wastewater Treatment Plant. Sample Location 1 is at the discharge of the high-solids centrifuge, Location 2 is on the horizontal screw conveyor, Location 3 is after a vertical screw conveyor, and Location 4 is just before the lime mixer. Location 5 is at the discharge end of the lime mixer, Location 6 is on the horizontal screw conveyors moving material to the storage bunkers, and Location 7 is at the point of discharge to the bunkers. Research results show that using screw conveyors before mixing adds

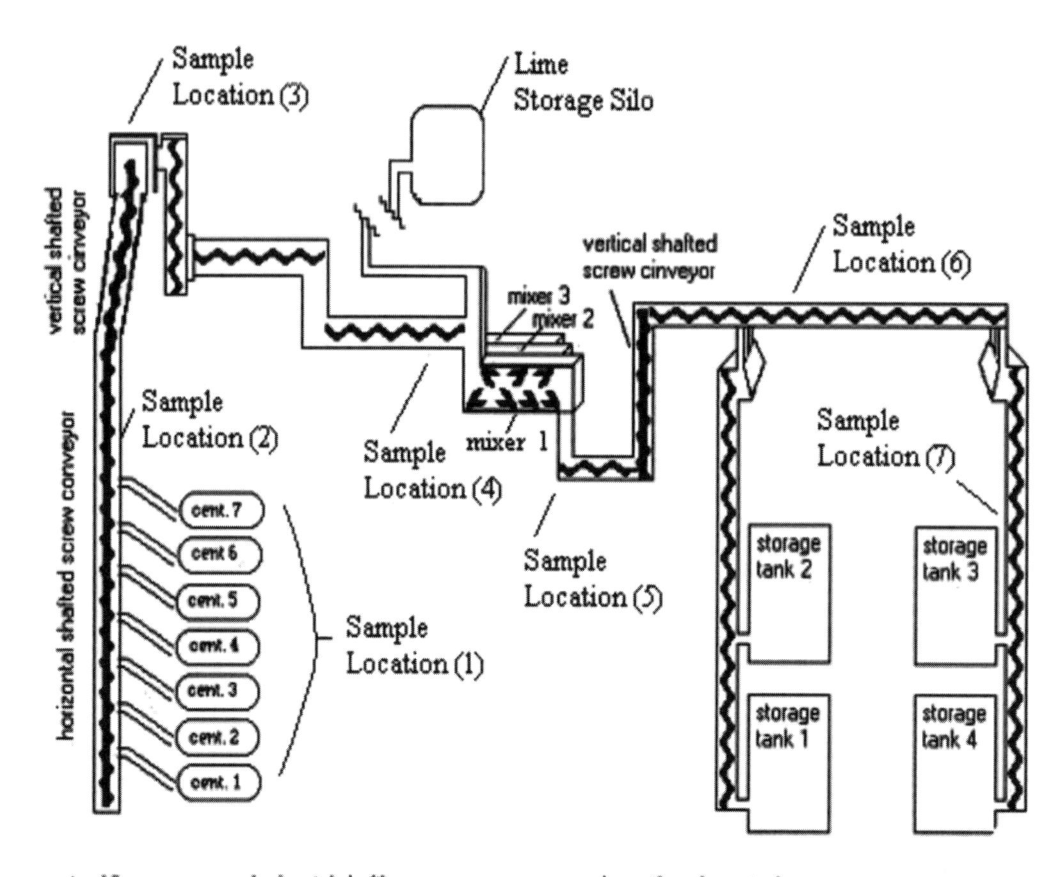

FIGURE 25.109 Plan view of conveyors and sample locations in the DCWASW lime stabilization system.

mixing energy (thereby changing the material's rheology), making it more difficult to stabilize. Using screw conveyors after mixing, however, adds more mixing energy and thereby reduces odors and improves product quality.

Unlimed material conveyed a longer distance requires more energy for proper stabilization. Staff grabbed unlimed material from four locations before mixing and subjected it to bench-scale mixing tests. Results show materials that have been conveyed longer (Location 4), consistently have more residual coliforms than those conveyed for a shorter distance (Location 2) (see Figure 25.110). This shows that conveyance changes the material's rheology and affects its ability to mix properly. Visual observations showed that the material changed from a crumbly consistency to a toothpaste consistency between Locations 1 and 4.

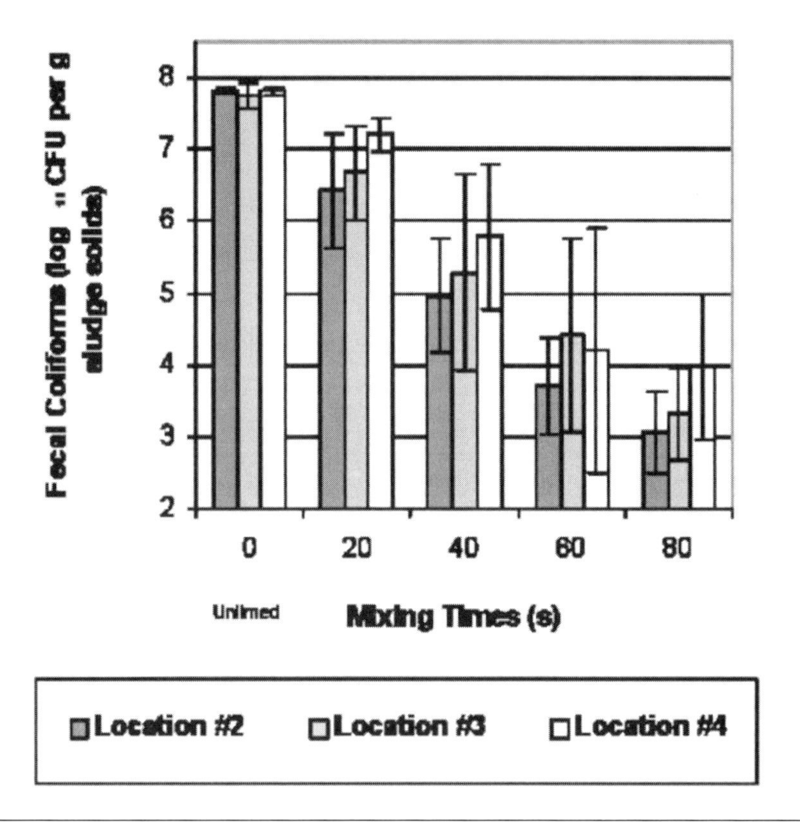

FIGURE 25.110 Effect of conveyance distance on solids stabilization.

When considering equipment changes (if no other intervention has helped), operators should compare the costs of replacing screw conveyors (with belt conveyors, which do not affect a material's rheology) to the cost of upgrading mixers. Existing mixers might be adequate for material that is not screw conveyed a long distance.

After mixing, screw conveyors can add mixing energy and improve product quality. Figure 25.111 shows that, the farther biosolids were conveyed, the less reduced sulfur it generated. These data also show the importance of sampling at the end of conveyor runs, rather than at the discharge end of the lime mixer. Fecal coliform samples from Locations 5 and 7 will bear strikingly different results, again showing that screw conveyors further stabilize biosolids after lime mixing.

5.7.4 *Class B Lime Stabilization Design Example*

To design a system to meet EPA Class B standards (<2,000,000 CFU/g fecal coliforms), a facility must adequately mix into a raw solids an appropriate amount of lime on a

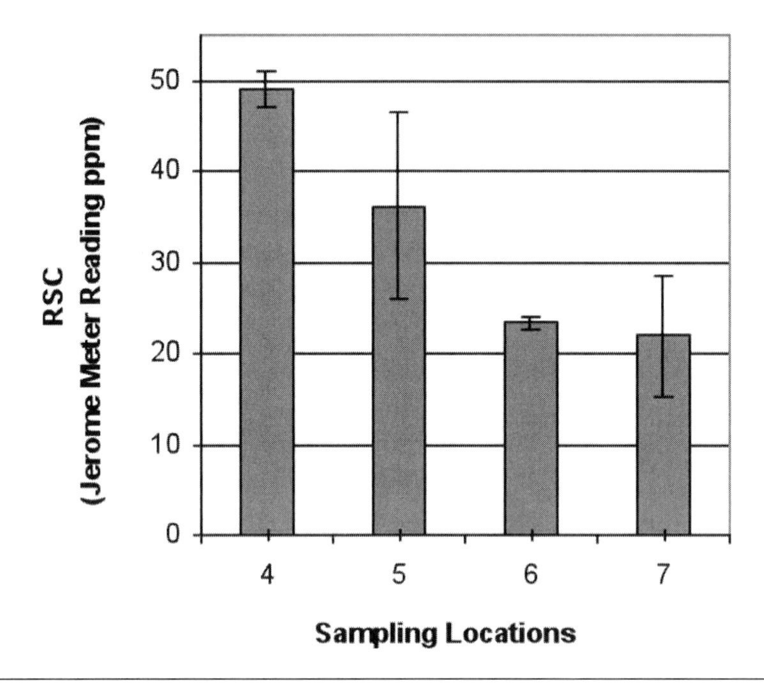

FIGURE 25.111 Effect of conveyance distance on odors.

consistent basis to produce a useable, low odor biosolids product. This requires a design for storage and conveyance of an appropriate amount of lime for the biosolids produced, and a mixing system with adequate mixing energy to match the material (which can vary considerably from plant to plant, depending on dryness, conveyance, and lime gradation). Lime dosing and lime mixing have been addressed in this MOP. The following example uses the concepts outlined in earlier sections of this manual.

5.7.4.1 Design Example—Part I

Design a solids stabilization system that meets Class B pathogen requirements using quicklime and a lime mixer. The wastewater treatment facility's peak production is 20 dry tonne/d of solids. Solids are dewatered by a belt filter press that can produce a cake containing 18% solids. The facility initially will produce biosolids 24 hours a day, 7 days a week, loading directly into trucks that each hold 21 wet tonne. So, it must be staffed round the clock. Suppose that eventually the facility expands its settling capacity and wants to reduce solids treatment system operations (and staffing) to 8 hours a day, 5 days per week. Assume the chosen mixer has a rated capacity of 100 kg/min, and provide about 15 to 25% extra mixing capacity (beyond the manufacturer-suggested rated capacity) in mixer and standby mixer equipment. Assume a lime dose of 15% on a dry weight basis.

5.7.4.1.1 Design a Lime Mixing System for 24/7 Operations

Using the average daily production of dewatered cake (7 days per week), design engineers first should calculate how much lime and how many mixers are needed:

$$20 \text{ tonne/d solids} \times 15\% \text{ lime} \times 1000 \text{ lb/tonne} = 3000 \text{ kg/d lime} \tag{25.54}$$

$$\begin{gathered}20 \text{ dry tonne/d/18\% dry cake solids} = 111 \text{ wet tonne/d (77 kg/min)},\\ \text{so two mixers are required to achieve the desired redundancy.}\end{gathered} \tag{25.55}$$

Design engineers should consider for sizing mixers with extra capacity to take into account changes in the solids' rheology characteristics during the project's design life. In this example, one mixer provides enough capacity for peak conditions, but the standby mixer is available for O&M considerations.

5.7.4.1.2 Design a Lime Mixing System for 8 hr/d, 5 d/wk Operations

After plant expansion and staff reduction, how much cake is produced during the 8-hour, 5 day-per-week operational shift? How much lime is required? How many mixers are required?

$$111 \text{ w tonne/d} \times 7/5 = 156 \text{ wet tonne/8-hour shift} \tag{25.56}$$

$$\begin{gathered}156 \text{ wet tonne} \times 18\% \text{ dry cake solids} \times 15\% \text{ lime} \times 1\,000 \text{ lb/tonne} =\\ 4\,200 \text{ kg lime per shift (9 kg/min)}\end{gathered} \tag{25.57}$$

$$156 \text{ wet tonne/8 hr/60 min} \times 1\,000 \text{ lb/ton} = 324 \text{ kg/min to mixer(s)} \tag{25.58}$$

$$\begin{gathered}324 \text{ kg/min} + 9 \text{ kg/min} = 333 \text{ kg/min solids and lime to mixers,}\\ \text{so four mixers are required at 83 kg/min per mixer,}\\ \text{plus one standby mixer for O\&M considerations.}\end{gathered} \tag{25.59}$$

5.7.4.2 Design Example—Part II

Several years into the life of the project, plant personnel decide to convert from belt filter presses to centrifuges, which produce a cake containing 24% solids. Before this conversion, plant operators test the mixers' adequacy using the calcium test method. Results showed that the bench-scale mixer provided a minimum standard deviation of 2.6 after 40 seconds of mixing, beyond which the standard deviation did not further improve (Figure 25.105). The full-scale mixer also achieved a standard deviation of 2.6 for solids dewatering with a belt filter press, indicating mixing was adequate.

Knowing that as solids content increases, mixing energy requirements also increase [a 3% increase in solids nearly doubles (100% more) the required mixing energy], staff tested the mixers again after the centrifuges were installed. The new results had a standard deviation of 8.4, indicating less-than-optimal mixing.

Personnel needed to determine the answers to the following questions:

- What is the flow to each of the four existing mixers?
- What is the effect of the higher solids on the material's rheology?
- Does a decrease in solids flowrate (because of the centrifuge installation) compensate for an increase in mixing energy requirement?

While the extra capacity originally designed into the mixing system was sufficient to account for changes in rheology in belt-pressed solids, it may not be enough for the drier cake produced by a centrifuge.

Calculations showed that the cake flowrate from a centrifuge is lower than that from a belt filter press:

$$156 \text{ wet tonne} \times 18\% \text{ belt press dry cake solids}/24\% \text{ centrifuge dry cake solids}/\ 8\text{ h}/60\text{ min} \times 1000\text{ kg/tonne}$$

$$= 244\text{ kg/min solids} + 9\text{ kg/min} \qquad (25.60)$$

$$= 253\text{ kg/min to four mixers, or } 63\text{ kg/min per mixer}$$

The flow to the mixers decreases by 25% (from 83 to 63 kg/min) with about 40% excess capacity (100 kg/min rated capacity), but mixing energy requirements also change:

$$24\% \text{ dry cake solids} - 18\% \text{ dry cake solids} = 6\% \text{ increase (or a 200\% increase in required mixing energy)} \qquad (25.61)$$

Knowing that inadequate mixing can increase odors and fecal coliforms, what can staff do to ensure that the existing mixers provide enough mixing energy to maintain the product quality of past production years?

If the mixers are left untouched, biosolids quality probably will decline (i.e., produce more fecal coliforms and odors) once the centrifuges are installed. Calcium test results confirm this assumption; they show that the standard deviation is not minimized after dewatering improvements.

Facility personnel can install weirs in the mixer to increase detention time. If this does not provide adequate mixing, they also can consider installing another mixer.

Finally, personnel may need to consider detuning the new centriguge to produce cake with lower solids content.

This example shows that changes in upstream rheology should be considered when sizing lime-stabilization equipment. In many cases, significant changes in dewatering equipment may require pilot-tests and changes in mixer design to compensate.

5.8 Product End-Use Considerations

Wastewater solids contain organic matter and plant nutrients, making them a valuable crop fertilizer and soil conditioner. However, adding alkaline material dilutes some plant nutrients and volatilizes the ammonia-nitrogen content. Also, the alkaline material adsorbs a substantial portion of mineralized organic nitrogen, further reducing the amount of nitrogen available to plants (Logan, 1990a). The net result may be a relatively low-grade fertilizer, but a good lime substitute and organic soil amendment.

That said, alkaline materials can be custom-blended with solids and other feedstocks (e.g., sand, topsoil, yard waste, and leaves) to produce a specific, marketable product. Some municipalities (e.g., Warren, Ohio) have done this successfully, creating a more publicly acceptable product with a lower pH. Such products may be called *artificial soils* (WEF, 1994a).

Applications for alkaline-stabilized biosolids include

- Agriculture (e.g., organic fertilizer, agricultural-lime substitute, or soil amendment);
- Horticulture (e.g., nurseries and sod farms);
- Residential lawns and gardens (e.g., manufactured organic topsoil blends);
- Bulk fill (e.g., slope stabilization and dike construction);
- Nonagricultural land application, land reclamation, or dedicated land disposal; and
- Landfill (e.g., disposal or daily, intermediate, final, and vegetative cover).

Each has particular quality requirements and standards. For example, the application rate of alkaline-stabilized biosolids for agronomic purposes may be limited based on calcium carbonate equivalence or alkalinity content, rather than its low plant-nutrient content. If the material will be used at landfills, most regulators require extensive testing and documentation first.

An advantage of an alkaline-stabilized biosolids compared to other biosolids (e.g., compost) is that it can partially or fully satisfy the liming requirements of many soils. Also, it may contain small amounts of plant nutrients. For example, most cement-plant kiln dust contains significant amounts of potassium and smaller amounts of trace nutrients. Some alkaline stabilization processes use mineral byproducts [e.g, cement-plant kiln dust, lime kiln dust (lime-plant kiln dust), and alkaline fly ash]. Nutrient content is biosolids-specific and should be carefully monitored. Biosolids also may contain regulated trace elements that should be carefully monitored to avoid exceeding regulatory limits.

For more information on biosolids uses, see Chapter 27.

6.0 REFERENCES

Abu-Orf, M. M.; Griffin, P. P.; Dentel, S. K. (2001) Chemical and Physical Pretreatment of ATAD Biosolids for Dewatering. *Water Sci. Technol*, **44** (10), 309–314.

Adams, G. M.; Witherspoon, J.; Card, T.; Erdal, Z.; Forbes, B.; Geselbracht, J.; Glindemann, D.; Hargreaves, R.; Hentz, L.; Higgins, M. J.; McEwen, D.; Murthy, S. (2004) *Identifying and Controlling the Municipal Wastewater Odor Environment Phase 2: Impacts of In-Plant Operational Parameters on Biosolids Odor Quality*; Report No. 00HHE5T; Water Environment Research Foundation: Alexandria, Virginia.

Agarwal, S.; Abu-Orf, M.; Novak, J. T. (2005) Sequential Polymer Dosing for Effective Dewatering of ATAD Sludges. *Water Res.*, **39**, 1301–1310.

Al-Ghusain, I.; Hamoda, M. F.; El-Ghany, M. A. (2004) Performance Characteristics of Aerobic/Anoxic Sludge Digestion at Elevated Temperatures. *Environ. Technol.*, **25**, 501–511.

American Society of Civil Engineers (1959) *Sewage Plant Design*; prepared by American Society of Civil Engineers and Water Pollution Control Federation.

American Society of Civil Engineers (1983) *A Survey of Anaerobic Digester Operations*; American Society of Civil Engineers: New York.

American Society of Heating, Refrigerating, and Air Conditioning Engineers (2005) *ASHRAD Handbook Fundamentals*, inch-pound edition; American Society of Heating, Refrigeration, and Air-Conditioning Engineers: Atlanta, Georgia.

American Water Works Association (1983) *Standard for Quicklime and Hydrated Lime*; AWWA B202-83: American Water Works Association: Denver, Colorado.

Arant, S.; Boden, R. (2000) Design and Startup of a Unique Auto-Heated Aerobic Digester. *Proceedings of the 73rd Annual Water Environment Federation Technical Exposition and Conference* [CD-ROM]; Anaheim, California, Oct 14–18; Water Environment Federation: Alexandria, Virginia.

Avallone, E. A.; Baumeister, T. (1996) *Marks' Standard Handbook for Mechanical Engineers*, 10th ed.; McGraw-Hill: New York.

Baier, U.; Zwiefelhofer, H. P. (1991) Effects of Aerobic Thermophilic Pretreatment. *Water Environ. Technol.* **3,** 56–61.

Barber, N. R.; Dale, C. W. (1978) Increasing Sludge Digester Efficiency. *Chem. Eng.*, **85** (16), 147–149.

Barker, J. C. (1996) *Crystalline (Salt) Foramtion in Wastewater Recycling Systems*; Publication EBAE 082-81; North Carolina Cooperative Extension Service: Raleigh, North Carolina.

Barnes, C.; Walker, S.; Anderson, W.; Papke, S. (2007) Implementation of a Two-Phase Anaerobic Digestion System. *Proceedings of the 21st Annual Water Environment Federation Residuals and Biosolids Conference*; Denver, Colorado, Apr 15–18; Water Environment Federation: Alexandria, Virginia.

Batstone, D. J.; Keller, J.; Angelidaki, I.; Kalyuzhnyi, S. V.; Pavlostathis, S. G.; Rozzi, A.; Sanders, W. T. M.; Siegrist, H.; Vavilin, V. A. (2002) The IWA Anaerobic Digestion Model No. 1 (ADM1). *Water Sci. Technol.*, **45** (10), 65–73.

Beall, S. S.; Jenkins, D.; Vidanage, S. A. (1998) A Systematic Analytical Artifact that Significantly Influences Anaerobic Digestion Efficiency Measurement. *Water Environ. Res.*, **70**, 1019.

Beals, J. L. (1976) Mechanics of Handling Lime Slurries. *Proceedings of the Int. Water Conference,* Pittsburgh, Pennsylvania, Oct 26–28; Engineers' Society of Western Pennsylvania.

Benefield, L. D.; Randall, C. W. (1980) *Biological Process Design for Wastewater Treatment*; Prentice-Hall Inc.: Englewood Cliffs, New Jersey.

Bitton, G.; Damron, B. L.; Edds, G. T.; Davidson, J. M. (1980) *Sludge Health Risks of Land Application*; Ann Arbor Science Publishers: Ann Arbor, Michigan.

Bird, E. B.; Stewart, W. E.; Lightfoot. E. N. (1960) *Transport Phenomena*; Wiley & Sons: New York.

Bowker, R. P. G.; Trueblood, R. (2002) Control of ATAD Odors at the Eagle River Water and Sanitation District. *Proceedings of the 2002 Water Environment Federation Odors and Toxic Air Emissions Conference*; Albuquerque, New Mexico, Apr 29–30; Water Environment Federation: Alexandria, Virginia.

Brinkman, D. G.; Voss, D. (1997) *Egg-Shaped Digesters: Are They All They're Cracked Up To Be?*; Operational Survey; Black and Veatch: Gaithersburg, Maryland.

Bryers, J. D. (1984) Structured Modeling of the Anaerobic Digestion of Biomass Particulates. *Biotechnol. Bioeng.*, **27**, 638–649.

Burd, R. S. (1968) *A Study of Sludge Handling and Disposal*; Publication No. WP-20-4; U.S. Department of the Interior, Federal Water Pollution Control Administration: Washington, D.C.

Burnham, J. C.; Hatfield, N.; Bennett, G. F.; Logan, T. J. (1992) *Use of Kiln Dust with Quicklime for Effective Municipal Sludge Treatment with Pasteurization and Stabilization with the N-Viro Soil Process*; Stand. Tech. Publication 1135; American Society for Testing and Materials: Philadelphia, Pennsylvania.

Buswell, A. M.; Neave, S. L. (1939) Laboratory Studies of Sludge Digestion. *Ill. State Water Surv. Bull.*, **30**. Institute for Natural Resource Sustainability, University of Illinois, Champagne, IL.

Byers, H. W.; Jensen, B. (1990) Stabilizing Sludge with Fly Ash-Sludge. Paper presented at Dep. Eng. Professional Development; University of Wisconsin: Madison, Wisconsin.

Camp, T. R.; Stein, P. C. (1943) Velocity Gradients and Internal Work in Fluid Motion. *J. Boston Soc. Civ. Eng.*, **203**.

Chang, Y. (1967) The Fungi of Wheat Straw Compost: Part II—Biochemical and Physiological Studies. *Trans. Br. Mycol. Soc.*, **50**, 667.

Chapman (1989) *Mixing in Anaerobic Digesters: State of the Art in Encyclopedia of Environmental Control Technology; Vol 3, Wastewater Treatment Technology;* Cheremisinoff, P. N., Ed.; Gulf Publishing Co.: Houston, Texas.

Chen, Y. C., Higgins, M. J.; Murthy, S. N.; Beightol, S. M. (2008) The Link between Odors and Regrowth of Fecal Coliforms after Dewatering. *Proceedings of the 22nd Annual Water Environment Federation Residuals and Biosolids Conference*; Philadelphia, Pennsylvania, Mar 30–Apr 2; Water Environment Federation: Alexandria, Virginia.

Chen, Y.; Higgins, M. J.; Murthy, S. N.; Maas, N. A.; Covert, K. J.; Toffey, W. E. (2006) Production of Odorous Indole, Skatole, p-Cresol, Toluene, Styrene, and Ethylbenzene in Biosolids. *J. Residuals Sci. Technol.*, **3** (4), 193–202.

Chen, Y.; Higgins, M. J.; Maas, N. A.; Murthy, S. N.; Toffey, W. E.; Foster, D. J. (2005) Roles of Methanogens on Volatile Organic Sulfur Compound Production in Anaerobically Digested Wastewater Biosolids. *Water Sci. Technol.*, **52**, 67–72.

Cheunbarn, T.; Pagilla, K. R. (1999). Temperature and SRT Effects on Aerobic Thermophilic Sludge Treatment. *J. Environ. Eng.*, **125** (7), 626–629.

Cheunbarn, T.; Pagilla, K. R. (2000). Aerobic Thermophilic and Anaerobic Mesophilic Treatment of Sludge. *J. Environ. Eng.*, **126** (9), 790–795.

Christensen, G. L. (1982) Dealing with the Never-Ending Sludge Output. *Water Eng. Manage.*, **129**, 25.

Christensen, G. L. (1987) Lime Stabilization of Wastewater Sludge. In *Lime for Environmental Uses*; Gutschick, K. A., Ed.; American Society for Testing and Materials: Philadelphia, Pennsylvania.

Christi, Y. (2003) Sonobioreactors: Using Ultrasound for Enhanced Microbial Productivity. *Trends in Biotechnol.*, **21** (2), 89–93.

Christy, R. W. (1992) Process and Mechanical Design Considerations for Sludge/Lime Mixing. *Proceedings of the 6th Annual Water Environment Federation Residuals Management Conference: Future Directions in Municipal Sludge (Biosolids) Management: Where We Are and Where We're Going*; Portland, Oregon, Jul 26–30; Water Environment Federation: Alexandria, Virginia.

Clements, R. P. L. (1982) Sludge Hygienization by Means of Pasteurization Prior to Digestion. *Disinfection of Sewage Sludge: Technical, Economic and Microbiological Aspects: Proceedings of a Workshop in Zurich*; Zurich, Switzerland, May 11–13; Bruce, A. M.; Havelaar, A. H.; Hermite, P. L., Eds.; D. Riedel Pub. Co.: Dordrecht, Switzerland; 37–52.

Conklin, A.; Stensel, H. D.; Ferguson, J. (2006) Growth Kinetics and Competition Between Methanosarcina and Methanosaeta in Mesophilic Anaerobic Digestion. *Water Environ Res.*, **78**, 486–496.

Counts, C. A.; Shuckrow, A. J. (1975) *Lime Stabilized Sludge: Its Stability and Effect on Agricultural Land*; EPA-670/2-75-012; Battelle Memorial Institute: Richland, Washington.

Dague, R. R. (1968) Application of Digestion Theory to Digester Control. *J. Water Pollut. Control Fed.*, **40**, 2021.

DeGarie, C. J.; Crapper, T.; Howe, B. M.; Burke, B. F.; McCarthy, P. J. (2000) Floating Geomembrane Covers for Odour Control and Biogas Collection and Utilization in Municipal Lagoons. *Water Sci. Technol.*, **42**, 291–298.

Daigger, G. T.; Bailey, E. (2000) Improving Aerobic Digestion by Prethickening, Staged Operation, and Aerobic–Anoxic Operation: Four Full-Scale Demonstrations. *Water Environ. Res.*, **72**, 260–270.

Daigger, G. T.; Ju, L. K.; Stensel, D.; Bailey, E.; Porteous, J. (2001) Can 3% SS Digestion Meet New Challenges? *Proceeding of the Aerobic Digestion Workshop*, Volume V; featured presentation sponsored by Enviroquip, Inc. (Austin, Texas) at the Water Environment Federation 74th Annual Exposition and Conference, Atlanta, Georgia, Oct 14–19.

Daigger, G. T.; Novak, J.; Malina, J.; Stover, E.; Scisson, J.; Bailey, E. (1998) Panel of Experts. *Proceedings of the Aerobic Digestion Workshop*, Volume II; Sponsored by Enviroquip, Inc.; Orlando, Florida, Oct 3.

Daigger, G. T.; Scisson, J.; Stover, E.; Malina, J.; Bailey, E.; Farrell, J. (1999) Fine Tuning the Controlled Aerobic Digestion Process. *Proceedings of the Aerobic Digestion Workshop*, Volume III; Sponsored by Enviroquip, Inc.; New Orleans, Louisiana, Oct 10.

Daigger, G. T.; Stensel, D.; Ju, L. K.; Bailey, E.; Porteous, J. (2000) Experience and Expertise Put to the Test. *Proceedings of the Aerobic Digestion Workshop*, Volume IV; Sponsored by Enviroquip, Inc.; Anaheim, California, Oct 15.

Daigger, G. T.; Yates, R.; Scisson, J.; Grotheer, T.; Hervol, H.; Bailey, E. (1997) The Challenge of Meeting Class B While Digesting Thicker Sludges, *Proceedings of the Aerobic Digestion Workshop*, Volume I; Sponsored by Enviroquip, Inc.; Chicago, Illinois, Oct 18.

Deeny, K.; Hahn, H.; Leonard, D.; Heidman, J. (1991) Autoheated Thermophilic Aerobic Digestion. *Water Environ. Technol.*, **3**, 65–72.

Dichtl, N. (1997) Thermophilic and Mesophilic (Two Stages) Anaerobic Digestion: Innovative Technologies for Sludge Utilization and Disposal. *J. Chartered Inst. Water Environ. Manage.*, **11**, 98–104.

Drury, D. D.; Lee, S.; Baker, C. (2002) Comparing Pathogen Reduction in Three Different Anaerobic Thermophilic Processes. *Proceedings of the 75th Annual Water Environment Federation Technical Exposition and Conference* [CD-ROM]; Chicago, Illinois, Sep 28–Oct 2; Water Environment Federation: Alexandria, Virginia.

Engineering–Science, Inc.; Black and Veatch (1991) *Technology Evaluation Report: Alkaline Stabilization of Sewage Sludge*; Report prepared for U.S. EPA; Contract No. 68-C8-0022; Work Assignment No. 01-08; U.S. Environmental Protection Agency: Washington, D.C.

Eschborn, R.; Higgins, M. J.; Johnston, T.; Toffey W.; Chen, Y. C. (2006) Philadelphia's Experience Using Static, Non-Aerated Curing to Produce Low Odor Biosolids. *Proceedings of the 20th Annual Water Environment Federation Residuals and Biosolids Management Conference*, Cincinnati, Ohio, Mar 12–15; Water Environment Federation: Alexandria, Virginia.

Eschborn, R.; Thompson, D. (2007) The Tagro Story—How the City of Tacoma, Washington, Went Beyond Public Acceptance to Achieve the Biosolids Program Words We'd Like to Hear: Sold Out. *Proceedings of the 21st Annual Water Environment Federation/American Water Works Association Joint Residuals and Biosolids Management Conference*; Denver, Colorado, Apr 15–18; Water Environment Fedearation: Alexandria, Virginia.

Farrah, S. R.; Bitton, G.; Zan, S. G. (1986) *Inactivation of Enteric Pathogens during Aerobic Digestion of Wastewater Sludge*; EPA-600/2-86-047; U.S. Environmental Protection Agency, Water Engineering Research Laboratory: Cincinnati, Ohio.

Farrell, J. B.; Smith, J. ; Hathaway, S. Dean, R. (1974) Lime Stabilization of Primary Sludge. *J. Water Pollut. Control Fed.*, **46**, 113.

Fergen, R. E. (1991) Stabilization and Disinfection of Dewatered Municipal Wastewater Sludge with Alkaline Addition. *Proceedings of the 5th Annual American Water Works Association/Water Pollution Control Federation Joint Residuals Management Conference*; Durham, North Carolina, Aug 11–14; Water Pollution Control Federation: Alexandria, Virginia.

Garber, W. F. (1982) Operating Experience with Thermophilic Anaerobic Digestion. *J. Water Pollut. Control Fed.*, **54**, 1170.

Gemmell, R.; Deshevy, R.; Elliott, M.; Crawford, G.; Murthy, S. (1999) Full Scale Demonstration of Dual Digestion: Thermodynamic and Kinetic Analysis. *Proceedings of the 72nd Annual Water Environment Federation Technical Exposition and Conference* [CD-ROM]; New Orleans, Louisiana, Oct 10–13; Water Environment Federation: Alexandria, Virginia.

Gemmell, R.; Deshevy, R.; Elliott, M.; Crawford, G.; Murthy, S. (2000) Design Considerations and Operating Experience for a Full Scale Dual Digestion System with Separate Sludge Thickening; *Proceedings of the 73rd Annual Water Environment Federation Technical Exposition and Conference* [CD-ROM]; Anaheim, California, Oct 14–18; Water Environment Federation: Alexandria, Virginia.

Ghosh, S.; Conrad, J. R.; Klass, D. L. (1975) Anaerobic Acidogenesis of Wastewater Sludge. *J. Water Pollut. Control Fed.*, **47**, 30.

Ghosh, S.; Henry, M. P.; Sajjad, A.(1987) *Stabilization of Sewage Sludge by Two-Phase Anaerobic Digestion*; EPA-7600/2-87-040; U.S. Environmental Protection Agency, Water Engineering Research Laboratory: Cincinnati, Ohio.

Ghosh, S.; et al. (1991) Pilot- and Full-Scale Studies on Two-Phase Anaerobic Digestion for Improved Sludge Stabilization and Foam Control. *Proceedings of the 64th Annual Water Pollution Control Federation Technical Exposition and Conference*; Toronto, Ontario, Canada, Oct 7–10; Water Environment Federation: Alexandria, Virginia.

Ghosh, S.; Buoy, K.; Dressel, L.; Miller, T.; Wilcox, G.; Loos, D. (1995) Pilot- and Full-Scale Studies on Two-Phase Anaerobic Digestion of Municipal Sludge. *Water Environ. Res.*, **67**, 206.

Golueke, C. G. (1977) *Biological Reclamation of Solid Waste*; Rodale Press: Emmaus, Pennsylvania.

Grady, C. P. L., Jr.; Daigger, G. T.; Lim, H. C. *(1999) Biological Wastewater Treatment*, 2nd ed.; Marcel Dekker: New York.

Gray, D. M.; Suto, P. J.; Chien, M. H. (2008) Producing Green Energy from Post-Consumer Solid Food Wastes at a Wastewater Treatment Plant Using an Innovative New

Process. *Proceedings of the Water Environment Federation Sustainability Conference*; Washington, D.C., June 22–25. Water Environment Federation: Alexandria, Virginia.

Great Lakes Upper Mississippi River Board of State Sanitary Engineering Health Education Services Inc. *(1997) Recommended Standards for Wastewater Facilities*; Great Lakes Upper Mississippi River Board of State Sanitary Engineering Health Education Services Inc.: Albany, New York.

Griffin, M. E.; McMahon, K. D.; Mackie, R. I.; Raskin, L. (2000) Methanogenic Population Dynamics during Start-Up of Anaerobic Digesters Treating Municipal Solid Waste and Biosolids. *Biotechnol. Bioeng.*, **57**, 342–355.

Gujer, W.; Zehnder, A. J. B. (1983) Conversion Process in Anaerobic Digestion. *Water-Sci. Technol.*, **15**, 127–167.

Haas O. (1984) *Demonstration of Thermophilic Aerobic-Anaerobic Digestion at Hagerstown, MD*; Grant S-805823-01-0; Final Report; EPA-600/S2-84-142; U.S. Environmental Protection Agency, Municipal Environmental Research Laboratory: Cincinnati, Ohio.

Han, Y.; Dague, R. (1996) Heat Control: Temperature-Phased Anaerobic Digestion Reduces Foaming and Produces Class A Biosolids Without the Odors. *Oper. Forum*, **13**, 19–23.

Hartman, R. B.; Smith, D. G.; Bennett, E. R.; Linstedt, K. D. (1979) Sludge Stabilization through Aerobic Digestion. *J. Water Pollut. Control Fed.*, **51**, 2353.

Haug, R. T.; LeBrun, T. I.; Tortorici, L. D. (1983) Thermal Pretreatment of Sludges—A Field Demonstration. *J. Water Pollut. Control Fed.*, **55**, 23–34.

Haug, R. T.; Stuckey, D. C.; Gossett, I. M.; McCarty, P. I. (1978). Effect of Thermal Pretreatment on Digestibility and Dewaterability of Organic Sludges, *J. Water Pollut. Control Fed.*, **50**, 73–85.

Haug, R. T. (1980) *Compost Engineering*; Ann Arbor Science Publishers: Ann Arbor, Michigan.

Hepner, S.; Striebig, B.; Regan, R.; Giani, R. (2002) Odor Generation and Control from the Autothermal Thermophilic Aerobic Digestion (ATAD) Process. *Proceedings of the 2002 Water Environment Federation Odors and Toxic Air Emissions Conference*; Albuquerque, New Mexico, April 29–30; Water Environment Federation: Alexandria, Virginia.

Higgins, A. J.; Chen, S.; Singley, M. E. (1982) Airflow Resistance in Sewage Sludge Composting Systems. *Trans. Am. Soc. Agric. Eng.*, **25** (4), 1010–1014, 1018.

Higgins, M.; Murthy, S.; Toffey, W.; Striebig, B.; Hepner, S.; Yarosz, D.; Yamani, S. (2002) Factors Affecting Odor Production in Philadelphia Water Department

Biosolids. *Proceedings of the 2002 Water Environment Federation Odors and Toxic Air Emissions Conference*; Albuquerque, New Mexico, April 29–30; Water Environment Federation: Alexandria, Virginia.

Higgins, M.; Yarosz, D.; Chen, Y.; Murthy, S. (2003) Mechanisms for Volatile Sulfur Compound and Odor Production in Digested Biosolids. *Proceedings of the 17th Annual Water Environment Federation/American Water Works Association Joint Biosolids and Residuals Conference*; Baltimore, Maryland, Feb 19–22; Water Environment Federation: Alexandria, Virginia.

Higgins, M. J.; Yarosz, D. P.; Chen, D. P.; Murthy, S. N.; Maas, N.; Cooney, J.; Glindemann, D.; Novak, J. T. (2006a) Cycling of Volatile Organic Sulfur Compounds in Anaerobically Digested Biosolids and Its Implications for Odors. *Water Environ. Res.*, **78**, 243–252.

Higgins, M. J.; Chen, Y. C.; Murthy, S. N.; Hendrickson, D. (2006b) *Examination of Reactivation of Fecal Coliforms in Anaerobically Digested Biosolids*; Report No. 03-CTS-13T; Water Environment Research Foundation: Alexandria, Virginia.

Higgins, M. J.; Chen, Y. C.; Novak, J. T.; Glindemann, D.; Forbes, R. H.; Erdal, Z.; Witherspoon, J.; McEwen, D.; Murthy, S.; Hargreaves, J. R.; Adams, G. (2008a) A Multi-Plant Study to Understand the Chemicals and Process Parameters Associated with Biosolids Odors. In *Environmental Engineer: Applied Research and Practice*; American Academy of Environmental Engineers: Annapolis, Maryland.

Higgins, M. J.; Chen, Y. C.; Murthy, S. N.; Hendrickson, D. (2008b) *Evaluation of Bacterial Pathogen and Indicator Densities After Dewatering of Anaerobically Digested Biosolids: Phase II and III*; Report No. 04-CTS-3T; Water Environment Research Foundation: Alexandria, Virginia.

Iacoboni, M. D.; Leburn, T. J.; Lingston, J. (1980) Deep Windrow Composting of Dewatered Sewage Sludge. *Proceedings of the National Conference of the Municipal and Industrial Sludge Composting Hazardous Materials Control Research Institute*; Silver Spring, Maryland; pp. 88–108.

Jacobs, A.; et al. (1992) Odor Emissions and Control at the World's Largest Chemical Fixation Facility. *Proceedings of the 6th Annual Water Environment Federation Residuals Management Conference: Future Directions in Municipal Sludge (Biosolids) Management: Where We Are and Where We're Going*; Portland, Oregon; Water Environment Federation: Alexandria, Virginia.

Jacobs, A.; Silver, M. (1990) Sludge Management at the Middlesex County Utilities Authority. *Water Sci. Technol.*, **22**, 93.

Jewell, W. J.; Kabrick, R. M. (1980) Autoheated Aerobic Thermophilic Digestion with Air Aeration. *J. Water Pollut. Control Fed.*, **52**, 512.

Jimenez, E. I.; Garcia, V. P. (1989) Evaluation of City Refuse Compost Maturity: A Review. *Biol. Wastes*, **27**, 115.

Jones, R.; Parker, W.; Khan, Z.; Murthy, S.; Rupke, M. (2008a) Characterization of Sludges for Predicting Anaerobic Digester Performance. *Water Sci. Technol.*, **57**, 721–726.

Jones, R.; Parker, W.; Zhu, H.; Houweling, D.; Murthy, S. (2008b) Predicting the Degradability of Waste Activated Sludge. *Proceedings of the 81st Annual Water Environment Federation Technical Exhibition and Conference* [CD-ROM]; Chicago, Illinois, Oct 18–22; Water Environment Federation: Alexandria, Virignia.

Kampelmacher, E. H.; van Noorle Jansen, L. M. (1972) Reduction of Bacteria in Sludge Treatment. *J. Water Pollut. Control Fed.*, **44**, 309.

Kayhanian, M.; Tchobanoglous, G. (1992) Computation and Importance of Carbon to Nitrogen (C/N) Ratios for Various Organic Fractions of Municipal Solid Waste. *BioCycle*, **33**, 58–60.

Keller, U. (1980) Klarschlammpasteurisierung in der Abwasserreinigungsanlage Altenrhein. Wasser, Energie, Luft. 72 Jahrgang. Heft 1/2 (side-by-side article in French).

Kelly, H. G.; Mavinic, D. S.; Trueblood, B.; Zhou, J.; Hystad, B.; Frese, H.; Cheshuk, J. (2003) Autothermal Thermophilic Aerobic Digestion Research Application and Operational Experience. *Proceedings of the 76th Annual Water Environment Federation Technical Exhibition and Conference*; Workshop W104—Thermophilic Digestion: Hot Update!; Los Angeles, California, Oct 11–15; Water Environment Federation: Alexandria, Virginia.

Kelly, H. G.; Melcer, H.; Mavinic, D. S. (1993) Autothermal Thermophilic Aerobic Digestion of Municipal Sludge: A One-Year Full-Scale Demonstration Project. *Water Environ. Res.*, **65**, 849.

Kelly, H. G. (1991) Autothermal Thermophilic Aerobic Digestion: A Two Year Appraisal of Canadian Facilities. *Proceedings of the American Society of Civil Engineers Environmental Engineering Specialty Conference;* Reno, Nevada, Jul 10–12; American Society of Civil Engineers: New York, New York.

Kester, G.; Schafer, P.; Gillette, B. (2008) Using Treatement Plant Digesters to Process Fats, Oils and Grease. *BioCycle*, **49**, 47.

Kim, H.; Murthy, S.; Peot, C.; Ramirez, M.; Strawn, M.; Park, C.; McConnell, L. (2003) Examination of Mechanisms for Odor Compound Generation during Lime Stabilization *Water Environ. Res.*, **75**, 121.

Knoll, K. H. (1964) Information Bulletin No. 13-20; Int. Research Group Refuse Disposal, U.S. Public Health Service, Rockville, Maryland.

Kopp, J.; Ewert, W. (2006) New Processes for the Improvement of Sludge Digestion and Sludge Dewatering. *Proceedings of the 11th European Biosolids and Organic Resources Conference*; Wakefield, United Kingdom, Nov 13–15; Chartered Institution of Water and Environmental Management: London, U.K.

Krugel, S.; Nemeth, L.; Peddie, C. (1998) Extending Thermophilic Anaerobic Digestion for Producing Class-A Biosolids at the Greater Vancouver Regional Districts Annacis Island Wastewater Treatment Plant. *Water Sci. Technol.*, **38** (8), 409–416.

Krugel, S.; Parella, A.; Ellquist, K.; Hamel, K. (2006) Five Years of Successful Operation–A Report on North America's First New Temperature Phased Anaerobic Digestion System at the Western Lake Superior Sanitary District (WLSSD). *Proceedings of the 79th Annual Water Environment Federation Technical Exposition and Conference* [CD-ROM]; Dallas, Texas, Oct 21–25; Water Environment Federation: Alexandria, Virginia.

Kumar, N.; Novak, J. T.; Murthy, S. N. (2006a) Sequential Anaerobic-Aerobic Digestion for Enhanced Volatile Solids Reduction and Nitrogen Removal. *Proceedings of the 20th Annual Water Environment Federation Residuals and Biosolids Management Conference*; Cincinnati, Ohio, March 12–14; Water Environment Federation: Alexandria, Virginia.

Kumar, N.; Novak, J. T.; Murthy, S. N. (2006b) Effect of Secondary Aerobic Digestion on Properties of Anaerobic Digested Biosolids. *Proceedings of the 79th Annual Water Environment Federation Technical Exhibition and Conference* [CD-ROM]; Dallas Texas, Oct 21–25; Water Environment Federation: Alexandria, Virginia.

Lawrence, A. W. (1971) Application of Process Kinetics to Design of Anaerobic Processes. In *Anaerobic Biological Treatment Processes*; Pohland, F. G., Ed.; Advances in Chemistry Series; American Chemical Society: Washington, D.C., 105.

Lawrence, A.W.; McCarty, P. L. (1969) Kinetics of Methane Fermentation in Anaerobic Treatment. *J. Water Pollut. Control Fed.*, **41**, 1–17.

Layden, N. M.; Kelly, H. G.; Mavinic, D. S.; Moles, R.; Bartlett, J. (2007) Autothermal Thermophilic Aerobic Digestion (ATAD) Part II: Review of Research and and Full-Scale Operating Experiences. *J. Environ. Eng. Sci.*, **6** (6), 679–690.

LeBrun, T. (1979) Memorandum to the LA/OMA Project on Status of *Aspergillus* Monitoring. Los Angeles County Sanitation Districts, California.

Lee, K. M.; Brunner, C. A.; Farrell, J. B.; Ealp, A. E. (1989) Destruction of Enteric Bacteria and Viruses during Two-Phase Digestion. *J. Water Pollut. Control Fed.*, **61**, 1421–1429.

Leicht, R. K.; Regan, J. T.; Toy, D. A. (1986) Refinery Hydrogen Sulfide Emissions Cut 99.9% with Chelated Catalyst. *Chem. Proc.*, **Aug**, 106–108.

Lewis, C. J.; Gutschick, K. A. (1980) *Lime in Municipal Sludge Processing*; National Lime Association: Washington, D.C.

Li Y. Y.; Noike T. (1992) Upgrading of Anaerobic Digestion of Waste Activated Sludge by Thermal Pre-Treatment. *Water Sci. Technol.*, **26** (3–4), 857–866.

Liptak, B. G. (1974) *Environmental Engineers' Handbook*; Chilton Book Co.: Radnor, Pennsylvania.

Logan, T. J. (1990) Chemistry and Bioavailability of Metals and Nutrients in Cement Kiln Dust-Stabilized Sewage Sludge. *Proceedings of the 4th Annual Water Environment Federation/American Water Works Association Joint Residuals and Biosolids Conference*; New Orleans, Louisiana; Water Pollution Control Federation: Washington, D.C.

Lue-Hing, C.; Zenz, D. R.; Kuchenrither, R., Eds. (1992) *Municipal Sewage Sludge Management: Processing, Utilization and Disposal*; Technomic Publishing Co. Inc.: Lancaster, Pennsylvania.

Mason, T. J.; Lorimer, J. P. (1988) *Sonochemistry: Theory, Applications and Uses of Ultrasound in Chemistry*; Ellis Horwood: Chichester, U.K.

Matsch, L. C.; Drnevich, R. F. (1977) Autothermal Aerobic Digestion. *J. Water Pollut. Control Fed.*, **49**, 296.

Maxwell, M. J.; et al. (1992) Impact of New Sludge Regulations on Aerobic Digester Sizing and Cost-Effectiveness. *Proceedings of the 65th Annual Water Environment Federation Exposition and Conference*; New Orleans, Louisiana, Sep 20–24; Water Environment Federation: Alexandria, Virginia.

McCarty, P. L.; Smith, D. P. (1986) Anaerobic Wastewater Treatment: Fourth of a Six-Part Series on Wastewater Treatment Processes. *Environ. Sci. Technol.*, **20** (12), 1200–1206.

McHugh, S.; Carton, M.; Mahony, T.; O'Flaherty, V. (2006) Methanogenic Population Structure in a Variety of Anaerobic Bioreactors. *FEMS Microbiol. Lett.*, **219**, 2297–2304.

Messenger, J. R.; de Villiers, H. A.; Ekama, G. A. (1993) Evaluation of the Dual Digestion System, Part 2: Operation and Performance of the Pure Oxygen Aerobic Reactor. *Water SA.*, **19** (3), 193–200.

Metcalf and Eddy, Inc. (1991) *Wastewater Engineering: Treatment, Disposal, Reuse*, 3rd ed.; Tchobanoglous, G., Ed.; McGraw-Hill: New York.

Metcalf and Eddy, Inc. (2003) *Wastewater Engineering: Treatment and Reuse*, 4th ed.; Tchobanoglous, G., Ed.; McGraw-Hill: New York.

Millner, P. D.; Marsh, P. B.; Snowden, R. B.; Parr, J. F. (1977) Occurrence of *Aspergillus fumigatus* during Composting of Sewage Sludge. *Appl. Environ. Microbiol.*, **34**, 6.

Morgan, M. T.; MacDonald, F. W. (1969) Tests Show MB Tuberculosis Doesn't Survive Composting. *J. Environ. Health*, **32**, 101.

Muller, C. D. (2006) Shear Forces, Floc Structure and their Impacts on Anaerobic Digestion and Bisolids Stability. Ph.D. disseration, Virginia Polytechnic Institute and State University, Blacksburg, Virginia.

Muller, C. D.; Novak, J. T. (2007) The Influence of Anaerobic Digestion on Centrifugally Dewatered Biosolids. *Proceedings of the 21st Annual Water Environment Federation Residuals and Biosolids Conference*; Denver, Colorado, April 15–18. Water Environment Federation: Alexandria, Virginia.

Murray, C. M.; Thompson, J. L. (1986) Strategies for Aerated Pile Systems. *BioCycle*, **6.**

Murray, K. C.; Tong, A.; Bruce, A. M. (1990) Thermophilic Aerobic Digestion: A Reliable and Effective Process for Sludge Treatment at Small Works. *Water Sci. Technol.*, **22**, 225.

Murthy, S. N.; Novak, J. T.; Holbrook, R. D. (2000a) Optimizing Dewatering of Biosolids from Autothermal Thermophilic Aerobic Digesters (ATAD) Using Inorganic Conditioners. *Water Environ. Res.*, **72**, 714–721.

Murthy, S. N.; Novak, J. T.; Holbrook, R. D.; Surovik, F. (2000b) Mesophilic Aeration of Autothermal Thermophilic Aerobically Digested Biosolids to Improve Plant Operations. *Water Environ. Res.*, **72**, 476–483.

Murthy, S. N.; Forbes, B.; Burrowes, P.; Esqueda, T.; Glindemann, D.; Novak, J.; Higgins, M.; Mendenhall, T.; Toffey, W.; Peot, C. (2002a) Impact of High Shear Solids Processing on Odor Production from Anaerobically Digested Biosolids. *Proceedings of the 75th Annual Water Environment Federation Technical Exposition and Conference* [CD-ROM]; Chicago, Illinois, Sep 28–Oct 2; Water Environment Federation: Alexandria, Virginia.

Murthy, S. N.; Peot, C.; North, J.; Novak, J.; Glindemann, D.; Higgins, M. (2002b) Characterization and Control of Reduced Sulfur Odors from Lime-Stabilized and Digested Biosolids. *Proceedings of the 16th Annual Water Environment Federation Residuals and Biosolids Conference*; Austin, Texas, Mar 3–6; Water Environment Federation: Alexandria, Virginia.

Murthy, S. N.; Higgins, M. J.; Chen, Y. C.; Toffey, W.; Golembeski J. (2003) Influence of Solids Characteristics and Dewatering Process on Volatile Sulfur Compound Production from Anaerobically Digested Biosolids. *Proceedings of the 17th Annual Water Environment Federation/American Water Works Association Joint Residuals and Biosolids Con-*

ference; Baltimore, Maryland, Feb 19–22;.Water Environment Federation: Alexandria, Virginia.

National Fire Protection Association (1993) *National Electric Code*; National Fire Protection Association: Quincy, Massachusetts.

National Fire Protection Association (1995) *Standard for Fire Protection in Wastewater Treatment and Collection Facilities*; NFPA 820; National Fire Protection Association: Quincy, Massachusetts.

National Lime Association (1988) *Lime: Handling, Application, and Storage in Treatment Processes*; Bulletin 213; National Lime Association: Arlington, Virginia.

Nielsen, H. B.; Angelidaki, I. (2008) Strategies for optimizing recovery of the biogas process following ammonia inhibition. *Bioresour. Technol.*, **99**, 7995–8001.

North, J. M.; Becker, J. G.; Seagren, E. A.; Ramirez, M.; Peot, C. (2008a) Methods for Quantifying Lime Incorporation into Dewatered Sludge. I: Bench-Scale Evaluation. *J. Environ. Eng.*, **134** (9), 750–761.

North, J. M.; Becker, J. G.; Seagren, E. A.; Ramirez, M.; Peot, C.; Murthy, S. N. (2008b), Methods for Quantifying Lime Incorporation into Dewatered Sludge. II: Field-Scale Application. *J. Environ. Eng.*, **134** (9), 762–770.

Novak, J. T.; Becker, H.; Zurow, A. (1977) Factors Influencing Activated Sludge Properties. *J. Environ. Eng.*, **103**, 815.

O'Rourke, J. T. (1968) Kinetics of Anaerobic Treatment at Reduced Temperatures. Ph.D. thesis, Stanford University, Palo Alto, California.

Oerke, D. W. (1999) Alkaline Stabilization of Biosolids Can Save Money, Space. *Water World*, **Mar**, 14–16.

Oerke, D. W; Rogowski, S. M. (1990) Economic Comparison of Chemical and Biological Sludge Stabilization Processes. *Proceedings of the 4th Annual Water Pollution Control Federation Specialty Conference on the Status of Municipal Sludge Management*; New Orleans, Louisiana; Water Pollution Control Federation: Washington, D.C.

Oerke, D. W.; Stone, L. A. (1991) Detailed Case Study Evaluation of Alkaline Stabilization Processes. *Proceedings of the 5th Annual American Water Works Association/Water Pollution Control Federation Joint Residuals Management Conference*; Durham, North Carolina, Aug 11–14; Water Pollution Control Federation: Washington, D.C.

Otoski, R. M. (1981) *Lime Stabilization and Ultimate Disposal of Municipal Wastewater Sludge*; EPA-600/S2-81-076; U.S. Envrionmental Protection Agency, Municipal Environmental Research Laboratory, Center for Environmental Research: Cincinnati, Ohio.

Pagilla, K. R.; Craney, K. C.; Kido, W. H. (1996) Aerobic Thermophilic Pretreatment of Mixed Sludge for Pathogen Reduction and *Nocardia* Control. *Water Environ. Res.*, **68**, 1093–1098.

Park, C.; Abu-Orf, M. M.; Novak, J. T. (2003) Predicting the Digestibility of Waste Activated Sludges Using Cation Analysis. *Proceedings of the 76th Annual Water Environment Federation Technical Exhibition and Conference*; Los Angeles, California, Oct 11–15; Water Environment Federation: Alexandria, Virginia.

Parkin, G. F.; Owen, W. F. (1986) Fundamentals of Anaerobic Digestion of Wastewater Sludge. *J. Environ. Eng.*, **112**, 5.

Parravicini, V.; Svardal, K.; Hornek, R.; Kroiss, H. (2008) Aeration of Anaerobically Digested Sewage Sludge for COD and Nitrogen Removal: Optimization at Large-Scale *Water Sci. Technol.*, **57**, 257.

Paulsrud, B.; Eikum, A. S. (1975) Lime Stabilization of Sewage Sludge. *Water Res.* (G.B.), **9**, 297.

Pavlostathis, S. G.; Gossett, J. M. (1988) Preliminary Conversion Mechanisms in Anaerobic Digestion of Biological Sludges. *J. Environ. Eng.*, **114**, 575–592.

Pavlostathis, S. G.; Gossett, J. M. (2004) A kinetic Model for Anaerobic Digestion of Biological Sludge. *Biotech. Bioeng.*, **28**, 519–1530.

Perry, R. H.; Green, D. W. (1997) *Perry's Chemical Engineers' Handbook,* 7th ed.; McGraw-Hill: New York.

Pitt, A. J.; Ekama, G. A. (1996) Dual Digestion of Sewage Solids Using Air and Pure Oxygen. *Proceedings of the 69th Annual Water Environment Federation Technical Exposition and Conference* [CD-ROM]; Dallas, Texas, Oct 5–9; Water Environment Federation: Alexandria, Virginia.

Poincelot, R. P. (1975) *The Biochemistry and Methodology of Composting.* Bulletin 754. Connecticut Agricultural Experiment Station: New Haven, Connecticut.

Popel, F. (1971a) Die Theoretischen und Praktischen Grunlagen der Flussig-Kompostierung Hockkonzentrierten Substrate. *Ausgearbeitet fur die Bad-ische Anilin-und Sodafabrik Ludwigshafen, Stuttgart,* November.

Popel, F. (1971b) Energieerzeugung Beim Biologischen Abbau Organischer Stoffe. *Gewaesser Abwaesser* (Ger.), **112**.

Portenlänger, G.; Heusinger, H. (1997) The Influence of Frequency on the Mechanical and Ramirez, A.; Malina, J. (1980) Chemicals Disinfect Sludge. *Water Sew. Works*, **127** (4), 52.

Reimers, R. S.; Little, M. D.; Englande, A. J.; Leftwich, D. B.; Bowman, D. D.; Wilkinson, R. F. (1981) *Parasites in Southern Sludge and Disinfection by Standard Sludge Treatment*; EPA-600/2-81-166; U.S. Environmental Protection Agency: Washington, D.C.

Reynolds, D. T.; Cannon, M.; Pelton, T. (2001) Preliminary Investigation of Recuperative Thickening for Anaerobic Digestion. *Proceedings of the 74th Annual Water Environment Federation Technical Exhibition and* Conference; Atlanta, Georgia, Oct 13–17; Water Environment Federation: Alexandria, Virginia.

Reynolds, T. D. (1973) Anaerobic Digestion of Thickened Waste Activated Sludge. *Proceedings of the 28th Purdue Industrial Waste Conference*; West Lafayette, Indiana, May 1–3: Purdue University: West Lafayette, Indiana.

Rubin, A. R. (1991) Agricultural Limitations and Criteria for Lime Stabilized Sludge (PSRP or PFRP). *Proceedings of the 5th Annual American Water Works Association/Water Pollution Control Federation Joint Residuals Management Conference;* Durham, North Carolina, Aug 11–14: Water Pollution Control Federation: Washington, D.C.

Salsali, H. R.; Parker, W. J. (2007) An Evaluation of 3 Stage Anaerobic Digestion of Municipal Wastewater Treatment Plant Sludges, *Water Practice*, **1**, 1-12.

Sanders, W. T. M.; Geerink, M.; Zeeman, G.; Lettinga, G. (2000) Anaerobic hydrolysis kinetics of particulate substrates. *Water Sci.Technol.*, **41**, 17–24.

Schafer, P., Wolfenden, A. (1982) Odor Control Features Make Lagoons an Acceptable Sludge Process *Sixth Mid-America Conference on Environmental Engineering Design, American Society of Civil Engineers*, Kansas City, Missouri, Junc.

Schafer, P. L.; Farrell, J. (2000) Performance Comparisons for Staged and High-Temperature Anaerobic Digestion Systems, *Proceedings of the Annual Water Environment Federation Technical Exposition and Conference* [CD-ROM]; Anaheim, California, Oct 17; Water Environment Federation: Alexandria, Virginia..

Schafer, P. L.; Farrell, J.; Newman, G.; Vandenburgh, S. (2002) Advanced Anaerobic Digestion Performance Comparisons. *Proceedings of the 75th Annual Water Environment Federation Technical Exposition and Conference* [CD-ROM]; Chicago, Illinois, Sep 28–Oct 2; Water Environment Federation: Alexandria, Virginia.

Schafer, P. L.; Trueblood, D.; Fonda, K.; Lekven, C. (2008) Grease Processing for Renewable Energy, Profit, Sustainability, and Environmental Enhancement. *Proceedings of the Water Environment Federation Biosolids Specialty Conference*, Philadelphia, Pennsylvania. April.

Schwinning, H. G.; Denny, K. J.; Hong, S. N. (1997) Experience with autothermal thermophilic aerobic digestion (ATAD) in the United States. *Proceedings of the 70th Annual*

Water Environment Federation Technical Exposition and Conference, Chicago, Illinois, Oct 18–22 Water Environment Federation: Alexandria, Virginia.

Scisson, J. P. (2006) ATAD, the far country: Improvements to the 3rd, 2nd Generation ATAD Yield Better than Expected Returns. *Proceedings of the Water Environment Federation Residuals and Biosolids Management Specialty Conference;* Cincinnati, Ohio, Mar 12–15; Water Environment Federation: Alexandria, Virginia.

Shell, G. L.; Boyd, J. L. (1969) *Composting Dewatered Sewage Sludge*; SW-12c; U.S. Department of Health, Education, and Welfare: Washington, D.C.

Sloan, D. (1992) Design and Process Considerations for Advanced Alkaline Stabilization with Subsequent Accelerated Drying Facilities. Paper presented at the 5th Annual International Conference on Alkaline Pasteurization Stabilization, Somerset, New Jersey.

Speece, R. E. (1972) Anaerobic Treatment. In *Process Design in Water Quality Engineering.* E.L. Thackston and W.W. Eckenfelder (Eds.), Jenkins Publishing Company, New York.

Spinosa, L.; Vesilind, P. A.; Eds. (2001) *Sludge into Biosolids. Processing, Disposal, and Utilization;* IWA Publishing, London, U.K.

Stege, K.; Bailey, E. (2003) Aerobic digestion Operation of Stockbridge Wastewater Treatment Plant, GA; *Georgia Water Pollution Control Association.*

Stensel, H. D.; Coleman, T. E. (2000) *Assessment of Innovative Technologies for Wastewater Treatment: Autothermal Aerobic Digestion (ATAD),* Preliminary Report, Project 96-CTS-1; U.S. Environmental Protection Agency: Washington, D.C.

Stone, L. A., et al. (1992) The Historical Development of Alkaline Stabilization. Paper presented at the Water Environment Federation Specialty Conference on Future Directions in Municipal Sludge (Biosolids) Management: Where We Are and Where We're Going, Portland, Oregon.

Stuckey, D. C.; McCarty, P. L. (1984) The Effect of Thermal Pretreatment on the Anaerobic Biodegradability and Toxicity of Waste Activated Sludge, *Water Res.,* **18** (11), 1343–1353.

Stutzenberger, F. J. (1971) Cellulase Production by *Thermomonospora curvata* Isolated from Municipal Solid Waste Compost. *Appl. Microbiol.,* **22**, 2, 147.

Torpey, W. N., et al. (1984) Effects of Multiple Digestion on Sludge. *J. Water Pollut. Control Fed.,* **56**, 62.

Tortorici, L.; Stahl, J.F. (1977) Waste Activated Sludge Research. *Proceedings of the Sludge Management, Disposal, and Utilization Conference,* Information Transfer, Inc., Rockville, Maryland.

U.S. Environmental Protection Agency (1975) *Lime Stabilized Sludge: Its Stability and Effect on Agricultural Land.* EPA-670/2-75-012, National Environmental Research Center; U.S. Environmental Protection Agency: Washington, D.C.

U.S. Environmental Protection Agency (1978) *Sludge Treatment and Disposal—Volume I;* EPA-625/4-78-012, Technology Transfer, U.S. Environmental Protection Agency: Washington, D.C.

U.S. Environmental Protection Agency (1979) *Process Design Manual for Sludge Treatment and Disposal;* EPA-625/1-79-011, U.S. Environmental Protection Agency: Cincinnati, Ohio.

U.S. Environmental Protection Agency (1982) *Guide to the Disposal of Chemically Stabilized and Solidified Waste;* SW-872, Office of Solid Waste Emergency Response, U.S. Environmental Protection Agency: Washington, D.C.

U.S. Environmental Protection Agency (1989) *1988 Needs Survey of Municipal Wastewater Treatment Facilities;* EPA-430/09-89-001; U.S. Environmental Protection Agency: Cincinnati, Ohio.

U.S. Environmental Protection Agency (1990) *Autothermal Thermophilic Aerobic Digestion of Municipal Wastewater Sludge;* EPA-625/10-90-007; U.S. Environmental Protection Agency: Cincinnati, Ohio.

U.S. Environmental Protection Agency (1992) *Control of Pathogens and Vector Attraction in Sewage Sludge;* EPA-625/R-92-013; Environmental Regulations and Technology; U.S. Environmental Protection Agency: Washington, D.C.

U.S. Environmental Protection Agency (1993) Standards for the Use or Disposal of Sewage Sludge. *Fed. Regist.*, **58**, 32.

U.S. Environmental Protection Agency (1999a) *Control of Pathogens and Vector Attraction in Sewage Sludge;* EPA-625/R-92-013, Environmental Regulations and Technology; U.S. Environmental Protection Agency: Washington, D.C.

U.S. Environmental Protection Agency (1999b) *Standards for the Use or Disposal of Sewage Sludge.* Code of Federal Regulations, Part 503, Title 40.

U.S. Environmental Protection Agency (2003) *Control of Pathogens and Vector Attractionin Sewage Sludge;* EPA-625/R-92013; U.S. Environmental Protection Agency: Cincinnati, Ohio.

U.S. Environmental Protection Agency (in press) *Alkaline Treatment of Municipal Wastewater Treatment Plant Sludge Technical Guide*; National Risk Management Research Laboratory; Office of Research and Development; U.S. Environmental Protection Agency: Cincinnati, Ohio.

Van Lier M. J. B.; Martin M. J. L. S.; Lettinga, M. G. (1996) Effect of Temperature on the Anaerobic Thermophilic Conversion of Volatile Fatty Acids by Dispersed and Granular Sludge. *Water Res.*, **30**, 199–207.

Verschueren, K. (1983) *Handbook of Environmental Data on Organic Chemicals*, 2nd ed.; Van Nostrand Reinhold Co.: New York.

Volpe, G.; Keaney, J.; Schlegel, P.; Tyler, C.; Carr, J.; Nagel, J. (2004) Large Egg-Shaped Digesters: Issues and Improvements. *Proceedings of the 77th Annual Water Environment Federation Technical Exhibition and Conference* [CD-ROM]; New Orleans, Louisiana, Oct 2–6; Water Environment Federation: Alexandria, Virginia.

Waksman, S. A.; Cordon, T. C. (1939) Thermophilic Decomposition of Plant Residues in Composts by Pure and Mixed Cultures of Microorganisms. *Soil Sci.*, **47**, 217.

Ward, A.; Stensel, H. D.; Ferguson, J.; Ma, G.; Hummel, S. (1999) Preventing Growth of Pathogens in Pasteurized Digested Sludge. *Water Environ. Res.*, **71**, 176.

Water Environment Federation (1994a) *Beneficial Use Programs for Biosolids Management*; Special Publication; Water Environment Federation: Alexandria, Virginia.

Water Environment Federation (1994b) *Safety and Health in Wastewater Systems*, 5th ed.; Manual of Practice No. 1; Water Environment Federation: Alexandria, Virginia.

Water Environment Federation (1995) *Wastewater Residuals Stabilization*; Manual of Practice No. FD-9; Water Environment Federation: Alexandria, Virginia.

Water Environment Federation (2007) *Operation of Municipal Wastewater Treatment Plants*, 6th ed.; Manual of Practice No. 11; McGraw-Hill: New York.

Water Environment Federation (2009) *An Introduction to Process Modeling for Designers*; Manual of Practice No. 31; Water Environment Federation: Alexandria, Virginia.

Water Pollution Control Federation (1987) *Anaerobic Sludge Digestion*; Manual of Practice No. 16; Water Pollution Control Federation: Washington, D.C.

Water Environment Research Foundation (2004) *Producing Class A Biosolids with Low-Cost, Low-Technology Treatment*; Report No.99-REM-2; Water Environment Research Foundation: Alexandria, Virginia.

Werker, A. G.; Carlsson, M.; Morgan-Sagastume, F.; Le M. S.; Harrison, D. (2007) Full-Scale Demonstration and Assessment of Enzymatic Hydrolysis Pretreatment for Mesophilic Anaerobic Digestion of Municipal Wastewater Treatment Sludge. *Proceedings of the 80th Annual Water Environment Federation Technical Exhibition and Conference* [CD-ROM]; San Diego, California, Oct 13–17; Water Environment Federation: Alexandria, Virginia.

Westphal, A.; Christensen, G. L. (1983) Lime Stabilization: Effectiveness of Two Process Modifications. *J. Water Pollut. Control Fed.*, **55**, 1381.

Wiley, J. S.; Westerberg, S. C. (1969) Survival of Human Pathogens in Composted Sewage. *Appl. Microbiol.*, **18**, 944.

Williams, T. O.; Forbes, Jr., R. H.; Wagoner, D. L.; Hahn, J. T. (2008) Control of Biosolids Cake Odors Using the New Biosolids Odor Reduction Selector Process. *Proceedings of the 22nd Annual Water Environment Federation Residuals and Biosolids Management Conference,* Philadelphia, Pennsylvania, Mar 30–Apr 2; Water Environment Federation: Alexandria, Virginia.

Willis, J.; Aiken, M.; Arnett, C.; Hull, T.; Matthews, J.; Schafer, P.; Sobsey, M.; Turner, B. (2003) Cost to Convert to Class A: Columbus Biosolids Flow-Through Thermophilic Treatment (CBTF3) as a Cost Effective Option. *Proceedings of the Water Environment Federation Technical Exhibition and Conference* [CD-ROM], Los Angeles, California, Oct 11–15; Water Environment Federation: Alexandria, Virginia.

Wilson, C. A.; Fang, Y.; Novak, J. T.; Murthy, S. N. (2008a) The Effect of Temperature on the Performance and Stability of Thermophilic Anaerobic Digestion. *Water Sci. Technol.*, **57**, 297–304.

Wilson, C. A.; Murthy, S. N.; Novak, J. T. (2008b) Laboratory Digestibility Study of Wastewater Sludge Treated by Thermal Hydrolysis. *Proceedings of the 22nd Annual Water Environment Federation Residuals and Biosolids Conference: Traditions, Trends & Technologies*; Philadelphia, Pennsylvania, March 30–April 2; Water Environment Federation: Alexandria, Virginia.

Zahller, J. D.; Bucher, R. H.; Ferguson J. F.; Stensel, H. D. (2005) Performance and Stability of Two-Stage Anaerobic Digestion. *Proceedings of the 78th Annual Water Environment Federation Technical Exhibition and Conference* [CD-ROM]; Washington, D.C., Oct 29–Nov 2, Water Environment Federation: Alexandria, Virginia.

Zhou, J.; Mavinic, D. S.; Kelly, H. G.; Ramey, W. D. (2002) Effects of Temperature and Extracellular Proteins on Dewaterability of Thermophilically Digested Biosolids. *J. Environ. Eng. Sci.*, **1**, 409–415.

Chapter 26

Thermal Processing

1.0 INTRODUCTION

The importance of municipal wastewater sludge processing and disposal has grown since the establishment of secondary treatment standards by the Clean Water Act of 1972 and subsequent construction of wastewater treatment plants that have generated increasingly larger volumes of sludge. At the same time, increasingly strict regulations governing wastewater solids disposal combined with the decreasing availability of disposal sites have limited disposal options and increased disposal costs.

Consequently, interest continues in thermal processing methods as a means of producing a marketable product or reducing sludge or biosolids volumes. Economic, environmental and socio-political analyses provide the best basis for deciding whether to use thermal processing methods. Such an analysis considers all available reuse and disposal options and their costs and revenues. Costs include process costs and the cost of further handling and management (hauling, tipping fees, and so on). Revenues include those generated by production of a marketable product (fertilizer, electricity, and so on). The selection of a thermal processing system should recognize new, rapid advancements in dewatering technology. Some of these advancements may eliminate the future need for, or economic feasibility of, some thermal processing alternatives. Further, the selection process must weigh the likely effect of present and future regulations on the thermal processing system, associated air pollution control equipment, and the reuse or disposal option. The process should also assess whether the options are flexible enough to accommodate likely regulation changes. Such forethought may prevent the installation of a system that becomes obsolete or uneconomical, or requires major modifications to comply with future regulations.

A sound analysis of thermal processing options requires realistic expectations for advanced technology. Some existing sludge-processing systems were planned with high expectations for advanced technology that have often not been realized. As a result, the best reuse or disposal option available for these plants might have been overlooked.

The approach for operating advanced technology systems for thermal processing differs from that for operating typical wastewater treatment plant systems. Successful operation of some thermal processing systems requires highly trained and skilled operators or process engineers to be available 24 hours per day, especially during initial start-up of the plant. Some municipalities may not offer a salary structure sufficient for such staff. A good analysis will account for these specialized staffing needs. A comprehensive cost/benefit analysis will help with selection of the best sludge-processing strategy for the wastewater treatment plant.

The design and procurement of a thermal processing system often depend on proprietary vendor information, which may not be readily accessible. As a result, a designer should avoid specifying equipment or processes in a manner that may unnecessarily restrict some vendors from bidding on the project. In some cases, however, restrictive specifications reflect good engineering judgment. An engineer must set basic requirements such as size of the unit, types of support equipment, and minimum standards for materials.

The following five sections describe the prevalent thermal processing methods for municipal wastewater sludge, including conditioning, wet air oxidation, drying, oxidation, vitrifcation and biogasification. It should be noted that since the last edition of this manual, conditioning technologies have generally be phased-out, although there are some promising emerging technologies under development. Vitrification and biogasification remains an emerging technology for wastewater solids management. The fifth section discusses the regulations that govern particulate and gaseous emissions and emission control technology.

2.0 THERMAL CONDITIONING

Thermal conditioning is the simultaneous application of heat and pressure to solids to enhance dewaterability without adding conditioning chemicals. During thermal conditioning, heat lyses the cell walls of microorganisms in biological solids, releasing bound water from the particles. This process further hydrolyzes and solubilizes hydrated particles in biological solids and, to a limited degree, organic compounds in primary solids. Conventional mechanical dewatering devices can then readily separate released water from particles as long as the solids has enough fibrous solids for cake structure.

The two basic modifications of thermal conditioning that were historically used in wastewater treatment were heat treatment and low-pressure oxidation. At one time, about 31 U.S. wastewater treatment plants were using heat treatment systems. About 78 U.S. plants used low-pressure oxidation systems. Few currently operate today, and vendors are not actively marketing these systems in the United States.

2.1 Slurry Carbonization

Slurry carbonization is an emerging thermal treatment process to improve the dewaterability of biosolids. The process has been tested using cake containing up to 20% solids (OCSD, 2003). If the feed cake contains more than 20% solids, it is diluted and macerated to 6-mm (0.25-in.) particles. (The feed solids concentration is limited by pumping requirements.) The resulting feed slurry then is pumped up to the required pressure setpoint [about 5 MPa (750 psi)] and passed through heat exchangers to raise its temperature to about 200 to 230°C (400 to 450°F) (see Figure 26.1). The slurry then enters a reactor, where thermal decomposition occurs. At this point in the process, organics break down, carbon dioxide gas separates from the solids, and chlorine converts to hydrochloric acid, which is neutralized via the slurry's inherent buffering strength. Chlorine (a precursor to dioxins and furans) also washes out of solids in the form of aqueous salts. These chemical changes greatly reduce the slurry's viscosity and increase its dewaterability. The slurry then passes through recovery heat exchangers, where residual heat is captured for use in heating the feed slurry. Then it is washed and dewatered in a centrifuge to a solids concentration between 50 and 55%. Depending on its intended use, the dewatered cake is either used or further dried to 95% solids. If fed undigested solids, slurry carbonization produces a material with a heating value of about 15 100 kJ/kg (6 500 Btu/lb) at 95% solids.

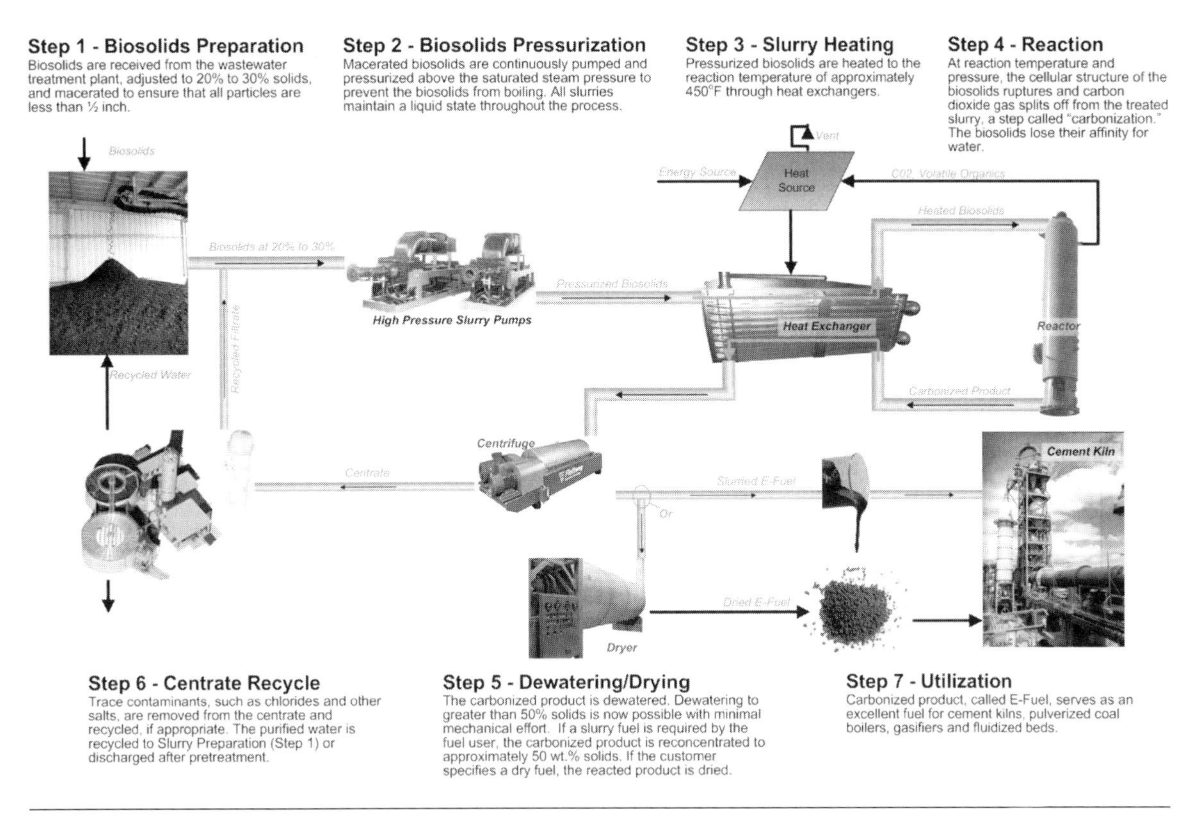

FIGURE 26.1 Schematic of slurry carbonization process (courtesy of Enertech).

The process also generates a wastestream high in ammonia and organics, which can be treated via membrane filtration, anaerobic digestion, or dissolved air flotation. The carbon dioxide and reactor exhaust are directed to a regenerative thermal oxidizer (RTO) to oxidize any volatile organic compounds (VOCs) in the offgases.

A 20-metric ton/d (22-ton/d) system in Japan treated slurried organic solid waste. A demonstration-scale facility in Atlanta, Ga., used residuals from several wastewater treatment plants.

In southern California, a proposed regional facility next to the Rialto Wastewater Treatment Plant is designed to process about 178 metric ton/d (196 dry ton/d) of solids from five municipalities and generate about 150 metric ton/d (167 ton/d) of product. The nearly 2.5-ha (6.2-ac) site is in a heavily industrialized area. The privately operated slurry carbonization unit is expected to have a footprint of about 30 m × 30 m (100 ft × 100 ft).

According to the private operator, slurry carbonization generates more than 1.98 times the energy it needs to treat undigested solids. A number of cement kilns within

80 km (50 mi) of the proposed site have expressed interest in using the product for fuel. Combustion tests show that its nitrogen oxide emissions are similar to those for coal, and its sulfur oxide and other emissions are lower, making it a cleaner burning fuel overall (EnerTech, 2001). Also, the remaining ash will be used in the cement process.

3.0 THERMAL DRYING

3.1 Overview of Technology

Thermal drying involves heating biosolids to evaporate water and thereby reduce the moisture content further than conventional mechanical-dewatering methods can achieve. Thermally dried biosolids cost less to transport because of the mass and volume reduction that occur; they also meet Class A pathogen standards and are more marketable than dewatered cake. However, thermal drying processes are more complex and expensive to operate than conventional dewatering methods. There also are safety concerns addressed later in this section.

Thermal drying has been practiced since the 1920s when the Milwaukee (Wis.) Metropolitan Sewerage District began producing and selling Milorganite®, its thermally dried biosolids fertilizer. For decades, only a few U.S. municipalities used this technology, but its popularity has grown since the early 1990s because of technology improvements, increasingly stringent regulations, and social factors affecting land application and landfilling. More than 50 thermal drying facilities were operating or under construction by 2009. Their throughputs range from less than 1 to more than 100 dry metric ton/d.

Thermally dried biosolids can be used as a fertilizer, soil conditioner, or biofuel. The use depends on the source of the solids, the pretreatment and drying systems used, and local conditions.

Some manufacturers call thermally dried biosolids "pellets", while others call them "granules". In this chapter, the term *granules* will be used.

3.2 Process Fundamentals

The drying rate depends on both the solids flow and evaporation rates. During drying, a temperature gradient develops from the heated surface inward, causing moisture to migrate from inside wet solids to the surface via diffusion, capillary flow, and internal pressures generated as the solids dry and shrink. As heat transfers from the machine to the solids, the solids' temperature rises and water evaporates from its surface. Other conditions that affect the process include the temperature, humidity, rate and direction of gas flow, exposed surface area, physical form of solids, agitation, detention time, and the support method used. Design engineers need to understand these conditions and their effects when investigating a solids' drying characteristics, choosing the correct dryer, and determining the optimal operating conditions.

3.2.1 Stages of Thermal Drying

There are three stages of thermal drying: warm-up, constant-rate, and falling-rate (see Figure 26.2).

3.2.1.1 Warm-Up Stage

During the warm-up stage, both the solids temperature and drying rates increase until they reach steady-state conditions. This stage typically is short and results in little drying.

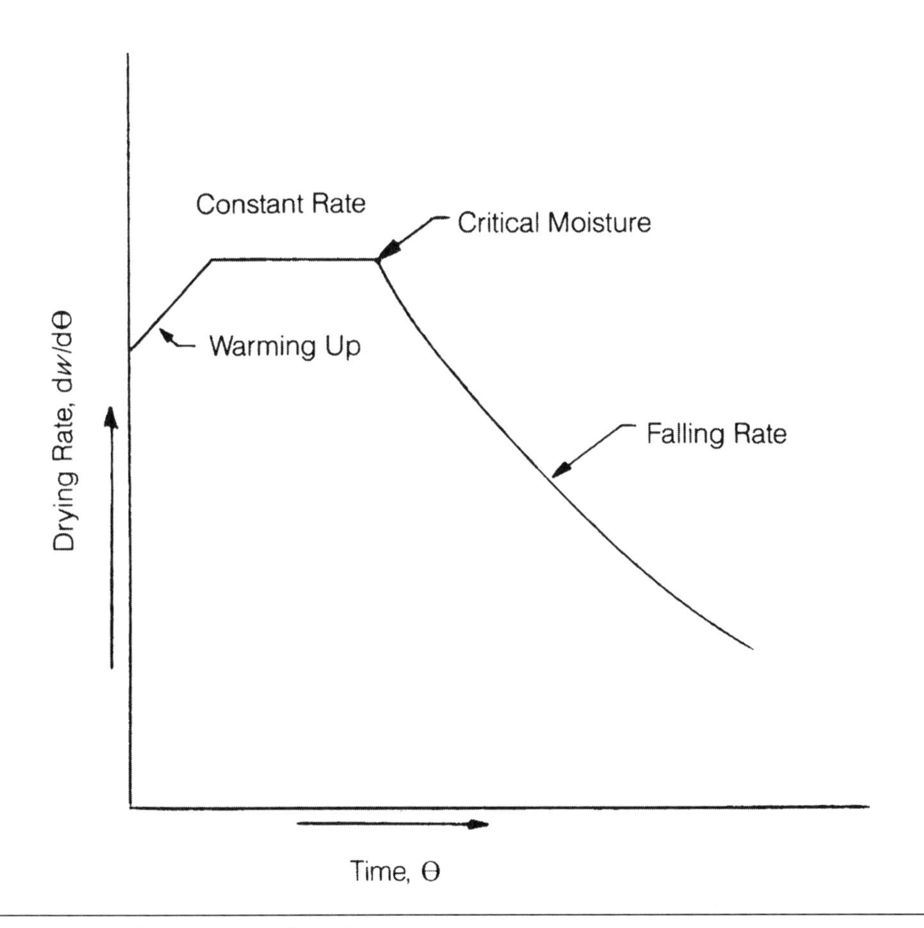

FIGURE 26.2 Three stages of drying.

3.2.1.2 Constant-Rate Stage

During the constant-rate stage, the surface of the solids remains saturated as interior moisture moves outward to replace the moisture that has evaporated. The drying rate depends on how much heat is transferred to the solids surface (the heat- or mass-transfer coefficient). It also depends on how much area is exposed to the drying medium, as well as temperature and humidity differences between the drying medium and the solids surface. This stage typically is the longest and when most drying occurs.

3.2.1.3 Falling-Rate Stage

During the falling-rate stage, moisture evaporates faster from the solids surface than it can be replaced by the moisture within the solids. So, the exposed surface is no longer saturated, solids do not transfer latent heat as rapidly as they receive sensible heat from the drying medium, and the temperature of the solids increases. This stage is when the drying rate decreases.

The drying rate at the transition point between the constant- and falling-rate stages is called *critical moisture.*

3.2.2 Heat-Transfer Methods

Dryers are classified based on the method (convection or conduction) predominantly used to transfer heat to wet solids. Although most systems actually use multiple heat-transfer methods, one method typically is more dominant than the others.

3.2.2.1 Convection

In convection (direct-drying) systems, the heat-transfer medium (e.g., hot gas) contacts the wet solids directly. Convection heat transfer is expressed mathematically as follows (U.S. EPA, 1979):

$$q_{\mathrm{conv}} = h_c A(t_g - t_s) \tag{26.1}$$

Where

q_{conv} = convective heat transfer [kJ/h (Btu/hr)];
h_c = convective heat-transfer coefficient [kJ/m^2·h/°C (Btu/hr/sq ft/°F)];
A = area of wetted surface exposed to gas [m^2 (sq ft)];
t_g = gas temperature [°C (°F)]; and
t_s = temperature at sludge–gas interface [°C (°F)].

Manufacturers have developed convective heat-transfer coefficients for their proprietary systems that they use to size the systems.

3.2.2.2 *Conduction*

Conduction (indirect-drying) systems transfer heat from the heat-transfer medium (e.g., hot oil or steam) to the solids via a material (e.g., a steel plate) that separates the solids from the medium. The mathematical expression for conduction heat transfer is as follows (U.S. EPA, 1979):

$$q_{\text{cond}} = h_{\text{cond}} A (t_m - t_s) \tag{26.2}$$

Where

q_{cond} = conductive heat transfer [kJ/h (Btu/hr)];
h_{cond} = conductive heat-transfer coefficient [kJ/m^2·h/°C (Btu/hr/sq ft/°F)];
A = area of heat-transfer surface [m^2 (sq ft)];
t_m = temperature of heating medium [°C (°F)]; and
t_s = temperature of solids at drying surface [°C (°F)].

The conductive heat-transfer coefficient is a composite term that includes the effects of both the solids' and the medium's heat-transfer surface films. Manufacturers have developed system-specific coefficients that they use to size their proprietary systems.

3.3 Process Description

As with most technologies, thermal dryers have continually evolved in the more than 75 years since they first were used to treat solids. Some systems have adapted to the industry's changing needs, while others have come and gone. Some systems ceased being viable technologies because

- Adverse commercial issues affected the manufacturers,
- The technology was overly complex,
- The technology could not produce a product that was easy to handle, or
- Full-scale installations performed poorly.

Below are basic descriptions of both established and emerging heat-drying processes.

3.3.1 Established Systems

3.3.1.1 *Convection Dryers*

Convection dryers that have been used successfully to dry municipal solids include the flash dryer, rotary drum dryer, and fluidized-bed dryer.

3.3.1.1.1 Flash Dryer

At one point, flash dryers were being used in as many as 50 U.S. facilities. As of 2008, however, only Houston, Texas, still used this technology at two of its wastewater treatment plants, and it was replacing one with a rotary drum dryer at press time. The reasons for its decline include safety concerns, high energy and O&M costs, and the primary equipment supplier's limited interest in the solids-treatment sector. Therefore, flash drying will not be further reviewed.

3.3.1.1.2 Rotary Drum Dryer

Rotary drum dryers have been used to treat solids since Milwaukee first installed them in the 1920s. In 2008, rotary drum dryers were in use or under construction at more than 20 U.S. and 75 European facilities.

Although some system components are manufacturer-specific, all rotary drum dryers have long, cylindrical steel drums that rotate on bearings (see Figure 26.3). Some have one drum through which solids make a single pass, while others have concentric drums to allow solids to make multiple passes through the drum (e.g., triple-pass systems). Hot gases dry solids and help move them through the system.

Some of the granules are recycled to the feed solids (dewatered solids), where a mixer blends them to reduce the feed material's moisture content to typically less than 35% so it is not "sticky" and has better handling characteristics. Actual moisture content depends on the dryer supplier and solids characteristics. Feed solids and hot furnace gases [400 to 650°C (750 to 1 200°F)] continuously enter the upper end of the drum dryer and are conveyed (typically concurrently) to the discharge end (see Figure 26.4). As they travel through the drum, axial flights along the slowly rotating interior wall pick up and cascade solids through the dryer. This thin sheet of falling solids directly contact the hot gases, which rapidly dry it.

In most systems, about 80% of furnace gases are recycled to enhance the system's thermal efficiency and help maintain an inert atmosphere. (See Section 3.5.4 on why inertization is important.) Recycling gas also reduces the volume of exhaust gases to be treated. Exhaust gases exit the dryer at 66 to 105°C (150 to 220°F) and travel to emissions-control equipment for particulate removal and odor control (Niessen, 1988).

Rotary drum dryers have been successfully used to dry both digested and undigested solids. In Europe, where granules typically become a biofuel, dryers often process undigested solids. In the United States, where the granules primarily are used as a fertilizer or soil amendment, the dryers typically process digested solids. Digestion can break down much of the fiber in raw solids that can cause operating problems for dryers. It also reduces the putrescible compounds that can cause odor problems in granules.

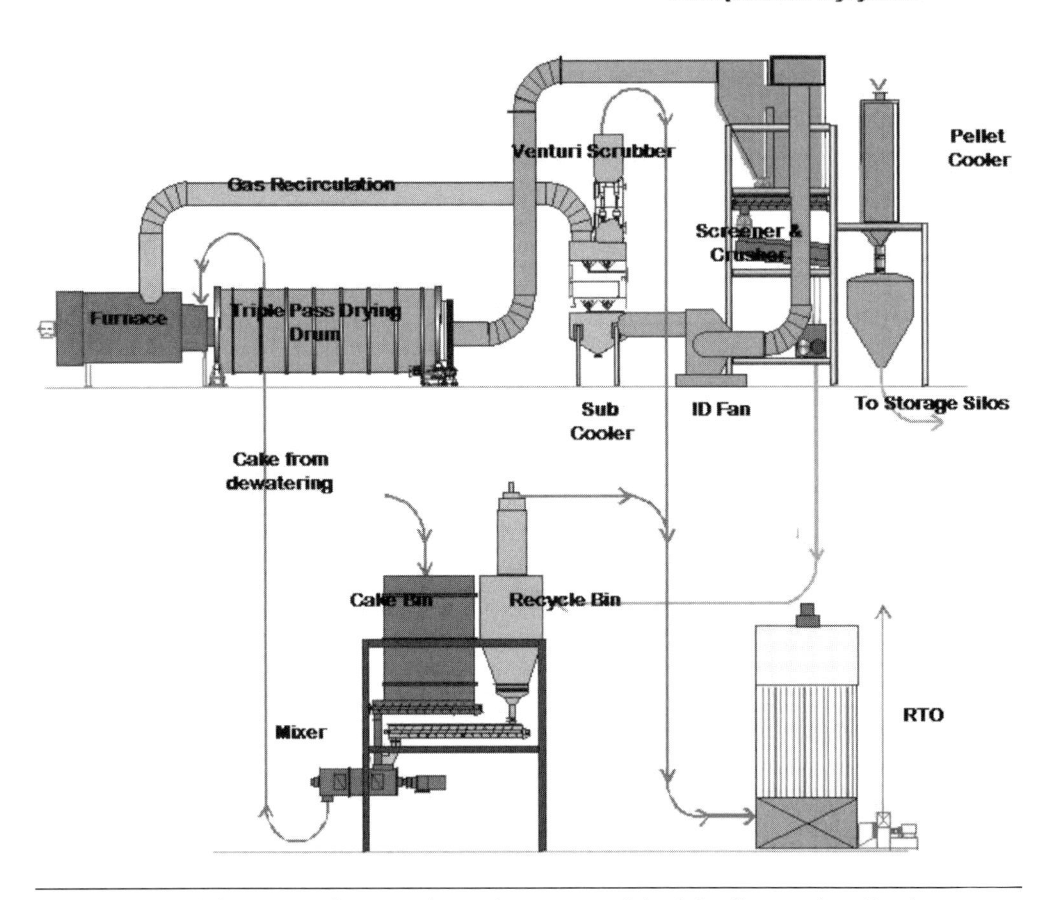

FIGURE 26.3 Schematic of rotary drum (courtesy of Andritz Separation, Inc.).

3.3.1.1.3 Fluidized-Bed Dryer

Fluid-bed technology first was applied to drying biosolids in Europe in the early 1990s; now, more than 30 European facilities use this technology. This application was first used in the United States in the late 1990s, and two U.S. systems were operating in 2008.

The main component of a fluidized-bed dryer is a stationary, vertical chamber segregated into three zones. The first zone is the wind box (gas plenum), which distributes hot fluidizing gases [typically air at about 85°C (185°F)] evenly through a bed of granules. The gases are forced or induced through the system by a blower arrangement in a closed-loop system.

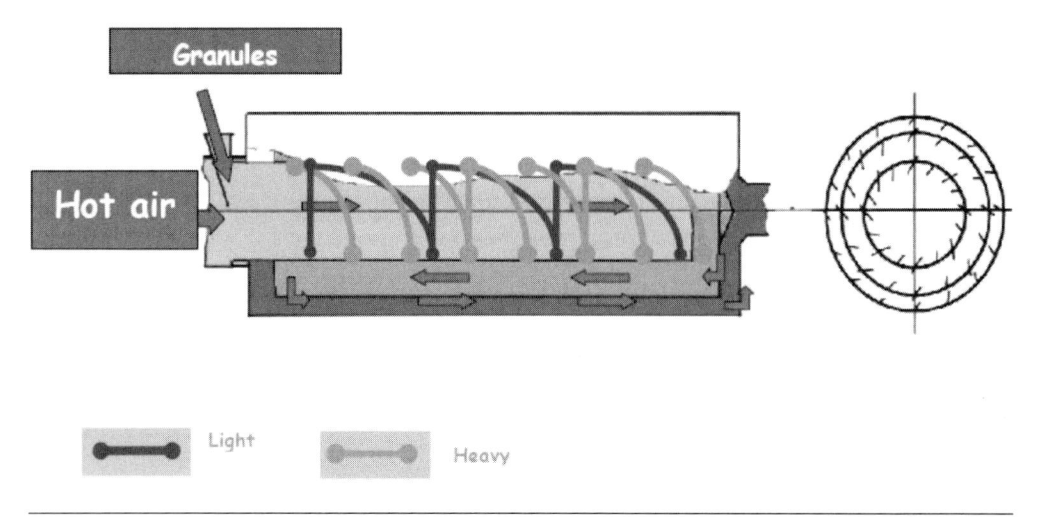

FIGURE 26.4 Schematic of solids movement through dryer (courtesy of Andritz Separation, Inc.).

The second zone is a heat exchanger. A heat-transfer medium (typically steam or hot oil) circulates through the heat exchanger tubes, while fluidizing gas keeps the granules in suspension, helps transfer heat from the tubes to the granules, and removes moisture from the system.

The third zone is a hood, where fluidizing gases exit the unit and most solids are separated from the gas flow. Exhaust gas then is sent to emissions-control equipment (e.g., cyclone separator and condenser), which removes particulate and moisture.

The method for introducing dewatered solids to the dryer is manufacturer-specific. Some systems pump solids into the dryer, where a rotating device cuts dewatered cake into small, irregular pieces that fall into the fluid bed. The fluidizing gas intimately contacts and highly intermixes the solids, resulting in high mass and heat transfer from gas to solids. Agitation in the fluidized bed removes rough edges, resulting in granules with a relatively uniform shape. Granules exit the dryer by overflowing from the chamber via an adjustable weir with a rotary airlock.

As with rotary drum systems, fluid-bed dryers will reduce the solids moisture content to less than 10%. These systems are not designed to operate as scalping (partial drying) systems.

Because of the low temperatures involved, fluid-bed systems are well suited for applications where energy recovery is possible. European systems have been operated off waste steam from garbage incineration plants. A relatively new facility at the North Shore Sanitation District in Waukegan, Ill., has used waste heat from a vitrification process.

3.3.1.2 Conduction Drying Systems

Over the last decade, conduction dryers have become more popular in the United States as more small and medium-size utilities have begun drying solids. The systems' simple materials-handling and emissions-control systems appeal to smaller utilities. Most facilities use the dryers to reduce the solids' moisture content to less than 10%, but some use these systems as scalping dryers to reduce the solids' moisture content to about 50%.

Conduction dryers that have been used successfully to dry solids include the paddle dryer, hollow-flight dryer, rotary chamber dryer, tray dryer, and pressure filter/vacuum dryer.

3.3.1.2.1 Paddle and Hollow-Flight Dryers

There are three types of hollow-flight dryers: paddle, disc, and rotary chamber. Paddle dryers have been used in industrial applications for many decades, but were first applied to drying biosolids in Japan in the late 1980s. Both paddle and disk dryers have been used to treat biosolids in the United States since the mid-1990s. In 2008, more than 30 U.S. facilities were building or operating paddle, disk, or rotary chamber dryers.

Both paddle and disc dryers consist of a stationary horizontal vessel (trough) and a series of agitators (paddles, discs, or flights) mounted on a rotating shaft (rotor). In some systems, the trough has a jacketed shell through which a heat-transfer medium (typically hot oil or steam) circulates. Rotary chamber dryers are similar, except that the trough's outer shell rotates and is heated by gas-fired burners. All of these systems have a hollow rotor and agitators, through which the heat-transfer medium circulates. The surfaces of the trough, agitators, and rotors conduct heat to the biosolids.

Dewatered biosolids typically are fed directly to the dryer by a positive-displacement pump or screw conveyor. (In these systems, feed solids typically are not mixed with recycled granules before treatment.) Granules exiting the dryer are irregularly shaped and vary in size because the agitation associated with paddles or discs can increase the concentration of fines in the product. Screens or conditioners can be used to produce a more uniformly sized product.

Rotor and agitator assemblies are fabricated of carbon steel, stainless steel, or a combination of both. They also can be made of various special alloys if highly corrosive elements are present. In some systems (primarily the paddle type), the agitators are arranged on the shaft to intermesh, which enhances the relative contact between heated surfaces and solids, thereby increasing the heat-transfer coefficient. They also are self-cleaning. In other systems, stationary agitator ploughs or breaker bars are placed between rotating agitators to improve mixing and prevent cake buildup on the agitator surface.

The temperature of the heat-transfer medium varies, but ranges from about 150 to 205°C (300 to 400°F). After transferring its available energy to solids, the heat-transfer medium typically is recirculated through the heating system. A high-pressure condensate return system can be used for steam to reduce energy consumption.

The water vapor concentration in the vessel creates an inert environment, but enough gas (air) is drawn through the dryer to remove evaporated water so it does not condense in the system. Instead, it is exhausted from the dryer and conveyed to a condenser. Noncondensable gasses then can be routed to the odor-control system. Some facilities use chemical scrubbers for odor control, while others convey the exhaust to the activated sludge process' aeration system. Thermal oxidation also can be used. The volume of exhaust gas involved is less than that produced by convection systems.

Paddle, disc, and rotary chamber dryers can be either continuous or batch fed. In continuous-feed systems, solids are conveyed through the dryer via the paddles' agitation of the bed, which causes volume displacement. Some systems use a weir at the discharge end of the dryer to help submerge the heat-transfer surface in the solids.

3.3.1.2.2 Tray Dryers

Tray dryers are similar to disc dryers, except that they are configured vertically and mix recycled granules with dewatered solids before feeding them to the dryer. The blended solids are fed through a top inlet in the vertical multistage dryer. The dryer has a central shaft with attached rotating arms that move solids from one heated stationary tray to another in a rotating zigzag motion until they exit at the bottom as granules. The rotating arms have adjustable scrapers that move and tumble solids in 20- to 30-mm layers over the heated trays.

After spiraling through the trays, solids exiting the dryer typically contain less than 10% moisture. These granules are screened and then either sent to a recycle bin for further processing or cooled to 30°C (90°F) and pneumatically conveyed to a storage silo.

Tray dryers have had limited growth in North America over the last decade. The Baltimore, Md., Back Water Wastewater Treatment Plant has used tray dryers to treat solids since 1995. A tray dryer at the Ashbridges Bay treatment plant in Toronto, Ontario, is scheduled to become operational in 2009. A tray dryer at the Stickney Wastewater Treatment Plant in Chicago began startup operations in 2008.

3.3.1.2.3 Pressure Filter/Vacuum Dryer

A pressure filter/vacuum dryer is a variation on a standard pressure filter that combines dewatering and drying in one process. First, the process uses the standard pressure cycle to mechanically dewater solids, which typically have been conditioned with lime and ferric chloride before entering the unit. After the pressure cycle, hot water

[80°C (180°F)] is circulated through the system to heat the chambers while a vacuum is drawn across the system to lower the boiling point of the water still in the solids. This is a batch process, which allows operators to control the level of dryness achieved.

To date, the pressure filter/vacuum dryer has not been widely used to treat solids; only two installations existed in 2009. Mountain City, Tennessee, has been using a pressure filter/vacuum dryer to treat solids since 2000. This utility uses the unit to produce solids containing 50% moisture, although it could dry solids further.

3.3.2 *Emerging Convection Dryers*

Emerging processes are those that have not been in full-scale, commercial use inside or outside the United States for at least 5 years. Belt and solar dryers are both emerging technologies. Both systems started out in Europe, have been used there for more than 5 years, and now are being used in the United States.

3.3.2.1 Belt Dryers

In belt dryers, solids are placed on a porous conveyor belt, and large volumes of heated gas are blown or drawn through them. The belt speed and configuration control the solids retention time. The belts are made of either steel mesh or synthetic material (similar to the filter media used in a belt filter press). The unit may have one or two belts, and its configuration depends on the manufacturer. The granules typically are less uniform in size than those from more complex systems because of the limited materials handling. Screens and conditioners can be used to improve uniformity.

The gas-management system is also supplier-specific. To remove moisture from the process, a fan typically removes a portion of the gas from the process air loop and sends it to a condenser. One supplier returns most of the scrubbed gas to the dryer, but diverts a portion to mix with combustion air. Other suppliers recommend that the scrubbed gas be treated by a biofilter or chemical scrubber to control odors before discharge. Recycling most of the process air improves the system's thermal efficiency and reduces the exhaust needing treatment.

The odor-control method is supplier-specific. Biofilters traditionally have been used in Europe to control odors from these systems, but chemical scrubbers also can be used. One supplier uses the furnace to combust exhaust gases.

Belt dryers use a relatively low-temperature gas [about 130 to 177°C (265 to 350°F)], so low-grade waste heat can be used as an energy source. This allows for a wide range of energy-recovery options. European facilities have used waste heat from gas-fired engines, municipal solid waste incinerators, and solids incinerators. One U.S. system scheduled for startup in 2009 plans to combust its granules to supply heat for the dryer. Other U.S. facilities were developing belt dryers with waste-heat options in mid-2009.

Belt dryers have been used to dry wood wastes and other homogeneous materials for decades in Europe. As of 2009, at least twenty systems were being built or operated in Europe, while one system was under construction and another was operating in the United States.

3.3.2.2 *Solar Dryers*

Solids have been dewatered naturally (e.g., sand drying beds) for decades, but such systems were susceptible to weather and typically produced solids containing about 70% moisture (for ease of disposal). In the 1990s, German researchers developed systems based on radiant (solar) energy and convective (air) drying theory that could produce solids containing no more than 10% moisture. The first full-scale facility began operating in Europe in the early 1990s; by 2008, more than 100 installations were in use there, and 10 facilities were being built or operated in the United States. The U.S. systems often have been designed to produce solids containing less than 25% moisture so they will comply with 40 CFR 503's vector attraction-reduction criteria.

Solar dryers typically consist of a concrete pad with low walls that is surmounted by a greenhouse-type structure. Unthickened, thickened, or dewatered solids are trucked or pumped onto the pad. A program monitors climatic variables (e.g., humidity and temperature) both inside and outside the structure, and adjusts fans and louvers to provide enough ventilation for drying. An electrically powered, mobile mixer tills the solids to expose more surface for evaporation. Solids may be spread in a relatively thin layer or arranged into windrows.

In the United States, this technology typically has been used by smaller treatment facilities. Some of the European installations, however, serve relatively large treatment facilities. For example, one system with a 2-ha (5-ac) footprint handles solids from a city with 650 000 inhabitants. Another is designed to dewater solids from a 4970-m^3/hr (31.5-mgd) facility.

Most of these dryers obtain their thermal energy solely from solar radiation, but about 20 use auxiliary heat (e.g., burned digester gas or waste heat from a power cogeneration facility) to increase drying performance. An Austrian facility burns its solar-dried solids onsite and uses this heat to speed up the drying process.

3.4 Process Design Guidelines

3.4.1 *Sizing Parameters*

3.4.1.1 *Evaporative Capacity*

Dryers are evaporation devices, so a design engineer's first step is to calculate how much water needs to be evaporated in a given time. The key parameters when calculating the desired evaporative capacity are the solids concentration of both feed solids and granules, as well as operating time. Evaporative capacity typically is expressed in terms of kilograms (pounds) of water to be evaporated per hour.

Then, designers should consult with manufacturers, who have established their systems' evaporative capacity based on heat-transfer calculations, lab tests, and demonstrated full-scale performance using various solids. A drying system may have multiple evaporative capacities based on the type of solids (e.g., undigested, digested, or combined primary and secondary), which affect heat-transfer coefficients and materials-handling characteristics.

Design engineers should keep in mind that the performance of the mechanical dewatering process substantially affects the size of a thermal dryer and its energy requirements. It typically is more cost-effective to remove water mechanically than thermally (to the extent possible).

3.4.1.2 Hours of Operation

Another key parameter in sizing thermal dryers are the acceptable hours of operation. Dryers have a fixed evaporative capacity, so the longer they operate each day, the more solids they can process. A dryer that operates 24 hr/d only needs one-third of the evaporative capacity to process the same amount of solids as a dryer that operates 8 hr/d. Because thermal dryers are expensive, many are sized to run almost continuously [e.g., 24 hours/day for 5 days each week (about 100 to 105 hours/week)] to fit with typical work schedules while leaving some down time for preventive maintenance. Also, running nearly continuously reduces equipment wear. Frequent heating and cooling cycles increase metal fatigue and, therefore, the rate of component failure.

3.4.1.3 Solids Residence Time

Solids residence time (SRT) depends on the type of dryer and the manufacturer's design. Manufacturers determine SRT based on system capacity and the heat transfer rate. System temperature does not affect SRT in thermal dryers; a rotary drum dryer and belt dryer may have similar SRTs even though their operating temperatures are significantly different. Some manufacturers also can use the dryer's SRT to demonstrate compliance with 40 CFR 503's pathogen-reduction criteria.

3.4.1.4 Operating Temperatures

The heat-transfer equations (presented earlier in this chapter) show that the heat-transfer rate depends on the temperature of the drying medium (e.g., gases, oil, or steam). When two similar systems are operating at different temperatures, the one with hotter temperatures typically will dry solids faster. Although operating temperature may affect the size of the unit, the heat-transfer rate is affected by several factors (e.g., area of exposed, wetted surface and the unit's heat-transfer coefficient), so temperature will not directly indicate the size.

3.4.1.5 Storage

Storage is an important but often overlooked element of a thermal drying system. It typically is needed both before and after the dryer. The system needs enough storage before the dryer to attenuate solids variations and allow for system shutdowns. It also needs enough storage after the dryer to handle changes in hauling schedules and seasonal fluctuations in market demand.

Granules often are stored in silos that are sealed from the outside environment. Because of the cost of silos, design engineers should size them carefully. Alternative use or disposal methods (e.g., landfilling) may be more cost-effective than significant storage capacity. Small and medium-sized facilities can use other storage methods (e.g., bulk bags or covered pads) as long as dust control and other safety issues are addressed. For more information on safety and solids storage, see Section 3.5.4.

3.4.2 Selection

3.4.2.1 Product Quality and Use

Design engineers should carefully consider the planned use for granules when selecting a thermal dryer. Some uses (e.g., high-end fertilizer) will require a higher quality product than others (e.g., soil amendment, bulk land-application, or biofuel). Although thermal dryers typically produce similar granules, design engineers should note the variations among suppliers, which may affect product use. For example, one belt-dryer supplier uses backmixing, while another extrudes material directly onto the belt, resulting in granules with different characteristics.

Utilities considering thermal drying can perform preliminary market/use analyses before selecting a system to determine which outlets would be most viable. (For more information on biosolids marketing and use issues, see Chapter 27.)

3.4.2.2 Processing Train Unit Capacity

When selecting a thermal drying system, engineers should consider the desired design capacity and the number of processing trains needed to supply it. Thermal dryers should be sized to provide an optimal balance among capital costs, redundancy, and space constraints. Design engineers also should carefully consider turndown capacity when sizing a system that is expected to accommodate significant increases in throughput capacity over time. Thermal dryers typically have limited turndown capability, so multiple trains may be needed to meet future processing needs or else the system may need shorter operating cycles initially to optimize throughput. (Throughput optimization also improves the system's thermal efficiency.)

Typically, it is more cost-effective to install one processing train rather than several to meet capacity requirements because this reduces space and ancillary support-system requirements and takes advantage of equipment-related economies of scale. However, one train may not meet the utility's needs for redundancy. To address this issue, numerous U.S. systems have one processing train (to reduce capital costs) and another dewatering and disposal option (e.g., landfilling) when the thermal dryer is out of service.

Convection systems (e.g., rotary drum and fluid bed) typically have larger single-train capacities than conduction systems (e.g., paddle, disk, and rotary chamber). Rotary drum and fluid bed systems typically can be sized to process between 2 000 to 11 000 kg H_2O/hr. Tray dryers also are considered large-capacity systems. Conduction systems, on the other hand, typically can be sized to process from less than 500 to 4 000 kg H_2O/h. The exception is belt dryers. One supplier has developed belt dryers with capacities comparable to those of rotary drum systems (although none had been installed in the United States as of 2009).

3.4.2.3 *Labor Requirements*

Labor requirements depend on the type of thermal dryer and its design. Some systems require more supervision and O&M staff skill sets than others. For example, rotary drum systems require continuous supervision and well-trained operators. Several fluid-bed and belt drying systems in Europe, on the other hand, operate in automatic mode with minimal supervision during off-shifts. A fluid-bed system at Houthalen, Belgium, operates 24 hours a day, 7 days a week in two 12-hour shifts. During one of these shifts, the dryer operates unattended in automatic mode.

Some municipal treatment plants hire contractors to handle solids drying and marketing operations. These private firms have the expertise for system O&M and product marketing.

3.4.3 *Utility Requirements*

3.4.3.1 *Electrical*

Thermal dryers may need substantial electrical power for the fans and solids-handling components. Requirements depend on system type and manufacturer. More complex systems will need more power because of their additional processing equipment.

3.4.3.2 *Thermal*

Thermal energy traditionally was supplied via the combustion of fossil fuels (e.g., natural gas and fuel oil); however, alternative fuels (e.g., digester biogas, landfill gas, and wood waste gasification) have become more common during the last decade, helping reduce or eliminate fuel costs. One thermal dryer under construction in 2009 was designed to use dried biosolids as its primary source of fuel.

Thermal energy requirements typically range from about 3 250 to 3 720 kJ/kg (1 400 to 1 600 Btu/lb) H_2O evaporated. System needs will depend on the type of thermal dryer and the manufacturer. Numerous factors affect energy efficiency, including many proprietary features that system owners and designers cannot control. Manufacturers can provide guaranteed energy-consumption factors based on tested operating conditions and feed solids characteristics.

One factor that owners can control to improve thermal efficiency is the mode of operation. Continuously operated systems with weekly startup and shutdown cycles will be more energy-efficient than those with daily startup and shutdown cycles. Also, operating close to design capacity will enhance overall thermal efficiency.

3.4.3.3 Water

Thermal dryers require substantial supplies of water to cool granules and remove moisture from process gas (via a condenser). Treatment plant effluent can be used for these purposes. Potable water may be required for some safety systems and emissions-control systems (e.g., demisters), but such demands typically are not significant.

3.4.3.4 Sidestreams

Design engineers should consider the effects of all thermal-dryer sidestreams on the wastewater treatment plant. For example, condensers produce odorous liquid sidestreams that contain both organic oils and ammonia. Wet scrubbers and other ancillary equipment also generate sidestreams.

3.4.3.5 Emissions and Odor Control

It is good design practice to enclose dryer equipment, as well as handling and storage areas, and vent these spaces to air pollution-control equipment. Odors typically are controlled via chemical scrubbing and thermal oxidation. Cyclone separators, wet scrubbers, bag houses, or a combination of these technologies can remove particulate from exhausted air.

3.5 Design Practice

3.5.1 Pre-Processing Equipment

Mechanical dewatering systems significantly affect thermal dryer requirements. It is more cost-effective to remove water mechanically (to the extent possible) than thermally. Slight improvements in dewatering-system performance can significantly reduce the overall size thermal dryers and their energy-consumption needs.

Consistent dewatering performance is also important. Large variations in feed solids concentrations can be problematic for some systems (e.g., drum dryers). The variability can affect the ratio of recycled granules to wet cake, thereby degrading the performance of drum dryers and subsequent processes. Batch-run systems (e.g., fluid-bed systems and some conduction systems) can better process feed cake with varying solids concentrations.

3.5.2 *Post-Processing Equipment*

Post-drying systems (e.g., screening, cooling, and dust control) depend on dryer type and manufacturer. These systems can fulfill processing requirements, improve granule-handling characteristics, and make the overall system safer. For example, screening makes the granules more uniform, and its reject can be recycled for blending with feed cake. Granules must be cooled before storage to reduce the risk of auto-oxidation (see Section 3.5.4). Dust-control systems (e.g., mineral-oil spray and mixing systems) agglomerate dust particles to improve granule quality and enhance system safety.

3.5.3 *Materials of Construction*

The choice of construction materials for a thermal dryer and its ancillary equipment depends on the characteristics of the solids being treated. Mild steel typically is appropriate for components that only touch granules, but stainless steel or other corrosion-resistant materials may be required for components that contact wet cake. Solids are abrasive, so design engineers should consider hard surfaces in areas especially prone to wear (e.g., agitators in indirect dryers) to avoid frequent equipment replacement. Also, all equipment should be airtight and well-insulated to minimize heat and efficiency losses.

3.5.4 *Safety*

Dried biosolids are combustible and, therefore, a fire hazard. Under certain conditions (even when wet), the material can *auto-oxidize*—generate enough heat via either biological or chemical processes to combust. A fire requires three components: a fuel source, oxygen, and an ignition source. So, fire-prevention measures for thermal dryers should focus on minimizing oxygen levels (inerting) and eliminating ignition sources. [The fuel source (granules) cannot be eliminated.]

Thermal dryers also have explosion hazards (e.g., combustible dust generated in solids-handling equipment, and CO and other combustible gases created during a smoldering fire). An explosion also requires three components: combustible dust or gases, oxygen, and an ignition source. So, explosion-prevention measures should focus on keeping combustible dust and gases below explosive concentrations, minimizing oxygen levels (inerting), and eliminating ignition sources.

Because design engineers cannot eliminate every condition that could lead to fires or explosions, they should provide mitigation measures (e.g., explosion venting) designed to limit potential damage to people and equipment, as well as isolation systems to prevent fires and explosions from spreading among drying-system components.

3.5.4.1 Dryer Equipment

To prevent fires and explosions in thermal drying systems, engineers should make sure the final design includes the following safety equipment:

- Gas-inerting systems in granule storage silos, solids-handling equipment, and the drying area that either provide continuous inerting or activate in response to incipient fire conditions;
- Water sprays or other suppression systems designed to quench fires;
- Some means of temperature control to keep temperatures from rising to unsafe levels;
- A system for cooling granules before storage to prevent auto-oxidation; and
- Ventilation systems for solids-handling areas to remove condensation and keep dust concentrations below explosive limits;
- Explosion venting in granule-storage areas and higher-risk solids-handling equipment (e.g., bucket elevators) to relieve the pressure from an explosion and minimize equipment damage; and
- Isolating devices (e.g., rotary valves) are provided between drying-system components—particularly areas with large amounts of granules (e.g., storage bins) or high dust concentrations (e.g., baghouses)—to prevent fires or explosions from spreading.

3.5.4.2 Dryer Operation

Because design engineers cannot eliminate all hazardous conditions associated with drying biosolids, thermal dryers must be operated to minimize such conditions. For example, controlling the solids' moisture content throughout the system is critical to eliminating auto-oxidation and excessive-dust hazards. In a solids-recycling system, granules that are too wet can provide the water needed for auto-oxidation to occur, as well as allow solids to build up inside equipment. However, granules that are too dry can form excessive amounts of dust. Maintaining steady-state operations and optimal moisture levels throughout the system can reduce such hazards.

3.5.4.3 Industry Standards

To ensure that proven safety measures are implemented, design engineers should consult the relevant industry standards. Those addressing the fire and explosion risks associated with drying biosolids include:

- *Standard for the Prevention of Fire and Dust Explosions from the Manufacturing, Processing, and Handling of Combustible Particulate Solids* (NFPA 654);
- *Standard on Explosion Prevention Systems* (NFPA 69);
- *Guide for Venting of Deflagrations* (NFPA 68); and
- *Standard for Fire Protection in Wastewater Treatment and Collection Facilities* (NFPA 820).

4.0 THERMAL OXIDATION

4.1 Process Fundamentals

Thermal oxidation systems totally or partially convert organic solids into oxidized products (primarily carbon dioxide and water) or else partially oxidize and volatilize organic solids via starved air combustion into products with a residual caloric value. *Thermal oxidation* refers to high-temperature oxidation in the presence of excess air. *Starved air combustion* is thermal oxidation that restricts airflow (oxygen) to produce three potentially energy-rich products: gas, oil or tar, and a char.

Although wastewater solids typically are organic, they only will sustain combustion without auxiliary fuel if enough water has been removed. Hence, dewatering should always precede thermal oxidation. Dewatered cake with 20 to 35% total solids can be burned with or without auxiliary fuels, depending on the temperature of the combustion air.

Combustion (*burning*) is the rapid combination of oxygen with fuel, resulting in the release of heat. The typically combustible elements of wastewater solids—carbon, hydrogen, and sulfur—chemically combine in organic solids as grease, carbohydrates, and protein. The combustible portion of the solids has a heat content about equal to that of lignite coal. Adding air provides oxygen to support the combustion. The products of complete combustion are carbon dioxide, water vapor, sulfur dioxide, and inert ash. Good combustion requires proper proportioning and thorough mixing of fuel and air and the initial and sustained ignition of the mixture. To achieve complete combustion, three fundamental principles must be followed: temperature, time, and turbulence (see Section 4.3).

The gaseous products of thermal oxidation include dry combustion gases, excess air, and water vapor from the moisture in the solids and the oxidation of hydrogen. The heat content of dry combustion gases is determined by multiplying the individual weight of each gas by its specific heat content at the exit temperature. The weight of moisture in the exit gases (resulting from the combustion of hydrogen) typically is combined with the weight of the moisture in the wet feed to calculate the total heat content of the total water vapor exiting in combustion gases. If supplemental fuel is used, the volumes and heat contents of stack gases resulting from fuel and solids combustion must be calculated separately.

The main objective of thermal oxidation is to reduce the quantity of solids requiring disposal. This process reduces the volume and weight of solids by about 95%, oxidizes or reduces toxics, can operate without consuming fossil fuels, and can produce power. However, its capital and O&M costs are higher than those for other solids use and disposal options, and its exhaust gases must be treated extensively to comply with increasingly stringent air emission regulations.

4.1.1 Solids Calorific Values

The composition and quantity of fuels to be burned are fundamental inputs for the design of any combustion system. The composition determines the fuel's calorific or heating value, and the fuel's quantity determines the required size of the unit.

A solids' *heating value* is the amount of heat that can be released per unit mass of solids. Its gross heating value—the prime indicator of combustion potential—depends on how much carbon, hydrogen, and sulfur it contains. Carbon burned to carbon dioxide has a gross heating value of 3.4×10^4 kJ/kg (1.46×10^4 Btu/lb). Hydrogen has a gross heating value of 1.44×10^5 kJ/kg (6.2×10^4 Btu/lb). Sulfur has a gross heating value of 1.0×10^4 kJ/kg (4 500 Btu/lb). So, any changes in the solids' carbon, hydrogen, or sulfur content will raise or lower its heating value.

Each fuel has a characteristic ("effective") heating value and a gross heating value. *Effective heat* (*available heat*) is the amount of heat released in a combustion chamber minus both the dry flue-gas loss and the moisture loss. It depends on the solids' characteristics and the combustion chamber's design and operating characteristics; it cannot be determined via a bomb calorimeter alone. Incinerators that need large amounts of excess air offer less available heat because of the energy needed to heat the excess air (see Figure 26.5) (North American Manufacturing Co., 1965; LACSD, 1988).

The typical heating values of wastewater solids obtained via bomb calorimetry are shown in Table 26.1 (U.S. EPA, 1979). Often, the combustibles in primary and WAS have a heat value of about 2.2×10^4 kJ/kg (9500 Btu/lb).The literature contains several methods for calculating the heating values of organic materials using proximate and ultimate analyses, but none of these—including the Dulong formula—accurately approximate the heating value of wastewater solids.

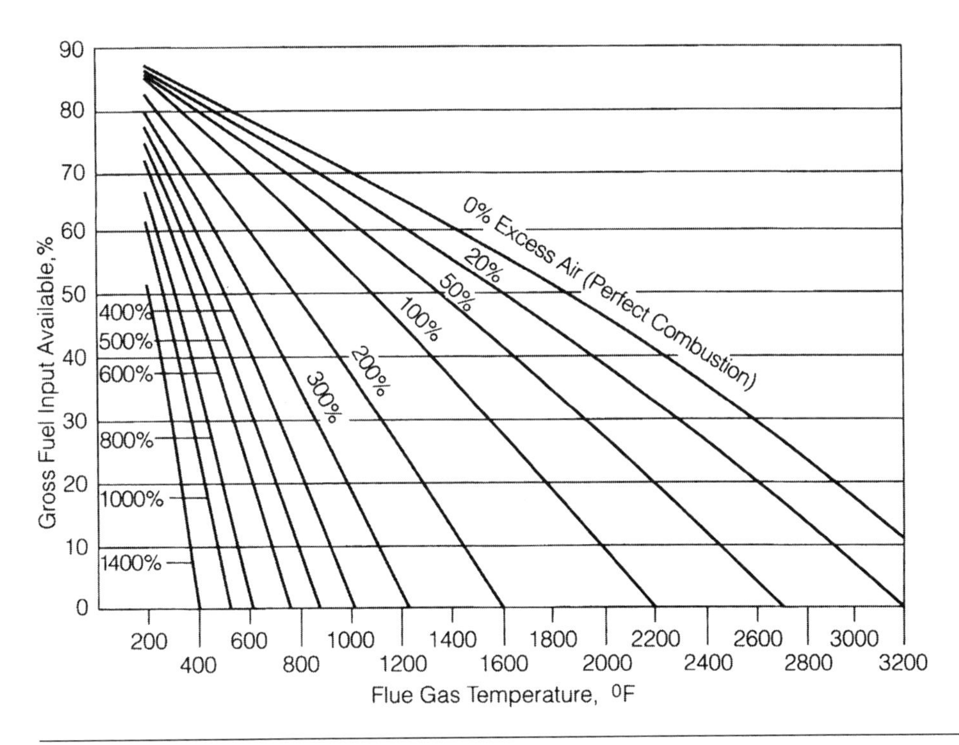

FIGURE 26.5 Generalized available heat chart.

4.1.2 *Oxygen Requirements*

The oxygen needed for combustion typically comes from air. Because air contains a large amount of nitrogen, thermal oxidizers need much more air than would be necessary if pure oxygen were used (see Table 26.2). For every kilogram of oxygen required, about 4.6 kg of air must be supplied.

TABLE 26.1 Representative heating values of some solids (courtesy of Infilco Degremont, Inc.).

Material	Combustible solids (%)	High heating value [kJ/kg (Btu/lb) of dry solids]
Grease and scum	88	38 816 (16 700)
Raw wastewater solids	74	17 678 (7 600)
Fine screenings	75	15 700 (6 750)
Ground garbage	75	13 095 (5 630)
Biosolids	60	12 793 (5 500)
Grit	30	9 304 (4 000)

TABLE 26.2 Theoretical air and oxygen requirements for complete combustion (courtesy of Infilco Degremont, Inc.).

	kg/kg of substance	
Substance	**Air**	**Oxygen**
Carbon	11.53	2.66
Carbon monoxide	2.47	0.57
Hydrogen	34.34	7.94
Sulfur	4.29	1.00
Hydrogen sulfide	6.10	1.41
Methane	17.27	3.99
Ethane	16.12	3.73
Ammonia	6.10	1.41

4.1.2.1 Stoichiometric Air

Insufficient quantities of oxygen lead to partial combustion and the resulting soot, carbon monoxide, and gaseous hydrocarbon emissions. Heat and material balances stoichiometrically indicate the amount of oxygen needed to completely combust solids, and design engineers typically should add 25 to 100% more air to ensure that the system has enough oxygen for complete combustion. However, they should not provide too much excess air and, thereby, waste fuel heating unneeded oxygen and inert nitrogen. Most multiple-hearth furnaces are designed to operate with 75 to 100% excess air. Fluidized-bed combustors typically operate with about 40% excess air (Dangtran et al., 2000). Turndown on a fluidized-bed combustor is limited by the air velocity required to fluidize the media bed, so fluidized-bed furnaces always must be operated near design solids loadings, even for short periods.

4.1.2.2 Oxygen Content

The design oxygen content of exhaust gases typically ranges from 6 to 8% on a dry weight basis (WEF, 1991). To control combustion, the exhaust gas' oxygen content must be measured. Two types of analyzers are used to measure this oxygen content: extraction and *in situ*. The extraction analyzer withdraws a sample of exhaust gas and then cools it, filters it, and analyzes it with a zirconium oxide element. The *in situ* analyzer uses a zirconium oxide element surrounded by a ceramic filter to directly measure the oxygen content of the exhaust gas.

4.1.3 Heat and Temperature

When fuel, oxygen, and an ignition source combine, the resulting combustion releases energy that is called heat when it transfers from the burning material to cooler solids and gases. This transfer is driven by the *temperature differential*—the difference in temperature between two objects.

A combustion system releases two types of heat: sensible and latent. *Sensible heat* is heat that, when gained or lost by a body, is reflected by a change in that body's temperature. This heat is computed by multiplying the materials's heat capacity (specific heat) by its temperature above some reference point [typically 0 or 15.5°C (32 or 60°F)] as follows:

$$Q_s = (M_{cp})(W)(T - t) \tag{26.3}$$

Where

Q_s = sensible heat [kJ (Btu)];
M_{cp} = mean specific heat [kJ/kg·°C (Btu/lb/°F)];
W = weight of material [kg (lb)];
T = temperature above a reference temperature [°C (°F)]; and
t = reference temperature [°C (°F)].

Examples of sensible heat include the heat content of ash and the heat required to raise the temperature of flue gases.

Latent heat is the heat required to evaporate moisture from solids. The latent heat of vaporization is a function of temperature; in combustion calculations, it is assumed to occur at a temperature of either 0 or 15.5°C (32 or 60°F).

A substance's theoretical flame temperature is:

$$T = [Q_s/(W \times M_{cp})] + t \tag{26.4}$$

This equation assumes all heat is recovered and used to elevate the temperature of combustion products; none is lost (e.g., radiation) to the combustion environment. In practice, however, theoretical flame temperatures are impossible to reach because several factors limit the temperature to a level somewhat below the calculated value. For example, when wastewater solids are burned, the theoretical flame temperature will be reduced by the following major heat losses to the combustion environment:

- Moisture in the stack gases (latent heat),
- Unburned carbon,
- Excess air,
- Radiation,
- The ash's heat content, and
- Heats caused by reaction or formation.

Combustion actually seeks a final equilibrium temperature at which the heat input from the fuel equals the heat losses. When self-supporting, this equilibrium temperature is called the *autogenous burning temperature.*

Wastewater solids dried to about 40% solids will ignite at less than 538°C (1 000°F). Operating temperatures greater than 760°C (1 400°F) are required to combust all organic material completely. Operating multiple-hearth furnaces at temperatures below 982°C (1 800°F) prevents the damage of furnace insulation and structural components, and the slagging of ash (U.S. EPA, 1979, 1983a).

4.2 Process Description

Three thermal oxidation processes are discussed in this section: fluidized-bed furnaces, multiple-hearth furnaces, and multiple-hearth enhancement processes.

4.2.1 *Established Technologies*

4.2.1.1 Fluidized Bed Incinerator

The principle of using a high-temperature gas to fluidize solid materials was first commercially developed by the oil-refining industry in a technology called fluidized-bed catalytic crackers. While fluidized-bed combustors only vaguely resemble catalytic crackers, they function similarly. In a fluidized-bed combustor, combustion air is passed through an enclosed space (fluidized-bed zone) in a way that sets all the particles in that zone in a homogeneous, boiling motion (i.e., fluidizes them) (see Figure 26.6). In this state, particles are separated from each other by an envelope of fluidizing gas (combustion air), thereby presenting more surface for a gas-to-solid reaction (e.g., air to carbon). The maximized surface area is what gives most fluidized-bed combustors their high thermal efficiency.

Treatment capacity, therefore, is a function of total reactor bed volume (although typically expressed as either fluidized-bed surface area or freeboard area) and the maximum gas velocity permitted because of particle entrainment.

At combustion equilibrium, the fluidized bed resembles a boiling liquid and obeys most hydraulic laws. A specially designed orifice plate disperses fluidizing gas throughout the fluidized-bed zone to ensure complete mixing. Temperature variations between any two spots in the fluidized bed typically will not exceed 6 to 8°C (10 to 15°F).

Dewatered cake is fed directly into the fluidized bed, which is preheated with auxiliary fuel to about 650°C (1 200°F) (WEF, 2009). In the bed, solids moisture evaporates, and the volatile fraction burns at temperatures of 760 to 816°C (1 400 to 1 500°F) (U.S. EPA, 1985b). The fluidized-bed medium is kept at combustion temperature via oxidation of organic material in the feed. Rather than flaming, the medium glows. Combustion is

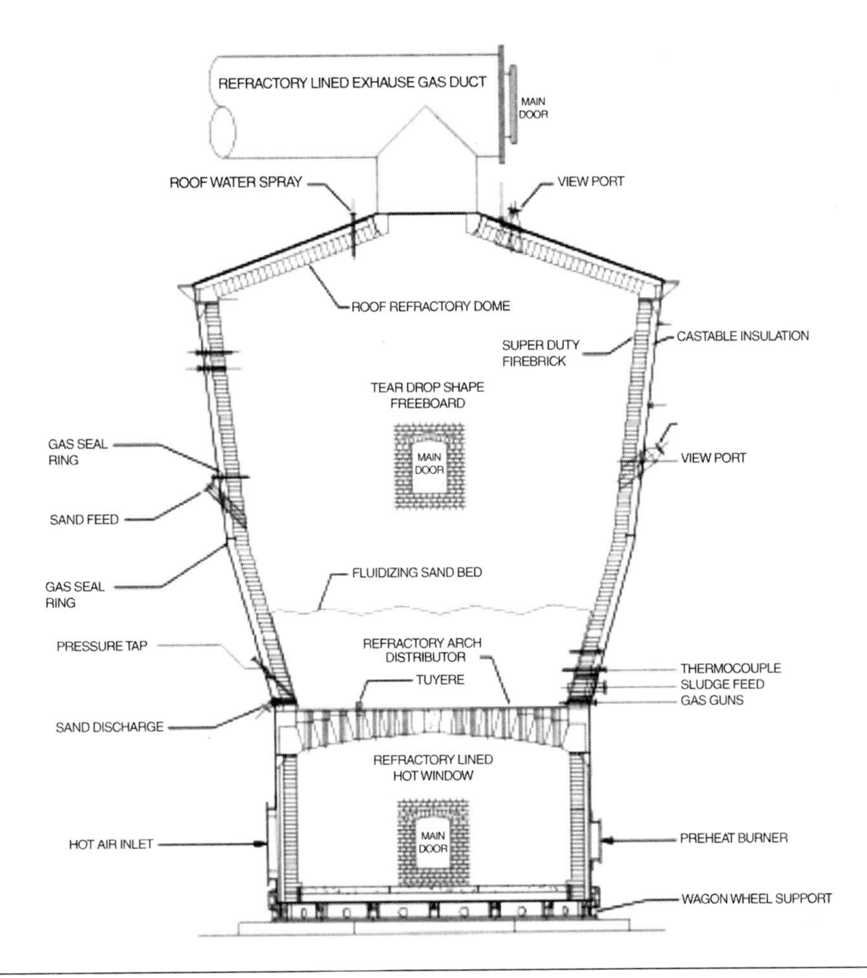

FIGURE 26.6 Typical section of a fluidized bed furnace (courtesy of Infilco Degremont, Inc.).

rapid, and the fluidized bed itself will contain negligible amounts of unburned organic matter. (Complete oxidation of organic matter is the first step in controlling air pollution.)

There typically is a temperature difference between the fluidized bed and the freeboard (disengaging zone) because a portion of VOCs will be combusted in the area above the bed, causing a temperature increase of 80 to 170°C (150 to 300°F). In other words, while the bed temperature typically ranges from 730 to 760°C (1 350 to 1 400°F), the freeboard temperature will be significantly hotter [840 to 900°C (1 550 to 1 650°F)].

Combustion gases and ash discharge from the top of the furnace. Entrained ash typically is separated from the combustion gases by a Venturi scrubber or another particulate-removal device.

Fluidized-bed combustors have thermally oxidized various wastewater solids successfully. They save several advantages:

- The smooth, liquid-like flow of particles allows continuous, automatically controlled operations with ease of handling;
- Solids can be circulated between two combustors, making it possible to transport vast quantities of heat produced or needed in large reactors;
- They are well-suited to large-scale operations;
- Heat- and mass-transfer rates between gas and particles are high;
- Heat exchangers inside fluidized bed combustors require relatively small surface areas because the heat-transfer rate between the combustor and an immersed object is high; and
- The combustor can be shut down and started up daily without a preheating period because the bed cools slowly [3 to 8°C/h (6 to 15°F/hr)].

There are two types of fluidized-bed combustors: hot wind-box and cold (warm) wind-box. The hot wind-box systems typically incinerate wastewater solids with typically low heat values that require intensive air preheating to minimize auxiliary fuel consumption. The cold wind-box system typically is used to burn heat-treated solids, dried solids, primary solids, scum, grease, and other materials (e.g., wood chips or sawdust) that can burn autogenously without heat recovery or with moderate heat recovery.

4.2.1.1.1 Hot Wind Box

The hot wind-box fluid bed is designed for maximum temperatures of about 650 to 980°C (1 200 to 1 800°F) for most applications. It is used when the wind-box air temperature is greater than about 400°C (750°F). A cross-section of a typical hot wind-box fluid bed is shown in Figure 26.6. The unit is a vertical steel shell made of carbon steel. The inside lining is made of refractory and insulating brick. The refractory lining is necessary because of the inside temperature of approximately 980°C (1800°F). The fluid bed is composed of four sections: the wind box, the bed support and air distributor, the bed, and the freeboard.

The lower section of the furnace is the *hot wind box*, a refractory-lined plenum in which hot combustion air is received. It serves as a distribution chamber for fluidizing air and a combustion chamber for the preheat burner. The wall of the wind box has openings for fluidizing air supply, preheat burner, observation port, and instrument ports.

The roof of the wind box serves as a bed support and air distributor. Typically called the *dome*, it can be made of refractory or metal alloys, depending on design requirements. Its refractory arch is constructed to be self-supporting, because of the special shape of the refractory elements used. The dome supports the weight of the unfluidized bed material and distributes fluidizing air via a number of air nozzles (typically called *tuyeres*). The tuyeres' special shape and material prevent sand drainage, provide uniform air distribution, and withstand furnace operating temperatures. Both dome and wind box are designed for about 980°C (1 800°F); the combustion air typically is preheated to about 675°C (1 250°F).

The bed (combustion zone) contains the fluidized mass of sand. Air from the distributor causes the sand to fluidize. The height of the fluidized sand layer depends on the amount of sand in the bed. An expanded bed of about 1.5 m (5 ft) typically is used to thermally oxidize wastewater solids. The side walls slope outwardly from the bottom of the bed to ensure that water vapor expands and to keep gas velocity within acceptance limits. The walls are also equipped with nozzles for solids and auxiliary fuel injections, and ports for various instruments.

The space above the bed is called the *freeboard* (*disengagement zone*). It acts as both a gas-retention and particle-separation chamber. To ensure complete combustion of any volatile hydrocarbons escaping from the bed, the freeboard must be large enough to provide about 6.5 seconds of gas-residence time at a minimum. It is typically 4.6 m (15 ft) high. The freeboard could be shaped as a cylinder or a teardrop. The cylinder typically is designed based on a gas velocity of 0.76 m/s (2.5 ft/sec). The teardrop is designed with the narrowest part at the bottom to maximize residence time and further reduce gas velocity. The gas velocity at the top of the teardrop is 0.64 m/s (2.1 ft/sec). An exhaust-gas duct is installed in the center of the roof dome to minimize gas bypassing, minimize dead zone, and maximize residence time in the freeboard. To minimize sand losses, gas velocity is designed to decrease in the freeboard so particles drop out of suspension.

4.2.1.1.2 Cold (Warm) Wind Box

A cold (warm) wind-box thermal oxidizer is used for solids that can be incinerated without heat recovery (or with moderate heat recovery). The air temperature in the wind box typically is limited to less than 400°C (750°F). The designs for hot and cold (warm) wind-box systems are similar, except that in cold wind-box systems,

- The wind box is not refractory lined;
- The bed support and air distributor can be a metal alloy plate, which typically is refractory lined to sustain the bed's high temperature; and
- The preheat burner is installed in the freeboard and angled downward to heat the top of the fluidized sand bed during startup.

4.2.1.2 Multiple-Hearth Furnace

A multiple-hearth furnace consists of a vertical, refractory-lined steel cylinder containing a series of stacked horizontal refractory shelves (hearths) (see Figure 26.7) (U.S. EPA, 1979, 1983b). A hollow cast-iron rotating shaft runs through the center of the hearths. Rabble arms are attached to this shaft above each hearth, and cooling air may flow through them. Metal blades (rabble teeth) on each arm may be angled in the direc-

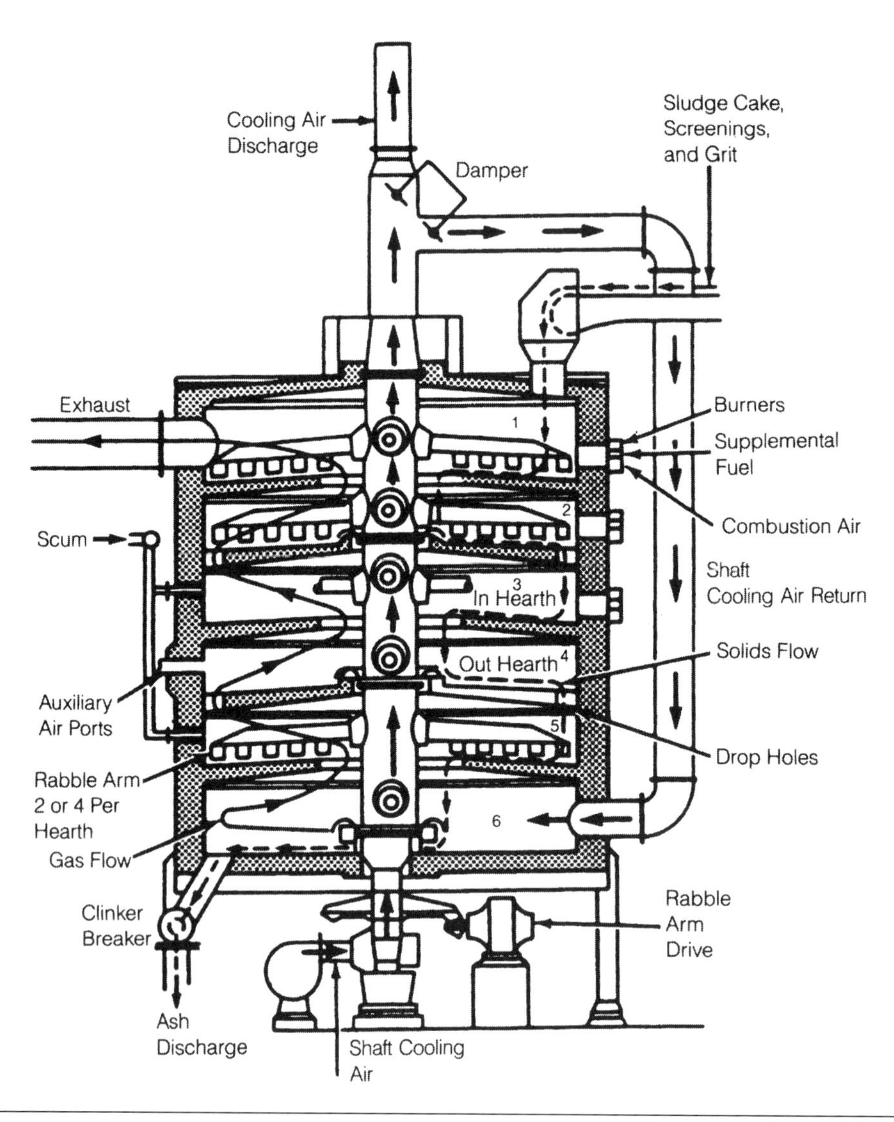

FIGURE 26.7 Typical cross-section of a multiple-hearth furnace.

tion of rotation (forward rabbling) or may be reversed (back rabbling). Back rabbling increases the cake's detention time and improves drying. Combustion air enters at the bottom of the hearth and circulates upward through drop holes in the hearths, countercurrent to solids flow. It exhausts through the top hearth to the waste-heat exchanger (if present) and then to the air pollution-control equipment.

Solids typically flow down through the furnace. This is not a simple vertical drop, however; they move outward and inward on successive hearths, dropping through either peripheral holes in a hearth ("out-hearth") or an annular opening at the center ("in-hearth"). The rabble arms and teeth sweep solids forward and radially several times on each hearth, resting after each plowing action. In addition to moving solids, the rabble teeth cut, furrow, turn, and open them, thereby exposing new surfaces for drying. So, the effective drying surface is estimated to be 130% of the hearth's actual area (Niessen, 1988; U.S. EPA, 1979).

A multiple-hearth furnace has three zones: drying, combustion, and cooling. Most of the water in solids evaporates in the upper hearths (drying zone). The heat-transfer mechanisms in this zone are about 15% convection and 85% radiation (Lewis and Lundberg, 1988). The temperature in the drying zone rises from 427 to 760°C (800 to 1 400°F) as solids move from hearth to hearth. In the central hearths (combustion zone), the solids' combustibles are burned at 760 to 927°C (1 400 to 1 700°F) (U.S. EPA, 1979, 1983a). Volatile gases and solids burn in the upper middle hearths, while fixed carbon burns in the lower middle ones. Incoming combustion air cools ash to between 93 and 204°C (200 and 400°F) in the lowest hearths (ash cooling zone). Ash exits through the bottom of the furnace (WEF, 1991).

Multiple-hearth furnaces have successfully treated various solids. They have minimum space requirements and are reliable, easy to operate, effective, reduce solids volume, and produce a sterile ash (U.S. EPA, 1983a). However, the complex systems have high capital costs, O&M requirements, and energy use, as well as air-pollution and odor problems (Dangtran et al., 2000).

4.2.2 *Emerging Technologies*

4.2.2.1 *Multiple-Hearth Furnace Innovations and Enhancements*

To help multiple-hearth furnaces meet the requirements of 40 CFR 503 (U.S. EPA, 1993) and other regulations, several enhancement have been developed. These include furnace modifications, a reheat-and-oxidation process and flue-gas recirculation.

4.2.2.1.1 Furnace Modifications

Furnace modifications include internal afterburners, hole modifications, burner improvements, and burner-control modifications. Engineers have installed internal afterburners in many furnaces by converting the top hearth into a "zero hearth" and feeding solids to the hearth below (Dangtran et al., 2000). Zero hearths are constructed by either removing the top hearth or using the top two hearths as the zero hearth. Solids then start treatment on the next hearth below; they are either fed via refractory-lined chutes or dropped directly onto the hearth. The zero hearth increases the residence time for products of incomplete combustion so they can burn out in furnace exhaust gases at elevated temperatures before they leave the furnace. Zero hearths typically are operated at 590 to 760°C (1 100 to 1 400°F); this heat is provided by burners (typically larger ones than the furnace originally used). Actual operating temperatures are site-specific and established to ensure that the Part 503 hydrocarbon limit is met. Zero-hearths provide better fuel efficiency than external afterburners.

Out-hearth drop holes have been enlarged to reduce slagging caused by the "blow torch" effect of high-velocity gases from one hearth contacting smoldering solids from the hearth above. However, larger holes can reduce effective hearth area and has met with mixed success.

Manufacturers have improved burner design to reduce burner tile and refractory slagging. One of these burners, which can operate on dual fuel, has been used in many furnaces (Nuss et al., 2008). Also, various improvements to burner controls have been implemented. One modification, called the "on-the-fly air : fuel ratio control", keeps airflow to the burner at the maximum rate and adjusts the fuel rate to match heat requirements. This maintains maximum turbulence for good solids combustion (O'Kelley et al., 2006).

Another furnace enhancement is installing variable-frequency drives on induced draft fans to replace dampers and improve furnace-draft control and electrical efficiency. Yet another improvement is replacing Venturi/impingement scrubbers with post quench/impingement multiple Venturi scrubbers to improve particulate removal efficiency and reduce power consumption.

4.2.2.1.2 Reheat and Oxidation Process

The reheat and oxidation process is a practical, economical means of producing high-quality exhaust gas while maintaining the conditions needed to combust wet solids (RHOX, 1989). In this process, a regenerative thermal oxidizer (RTO) equipped with a low-NO_X burner is installed downstream of the scrubber to reduce emissions of THCs, carbon monoxide, dioxin, and furan (see Figure 26.8). The regenerative thermal oxidizer uses its high heat-transfer efficiency to raise the scrubber's exhaust gas temperature from about 38°C (100°F) to about 110 to 140°C (230 to 280°F).

The regenerative thermal oxidizer is more fuel-efficient than a conventional external or internal afterburner. In a conventional afterburner, the mass load through

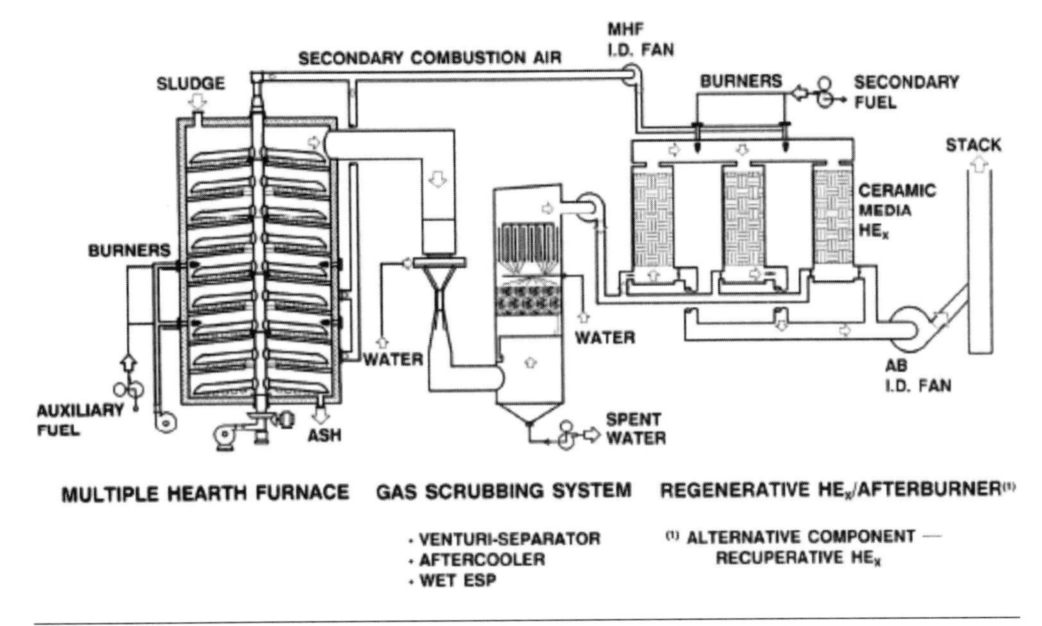

FIGURE 26.8 Reheat and oxidation process flow sheet for a multiple-hearth furnace with regenerative heat exchanger (RHOX, 1989).

the afterburner includes the water evaporated in the multiple-hearth furnace. So, if the furnace exhausts gas that is about 480°C (900°F) and the afterburner temperature is about 675°C (1 250°F), the temperature of the entire mass of exhaust gas has to be raised about 195°C (350°F). The regenerative thermal oxidizer, on the other hand, only has to raise the temperature of the saturated gas, which contains relatively little water, from about 60 to 140°C (140 to 280°F)—a difference of about 80°C (140°F). Also, the mass loading of scrubber exhaust gas is about 70% of the mass loading of furnace exhaust gas.

The exhaust gas must have low particulate concentrations to avoid fouling the ceramic surfaces of the heat exchangers. So, existing scrubbers may have to be augmented by a wet ESP or replaced by a scrubber.

4.2.2.1.3 Flue Gas Recirculation

In flue-gas recirculation, ducting and fans recirculate flue gas from the top hearth to one near the bottom (see Figure 26.9). Typically, two sets of ducts and fans are provided. The temperature and flowrate of recirculated gas is measured by a flow meter in the duct. The flowrate is controlled by dampers or a variable-speed fan. Cooling air is provided to

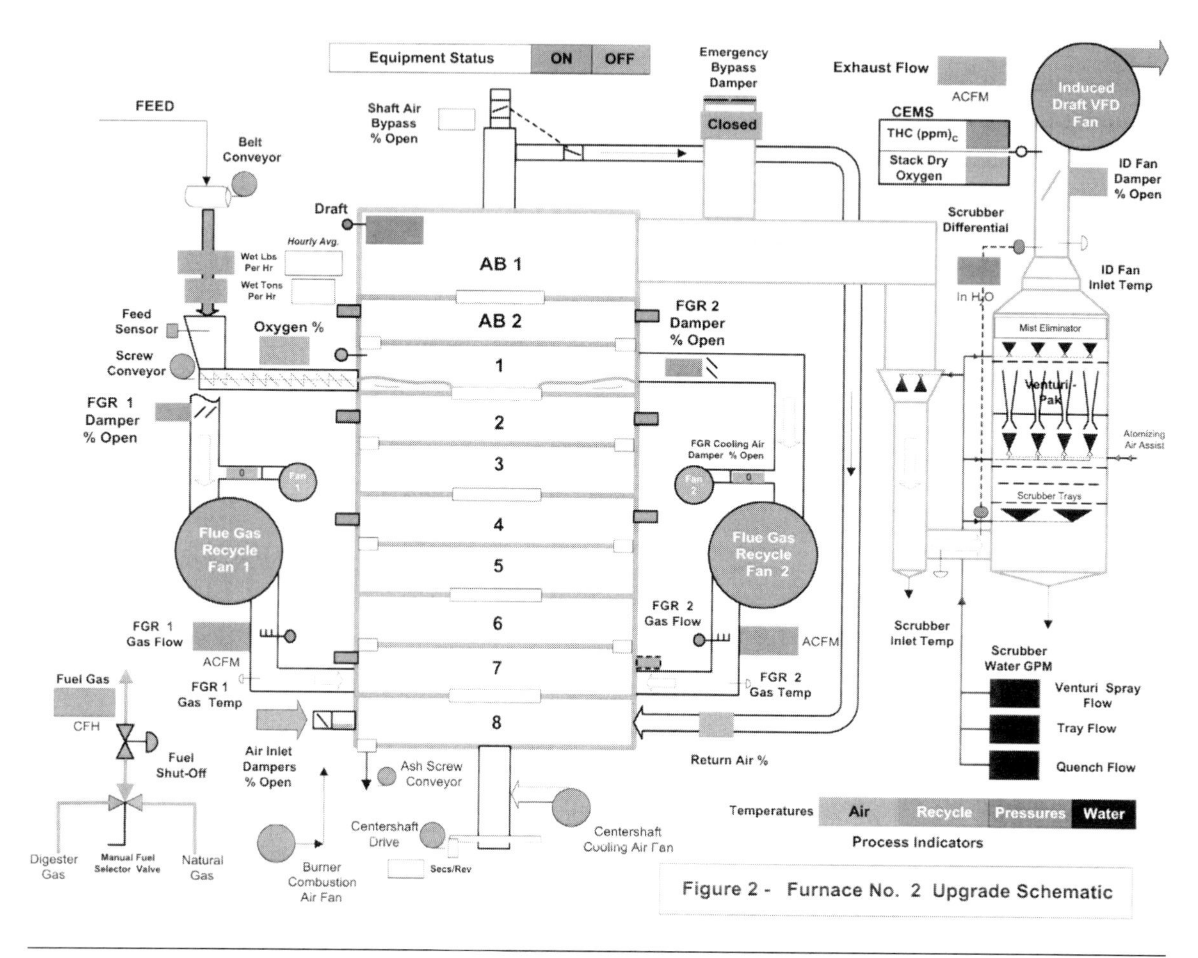

FIGURE 26.9 Schematic of a multiple-hearth furnace with flue gas recirculation (Porter and Mansfield, 2002).

the flue-gas recirculation system to prevent overheating. Flue-gas recirculation offers several benefits (Sapienza et al., 2007):

- More stable furnace operations,
- Lower NO_X emissions (elimination of yellow plume),
- Less slag formation in the multiple-hearth furnace,
- Higher throughput capacity (because of less downtime due to slag removal),
- Reduced THC emissions (because of more stable furnace operations), and
- Complete ash burnout (because recirculated gas raised the temperature in the lower hearths).

4.3 Process Design Guidelines

Because thermal oxidizers should produce minimal air-pollution emissions, fluidized-bed incinerators typically are preferred to multiple-hearth furnaces. For the past few decades wastewater utilities have installed far more fluid-bed incinerators than multiple-hearth furnaces, and even have replaced multiple-hearth units with fluid-bed ones. So, this section focuses on fluid bed incinerators.

4.3.1 *Furnace Sizing Parameters*

Combustion (*burning*) is the rapid union of oxygen with carbon, hydrogen, and sulfur. Complete combustion primarily depends on temperature, time, and turbulence.

4.3.1.1 Temperatures

Feed solids should be introduced to the fluidized bed when the bed's operating temperature exceeds the material's ignition temperature. Ignition temperatures vary. Some compounds (e.g., chlorinated compounds) require temperatures hotter than 980°C (1 800°F) to oxidize. Wastewater solids typically can be completely combusted when the bed temperature ranges from 650 to 760°C (1 200 to 1 400°F) and the freeboard temperature is between about 815 and 870°C (1 500 and 1 600°F).

The bed's huge thermal inventory ensures that feed solids quickly reach the bed temperature, at which point organics volatilize and most VOCs oxidize, releasing heat and maintaining the thermal inventory. The small amount of fixed carbon in solids typically is combusted in the bed. The bed temperature is stable because the large heat inventory and the ease of automatic control ensure that the necessary combustion temperature is maintained.

4.3.1.2 Time

Combustible materials must have sufficient time to react. Fluid-bed thermal oxidizers are designed to allow enough time for feed solids and any auxiliary fuels to react with oxygen in fluidizing air. The bed is designed to completely disintegrate feed solids in a fraction of second and combust some of its volatiles to keep temperatures above 650°C (1 200°F).

Residual volatiles finally burn out in the elevated temperature zone above the bed (the freeboard). With temperatures between 840 and 900°C (1 550 and 1 650°F), the freeboard is designed to completely combust any volatiles that escape from the bed. Typically, gas residence times are 2 to 3 seconds in the bed and 6 to 7 seconds in the freeboard. Although high combustion efficiencies can be achieved at lower freeboard residence times, design engineers should provide enough disengagement height [5 to 6 m (15 to

18 ft)] in the freeboard to reduce sand carryover in exhaust gases, and doing this lengthens freeboard residence times.

4.3.1.3 Turbulence

To optimize combustion (operate efficiently with low excess air), combustible materials and fluidizing air must be mixed so enough of the combustible material's surface area will contact oxygen and react. Turbulence better distributes feed solids in the bed and exposes every particle of the fluid bed to fluidizing air. The highly agitated hot sand quickly fragments feed material into small particles, which in turn are quickly heated to volatilization temperature without lowering bed temperature (because of the sand bed's large heat inventory).

Without turbulence, feed material will be poorly distributed and more volatilized organics will reach the freeboard before being oxidized in the bed. This phenomenon can lead to excessive over-bed burning and, subsequently, higher emissions of hydrocarbon volatiles and other products of incomplete combustion.

A certain amount of freeboard burning is inescapable; however, freeboard-to-bed differentials should not exceed 140 to 170°C (250 to 300°F). The goal should be less than 110°C (200°F) unless temperatures higher than 840°C (1 550°F) are required to completely combust VOCs.

4.3.1.4 Space Velocity

When designing fluid-bed systems, the selection of bed material is critical because particle size directly affects fluidization quality (see Figure 26.10). Thermal oxidizers treating wastewater solids typically use a sand-like material with a median size of 550 μm (30 mesh). At bed operating conditions, such particles have a minimum fluidization gas velocity (U_{mf}) of 0.33 m/s (1 ft/sec). The optimal bed fluidizing gas velocity is between 2.5 and 3 U_{mf}, so the particles' gas spatial velocity (fluidization gas superficial velocity) would be 0.75 to 1 m/s (2.5 to 3 ft/sec). When sizing the bed, design engineers should use gas velocities corrected to bed temperature and pressure. Bed velocities greater than 1.0 m/s (3.0 ft/sec) are not recommended for solids combustion because fluidization is less stable, and bed losses, caused by entrainment, are higher. If higher velocities are used, coarser bed material is required.

Freeboard gas velocity typically ranges from 0.76 to 0.64 m/s (2.5 to 2.1 ft/sec). To minimize sand loss, freeboard gas velocity must be kept as low as possible.

4.3.1.5 Evaporation/Heat Release Limitations

Fluidized-bed furnaces typically burn up to 552 kJ/m^2·s (1.75 × 10^5 Btu/hr/sq ft) of freeboard area. High-velocity fluidized-bed combustors can process up to 1 300 kJ/m^2·s

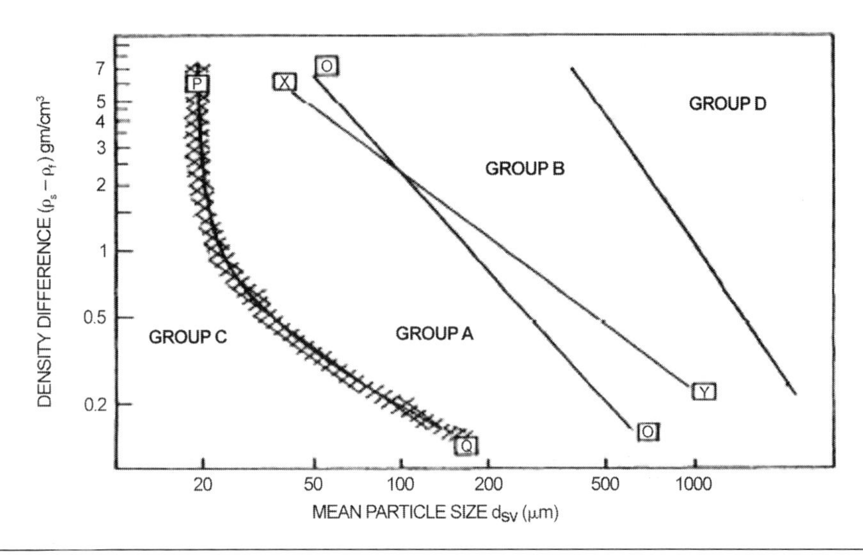

FIGURE 26.10 Geldart classification of powders (Geldart, 1973).

(4.0×10^5 Btu/hr/sq ft) in some installations. Coal-burning boilers, on the other hand, only consume 130 kJ/m^2·s (4.0×10^4 Btu/hr/sq ft) of grate.

The feed material's water content significantly affects thermal oxidizer operations. Water has no heating value but requires a large amount of heat to be vaporized and heated to the oxidizer's operating temperature. This heat must be supplied by combustible materials in the feed solids or by auxiliary fuel.

Changes in water content will affect the thermal oxidation system's operating temperature and capacity. When feed solids are wetter than the design condition, more auxiliary fuel will be required to maintain bed temperature, reducing system capacity. To maintain design capacity and optimize operations, the percentage of total solids in feed cake should be monitored twice per day.

4.3.1.6 Sufficient Air

A fluid-bed thermal oxidizer receives oxygen in the form of fluidizing (combustion) air, which must be supplied at amounts slightly greater than that theoretically required for complete combustion. Operators typically monitor excess air levels measured as the percent of free oxygen in the air released to the atmosphere. Depending on feed solids and combustion temperature, the oxygen content in exhaust gases should be at least 4% by volume on a dry basis (about 2% on a wet basis) just before the gas enters the scrubber.

For efficient operations, the oxygen content in exhaust gases should not exceed 5% (on a wet gas basis). Anything more than 5% will quench (cool) the bed and reduce overall efficiency.

4.3.1.7 Summary

If thermal oxidizer operations violate one or more of the six principles of combustion, then combustion will be poor or incomplete. The following criteria indicate that a fluidized-bed system is being operated inefficiently and, therefore, ineffectively:

- Excess oxygen below prescribed minimum,
- Black and smoky exhaust gases,
- High residual carbon content in bed material,
- High carbon content (black color) in scrubber effluent, and
- Excessive freeboard-to-bed differential temperatures.

4.3.2 Selection and Sizing

Selecting and designing a thermal oxidation system to treat solids is a complex, technical, and highly specialized task. A computer can best handle the iterative heat-transfer calculations. Design procedures include empirical methods that rely on extensive pilot-plant kinetic data. Such information typically is regarded as proprietary and may be patented by the manufacturers who typically perform the mass and energy balance and design calculations after receiving input data, conditions, and functional specifications from the operating agency (see Table 26.3).

A thermal oxidation system should be able to burn dewatered cake and concentrated scum to an inert ash without emitting smoke, flash, or objectionable odors. The U.S. Environmental Protection Agency also typically requires that thermal oxidation systems:

- Combust dewatered cake and concentrated scum satisfactorily within an exhaust gas temperature range of 427 to 760°C (800 to 1 400°F);
- Burn dewatered cake and concentrated scum to a sterile ash that contains no more than 5% combustibles;
- Produce flue gas that contains between 6 and 10% oxygen;
- Will not discharge exhaust gases hotter than 67°C (120°F) to the atmosphere;
- Will not discharge exhaust gases whose opacity exceeds 20% to the atmosphere; and
- Will not emit more than 0.65 g/kg of feed solids (1.3 lb/dry ton of feed solids) of particulate to the atmosphere

4.3.2.1 Furnace Selection

When selecting a furnace, design engineers should consider plant size, solids composition, air emission regulations, plant maintenance, sustainability, and labor requirements. A fluidized-bed furnace typically is feasible for plants with flows more than

TABLE 26.3 Typical parameters for selecting and designing a thermal destruction system for municipal solids(courtesy of Infilco Degremont, Inc.).

Dewatered solids characteristics
- Description of the feed, including its nature, material handling, pumpability and hazardous constituents POHCs if applicable.
- Furnace design capacity [kg DS/h (lb DS/hr)]
- Moisture content (%)
- Volatile solids (%)
- Heating value [kcal/kg (Btu/lb) volatile]
- Bulk density [kg/m^3 (lb/cu ft)]
- Ultimate analysis (volatile basis): concentrations of the following volatiles: C, H, O, N, S, Cl.
- Trace metals: concentration of regulated metals in dry solid in ppm weight

Solids ash chemical compositions, including soluble and insoluble parts of the following elements: Al, Na, K, P, Mg, Ca, Fe, Si, S

Auxiliary fuel characteristics
- Description of the fuel, including its nature, viscosity (if liquid), material handling, pumpability
- Heating value [kcal/kg (Btu/lb) of the fuel]
- Ultimate analysis (volatile basis): concentrations of the following volatiles: C, H, O, N, S, Cl.
- Moisture content (%)

If steam is desired, the following are required:
- Steam temperature [°C (°F)]
- Steam pressure [barg (psig)]

Air pollution-control system
- Maximum pollutant emissions
- Maximum or minimum allowable flue gas temperature

Process water analysis (e.g. flowrate, temperature, alkalinity, and TSS)

Environmental conditions [e.g., seismic zone, elevation, design air temperature and moisture, and design windload (if outdoor)]

3.79×10^4 m^3/d (10 mgd); a multiple-hearth furnace typically is economical when plant flows exceed 7.57×10^4 m^3/d (20 mgd) (U.S. EPA, 1985b). That said, site-specific factors (e.g., plant location, land available for solids reuse or disposal, and air emission regulations) may make the other option more attractive. Fluidized-bed furnaces typically are more economical if high-temperature combustion of exhaust gases is required, because multiple-hearth furnaces require afterburners to achieve high temperatures. A fluidized-bed furnace has lower capital and O&M costs (amortized to 20 years) than a multiple-hearth furnace equipped with RTO (Dangtran et al., 2000).

Since 1988, 53 fluid bed systems and one multiple-hearth system have been installed at North American wastewater treatment facilities. Of the fluid bed installations, 18 replaced existing multiple-hearth furnaces.

4.3.2.2 Equipment Sizing

Design engineers must size thermal oxidation systems based on realistic design loadings. Using overly conservative peaking factors for solids loadings and flows results in oversized systems that fail to operate efficiently during the initial years, when solids quantities and contents are lower than design values. Oversized systems must be operated intermittently, which increases auxiliary fuel costs (due to repeated shutdowns and startups, or maintaining a constant furnace temperature in standby mode). To avoid such problems, design engineers should consider furnace systems that can be modified incrementally to increase capacity, or multiple units that can enter service incrementally as solids production increases.

4.3.2.1 Effect of Dewatered Solids Characteristics on Selection and Sizing

4.3.2.1.1 Moisture Content

Wastewater solids with high moisture contents have complicated solids processing at older plants, particularly those constructed before 1980 (U.S. EPA, 1985b). High moisture contents reduce a furnace's equivalent dry solids throughput capacity and require more auxiliary fuel to evaporate water before and during combustion. Treatment plants that dewater solids via efficient belt presses, recessed plate filter presses, or high-solids centrifuges produce drier cakes. When designing thermal oxidation systems, engineers should consider replacing existing, inefficient dewatering systems with more effective ones.

4.3.2.1.2 Solids Composition

Feed cake properties that affect thermal oxidizers include solids content, the percentage of combustibles (volatile solids plus fixed carbon), the heating value of combustibles, and the presence of chemicals that react endothermically (e.g., lime). Because treatment plant screenings tend to clog feed mechanisms, they are ground or shredded before thermal oxidation. Wastewater solids with high percentages of volatile solids (e.g., grease and scum) have high heating values that can cause localized heat release and slagging, resulting in clinkering and air emissions associated with incomplete combustion. A fluidized-bed furnace can accommodate grease, scum, and other high-caloric materials better than a multiple-hearth furnace because it provides better contact with combustion air.

Grit and chemical precipitates (e.g., lime and ferric chloride) contain a large percentage of inert materials with low heating values and high ash production, so they should not be thermally oxidized unless necessary for odor control. Also, because the heating value of combustibles affects the amount of auxiliary fuel required for complete combustion, engineers should examine ranges of expected values to ensure that the thermal oxidation system will meet present and future needs.

4.3.2.1.3 Inorganic Sludge Conditioning

Conditioning solids with inorganic chemicals can increase O&M costs, fuel consumption, and corrosion. Inert conditioning chemicals produce more ash and promote the formation of metal salts, resulting in slagging and clinkering; this can increase O&M costs. Adding inert conditioning chemicals with low heating values or an endothermic lime reaction probably will increase fuel consumption. Polymers may cause uneven burning in a multiple-hearth furnace (unlike a fluidized-bed furnace, because of its turbulence and huge heat reservoir), leading to the formation of combustion-resistant solids balls. Ferric chloride can produce a chlorine-laden exhaust gas that is extremely corrosive to steel surfaces at high temperatures. To reduce these adverse effects, design engineers should optimize the use of conditioning chemicals, improve the dewatering process, use polymers rather than lime or metal salts, and use ferric sulfate rather than ferric chloride.

Design engineers also should evaluate the solids' chemical composition. Sodium and potassium chlorides have low melting points, so large quantities of them can lead to vitrification of bed media [i.e., the bed media can become sticky, agglomerates (clinkers) can form, bed materials segregate, and eventually the bed de-fluidizes]. Iron, phosphorus, and chlorides can lead to the deposition of iron oxides. Iron oxide scaling can obstruct the exhaust gas duct, potentially resulting in excessive backpressure, operating difficulties, and system shutdown. These problems can be eliminated by chemical addition (Dangtran et al., 1999).

Kaolin clay (a mixture of hydrous aluminum silicates) can neutralize sodium and potassium. It typically is available as a fine powder and is a convenient source of both silicate oxide and aluminum oxide, which react with sodium and potassium chlorides to form crystalline sodium and potassium aluminum silicates. These silicates have a melting point above 1 093°C (2 000°F).

Lime can convert iron phosphate into iron oxides in the sand bed at bed temperature. This conversion prevents iron from forming gaseous iron chloride, which can precipitate and form scales in the freeboard and exhaust gas duct.

To calculate the appropriate dose(s) of such chemical additives, design engineers must perform a complete ash analysis (both soluble and total concentrations of the components). For more information on this analysis, see "Control Problem Waste Feeds in Fluid Beds" (Dangtran et al., 1999).

4.3.2.2 Fuel Optimization

One of the most important criteria in designing a thermal oxidation system is minimizing fuel consumption. Engineers calculate auxiliary fuel needs based on heat and mass balances; the result depends on the feed cake's heat content [specifically, its solids content (percent dry solids)] and the combustion air's heat content (temperature). The feed cake's solids content depends on the dewatering equipment used and the quantity of polymer used as a dewatering aid. The combustion air's temperature depends on how intense heat recovery is in the heat exchanger. Typically, a heat exchanger can recover up to about 50% of the flue gas' available enthalpy to preheat combustion air to about 675°C (1 250°F).

A theoretical curve of supplementary fuel consumption is presented in Figure 26.11. The calculations were based on a combustion gas temperature of 843°C (1 550°F) and a throughput capacity of 454 kg/h (1 000 lb/hr) of dry solids. The feed cake contains 75% volatile solids and a high heating value of 23 260 kJ/kg (10 000 Btu/lb) of volatile solids (values typical for wastewater solids).

The auxiliary fuel requirement drops as the solids content and combustion air temperature increase. When the wind-box temperature is 648°C (1 200°F), feed cake containing 27% solids burns autogenously (i.e., is thermally self-supporting).

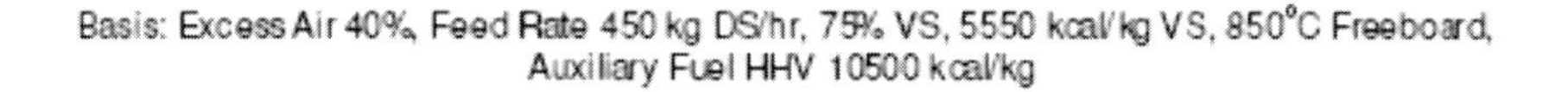

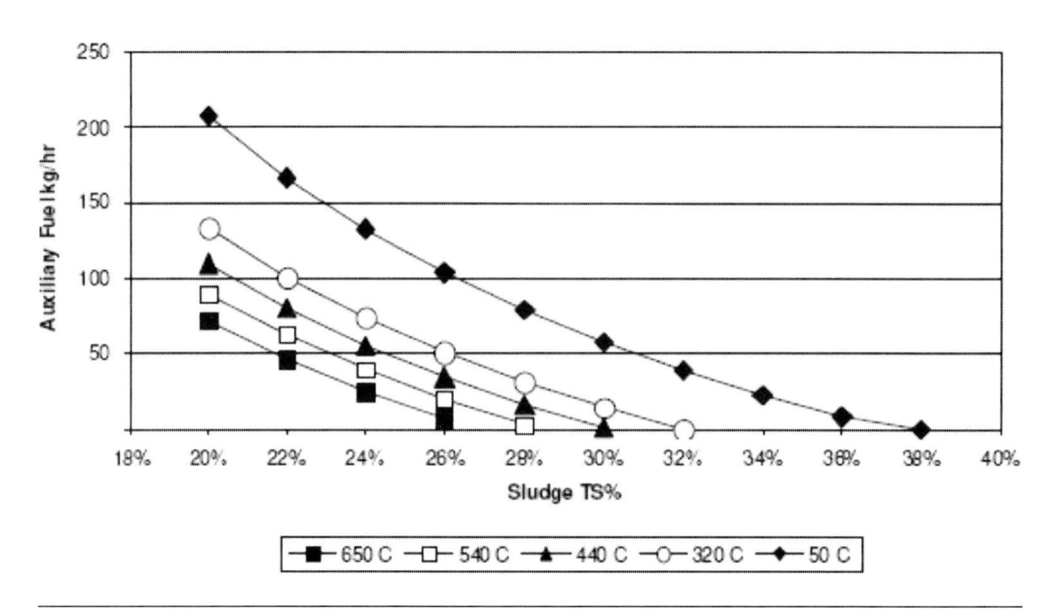

FIGURE 26.11 Theoretical fuel requirement versus cake solids content at various combustion air temperatures (courtesy of Infilco Degremont, Inc.).

Meanwhile, to lower polymer consumption and minimize NO_X emissions [which increase as dry solids increase, according to Dangtran and Holtz (2001)], a thermal oxidizer typically is designed based on autogenous combustion at maximum air temperature and minimum dry solids content.

4.3.2.3 *Air Emission Objectives*

If inadequately designed or operated, thermal-processing facilities for wastewater solids can contribute significantly to air pollution. Two categories of air contaminants associated with thermal processing are odors and combustion emissions. Engineers should design thermal oxidation systems to generate no odors and to meet the requirements of the air permit. Odors are controlled by keeping the furnace's exhaust temperatures above the odor-destruction threshold temperature. This will be site specific. Combustion emissions are controlled by maintaining good combustion efficiency (see Section 4.3.2.2) and by installing appropriate air pollution-control equipment (see Section 7.0).

4.3.2.4 *Mass and Energy Balance*

A *mass balance* applies the law of conservation of mass to the analysis of physical systems. By accounting for material entering and leaving a system, design engineers can identify mass flows that might have been unknown or difficult to measure otherwise. Similarly, an *energy balance* quantifies the energy used or produced by a system. However, while a system may have a closed mass balance, its energy balance may be open. Also, while it is possible for a system to have more than one mass balance, it can have only one energy balance.

Mass and energy balances typically are performed via computer models.

Suppose a fluidized bed furnace is designed to handle 105 dry metric ton/d (115.5 dry ton/d) of solids. The feed cake contains 28% total solids and 68% volatile solids. The volatile solids consist of 55.83% carbon, 7.52% oxygen, 8.13% hydrogen, 27.10% nitrogen, 1.42% sulfur, and 0.1% chlorine. The volatiles high heating value (HHV) is 5 560 kcal/kg (10 007 Btu/lb).

A summary of the mass and energy balances calculated for this system are presented in Table 26.4. All enthalpies are based on a reference temperature of 0°C (32°F). Feed cake provides about 74% of the inlet heat required to sustain combustion; the remaining 26% is provided by preheated air. With a hot wind-box temperature of 654°C (1 210°F) and a feed material containing 28% total solids, no auxiliary fuel is required during normal operations. Most of the exhaust-gas heat is in the water vapor (63%). The rest is in the dry combustion gas (34%), ash (1%), and heat-loss (2%).

TABLE 26.4 Example of a mass and energy balance for a fluid bed furnace (courtesy of Infilco Degremont, Inc.).

Mass and energy input	kg/h (lb/hr)	kcal/h (Btu/hr)	Heat (%)
Cake solids	4 375 (9 645)	16 560 068 (65 710 349)	73%
Cake water	11 250 (24 802)	236 250 (937 440)	1%
Aux fuel	0 (0)	0 (0)	0%
Quench water	33 (74)	701 (2 781)	0%
Purge air (dry)	1 085 (2 392)	12 931 (51 309)	0%
Fluidizing air (dry)	32 630 (71 931)	5 405 955 (21 450 830)	24%
Air moisture	635 (1 400)	574 696 (2 280 394)	2%
Sand	51 (112)	236 (935)	0%
Total in	**50 059** **(110 360)**	**22 790 836** **(90 434 038)**	**100%**

Mass and energy output	kg/h (lb/hr)	kcal/h (Btu/hr)	Heat (%)
Dry flue gases	34 516 (76 095)	7 642 654 (30 326 049)	34%
Water vapor	14 095 (31 073)	14 398 243 (57 132 228)	63%
Ash	1 400 (3 086)	261 800 (1 038 822)	1%
Sand	51 (112)	9 537 (37 843)	0%
Heat losses		478 603 (1 899 095)	2%
Total out	**50 062** **(110 367)**	**22 790 836** **(90 434 038)**	**100%**

4.3.2.5 Operating Schedule and Redundancy

Thermal equipment typically should not be heated and cooled frequently because cycling temperatures increase the material's fatigue and rate of failure. However, fluidized-bed furnaces have minimal heat loss during shutdown [around 5°C/h (10°F/hr)] because of their thick refractory insulation and mass of granular sand. The hot granular bed material can act as a thermal reservoir, retaining the heat during shutdown, so the system can be operated intermittently. Existing fluidized-bed furnace installations have varied operating schedules, ranging from 24 hr/day in some plants to 16 hours/day or 5 days/week. This allows downtime for preventive maintenance without exposing too much of the refractory layer to temperature cycling.

Redundancy is not common at fluidized-bed furnace systems.

4.3.2.6 Ash Handling

Thermal oxidation plants generate two types of ash—wet and dry—so there are two types of ash-handling systems, which are briefly described below. For more information on ash-handling systems, see *Incineration Systems* (WEF, 2009).

4.3.2.6.1 Wet Ash-Handling System

Wet ash-handling systems are hydraulic and remove ash as a slurry. They typically are used in fluid-bed systems with wet scrubbers because most of the ash removed from the process train is wet. In fluid-bed thermal oxidizers, ash slurry is drained from the bottom of the wet scrubber and conveyed (via pump or gravity) to a lagoon or to mechanical thickening and dewatering equipment.

4.3.2.6.2 Dry Ash-Handling System

Dry ash-handling systems may be mechanical or pneumatic. Pneumatic systems may be either pressure or vacuum and either dilute- or dense-phase. They are suitable for fluid-bed thermal oxidizers with waste-heat boilers, fabric filters, or electrostatic precipitators because most of the captured fly ash is dry. Bottom ash systems may need a grinder to ensure that ash can be transported effectively and protect downstream conveyance equipment. Dry ash typically is kept in storage bins until it can be hauled offsite. While being discharged from bins into disposal trucks, the ash typically is conditioned (wetted with water) to reduce fugitive dust during loading. Then, it is transported to the ultimate use or disposal site.

4.3.3 Electricity Requirements

Suppose a fluidized-bed incinerator treats 100 dry metric ton/d under positive pressure, with primary heat recovery for air preheat and a Venturi scrubber. It would have an operating horsepower of 684 kW (917 hp) and, therefore, consume 164 kW/dry metric ton of electricity (see Table 26.5). The fluidizing air blower would use the most power.

Suppose instead that this system included superheated steam production via a water-tube waste-heat boiler and associated boiler feed-water, condensate, and water treatment pumps (see Table 26.6). The boiler also had a bottom screw to remove ash

TABLE 26.5 Horsepower list for a 100-dry metric ton/d fluidized bed furnace with push system, primary heat exchanger, and wet scrubber (courtesy of Infilco Degremont, Inc.).

Equipment	Time operation (%)	Motor horsepower	Average operating horsepower
Air			
Fluidizing blower	100	700	626
Preheat burner air blower	0	20	0
Oil injection blower	100	20	17
Gas injection blower	100	20	17
Sand air compressor	20	15	2.5
Instrument air compressor	100	40	34.5
Oil			
Preheat burner pump	0	2	0
Bed injection pump	10	2	0.1
Water			
Roof spray pump	20	20	3.4
Scrubber booster pump	10	20	1.7
Solids			
Hopper sliding frame	100	20	17
Hopper extraction conveyor	100	20	17
Feed pump	100	200	177
Pipe lube pump	100	5	3.7
Total operating horsepower			917 hp (684 kW)

from the economizer and boiler, as well an induced draft fan to keep negative pressure in the boiler and avoid ash leakage. Then, it would use 50% more electricity (244 kW/dry metric ton). However, if some of the generated steam were used in a turbine to drive the fluidizing air blower, its electricity consumption would drop to 141 kW/dry metric ton.

4.4 Design Practice

4.4.1 Feed Equipment and Systems

Continuous, even transport and distribution of feed cake to the furnace is important. Past practice was to use screw extrusion feeders for dry cake and progressing cavity pumps for wet cake. Now, hydraulic piston or progressing cavity pumps typically are used to convey cake from the dewatering equipment to the furnace. A piston pump is preferred because of its flexibility (insensitivity to feed solids quality).

TABLE 26.6 Horsepower list for a 100-dry metric ton/d fluidized bed furnace with push system, primary heat exchanger, and wet scrubber (courtesy of Infilco Degremont, Inc.).

Equipment	Time operation (%)	Motor horsepower	Average operating horsepower
Air			
Fluidizing blower	100	600	536
Induced draft fan	100	500	447
Oil injection blower	100	20	17
Gas injection blower	100	20	17
Sand air compressor	20	15	2.5
Instrument air compressor	100	40	34.5
Oil			
Preheat burner pump	0	2	0
Bed injection pump	10	2	0.1
Water			
Roof spray pump	20	20	3.4
Scrubber booster pump	10	20	1.7
Solids			
Hopper sliding frame	100	20	17
Hopper extraction conveyor	100	20	17
Feed pump	100	200	177
Pipe lube pump	100	5	3.7
Steam			
Boiler feed water pump	100	75	66
Boiler sootblowers	40	0.17	0.2
Chemical feed pump	100	1.5	0.8
Chemical feed mixer	20	1.5	0.2
Condensate pump	100	15	12
Dearator recirc. pump	100	7.5	6
Water treat feed pump	100	7.5	6
Water treat conc pump	100	.75	0.3
Ash screw conveyor	100	1	0.4
Total operating horsepower			1365 hp (1 018 kW)

Two types of solids-feeding systems can be found in the literature: over-bed and in-bed. In over-bed feeding, solids either drop via gravity or are air sprayed onto the bed from the freeboard sidewall or furnace roof. In in-bed feeding, solids are conveyed at high pressure directly into the sand bed about 1.2 m (4 ft) below the bed surface.

Over-bed feeding is simple but prone to bypassing. Solids can end up unburned in the exhaust, leading to excessive fuel use in the bed, explosions in the freeboard, or explosions in the heat-recovery and air pollution-control equipment. This method typically is used in other applications (e.g., fluid-bed boilers burning coal or another solid-waste fuel); a cyclone at the exhaust end of the furnace recycles unburned carbon back to the bed.

In-bed feeding typically is used for thermal oxidation of solids because the combustion process is slower and in two stages (evaporation and combustion). The feeding location [1.2 m (4 ft) under the bed surface or 30 cm (1 ft) above the distributor] ensures that the maximum possible SRT is obtained before solids particles reach the bed surface, where carryover may occur. Solids should release their maximum energy to the sand bed, to counteract the quenching effect due to water evaporation.

The number of feed points needed to evenly distribute solids throughout the bed depends on the furnace's diameter. Feed ports typically are arranged in pairs (e.g., two or four).

4.4.2 Process Train Equipment

Fluid-bed thermal oxidation systems could be located indoors or outdoors. Puerto Nuevo, Puerto Rico, has an outdoor system that consists of a hot wind-box fluid bed, a heat exchanger to preheat combustion air to about 675°C (1 250°F), a quench section followed by cooling tray and multiple Venturi scrubber, a wet electrostatic precipitator, and a stack (see Figure 26.12).

The storage and feed system typically consists of a live-bottom bin and piston pumps. The thermal oxidation process could be either a hot or cold wind-box. The heat-recovery system could consist of heat exchangers (to preheat combustion air and suppress plumes), a waste-heat boiler (to generate steam), or both. The air pollution-control system could be either a dry-ash or wet-ash system.

Common practice in the United States has been to provide a fully positive combustion and gas-cleaning system, from the windbox to the stack. European practice typically involves a "push and pull" concept, in which the reactor's freeboard is under negative pressure and the zero-pressure point is in the top portion of the bed. There are advantages to each approach.

The fully pressurized system, in which a fluidizing blower provides total pressurization and gas flow, is a simpler and less costly design. It probably is best used in situations where the scrubbing systems have high headloss [e.g., a Venturi scrubber and tray cooler producing pressure differentials of up to 10 kPa (40 in.) of water column]. A disadvantage of this system is that the freeboard and top feed systems must be completely sealed to prevent hot gases and ash from discharging into the combustion working area. So, the reactor can only be fed at the top.

FIGURE 26.12 Puerto Nuevo, Puerto Rico's fluid bed incinerator system, which processes 2 415 kg/h (5 325 lb/h) of dry solids (courtesy of Infilco Degremont, Inc.).

The push-and-pull system initially was used for top-feed fluidized-bed reactors. Controlled air leakage can help cool the feeder without disrupting fluidization if a fluidizing air blower provides the air. The induced draft fan must be paced with the fluidizing air blower and the amount of air leakage into the reactor freeboard to maintain a negative pressure of about 0.5 kPa (2 in.) of water column. Design engineers should keep in mind that fluidized-bed reactors used to combust solids are prone to bed surging, which causes pulsations in freeboard and bed pressure. These, in turn, can cause backflow unless the feeder is adequately sealed.

A typical process flow diagram of a fluid-bed thermal oxidation system is shown in Figure 26.13. Solids are dewatered by belt filter presses and then transported by piston pumps to the furnace, where they enter through two or four feed ports. This thermal oxidation system is a wet-ash system with a hot wind-box furnace and heat recovery via heat exchangers. The hot wind-box furnace has a refractory arch that supports the sand bed and evenly distributes air. Makeup sand may be fed to the bed pneumatically during operation, if required. The reactor has an expanded freeboard so the larger particles can decelerate to minimize sand carryover and maximize carbon burnout. The

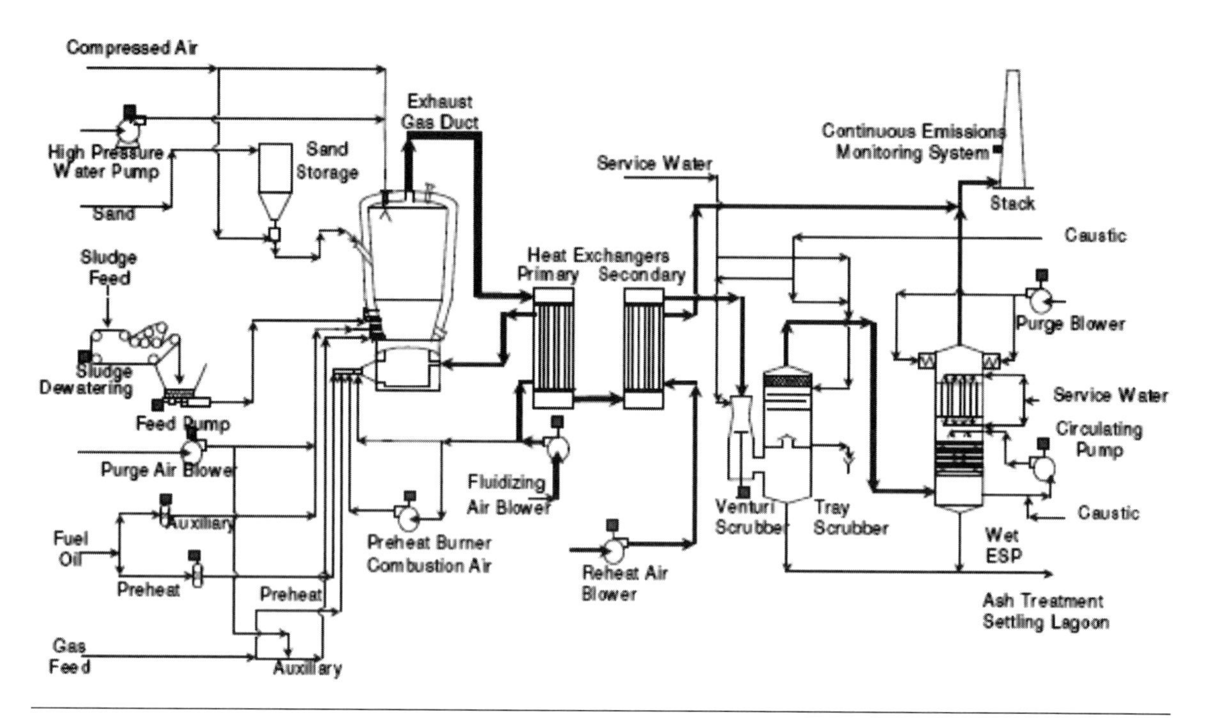

FIGURE 26.13 Typical process flow diagram for a wet ash system (courtesy of Infilco Degremont, Inc.).

freeboard operates at a design temperature of about 843°C (1 550°F). No. 2 fuel oil or natural gas is used as auxiliary fuel during startup and operation as needed. To minimize its use, fluidizing air is preheated to about 675°C (1 250°F) in an external tube-and-shell heat exchanger heated by the reactor's exhaust flue gas.

The air pollution-control system includes a Venturi scrubber followed by a tray tower. A wet electrostatic precipitator could be used to eliminate submicrometer particulate matter. The Venturi scrubber, which has a high pressure drop, removes ash and fine sand particles from flue gas, creating an ash slurry that is sent (via pump or gravity) to an outdoor ash-settling lagoon for dewatering. Dry ash (about 50% total solids) is removed from the drying lagoon about once per month, depending on the size of the lagoon. Meanwhile, hot air [260°C (500°F)] can be added to the stack gas to suppress plumes. This air is preheated in a secondary heat exchanger using exhaust flue gas from the primary heat exchanger.

In the wet-ash system, acid gases (e.g., SO_2 and HCl) are removed by water in the Venturi scrubber and cooling tray. These gases are soluble in water, which means that up to 95% of the acids can be removed by effluent alone. To meet stricter limits, a caustic solution can be added to the cooling tray to further reduce acid gas concentrations. Mercury and dioxins can be removed from flue gas via an activated carbon adsorption system installed before the stack.

More than 90% of North American installations currently use the wet-ash system because of its simplicity and the availability of effluent and space at the treatment plant. However, other types of heat recovery (e.g., waste heat boilers and dry air pollution-control systems) could be used. In the dry-ash system (Figure 26.14), the flue gas temperature has to be in the range of 150 to 205°C (300 to 400°F) before entering a fabric filter (or dry electrostatic precipitator). A waste heat boiler or an economizer can be installed between the fluid bed or heat exchanger and the air pollution-control system to generate steam or hot water. To remove acid gases, mercury, and dioxins, chemical sorbent can be injected into the duct or into a reactor chamber installed between the heat-recovery and air pollution-control systems.

4.4.3 *Fans and Blower Equipment*

A thermal oxidation system needs three types of air: fluidizing air, purge air and atomizing air.

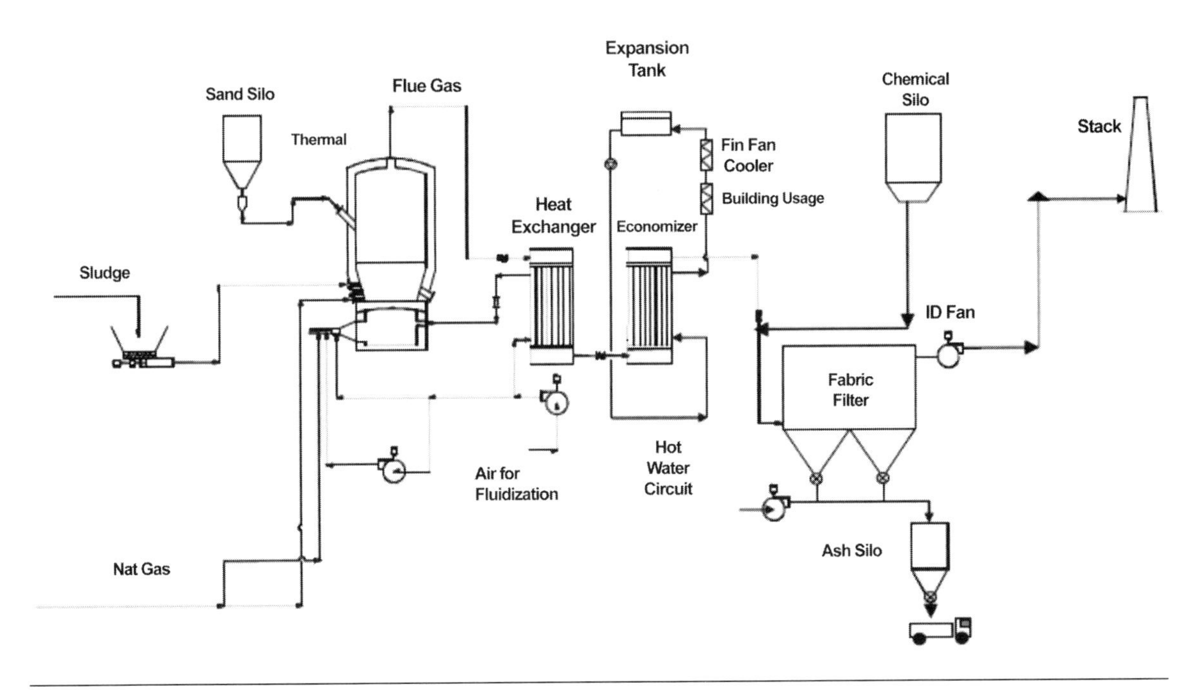

FIGURE 26.14 Typical process flow diagram for a dry ash system (courtesy of Infilco Degremont, Inc.).

4.4.3.1 Fluidizing Air

To calculate the amount of fluidizing (combustion) air needed, design engineers typically need to know the maximum amount of solids and fuel heat anticipated and add an average excess air level of 30 to 50%. They also typically add a 10 to 15% peaking (safety) factor.

Fluidizing air typically is supplied by a multi-stage centrifugal blower. It typically is made of stainless steel or aluminum, and designed with several stages to provide the required outlet pressure. Thermal oxidation capacity is regulated by the amount of fluidizing air entering the wind-box. This airflow is controlled by a damper on the blower inlet side and measured by a flow meter.

If the thermal oxidizer uses a wet-ash system that is entirely under positive pressure, then the air pollution-control system must be completely airtight. If it uses a dry ash system under both positive and negative pressure (i.e., a push–pull system), then an induced draft fan is needed to ensure that negative pressure is maintained in the air pollution-control system (e.g., fabric filters) and avoid leaking dry ash. Induced draft fans typically are heavy-duty industrial centrifugal fans made of stainless steel or higher grades of alloy (e.g., Hastelloy C). The fan typically is controlled by a variable-speed drive or variable-inlet vane damper.

4.4.3.2 Purge Air

Purge air is used to keep all thermal-oxidizer ports, ductwork expansion joints, and ductwork pressure taps cool and free of sand and ash. It can be either high-pressure or low-pressure air. For example, a compressed air system can supply high-pressure air to keep roof spray nozzles and pressure ports cool and free of deposits. A fluidizing air blower can provide air for the annular sleeve of all ports, including site ports; bed oil guns; sand inlet and outlet; solids inlet; and roof sprays.

4.4.3.3 Atomizing Air

If the auxiliary fuel is fuel oil, then about 0.48 to 0.55 bar (7 to 8 psi) of atomizing air at is required. Atomizing air can be provided by an injection-purge air blower, which typically is a lobe-type blower.

4.4.4 Other Auxiliary Equipment

4.4.4.1 Auxiliary Fuel System

Auxiliary fuel is used in a preheat burner during startup operations. It also is directly injected (via fuel guns) into the fluidized sand bed during normal operations. A wide variety of fuels (e.g., coal, sawdust, and digester gas) can be used, as long as the fuel can be fed to the system in a reliable, controlled manner. The most commonly used auxiliary fuels are natural gas and No. 2 fuel oil.

4.4.4.2 Preheat Burner

When starting up a cold thermal oxidizer, operators need a preheating method (e.g., a burner that heats the fluid bed to a temperature at which fuel injected into the fluidized bed will ignite). In a hot wind-box system, preheating is done by a standard industrial oil (or gas) burner placed in the sidewall of the wind box. Fluidizing air from the heat exchanger (preheater) is thus heated by mixing with the hot gases from the burner, and the resulting hot air then fluidizes the sand bed. Heat from the air is then intimately transferred to the fluidizing sand.

In a cold (warm) wind-box system, the preheat burner is in the freeboard. It transfers heat from the gas to the bed less efficiently, so more time and fuel are needed to get the unit to operating temperatures. The preheat burner gets its air supply from the outlet side of the fluidizing air blower. The air pressure is about 0.13 bar (2 psi) above the pressure inside the wind box.

Preheat burners should come with the burner-management package required by process and local insurance regulations. Operators manually adjust the fuel's flowrate from the control panel. Flow adjustments are based on required temperature and firing rates.

4.4.4.3 Bed Fuel Injection

During startup and other modes of operation when there are not enough solids for autogenous combustion, an auxiliary fuel (e.g., No. 2 fuel oil, natural gas, or digester gas) must be added to the system. It can be added via the wind-box preheat burner or via guns directed into the fluidized sand bed.

Fuel oil can be injected into a working fluidized bed via parallel, positive-displacement rotary-gear pumps driven by a common variable-speed drive. An automatic temperature-control loop senses bed temperature and adjusts the pumping speed accordingly. Oil is injected by oil guns installed at the periphery of the bed about 30 cm (1 ft) above the distributor. To prevent fouling, the oil is mixed with purging air at the supply end of the gun, and blown into the bed in the form of a coarse mist. Purge air is provided by a blower. Fluidizing air can be used to cool and continuously purge the gun sleeves. Each oil-injection nozzle can be shut down and withdrawn for cleaning and inspection during system operations or shutdown periods.

Natural gas is directly injected into the bed via gas guns. To maximize in-bed combustion (and avoid overbed burning), the gas must be distributed homogenously throughout the sand. The guns must be located in the bottom against the distributor and deliver gas throughout the sand bed in different lengths. For safety reasons, gas should only be fired into beds with temperatures higher than 730°C (1 350°F), so some treatment plants use the wind-box gas burner to control the bed temperature instead.

It is important to maintain airflow in oil or gas guns when they are inserted into the bed, so the air-supply line should be equipped with a flow indicator.

4.4.4.4 Water System

To protect the heat exchanger from excessive temperatures, the fluidized bed has water quench spray nozzles that are installed through the roof of the reactor above the freeboard. The high-pressure nozzles [about 26.7 bar (300 psi)] create a fine water mist that will evaporate and quench exhaust gas as fast as possible. The nozzles are used in sequence, depending on the temperature of the gas at heat-exchanger inlet. Evaporation occurs close to the freeboard exhaust-gas duct to limit the cooling effect to flue gas leaving the thermal oxidizer.

The spray nozzles have very small orifices and swirl grooves. To maintain clear internal passages and cool them when not in use, the nozzles are purged with air from the compressed-instrument air system at a pressure of about 4.1 bar (60 psi). Both air and water supplies have check valves to prevent one medium from backflowing into the pipes of the other.

The water supply system also includes a water pump, pressure regulator, filter, and a relief valve.

4.4.4.5 Sand System

The fluid bed typically is filled with a sand-like material to a static depth of 0.9 m (3 ft). During operations, fluidizing air will expand the bed material to about 1.5 m (5 ft). The sand will become abraded over time, and makeup sand will be required. Makeup sand typically is pneumatically fed to the furnace by dense-phase conveyors during normal operations. (The reactor typically is equipped with a pneumatic sand makeup system and a bed drain system.)

Because the furnace is sized based on gas flowrate, the hydrodynamics of the fluidized layer depends on the size and density of the media. The media should have a bulk density of about 1 601 kg/m^3 (100 lb/cu ft) and a particle size distribution similar to that shown in Table 26.7.

It must be angular, dry, and free of sodium and potassium. It also must not grind into fines at 871°C (1 600°F) or fuse at 982°C (1 800°F).

Two types of fine bed media can be used: silica or olivine sands. The choice of media depends on feed solids quality. Silica sand is less expensive but more abrasive than olivine sand. Olivine sand is not desirable when

- Feed solids contain grit, which can build up in the bed until it is too deep and excess material must be removed, or

Table 26.7 Typical sand particle size analysis (courtesy of Infilco Degremont, Inc.).

Particle size [μm (U.S. Mesh)]	Distribution (%)
2 380–841 (8–20)	0 to 20
841–500 (20–30)	10 to 30
500–350 (30–40)	20 to 25
350–295 (40–50)	20 to 25
295–210 (50–70)	0 to 5

- Feed solids contain high alkali metals, which can accumulate on the bed and lead to low melting eutectics.

4.4.5 Materials of Construction

The choice of construction materials for fluidized-bed furnace and ancillary equipment depends on the composition of both the solids and flue gas. Although mild steel typically is appropriate, stainless steel or special alloy steel are highly recommended for equipment that directly contacts the flue gas. The furnace's shell is carbon steel ASTM-A36. Design engineers should carefully consider the thickness of this steel, which can range from 0.375 to 0.5 in., depending on the unit's size.

Refractory materials are selected in accordance with three basic criteria:

- The material's interior face must be strong and wear-resistant enough to withstand temperature and abrasion in the three zones of the vessel,
- The material must be backed by an insulating material that can sustain solids temperatures within accepted norms, and
- The refractory layers must be designed and installed to comply with gas-tight parameters.

The third criterion is particularly critical in the fluidized-bed zone of a sand bed reactor, where the pressure is highest. If this zone is not constructed to be gas-tight, gas leaks can occur and form pockets behind the refractory layer. The pockets fill with combustible mixtures that can ignite and cause "hot spots" on the shell. Both the reactor and any accompanying nozzles should be fitted with gas baffle rings. Also, the entire vessel's refractory lining should be designed with enough expansion provisions to accommodate the thermal stresses developed by the shell or refractory layer.

An air preheater typically has a severe operating environment: hot, corrosive, and erosive. The material used for the heat-exchanger tubes depends on the concentration of hydrochloric acid in the flue gas. Type 300, Alloy 20, and Alloy 625 stainless steel have been used in past projects. Experience indicates that stainless steels have considerably more problems when chloride levels in the flue gas reach about 100 ppm; such problems become progressively worse as chloride levels further increase. Intermediate alloys (e.g., Alloy 20, 800H, and Type 825) work when chloride levels are above 100 ppm. Alloy 625 is required if chloride levels exceed 1 000 ppm.

Hot tube sheets follow the same pattern and should be compatible with welding the tubes to the sheets. The lower tube sheet typically is made of carbon steel. It is sufficiently cool and protected from flue gas exposure by refractory and insulation layers, so failures there are rare.

Expansion joints now typically are furnished with Alloy 625 bellows, whose high nickel content effectively resists corrosion cracking from chloride stress. Although the alloy is costly, the thin bellows only use a small amount, and the extra cost is easily justified when compared to the cost of a typical heat exchanger and the longer life it has when this alloy is used.

4.4.6 *Process Control*

The control system is designed with alarms and interlocks to ensure safe operation. Interlocking typically is based on the following fundamental philosophies. No combustion operation can be started until the various safety checks (e.g., airflow rates as per design condition, water flowrates to Venturi scrubber) are clear. The control system typically consists of PLC and personal computer with screen control monitor as interface. All process information recorded by the instrument and control equipment is displayed on the operator's graphics computer screen for plant monitoring.

For the safety of the operation, the plant is fitted with temperature elements (thermocouples) that give control signals to various combustion-control loops associated with thermal oxidizer operation. The thermocouples also determine the bed temperature span, which is an indication of fluidization quality. The thermal oxidizer also is fitted with pressure taps. Differential pressures in the bed indicate bed height, and are also used to monitor the quality of fluidization. A wide span of bed-pressure differential typically indicates a well-fluidized bed. Thermocouples and pressure taps also are used in the heat recovery and air pollution-control systems. Water flow and airflow are measured by a mass flow meter.

Fluidized bed exhaust is supplied with an oxygen sampling and monitoring system to help operators monitor combustion and function as a source of interlocks and alarm.

The operation of the fluid bed and its performance depend on the feed rates of the three major flows (air, solids, and supplemental fuel) to the thermal oxidizer. While airflow can be constant, the solids composition—especially solids content—and, therefore, cake feed rate varies with time. Variations in solids feed rate is the major reason operators must observe and control the process continuously. The control is simple and based on only two parameters—temperature and excess air (or oxygen)—that are continuously monitored.

4.4.7 *Safety*

When designing safety measures for a thermal oxidation system, the first priority is to protect plant personnel, contractors, and visitors. The second priority is to protect the thermal oxidation system itself (i.e., its equipment and structures). Also, while the following safety issues are particular to thermal oxidation, this system cannot be considered separately from overall plant safety. Adequate thermal oxidation safety procedures depend on a strong overall plant safety program, which creates a culture that emphasizes safety in all decisions and procedures.

4.4.7.1 *Regulations, Codes and Standards*

The design and operation of thermal oxidation systems are governed by numerous federal, state, and local regulations, codes, and standards. Some are guidelines, while others are enforceable. In all cases, engineers should consider the underlying safety principles and incorporate them into the design.

4.4.7.1.1 Occupational Safety and Health Standards

The Occupational Safety and Health Act established general national *Occupational Safety and Health Standards* (29 CFR 1910), which apply directly to privately owned wastewater treatment facilities. It is common practice for publicly owned treatment facilities to meet these standards also. Some of the items that apply to thermal oxidation systems include

- Walking–working surfaces,
- Means of egress,
- Occupational health and environmental control (ventilation and noise exposure),
- Personal protective equipment,
- Permit for entry to confined spaces,
- Control of hazardous energy (lockout/tagout),
- Fire protection,
- Machinery and machine guarding,
- Electrical, and
- Fall protection.

4.4.7.1.2 Building, Fire, and Mechanical Codes

Local codes typically include safety requirements that apply to thermal oxidation systems. Code requirements are location-specific, but the model codes listed below illustrate the types of codes that would apply to thermal oxidation systems:

- *International Building Code,* which addresses egress requirements for thermal oxidizer rooms;
- *International Fire Code,* which addresses thermal oxidizer requirements related to fires and means of egress for thermal oxidizer rooms; and
- *International Fuel Gas Code,* which addresses commercial–industrial thermal oxidizers constructed and installed in accordance with NFPA 82, gas-piping installation requirements (e.g., sizing, materials, support, shutoff valves, and flow controls).

These model codes were issued by the International Code Council.

4.4.7.1.3 National Fire Protection Association

The National Fire Protection Association (NFPA) maintains codes and standards that can apply to various aspects of fire and explosion safety for thermal oxidation equipment:

- *Flammable and Combustible Liquids Code* (NFPA 30), which applies to the storage, handling, and use of liquid fuels, including fuel oil used as auxiliary fuel for thermal oxidizers;
- *Standard for the Installation of Oil-Burning Equipment* (NFPA 31), which applies to the installation of stationary oil-burning equipment and appliances;
- *National Fuel Gas Code* (NFPA 54), which addresses safety issues (e.g., piping materials, operating pressures, over- and under-pressure protection, need for shutoff valves, and combustion air) relating to the design, sizing, and installation of gaseous fuel systems for thermal oxidizers that use natural gas, propane, or other similar fuel;
- *Standard on Thermal Oxidizers and Waste and Linen Handling Systems and Equipment* (NFPA 82), which applies to the installation and use of thermal oxidizers; [Although seemingly directed at thermal oxidizers burning solid waste, the explanatory material notes that there are many types of thermal oxidizers burning a wide range of wastes, so the standard is not intended to address all design details for each thermal oxidation technology. However, it includes many requirements that seem applicable to solids thermal oxidizers (e.g., requirements for auxiliary fuel, air for combustion and ventilation, thermal oxidizer design, placement, and clearances).]
- *Standard for Ovens and Furnaces* (NFPA 86), which applies to ovens, dryers, or furnaces that process industrial materials at about atmospheric pressure; [Although it contains no direct reference to combustion of wastewater solids, this standard typically is applied to municipal thermal oxidizers and used to specify the safety equipment and practices (e.g., purging requirements) associated with using fuel-firing burners.]

- *Fire Protection in Wastewater Treatment and Collection Facilities* (NFPA 820), which establishes minimum requirements for preventing and protecting against fire and explosions in wastewater treatment facilities. This code addresses hazard classifications for specific processes (e.g., thermal oxidizers) and addresses the buildings in which they are housed as follows:
  - Does not require ventilation in the thermal oxidizer area because ventilation typically is provided for other purposes (e.g., heat removal);
  - Requires an unclassified area for electrical equipment;
  - Requires limited-combustion, low-flame-spread, or noncombustible building materials; and
  - Requires a fire-suppression system (e.g., sprinklers).

4.4.7.1.4 Insurance and Other Industry Standards

Insurance standards often are more stringent than local codes or NFPA regulations and should be considered when selecting and designing safety provisions for thermal oxidation systems, especially when particular hazards are present. When designing the fuel valve train and burner safety systems, for example, engineers should identify the relevant insurance standards, including those issued by Industrial Risk Insurers (IRR) or Factory Mutual (FM), and industry standards, including those issued by Underwriters Laboratories (UL). They also should identify the wastewater treatment facility's insurance carrier and its requirements to ensure that the design will meet them.

4.4.7.2 Thermal Oxidizer Safety Considerations

Thermal oxidation involves high temperatures, fuel supply and combustion, and solids combustion. They should be designed and operated with adequate safety features and procedures to address the related hazards.

4.4.7.2.1 Hot Equipment Surfaces, Personnel Protection

Thermal oxidizers and some of their exhaust breeching are refractory-lined (insulated) but must operate at surface temperatures above 60°C (140°F). The reactor walls are often designed to keep shell temperatures above 100°C (212°F) to prevent condensation from exhaust leaks to the inside wall of the metal. To protect facility personnel against burns from such hot surfaces,

- They should have protective gear (e.g., gloves, clothing, and eye shields) while operating or servicing a hot thermal oxidizer;
- Ducts and equipment surfaces should be insulated, where possible, to keep surface temperatures at 60°F (140°F) or lower while operating in an ambient temperature of 32°C (90°F) or higher; and

- Barriers (e.g., expanded metal shields or barrier fences with locked gates) should be used to protect personnel from contact with hot ducts and equipment, and prevent unauthorized access.

4.4.7.2.2 Fuel Safety Provisions

Thermal oxidizers use an auxiliary fuel (e.g., oil or natural gas) to heat the unit during startup and to supplement operations when needed. Fuel supply and combustion systems should include the following safety features:

- Supplemental fuel-supply piping whose size, materials, configuration, support, shutoff valves, and pressure and flow controls are in accordance with applicable standards; and
- Supplemental fuel-safety systems whose components (e.g., gas conditioner, burner, pilot, ignition, flame monitor, combustion air-pressure monitor, fuel-pressure control and monitor, emergency-fuel shutoff, venting, and purging equipment) are in accordance with applicable standards.

4.4.7.2.3 Fire and Explosion Protection

Fluid-bed thermal oxidizers have a good safety record in preventing fire and explosions. When designing thermal oxidizers, engineers should make sure that

- Solids-handling systems are configured to reduce the risk of spills and accumulation of solids that could dry and produce combustible dust;
- The preheat burner is purged before startup;
- Instrumentation (e.g., temperature, pressure, and feed monitors) is adequate;
- All equipment is sized and controlled to ensure proper combustion conditions;
- Flue-gas ductwork and equipment are designed to prevent exhaust gas from leaking into the building;
- Dewatered cake combusts under slightly positive pressure in a reactor vessel that meets the applicable structural and welding standards for the pressures to be encountered;
- Natural gas, fuel oil, or other fuels cannot accumulate in the reactor—particularly when cooled—because appropriate interlocks, equipment (e.g., block and bleed valves on fuel-supply lines and removable oil lances), and operating practices are in place;
- The reactor's refractory layer is installed to prevent pockets from developing between it and the shell (such pockets could allow combustible gases to accumulate and cause minor explosions, or permit condensate to collect on the interior of the shell and promote corrosion).

If the thermal oxidizer is a multiple-hearth furnace, design engineers also should make sure that:

- The system has an emergency bypass damper and ductwork to vent combustion gases directly to atmosphere, if the power fails and
- The thermal oxidizer is purged before the burners are started.

4.4.7.2.4 Hazard and Operability (HAZOP) Reviews

Some North American and European industries use hazard and operability (HAZOP) reviews to identify major operability problems and significant hazards to health, safety, and the environment in a facility design. Once potential problems have been identified, they must be resolved, mitigated, or eliminated via design changes.

HAZOP reviews also are beginning to be used for North American solids-processing systems. This is a systematic, structured method for identifying potential safety problems and taking appropriate mitigation measures to reduce the risks. HAZOP reviews can be performed during all stages of design, from conceptual design through completion of final contract documents. They also may be used on existing facilities. Ideally, though, the review should be timed so design changes can be made before construction begins.

HAZOP reviews typically proceed as follows:

- The HAZOP review should be facilitated by an individual qualified in HAZOP reviews. The HAZOP review team should include, as a minimum, process design engineers, discipline design engineers, operators, construction engineers, safety engineers, and an owner's representative.
- The team determines what will be reviewed (e.g., the entire system, a part of it, or specific items of equipment) and gather the drawings (e.g., process and instrumentation diagrams) needed to identify nodes for study and chart the review's progress.
- The team selects parts of the facility (nodes) for review, and reviewers evaluate them using parameters (e.g., flow, temperature, pressure, and level) and deviation guidewords (e.g., more, less, obstructed, reverse, higher, or lower) to describe the causes and effects of identified deviations;
- Reviewers discuss the consequences of deviations; identify safeguards (anything that could prevent or alleviate the consequences); request more information, if necessary; and recommend appropriate modifications (e.g., design changes).
- The review is complete when all nodes have been examined.

After completing the review, engineers evaluate the identified safeguards to determine which should be incorporated into the design. They must follow up on reviewers' recommendations to ensure that concerns are mitigated. If design changes are recommended, a follow-up HAZOP review should be considered.

The primary benefit of HAZOP reviews is that they raise more awareness among designers, owners, equipment suppliers, and operators of the potential risks associated with process operations. They force all parties to have a thorough discussion of all facility functions, helping them understand facility operations and related safety issues before construction is completed.

4.4.8 Heat Recovery and Use Opportunities

Hot flue gas is the chief source of recoverable energy in thermal oxidation systems. This gas contains most of the heat energy added to the system by feed cake, auxiliary fuel, and combustion air.

Heat-transfer technology is either direct or indirect. In *direct heat-transfer processes,* the heat source comes in direct contact with the material being heated. In *indirect heat-transfer processes,* a physical barrier separates the heat source from the material being heated. A heat exchanger that uses a thermal oxidizer's flue gases to preheat combustion air is an indirect heat-transfer process. Most municipal thermal oxidation systems use indirect heat transfer.

The following heat-recovery methods typically are used with thermal oxidation systems:

- A *recuperative air preheater* is a shell-and-tube heat exchanger in which hot flue gases indirectly preheat fluidizing air. This helps maintain autogenous operation.
- A *secondary heat exchanger* is a shell-and-tube heat exchanger used to cool flue gas before it enters the air pollution-control equipment. The recovered heat typically is used to raise the temperature of cleaned exhaust gases before they are discharged to the stack to suppress stack plumes.
- A *waste-heat recovery boiler* is a fire- or water-tube boiler used to cool flue gases; it can be installed after the furnace or a recuperative air preheater. The heat it recovers is transferred to a heat-transfer medium (e.g., water, steam, and thermal fluid) for external use. Water or steam can be used to heat processes or buildings, pre-dry solids, drive steam turbines, or generate steam in steam turbine/generator sets. Thermal fluid can be used to heat processes or buildings.

If waste-heat recovery boilers will produce steam, design engineers should select a unit that can operate at saturation pressures or superheated temperatures. This process has been installed as part of fluid-bed incinerator systems (Quast, 2006) and retrofitted to multiple-hearth furnace systems (DiGangi et al., 2008).

For more information on heat recovery, see *Wastewater Solids Incineration Systems* (WEF, 2009).

5.0 VITRIFICATION

Vitrification—a thermal process for converting minerals into glass—has a well-established track record in other industries (e.g., glass furnaces in the glass manufacturing industry and slagging furnaces in coal-fired power generation) but is new to the solids treatment field.

Wastewater solids (e.g., from paper mills and municipal treatment plants) have some characteristics in common with glass-manufacturing and power-generation wastes that play two important roles in vitrification. First, the organic fraction provides the thermal energy required to complete vitrification. Secondly, the mineral fraction (ash, clays, and mineral fillers) melts into a glass aggregate that has multiple construction and industrial applications.

Japan has vitrified wastewater solids for a couple of decades, but until recently, commercial-scale vitrification of solids was limited to hazardous waste applications in the United States. In 1998, the Fox Valley Glass Aggregate Plant in Neenah, Wis., began vitrifying industrial wastewater residuals. At press time, the plant vitrified about 270 000 metric ton/a (350 000 ton/yr) of wastewater solids from local paper mills into about 45 000 metric ton/a (50 000 ton/yr) of glass aggregate that is sold and used locally. The company that runs the Fox Valley plant has since developed a second-generation vitrification technology applicable for individual onsite use. However, this unit was still in the startup phase as of 2009.

6.0 BIOGASIFICATION

A *gasification* process uses heat, pressure, and steam to convert solids into a gas composed primarily of carbon monoxide and hydrogen (see Figure 26.15) (CIWMB, 2001). Variations in operating temperatures and pressures affect the byproducts, which may be a syngas, char or slag, oils, and reaction water. Operating temperatures may be in the range of 815° to 1 815°C (1 500 to 3 300°F) and pressures may be up to 200 kPa (400 psi).

Process dynamics and products depend on the type of feed used. Pilot testing typically is required to determine the yields of offgases and residues. The process has proven to be expensive when treating wastewater solids because of the low calorific value and high moisture content of the feed cake, which must be heat-dried before gasification. Also, the resulting syngas typically has a low heating value [about 4 000 to

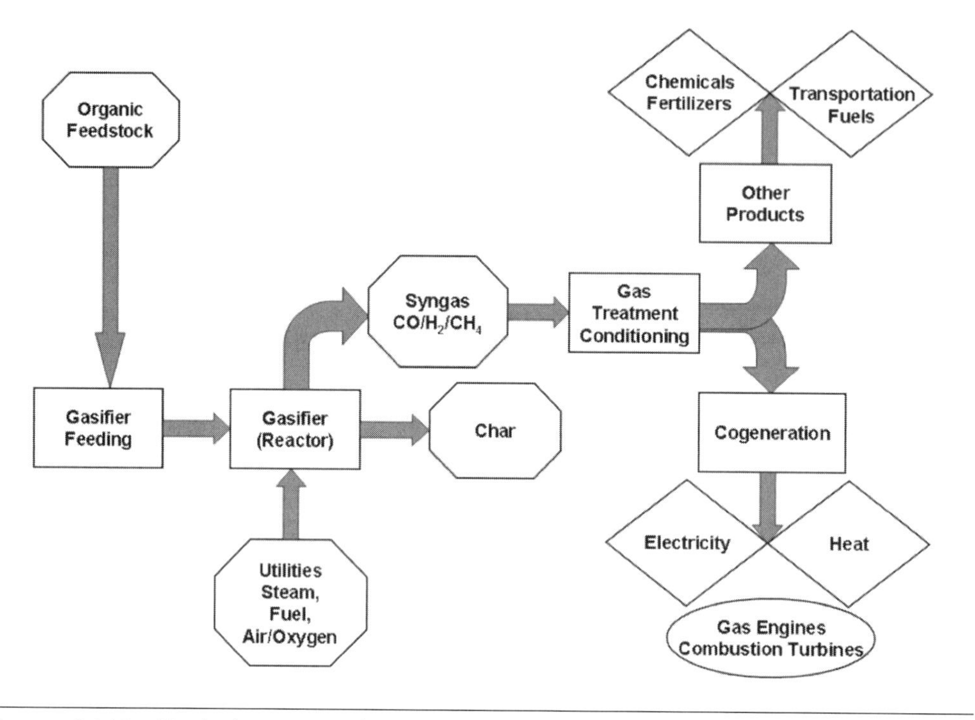

FIGURE 26.15 Typical process schematic for gasification.

8 000 kJ/m^3 (105 to 210 Btu/cu ft)] and must be combined with higher-quality fuels (e.g., natural gas) before it can be used. Likewise, the char and oils have less heating value than those produced in a pyrolysis system because the organics only partially combust.

Gasification systems with high-calorific-value feed stocks (e.g., wood wastes) have been used more widely in Europe and Asia. Compared to incineration, gasification provides better control of air emissions, is considered an energy-recovery technology because it produces items with energy value, and does not have the negative public perception that incineration has. (Incineration is often considered a disposal technology.)

As of 2008, however, no full-scale biosolids-gasification systems had been operating for 5 years or longer.

7.0 EMISSIONS CONTROL

7.1 Odors

The most noticeable form of air pollution, odors are the reason for about half of all citizen complaints to local air pollution-control agencies. Thermal-processing odors are particularly offensive, so design engineers must make every effort to minimize them. For specific information on odor control, see Chapter 7 and various manuals on the subject [e.g., *Control of Odors and Emissions from Wastewater Treatment Plants* (WEF, 2004)].

7.2 Combustion Emissions

The quantity and quality of combustion emissions from thermal processes depend on the thermal processing method used and the composition of the solids and auxiliary fuel. Completely combusted solids produce carbon dioxide, sulfur dioxide, water vapor, hydrogen chloride, and hydrogen fluoride. The exhaust gases will contain nitrogen, excess oxygen, and an inert ash that contains heavy metals, some of which (e.g., mercury) are volatile and present in both solid and gaseous forms.

In real life, however, combustion is never complete, so thermal processes release both nitrogen oxides and products of incomplete combustion. Such products include carbon monoxide, volatile organic compounds, and polycyclic organic matter.

7.2.1 Carbon Monoxide

Carbon monoxide is a product of incomplete combustion that occurs when carbon in the feed cake partially oxidizes with oxygen in the combustion air. It is the result of one or more of the following combustion-system deficiencies: inadequate temperatures, inadequate residence time for combustion gases, or inadequate mixing or turbulence (needed to bring combustion gases in dynamic contact with oxygen in the air supply).

In general, multiple-hearth and fluidized-bed furnaces have markedly different combustion environments. In a multiple-hearth furnace, the feed cake dries on the upper hearths before being combusted on the middle hearths. While this arrangement is efficient (using the heat from combusting dried cake to dry incoming wet feed cake), it releases partially oxidized combustion gases and products of incomplete combustion from the upper hearths of the furnace when dried feed cake begins burning. The slow, stratified flow of combustion gases (i.e., lack of turbulence) in this part of the furnace leads to emission of high levels of CO and products of incomplete combustion.

Carbon monoxide emissions from a multiple-hearth furnace without an afterburner can range from 900 to 2 700 mg/Nm^3dv_{11} (1 000 to 3 000 $ppmdv_7$); the typical CO mass emission rate is 15.5 g/kg (31 lb/dry ton) of solids incinerated. High CO and VOC emissions are one reason why the use of multiple-hearth furnaces has declined steadily in recent decades. A high-temperature afterburner can control such emissions. Multiple-hearth furnaces that have been retrofitted with top-hearth ("zero-hearth") afterburners have significantly lower CO and VOC emissions. However, afterburners also increase fuel requirements.

A fluidized-bed furnace is a completely mixed, highly turbulent system in which drying and combustion take place concurrently and in a matter of seconds. Turbulence provides complete, intimate contact of feed cake, volatilized gases, and oxygen in the fluidizing air. As hot combustion gases rise from the bed, they enter the freeboard, which provides a long residence time that allows CO and other volatilized organics to fully burn out. Carbon monoxide emissions from an fluidized-bed furnace are invariably less than 45 mg/Nm^3dv_{11} (50 $ppmdv_7$) and often less than 9 mg/Nm^3dv_{11} (10 $ppmdv_7$). Mass emission rates are typically less than 0.5 g/kg (1.0 lb/dry ton) of solids incinerated.

State regulators typically require a new facility to meet a CO emission limit of 90 mg/Nm^3dv_{11} (100 $ppmdv_7$). A fluidized-bed furnace can easily meet this limit. A multiple-hearth furnace would require an afterburner operating at a minimum of 816°C (1 500°F) to meet this standard.

7.2.2 *Volatile Organic Compounds*

Volatile organic compounds (e.g., CO) are products of incomplete combustion that occur when organic matter in feed cake vaporizes and volatilized compounds partially oxidize. Incomplete combustion happens when the temperature, residence time, and/or mixing in the thermal oxidizer is insufficient.

Chemically, a wide variety of compounds [e.g., straight and branched chain aliphatic hydrocarbons (methane, ethane, acetylene, etc.); oxygenated hydrocarbons (acids, aldehydes, ketones, etc.); chlorinated hydrocarbons (perchloroethylene, trichloroethane, etc.); and saturated and unsaturated ring compounds (benzene, toluene, phenols, etc.)] are volatile and organic. Such compounds are regulated under 40 CFR 503 (U.S. EPA, 1993), which requires that thermal oxidizers emit them at rates less than 100 ppm as propane on a dry volume basis corrected to 7% oxygen [i.e., 100 $ppmdv_7$ (140 mg/Nm^3dv_{11})].

Because multiple-hearth and fluidized-bed furnaces have different combustion conditions, the VOC emissions from each are quite different. The upper drying hearths of a multiple-hearth furnace typically will be hot enough to volatilize organic compounds but not fully oxidize them. Also, VOC emissions from a multiple-hearth furnace depend on the cake's feed rate and combustion characteristics (percent solids, percent volatile solids, and heating value), as well as the furnace's operating conditions (hearth temperatures, excess air, and burner firing rates on different hearth levels). Some multiple-hearth furnaces can meet the 140-mg/Nm^3dv_{11} (100-$ppmdv_7$) standard without an afterburner. Maintaining a top hearth temperature of at least 593°C (1 100°F) typically is important for achieving the standard (Waltz, 1990; Baturay, 1990).

In a fluidized-bed furnace, temperatures and turbulence are high, and total hydrocarbon emissions are low [typically less than 14 mg/Nm^3dv_{11} (10 $ppmdv_7$) as propane].

7.2.3 *Polycyclic Organic Matter*

Polycyclic organic matter [e.g., polychlorinated biphenols (PCBs), polychlorinated dibenzo-p-dioxin (PCDD), and polychlorinated dibenzo furan (PCDF)] is a subset of VOCs that is of particular concern to regulators because of its potentially high health-risk effects. The U.S. Environmental Protection Agency does not have specific emission limits for these pollutants, but some state regulatory agencies have included emission criteria for a few of these compounds in permits for new thermal oxidizers. The U.S. Environmental Protection Agency's *Compilation of Emission Factors* (AP-42) indicates that both multiple-hearth and fluidized-bed furnaces emit low levels of these pollutants [in the 0.5×10^{-7} to 0.5×10^{-9} g/kg (1×10^{-7} to 1×10^{-9}) range] (U.S. EPA, 1998). To minimize emissions of polycyclic organic matter, system design should maximize the thermal oxidizer's combustion efficiency. Multiple-hearth furnaces may need high-temperature afterburners if more control of polycyclic organic matter is necessary. (This would not be necessary for a fluidized-bed furnace.)

7.2.4 *Nitrogen Oxides*

Nitrogen oxides (e.g., nitrogen dioxide, nitrous oxide, and nitric oxide) are formed by the oxidation of nitrogen during combustion. The two types of nitrogen oxides are thermal nitrogen and fuel nitrogen. Thermal nitrogen is formed by atmospheric nitrogen in the combustion air. Fuel nitrogen is formed by the chemical oxidation of fuel- or solids-bound nitrogen.

7.3 Emission Regulations

Federal, state, and local governments all promulgate air pollution regulations, and they require emission sources to have permits to operate. For information on permitting thermal processes, see Chapter 7. The rest of this section discusses regulations applicable to thermal processes.

The Clean Air Act (CAA) of 1970 gave U.S. EPA the responsibility and authority to establish a nationwide program to abate air pollution and enhance air quality, and the 1990 amendments strengthened that authority. State governments have primary responsibility for implementing the program under federal supervision, and each state has developed its own plan for achieving and maintaining the National Ambient Air Quality Standards (NAAQS). State governments often delegate the responsibility and authority for implementing the plan to local pollution-control agencies.

The U.S. Environmental Protection Agency promulgated *Standards for the Use or Disposal of Sewage Sludge* (40 CFR 503) in 1993. This rule, which falls under the Clean Water Act (CWA), includes emission standards and management practices for solids incinerators.

There are two types of air pollution regulations: limiting and administrative. Limiting regulations establish specific limits for specific air pollutants. Such regulations include the New Source Performance Standards, National Emission Standards for Hazardous Air Pollutants, local prohibitory rules, and source-specific standards. Administrative regulations establish requirements for gathering data (i.e., source testing and ambient monitoring), controlling the growth of emission sources, and using pollution-control technologies. Such regulations include prevention of significant deterioration (PSD) regulations for attainment areas (areas that meet NAAQS) and new-source review regulations for nonattainment areas (areas that do not meet NAAQS).

7.3.1 New Source Performance Standards

Under federal standards for wastewater treatment plants (40 CFR 60, Subpart O), solids incinerators with capacities larger than 1 000 kg/d (2 205 lb/d) on a dry basis are limited to particulate discharges of 0.65 g/kg (1.30 lb/ton) of dry solids input, and to an opacity of 20%. State standards already in effect may limit particulate emissions to less than 0.2 g/kg (0.4 lb/ton) of dry solids input.

7.3.2 National Emission Standards for Hazardous Air Pollutants

The U.S. Environmental Protection Agency established National Emission Standards for Hazardous Air Pollutants to address listed hazardous pollutants (e.g., carcinogens, mutagens, and toxicants) emitted from specific new and existing sources. For example, they must emit no more than 10 g of beryllium over a 24-hour period, according to 40 CFR 61, Subpart C. Also, the beryllium content of ambient air near the source must be averaged over a 30-day period. Solids incinerators must emit no more than 3 200 g of mercury over a 24-hour period, according to 40 CFR 61, Subpart E.

7.3.3 Prevention of Significant Deterioration

These regulations are designed to prevent the degradation of existing air quality beyond an allowable increment. For example, new stationary sources of a regulated pollutant that are located in a pollutant attainment area and emit more than 620 kg/d (250 ton/yr) of that pollutant are subject to PSD regulations (40 CFR 61, Subpart A, Section 52.21). Modifications of major sources also are subject to PSD if they emit more than the specified minimum amount of any pollutant.

7.3.4 *New Source Review*

The Clean Air Act Amendments of 1977 require all states to establish a permit program for major new, modified, or reconstructed stationary sources in nonattainment areas. The goal is to prevent potential new sources of pollutants from increasing the net air pollution or delaying the area's ability to comply with NAAQS.

To obtain a permit in a nonattainment area, a major source must use the best available control technology that will produce the lowest achievable emission rate, demonstrate a net air quality improvement by offsetting new emissions with emission reductions from other combustion sources in the area (emission-reduction credits), and certify that all other similarly regulated sources are in compliance with all applicable emission regulations.

Sales of emission-reduction credits have developed into a full-fledged commercial "trade" in the past 5 years.

7.3.4.1 Clean Air Amendments of 1990

The Clean Air Amendments of 1990 expand the scope of the 1970 Clean Air Act and 1977 amendments. For example, they divide nonattainment areas into categories based on the severity of the nonattainment and set category-specific timetables for attainment.

The 1990 amendments also directed U.S. EPA to change its approach to regulating air toxics. Rather than analyzing data for a chemical emitted from a specific source and setting standards to protect human health, the amendments require all *major sources*—facilities that emit at least 25 kg/d (10 ton/yr) of one of the 189 chemicals listed in the 1990 amendments or whose emissions of all of these chemicals total 60 kg/d (25 ton/yr) or more—to install maximum achievable control technology (MACT). *Maximum achievable control technology* is the control technology used by the cleanest 12% of facilities in the same source category. Once these controls are in place, U.S. EPA plans to review scientific data to assess the health risk of each chemical and issue new standards if assessment results indicate that MACT did not provide an ample margin of safety for the people most exposed to high-risk emissions.

The 1990 amendments also expand U.S. EPA's authority to enforce the provisions of the CAA and its amendments.

7.3.5 *Standards for the Use and Disposal of Sewage Sludge*

Federal standards (40 CFR 503, Subpart E) limit emissions of beryllium, mercury, lead, arsenic, cadmium, nickel, and chromium from solids incinerators and prescribe an operational standard of <100 ppm for total hydrocarbon (THC) or CO. The rule also prescribes management practices, including continuous monitoring of THC or CO, oxygen, moisture, solids feed rate, and furnace temperature. In addition, it prescribes the frequency of monitoring, recordkeeping, and reporting.

7.3.6 Local Prohibitory Rules

Local air-quality agencies also can promulgate various regulations to limit specific air pollutants, thereby protecting local air quality and public health. For example, the South Coast Air Quality Management District of California imposes a "nuisance rule" prohibiting emissions from causing injury or being a detriment, nuisance, or annoyance to the public.

7.3.7 Future Regulations

In addition to existing regulations, designers and operators of solids-processing facilities should be aware of pending air pollution regulations. As public concerns over toxic air pollutants grow, air pollution-control agencies are developing regulations to limit toxic air pollutants from new and modified sources. Solids-processing facilities have been identified as potential sources of toxic air pollutants, so efforts to permit new processing facilities or modify existing ones will face scrutiny under toxic air pollutant regulations.

In the Clean Air Act Amendments (CAAA) of 1990 (see Section 7.3.4.1), solids incinerators were listed as source categories under both Section 112 and Section 129. In January 1997, U.S. EPA indicated in the *Federal Register* that biosolids incinerators would be delisted from Section 112 and regulated under Section 129's Other Solid Waste Incinerators category. Then on Nov. 15, 2000—based on recommendations and data analysis results submitted by the Industrial Combustion Coordinated Rulemaking (ICCR) advisory committee under the Federal Advisory Committee Act—U.S. EPA issued its final rulemaking that solids incinerators would not be regulated as a category under Section 129 of the CAA.

Confirming its earlier announcement of January 1997, U.S. EPA announced changes to the MACT standards that affect biosolids incinerators favorably. Section 112 (c) of the CAAA of 1990 had listed sewage sludge incinerators under the Waste Treatment and Disposal category, so U.S. EPA had to develop MACT standards for publicly owned treatment works (POTWs). After evaluating all the emissions information available, including testing done since the initial listing in June 1992, U.S. EPA concluded that sewage sludge incinerators do not have any sources that could emit HAP at a level approaching major source levels, so the agency deleted such incinerators from the Section 112 source category list (effective Feb. 12, 2002) (see 67 *Fed. Reg.* 6521, 6523; Feb. 12, 2002). In June 2006, U.S. EPA published a notice reconsidering the issue of whether sewage sludge incinerators should be excluded from regulation under Section 129, and on Jan. 22, 2007, the agency confirmed its earlier decision to exclude them.

In June 2002, U.S. EPA listed sewage sludge incinerators under Sections 112(c)(3) and 112(k)(3)(B)(ii) as an area source category, and the agency must promulgate standards for such incinerators in June 2009. Sources regulated under the area source category typically are required to use generally acceptable control technology (GACT), which less stringent than the maximum achievable control technology (MACT) that is mandatory for sources regulated under Section 129 and Section 112. The expectation is that U.S. EPA will require sewage sludge incinerators to use GACT, and the effect on existing control technologies would be negligible. The emissions-control technologies discussed in this chapter are expected to be suitable for GACT.

In December 2007, the National Association of Clean Water Agencies (NACWA) issued an alert to its members concerning a decision made by the D.C. Circuit Court of Appeals. In June 2007, the Court required U.S. EPA to regulate some incineration sources that previously had been excluded from Section 129. The association's opinion is that U.S. EPA has interpreted the ruling to mean that sewage sludge incinerators must be regulated under Section 129. If U.S. EPA issues such regulations, sewage sludge incinerators would, at minimum, be subject to numeric emissions limits for particulate, sulfur dioxide, NO_X, hydrogen chloride, carbon monoxide, cadmium, mercury, lead, dioxins, and furans. New incinerators would be required to meet best control technology limits, and existing ones would be limited to the average emissions levels achieved by the best performing 12% of units in the category. If MACT are required in future, the emission technologies evaluated in this chapter would need to include mercury controls, NO_X controls and hydrochloric acid controls. Such technologies currently are available and used commercially. As of 2009, this pending regulatory issue has not been resolved.

7.3.8 Source-Specific Standards

Air emissions can be divided into criteria pollutants and noncriteria pollutants. *Criteria pollutants* are air contaminants for which U.S. EPA has established NAAQS (e.g., nitrogen, carbon monoxide, sulfur oxides, reactive organic gases, particulate, and lead). *Noncriteria pollutants* are air contaminants for which U.S. EPA has not yet established NAAQS (e.g., hydrogen chloride, hydrogen sulfide, vinyl chloride, trace metals, polynuclear aromatic hydrocarbons, dioxins, and polychlorinated biphenyls).

Air contaminant concentrations can be determined via emission factors or source testing. Emission factors for criteria pollutants from fuel combustion are well established and readily available from combustor manufacturers and other references. However, emission factors for noncriteria pollutants or solids combustion are not readily available; source testing often is required.

7.3.9 *Nonregulated Emissions (Greenhouse Gases)*

Greenhouse gases are compounds that contribute to global warming; they currently are unregulated. The four principal greenhouse gases of concern are carbon dioxide (CO_2), methane (CH_4), nitrous oxide (N_2O), and halocarbons (a group of gases containing fluorine, chlorine, and bromine). Of these, the first three can be emitted by thermal processes, the fuel used to operate them, or the generators that provide the electricity to operate the equipment. The *global warming potential* (intensity factor) is a measure of a given substance's radiative effect compared to CO_2, integrated over a given horizon, for example, 100 years. Table 26.8 shows 100-year estimates of global warming potentials for greenhouse gases associated with fossil fuel combustion.

To estimate greenhouse gas emissions, design engineers should determine a source's emission rate of such gases, use the global warming potential to convert measurements of non-CO_2 gases into CO_2 equivalents, and calculate how many metric tons of CO_2 equivalents are emitted annually. Greenhouse gas emissions are classified as direct or indirect. Direct emissions include those produced when natural gas is combusted in a dryer, incinerator, or boiler. Indirect emissions include those associated with the electricity generated to operate a thermal process.

When estimating greenhouse gas emissions, design engineers need to distinguish between anthropogenic effects and biogenic sources. *Anthropogenic effects* are processes or materials that occur as a result of human activities. *Biogenic sources* are processes or materials that occur as a result of biological or ecological activities. For example, fossil fuel combustion is considered an anthropogenic effect, and experts believe that the greenhouse gases emitted during this process contribute to global warming. Wastewater solids, on the other hand, are considered biogenic substances whose CO_2 emissions do not contribute to global warming because the carbon is already part of the active global carbon cycle. The *active global carbon cycle* is the biogeochemical cycle by which carbon is exchanged among the Earth's biosphere, pedosphere, geosphere, hydrosphere, and atmosphere.

Design engineers can minimize greenhouse gas emissions from thermal processes by selecting and designing processes that are process- and energy-efficient. Below are

TABLE 26.8 Comparison of 100-year global warming potential (GWP) estimates from the IPCC's second, third, and fourth assessment reports (IPCC, 1995, 2001, 2007).

Gas	1996 IPCC GWP	2001 IPCC GWP	2007 IPCC GWP
Carbon dioxide (CO_2)	1	1	1
Methane (CH_4)	21	23	25
Nitrous oxide (N_2O)	310	296	298

some of the direct and indirect emissions they should take into account when designing thermal processes.

7.3.9.1 Thermal Drying

Thermal drying typically emits methane, which may volatilize when solids are heated. The process also emits CO_2 via the combustion of fossil fuels (e.g., natural gas) to heat solids and evaporate moisture.

7.3.9.2 Incineration

Incineration typically emits methane and nitrous oxide when organic solids are incompletely combusted. The process also emits CO_2 via the combustion of fossil fuels (e.g., natural gas) used both as an auxiliary fuel and to start up the incinerator after shutdowns.

7.3.9.3 Indirect Emissions Associated with Thermal Processes

Indirect emissions associated with thermal processes typically include CO_2 equivalents emitted by the generators that produce the electricity used to power the processes. They also include CO_2 equivalents emitted during transportation of dried solids or ash from the facility to its intermediate or final destination.

7.4 Air Pollution Control Methods and Technology

To design an effective air pollution control system, engineers need the following data:

- Contaminated gas characteristics (flowrate, temperature, pressure, composition, and moisture content);
- Particulate characteristics (resistivity, flowability, and emission limits);
- Emission parameters (concentration, solubility, absorbability, combustibility, and emission limits); and
- Site-specific constraints (availability of power, fuel, water, and space; aesthetic considerations; noise restrictions; and budget).

Table 26.9 provides a summary of methods and technologies used to control emissions from incinerators. Below are pollutant-specific introductions to available control technologies.

7.4.1 Particulate and Metals

Particulate—the predominant air contaminant in thermal processing—includes both solid particles and liquid droplets (excluding uncombined water) that are swept along by the gas stream or formed via condensation of flue gas. It can be enriched with

TABLE 26.9 Summary of emission controls applied to biosolids incinerators.*

Emission control components	Function	Applicability
Incinerator combustion–design temperature, residence time and turbulence	Complete combustion–conversion of organics to carbon dioxide and water and sulfur to sulfur dioxide. Minimize formation of carbon monoxide, methane, non-methane VOCs, dioxins/furans and NO_x	Applied to both FBI and MHF
Injection of urea or ammonia to flue gas	Reacts with NO_x gases	Typically applied to FBI
Quench	Cools gases and assists agglomeration	Applied to both FBI and MHF
Venturi (in combination with tray cooling tower)	Removes course and fine particulate, heavy metals, acid gases—wet (>1 μm)	Applied to both FBI and MHF
Packed tower (in combination with quench)	Removes acid gases (may require caustic addition for pH control)	Typically on applied to FBI
Wet ESP (in combination with venturi/tray cooling tower)	Removes fine particulate, heavy metals-wet (<1 μm), removes moisture droplets	Applied to both FBI and MHF
Multiple venturi post quench (in combination with quench)	Removes fine particulate, heavy metals acid gases-—wet (<1 μm)	Applied to both FBI and MHF
Dry ESP	Removes course and fine particulate, heavy metals—dry (>1 μm)	Typically applied to FBI (only applications in Europe)
Baghouse (fabric filter)	Removes course and fine particulate, heavy metals—dry (<1 μm)	Typically applied to FBI
Conditioning tower for carbon injection	Adsorbs mercury for removal in dry ESP and baghouse. Also removes dioxins/furans	Typically applied to FBI
Spray dryer/absorber for alkali injection	Absorbs acid gases for removal in dry ESP and baghouse.	Typically applied to FBI (only applications in Europe)
Carbon adsorber (in conjunction with all emission control components)	Adsorbs mercury (and dioxins/furans)	Typically applied to FBI (downstream of wet scrubbing system)
RTO (after particulate control)	Thermally oxidizes organics (THCs and dioxins/furans), plume suppression	Applied to MHF
Exhaust gas reheat	Plume suppression	Typically applied to FBI

* ESP = electrostatic precipitator; FBI = fluidized bed incinerator; MHF = multiple-hearth furnace; RTO = regenerative thermal oxidation; THC = total hydrocarbons; and VOC = volatile organic compound.

volatile trace metals (e.g., cadmium, lead, and zinc). Most particles are smaller than 2 m; volatile elements primarily can be found in the submicron particles.

Particulate can be removed from gas streams by mechanical collectors, wet scrubbers, fabric filters, or electrostatic precipitators. The choice of technology depends on the nature of the particulate, gas-stream conditions, and emission limits. Figures 26.16 and 26.17 compare the effectiveness of various particulate control equipment.

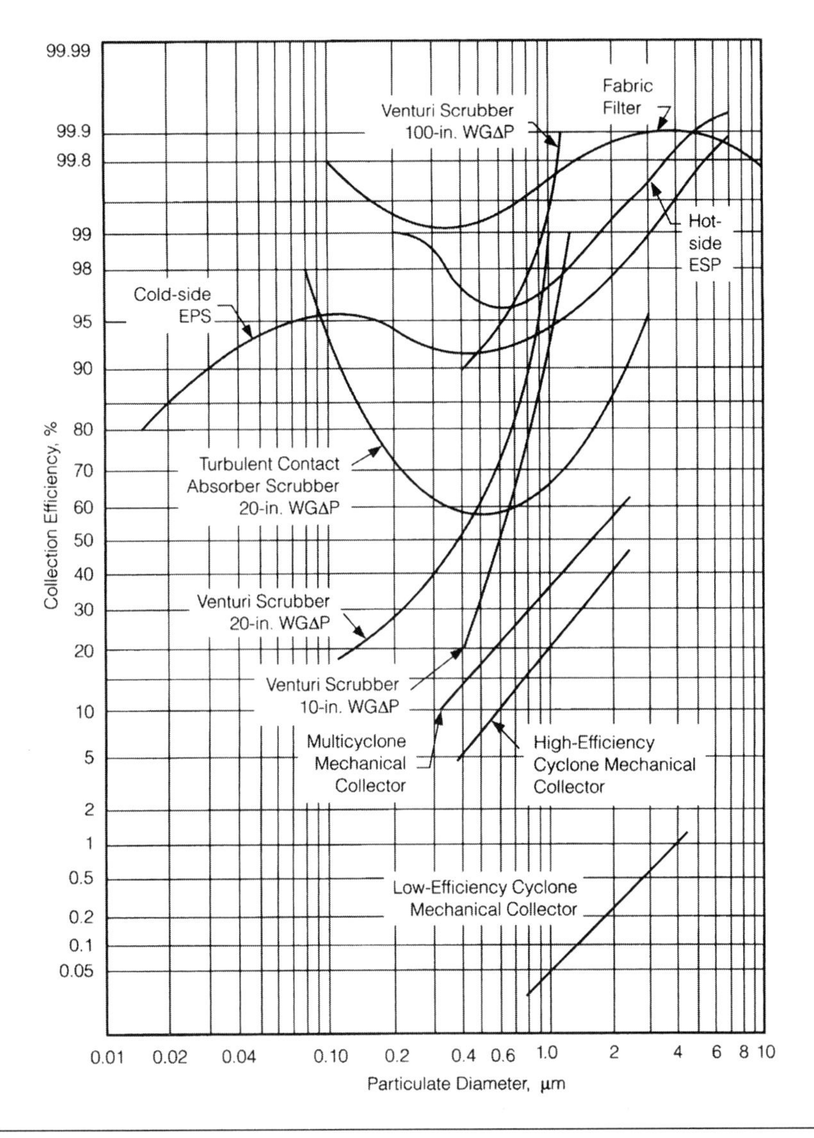

FIGURE 26.16 Efficiency curves for conventional air pollution control equipment.

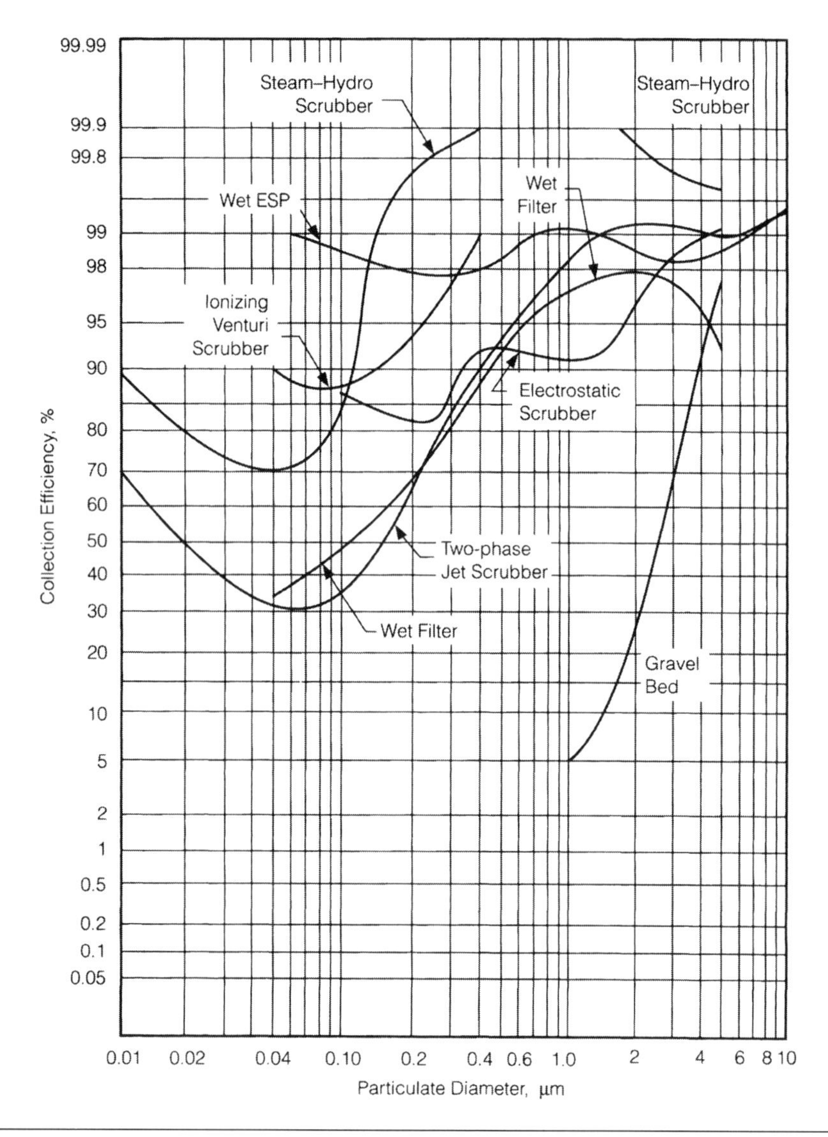

FIGURE 26.17 Efficiency curves for novel air pollution control equipment.

7.4.1.1 Mechanical Collectors

Mechanical collectors use inertia to separate particles from gas. They are less expensive than other particulate-control devices, but their collection efficiency is also lower. They typically are used as a pretreatment step for other particulate-control devices.

There are three types of mechanical collectors: settling chambers, impingement separators, and cyclone separators. The settling chamber is the simplest mechanical collector. The low gas velocity through the chamber }{ss than 3 m/s (10 ft/sec)] allows heavy particles to settle. Pressure drop across the chamber is negligible, however, and particle separation is limited.

Impingement separators direct gas against collecting bodies, where particles lose momentum and drop out of the gas. Typically designed with pressure drops between 2 and 40 mm (0.1 and 1.5 in.) of water column, they remove particles larger than 10 μm.

Cyclone separators are the most widely used mechanical collectors. Gas enters the device tangentially at the top of a cylindrical shell and then spirals downward toward the narrowest part of a conical section. Particles exit the bottom via an airlock, while the gas is directed back up through the center of the vortex and discharged from the top. Cyclones have top diameters ranging from 0.7 to 3 m (2 to 10 ft) and can remove 85% of particles as small as 10 μm at pressure drops between 10 and 80 mm (0.5 and 3 in.) of water column. Smaller-diameter cyclones have higher removal efficiencies than larger-diameter ones with the same pressure drop.

A multi-tube collector (multiclone) uses many small cyclones in parallel to improve the removal efficiency. Units with tube diameters of 100 to 300 mm (4 to 12 in.) can remove 95% of particles as small as 5 μm.

7.4.1.2 Wet Scrubbers

Wet scrubbers use a liquid (typically water) to separate dusts or mists from gas streams. They are the most widely used emission-control equipment for solids incinerators. Wet scrubbers can handle hot gases with high moisture contents. They also remove water-soluble air contaminants (e.g., hydrogen chloride, sulfur dioxide, and ammonia).

The simplest type of wet scrubber is a spray tower. In this device, gas enters at the bottom and rises to the top, where liquid is sprayed continuously to contact the gas. As liquid droplets fall and collide with the rising gas stream, they capture particles, removing them from the gas. Spray towers typically contain packing material (packed towers) or trays with bubble caps (tray towers) to increase the mass-transfer surface area. Although pressure drop through the spray tower is low, particulate-removal efficiencies are less than 50%, so these devices are typically used in combination with Venturi scrubbers for particulate removal or for chemical scrubbing.

Cyclone scrubbers use centrifugal force to increase the momentum of particle-and-liquid-droplet collisions. They can remove 90% of particles as small as 5 μm at pressure drops between 50 and 150 mm (2 and 6 in.) of water column.

The most widely used particulate scrubbers are Venturi scrubbers. In this device, the gas stream is accelerated across a Venturi (orifice), where liquid is sprayed and then turbulently mixed with the gas at the throat. The high turbulence promotes collisions between liquid droplets and particulates, thereby capturing small particles. The Venturi scrubber's particulate-removal efficiency is directly proportional to the power input (pressure drop). At a pressure drop of 250 mm (10 in.) of water column, the device can remove about 90% of particles as small as 1 μm. As the pressure drop increases to 500 mm (20 in.) of water column, it removes up to 98% of such particles. Variable-throat Venturi scrubbers allow efficient operations at a range of flowrates. Multiple-Venturi scrubbers have a quench and tower section followed by multiple Venturis. The tower section removes larger particulate, and the multiple Venturis remove smaller particles. Multiple-Venturi scrubbers have particulate-removal efficiencies that are similar to those of traditional Venturi scrubbers, but at lower pressure drops.

7.4.1.3 Fabric Filters

Fabric filters (bag houses) collect solid particles by passing dust-laden gas through a filter medium or fabric that most particles cannot penetrate. As the dust layer accumulates, the pressure drop increases, so accumulated dust must be removed periodically. Operators can clean bags via mechanical shaking, pulse jet, or reverse airflow.

Fabric filters can remove more than 99% of particles down to submicron sizes with pressure drops typically between 50 and 100 mm (2 and 4 in.) of water column. These devices typically are suitable for solids incinerators when the gas temperature can be reduced reliably to about 149 to 177°C (300 to 350°F) before gas enters the bag house. A heat-recovery boiler can be used for this purpose.

7.4.1.4 Electrostatic Precipitators

In a dry electrostatic precipitator, in exhaust gases pass through large chamber, where a negative charge is imparted to particulate. It then is attracted to positively charged collector plates that parallel the gas flow through the chamber. Collected particles are removed periodically via vibration, rapping, or rinsing.

Some precipitators use pipes as collectors rather than plates. These units remove liquid droplets and particulate fumes, as well as solid particles.

Wet electrostatic precipitators are similar to dry ones except that they contain a washing mechanism to counteract the buildup of volatile or particulate matter on the plates. They also typically perform better than dry ones when treating certain types of pollutants, removing 99% or more of particulate with negligible pressure drops. In many incineration systems, wet electrostatic precipitators have been installed downstream of wet scrubbers to help the system comply with 40 CFR 503.

7.4.2 Nitrogen Oxides

Two types of nitrogen-control technologies are combustion modification and flue gas treatment.

7.4.2.1 Combustion Modification

Systems can be designed to reduce the formation of nitrogen oxides through operation at reduced combustion temperature and lower amounts of oxygen. There are five principal methods for modifying combustion to limit nitrogen oxide formation: low excess air, staged air combustion, staged fuel combustion, low-nitrogen burners, and flue gas recirculation. The simplest option is operating burners with low excess air. Doing this reduces nitrogen oxides by about 20%. The degree of control is constrained by increases in CO emissions.

Staged air combustion can reduce the formation of nitrogen oxides by up to 70%. In this option, combustion occurs in two stages: In Stage 1, fuel is burned with insufficient air, and in Stage 2, more air is mixed with the fuel to complete combustion. Rotary kilns and fluidized-bed combustors often are designed with staged air combustion.

In staged fuel combustion, the first stage of combustion occurs with insufficient fuel, and the cooling effect of excess air suppresses the formation of nitrogen oxides. The rest of the fuel is injected and burned in the second stage. Here, inerts from the first stage depress peak temperatures and reduce oxygen concentration, thereby reducing the formation of nitrogen oxides. Reserving about 20% of the fuel for the second stage can reduce nitrogen oxide formation by 50 to 70%.

Low-nitrogen burners limit the exposure of fuel to oxygen in the immediate flame zone. As a result, fuel and air mix gradually, reducing peak flame temperatures. This option can reduce nitrogen oxide formation by 30 to 50%.

Recirculating some of the flue gas as inlet combustion air lowers both the peak flame temperature and the formation of thermal nitrogen. Recirculating up to 15% of flue gas can reduce nitrogen oxide formation by 50%.

7.4.2.2 Flue Gas Treatment

Nitrogen oxide emissions can be reduced by injecting the appropriate reducing agents to the post-combustion region (flue), where they reduce flue-gas nitrogen to molecular nitrogen. Three types of flue-gas treatment are the thermal denitrification process, urea injection process, and selective catalytic reduction process. In the thermal denitrification process, ammonia is injected into an incinerator heated to between 871 and 1 093°C (1 600 and 2 000°F) to reduce nitrogen oxides. This process can reduce nitrogen oxide emissions by 35 to 70%, depending on flue gas temperature, residence time, degree of mixing, and the ammonia-to-nitrogen mole ratio.

The urea injection process is similar to thermal denitrification, except that it uses urea as the reducing agent. Urea breaks down to form CO and ammonia, which reacts with nitrogen oxides to form nitrogen and water. This process also can reduce nitrogen oxide emissions by 35 to 70%, depending on flue gas temperature, residence time, degree of mixing, and the ammonia-to-nitrogen mole ratio.

In the selective catalytic reduction process, a catalyst bed operates at lower flue gas temperatures, allowing ammonia to react with nitrogen. It can treat flue gases with temperatures of 316 to 427°C (600 to 800°F) and reduce nitrogen oxide levels by 90% or more. However, the process is expensive and the catalyst can be adversely affected by trace metals in flue gas.

7.4.3 Acid Gases, Including Sulfur Oxides

Acid gases include sulfur oxides, hydrogen chloride, hydrogen fluoride, and hydrogen bromide. The emission concentrations of such gases is a direct function of the sulfur, chloride, fluoride, and bromide levels in the fuel and solids to be combusted. Acid gas emissions can be reduced via fluidized-bed combustion, wet scrubbers, and dry scrubbers. In fluidized-bed combustion, fuel is combusted in a bed of granular limestone (dolomite). Acid gases react with limestone to form solid calcium compounds, which are removed with the ash. Fluidized-bed combustion also has relatively low nitrogen oxide emissions because combustion temperatures are moderate [816 to 927°C (1 500 to 1 700°F)].

Wet scrubbers remove both particulate and acid gases from flue gas. They reduce acid gases by diffusing them into liquid droplets. Removal efficiency may be improved by adding an alkaline reagent.

In dry scrubbers, a dilute slurry of lime or another reagent is sprayed into the flue gas stream. The water fraction of the slurry instantly evaporates, while the atomized reagent reacts with acid gases to form sulfate, sulfite, and chloride precipitates. These precipitates then are removed by particulate-control devices.

7.4.4 Mercury Control

Flue gas may contain two types of mercury: elemental (ionic) and oxidized (oxides, chlorides, etc.). Wet scrubbers absorb and remove some of the oxidized form of mercury. If scrubber underflow is returned to the treatment plant, however, the mercury it contains will enter treatment plant flows and could affect effluent concentrations.

Gaseous mercury emissions can be reduced through adsorption onto activated carbon. This is achieved via static bed adsorbers or flue gas injection. Static bed adsorbers are multi-compartment vessels located downstream of air-pollution control equipment. The compartments contain media—typically a combination of inert material (e.g., alumino-silicates and activated carbon) that may be impregnated with sulfur or other chemicals—that provide surface area for mercury to adsorb to as flue gases pass through. Afterward, the flue gas passes through a conditioning chamber where the humidity and temperature are controlled.

In flue gas injection, activated carbon is injected into a reaction chamber, where it mixes with flue gas and mercury is transferred to the carbon. This process is used as a pretreatment step for baghouses or dry electrostatic precipitators, which then remove mercury-laden carbon from the gas. Baghouses require less carbon than electrostatic precipitators because the carbon trapped on the bag's dust layer continues to capture mercury.

7.4.5 Stack-Gas Reheat

While an occasional steam plume may not hurt the environment, it can hurt public perception of solids combustion. Eliminating such plumes is important for widespread public acceptance of solids combustion.

In an efficient gas-cooling system, exhaust gases typically are not more than 11°C (20°F) above the inlet water temperature [about 20 to 40°C (70 to 110°F)] and, therefore, will not produce a steam plume under most atmospheric conditions. Reheating stack gas, however, can completely eliminate the steam plume and better disperse gases to the atmosphere.

Wastewater treatment plants typically have short exhaust stacks and other building and topographical conditions that limit the dispersal of gases to the atmosphere. Relatively dry, warm gases rise more rapidly and disperse better.

There are three methods for raising the temperature of exhaust gases from between 20 and 40°C (70 and 100°F) up to 90 to 150°C (200 to 300°F): a fuel-fired afterburner, indirect reheating with preheated air using furnace offgases (see Figure 26.18), and direct reheating via furnace offgases (see Figure 26.19). However, treatment plant personnel are reluctant to use fuel-fired afterburners because of the fuel costs. (They use natural gas to avoid secondary contamination of stack gases.)

Both furnace offgas methods are effective and require exhaust gases at temperatures of 430 to 540°C (800 to 1 000°F). However, the direct heating method is more thermally efficient. In the indirect method, primary exhaust gases are passed through a secondary heat exchanger, where they transfer heat to an ambient airflow, which then is discharged to the stack via a low-pressure blower. The direct method does not involve this blower.

8.0 DESIGN EXAMPLE

These examples have been selected to illustrate how design engineers select and size both thermal dryers and incinerators for a wastewater treatment plant that produces digested and dewatered solids. The incineration example also illustrates how to select and size an incinerator for dewatered raw solids.

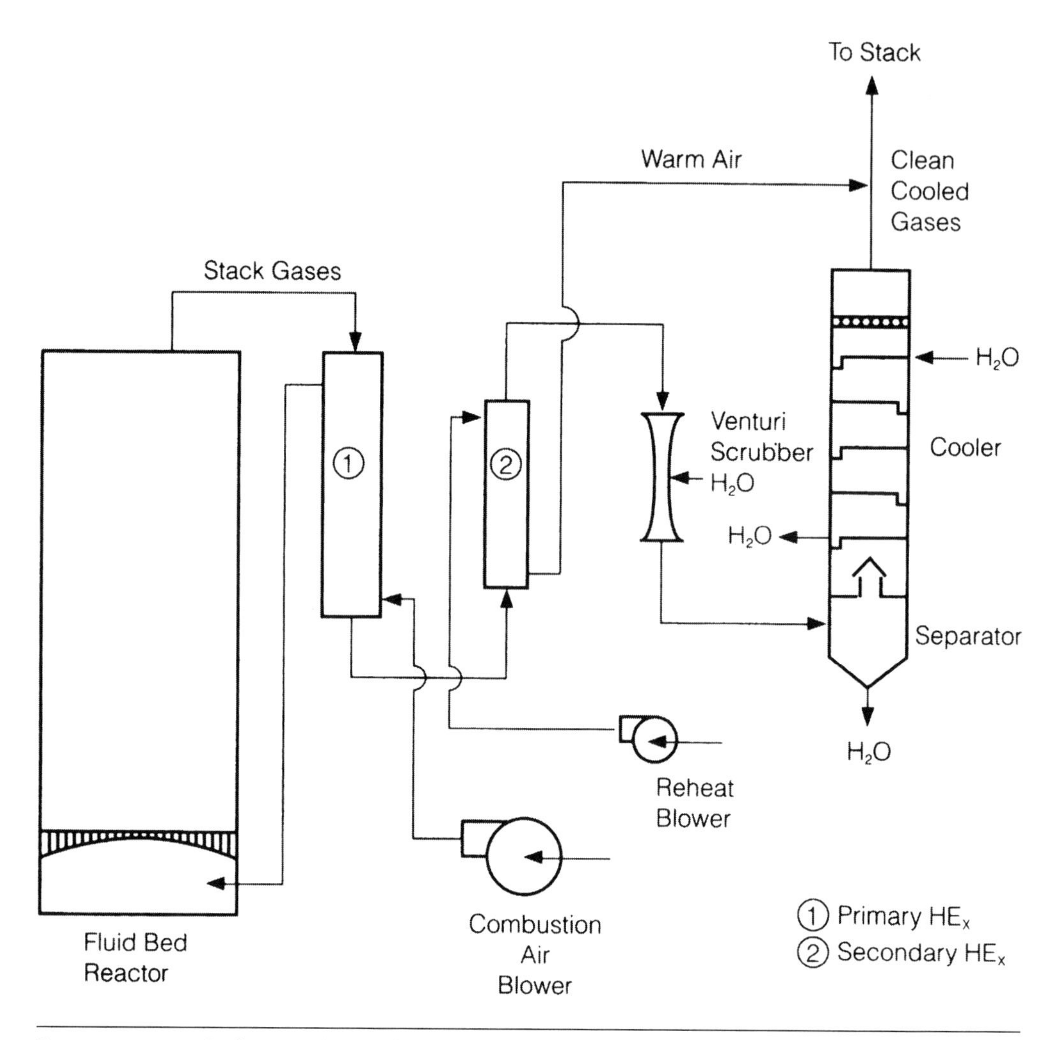

FIGURE 26.18 Indirect reheat of stack gases.

In these examples, the treatment plant produces 50 metric ton/d (55 ton/d) (dry solids basis) of digested and dewatered solids at 25% solids. Peak month production is 65 metric ton/d (72 ton/d) dry solids. The equivalent dewatered raw solids production is 75 metric ton/d (dry solids basis) at 28% solids. Solids characteristics and other relevant parameters are provided in each example.

8.1 Thermal Drying

Note, many parameters vary by system type and manufacturer. Confirm specific parameters with manufacturer during design.

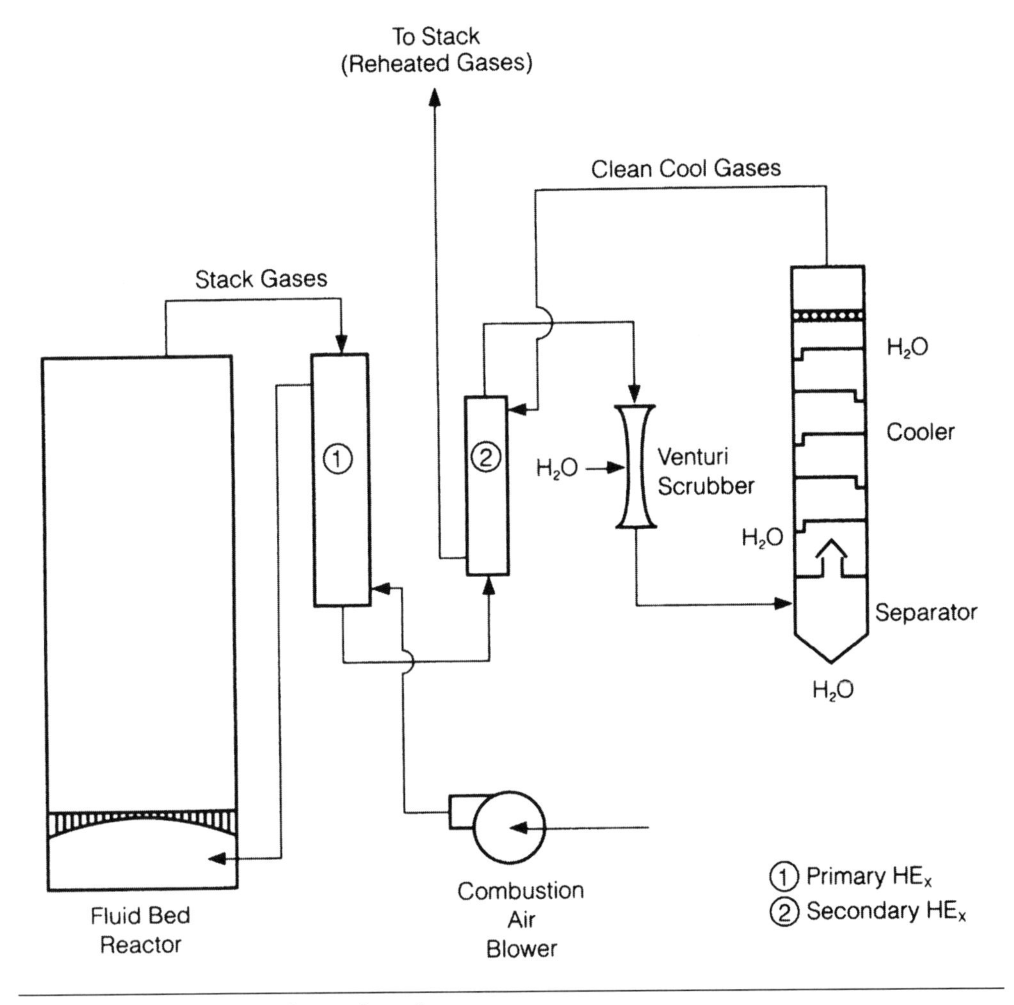

FIGURE 26.19 Direct reheat of stack gases.

Input parameters are based on approximate values for a rotary drum system using a regenerative thermal oxidizer for odor control.

8.1.1 Given Information

Anaerobically digested solids		
Annual average production	50 metric ton/d	55 dry ton/d
Maximum month production	65 metric ton/d	72 dry ton/d
Cake solids concentration (total solids)	25%	

8.1.2 *Procedure*

(1) Establish hours of operation and redundancy requirements.
(2) Establish throughput requirements and product dryness.
(3) Calculate required evaporation rate and thermal energy requirement.
(4) Determine capacity for one drying train and confirm number of dryer trains.
(5) Establish annual thermal and electrical energy requirements.
(6) Establish storage requirements (daily or weekly) and calculate storage capacity requirements.

8.1.3 *Assumptions*

Establish hours of operation. Drying systems are best suited for continuous operation for extended periods to improve thermal efficiency and reduce equipment wear due to frequent starts and stops. They typically operate continuously throughout the week and allow downtime on weekends.

Sizing considerations:

- Redundancy often is not provided because of high capital cost, but unit capacity can affect this decision.
- In many cases, the system can be sized conservatively to allow adequate time for preventive maintenance. Design engineers should consider whether landfilling is available as a backup option.
- In this example, size is based on processing maximum monthly production in a 5-day work week, 24 hours/d.
- Assume two units will be provided to meet total drying capacity requirements. This will provide 50% redundancy.

Drying trains installed	2	
Actual hours of operation per week	103 hr/wk	
Annual operation	52 wk/yr	
Dried product's solids concentration (total solids)	92%	
Dried product's bulk density *(Varies by product type. Contact manufacturer for correct value for system considered)*	729 kg/m^3	45 lb/cu ft

Heat energy consumed to evaporate water *(Varies by system. Contact manufacturer for correct value for system considered)*	3.72 MJ/kg H_2O	1 600 BTU/lb H_2O
Electrical energy consumed by the system *(Varies by system. Contact manufacturer for correct value for system considered)*	85 kW/dryer capacity in metric ton H_2O/h	
Required days of storage volume at annual average conditions required in silos	10 days	

8.1.4 Calculations

Determine total evaporation load during the week at maximum monthly design conditions.

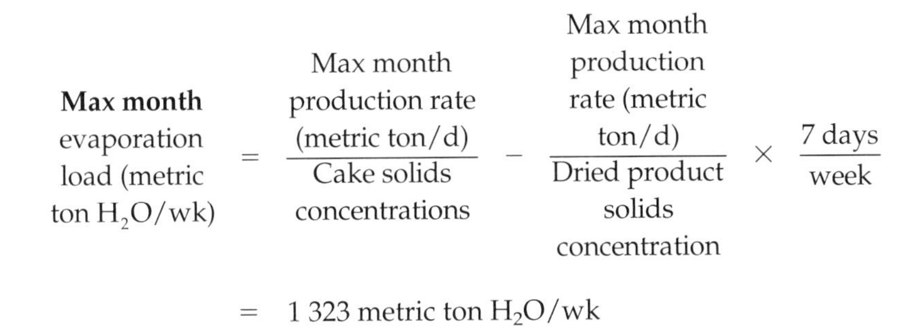

$$\textbf{Max month}\text{ evaporation load (metric ton }H_2O\text{/wk)} = \left(\frac{\text{Max month production rate (metric ton/d)}}{\text{Cake solids concentrations}} - \frac{\text{Max month production rate (metric ton/d)}}{\text{Dried product solids concentration}}\right) \times \frac{7\text{ days}}{\text{week}}$$

$$= 1\,323 \text{ metric ton } H_2O\text{/wk}$$

$$\textbf{Annual average}\text{ evaporation load (metric ton }H_2O\text{/wk)} = \left(\frac{\text{Annual average production rate (metric ton/d)}}{\text{Cake solids concentration}} - \frac{\text{Annual average production rate (metric ton/d)}}{\text{Dried product solids concentration}}\right) \times \frac{7\text{ days}}{\text{week}}$$

$$= 1\,018 \text{ metric ton } H_2O\text{/wk}$$

Determine hourly evaporation rate at maximum monthly conditions to establish drying capacity.

$$\begin{array}{c}\text{Evaporation}\\ \text{rate for max}\\ \text{month load}\\ (\text{kg } H_2O/\text{h})\end{array} = \frac{\begin{array}{c}\text{Weekly}\\ \text{evaporation}\\ \text{load (metric ton}\\ H_2O/\text{wk)}\end{array}}{\begin{array}{c}\text{Hours of}\\ \text{operation/week}\end{array}} \times \frac{1\,000 \text{ kg } H_2O}{\begin{array}{c}\text{Metric ton}\\ H_2O\end{array}}$$

$$= \boxed{\begin{array}{c}12\,845 \text{ kg } H_2O/\text{h}\\ \textit{This is required}\\ \textit{total drying}\\ \textit{evaporation capacity.}\end{array}} \quad \text{or} \quad 28\,310 \text{ lb } H_2O/\text{hr}$$

Determine capacity for one drying train.

$$\begin{array}{c}\text{One dryer's}\\ \text{evaporative}\\ \text{capacity}\\ (\text{kg } H_2O/\text{h})\end{array} = \frac{\begin{array}{c}\text{Design}\\ \text{evaporation rate}\\ (\text{kg } H_2O/\text{h})\end{array}}{\begin{array}{c}\text{Number of}\\ \text{drying trains}\end{array}}$$

$$= \boxed{\begin{array}{c}6\,422 \text{ kg } H_2O/\text{h}\\ \textit{This is required}\\ \textit{evaporative capacity for}\\ \textit{one drying train.}\end{array}} \quad \text{or} \quad 14\,155 \text{ lb } H_2O/\text{hr}$$

Determine hours of operation given this capacity and processing annual average throughput.

$$\begin{array}{c}\text{Hours of}\\ \text{operation at}\\ \text{annual average}\\ \text{capacity}\\ (\text{hr/wk})\end{array} = \frac{\begin{array}{c}\text{Annual average}\\ \text{evaporation}\\ \text{load (metric}\\ \text{ton } H_2O/\text{wk)}\end{array}}{\begin{array}{c}\text{Drying capacity}\\ (\text{kg } H_2O/\text{h})\end{array}} - \frac{1\,000 \text{ kg } H_2O}{\begin{array}{c}\text{Metric ton}\\ H_2O\end{array}}$$

$$= 80 \text{ hr/wk}$$

Therefore, the units would only need to operate 3 to 4 days per week at average conditions. May require biosolids storage upstream to allow for optimum dryer operations.

Note, if initial biosolids quantity is significantly less than design quantity, the designer may want to consider staging installation. Dryers are not thermally efficient if operated at throughput rates significantly lower than design capacity. Designer should consult with manufacturer to determine optimum turndown capacity.

Determine annual heat energy requirements using annual average biosolids production.

$$\text{Annual heat energy required (MJ/a)} = \frac{\text{Annual average evaporation load (metric ton } H_2O\text{/wk)}}{1\,000 \text{ kg } H_2O\text{/metric ton } H_2O} - \text{Heat energy consumed to evaporate water (MJ/kg } H_2O \times \text{Weeks per Year}$$

$$= 196\,885\,384 \text{ MJ/a} \quad \text{or} \quad 186\,621 \text{ MMBTU/yr}$$

Determine approximate annual electrical energy demand.

NOTE: This parameter varies significantly per system type and manufacturer. Designer should consult manufacturer.

$$\text{Annual electrical energy required (kWh/a)} = \frac{\text{Total drying evaporation capacity (kg } H_2O\text{/h)}}{1\,000 \text{ kg } H_2O\text{/metric ton } H_2O} - \text{System electrical energy requirement (kW/metric ton } H_2O\text{/h)} \times \text{Hours/Year of operation}$$

$$= 5\,847\,789 \text{ kWh/a}$$

8.1.5 Mass Balance

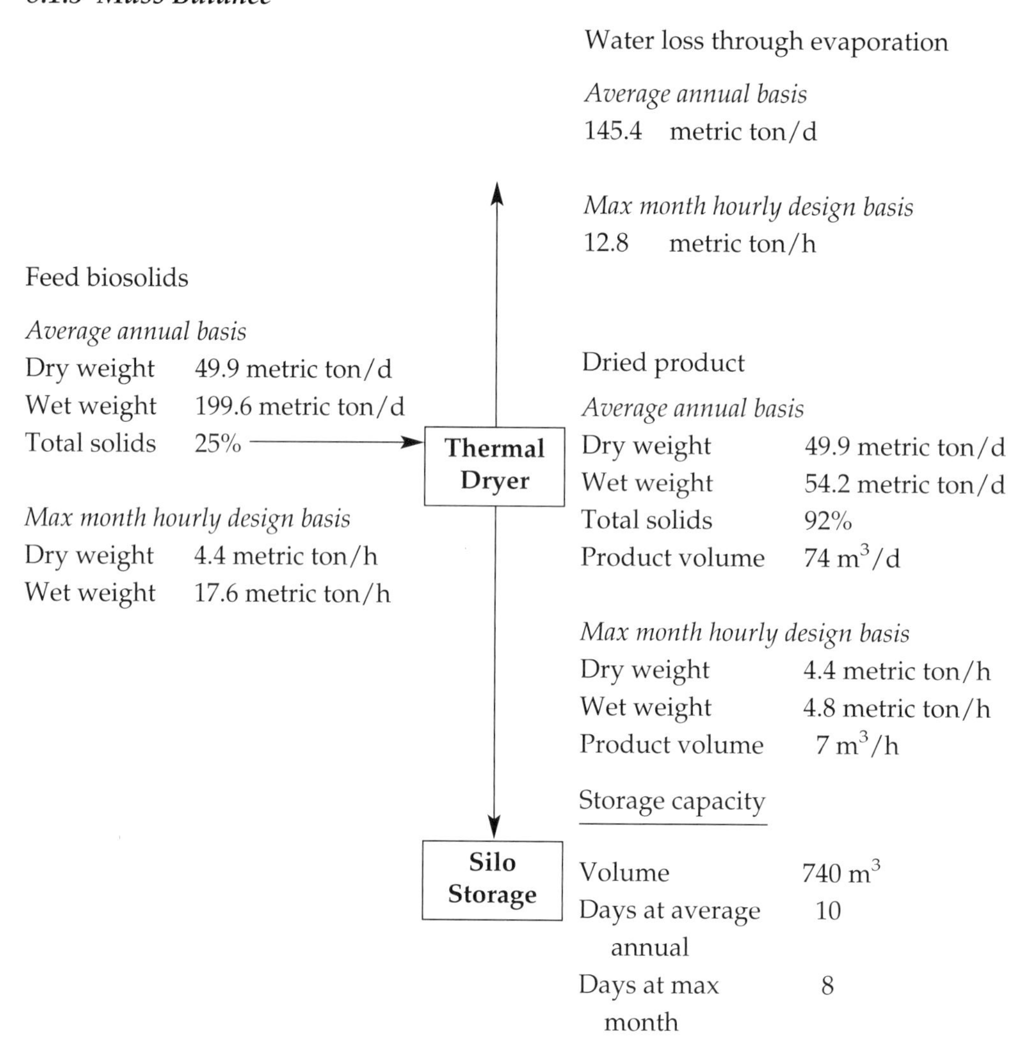

8.2 Incineration

Fluid-bed thermal oxidation can be used to dispose of both digested and undigested solids (raw solids or typical mixture of primary and secondary). Typically, digestion can reduce solids production by up to 33%, so it can reduce the size of the incinerator.

However, incinerating digested solids also has disadvantages. Compared to raw solids, digested solids have a lower volatiles content and, therefore, a lower heat content. Furthermore, dewatering digested solids is more difficult than dewatering raw solids.

TABLE 26.10 Solids feed rates.

Component	Raw solids	Digested solids
Dry metric ton/d	75	50
Wet metric ton/d	268	200
Total solids	28%	25%

In the following example, two sizes of fluidized-bed incinerators are shown. The first, which is incinerating raw sludge, has a capacity of 75 dry metric ton/d (82.5 ton/d) for raw solids. The second, which is incinerating digested solids, has a capacity of 50 metric dry ton/d (55 ton/d). Raw and digested solids characteristics, including proximate analysis, ultimate analysis, and heating values are presented in Tables 26.10 through 26.12.

8.2.1 *Design Data*

A centrifuge decanter is used to dewater solids. The raw and digested solids cake contains 28% and 25% solids, respectively.

Raw solids have a higher combustible content and heat value than digested solids. We use 74% and 61% as combustible contents for raw and digested solids, respectively. The combustible heating values are 26 050 kJ/kg (11 200 Btu/lb) for raw solids, and 23 260 kJ/kg (10 000 Btu/lb) for digested solids.

Three alternative cases are developed as follows:

(1) Case No. 1: Incineration of raw solids (or mixture of primary and secondary)
(2) Case No. 2: Incineration of digested solids, where biogas is used as auxiliary fuel
(3) Case No. 3: Incineration of digested solids, where dryer is used before incineration.

In each case, maximum heat is removed from the incinerator exhaust gas to preheat combustion air and generate steam. Hot air up to 665°C is generated in a tube and shell

TABLE 26.11 Proximate analyses and heating values.

Component	Raw solids	Digested solids
Ash	26.0%	39%
Volatiles	70.5%	60.0%
Fixed carbon	3.5%	1.0%
Total	100%	100%
HHV (kcal/kg volatile solids)	26 050	23 260

TABLE 26.12 Ultimate analysis of primary and digested solids.

Volatile composition	Raw solids	Digested solids
Carbon	55.8%	53.3%
Hydrogen	8.1%	7.7%
Oxygen	27.5%	29.7%
Nitrogen	7.4%	7.1%
Sulfur	1.2%	2.2%
Total	**100%**	**100%**

gas to air heat exchanger and is used as combustion and fluidizing air to minimize auxiliary fossil fuel consumption.

In the first two cases (Nos. 1 and 2), superheated steam at 3500 kPa (507 psi) and 370°C (700°F) is generated in a water-tube waste-heat boiler. The steam will be used in a steam turbine generator to produce electricity (see Figure 26.20a). In these two cases, a push–pull system with a force-draft fluidizing air blower and an induced draft fan is used to maintain negative pressure in the boiler and to minimize risk of ash leakage.

In Case No. 3 where sludge is pre-dried, saturated steam at 850 kPa (125 psi) is generated in a fired-tube boiler and used in a dryer to evaporate enough water so the incinerator can operate autogenously (see Figure 26.20b). Because of the incinerator's size and the amount of steam used for pre-drying, no electricity is produced. With low-pressure steam, a fire-tube boiler can be used instead of a water-tube boiler. So, a push system can be conveniently used in this case without risk of ash leakage.

In all cases, a Venturi scrubber is used to remove particulate, heavy metals, and acid gases.

8.2.2 *Procedure*

(1) Establish hourly throughput requirements and feed parameters.
(2) Perform mass and heat balances. [These are complex and difficult to perform manually. They can be performed by incinerator vendors or via commercially available models. A description of how to perform mass and heat balances is available in the literature (WEF, 2008).]
(3) Use results of mass and heat balances to select and size equipment. Designers should consult incinerator vendors, who are experienced in selecting and sizing of equipment.

Results of calculations follow (see Tables 26.13 through 26.16).

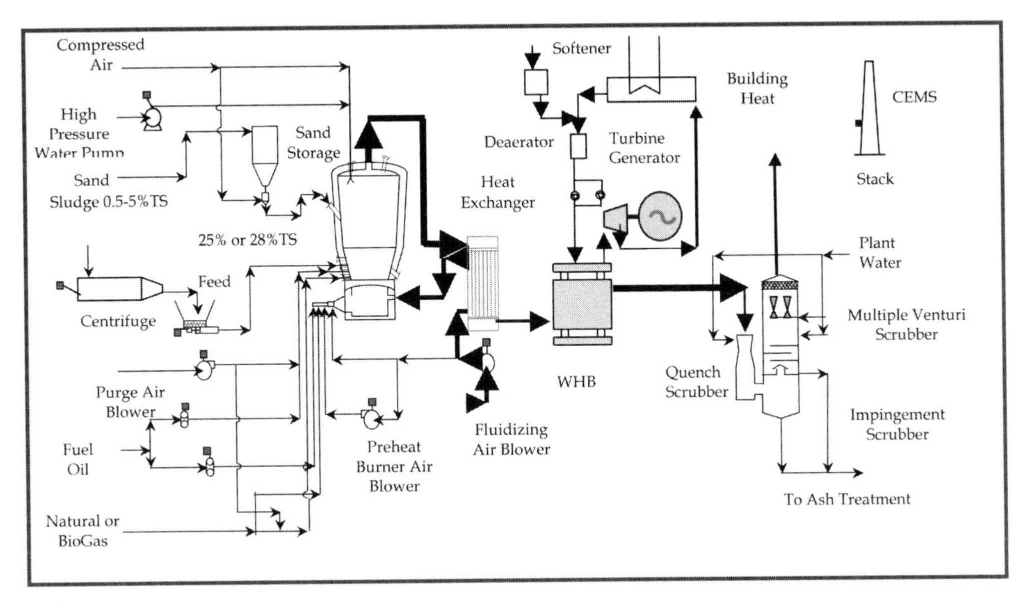

(a)

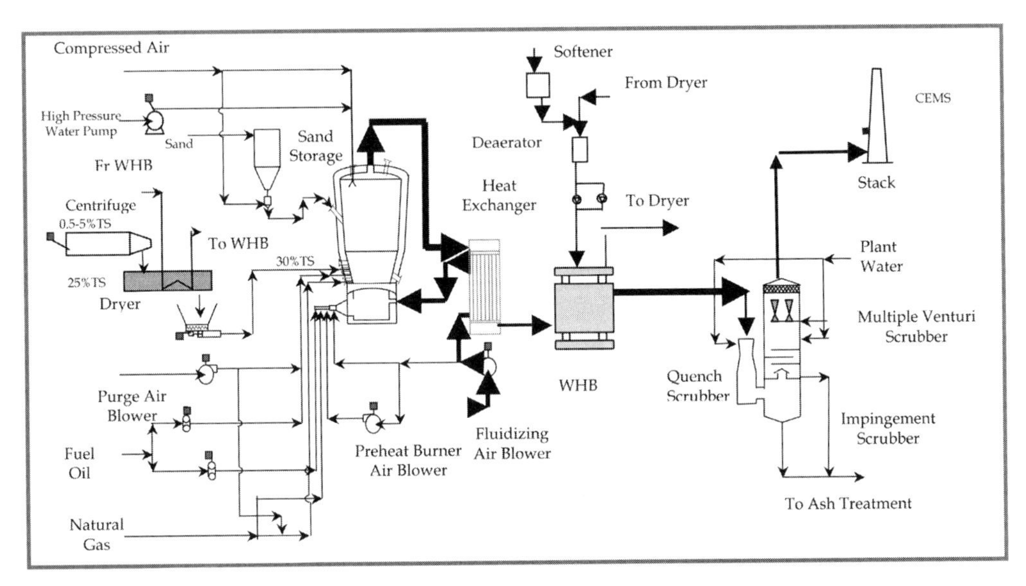

(b)

FIGURE 26.20 (a) Push–pull system with water tube boiler for high-pressure superheated steam production and (b) push system with fired-tube boiler for low-pressure saturated steam production.

TABLE 26.13 Case 1: Incineration of raw solids (cake containing 28% total solids).

	kg/h	kJ/h
Waste solid	3 125	60 292 323
Waste water	8 036	706 388
Aux fuel	0	0
Quench water	12	1 059
Purge air	1 063	53 058
Fluidizing air	25 074	8 115 775
Air moisture	492	1 519 723
Sand	39	753
Total in	**37 841**	**70 689 078**
	kg/h	**kJ/h**
Dry gases	26 764	24 809 974
Water vapor	10 226	43 728 111
Ash	813	636 012
Sand	39	30 528
Heat losses	0	1 484 456
Total out	**37 841**	**70 689 078**

TABLE 26.14 Case 2: Incineration of biosolids using biogas (cake containing 25% total solids).

	kg/h	kJ/h
Waste solid	2 083	29 594 225
Waste water	6 250	549 413
BioGas	147	7 892 498
Quench water	0	0
Purge air	1 063	53 058
Fluidizing air	15 410	10 591 321
Air moisture	310	1 159 472
Sand	25	486
Total in	**25 290**	**49 840 475**
	kg/h	**kJ/h**
Dry gases	16 846	15 609 632
Water vapor	7 607	32 528 623
Ash	813	636 012
Sand	25	19 570
Heat losses	0	1 046 638
Total out	**25 290**	**49 840 475**

TABLE 26.15 Case 3: Incineration of biosolids using dryer (cake containing 30% total solids).

	kg/h	kJ/h
Waste solids	2 083	29 594 225
Waste water	4 861	1 220 918
Biogas	0	0
Quench water	105	26 409
Purge air	1 063	53 058
Fluidizing air	12 452	8 557 750
Air moisture	255	946 819
Sand	19	1 051
Total in	**20 838**	**40 475 049**
	kg/h	**kJ/h**
Dry gases	13 905	12 882 834
Water vapor	6 102	26091 376
Ash	813	636 012
Sand	19	14 873
Heat losses	0	849 955
Total out	**20 838**	**40 475 049**

8.2.2.1 Heat and Mass Balances

For raw solids at 28% total solids, the process is autogenous with a wind-box air temperature at 326°C. The incinerator's thermal capacity is 70.7×10^6 kJ/h.

At 25% total solids and combustion of digested solids with a maximum wind-box air temperature of 650°C, the process still requires 7.9×10^6 kJ/h of biogas. As the plant capacity is reduced to 50 dry metric ton/d, the plant thermal capacity is also reduced to 49.8×10^6 kJ/h (47.2 MM Btu/hr) from 70.7×10^6 kJ/h (67 MM Btu/hr) for raw solids.

In this case, a disc dryer is used to thermally dewater centrifuge cake. The dryer is designed to raise the solids content of digested solids from 25 to 30% total solids. With less water in the solids than in Case No. 2, the incinerator's thermal capacity is reduced from 49.8×10^6 kJ/h (47.2 MM Btu/hr) to 40.5×10^6 kJ/h (38.5 MM Btu/hr). Among the three cases, this case results in the smallest incinerator.

8.2.2.2 Other Results and Conclusions

The calculations have shown that with raw solids, the fluid bed is larger with more exhaust-gas flowrates. Therefore, more heat can be recovered from exhaust gases in Case No. 1 (31×10^6 kJ/h as opposed to 21 and 17×10^6 kJ/h for Case Nos. 2 and 3, respectively).

TABLE 26.16 Other results and conclusions.

	Case 1 raw	Case 2 digested using biogas	Case 3 digested using dryer
Dry feed rate (metric ton/d)	75	50	50
Volatile solids	74%	61%	61%
Total solids	28%	25%	30%
Incinerator			
OD freeboard (m)	9.26	7.82	6.67
Diameter distributor (m)	5.26	4.09	3.51
Operating weight (tonne)	296	215	173
Windbox temperature (°C)	316	649	649
Combustion airflow rate (kg/h)	25 546	15 701	12 686
Flue gas flowrate (kg/h)	36 990	24 453	20 007
Thermal capacity ($\times 10^6$ kJ/h)	71	50	42
HE capacity ($\times 10^6$ kJ/h)	7	10	8
WHB capacity ($\times 10^6$ kJ/h)	24	11	9
Steam generation (kg/h)	9 216[a]	4 140[a]	4 034[b]
Steam use (kg/h)	9 216[a]	4 140[a]	1 935[b]
Net steam production (kg/h)	0[a]	0[a]	2 099[b]
Venturi scrubber gas (m^3/h)			
Dry gas rate (kg/h)	26 764	17 556	13 905
Water vapor (kg/h)	13 906	7 607	6 102
Water vapor condensation (kg/h)	11 833	6 464	5 304
Auxiliary fuel ($\times 10^6$ kJ/h)	—	8	—
Venturi scrubber plant effluent water rate (m^3/h)	228	171	171
Electricity consumption[c] (kW-h/h)	968	692	759
Estimated electricity production (kW-h/h)	1 551	697	—
Net production (kW-h/h)	583	5	(759)

[a] Superheated steam (3 500 kPag and 370°C).
[b] Saturated steam (850 kPag).
[c] Not including dewatering equipment.

Furthermore, because raw solids have a higher heat value, the fraction of heat needed to be recovered in the air preheater is also minimal (23% in case 1, versus over 47% for the two other cases). Therefore, more steam can be generated.

With more steam generated in Case No. 1, more electricity can be produced (1.55 MW-h/h for Case 1, versus Case 2 with 0.7 MW-h/h). Electricity generation in Case Nos. 1 and 2 is large enough to cover all electricity needed in the incinerator. A net electricity production of 37% (0.6 MW-h/h) is found in Case 1 with raw solids.

Case No. 3, with 4034 kg/h of 850-kPa saturated steam generated, 1935 kg/h is used in the dryer and the remaining 2099 kg/h can be used as building heat. This represents a net steam production of about 52%.

9.0 REFERENCES

Baturay, A. (1990) *Case Studies of Total Hydrocarbon Emissions (THC) From Multiple Hearth Sewage Sludge Incinerators and THC Reduction Strategies;* Prepared for the Association of Metropolitan Sewerage Agencies Incinerator Workgroup Meetings; New Orleans, Louisiana; Association of Metropolitan Sewerage Agencies: Washington, D.C.

California Integrated Waste Management Board (2001) *Conversion Technologies for Municipal Residuals;* a background primer prepared for the Conversion Technologies for Municipal Residuals Forum; Sacramento, California, May 3–4; California Integrated Waste Management Board: Sacramento, California.

Dangtran, K.; Jeffers, S.; Mullen, J. F.; Cohen, A. J. (1999) Control Problem Waste Feeds in Fluid Beds. *Chem. Eng. Prog.*, May.

Dangtran, K.; Mullen, J. F.; Mayrose, D. T. (2000) A Comparison of Fluid Bed and Multiple Hearth Biosolids Incineration. *Proceedings of the 14th Annual Water Environment Federation Residuals and Biosolids Management Conference*; Boston, Mass., Feb; Water Environment Federation: Alexandria, Virginia.

Dangtran, K.; Holst, T. (2001) Minimization of Major Air Pollutants from Sewage Sludge Fluid Bed Thermal Oxidizers. *Proceedings of the 74th Annual Water Environment Federation Exposition and Conference*; Atlanta, Georgia, Oct 13–17; Water Environment Federation: Alexandria, Virginia.

DiGangi, D.; Melchiori, E.; Habetz, D. (2008) The Beneficial Reuse of Sludge Incinerator Exhaust Gases to Produce Renewable Energy. *Proceedings of the 81st Annual Water Environment Federation Technical Exhibition and Conference* [CD-ROM]; Chicago, Illinois, Oct 18–22; Water Environment Federation: Alexandria, Virginia.

EnerTech Environmental (2001) *Phase II SBIR Final Technical and Commercialization Report*; NSF Award Number DMI-9983559; National Science Foundation, Arlington, Virginia; Feb.

Geldart, D. (1973) Types of Gas Fluidization, *Powder Technol.*, **7**, 285–292.

Intergovernmental Panel on Climate Change (1995) *Climate Change 1995: The Science of Climate Change,* Published for the Intergovernmental Panel on Climate Change; Cambridge University Press: New York.

Intergovernmental Panel on Climate Change (2001) *Climate Change 2001: The Scientific Basis,* Published for the Intergovernmental Panel on Climate Change; Cambridge University Press: New York.

Intergovernmental Panel on Climate Change (2007) *Climate Change 2007: The Physical Scientific Basis,* Published for the Intergovernmental Panel on Climate Change; Cambridge University Press: New York.

Lewis, F. M.; Lundberg, L. A. (1988) Modifying Existing Multiple-Hearth Incinerators to Reduce Emissions. *Proceedings of the National Conference on Municipal Sewage Treatment Plant Sludge Management*; Palm Beach, Florida, June; Hazardous Materials Research Institute.

Los Angeles County Sanitation Districts (1988) *1988 Survey of Thermal Conditioning and Wet Air Oxidation Facilities*; Los Angeles County Sanitation Districts: Whittier, California.

Niessen, W. R. (1988) Thermal Processing of Wastewater Treatment Plant Sludges. *Proceedings of the National Conference on Municipal Sewage Treatment Plant Sludge Management*; Palm Beach, Florida, June; Hazardous Materials Research Institute.

North American Manufacturing Co. (1965) *North American Combustion Handbook*; North American Manufacturing: Cleveland, Ohio.

Nuss, S.; Persinger, D.; Brunner, T.; Netzel, J. (2008) Performance of Instrumentation and Control Upgrades to Multiple Hearth Furnace in Anchorage, Alaska. *Proceedings of the 81st Annual Water Environment Federation Technical Exhibition and Conference* [CD-ROM]; Chicago, Illinois, Oct 18–22; Water Environment Federation: Alexandria, Virginia.

O'Kelley, S.; Williamson, R.; Lewis, F. M. (2006) Who Says you Can't Teach Old Dogs New Tricks? *Proceedings of the 20th Annual Water Environment Federation Residuals and Biosolids Management Conference*; Cincinnati, Ohio, March; Water Environment Federation: Alexandria, Virginia.

Orange County Sanitation District (2003) *Long-Term Biosolids Master Plan*; Job No. J-40-7; Orange County Sanitation District: Huntington Beach, California.

Quast, D. (2006) Energy Efficiency Improvements for the Metropolitan Wastewater Treatment Plant Solids Processing Facility. Paper presented at the Conference on the Environment, a Joint Conference by the Minnesota Section of the Central States Water Environment Association and the Air and Waste Management Association; Bloomington, Minnesota, November.

Porter, J.; Lill, W.; Mansfield, W. (2002) Reviewing Multiple Hearth Furnaces: The Atlanta Experience. *Proceedings of the 16th Annual Water Environment Federation*

Biosolids Conference [CD-ROM]; Austin, Texas, Feb 27–29; Water Environment Federation: Alexandria, Virginia.

RHOX International, Inc. (1989) *RHOX Process*; Technical Bulletin MHF-1; RHOX International Inc.: Salt Lake City, Utah.

Sapienza, F.; Walsh, T.; Sangrey, K.; Barry, L. (2007) Upgrade of UBWPAD's Multiple-Hearth Furnace Sludge Incinerators. *Proceedings of the Annual Water Environment Federation/American Water Works Association Joint Residuals and Biosolids Management Conference*; Denver, Colorado, April; Water Environment Federation: Alexandria, Virginia.

U.S. Environmental Protection Agency (1979) *Process Design Manual for Sludge Treatment and Disposal*; EPA-625/1-79-011; U.S. Environmental Protection Agency: Washington, D.C.

U.S. Environmental Protection Agency (1983a) *Assessment of Sludge Processing Problems at Selected POTWs*; U.S. Environmental Protection Agency, Office of Water Management: Cincinnati, Ohio.

U.S. Environmental Protection Agency (1983b) *Municipal Wastewater Sludge Combustion Technology*; EPA-625/4-85/015; U.S. Environmental Protection Agency: Washington, D.C.

U.S. Environmental Protection Agency (1985b) *Multiple-Hearth and Fluid Bed Sludge Incinerators: Design and Operational Considerations*; EPA-430/9-86-002; U.S. Environmental Protection Agency: Washington, D.C.

U.S. Environmental Protection Agency (1993) Standards for the Use or Disposal of Sewage Sludge. *Code of Federal Regulations*, **40**, 503, Subpart E.

U.S. Environmental Protection Agency (1998) Compilation of Air Pollutant Emission Factors, Volume I: Stationary Point and Area Sources; AP-42; Section 2.2; U.S. Environmental Protection Agency, Office of Air Quality Planning and Standards: Research Triangle Park, North Carolina.

Waltz, E. W. (1990) *Technical Discussion of Proposed EPA Hydrocarbon Regulation for Sludge Incinerators—Charts and Graphs*; Prepared for the Association of Metropolitan Sewerage Agencies Incinerator Workgroup Meetings; New Orleans, Louisiana; Association of Metropolitan Sewerage Agencies: Washington, D.C.

Water Environment Federation (1991) *Sludge Incineration: Thermal Processing of Wastewater Treatment Residues*; Manual of Practice FD-19; Water Environment Federation: Alexandria, Virginia.

Water Environment Federation (2004) *Control of Odors and Emissions from Wastewater Treatment Plants*; Manual of Practice No. 25; McGraw-Hill: New York.

Water Environment Federation (2009) *Wastewater Solids Incineration Systems*; Manual of Practice No. 30; McGraw-Hill: New York.

10.0 SUGGESTED READINGS

Baturay, A.; Bruno, J. M. (1990) Reduction of Metal Emissions from Sewage Sludge Incinerators with Wet Electrostatic Precipitators. Paper presented at 83rd Annual Meeting of the Air Waste Management Association; Pittsburgh, Pennsylvania, June.

Borghesi, J.; Burrowes, P.; Voth, H.; Flood, R. (2002) A State-of-the-Art Fluid Bed Incineration Process to Meet the Solids Processing Needs of the Twin Cities. *Proceedings of the 16th Annual Water Environment Federation Residuals and Biosolids Management Conference*; Austin, Texas; February; Water Environment Federation: Alexandria, Virginia.

Burrowes, P.; Brady, P. (2005) Development in Emissions Technology for Incinerators. *Proceedings of the 78th Annual Water Environment Federation Technical Exhibition and Conference* [CD-ROM]; Washington, D.C., Oct 29–Nov 2; Water Environment Federation: Alexandria, Virginia.

Burrowes, P.; Borghesi, J.; Quast, D. (2007) The Twin Cities Sludge-to-Energy Plant Reduces Greenhouse Gas Emissions. *Proceedings of the 80th Annual Water Environment Federation Technical Exhibition and Conference* [CD-ROM]; San Diego, California, Oct 13–17; Water Environment Federation: Alexandria, Virginia.

Haug, R. T.; et al. (1983) Thermal Pretreatment of Sludges: A Field Study. *J. Water Pollut. Control Fed.*, **55**, 23.

Los Angeles County Sanitation Districts (1977) *Assessment of Existing and Past Sludge Product Marketing Experiences*; Technical Report 9; Prepared for Los Angeles/Orange County Metropolitan Area Project; Los Angeles County Sanitation Districts: Whittier, California.

Los Angeles County Sanitation Districts (1978) *Carver Greenfield Process Evaluation: A Process for Sludge Drying*; Prepared for Los Angeles/Orange County Metropolitan Area Project; Los Angeles County Sanitation Districts: Whittier, California.

Marshall, D. W.; Gillespie, W. J. (1974) Comparative Study of Thermal Techniques for Secondary Sludge Conditioning. *Proceedings of the 29th Purdue Industrial Waste Conference*; West Lafayette, Indiana; Purdue University: West Lafayette, Indiana.

Metcalf and Eddy, Inc. (1991) *Wastewater Engineering: Treatment, Disposal, Reuse*, 3rd ed.; Tchobanoglous, G., Ed.; McGraw-Hill: New York.

Metro Wastewater Reclamation District (2006) *Biosolids Management Program/Facility Study PAR 880*; Metro Wastewater Reclamation District: Denver, Colorado.

Morton, E. L. (2006) A Sustainable Use for Dried Biosolids. *Proceedings of the 79th Annual Water Environment Federation Technical Exhibition and Conference* [CD-ROM]; Dallas, Texas, Oct 21–25; Water Environment Federation: Alexandria, Virginia.

Osaka Gas Engineering Co. Ltd. (1983) *Osaka Gas Sludge Melting Process Technical Notes*; Osaka Gas Engineering Co. Ltd.: Osaka, Japan.

Oxidyne Corp. (1987) *An Introduction to Oxidyne Technology*; Technical Bulletin M851; Oxidyne Corp.: Houston, Texas.

Perry, R. H.; Green, D. W., Eds. (1984) *Perry's Chemical Engineers' Handbook*; McGraw-Hill: New York.

U.S. Environmental Proctection Agency (1978) *Effects of Thermal Treatment of Sludge on Municipal Wastewater Treatment Cost*; EPA-600/2-78-073; U.S. Environmental Proctection Agency, Office of Water Management: Cincinnati, Ohio.

U.S. Environmental Protection Agency (1985) *Heat Treatment/Low Pressure Oxidation: Design and Operational Considerations*; EPA-430/9-85-001; U.S. Environmental Protection Agency: Washington, D.C.

U.S. Environmental Protection Agency (1987a) *Dewatering Municipal Sludges Design Manual*; EPA-625/1-87-014; U.S. Environmental Protection Agency, Office of Water Management: Cincinnati, Ohio.

U.S. Environmental Protection Agency (1987b) *Aqueous-Phase Oxidation of Sludge Using the Vertical Reaction Vessel System*; EPA-600/2-87-022; U.S. Environmental Proctection Agency, Office of Water Management: Cincinnati, Ohio.

Zimpro, Inc. (1978) *Characteristics of Thermally Treated Sewage Sludge*; Technical Bulletin 2302-T; Zimpro, Inc.: Rothschild, Wisconsin.

Zimpro, Inc. (1984) *Product Manual Sludge Management Systems 300*; Zimpro, Inc.: Rothschild, Wisconsin.

Chapter 27

Use and Disposal of Residuals and Biosolids

1.0 INTRODUCTION

The methods for using and disposing of wastewater solids vary considerably. The choice heavily depends on the type of solids involved. Solids products—and approaches to manage them include:

- Sludge. If dewatered, raw primary and secondary solids (or sludge) can be landfilled or incinerated. Most other use and disposal options—particularly beneficial uses (e.g., land application)—require that solids first be treated to meet the U.S. Environmental Protection Agency's (U.S. EPA's) requirements in 40 CFR 503, *Standards for the Use or Disposal of Sewage Sludge* (also called *Part 503*).
- Biosolids. *Biosolids* are any solids that have been stabilized to meet the criteria in the Part 503 regulations and, therefore, can be beneficially used. (For more information on Part 503, see Chapter 20.) There are a wide variety of stabilization processes, which produce differing types of biosolids (e.g., liquid or dewatered biosolids, compost, heat-dried biosolids, and alkaline-stabilized biosolids). Most of these products can be land-applied; some are suitable for commercial marketing and distribution.
- Ash. Ash is a product of incineration. Ash historically was landfilled, but in recent years, there has been more emphasis on finding beneficial uses for this material (e.g., as landfill cover, a soil amendment, an ingredient in concrete, a fine aggregate in asphalt, a flowable fill material, and an additive in brick manufacturing).

However, no matter which use or disposal option is selected, the project team should review the *National Manual of Good Practice for Biosolids* (NBP, 2005) for guidance on developing and implementing biosolids management practices that emphasize environmental stewardship and strong community relations. Published by the National Biosolids Partnership (NBP), this manual is the foundation of NBP's Environmental Management System (EMS) program (NBP, 2009).

Also, water treatment residuals can be co-disposed or beneficially used with biosolids. For more information on managing water residuals, see *Water Treatment Principles and Design* (MWH, 2005).

2.0 LAND APPLICATION

Land application is the practice of adding biosolids to land for beneficial purposes (e.g., to promote crop growth, to promote forest growth, and to reclaim former mining sites and other disturbed land). In these applications, both plants and soil benefit from the nutrients and organic matter in biosolids.

The popularity of land application has grown dramatically over the last 30 years. Rising disposal costs and encouragement from U.S. EPA have substantially increased the number of facilities undertaking land-application programs. According to a 2004 survey conducted by the New England Biosolids and Residuals Association (NEBRA), 55% of the biosolids generated nationwide were "applied to soils for agronomic, silvicultural, and/or land restoration purposes, or were likely stored for those purposes", and 74% of land-applied biosolids were used for agricultural purposes (NEBRA, 2007).

The NEBRA survey also noted the term *beneficial use* historically referred to "biosolids that are applied to soils to take advantage of the nutrients and organic matter they contain." In the future, it noted, this definition may be too narrow as biosolids are used in ways that provide other benefits (e.g., energy).

2.1 Regulatory Considerations

Federal and state regulations establish controls for the land application of biosolids. Some activities (e.g., the selection and management of land-application sites) also can be governed on a local level.

2.1.1 Federal Requirements

At the federal level, biosolids are regulated under Section 405(d) of the Clean Water Act. Specific criteria are set forth in 40 CFR 503 (U.S. EPA, 1993a). (For more information on Part 503, see Chapter 20.)

2.1.2 *State Requirements*

Some states have enacted regulations more stringent than Part 503. Wastewater treatment professionals should consult state regulatory requirements when assessing the feasibility of land application or planning a land-application program.

State requirements vary. However, an increasing number of states require nutrient management planning and some states (e.g., Virginia) also may require conservation planning (Evanylo, 1999). In addition, some states require permits for land-application sites.

2.1.2.1 Nutrient Management Planning

Land-applied biosolids supply nutrients (e.g., nitrogen, phosphorus, and micronutrients) to plants. However, too much of any nutrients can lead to water quality issues. Excess nitrogen is associated with groundwater concerns, for example, and excess phosphorus is associated primarily with surface water concerns. [For more information on nutrients and biosolids, see *Comparing the Characteristics, Risks and Benefits of Soil Amendments and Fertilizers Used in Agriculture* (Moss et al., 2002) and the *National Manual of Good Practice for Biosolids* (NBP, 2005).]

Part 503 requires that biosolids be applied at the agronomic rate for nitrogen (i.e., the rate needed to meet a given crop's nitrogen requirements), and many state regulations share this requirement. However, increasingly more states are requiring that land-appliers manage phosphorus as well. Phosphorus-based management can quadruple the area needed for land-application programs. Some of the states that require phosphorus-based management rely on the Nutrient Management Standard (Code 590) issued by the U.S. Department of Agriculture's (USDA's) Natural Resources Conservation Service (NRCS) (USDA-NRCS, 1999). Where used, this national standard (called the *P Index*) typically is modified by states to account for local conditions.

2.1.2.2 Site Permits

Site permitting programs typically are administered by state agencies, although local entities can be involved in some states. The permits contain the issuing agency's specific requirements for land application, which are designed to ensure that the biosolids are being beneficially used, not disposed. Permit requirements also are typically designed to protect public health, surface water, and groundwater, as well as address the aesthetic concerns (e.g., odor) of the land-application sites' neighbors.

For example, most state programs impose separation distances (setbacks) between the areas receiving biosolids and adjacent site features (e.g., developments, wells, dwellings, surface water, wells, roads, and rights of way). Setbacks are based primarily on the potential for surface runoff, leaching, and aesthetic concerns (see Table 27.1).

TABLE 27.1 Representative setback requirements for land application.

Feature	Setback distance	
	ft	m
Public road	0–50	0–15
House	20–500	6–152
Well	100–500	30–152
Surface water	25–300	8–91
Property line	None listed–100	None listed–30
Intermittent stream	10–200	3–61

Many states will reduce the setback requirements if the biosolids are immediately incorporated into the soil or applied via subsurface injection (Forste, 1996).

2.2 Project Planning

A successful land-application program is based on the careful consideration of multiple factors [e.g., the biosolids to be applied, the suitability of proposed land-application sites, and the estimated application rates (which determine the amount of land required)].

2.2.1 Biosolids Characteristics and Suitability

Regulations and crop needs typically determine which biosolids characteristics [e.g., pathogen content, vector attraction, and nutrient content (typically nitrogen, phosphorus and/or calcium)] are most important when developing a land-application program (see Table 27.2). Site conditions and economics also may determine whether the biosolids should be applied in liquid or cake form.

TABLE 27.2 Typical biosolids physical characteristics (Logan, 2008; Lue-Hing et al., 1998).

Property	Class A alkaline stabilized	Compost	Liquid	Cake	Heat-dried
Total solids (%)	30–65	58	4	22	95
Volatile solids (% of total solids)	12–20	60	61	60	67
Bulk density [kg/m^3 (lb/cu ft)]	1 040–1 200 (65–75)	720 (45)	960 (60)	880 (55)	530–740 (33–46)
Organic content (%)	12–20	50–75	60–75	60–75	78
pH	—	—	—	—	6.4–8.0

2.2.1.1 Physical Characteristics

The textures and particle sizes of biosolids are as varied as the processes used to create them. Heat-dried biosolids can be granular; the irregular or spherical particles are about the size of a small ball bearing. Composts have a texture similar to that of peat, but depending on how they were made, may contain some wood fibers or wood chips. Alkaline-stabilized materials may be clay-like or a soil-like, depending on moisture content and other factors. Aerobically and anaerobically digested biosolids may be liquid or dewatered into a clay-like material.

2.2.1.2 Pathogens and Vectors

Biosolids' pathogen content depends on the level of treatment they have received (see Table 27.3). Raw primary solids contain the most pathogens, while biosolids that meet Part 503's Class A standards have pathogen concentrations below detection limits. Concentrations of specific pathogens (e.g., helminths) depend on the sources in a sewershed, as well as the climate.

Before they can be land-applied, biosolids must meet Class A or B pathogen-reduction standards (as well as relevant vector attraction reduction and pollutant concentrations). Land-application sites that receive Class B biosolids have additional restrictions (e.g., waiting periods before specific crops can be grown) (U.S. EPA, 1994a).

Part 503 requires that Class B biosolids contain less than 2 million colony-forming units (CFU) [most probable number (MPN)] of fecal coliforms per gram of dry biosolids. Class A biosolids must contain less than 1 000 MPN/g of fecal coliforms and less than 3 MPN/4g of Salmonella bacteria and meet one of six alternatives. One

TABLE 27.3 Typical pathogen and pathogen-indicator concentrations in biosolids (Lue-Hing et al., 1998).

Pathogen	Class A alkaline stabilized	Class A composted	Class A heat-dried	40 CFR 503 Class A standards
No. of plants in sample	5	4	5	NA
Fecal coliform (MPN/g mean)	3	76	6	<1 000
Fecal coliform (MPN/g median)	1	506	8	NA
Salmonella sp. (MPN/4g mean)	2	2	0	<3
Salmonella sp. (MPN/4g median)	2	2	0	NA
No. of plants in sample	55	26	41	NA
Fecal coliform (MPN/g mean)	60	104 600	6 521	<2 million
Fecal coliform (MPN/g median)	9 600	472 600	16 071	NA
Salmonella sp. (MPN/4g mean)	2 070	NA	NA	NA
Salmonella sp. (MPN/4g median)	4 000	NA	NA	NA

alternative involves testing for enteric viruses and helminth ova (U.S. EPA, 2003). Wastewater treatment plants typically handle bacterial testing onsite, but virus and helminth ova testing may have to be done by commercial labs with the appropriate expertise.

Land-applied biosolids also must meet vector attraction reduction (VAR) requirements via extensive drying, volatile solids reduction (VSR), oxygen uptake rate, time and temperature, or pH increases (U.S. EPA, 2003). Eight VAR options are based on process parameters or testing; two are based on land management (e.g., soil incorporation or injection).

For details on pathogen and vector attraction reduction requirements, methodologies, monitoring, and testing, see *Environmental Regulations and Technology: Control of Pathogens and Vector Attraction in Sewage Sludge* (also called the "White House document") (U.S. EPA, 2003).

2.2.1.3 Metals

Biosolids contain a wide range of chemical elements, many of which are trace elements, also called *heavy metals*. This is a misnomer because the density of some trace elements (e.g., boron and arsenic) is too low to be considered a heavy metal. Likewise, not all trace elements behave as metals in the environment. The ones with metallic chemical behavior are those that are cations in the environment [e.g., copper (Cu^{2+}), cadmium (Cd^{2+}) and lead (Pb^{2+})]. Other trace elements are anions in the environment [e.g., arsenic (AsO_4^{3-}), molybdenum (MoO_4^{2-}) and selenium (SeO_4^{2-})]. This distinction is important because cationic trace elements are more bioavailable when soil pH is low, while anionic trace elements are more bioavailable when soil pH is high.

When developing 40 CFR 503, U.S. EPA conducted a risk assessment on a number of trace elements in biosolids and developed a *Hazard Index*, which considered various pathways of exposure (e.g., direct ingestion of biosolids, plant uptake, and uptake by soil organisms) from trace elements in land-applied biosolids (U.S. EPA, 1995a). The agency eliminated two trace elements (fluorine and iron) based on the Hazard Index. Then, regulators developed acceptable concentrations for the remaining 10 elements based on a pathway analysis. The *ceiling concentration* is the maximum safe concentration for biosolids land-applied at agronomic rates. A lower concentration also was established for what was later called *exceptional quality (EQ) biosolids* {this term refers to biosolids that meet Class A pathogen-reduction requirements, meet VAR requirements [503.33(a)(1) through (8)], and have low concentrations of regulated pollutants (503.13, Table 3). The intent of the concentrations for EQ biosolids is that these levels would be safe no matter how much biosolids were applied to land.

Part 503 originally regulated 10 trace elements: arsenic, cadmium, chromium, copper, lead, mercury, molybdenum, nickel, selenium, and zinc. Lawsuits filed against U.S.

EPA resulted in the removal of chromium as a regulated metal and the removal of the EQ limit for molybdenum.

When Part 503 was promulgated, there were a number of municipal wastewater treatment plants whose biosolids failed to meet the EQ limits for one or more elements (see Table 27.4). Most, however, met the ceiling concentrations, so their biosolids could be land-applied but there was a limit on the cumulative loading to a given site (U.S. EPA, 1995a). By the late 1990s, industrial pretreatment was so successful that most treatment plants could meet EQ limits. Today, trace elements in biosolids are no longer considered an important limitation in land-application programs.

2.2.1.4 Nutrients

Biosolids contain various concentrations of macro and micro nutrients.

As shown in Table 27.5, the nitrogen content in biosolids ranges from 1.0% (dry weight) for Class A alkaline-stabilized biosolids to up to 6.0% (dry weight) for heat-dried biosolids. Primary and waste activated solids mostly contain organic nitrogen (in the form of protein) while up to a third of total nitrogen in aerobically and anaerobically digested biosolids is ammonia. Biosolids contain very little if any nitrate.

The total phosphorus content in biosolids typically ranges from 0.9 to 3.1% (dry weight). This includes both organic phosphorus and various forms of inorganic phosphorus. Municipal treatment plants that use iron and aluminum salts for phosphorus removal will have solids that contain inorganic phosphorus in the form of iron and aluminum phosphates. Treatment plants that use lime for dewatering or disinfection will have solids that contain inorganic phosphorus in the form of calcium phosphates.

TABLE 27.4 Typical metals concentrations in biosolids (mg/dry kg) (Lue-Hing et al., 1998).

Metal	Class A alkaline stabilized	Compost	Liquid	Cake	Heat-dried	40 CFR 503 Table 3 limits
No. of treatment plants in sample	7	10	48	117	5	NA
Arsenic	5.79	5.39	7.78	13.86	5.58	41
Cadmium	3.07	4.64	4.90	7.03	8.94	39
Chromium	50.3	72.7	62.0	119.4	152.6	NA
Copper	176	317	448	559	472	1 500
Lead	62.0	80.4	74.8	128.6	93.3	300
Mercury	0.73	1.94	2.60	2.01	1.60	17
Molybdenum	8.20	13.8	11.0	15.3	17.9	75
Nickel	38.8	26.5	33.5	69.3	35.0	420
Selenium	2.43	3.73	5.86	6.06	8.32	100
Zinc	878	878	807	886	906	2 800

TABLE 27.5 Primary nutrient concentrations in biosolids (adapted from Moss et al., 2002; reprinted with permission from the Water Environment Research Foundation).

Nutrient content (as % of dry weight)	Class A alkaline stabilized	Compost	Liquid	Cake	Heat-dried
Nitrogen (N)	1.0	2.8	5.3	4.1	6.0
Total phosphorus (P)	0.4	1.7	2.2	2.2	3.1
Potassium (K)	0.3	0.3	0.3	0.2	0.3

Iron, aluminum, and calcium phosphates are relatively insoluble, which will affect the buildup of available phosphorus in soils receiving repeated applications of biosolids.

Biosolids also contain small amounts of potassium—typically 0.1 to 0.3% (dry weight)—and are not a significant source of potassium for crops. Most of the potassium in biosolids is in a water-soluble form.

In addition, biosolids contain various concentrations of other macro and micro nutrients [e.g., calcium, magnesium, sulfur, iron, manganese, as well as the regulated elements (copper, molybdenum, selenium, zinc)]. Some plant species also require nickel for growth. The plant availability of these nutrients depends on the major chemical phases in the biosolids. Elements like copper and zinc are known to be complexed with the biosolids organic matter, and most sulfur in biosolids is in the form of protein. On the other hand, iron and manganese exist as relatively insoluble oxides.

2.2.1.5 *Other Constituents*

Biosolids contain myriad organic and inorganic chemical compounds. Some are discharged into the collection system as household chemicals (e.g., surfactants and conditioning agents used in detergents) and very low levels of medicines (e.g., birth-control chemicals) and concentrate in biosolids. Biosolids also contain natural materials (e.g., silica and aluminosilicate clays) that enter the collection system as sediments.

In the case of alkaline-stabilized biosolids, lime and lime-containing chemicals (e.g., alkaline coal ash, cement kiln dust, and lime kiln dust) are added to biosolids for disinfection. The lime reacts with water in biosolids to form calcium hydroxide, which has a pH of 12 to 12.5 and adds soil liming value to the biosolids. Alkaline-stabilized biosolids often are used as a substitute for agricultural limestone.

2.2.2 *Site Suitability*

After establishing that the biosolids are suitable for land application, program staff must determine if suitable land is available.

2.2.2.1 *Objectives*

The objective of a site evaluation is to select application sites that not only meet the technical requirements of land application but also balance prevailing economic and social constraints. When determining how much land is needed for a land-application program, it is important to note that adequate acreage must be available during all but the most inclement weather conditions. Site availability is determined by local farming practices, and there are periods of the year when certain types of cropland are unavailable for the application of biosolids. Therefore, to ensure land availability, it is advisable to maintain two to three times the acreage actually needed in any given year to accommodate all of the biosolids produced. Typically, storage capacity at the treatment plant or in the field will also be necessary at some time during the year.

2.2.2.2 *Resources*

Soil surveys are a useful source of information for initially determining a site's suitability for land application. The surveys provide detailed soil maps on a photographic background, a soil description by series and mapping unit, data on the drainage and agronomic properties of soils, and interpretive tables. Soil surveys are prepared by the USDA's NRCS in cooperation with agricultural experiment stations and local government units. They are available from local NRCS offices or online (USDA-NRCS, 2009). The soil surveys are used extensively to identify potential sites that meet the regulatory and agronomic requirements for land application.

2.2.2.3 *Site Evaluation Criteria*

While selection criteria for land-application sites vary from state to state, the *National Manual of Good Practice for Biosolids* discusses a variety of factors that should be considered in any land-application program (NBP, 2005). These selection criteria are summarized below.

Soil surveys contain essential information for determining whether a field has appropriate soil characteristics for biosolids application (e.g., soil texture, erodibility, soil drainage characteristics, and slopes). Favorable soil characteristics are discussed in Appendix C of the *National Manual of Good Practice for Biosolids* (NBP, 2005). U.S. Geological Survey (USGS) quadrangle maps are also helpful during preliminary planning and screening to estimate slope, topography, depressions or wet areas, rock outcrops, drainage patterns, and water table elevations. The following topographical and soil characteristics are unfavorable for biosolids application:

- Steep areas with sharp relief;
- Undesirable soil conditions (sandy soils, shallow, highly erodible, or poorly drained);

- Environmentally sensitive areas (e.g., intermittent streams, ponds);
- Rocky, non-arable land;
- Wetlands and marshes; and
- Areas bordered by surface waterbodies without appropriate setback areas.

Onsite verification of candidate sites also is recommended, particularly for small parcels of land that are not adequately represented on surveys or maps.

In addition to topography and soil characteristics, any preliminary site evaluation should assess water quality-based requirements at each farm (e.g., buffer zones, conservation planning, and nutrient management). Part 503 requires a buffer of at least 10 m (33 ft) to surface waters, and states often have additional buffer requirements to surface waterbodies, water supply wells, property lines, outcroppings, or public roads. Because the buffer requirements limit the actual land available for application, factoring in these limitations may make some sites unsuitable for land application.

Some states require that farms implement conservation plans to control soil erosion and nutrient management plans to limit excess nutrient transportation to water resources. Certain conservation measures (e.g., no-till or residue management) and nutrient management issues (e.g., fields with a long history of land-applied manure or biosolids) could limit a site's suitability for land application.

Site criteria important to the economic viability of a land-application program include the ease of access to the farm (e.g., road restrictions or traffic limitations) and the hauling distance or travel time (and therefore cost).

2.2.2.4 *Attitudes*

Assessing the public acceptance of and attitudes towards land application is an important part of the site evaluation process. The attitudes of stakeholders (e.g., neighbors) can be critical to the success of a land-application program, and discussions with farmers or landowners beforehand often can provide key insights in this regard.

Farmer and landowner attitudes are critical to program success as well. According to the *National Manual of Good Practice for Biosolids* (NBP, 2005), the landowner or farmer "must be confident about the benefits of biosolids and be willing to accept and comply with all of the . . . regulatory requirements, as well as meet biosolids program needs."

2.3 Design and Implementation

2.3.1 *Application Rates*

When determining application rates for biosolids, wastewater treatment professionals must consider metals content, nutrient content (nitrogen and occasionally phosphorus), and for liming applications, calcium content as well. The application rate selected will be the lowest of those calculated for each appropriate parameter.

2.3.1.1 Metals

Part 503 regulates the loading rates of the elements identified in Section 2.2.1.3. Biosolids whose metal and heavy metal concentrations exceed the EQ limit but are lower than ceiling concentrations can be land-applied up to the cumulative loading limit (EPA, 1993a) defined in Part 503. There also is a maximum annual loading rate. Arsenic, for example, has an EQ limit of 41 mg/kg, a ceiling concentration of 75 mg/kg, an annual loading limit of 2.0 kg/ha/yr, and a cumulative loading limit of 41 kg/ha. Originally, all of the cumulative loading limits and EQ limits in Part 503 were equal, except for selenium, whose EQ limit was 36 mg/kg and cumulative loading limit was 100 kg/ha. Later Part 503 was amended to increase the EQ limit for selenium to 100 mg/kg.

Several states have lower loading limits than Part 503, and some also regulate other elements (e.g., thallium in New York).

2.3.1.2 Nutrients

Part 503 restricts the land-application rate on agricultural land based on the crop's nitrogen requirements and the available nitrogen in biosolids. A given crop's nitrogen requirement is location-specific; the data are readily obtained from state agricultural extension publications. The available nitrogen in biosolids is the sum of its ammonia-nitrogen content (which is assumed to be 100% plant available) and a percentage of its organic nitrogen content. Organic nitrogen becomes plant available by converting to ammonia and then nitrate via mineralization by soil bacteria. A typical mineralization rate for aerobically or anaerobically digested biosolids is 30% in the first year of application. A lower percentage is assumed for compost and alkaline-stabilized biosolids. If biosolids are not incorporated or injected (e.g., at no-till sites), a percentage of the ammonia-nitrogen (typically 50%) is assumed to be lost via volatilization. For specific details on calculating available nitrogen in biosolids, see state biosolids regulations and recommendations.

The phosphorus in biosolids has variable plant availability (see Section 2.2.1.4), which is assessed by evaluating the effect of biosolids application rates on soil test phosphorus. Unlike nitrogen, phosphorus accumulates in soil, so repeated biosolids applications will increase soil test phosphorus concentrations. When biosolids are land-applied at nitrogen rates, its phosphorus content far exceeds the crop's phosphorus requirement, leading to large phosphorus buildups in some soils. Soil test phosphorus is strongly correlated with phosphorus in surface runoff, and states are beginning to regulate phosphorus applications to agricultural soils to protect water quality.

The Natural Resources Conservation Service has developed a soil model for phosphorus applications called the *Phosphorus Index*, and some states have adopted various forms of it to regulate fertilizer, manure, and biosolids application rates. As used in some states, the Phosphorus Index includes a factor that reflects the bioavailability of the material being applied. For example, biosolids with large iron, aluminum, and calcium contents will have lower phosphorus bioavailabilities than aerobically or anaerobically digested biosolids (although this difference is not always reflected in Phosphorus Index calculations).

Because phosphorus-based management requirements are state-specific, planners should contact state regulators to define potential requirements.

2.3.1.3 Calcium

Alkaline-stabilized biosolids contain various concentrations of calcium in the form of calcium hydroxide and calcium carbonate. When lime is added to biosolids, it picks up water from the biosolids and is rapidly converted to calcium hydroxide. Over a longer period of time (weeks and months) calcium hydroxide is converted to calcium carbonate by interacting with carbon dioxide in the atmosphere. Class B alkaline-stabilized biosolids contain between 2 and 5% lime or its equivalent, while Class A alkaline-stabilized biosolids contain between 12 and 30% lime or its equivalent, depending on which Class A alkaline-stabilization process is used.

The high content of lime in alkaline-stabilized biosolids makes these materials excellent substitutes for agricultural limestone (calcium carbonate or dolomite), which is applied to soils that are naturally acidic or have become so as a result of chemical fertilizer and organic matter. The land-application rates are based on the results of tests that determine the soil's lime requirement and the liming value of alkaline-stabilized biosolids.

Soil pH is not an effective measurement of soil acidity, and separate tests must be conducted to measure soil acidity. The lime-requirement soil test is state-specific; state agricultural extension services provide recommendations on test methods and the lime requirements of various crops. The results of the lime-requirement soil test typically include soil acidity and a recommended application rate (in metric ton/ha or ton/ac) of agricultural limestone required to neutralize soil acidity to a target pH value.

The test to determine the liming value of biosolids is based on acid-neutralizing capacity. Results are expressed as calcium carbonate equivalency (CCE), which represents the percentage of pure calcium carbonate, on either a wet weight or a dry weight basis. Standard tests to determine CCE are the ASTM International Standard C-25-06, *Standard Test Methods for Chemical Analysis of Limestone, Quicklime, and Hydrated Lime* (ASTM, 2006), and one derived from American Water Works Association Standard

B202-07 (AWWA, 2008). The alkaline-stabilized biosolids' CCE is used to adjust the lime requirement. For example, an alkaline-stabilized biosolids with a CCE of 25% would have to be applied at a rate of 8 metric ton/ha (3.6 ton/ac) to satisfy a lime requirement of 2 metric ton/ha (0.90 ton/ac) of agricultural limestone.

2.3.1.4 Design Examples

2.3.1.4.1 Nutrient and Metals-Based Land Application

A wastewater treatment plant plans to surface apply biosolids to hay crops. The facility produces 3 dry metric ton/d of liquid, aerobically digested biosolids. The biosolids will not be incorporated. The fields have not received soil amendments or fertilizers in the past. The aerobically digested biosolids have the following characteristics:

- Total nitrogen content = 3%
- Ammonia-nitrogen content = 2%
- Nitrate content = 0%
- Total phosphorus content = 1%
- The limiting metal is copper at 400 mg/dry kg solids

(1) Determine the land-application rates and required acreages on a) a nitrogen-limiting basis and b) a phosphorus-limiting basis. Assume that the annual crop requirement is 235 kg nitrogen/ha·a (210 lb nitrogen/ac/yr) and 73 kg phosphorus/ha·a (65 lb phosphorus/ac/yr).
(2) Determine the metal-limiting land-application rate.
(3) Which is more limiting: crop nutrient requirements or metals limits?

2.3.1.4.1.1 SOLUTION (1A)

First, the plant-available nitrogen (N_{PA}) in biosolids can be calculated as follows:

$$N_{PA} = 1\,000[(NH_3)K + NO_3 + N_0 f] \quad (27.1)$$

Where

N_{PA} = plant-available nitrogen (kg/metric ton dry solids);
NH_3 = percent ammonia-nitrogen in biosolids (as a decimal);
K = volatilization factor for ammonia;
NO_3 = percent nitrate in biosolids (as a decimal);
N_0 = percent organic nitrogen in biosolids (as a decimal); and
f = mineralization factor (conversion of organic nitrogen to ammonium-nitrogen).

K depends on the application method used (see Table 27.6).

TABLE 27.6 Volatilization factor (K) for ammonia (U.S. EPA, 1994b).

If solids are	K factor is
Liquid and surface applied	0.5
Liquid and injected into the soil	1.0
Dewatered and applied in any manner	1.0

Because the biosolids will not be incorporated, K is assumed to be 0.5 (50% loss of ammonia-nitrogen). The mineralization factor f can be determined via Table 27.7. Because this is the first year of application and aerobically digested biosolids are being applied, f can be assumed to be 0.3. Therefore,

$$N_{PA} = 13 \text{ kg nitrogen/metric ton dry solids (26 lb nitrogen/ton dry solids)}$$

Next, the nitrogen-limiting biosolids loading (R_N) can be calculated as follows:

$$R_N + \frac{U_N}{N_{PA} + N_{PM}} \tag{27.2}$$

Where

U_N = annual crop requirement for nitrogen,
N_{PA} = plant-available nitrogen from this year's sludge application, and
N_{PM} = plant-available nitrogen from mineralization of all previous applications.

TABLE 27.7 Mineralization factor (f) for nitrogen in various types of biosolids (U.S. EPA, 1994b).

Time after sludge application (year)	Percent of organic nitrogen mineralized from stabilized primary solids and WAS	Percent of organic nitrogen mineralized from aerobically digested biosolids	Percent of organic nitrogen mineralized from anaerobically digested biosolids	Percent of organic nitrogen mineralized from composted biosolids
0–1	40	30	20	10
1–2	20	15	10	5
2–3	10	8	5	3
3–4	5	4	3	3
4–5	3	3	3	3

As mentioned above, U_N is 235 kg N/ha·a (210 lb N/ac/yr). Also, N_{PM} is zero because this is the first year of application. Therefore,

$$R_N = 18 \text{ metric ton dry solids/ha·a (8.0 ton dry solids/ac/yr)}$$

The total land required on a nitrogen-limiting basis is then

$$= (3 \text{ metric ton solids/d}) \times (365 \text{ days})/(18 \text{ metric ton solids/ha·a})$$
$$= 61 \text{ ha (150 ac)}$$

2.3.1.4.1.2 SOLUTION (1B)

First, the total phosphorus percentage must be converted to the percentage of plant-available phosphorus (P_2O_5). This is done as follows:

$$P_2O_5 = \text{total phosphorus (\%)} \times 2.29$$
$$= 22.9 \text{ kg } P_2O_5\text{/dry metric ton solids (45.8 lb } P_2O_5\text{/dry ton solids)}$$

The phosphorus-limiting biosolids loading (R_P) can be calculated using a formula analogous to that for nitrogen. In this case, the phosphorus crop requirement (U_P) is 73 kg phosphorus/ha·a (65 lb phosphorus/ac/yr), all P_2O_5 is plant-available (P_{PA}), and the plant-available phosphorus from previous applications is zero because this is the first year of biosolids application. Thus,

$$R_P = 3.2 \text{ metric tons dry solids/ha·a (1.4 ton dry solids/ac/yr)}$$

The total land required on a phosphorus-limiting basis is then

$$= (3 \text{ metric ton/d}) \times (365 \text{ days})/(3.2 \text{ metric ton/ha·a})$$
$$= 344 \text{ ha (851 ac)}$$

2.3.1.4.1.3 SOLUTION (2)

Because copper is the limiting metal, the lifetime biosolids application rate is calculated based on the maximum lifetime cumulative loading for copper [1 500 kg/ha (1 340 lb/ac)] listed in 40 CFR 503 Table 2. The biosolids contain 400 mg copper/kg solids, which is equivalent to 0.4 kg copper/metric ton dry solids (0.8 lb copper/ton dry solids). Therefore, the metal-limiting solids loading rate is 3750 metric ton dry solids/ha (1 675 ton dry solids/ac).

2.3.1.4.1.4 SOLUTION (3)

A comparison of the nitrogen-, phosphorus-, and metal-limiting solids loading rates calculated above shows that crop nutrient requirements are the limiting factors when determining appropriate biosolids loading rates:

- Nitrogen-limiting rate = 18 metric ton dry solids/ha·a (8.0 ton dry solids/ac/yr),
- Phosphorus-limiting rate = 3.2 metric ton dry solids/ha·a (1.4 ton dry solids/ac/yr), and
- Metal-limiting rate = 3750 metric ton dry solids/ha (1675 ton dry solids/ac).

2.3.1.4.2 Alkaline Stabilized Biosolids Land Application

An agricultural operation finds that its soil has become more acidic over the years, and it wants to apply Class A alkaline-stabilized biosolids to raise the soil's pH. The biosolids' average solids content is 35%. A laboratory tested the soil's acidity and gave a liming recommendation of 2 metric ton/ha of agricultural limestone (0.9 ton/ac). The biosolids' CCE is 60% on a dry weight basis. Determine the appropriate land-application rate for the alkaline-stabilized biosolids.

The laboratory provided a recommended limestone application rate assuming that the agricultural limestone has 100% purity as calcium carbonate. Because biosolids do not have 100% purity, the application rate must be adjusted based on the biosolids' CCE. But first, the CCE on a dry weight basis must be converted into a wet weight basis:

= (0.60 kg as $CaCO_3$/kg dry biosolids) × (0.35 kg dry biosolids/kg wet biosolids)
= 0.21 kg as $CaCO_3$ per kg of wet biosolids
= 21% on a wet weight basis

Then, the limestone application rate can be used to determine the biosolids application rate:

= (2 metric ton limestone/ha)/(0.21 metric ton limestone/metric ton wet biosolids)
= 9.5 metric ton/ha (4.3 ton/ac)

It is important to keep in mind that regulations require the biosolids application rate to be based on nitrogen (and phosphorus, if the state has phosphorus limitations), which should be calculated according to crop requirements to ensure that loading limits are not exceeded. If the nutrient-based loading rate is less than the CCE rate, then program staff should use the nutrient-based rate and add supplemental limestone, or else complete the liming over a second year.

2.3.2 *Recommended Land Area*

Topographic and soil characteristics may limit the areas where biosolids can be applied. Other factors (e.g., buffer zones; setbacks from waterbodies; wells and other water sources; roads, treed areas; buildings and other structures) can limit the amount of land suitable for land application on a given property. A careful review of maps and a field visit are recommended to locate these areas and quantify the extent to which they will diminish lands available for biosolids applications. Care should be taken to ensure that the remaining site area meets application needs.

Land availability also is determined by local farming practices. There may be times when certain types of cropland are unavailable for the land application. Therefore, two to three times the calculated amount of usable land may be needed to meet land-application needs year-round. Storage can mitigate this need.

2.3.3 *Field Storage*

Many land-application programs may be subject to seasonal, weather, or other factors that temporarily limit the ability to land-apply material and, therefore, need short-term storage. Biosolids cake, alkaline-stabilized biosolids, compost, or heat-dried biosolids can be stored for up to 2 years in the field (depending on local regulations), in either stockpiles or constructed storage facilities. For details on properly siting, managing, and operating stockpiles, see the *Guide to Field Storage of Biosolids* (U.S. EPA, 2000). For information on designing constructed storage facilities, see the *Guide* and Table 27.8.

2.3.4 *Odor Management*

Odors are frequently cited as the basis for opposing land-application programs. One approach to minimizing the potential for offsite odors is maximizing the buffers between land-application fields and the public. As many agencies have found, however, development can encroach on previously remote sites, reducing buffers and increasing the potential for odor complaints. So, odor-management plans cannot be based on setbacks and buffers alone. The *National Manual of Good Practice for Biosolids* (NBP, 2005) recommends a comprehensive odor-management approach that addresses liquid and solids treatment at the wastewater treatment plant, transportation, site storage, field operations, and community relations. Addressing potential odors (and their effects) at each "critical control point" is essential to effectively manage odors at land-application sites.

2.3.5 *Land-Application Equipment and Methods*

Biosolids typically are land-applied as a cake or as liquid (exceptions include some Class A alkaline-stabilized biosolids, heat-dried biosolids, and composts). Manure spreaders are used to land-apply biosolids cake and, depending on the crop, discs may be used to

TABLE 27.8 Key design concepts for constructed biosolids storage facilities (U.S. EPA, 2000).

	Liquid/thickened	Dewatered/dry biosolids facilities		Liquid/thickened
	1–12% solids	*12–30% solids/>50% solids (dry)*		*1–12% solids*
Issue	**Lagoons**	**Pads/basins**	**Enclosed buildings**	**Tanks**
Design	Below ground excavation. Impermeable liner of concrete, geotextile, or compacted earth.	Above ground, impermeable liner of concrete, asphalt, or compacted earth	Roofed, open-sided or enclosed. Flooring: concrete, asphalt, or compacted earth	Above or below ground, concrete, metals or prefab. If enclosed—ventilation needed
Capacity	Expected biosolids volume + expected precipitation + freeboard	Expected biosolids volume, unless precipitation is retained; then, biosolids volume + expected precipitation + freeboard	Expected biosolids volume	Enclosed: expected biosolids volume. If open-top—expected biosolids volume + expected precipitation + freeboard
Accumulated water management	Pump out and spray irrigate or land apply the liquid, haul to WWTP, or mix with biosolids	Sumps/pumps if facility is a basin for collection of water for spray irrigation, land apply or haul to a WWTP	Roof and gutter system, enclosure, or up-slope diversions	Decant and spray irrigate, land apply or haul to WWTP or mix with biosolids in tank
Runoff management	Diversions to keep runoff out of lagoon	Diversions to keep runoff out of site, curbs and/or sumps to collect water for removal or down-slope filter strips or treatment ponds	Enclosure or up-slope diversions	Prevent gravity outflows from pipes and fittings. Diversions for open, below ground tanks
Biosolids consistency	Liquid or dewatered—removal with pumps, cranes or loaders	If no side-walls, material must stack without flowing	Material must stack well enough to remain inside	Liquid or dewatered biosolids. If enclosed, material must be liquid enough to pump.
Safety	Drowning hazard—post warnings, fence, locked gates and rescue equipment on site	Drowning hazard—post warnings, fences, locking gates, and rescue equipment on site	Post "No Trespassing", signs, remote location, lock doors, gates and fences	Posted warning, locking access points, e.g., use hatches, controlled access ladders, and confined space entry procedures to access

incorporate the biosolids into the soil. Liquid biosolids can applied to the soil surface via a tanker spreader or spray irrigation system; they also can be injected into the soil. For a comparison of application methods, along with method details and equipment, see the *National Manual of Good Practice for Biosolids* (NBP, 2005). The manual also discusses calibration needs, which are critical to ensure that the proper application rate is not exceeded.

2.3.6 *Biosolids Transportation*

When selecting transportation routes between a wastewater treatment plant and a land-application site, design engineers must consider odor and other nuisance factors (e.g., traffic and noise). Clean, covered trucks will help address odor concerns, as will minimizing waiting in line at the site. With respect to traffic, the ability of selected roads to accommodate the weight, width and turning radii of hauling vehicles must be considered in addition to nuisance concerns. The *National Manual of Good Practice for Biosolids* (NBP, 2005) should be consulted for additional discussions of biosolids transportation issues.

2.3.7 *Public Acceptance*

While farmers typically accept that land-applied biosolids function as a fertilizer, liming agent, or soil conditioner, concerns about public health and environmental safety remain. With this in mind, the National Biosolids Partnership (NBP) created an environmental management system (EMS) for beneficially used biosolids. The system is based on the premise that education and outreach, coupled with exemplary biosolids programs, are critical to gaining public acceptance. The partnership also has developed a set of planning and management documents that help wastewater utilities develop and improve their overall programs—especially their outreach programs.

The planning tools and guidance provided by NBP include comprehensive guidance on developing an environmentally sound biosolids management program. These documents, which include the *National Manual of Good Practice for Biosolids*, can be found on NBP's Web site (NBP, 2009). Wastewater treatment plants also can gain public acceptance of their beneficial use programs by active participation in state and regional WEF-sponsored biosolids programs. These organizations promote sound biosolids management and actively communicate with the public.

3.0 LAND RECLAMATION AND OTHER NONAGRICULTURAL USES

In addition to agriculture, biosolids can be used for land reclamation, silviculture and other nonagricultural purposes. When developing programs for such uses, design engineers typically should consider many of the recommendations noted above for agricultural programs. However, the application rate and frequency will be site-specific. For example, biosolids used to reclaim land may be applied only once (or infrequently) at a rate of 50 to 150 dry metric ton/ha (NBP, 2005). Biosolids also may be applied infrequently to silviculture sites, but the application rate will be between those for agricultural and reclamation sites (NBP, 2005).

Nonagricultural applications may be subject to both Part 503 and other state and federal requirements. For more information on the regulatory and other issues that should be considered for these programs, see the *National Manual of Good Practice* (NBP, 2005).

4.0 LANDFILLING

Landfilling is an option for disposing of wastewater residuals. This section presents information on planning, designing, constructing, monitoring, and closing such landfills. Important aspects include

- Landfill siting and capacity needs;
- Liner and leachate-collection system design;
- Surface water control;
- Landfilling methods;
- Daily, intermediate, and final covers;
- Monitoring requirements;
- Gas migration control, collection, and reuse;
- Covers and cap systems; and
- Landfill closure and reuse.

4.1 Regulatory Considerations

There are two types of landfill sites that typically accept solids. One is a *monofill*—a landfill that only accepts stabilized or unstabilized municipal wastewater solids. The other is a *co-disposal landfill*, which accepts both residuals and municipal solid waste.

4.1.1 *Monofills*

The U.S. Environmental Protection Agency regulates monofill design and operation under Subpart C of 40 CFR 503, which governs surface disposal of wastewater solids. Part 503 standards include general requirements; pollutant limits; management practices; pathogen-reduction and VAR alternatives; and monitoring, recordkeeping, and reporting requirements. (For more information on 40 CFR 503, see Chapter 20.)

Subpart C addresses two types of monofill operations. Unstabilized solids must adhere to VAR Alternative 11 and must be covered with soil or another material at the end of each operating day. Stabilized solids, on the other hand, must meet one of the VAR alternatives and do not have to be covered each day. Monofills for stabilized solids operate more like land-application sites with exceptionally high application rates. Further guidance on monofills for raw solids are covered in this section (Section 4), while monofills for stabilized solids (also called *dedicated land disposal* sites) are addressed in Section 5.

4.1.2 Co-Disposal Landfills

The U.S. Environmental Protection Agency regulates co-disposal landfill design and operation under 40 CFR 258, which governs the disposal of municipal solid waste (e.g., household wastes) (U.S. EPA, 1991). Because wastewater solids typically are a small percentage of the waste at such sites, Part 258 regulations are not referenced in detail in this section.

The disposal rate partially depends on the residuals' solids content. At co-disposal landfills, solids typically are spread in the active area (they part of the landfill accepting waste) and mixed with incoming solid waste to ensure that the material has acceptable handling characteristics. For example, a cake containing 20% solids might be mixed with solid waste at a 4:1 ratio [i.e., 4-metric ton of solid waste to 1-metric ton of solids (wet ton basis)]. Residuals with lower solids contents would need to be mixed with more solid waste. The mixing process (also called *bulking operation*) typically depends on the type and quantities of wastes delivered to the landfill.

Solids delivered to a co-disposal site must not contain any free liquids, as defined via the paint filter liquids test (U.S. EPA, 1995b). Dewatered cakes containing 20% solids typically pass this test.

4.2 Planning

The first step in planning a new landfill is to identify a site that meets the related design, regulatory, and cost requirements. Project teams typically evaluate and compare several prospective sites before selecting one. The siting process often involves gathering input from both the public and regulators. It may take several years to identify, design, permit, and construct a new landfill.

4.2.1 Siting

A monofill must meet the siting criteria in Part 503 (U.S. EPA, 1994a). Such criteria include:

- The monofill cannot be likely to adversely affect a threatened or endangered species;

- It cannot restrict the flow of a base flood (i.e., a 100-year flood event);
- It must not be in a geologically unstable area;
- It must be at least 60 m from a fault area that experienced displacement in Holocene time;
- If regulators permit the monofill to be located in a seismic impact zone, then it must be designed to resist seismic forces (a *seismic impact zone* is an area where the ground-level rock has at least a 10% probability of accelerating horizontally more than 0.10 g once in 250 years.); and
- The monofill cannot be located in a wetland (unless a special permit is obtained).

Some states may have additional siting criteria (e.g., setbacks from property lines, public or private drinking water wells, surface drinking water supplies, and buildings or residences). Design engineers should use these criteria to screen potential sites and select one that meets all applicable criteria.

Design engineers then need to determine the landfill footprint (where actual disposal activities occur) within the site boundaries. This footprint is determined based on proximity to sensitive receptors (e.g., wetlands, residences, water bodies, property boundaries, and roads). This footprint is surrounded by dikes constructed to both structurally support fill material and contain surface runoff. Dikes are made of soil compacted to a specified strength. The footprint also should include special working areas for use in inclement weather or other contingency operations.

Design engineers also should establish appropriate buffer distances between landfill footprint and sensitive receptors. Buffers are site-specific, based on regulations or general guidelines for mitigating adverse environmental effects. The landfill design also must accommodate support facilities, which may include:

- Access roads,
- Administrative offices and employee facilities,
- Equipment storage and maintenance areas,
- Stockpiling areas,
- Utilities,
- Fencing,
- Lighting,
- Truck-washing facilities,
- Leachate storage and pumping stations,
- Monitoring wells, and
- Stormwater detention basins.

4.2.2 *Landfill Capacity Needs*

When selecting a new site, the landfill footprint and site geometry must provide enough capacity for its entire operational life, which should be at least 20 years. A landfill's operational life is affected by many variables (e.g., the solids production rate, the volume consumed by liners and capping systems, and the volume consumed by bulking soils and cover materials). The volume lost to liner and cover systems is easy to calculate and is based on the thickness of these layers over the landfill area. The volume consumed by cover and bulking material depends on the characteristics (solids content, bulk density, etc.) of these materials. Solids must be mixed with a certain amount of bulking material (e.g., soil) to improve their strength and handling characteristics. Daily and intermediate cover material may consume 20% of the overall landfill volume. Design engineers need to conduct actual capacity assessments for each site. Assessment results will depend on the type of materials to be landfilled, the proposed method of landfill operation, cover requirements, and other factors.

4.3 Landfill Design

4.3.1 *Regulatory Requirements for Liners*

Part 503 does not require that all monofills (or Dedicated Land Disposal, or DLD, sites) have liner systems. The need for a liner is based on the solids' pollutant concentrations; U.S. EPA allows for disposal without a liner if the solids' arsenic, chromium, and nickel concentrations are below the pollutant limits. The limits are based on the distance between the monofill footprint and the property line (see Table 27.9). If a liner is

TABLE 27.9 Pollutant concentration limits for unlined monofills (U.S. EPA, 1994a).

Location in the Part 503 rule	Distance from the boundary of active biosolids unit to surface disposal site property line, m	Pollutant concentration*		
		Arsenic, mg/kg	Chromium, mg/kg	Nickel mg/kg
Table 2 of section 503.23	0 to less than 25	30	200	210
	25 to less than 50	34	220	240
	50 to less than 75	39	260	270
	75 to less than 100	46	300	320
	100 to less than 125	53	360	390
	125 to less than 150	62	450	420
Table 1 of section 503.23	Equal to or greater than 150	73	600	420

* Dry-weight basis (basically, 100% solids content).

not used, a sampling and analysis program must be established in accordance with Part 503.

If the solids exceed the pollutant concentration limits, the project team may be able to obtain a site-specific permit from regulators that allows the monofill to be unlined. To do this, the team must demonstrate that site conditions vary significantly from the criteria U.S. EPA used to derive pollutant concentrations. Otherwise, U.S. EPA requires that the monofill be lined.

Increasingly, professionals are considering it best engineering practice to design all monofills with linings to prevent the escape of contaminants, regardless of their concentrations. Many states' regulations for monofills have more stringent requirements than Part 503, including the need for a liner system.

4.3.2 Liner Design

Liner systems are used to contain waste in landfills and prevent the migration of leachate constituents out of the landfill (U.S. EPA, 1993b). Three types of materials typically are used: low-permeability-soil (clay) liners, geosynthetic clay liners, and geomembrane liners. A combination of these materials may be used, depending on regulatory or site-specific conditions. Geosynthetic clay liners, which are substitutes for low-permeability soil liners, are manufactured of bentonite clay supported by geotextiles held together by needling, stitching, or chemical adhesives. Geomembranes typically are placed on top of a clay layer (or other low-permeabilty layer) to form a composite lining.

Part 503 requires that a monofill liner have a maximum hydraulic conductivity (water vapor transmission) of 1×10^{-7} cm/s. The permeability of a geosynthetic clay liner is about 1×10^{-10} cm/s. Geomembrane liners have an average hydraulic conductivity of about 1×10^{-14} cm/s.

For detailed guidance on designing landfill lining systems, see U.S. EPA's *Solid Waste Disposal Facility Criteria Technical Manual* (U.S. EPA, 1993b) and *Waste Containment Systems, Waste Stabilization and Landfills: Design and Evaluation* (Sharma and Lewis, 1994).

4.3.3 Leachate Collection System

As water percolates through a monofill, it dissolves ("leaches") various solids constituents and becomes polluted. One of the most important considerations in the design, operation, and long-term care of a landfill is leachate collection and management. A leachate-collection system should minimize the leachate's hydraulic head on the primary liner during landfill operations; it should be able to maintain a leachate head of less than 0.3 m (1 ft). The collection system also should remove leachate from the landfill through the post-closure monitoring period.

The leachate collection system consists of a drainage layer (e.g., sand or a geonet), leachate collection pipes, a sump or a series of sumps, and pumps to transport leachate to an onsite treatment system or to a sanitary sewer. Cleanouts should be provided in the collection-system piping. The design and layout of the collection system must be compatible with the phased development of the monofill. Each phase's liner and collection system should be installed concurrently. Also, later segments of the collection system should easily connect with previously installed piping.

The leachate-control system should include facilities to monitor leachate leaks at the base of the landfill and to withdrawal leachate and, therefore, prevent buildup that would promote leachate migration from the landfill.

For detailed guidance on designing leachate collection and removal systems, see U.S. EPA's *Solid Waste Disposal Facility Criteria Technical Manual* (U.S. EPA, 1993b) and *Waste Containment Systems, Waste Stabilization and Landfills: Design and Evaluation* (Sharma and Lewis, 1994).

4.3.3.1 Leachate Quantity

The amount of leachate generated in both the active and closed areas of a landfill depends on the amount of infiltration from the landfill surface. Various mathematical models are available to predict the amount of leachate produced during landfill operations. One commonly used model is U.S. EPA's Hydrologic Evaluation of Landfill Performance (HELP) model, which estimates the amounts of surface runoff, subsurface drainage, and leachate that may be expected in both the active and closed areas (U.S. EPA, 1994c). One advantage of this model is that it allows users to differentiate layers within the landfill (e.g., topsoil, sand drainage, barrier soil, and waste layer).

Many default values are available for various liner and leachate-collection components. Required model input parameters (e.g., porosity, field capacity, wilting point, and saturated hydraulic conductivity) can be determined by testing solids samples from the treatment plant. For example, Hundal et al., (2005) found that biosolids from the Chicago area had a hydraulic conductivity (saturated potassium) of 67 to 118 mm/h (17 to 30 in./hr); 54 to 74% porosity; and 18 to 25% plant-available water.

4.3.3.2 Leachate Quality

Biological decomposition is the primary solids-degradation mechanism that causes contaminant leaching (see Table 27.10). Biological decomposition begins aerobically, but as oxygen is depleted, aerobic microorganisms give way to anaerobes. Because anaerobic decomposition is slow, organic contaminants may take several decades to degrade. Leachate production may continue for years, but the leachate's strength gradually decreases. Collected leachate must be treated onsite or transported to the wastewater treatment plant.

TABLE 27.10 Average leachate values for solids-only test cells (U.S. EPA, 1995b).

Parameter	Concentration
Chemical oxygen demand, mg/L	2 258
Total organic carbon, mg/L	737
pH	6.2
Volatile acids	1 213
Volatile solids	5 555

4.3.4 *Surface Water Control*

Adequate drainage is essential to good monofill operations. If precipitation-related surface water is not controlled, it ponds on the liner, increases leachate generation, and causes operating problems at the working face and cover-material storage areas. To control surface water, design engineers should consider vegetation, compaction, drainage structures, holding ponds, the type and thickness of cover material, and modifications to surface slope and slope length.

Surface slope and slope length determine the expected degree of erosion. The top surface grade should be between 2 and 5% to promote runoff, inhibit ponding, and minimize soil erosion by keeping flow velocities relatively low. Side slopes should be a maximum of three horizontal to one vertical (3:1) and require more care in seeding and runoff protection.

To the maximum extent possible, run-on should be diverted from the landfill footprint. Runoff water from active landfill areas (which may have contacted solids or solids-contaminated soils) must be drained to the leachate system and treated as leachate, or stored separately, tested, and treated to ensure that it meets discharge standards. Meanwhile, U.S. EPA requires that surface water runoff be collected and disposed of in accordance with National Pollutant Discharge Elimination System (NPDES) requirements. Both runoff and run-on collection systems must be designed to handle a 25-year, 24-hour rain event to ensure that contaminants are not released to the environment.

4.4 Groundwater Monitoring Requirements

Both Part 503 and Part 258 require that a landfill not contaminate an aquifer. Under Part 503, a monofill has contaminated an aquifer if it causes the aquifer's nitrate concentration to exceed the maximum contaminant level (MCL), which is 10 mg/L. To avoid this situation, most states require that the groundwater at monofill sites be monitored. In fact, state or local regulators typically require routine sampling for several parameters.

When designing a groundwater monitoring network, engineers should start by evaluating the site's geological and hydrogeological conditions (e.g., groundwater elevations and flow rate, and soil and bedrock conditions). Groundwater quality should be monitored at upgradient locations to provide background data. Downgradient wells are used to determine whether and how the landfill has affected groundwater quality. The number and type of wells is site-specific, and may include overburden wells at various depths and bedrock wells.

For more guidance on the selection and placement of groundwater monitoring wells, monitoring procedures, data analysis, and recordkeeping, see U.S. EPA's *Solid Waste Disposal Facility Criteria Technical Manual* (U.S. EPA, 1993b) and *Waste Containment Systems, Waste Stabilization and Landfills: Design and Evaluation* (Sharma and Lewis, 1994).

4.5 Landfill Gas Management

Landfill gas primarily consists of methane and carbon dioxide. Methane is a combustible gas that is explosive at atmospheric concentrations of 5 to 15%. Under stable anaerobic conditions, landfill gas can contain between 45 and 55% of methane. The remaining 45 to 55% primarily consists of carbon dioxide, along with small amounts of hydrogen, oxygen, nitrogen, and traces of other gases.

Because of the potential explosion hazard, U.S. EPA has established monitoring requirements for methane gas at landfills. Methane's lower explosive limit (LEL) is 5% (by volume). (A *lower explosive limit* is the concentration above which a mixture of air and combustible gas is explosive.) Federal regulations governing landfills mandate that the methane concentration shall not exceed 25% of the LEL in onsite structures and shall not exceed the LEL (in subsurface soils) at the property line. Methane must be monitored throughout the landfill's operating life and for 3 years after it closes.

A landfill design also should include an active or passive gas-collection system. For guidance on designing both systems, see U.S. EPA's *Solid Waste Disposal Facility Criteria Technical Manual* (U.S. EPA, 1993b) and *Waste Containment Systems, Waste Stabilization and Landfills: Design and Evaluation* (Sharma and Lewis, 1994).

The collected gas (called *landfill gas* or *biogas*) typically is either flared or used as fuel, where practical. Increasingly, landfill gas is seen as a renewable resource, and efforts to use it are increasing. For information on potential uses for biogas, see Chapter 25. (While the chapter discusses uses for digester gas, the options also apply to landfill gas.)

4.6 Landfill Closure

4.6.1 Closure Plan

Part 503 requires that a surface disposal site (monofill or DLD) owner or operator submit a closure plan up to 180 days before closing the facility. This plan should include a description of closure and post-closure activities. It should document how the leachate collection system will be maintained for at least 3 years after closure and outline the methane monitoring program, which also must occur for 3 years after closure.

If ownership of the disposal site changes after closure, the new owner must be notified in writing that a surface disposal site was operated on the property.

4.6.2 *Cap System*

A *landfill cap* is a multi-layered cover for the piled wastes that is added just before a landfill is closed. It designed to prevent rainwater from infiltrating the waste, thereby minimizing leachate production. The layers of this cap typically include a

- Subgrade layer, which is used to contour the landfill and provide a base for the other layers;
- Gas-control layer, which transports gas from under the barrier layer to a venting system;
- Hydraulic barrier layer, which limits the water infiltrating the landfilled waste;
- Drainage layer, which collects and transports the water that percolates into the final cover from the surface;
- Biotic layer, which protects the hydraulic barrier layer from biointrusions by animals or plants;
- Filter layer, which prevents particles of a finer material (i.e., the surface layer) from migrating into a coarser material (i.e., the drainage layer); and
- Surface layer, which may be either a soil that can support vegetation or an armored protection layer.

These layers perform important complementary functions (see Figure 27.1). For example, the vegetative layer helps prevent erosion by promoting the growth of plants, whose root structure anchors the soil.

The first step in selecting the layers of a landfill cap is to ensure that the proposed cap complies with state and federal regulations (see Figure 27.2). Although the layers and their functions are relatively standard in current engineering practice, each layer's thickness and performance standards may depend on state or local regulations.

Also, when designing a landfill cap, it is important to consider the strength of the underlying residuals. The geotechnical properties of solids depend on several factors (e.g., the degree and method of dewatering, the polymers added, and age) (see Table 27.11). Such properties should be tested as part of landfill design.

For more guidance on designing a landfill cap, consult U.S. EPA's *Solid Waste Disposal Facility Criteria Technical Manual* (U.S. EPA, 1993b) and *Waste Containment Systems,*

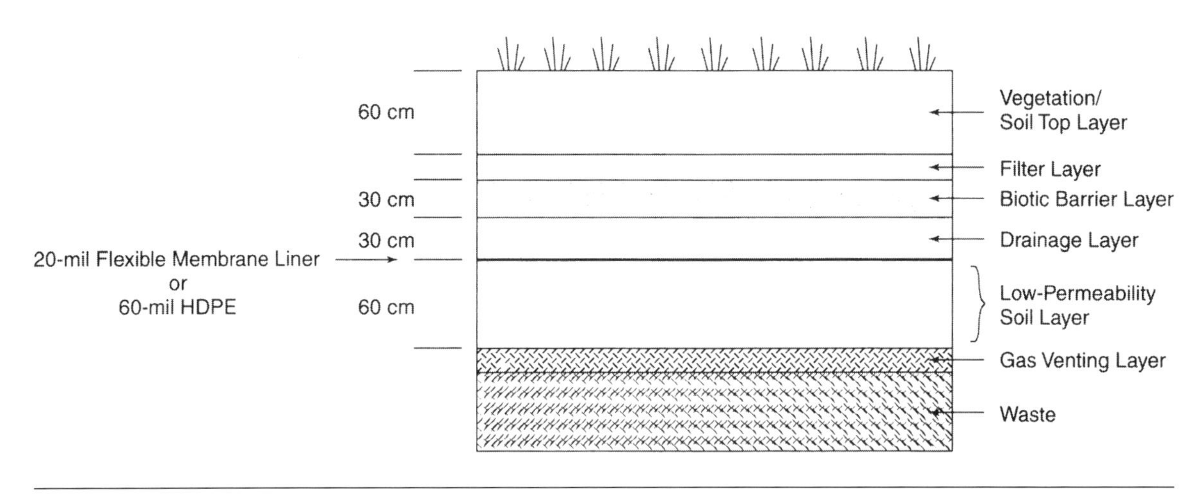

FIGURE 27.1 Example of various landfill cap components.

Waste Stabilization and Landfills: Design and Evaluation (Sharma and Lewis, 1994) provides guidance on selection and design of final cover layers.

4.6.3 Post-Closure Use Identification

When designing landfills, engineers should consider potential uses for the site once landfill operations cease and select cover material, grading, monitoring, and stormwater management options that are compatible with such uses (when possible). A former

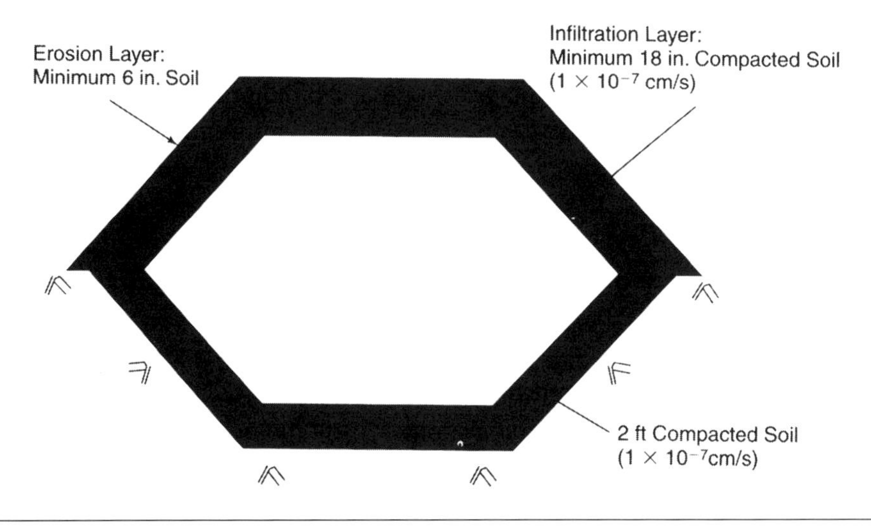

FIGURE 27.2 Minimum requirements for a landfill cap.

TABLE 27.11 Geotechnical properties of dewatered solids in Chicago area.

Parameter	Value
Bulk density	0.6–0.8 g/cm^3
Permeability (saturated potassium)	67–118 mm/h (17–30 in./hr)
Atterberg limits	LL = 71–119% PL = 54–85% PI = 17–53% Class = OH
Standard proctor test	Maximum dry density = 800–1090 kg/m^3 (50–68 lb/cu ft) Opt. moisture content (OMC) = 37–64%
Modified proctor test	Maximum dry density = 830–1 150 kg/m^3 (52–72 lb/cu ft) OMC = 31–64%
Compression index (Cc)	0.26–0.5
Recompression index (Cr)	0.03–0.1
Secondary consolidation index (C α)	0.02
Cohesion (C)	Triaxial CU test: Total strength: C = 0–40 kPa Effec. strength: C = 0–30 kPa Triaxial UU test: C = 0–20 kPa
Internal friction angle (α)	Triaxial CU test: Total strength: α = 21°–30° Effec. strength: α = 32°–41° Triaxial UU test: α = 32°–38°
Unconfined compressive strength (Qu) test	Qu = 23–126 kPa Strain at failure = 4.9–5.2%

landfill can become either a passive- or active-use area. Most become recreational and open-space areas (e.g., athletic fields, game courts, golf courses, playgrounds, and picnic areas).

5.0 DEDICATED LAND DISPOSAL

Dedicated land disposal is the process of applying municipal solids to land for disposal purposes. The equipment and process involved are similar to those used in land application, but the application rates typically are far above agronomic rates and no crop is grown. In effect, the site involved is a landfill that only accepts large quantities and heavy loadings of solids, and it is regulated as landfill.

5.1 Regulatory Considerations

Part 503 regulations apply to all surface disposal practices, including dedicated land disposal (and monofills, as discussed previously). The regulation is divided into several subparts, including general requirements, pollutant limits, management practices, and pathogen and vector-attraction controls.

5.1.1 General Requirements

The general requirements that apply to surface disposal of municipal solids include:

- Compliance with all applicable Part 503 requirements;
- Closure by 1994 of active units within 60 m of a fault with displacement in Holocene time, in an unstable area, or in a wetland, unless authorized by the permitting authority; and
- The need for closure and postclosure plans at least 180 days before closing any active units.

Also, site owners must provide written notification to the subsequent owner that municipal solids were placed on the land.

5.1.2 Pollutant Limits

Surface disposal sites that use liners and leachate collection systems do not have pollutant-concentration limits because the pollutants that seep from solids will be collected in the leachate and treated, as necessary, to avoid a pollution problem. For the site liner to qualify, it must have a hydraulic conductivity of 1×10^{-7} cm/s.

Surface disposal sites without liner and leachate collection systems have maximum allowable concentration limits for arsenic, chromium, and nickel (see Table 27.9) because these pollutants are the most likely to leach to groundwater, causing it to exceed the MCL. Different limits for these pollutants can be developed if a site-specific assessment (specified by the permitting authority) demonstrates that the site has different parameters than the ones U.S. EPA used to establish maximum allowable concentration limits.

5.1.3 Nitrate Contamination

Part 503 specifies that surface disposal operations cannot cause groundwater to exceed the MCL for nitrate (or exceed the existing concentration if it already is above the MCL). Groundwater-monitoring results or a statement from a qualified groundwater scientist can be used to demonstrate compliance with this requirement.

5.1.4 *Other Management Practices*

Other management practices that DLD facilities must follow include:

- Active disposal sites shall not be located within 60 m of a Holocene-period fault or in a wetland, unless authorized by the permitting authority; when located in a seismic impact zone, an active disposal site shall be designed to withstand the maximum-recorded horizontal ground-level acceleration.
- Surface runoff from a 25-year, 24-hour storm event shall be controlled in accordance with an NPDES permit.
- Active disposal sites shall not restrict base-flood flows, adversely affect threatened or endangered species, or be located in a structurally unstable area.
- Active surface disposal sites with a liner and leachate-collection systems shall operate and maintain them, and dispose of the collected leachate in accordance with applicable requirements for as long as the site is active and for 3 years after it is closed.
- No crops shall be grown nor animals grazed on surface disposal sites unless the permitting authority specifically authorizes such activities based on site-specific management practices.
- Public access to the site is restricted during operations and for 3 years after the site is closed.

5.1.5 *Pathogen and Vector Attraction Reduction Requirements*

Solids that will be placed in a surface disposal site must meet one of the Class A or Class B pathogen-control alternatives, unless they will be covered with soil or another material every day. They also must meet one of the first 11 vector attraction reduction options listed in Part 503.

5.2 Planning

The planning requirements for a new dedicated land disposal site are extensive. Significant hydrogeologic and soils investigations are required. Regulatory requirements and stormwater-management logistics also will affect site selection. Application rates will determine the size of the disposal site.

5.2.1 *Groundwater Protection*

Water that percolates from the soil of a surface disposal site (because of applied biosolids and precipitation) contains dissolved salts and products of decomposition.

The constituents of concern typically are nitrate and total dissolved salts (e.g., chloride); metals remain in the soil unless its pH is low. Design engineers need to carefully evaluate the fate of percolated constituents. Also, site-specific soil and hydrogeologic evaluations are mandatory.

Many sites need a relatively impermeable layer above the groundwater. This layer, which should be at least a few meters deep, collects percolated liquid. As the liquid turns anaerobic, much of the nitrate is denitrified and lost as nitrogen gas. If the percolation is minimal and shallow, most of the water may migrate back up to the surface and evaporate during the following year. Because the rule requires that surface-disposal operations not cause nitrate concentration in groundwater to exceed the MCL (or the existing concentration if it already exceeds the MCL), groundwater monitoring wells may need to be installed to verify compliance with this requirement.

Design engineers should evaluate lining systems and percolation controls based on biosolids and soil characteristics, site geology, climate, and application rates. Liner and leachate collection system design requirements are addressed in Section 4.

5.2.2 Site Selection

Design engineers also should consider the following when locating a dedicated land disposal site:

- Site runoff typically is considered wastewater, which needs to be returned to the treatment plant.
- The soil's pH must be 5.5 or higher to ensure that metals remain in the soil.
- A site susceptible to flooding or whose runoff could affect wetlands is not a good candidate for dedicated land disposal.
- Flat terrain is preferred because soil slope may affect site operations. A biosolids slurry requires limited slope, while dewatered cake can be applied on more variable slopes. Also, because runoff is an issue, dedicated land disposal typically is not considered on anything resembling hilly terrain.
- Buffer zones of about 150 to 300 m (500 to 1000 ft) have been suggested to minimize odor and dust complaints from the neighbors.

5.2.3 Application Rates

All land-application programs are "aerobic systems" at the surface incorporation layer. In this case, however, the goal is not crop or vegetation production but rather maximizing the aerobic stabilization of biosolids by soil bacteria. The application rates typically are much larger than agronomic rates and primarily limited by the evaporation rate of biosolids-related moisture from the soil.

Application rates typically range from 30 to 250 dry metric ton/ha·a (22 to 110 dry ton/ac/yr). Assuming that 25% of biosolids are lost to long-term products of decomposition, then these rates would increase soil depth by about 2.5 to 13 mm/yr (0.1 to 0.5 in./yr). If the site operates for 50 years, then total loading would be 1500 to 12 500 dry metric ton/ha. Bacteria in the soils can readily handle these application rates as long as enough water is evaporated and percolated to allow the aerobic microbe population to thrive. (Aerobic biological activity is about one to two orders of magnitude faster than anaerobic biological activity.)

The application rate typically depends on the evaporation rate, which affects how soon and how often equipment can be moved onto the site. If a slurry (6% solids) is applied at the above loading rates, it adds 8 to 38 cm (0.25 to 1.25 ft) of water to the site each year. This much water can be evaporated from DLDs in many (but not all) regions of the United States. Evaporation rates depend on temperature, humidity, wind, and rainfall. Because evaporation is a critical component of the operation, areas with limited evaporation seasons may find dedicated land disposal difficult to implement or may need to restrict application rates.

5.3 Design and Implementation

Engineers need a significant amount of site data to design and implement a dedicated land disposal site, including biosolids, hydrologic, topographic and meterological and other information.

5.3.1 General Considerations

Solids must be stabilized before they are applied to these sites. Subsurface injection is the preferred application method because it minimizes odor, minimizes vector problems, and ensures consistent application rates. Other general considerations include

- The biosolids, soil, and groundwater must be monitored.
- Traffic controls (e.g., hours of use, trucks per hour, and proper timing) may be required.
- Dust from the site and from access and internal roads needs to be controlled.

5.3.2 Biosolids Storage

Dedicated land disposal typically is a seasonal operation, so biosolids must be stored during the off seasons. Storing biosolids for several months in an environmentally acceptable manner with limited odor is often difficult.

5.3.3 *Groundwater Protection*

Extensive soil and geologic data need to be collected to determine how the site will react to heavy biosolids applications. This data-collection effort often involves soil borings, groundwater monitoring wells, surface and near-surface soil collection, and laboratory analysis. The primary goals are to

- Ascertain whether the site has a natural liner (impermeable soil);
- Determine if a supplemental lining is needed and how to implement it;
- Create an appropriate leachate-collection and -handling system, if necessary; and
- Discover likely limitations to loading rates.

5.3.4 *Biosolids Application Rates*

Design engineers need to calculate the application rate so the total site area requirements can be determined (see Table 27.12). Application rates are a function of the material applied; its water content; the seasonal evaporation rate; and the methods used to apply biosolids and subsequently aerate or mix them into the soil.

5.3.5 *Application Methods*

Biosolids can be applied in slurry or dewatered form. Dewatered cake allows for higher application rates (because of its lower moisture content) but can be difficult to spread over the entire site in a consistent layer. Slurries need high evaporation rates but can be pumped and are easier to spread. They also can be injected just below the surface by mobile equipment specifically designed for this purpose.

6.0 ASH USE AND DISPOSAL

Ash is the product of solids incineration, essentially consisting of the noncombustible portions of the feed material. There are numerous methods and equipment available for handling ash and various uses for ash. Handling methods and the final destination or use of material is often site-specific. Ash-handling equipment—especially conveyance—can be the most troublesome subsystem associated with incinerators.

TABLE 27.12 Typical design practices for several dedicated land disposal facilities.

Facility	Size mt/a*	Liquid or cake	Crop grown
Evanston, Wyoming	295	Cake	Grasses
Rapid City, South Dakota	1 222	Liquid	—
Colorado Springs, Colorado	8 939	Liquid	None

* 1996 data.

Multiple-hearth furnaces and fluid bed incinerators use either wet or dry systems for ash conveyance. Ash is abrasive and sometimes non-uniform, making it difficult to convey as a bulk material, and conveyance systems must be designed accordingly. The design of such systems is not covered in this chapter, but is detailed in the WEF Manual of Practice titled *Incineration Systems* (WEF, 2009). That document also covers another critical aspect of ash handling: storing wet and dry ash.

6.1 Regulatory Considerations

Regulations vary from state to state. Some states do not regulate ash; others treat it as a waste. Design engineers should check local, state, and federal regulations before determining whether to use or dispose incinerator ash. Each facility should conduct its own research on local, state, and federal regulations with regards to ash disposal and reuse.

Some landfills will require a toxicity characteristic leaching potential (TCLP) test or a pH test before accepting ash.

6.2 Use and Disposal Options

Biosolids incinerator ash is similar in physical and chemical characteristics to coal ash and can be used in applications where coal ash is permitted. Incinerator ash may be acidic, neutral, or alkaline, depending on the solids characteristics. Also, incinerator ash will contain more phosphorus than coal ash. While ash has historically been landfilled, interest in beneficial use of this material is increasing (see Table 27.13) (Dominak et al., 2005). Beneficial use options for ash include

- Landfill—Ash can be used as a landfill cover or blended with soil and used as a cover.
- Fill material—ash can be used as fill material for excavations. One utility, for example, has a contractor using the material to fill old solids lagoons. The material also can be used as a flowable fill.
- Soil amendment—in some areas (e.g., those with high clay soils), incinerator ash may be used as a soil additive that produces a soil that handles more easily, allows better drainage and airflow, and includes some valuable minerals.
- Concrete fly ash—ash has been used as a fly ash substitute in concrete mixes.
- Asphalt additive—the ash has been used as a mineral filler and fine aggregate in asphalt mixes. New Jersey's Department of Solid and Hazardous Waste (NJDEP) has permitted the use of incinerator ash for this purpose.

Additional valued-added products derived from ash are discussed later in this chapter.

TABLE 27.13 Results of an ash survey conducted by the Northeast Ohio Regional Sewer District (Dominak et al., 2005).

Wastewater treatment agency	State	City	Plant name	Ash handling method	Moisture content	Ash generated per year	Disposal method: Landfill	Disposal method: Beneficial reuse	Disposal cost	Description of beneficial reuse and/or landfilling
Central Contra Costa Sanitary District	California	Martinez	Main	Wet	65%	5110 wet ton[b]	X	X	\$21/ton	Most ash is beneficially reused as additives in landfill cover materials and brick making. Some ash is, however, disposed of in a commercial MSWLF.[f]
Albany County Sewer District	New York		Albany County	Wet	30–50%	8 000 cu yd[c]		X	No cost	Ash is blended with compost in a 50/50 mix and used as final landfill cover at no cost to Albany County. In return, Albany County accepts and treats landfill leachate.
Allegheny County Sanitary Authority	Pennsylvania		Alcosan	Dry	5–15%	6 500 dry ton	X		\$17–\$19/ton	Water is added to dry ash to control emissions when offloaded into trucks. Costs provided by Alcosan are for tipping fees only.
Metropolitan Council Environmental Service	Minnesota	St. Paul	Metro	Dry	NA[a]	15 000 dry ton		X	\$25.75/ton	Ash is used as a raw material in the manufacture of Portland cement.
Metropolitan Council Environmental Service	Minnesota	St. Paul	Seneca	Dry	NA	1 800 dry ton		X	\$25.75/ton	Ash is used as a raw material in the manufacture of Portland cement.
City of Canton Water Pollution Control Center	Ohio	Canton	City of Canton	Dry	NA	1 820 dry ton	X		\$210 per roll-off box plus taxes	Ash is disposed of at a commercial MSWLF. Ash is removed from the plant in 30-cu yd roll-off boxes.
Kansas City Water Services	Missouri	Kansas City	Blue River	Wet	50%	3500 dry ton	X		\$20/ton	Ash is disposed of at a commercial MSWLF after it is dewatered in a storage lagoon. Kansas City may shift to a dry ash system and close the ashfilled lagoon as is.

(*continued*)

TABLE 27.13 Results of an ash survey conducted by the Northeast Ohio Regional Sewer District (Dominak et al., 2005) (*continued*).

Wastewater treatment agency	State	City	Plant name	Ash handling method	Moisture content	Ash generated per year	Disposal method		Disposal cost	Description of beneficial reuse and/or landfilling
							Landfill	Beneficial reuse		
Upper Blackstone Water Pollution	Massachusetts		Upper Blackstone	Dry	NA	7 000–9 000 dry ton	X		?	Ash is disposed of at a publicly owned treatment works facility where screenings and grit are also disposed.
Narragansett Bay Commission			Field's Point	Wet	25%	2 341 wet ton		X	$16/ton	Ash is mixed with soil and used as a cover material at a quasipublic owned monofill
Hampton Roads Sanitation District	Virginia		Army Base	Dry	NA	1 300 dry ton	X	X	$46/ton (if BR[d]) $55/ton (if landfilled)	Water is added to dry ash to control emissions when offloaded into trucks. Some ash is disposed of in a commercial MSWLF; some ash is beneficially reused as select fill.
Hampton Roads Sanitation District	Virginia		Boat Harbor	Dry	NA	3 100 dry ton	X	X	$46/ton (if BR) $55/ton (if landfilled)	Water is added to dry ash to control emissions when offloaded into trucks. Some ash is disposed of in a commercial MSWLF; some ash is beneficially reused as select fill.
Hampton Roads Sanitation District	Virginia	Chesapeake/ Elizabeth		Dry	NA	2 500 dry ton	X	X	$46/ton (if BR) $55/ton (if landfilled)	Water is added to dry ash to control emissions when offloaded into trucks. Some ash is disposed of in a commercial MSWLF; some ash is beneficially reused as select fill.
Hampton Roads Sanitation District	Virginia		Virginia Initiative	Dry	NA	4 500 dry ton	X	X	$46/ton (if BR) $55/ton (if landfilled)	Water is added to dry ash to control emissions when offloaded into trucks. Some ash is disposed of in a commercial MSWLF; some ash is beneficially reused as select fill.

Hampton Roads Sanitation District	Virginia		Williams-burg	Dry	NA	4000 dry ton	X	X	$46/ton (if BR) $55/ton (if landfilled)	Water is added to dry ash to control emissions when offloaded into trucks. Some ash is disposed of in a commercial MSWLF; some ash is beneficially reused as select fill.
City of Palo Alto	California	Palo Alto	Palo Alto	Dry	NA	1 460 dry ton		X	?[e]	Water is added to dry ash to control emissions when offloaded into trucks. 500+ ton of ash is land applied; the rest is used as landfill cover.
City of Columbus Department of Public Utilities	Ohio	Columbus	Jackson Pike	Wet	35–65%	3 102 dry ton	X		$30–$38/ton	Ash is disposed of at a commercial MSWLF approximately 20 mile[g] from the plant.
City of Columbus Department of Public Utilities	Ohio	Columbus	Southerly	Wet	35–65%	4 458 dry ton	X		$30–$38/ton	Ash is disposed of at a commercial MSWLF approximately 20 mile from the plant.
Green Bay Metropolitan Sewerage District	Wisconsin	Green Bay	Green Bay	Dry	NA	3 766 wet ton	X		$26.77/ton	Water is added to dry ash to control emissions when offloaded into trucks. After wetting, ash has moisture content of 25–27%. Ash is disposed of at a county-owned MSWLF.
Buffalo Sewer Authority	New York	Buffalo	Buffalo	Dry	NA	6 100–6 500 dry ton	X		$27.40/ton	Ash is disposed of at a commercial MSWLF.
City of Youngstown	Ohio	Youngs-town	Youngs-town	Dry	NA	1065 dry ton	X		$15.24/ton	Water is added to dry ash to control emissions when offloaded into roll-offs. Plant transports ash to a MSWLF (Mahoning Landfill) in New Middletown, Ohio.

[a] NA = not applicable.
[b] ton × 0.9072 = Mg.
[c] cu yd × 0.7646 = m^3.
[d] BR = beneficially reused.
[e] ? = information not available.
[f] MSWLF = municipal solid waste landfill.
[g] mile × 1.609 = km.

Use options for incinerator ash tend to be site-specific. Utilities should pursue all avenues available for using ash. State departments of transportation should be contacted to determine requirements for using incinerator ash as a fly ash substitute in concrete or as a mineral filler or fine aggregate substitute in asphalt mixes. If the incinerator ash is approved for use in mix designs by the state department of transportation, then a considerable market can be opened for incinerator ash.

7.0 DISTRIBUTION AND MARKETING

In many parts of North America, the environment for land-based management of Class B biosolids has become more challenging. As farming areas are developed into residential and commercial areas, biosolids producers must transport their product farther and farther away to reach the rural environments necessary to support Class B agricultural application. Not only are the transportation-related energy costs becoming increasingly burdensome, but also the perceived imposition of urban and suburban "wastes" on rural communities can become a public relations issue for biosolids managers.

As a consequence, many wastewater treatment plants are evaluating techniques to produce a Class A biosolids product, which can be marketed locally or managed differently than Class B biosolids. Class A biosolids are essentially free of all pathogens, so the product can be distributed to the general public. EQ biosolids can be distributed as commercial products with even fewer limitations imposed by Part 503.

When contemplating a Class A process, it is important to determine how the product will be used. The costs associated with upgrading to a Class A process typically are not warranted if the material has no outlet. Also, once a market for Class A material is identified, developing that market will take time. That said, most of the classic Class A products can be successfully marketed locally, and the marketing effort can contribute a revenue stream that may offset some of the biosolids processing costs. Often market development requires 3 to 5 years of effort before the revenues might offset marketing costs.

7.1 Value-Added Products

The following biosolids products typically are marketed and distributed to various agricultural, horticultural, commercial landscape, and residential markets.

7.1.1 Compost

The public already considers compost to be an organic-based product that is a useful soil conditioner. Thus, biosolids compost producers have a less difficult time introducing their product to the marketplace. However, because many other organic byproducts also are composted (e.g., animal wastes, yard or green waste, and food wastes) under different regulatory requirements, biosolids compost may have more competition than other biosolids-derived products.

If biosolids generators produce a good-quality compost and commit the resources needed to develop and sustain the market, there typically is a strong demand for the product. Biosolids compost is successfully marketed throughout North America by generators ranging in size from the Washington (D.C.) Suburban Sanitary Commission and Los Angeles County to Davenport, Iowa, and the Hampton Roads (Virginia) Sanitation District. Each of these producers developed and exploited local markets for bulk and bagged compost through very active long-term marketing programs.

7.1.2 Heat-Dried Product

Heat dryers produce a biosolids product that can meet all U.S. EPA and most state requirements for unlimited distribution as a soil conditioner or fertilizer. Most drying systems are designed to meet the time and temperature requirements for Class A pathogen reduction. The dryers typically also meet VAR standards simply because they routinely dry biosolids to 90% solids or greater. Also, heat drying does not concentrate pollutants in the biosolids; the dry weight concentrations in both feedstock (dewatered cake) and product (heat-dried biosolids) will be nearly identical. So, if the dewatered cake met the pollutant concentrations in 40 CFR 503's Table 3, the heat dryers routinely produce EQ biosolids, which may be suitable for marketing as a soil conditioner or fertilizer.

The potential for producing fertilizer with a substantial nutrient content often is what makes heat drying attractive. Treatment plant managers look at examples like the Milwaukee Metropolitan Sewerage District (MMSD), which has produced and sold a heat-dried biosolids since about 1926, and at Houston, Texas, which has produced and distributed heat-dried biosolids since the 1950s. Both organizations developed successful markets for their products, and as a result, sales revenues play an important role in their budgets. However, such successes must be put into context; Milwaukee and Houston invested many years in developing and servicing their markets. New heat-drying facilities are unlikely to fully emulate these marketing programs until they have been operating for a number of years. Likewise, as the number of heat-drying facilities increase, there is likely to be a downward pressure on the prices of these products. Further, dryers require a lot of power, and increasing energy costs may negatively affect the economic feasibility of this option.

7.1.3 Advanced Alkaline Stabilized Products

Class A alkaline-stabilized products can vary substantially, depending on the process used to generate them. The generic lime process, as well as several proprietary processes (which use relatively low doses of lime and supplemental heat to meet Class A time and temperature requirements), create products that typically have final solids contents in the range of 30 to 40%, depending on initial cake solids. One proprietary process also dries the biosolids, producing a product with a solids content of 60%. Another process, which uses significant quantities of cement kiln dust and other waste alkaline products, typically produces a material containing 55 to 65% solids.

Alkaline-stabilized products typically are land-applied as substitutes for agricultural limestone (see Section 2.3.1.3). They are also excellent aids for reclaiming acidic soils and spoils. Alkaline-stabilized products with higher solids contents can be blended with native soils, composts, spent foundry sand, or dredge spoil to make marketable topsoil.

7.1.4 Ash-Derived Products

In addition to the ash use options discussed earlier, the following value added-products can be generated from ash.

- Brick—ash has been used in brick manufacturing by various utilities quite successfully. The Japanese have used incinerator ash to make water-permeable bricks. Brick manufacturers typically require large quantities of ash at a time. Such quantities could be obtained by emptying a lagoon and incinerating the solids.
- Worm castings (vermiculture)—one of the more innovative uses of incinerator ash is in vermiculture process. The ash is blended with food waste, and then worms are added. The worms are separated from the mixture once enough time has passed, and the remaining material is used as a soil amendment.
- Lightweight glass aggregate—a proprietary technology manufactures lightweight glass aggregates by burning a mixture of dewatered biosolids and fly ash.

7.1.5 Other Products

Any number of processes can create products that meet Class A and VAR standards. Some of these (e.g., vermicomposting and air drying) rely on Class A Alternatives 3 and 4 (see 40 CFR 503.32), which involve sampling and analyzing biosolids for helminth ova and enteric viruses. The specific processes may not have been approved by U.S. EPA as Class A alternatives, but the regulations state that a product is safe for distribution if monitoring demonstrates that it has low pathogen densities. That said, U.S. EPA and a number of states have expressed concern about the use of Alternatives 3 and 4 to meet Class A requirements. Generators considering such approaches should be aware of this and monitor the rulemaking activities of U.S. EPA and their state agency.

Other processes may meet the regulatory criteria for distribution and marketing as EQ biosolids. Technically, any biosolids that meet EQ criteria are suitable for marketing, but generators considering such processes should realize that merely achieving EQ status does not guarantee a market for the product. For example, some advanced digestion techniques may produce EQ biosolids, but if they are simply mechanically dewatered before use, then their appearance, handling properties, and odor may not be remarkably different from a Class B dewatered cake. There is little potential for marketing this material without further processing. A good general rule is that products must have high solids content—typically more than 50%—to be marketable.

Finally, there are emerging technologies (e.g., those aimed at manufacturing a biosolids-derived fuel or an aggregate component of building materials or cement) that may produce EQ biosolids or a marketable product. For more information on these processes, see *Emerging Technologies for Biosolids Management* (U.S. EPA, 2006). At press time, most of these processes had not been commercially proven, so it remains to be seen if they will produce a marketable product.

7.2 Regulatory Considerations

Marketed biosolids must meet 40 CFR 503 criteria, as well as state regulations.

7.2.1 *Federal Regulations*

All biosolids-derived products that will be distributed as a soil amendment or fertilizer are regulated under 40 CFR 503. Biosolids that meet Class A pathogen-reduction and VAR standards but do not meet Table 3 limits can still be marketed; however, their applications are subject to cumulative pollutant loadings. Such products probably have limited "marketability", although they remain suitable for broad-acre general land application.

7.2.2 *State Regulations*

State regulatory programs that affect the distribution and marketing of a biosolids-derived product include solid waste and water quality regulations, as well as some different regulatory controls.

Many states had solids or biosolids regulations in place before 40 CFR 503 was promulgated. The Part 503 provisions that facilitate the distribution and marketing of EQ biosolids may not have existed in those state regulations, so there was a lag time between 1993, when Part 503 was promulgated, and the year states adopted similar language (as agencies went through the processes necessary to modify their regulations). However, by 2008, most states had adopted the Part 503 provisions that facilitate distribution and marketing. Some states' provisions are somewhat different or more restrictive than Part 503, so any entity considering producing EQ biosolids should carefully evaluate state regulations in the areas where the product will be distributed to determine how those regulations may affect distribution costs.

Most states also have rules controlling how products with fertilizer, soil conditioning, or soil amendment (e.g., raise soil pH) value are distributed. These regulations may not be designed to control how these products are placed in the environment, but rather to ensure that financial transactions are based on real value for dollars charged. Therefore, state regulations controlling the sales and distribution of such products typically require certain guarantees of analysis (e.g., product contains at least 4% total nitrogen), efficacy of performance (e.g., product has a CCE of 55%), or other attributes (e.g., compost Salt Index is less than 2). These guarantees typically are embodied in a product label that must be approved by a state agency (e.g., the state department of agriculture) before the product can be sold. Many states also require producers or distributors of such products to be licensed or registered and typically require periodic payment of a *tonnage tax*—a fee collected on each ton of distributed product that typically pays for product inspections and analyses.

There may be fines and penalties for being out of compliance with the state fertilizer/soil amendment rules. For example, distributing a product that is dramatically different from the claims made on the product label can result in a fine or a "stop sales" order. So, entities planning to distribute and market biosolids-derived products should develop a sound database of their product's claimed attributes before labeling it for distribution. The label guarantees should be based on statistically reliable determinations of the attributes.

7.3 Marketability Criteria

7.3.1 Product Quality

The markets for each biosolids-derived product typically have a set of benchmarks that a product must meet before the market will be interested.

7.3.1.1 Compost

The product quality benchmarks for compost have been established by the U.S. Composting Council (USCC) and are included in their *Field Guide to Compost Use* (USCC, 2001). These parameters include (in addition to regulated pollutants): pH, soluble salts (salinity), nutrient content (N-P-K), water-holding capacity, bulk density, moisture content, organic matter content, particle size, growth screening, and stability.

7.3.1.1.1 pH

Most compost has a pH between 6 and 8. Adding compost can affect the pH of growing media; pH is adjusted by adding such materials as lime (alkaline) and sulfur (acidic).

7.3.1.1.2 Soluble Salts

The soluble salts concentration is the concentration of total soluble ions in a solution. Most plant species have a salinity tolerance rating, and maximum tolerable quantities are known. Excess soluble salts can cause phytotoxicity. Another measure of this parameter is the *Salt Index*, which is determined in the lab and compared to a standard (ammonium sulfate = 100). Products with a Salt Index well below 100 typically are not problematic. Ferric chloride conditioning may be one source of excessive salinity levels.

7.3.1.1.3 Nutrient Content

Nitrogen, phosphorus, and potassium are the three macronutrients required by plants in the greatest amounts, and therefore, the fertilizer components of greatest interest. These three nutrients are measured and expressed on a dry weight basis as a percentage (%) of the total dry mass; nitrogen typically is broken down in terms of inorganic and organic or slow-release forms, and both phosphorus and potassium are converted to the fertilizer forms of phosphoric acid (P_2O_5) and potash (K_2O). Laboratory determinations of extractable P_2O_5 and K_2O are necessary to determine plant-available fertilizer.

7.3.1.1.4 Water-Holding Capacity

Water-holding capacity is a measure of a given volume of compost's ability to hold water under 101.325 kPa (1 atm) of pressure, measured as a percent of dry weight. It indicates the potential benefit of reducing the required irrigation frequency, as well as gross water requirements. The water-holding capacity should be known to allow users to monitor or estimate the compost's effect on their crop-watering regime and growing media.

7.3.1.1.5 Bulk Density

Bulk density is the weight per unit volume of compost. It is used to convert compost application rates from metric tonnage to cubic meters. In a field application, cubic meters per hectare then would be extrapolated to express an application rate represented as a depth (e.g., 30-mm application rate). Bulk density also is used to determine the volume of compost that may be transported on a given occasion, taking into account that most vehicles have a specific maximum gross weight that may not be legally surpassed.

7.3.1.1.6 Moisture Content

Moisture content is the measure of the amount of water in a compost product, expressed as a percent of total solids. The moisture content of compost affects its bulk density, and, therefore, transportation costs. Moisture content also affects product handling. Dry compost can be dusty and irritating to work with, while wet compost can be heavy and clumpy, making it difficult to apply and more expensive to transport. The ideal moisture content for compost is between 35 and 45%.

7.3.1.1.7 Organic Matter Content

Organic matter content is a measure of organic carbon-based materials in compost. It typically is expressed as a percentage of dry weight.

7.3.1.1.8 Particle Size

Particle size is a measure (in percent) of how much of the product (sample) will pass through a certain sized screen opening. The particle size often dictates the product's use. For example, coarse compost is best suited for land reclamation and erosion control, while finer composts are used for topsoil blending and top dressing of turf. For most applications, merely specifying the product's maximum particle size (or the screen size through which it passes) is sufficient. However, some applications (e.g., potting/nursery media component), require a particular particle size distribution. *Particle size distribution* measures the amount of compost within a specific particle size range. It is determined by pouring a sample of compost through a series of sieves (screens), each having smaller openings than the one before. Results are expressed as the percent (by weight) of sample retained by each sieve size.

7.3.1.1.9 Growth Screening

The growth screening test is an indicator of the presence of phytotoxic substances (e.g., volatile fatty acids, alcohol, soluble salts, heavy metals, or ammonia). These substances may cause delayed seed germination, seed or seedling damage or death, or plant damage or death. Growth-screening tests include germination, root elongation, and pot tests; they are not intended to identify which growth inhibitor caused the poor growth response. Also, a product that passes initial growth-screening tests may fail later if improperly stored. Specific growth inhibitors (e.g., volatile fatty acids and alcohol) may form in compost stored under anaerobic conditions.

7.3.1.1.10 Stability

Stability is a measure of the level of biological activity in compost under a given set of conditions. Unstable compost consumes nitrogen and oxygen in significant quantities to support biologic activity and generates heat, carbon dioxide, and water vapor. Stable compost consumes almost no nitrogen and oxygen and generates almost no carbon dioxide or heat. Unstable compost demands nitrogen; it can cause nitrogen deficiency and be detrimental to plant growth, even killing plants in some cases. If stored, and left unaerated, unstable compost can become anaerobic and emit nuisance odors.

7.3.1.2 Heat-Dried Biosolids

Heat-dried biosolids typically should meet most of the physical, chemical, and microbiological regulatory criteria necessary to allow distribution of the product. Some of the important criteria include particle size, durability (hardness), dust, odor, bulk density, nutrients, Salt Index, and heating value.

7.3.1.2.1 Particle Size

As with compost, a measure of the size of the particles of dried biosolids is useful. Many biosolids drying processes will screen the final product as an important process step; this allows the producer to specify the size of the particles that will be managed merely by selecting a specific screen size or sizes. Preferred particle size may vary from market to market. For example, users seeking a standard-grade fertilizer might prefer a product with a particle size ranging from 2 to 4 mm in diameter, while a golf course superintendent seeking to use dried biosolids on tees and greens might prefer a smaller product (typically between 0.3 and 0.8 mm).

Particle size is an important consideration for a number of reasons. If the dried biosolids will be blended with other fertilizer ingredients, then it will be less likely to segregate after blending if it is about the same size as those other ingredients. Particle size also may affect the even distribution of a product. To an extent, particle size even affects the rate at which dried biosolids release nutrients (i.e., a smaller particle should release nutrients more quickly than a larger one).

Fertilizer users also may be interested in variations on the particle size determination. The *size guide number* (SGN) is a measure of the average size of granules in a fertilizer. To calculate SGN, product is screened through a nested series of screens with different-sized openings, screeners determine which screen opening size (in millimeters) retains 50% (by weight) of the material, and they then multiply that opening size by 100. In other words, a product with an SGN of 200 would have an average particle size of 2 mm.

Another physical measurement that may be important to fertilizer blenders is the Uniformity Index. In general, the *Uniformity Index* is the ratio of the size of the small particles (fines) to the large particles (coarse). A lower Uniformity Index indicates a broad particle size distribution; a higher one indicates a narrow distribution (i.e., a Uniformity Index of 100 would mean that all particles are the same size). Users typically prefer a higher Uniformity Index.

7.3.1.2.2 Durability (hardness)

Because heat-dried biosolids often are handled somewhat roughly (conveying, dumping, and moving with front-end loaders), the particles' durability (resistance to degradation) is important. Less durable particles will easily fracture, creating finer particles and dust.

There is no universally accepted test method for hardness. The Fertilizer Institute offers one test procedure. It involves placing selected fertilizer particles onto a ratcheting hardness tester, which exerts pressure on the particle and records the "weight" at which the particle crumbles as its "crushing strength". According to this method, urea pills have a crushing strength between 9 and 13 Newtons (2 and 3 lb), and ordinary super-phosphate has a crushing strength of 20 to 31 Newtons (4.5 to 7 lb). One could anticipate that dried biosolids with a crushing strength greater than 20 N (4 to 5 lb) should be reasonably durable.

7.3.1.2.3 Dust

A heat-dried biosolids' dust content may be the most important physical parameter when assessing whether the material is well-suited for a particular use; a dusty product tends to be poorly received by users. Many users and intermediaries demand that heat-dried biosolids contain virtually no dust.

One aspect of wastewater treatment that affects this characteristic is the degree to which grit, fine fibers, and hairs are kept out of the biosolids. These items tend to diminish the durability of heat-dried biosolids so that they are more susceptible to fracturing during handling and transfer, resulting in many fine particles. If the feedstock contains a substantial fraction of raw primary solids or the headworks equipment is not as efficient as possible, then the product typically will contain extraneous materials and could easily fracture. As a general rule, undigested solids typically produce a dustier heat-dried biosolids than biosolids that are digested prior to heat-drying.

Moreover, some drying technologies produce biosolids more prone to dust than others. Biosolids particles that are more angular tend to break during handling, creating fine particles and dust. More spherical particles typically are less likely to do so.

That said, most heat-drying technologies produce some dust and so many facilities install screening systems to remove fine particles or add dust-suppression oils to the product.

7.3.1.2.4 Odor

Most heat-dried biosolids emit a musty, earthy odor with overtones of ammonia and—depending on the system—sometimes a burnt smell. Anecdotally, heat-dried biosolids created from well-digested feedstock tend to be perceived as less odorous than those produced from raw solids. In most applications, odor does not limit the heat-dried biosolids' suitability for beneficial use. Occasionally, comments about product odor are heard after application and rewetting (e.g., after rainfall or irrigation). As with other biosolids, if heat-dried biosolids are surface applied at a high rate without subsequent incorporation, one can anticipate that some odor will emanate. The intensity of the odor will be a function of the application rate, but in most cases, the odor should dissipate over time. It is prudent, if such applications are planned, to select sites where fewer sensitive receptors might note the odors.

7.3.1.2.5 Bulk Density

Bulk density is a measure of the mass of a product per unit volume (e.g. pounds per cubic yard). This measurement is important in managing heat-dried biosolids for a number of reasons.

The bulk density of dried biosolids typically ranges from 560 to 720 kg/m^3 (35 to 45 lb/cu ft). However, the product's actual density cannot be determined until after the drying system is in operation, so design engineers will have to select a reasonable estimation when designing product handling and storage systems. Product-management issues may arise if storage-system designers based the work on a high bulk density and the actual product has a much lower density. Also, a product with low bulk density may fill up a transport unit before the maximum allowable weight is met, thereby increasing delivery costs.

If heat-dried biosolids will be bagged, either as a standalone fertilizer or as a component of blended dry fertilizers, its bulk density becomes important in bag sizing. Low bulk density can be especially problematic if the material is replacing a denser filler in a blended fertilizer; the manufacturer may need to purchase new bags to accommodate the lighter material.

7.3.1.2.6 Nutrients

The nutrient content of heat-dried biosolids typically is its chief value to prospective users. Users are interested in three types of nutrients: primary nutrients (nitrogen, phosphorous, and potassium); secondary nutrients (calcium, magnesium, and sulfur); and micronutrients (boron, chlorine, copper, iron, manganese, molybdenum, and zinc).

In most states, the state department of agriculture (or a similar agency) requires that all fertilizers (e.g., heat-dried biosolids) must be registered and labeled. Most states require that percentages of primary plant nutrients be guaranteed [typically presented on a fertilizer bag as % N–% P_2O_5–% K_2O (e.g., 10-10-10)]; they also will accept guarantees of secondary or micronutrients. These guarantees become regulatory standards, and if a product fails to provide the promised amount of nutrients, the generators can be subjected to fines, "stop sales" orders, or both.

Heat drying typically preserves the nutrient content of the feedstock solids, except for nitrogen. Typically, heat drying removes all but a small portion of the soluble ammonia-nitrogen from dewatered biosolids. It does not remarkably change the dry weight concentrations of organic nitrogen or most of the other inorganic elements.

Utilities typically measure biosolids' phosphorous content on an elemental basis. However, the fertilizer industry assesses a product's phosphorus value based on available phosphoric acid, which is determined by citric acid attraction.

Also, the fertilizer industry typically does not measure potassium in its elemental form, but rather as K_2O. A biosolids' total potassium content (dry weight basis) can be converted to K_2O by multiplying by 1.2. The fertilizer industry measures potassium via water extraction. Typically, potassium remains soluble through the wastewater treatment processes and does not accumulate in biosolids to any great extent. Most heat-dried biosolids producers do not guarantee K_2O content.

Secondary plant nutrients are typically important on a regional basis (e.g., calcium and magnesium may be important in areas with acidic soils) or for specific crops or cropping rotations. If guaranteed as a component of heat-dried biosolids, these elements typically are reported as the percent concentration of the element (dry weight basis). Most states have minimum concentration requirements for such guarantees.

Compared to the major elements, plants require small amounts of the micronutrients. Micronutrients are unique among the essential elements because their deficiency frequently is associated with a combination of crop species and soil characteristics. Copper, molybdenum, and zinc also are regulated pollutants under Part 503. As with the secondary nutrients, these elements typically are reported as the percent concentration of the element (dry weight basis), and most states have minimum concentration requirements for such guarantees.

The fertilizer components of heat-dried biosolids typically are either incorporated in the organic matter or precipitated as somewhat insoluble compounds in the solid matrix. Thus, most of the biosolids' fertilizer value is as a "slow-release" form of the nutrients. What this really means, in terms of marketing heat-dried biosolids, is that most of the fertilizer components are not released until naturally occurring populations of soil microbes break down the organic matter. This typically occurs when soil temperatures are above about 52°F and soil moisture is adequate.

7.3.1.2.7 Salt Index

Soluble salts are the concentration of dissolved ions in an extract of the water held within and by the soil matrix. In some parts of the country, soils may accumulate excessive concentrations of soluble salts, especially during summer, because of increased transpiration, insufficient leaching of salts from the soil solution, and overfertilization. Extremely high salt concentrations may inhibit water uptake and lead to chlorosis, foliage scorching, premature leaf drop, dieback, and inhibition of growth (frequently called *burning*). The *Salt Index* is a measure of a fertilizer's ability to cause these effects in plants. A fertilizer with a high Salt Index is more likely to lead to salt accumulation and desiccation injury than a fertilizer with a low Salt Index. High Salt Index fertilizers tend to cause water to move out of, not into, root cells.

Specifically, the Salt Index compares a fertilizer's potential to prevent water uptake by roots with that of an equivalent weight of the standard, sodium nitrate. Sodium nitrate, which has a Salt Index value of 100, was chosen as the standard because it is 100% water soluble and was a commonly used nitrogen fertilizer when the Salt Index was first proposed in 1943. Fertilizers with Salt Index values greater than 100 are more likely to prevent roots from absorbing water than sodium nitrate does. Fertilizers with Salt Index values greater than 20 should be worked into the soil before planting, or else applied to the soil surface and then sufficiently irrigated to reduce the potential of salt injury ("fertilizer burn").

Heat-dried biosolids are expected to have relatively low Salt Indices, but testing for this parameter and having the data available may facilitate conversations with prospective users.

7.3.1.2.8 Heating Value

An emerging market for heat-dried biosolids is as a biomass fuel in industrial applications. In Europe and other parts of the world, biosolids are combusted as a fuel in industrial processes (e.g., cement manufacturing). Biomass fuels are characterized by "proximate and ultimate analyses." The "proximate" analysis includes moisture content, volatile content (when heated to 950°C), the remaining free carbon at that point, the ash (mineral) in the sample, and the high heating value (HHV) based on complete combustion of the sample to carbon dioxide and water. The "ultimate" analysis gives the fuel's composition in weight percentage of carbon, hydrogen, and oxygen (the major components), as well as sulfur and nitrogen (if any).

Typical heating values for dried biosolids are anticipated to range from 12 000 to 28 000 kJ/kg (5 000 to 12 000 Btu/lb). Digested biosolids have about half the HHV of primary solids. In comparison, coal's HHV might be on the order of 30 000 kJ/kg (13 000 Btu/lb), while woody biomass might have an HHV around 19 000 kJ/kg (8 000 Btu/lb).

7.3.2 *Product Consistency*

Paying customers expect to get the same product and performance with each bag, batch, or load they buy. Therefore, biosolids generators should endeavor to produce the most reliable, consistent product possible.

Because the vagaries of seasons, flow, and process control naturally will result in some inconsistencies in product quality, generators should develop as much representative data as reasonably practical so they can show customers that each product characteristic remains within an expected range. This may involve more frequent sampling and analyses than required under Part 503, especially in the early years of a marketing program, so a database can be developed and statistically reviewed.

7.4 Identifying and Developing Markets

It is rare for generators of a new Class A biosolids product to find a fully developed market waiting for it.

Before investing in a new Class A facility, agencies should carefully assess local and regional markets for the products to be produced, use feedback from that assessment to finalize the design, and then endeavor to maintain communications with those markets once the process is selected and construction commences. If the marketplace is aware that a new product is coming and has some understanding of it, customers will be more receptive when the agency's new product is unveiled.

Doing this may shorten the period required to develop full market opportunities for the new product; however, biosolids generators still should expect at least 3 years to elapse before product revenues begin to meaningfully offset the costs of marketing. Likewise, organizations contemplating a new biosolids process should not presume that revenues from product sales will do more than slightly reduce O&M costs; rather, they should assume that the biosolids management program will remain a cost in the budget.

That being said, there are markets for various biosolids-derived products, and such marketing can support an agency's overall mission of producing clean water while limiting the total costs of residuals management.

7.4.1 *Typical Markets*

Markets are product-specific. Compost typically is sold as a soil conditioner or amendment to add organic matter to the soil or improve the soil's physical properties. Composting tends to reduce the biosolids' total nitrogen content, thus reducing its usefulness as a fertilizer; however, with fertilizer prices at historical highs in 2008, even a minor amount of nitrogen or phosphorus content might improve a compost's value.

Table 27.14 lists typical uses for compost, and the markets associated with each use. Each market has particular needs and preferences. For example, customers that use compost as a seed-germination medium are particularly sensitive to phytotoxicity and salinity. Those that use it as a top dressing are sensitive to particle size. In landfill cover, land reclamation, and erosion-control applications, coarser compost may be preferred because it will be more erosion-resistant. Meanwhile, texture and residual odor are key concerns in retail or homeowner applications.

Compost typically is sold in bulk to landscapers, soil blenders, and site-specific end users who may want to spread the product over a larger site (e.g., parks and ball fields). Some agencies also provide bulk compost on consignment to home and garden stores, which in turn allow customers to purchase in small bulk lots. Over time, as the market becomes more familiar with the product, some agencies will install bagging systems or contract with local baggers and begin to sell the end product in 18-kg (40-lb) bags or similarly sized containers. Bag sales typically are directed at the homeowner or smaller users, but it would not be inconceivable that other municipal agencies, golf courses and the like might buy compost in bags.

Unlike compost, heat-dried biosolids retain most of their initial nitrogen content and, therefore, offer somewhat more fertilizer value. A typical dried biosolids with 5% total nitrogen, in mid-2008, might have the equivalent nitrogen fertilizer value of about $90/metric ton (not taking into account anything other than the total nitrogen concentration). Some portion of that will be "slow-release," which represents both a market enhancement and detriment. To customers seeking release of nutrients over time, this

TABLE 27.14 Typical uses and markets for compost.

Uses	Markets
Potting and horticulture mixes	Greenhouses, nurseries, and retail distribution
Soil replacement	Field nurseries and sod farms
Blending to produce topsoil	Topsoil blenders and landscape material suppliers
Turf establishment	Landscape and site contractors, public sector*, and retail distribution
Top-dressing of turf	Gold courses, institutions, public sector, and retail distribution
Amendment of sandy or clay soils	Agriculture, also soil replacement, topsoil, turf establishment, and land reclamation markets
Land reclamation	Landfill operators, mine and gravel pit operators, landscape and site contractors
Landfill cover	Landfill operators
Private gardens	General public

* Public sector may include municipal, county, state, and federal agencies (such as parks departments, highway and public works departments, and airports).

is a valuable feature, but if a prospective customer is looking to provide all the nitrogen needs for a crop to be grown, the slow release property may reduce his or her perceived value of the product.

Heat-dried biosolids typically are sold as a component of dried, blended complete-analysis fertilizers (i.e., a fertilizer that offers nitrogen, phosphorus, and potassium). If the fertilizer is a low analysis (e.g., 10-10-10) the blender cannot construct that product merely with chemical fertilizer resources. Thus, the blender typically will add some amount of inert filler to complete the blend; if biosolids can be substituted for this product, the blender can both avoid the cost of filler and potentially reduce their purchase of other fertilizers necessary to complete the blend.

Heat-dried biosolids also can be sold as a standalone product both in bulk and in bags. Most "local" users (e.g., golf courses, parks departments, and turf managers) can often use product in bulk bags [about 450 kg (1000 lb) each]. Smaller bags [18 to 20 kg (40 to 50 lb)] can be distributed to homeowners and local users.

An emerging bulk user of dried biosolids is the cement industry. Cement kilns may burn a number of fuels—including "waste" products—so the industry is open to using dried biosolids as another fuel. Each cement company and kiln location probably will have a different perception of this practice and/or local limits on what can be burned. Nonetheless, if heat-dried biosolids can be demonstrated to contribute useful heat, then kiln operators (and other biomass-to-energy processes) may be interested in the product.

7.4.2 *Regional and Seasonality Issues*

The "peak" seasons for using biosolids-derived products vary throughout the United States. As a general rule, the two primary peaks are spring and fall. These peaks may be extended in areas with longer growing seasons, but agencies typically should plan on slow sales during some portion of the year. In the southwestern United States, for example, sales of biosolids-derived products may be robust from February to April, curtailed during the hot summer months, and peak again from September through December. Conversely, in the northeastern United States, spring sales may peak from April through June, and fall sales may rise in late August and remain strong through October.

A market assessment can help define the selling seasons. Agencies should use such an assessment when developing a product storage (inventory) strategy.

Although most biosolids-derived products can be used in most of the United States, agencies should not anticipate that sales of alkaline-based products will be robust in areas where soil pH levels are well above neutral. Folks in such areas just do not buy much liming agent. That is not to say that there is no market for such products there, but the value of the products may be limited unless they also possess other marketable attributes.

7.4.3 Distribution Approaches

Generators typically have two primary distribution approaches: They can have staff market a biosolids-derived product, or they can contract a third party to do it. If generators decide to market the product themselves, then the staff assigned to this work should have some experience with or expertise in the markets to be developed and served. Generators may find that they need to hire staff specifically for this purpose.

If distribution and marketing are contracted, the agency has a number of approaches to consider, ranging from using a broker to hiring a contractor to remove and manage the product on a daily basis. A *broker* typically is someone who adds the biosolids-derived product to a suite of products they market (similar to a manufacturer's rep). Brokers work hard to sell product and receive a commission when they do so, but typically there is no guarantee that a specific quantity of product will be distributed.

Some agencies enter into contracts with service providers that remove and manage product daily (similar to the companies that manage Class B biosolids). These companies are responsible for finding locations to distribute the material. Generators typically see little revenue from such contracts; in fact, they may have to pay the company for the service. The primary benefits of this approach are reduced staff costs and storage requirements.

An intermediate approach is a contract with a third party that will develop and service markets for the product. In this case, an agency may provide financial support until the market begins to realize the value of the product. Such contracts typically are at least 5 years long so contractors have an opportunity to realize the fruits of their early labors.

7.4.4 Distribution Methods

Various customers desire product delivery in various ways. Although a market assessment will indicate how the market prefers delivery, most agencies begin with a bulk distribution program until market demand is established. Once an agency has a reliable, consistent set of outlets for the bulk product, they can begin investigating the value of bagging. Over time, sales of bagged products may become a useful income stream, but that determination is better made when an agency has experience in the marketplace.

If an agency decides that sales of bagged product are viable, there are essentially two ways to implement the program:

- Invest in the equipment and labor necessary to bag product or
- Contract with a third party to bag, pallet, and wrap product.

In either approach, the agency will have to develop a bag label (which typically must be approved by state regulators) and purchase bags.

8.0 REFERENCES

American Water Works Association (2008). *Quicklime and Hydrated Lime*; B202-07; American Water Works Association: Denver, Colorado.

ASTM International (2006) *Standard Test Methods for Chemical Analysis of Limestone, Quicklime, and Hydrated Lime*; C25-06; ASTM International: Philadelphia, Pennsylvania.

Dominak, R. P. (2005) *Long Term Residuals Management Plan for the Northeast Ohio Regional Sewer District*; Northeast Ohio Regional Sewer District: Cleveland, Ohio.

Evanylo, G. K. (1999) *Agricultural Land Application of Biosolids in Virginia*; Publication No. 452-303; Virginia Cooperative Extension: Blacksburg, Virginia.

Forste, J. (1996) Land Application. In *Biosolids Treatment and Management*; Girovich, M., Ed.; Marcel Dekker, Inc.: Monticello, New York.

Hundal, L.; Cox, A.; Granato, T. (2005) Promoting Beneficial Use of Biosolids of Chicago: User Needs and Concerns. Paper presented at WEF Innovative Uses of Biosolids and Animal and Industrial Residuals Conference; Chicago, Illinois, June 29–July 1, 2005.

Logan, T. (2008) Personal communication. July.

Lue-Hing, C.; Tata, P.; Granato, T. A.; Pietz, R.; Johnson, R.; Sustich, R. (1998) *Sewage Sludge Survey*; Association of Metropolitan Sewerage Agencies: Washington, D.C.

Moss, L.; Epstein, E.; Logan, T. (2002) *Comparing the Characteristics, Risks and Benefits of Soil Amendments and Fertilizers Used in Agriculture*; Report 99-PUM-1; Water Environment Research Foundation: Alexandria, Virginia.

MWH Global, Inc. (2005) *Water Treatment Principles and Design*, 2nd ed.; Wiley & Sons: New York.

National Biosolids Partnership Home Page. http://www.biosolids.org (accessed May 2009).

National Biosolids Partnership (2005) *National Manual of Good Practice for Biosolids*; National Biosolids Partnership: Alexandria, Virginia.

Northeast Biosolids and Residuals Association (2007) *A National Biosolids Regulation, Quantity, End Use and Disposal Survey*; Northeast Biosolids and Residuals Association: Tamworth, New Hampshire.

Sharma S. D.; Lewis H. P. (1994) *Waste Containment Systems, Waste Stabilization and Landfills: Design and Evaluation*; Wiley & Sons: New York.

U.S. Composting Council (2001) *Field Guide to Compost Use*; U.S. Composting Council: Rokonkoma, New York.

U.S. Department of Agriculture Natural Resources Conservation Service (1999) *Nutrient Management Code 590*; Conservation Practice Standard Publication; U.S. Department of Agriculture, Natural Resources Conservation Service: Madison, Wisconsin.

U.S Department of Agriculture, Natural Resources Conservation Service, Web Soil Survey Home Page. http://websoilsurvey.nrcs.usda.gov (accessed May 2009).

U.S. Environmental Protection Agency (1991) Solid Waste Disposal Facility Criteria, Final Rule, Part II. *Code of Federal Regulations*, Parts 257 and 258, Title 40.

U.S. Environmental Protection Agency (1993a) Standards for the Use or Disposal of Sewage Sludge. *Code of Federal Regulations*, Parts 405(d) and 503, Title 40

U.S. Environmental Protection Agency (1993b) *Solid Waste Disposal Facility Criteria Technical Manual*; EPA 530R93017; U.S. Environmental Protection Agency: Washington, D.C.

U.S. Environmental Protection Agency (1994a) *Plain English Guide to the EPA Part 503 Rule*; EPA 832R93003; U.S. Environmental Protection Agency: Washington, D.C.

U.S. Environmental Protection Agency (1994b) *Guidance for Writing Permits for the Use or Disposal of Sewage Sludge*; U.S. Environmental Protection Agency: Washington, D.C.

U.S. Environmental Protection Agency (1994c) *The Hydrologic Evaluation of Landfill Performance Model (HELP)*; Users Guide Version 3; EPA600R94168a; U.S. Environmental Protection Agency: Washington, D.C.

U.S Environmental Protection Agency (1995a) *A Guide to the Biosolids Risk Assessment for the EPA Part 503 Rule*; EPA832B93005; U.S. Environmental Protection Agency: Washington, D.C.

U.S. Environmental Protection Agency (1995b) *Process Design Manual: Surface Disposal of Sewage Sludge and Domestic Septage*; EPA 625R95002; U.S. Environmental Protection Agency: Washington, D.C.

U.S. Environmental Protection Agency (2000) *Guide to Field Storage of Biosolids*; EPA832B00007; U.S. Environmental Protection Agency: Washington, D.C.

U.S. Environmental Protection Agency (2003) *Environmental Regulations and Technology: Control of Pathogens and Vector Attraction in Sewage Sludge*; EPA 625R92013; U.S. Environmental Protection Agency: Washington, D.C.

U.S. Environmental Protection Agency (2006) *Emerging Technologies for Biosolids Management*; EPA 832R06005; U.S. Environmental Protection Agency: Washington, D.C.

Water Environment Federation (2009) *Wastewater Solids Incineration Systems;* Manual of Practice No. 30; McGraw-Hill: New York.

Glossary

abiotic The nonliving elements in the environment.

absorption Assimilation of molecules or other substances into the physical structure of a liquid or solid without chemical reaction.

accuracy A measure of how closely the model matches observed data.

acetic acid Chemical produced by fermentation that is used as a carbon source for biological processes. Chemical formula is CH_3COOH.

acetogenesis The metabolic process that converts volatile acids to acetate, the primary substrate of acetoclastic methanogenesis.

acid (1) A substance that can react with a base to form a salt. (2) A substance that can donate a hydrogen ion or proton.

acidity The capacity of an aqueous solution to neutralize a base.

acidogenesis The formation of organic substrates to short-chain volatile fatty acids.

ACR Air volume-to-filter cloth surface area ratio in a baghouse.

actinomycetes Microorganisms typically found in compost; they have characteristics of both bacteria and fungi.

activated carbon A highly adsorbent form of carbon used to remove dissolved organic matter from water and wastewater, or to remove odors and toxic substances from gaseous emissions.

activated charcoal See *activated carbon.*

activated sludge The biologically active solids in an activated sludge process.

activated sludge process A biological wastewater treatment process in which a mixture of wastewater and biologically enriched sludge is mixed and aerated to facilitate aerobic decomposition by microbes.

activation energy The energy required to initiate a process or reaction.

admixture (1) A material or substance added during mixing. (2) A substance other than cement, aggregate, or water that is mixed with concrete.

adsorption The process of transferring molecules of gas, liquid, or a dissolved substance to the surface of a solid, where it is bound by chemical or physical forces.

adsorption capacity The maximum amount of contaminant an adsorptive media (e.g., activated carbon) can collect.

advanced oxidation processes Processes using a combination of disinfectants, (e.g., ozone and hydrogen peroxide) to oxidize toxic organic compounds into a nontoxic form.

advanced secondary treatment Secondary wastewater treatment with enhanced solids separation.

advanced wastewater treatment Treatment processes designed to remove pollutants that are not adequately removed by conventional secondary treatment processes.

aeration The addition of air or oxygen to water or wastewater, typically by mechanical means, to increase dissolved oxygen levels and maintain aerobic conditions.

aerator A device used to introduce air or oxygen to water or wastewater.

aerobe An organism that requires free oxygen for respiration.

aerobic Condition characterized by the presence of free oxygen.

aerobic digestion Solids stabilization process involving direct oxidation of biodegradable matter and oxidation of microbial cellular material.

agar A gelatinous substance extracted from red algae, typically used to culture microorganisms.

agar plate A circular glass plate, containing agar or another nutrient medium, used to culture microorganisms.

agglomeration Coalescence of dispersed suspended matter into larger flocs or particles.

agronomic rate The annual whole biosolids application rate (dry weight basis) is designed to do the following:

(1) Provide the amount of nitrogen needed by the food crop, feed crop, fiber crop, silvicultural crop, horticultural crop, or vegetation grown on the land.

(2) Minimize the amount of nitrogen that passes below the root zone of the crop or vegetation and enters the groundwater.

air changes per hour (ACH) A measure of the ventilation rate of an enclosed volume.

air diffuser A device designed to transfer atmospheric oxygen into a liquid.

airlift pump A device for pumping liquid by injecting air near the bottom of a riser pipe submerged in the liquid to be pumped, which lowers the specific gravity of the fluid mixture and allows it to rise up the riser pump.

air pollutant concentration The measure of a pollutant in the ambient air. The units may be expressed as mass per unit volume or as a molar ratio.

air scour The agitation of granular filter media with air during the filter backwash cycle.

air stripping The process of removing volatile and semivolatile contaminants from liquid by passing air and liquid countercurrently through a packed tower.

aliquot The amount of sample used for analysis.

alkali A substance with highly basic properties.

alkaline Water containing enough alkalinity to raise the pH above 7.0.

alkaline stabilization The process by which lime or other alkaline materials are added to solids to raise the pH above 12 for 2 hours to reduce pathogens.

alkalinity The ability of a water to neutralize an acid via the presence of carbonate, bicarbonate, and hydroxide ions.

alpha factor The ratio of oxygen-transfer coefficients for water and wastewater at the same temperature and pressure; used in sizing aeration equipment.

alternative system (onsite wastewater treatment) An onsite wastewater treatment system that is not a conventional system (as described by local regulatory code).

alum Common name for aluminum sulfate, frequently used as a coagulant in water and wastewater treatment. Chemical formula is $Al_2(SO_4)_3 \cdot 14H_2O$.

aluminum sulfate See *alum.*

amendment Organic material or bulking agent (e.g., wood chips or sawdust) added to municipal solids in a composting operation to promote uniform air flow.

ammonia A compound of hydrogen and nitrogen that occurs extensively in nature. Chemical formula is NH_3.

ammonia-nitrogen The quantity of elemental nitrogen present in the form of ammonia and the ammonium ion.

ammonification Bacterial decomposition of organic nitrogen to ammonia.

ammonium ion A form of ammonia found in solution; the ion NH_4^+.

amoeba A single-celled protozoan microbe. Also ameba.

amoebic dysentery A form of dysentery caused by a protozoan parasite, typically resulting from poor sanitary conditions and transmitted by contaminated food or water. Also amebic dysentery.

anaerobe An organism that can thrive in the absence of free oxygen, nitrate, and nitrite.

anaerobic Condition characterized by the absence of free oxygen and other electron receptors such as nitrate and sulfate. As a process, anaerobic implies the active presence of strictly anaerobic organisms. As an activated sludge process component, anaerobic implies the active presence of anaerobic and facultative organisms.

anaerobic digestion Solids stabilization process operated specifically without oxygen in which much of the organic wastefeed is converted to methane and carbon dioxide.

anion A negatively charged ion that migrates to the anode when an electrical potential is applied to a solution.

anionic flocculant A polyelectrolyte with a net negative electrical charge.

anionic polymer A polyelectrolyte with a net negative electrical charge.

annual whole solids application rate The maximum amount of biosolids (dry weight basis) that can be applied to a unit area of land during a 365-day period.

anode The positive electrode via which current leaves an electrolyte solution.

anodic protection Electrochemical corrosion protection achieved via the use of an anode with a higher electrode potential than the metal to be protected.

anoxic Condition characterized by the absence of free oxygen.

anthracite A hard, black coal containing a high percentage of fixed carbon and a low percentage of volatile matter; it burns with little or no smoke.

anthropogenic compounds Compounds created by human beings; they often are relatively resistant to biodegradation.

aqueous chlorine Chlorine or chlorine compounds dissolved in water; often mistakenly called "liquid chlorine."

arc screen Type of coarse screening device.

arc-flash The release of electrical energy at the point of a fault or short-circuit.

Archea The phylogenic class of prokaryotes that contain methanogens.

Archimedes' screw pump See *screw pump.*

area sources An open surface or basin that emits air pollutants.

aerated SRT Solids retention time within the aerated portion of an activated sludge tank. See *solids retention time.*

anammox Biological anaerobic ammonium oxidation process catalyzed by specialized planctomycete bacteria, whereby ammonia and nitrite are consumed to produce primarily nitrogen gas.

as-built drawings A copy of the original plans and specifications prepared for construction that have been corrected to reflect how a facility was actually built or installed.

ash The nonvolatile inorganic solids that remain after incineration.

aspergilosis A repertory infection, growth, or allergic response caused by the *Aspergillus* fungus.

aspirating aerator Aeration device using a motor-driven propeller to draw atmospheric air into the turbulence caused by the propeller and, thereby, form small bubbles.

aseptic The state of being free of pathogens.

at-grade alternative Above-grade soil treatment area designed and installed using suitable imported soil material for fill so some part of the infiltrative surface is located at the original ground.

atmospheric pressure The force exerted by the weight of the atmosphere above the point of measurement.

autooxidation A self-induced oxidation process.

autothermal thermophilic aerobic digestion An aerobic digestion process in which the microbes generate enough heat to maintain temperatures in the thermophilic range. When maintained for enough time to meet 40 CFR 503 requirements, the process results in a biosolid that is relatively pathogen free.

autotrophic membrane bio-reactor Biological reactor that uses low-pressure membranes to deliver hydrogen gas as an electron donor to autotrophic bacteria, which reduce nitrate or other oxidized contaminants.

available chlorine A measure of the total oxidizing power of chlorinated lime, hypochlorites, and other materials used as a source of chlorine (as compared with the measure of elemental chlorine).

available short-circuit current The amount of electrical current that will flow in an electrical system during a fault (short-circuit).

average day, maximum month The average daily flow or mass of a constituent during the month of maximum measurements for that constituent.

average daily flow The total flow past a physical point over a period of time divided by the number of days in that period, typically taken to mean a yearly (annual) average.

average flow The arithmetic average of flows measured at a given point, typically taken to mean a yearly (annual) average.

averaging periods The unit of time over which a measurement is taken.

autogenous combustion Burning that occurs when the heat of combustion of a wet organic material or solids is sufficient to vaporize the water and maintain combustion without auxiliary fuel.

autogenous temperature Equilibrium temperature in solids combustion wherein the heat input from the fuel equals the heat loss, and combustion is self-supporting.

autothermal thermophilic aerobic digestion A biological digestion system that converts soluble organics to lower-energy forms via anaerobic, fermentative, and aerobic processes at thermophilic temperatures.

autotroph Organism that derives its cell carbon from carbon dioxide.

axial flow The flow of fluid in the same direction as the axis of symmetry of a tank or basin.

axial flow pump A type of centrifugal pump in which fluid flow remains parallel to the flow path and develops most of its head by the lifting action of the vanes.

bacilli Rod-shaped bacteria.

backflow flow Liquid flowing opposite to the desired direction in a water distribution system; it may result in contamination from a cross-connection.

backflow prevention device Device used to prevent cross-connection (backflow) of nonpotable water into a potable water system.

backmixing Mixing thermally dried product with dewatered solids before the mix enters the dryer. Also, the dispersion of mixed liquor from the outlet end of an aeration tank towards the inlet, whereby nitrite and nitrate are subject to denitrification.

backwash A high-rate reversal of flow for the purpose of cleaning or removing solids from a filter bed or screening medium.

backwash rate The flowrate used during backwash operations.

backwash reject water ratio The ratio of backwash water used relative to the amount of total water filtered.

backwater Increased water surface elevation in a channel upstream of the reference point.

bacteria A group of universally distributed, rigid, essentially unicellular microscopic organisms lacking chlorophyll. They perform a variety of biological treatment processes (e.g., biological oxidation, solids digestion, nitrification, and denitrification). Some are pathogenic.

bacteriophage Viruses that infect bacteria.

baffle An obstructing device or plate used to provide even flow distribution or to prevent short-circuiting. Also, a wall or partition used to create zones within a basin.

baghouse An air-cleaning device that removes particulate from an airstream via filters. The filters are cleaned internally within the device.

ball valve A valve using a rotating ball with a hole through it that allows flow straight through when the ball is in the open position.

Band screen A type of fine screening device.

Bar A type of screening media.

base (1) A substance that can accept a proton. (2) A substance that can react with an acid to form a salt. (3) An alkaline substance.

baseline A sample used as a comparative reference point when conducting further tests or calculations.

basicity factor Factor used to determine the neutralization capabilities of alkaline reagents used to treat acidic wastes.

Bay Area Sewage Treatmen Emissions model A mass emissions model that predicts the fate of compounds in the liquid phase of a wastewater treatment plant and predicts mass emission rate into the air phase.

batch reactor A reactor that treats a specific volume (batch) of liquid or solids at a time; it is designed to completely mix its contents and ensure that every drop or particle receives the same degree of treatment.

bed depth The depth of media in a vessel.

Beggiatoa Filamentous microbe, typically associated with solids bulking, that thrives in low dissolved oxygen levels and/or high sulfide levels.

belt conveyor A device used to transport material; it consists of a continuous-loop belt that revolves around head and tail pulleys.

belt filter press See *belt press*.

belt press A device that uses a series of porous moving belts revolving over a series of pulleys to drain water from solids.

belt thickener A mechanical solids-processing device that uses a revolving horizontal filter belt to thicken solids before dewatering and use or disposal.

bench test A small-scale laboratory test or study used to determine whether a technology is suitable for a particular application.

benthic Relating to the environment at the bottom of a body of water.

bentonite Colloidal claylike mineral that can be used as a coagulant aid in water treatment systems. Sometimes used as the earth component or soil amendment when constructing a pond or landfill liner because of its low permeability.

berm A horizontal earthen ridge or bank.

best available control technology The best technology, treatment techniques, or other means available after considering field (not solely laboratory) conditions.

bicarbonate A chemical compound containing an HCO_3 group.

bicarbonate alkalinity Alkalinity caused by bicarbonate ions.

biflow filter Granular media filter characterized by water inflow from both the top and the bottom to a collector in the center of the filter bed.

bilharzia A waterborne disease; also known as schistosomiasis.

binary fission Form of asexual reproduction in some microbes in which the parent organism splits into two independent organisms.

bioaccumulative A characteristic of a chemical whose rate of intake into a living organism is greater than the organism's rate of excretion or metabolism.

bioaerosol A mist that could transport biological material.

bioassay An analytical method in which an organism's response to biological treatment or an environment is measured by changes in biological activity.

biochemical oxidation Oxidative reactions initiated by biological activity and resulting in the chemical combination of oxygen with organic matter.

biochemical oxygen demand A standard measure of wastewater strength that quantifies the oxygen consumed by microorganisms in the fluid within a stated period of time (typically 5 days at 20°C). See *BOD* and *cBOD*.

biocide A chemical used to inhibit or control the population of troublesome microbes.

biodegradable Capable of undergoing biological decomposition.

biofilm An accumulation of microbial growth on the surface of an object.

biological filter A bed of sand, stone, or other media on which a layer of microorganisms grow. (1) A wastewater system in which microbes metabolically break down the complex organic materials into simple, more stable substances. (2) A fixed-media air-treatment system in which microorganisms on the media surface oxidize contaminants.

biofilter See *biological filter*.

biofuel A combustible organic material (e.g., biosolids) that can be used as a substitute for fossil fuel.

biogas The gases produced via the anaerobic decomposition of organic matter.

biological denitrification An anoxic process in which microorganisms transform nitrate-nitrogen into inert nitrogen gas; an electron donor drives the reaction.

biological oxidation A process in which living organisms oxidize organic matter into a more stable or mineral form.

biological process Any process in which microorganisms metabolically break down complex organic materials into simple, more stable substances.

biological synthesis yield Mass of solids produced through biological growth divided by the mass of substrate removed (typically BOD or COD).

biomass The mass of microorganisms in a biological treatment process.

biomass concentration (M/L^2) A measure of the amount of biomass available to treat wastewater or solids. It is the mass of biomass removed from a known quantity of carriers divided by the specific surface area provided by the carriers.

biomass density (M/L^3) The biomass concentration divided by the measured or estimated biofilm thickness (L).

bioreactor An vessel in which microorganisms suspended in a liquid metabolically break down complex organic materials into simple, more stable substances. The vessel also may include a variety of attached devices used to control the reaction.

bioscrubber An air-treatment system in which contaminants are oxidized by microorganisms.

biosolids Solids that have been removed from wastewater and stabilized (e.g., digested or composted) to meet the criteria in the U.S. Environmental Protection Agency's (U.S. EPA's) 40 CFR 503 regulations and, therefore, can be beneficially used.

biota All living organisms in a system.

bioturbation The displacement and mixing of sediment particles by benthic organisms. Such activity in the mixing zone may help disperse any contaminants remaining in effluent and increase the exchange of oxygen and nutrients between the sediment and the water.

biotower A generic term for in-vessel biological air treatment. Can be either a bioscrubber or biotrickling filter. See *biological filter.*

biotrickling filter A biological air-treatment technology in which microorganisms attached to a fixed media oxidize contaminants in the gas phase. Nutrients are often add to the liquid that continuously trickles over the media to ensure that the microbes have enough substrate.

black box model See *statistical model.*

blinding The reduction or cessation of flow through a filter caused by solids obscuring or filling the openings in the filter media.

blower A device that conveys air at pressures up to 103 kPa (15 psi); it has many uses at a wastewater treatment plant (to provide dissolved oxygen in aerobic treatment processes, to provide fresh air in enclosed spaces, etc.).

BOD_5 Five-day uninhibited biochemical oxygen demand.

booster pump A pump used to raise the pressure of the fluid on its discharge side.

bound water (1) Water strongly adsorbed to or absorbed by colloidal particles. (2) Water associated with the hydration or crystalline compounds.

breakpoint chlorination A treatment approach in which chlorine is added to the disinfection process until all chlorine demand has been oxidized. Further addition will result in a chlorine residual.

brine-recovery reverse osmosis A reverse osmosis process used to concentrate the brine generated by another reverse osmosis process. Also called *secondary reverse osmosis.*

Brinell hardness number A measure of the indentation hardness of materials.

broad-crested weir A weir with a substantial crest width in the direction parallel to the direction of water flowing over it.

bromine A halogen used as a water disinfectant in combination with chlorine as a chlorine-bromide mixture.

brownfield Currently occupied or previously developed industrial site.

brush aerator Mechanical aeration device most frequently used in oxidation ditch wastewater treatment plants; consists of a horizontal shaft with protruding paddles that is rapidly rotated at the water surface. Also called a *rotor.*

British thermal unit (Btu) The quantity of heat required to raise the temperature of 1 lb of water from 60 to 61°F at a constant pressure of 1 atm.

budget estimate A capital cost estimate used to establish the owner's budget. It should be prepared based on flow sheets, layouts, and details about major equipment quantity, type, and sizing (at a minimum).

buffer zone The distance between a wastewater or solids management process and an environmentally or socially sensitive area (e.g., a waterbody, the habitat of an endangered species, or a residential area).

bulk density The density-to-volume ratio for a solid, including the voids in the bulk material.

bulk-specific surface area (L^2/L^3) The total surface area provided per bulk unit volume of carriers. This value is reported by the manufacturer and may or may not include a reduction to account for the portion of the carrier element not conducive to biofilm attachment because of scouring resulting from collisions with other carriers in the reactor (i.e., unprotected area).

bulk-specific volume **(L^3/L^3)** The volume displaced by carriers per bulk unit volume of media; determined by the bulk density of the carriers divided by the specific gravity of the carrier material.

bulking A phenomenon at activated sludge plants in which solids do not readily settle or concentrate because of the predominance of filamentous organisms.

bulking agent In composting, a carbon-rich material added to a solids mixture to increase porosity, solids content, and carbon-to-nitrogen ratio.

butterfly valve A valve with a stem-operated disk that is rotated to be parallel to the liquid flow when opened and perpendicular to the flow when closed.

bypass A channel or pipe arranged to divert flow around a tank, treatment process, or control device.

Bypass screens Devices used for emergency screening.

cake Dewatered material from a filter press, centrifuge, or other dewatering device; it is now thick enough to be handled as a solid, not a liquid.

cake filtration A classification for filteration technology that removes solids on the entering face of the granular media.

cake solids The percentage of suspended solids in the dewatered cake.

calcining Exposing an inorganic compound to a high temperature to alter its form and drive off a substance that originally was part of the compound.

calcium carbonate A white, chalky substance that is the principal source of water hardness and scale. Chemical formula is $CaCO_3$.

calcium carbonate equivalent A convenient unit of exchange (mg/L $CaCO_3$) for expressing all ions in water by comparing them to calcium carbonate. Calcium carbonate has a molecular weight of 100 and an equivalent weight of 50.

calcium hydroxide A compound used as a solids conditioner to improve the thickening and/or dewatering properties of solids; also used in alkaline stabilization processes. Commonly called *slaked lime, hydrated lime,* or *pickling lime*. Chemical formula is $Ca(OH)_2$.

calcium hypochlorite A chlorine compound frequently used as a water or wastewater disinfectant. Chemical formula is $Ca(OCl)_2$.

calcium oxide Compound used as a solids conditioner to improve the thickening and/or dewatering properties of solids; also used in alkaline stabilization processes. Commonly known as *burnt lime, lime,* or *quicklime*. Chemical formula is CaO.

capillary force The attraction of water to soil particle surfaces; it increases as soil pore size decreases.

capture efficiency The percentage of the solids fed to a dewatering device that is removed with the cake.

carbon An element present in many inorganic and all organic compounds.

carbon dioxide A nonflammable, colorless, odorless gas formed in animal respiration, as well as the combustion and decomposition of organic matter. Chemical formula is CO_2.

carbon dioxide equivalent (CO_2^e) A measure of how much global warming a given type and amount of greenhouse gas may cause, using carbon dioxide as a baseline.

carbon footprint A measure of the amount of carbon dioxide produced by a person, organization, or location at a given time.

carbonaceous biochemical oxygen demand (CBOD) The portion of biochemical oxygen demand that consumes oxygen via carbon oxidation; typically measured after a sample has been incubated for 5 days. Also called *first-stage biochemical oxygen demand*.

carbon adsorption The use of powdered or granular activated carbon to remove refractory and other organic matter from water.

carbonation The diffusion of carbon dioxide gas throughout a liquid.

catenary screen A type of coarse screening device.

cathode The negative electrode where the current leaves a electrolyte solution.

cathodic protection Electrochemical corrosion protection achieved by imposing an electrical potential to counteract the galvanic potential between dissimilar metals, which otherwise would lead to corrosion.

cation A positively charged ion that migrates to the cathode when an electrical potential is applied to a solution.

cationic flocculant A polyelectrolyte with a positive electrical charge.

cationic polymer A polyelectrolyte with a net positive electrical charge.

cavitation The formation of a partial vacuum in a flowing liquid as a result of the separation of liquid's parts. The collapse of these parts may cause pitting on metal surfaces inside the pump such as the impeller and casing.

cell yield The mass of cells produced per the mass of substrate consumed.

Celsius The international name for the centigrade temperature scale, in which the freezing and boiling points of water are 0 and 100°C, respectively, at a barometric pressure of 760 mm of mercury. Also called *centigrade*.

centigrade A thermometer temperature scale, in which the freezing and boiling points of water are 0 and 100°C, respectively, at a barometric pressure of 760 mm of mercury. Also called *Celsius*. To convert temperatures from centigrade to Fahrenheit, multiply by 1.8 and add 32.

centrate The liquid remaining after solids have been removed via a centrifuge.

centrifuge A dewatering device the relies on centrifugal force to separate particles of varying density (e.g., water and solids).

centrifugal pump A pump with a high-speed impeller that relies on centrifugal force to convert velocity into head pressure. Flow enters in front of the impeller and flows peripherally out of the pump and into the piping system. The amount of flow varies based on pressure.

chain-and-flight collector A mechanism for collecting solids in rectangular sedimentation basins or clarifiers.

chain-driven screen A type of coarse screening device.

channel (1) A perceptible natural or artificial waterway that contains moving water or forms a connecting link between two bodies of water or two water-bearing structures. (2) The deep portion of a river or waterway where the main current flows. (3) The part of a waterbody deep enough to be used for navigation in an area otherwise too shallow for shipping.

channeling A situation in processes with packed material that occurs when water finds furrows or channels through the media in which it can flow without contacting—and reacting with—the microorganisms.

check valve A valve that opens in the direction of normal flow and closes when flow reverses.

chemical coagulation A process in which an inorganic chemical is added to wastewater to destabilize and aggregate colloidal and finely divided suspended matter.

chemical conditioning Mixing chemicals with solids before dewatering and/or thickening to improve the solids separation characteristics. Typical conditioners include polyelectrolytes, aluminum and iron salts, and lime.

chemical dose A specific quantity of chemical applied to a specific quantity of fluid or solids for a specific purpose.

chemical equilibrium A condition in which there is no net transfer of mass or energy between system components. It occurs in a reversible chemical reaction when the rate of the forward reaction equals the rate of the reverse reaction.

chemical equivalent The weight in grams of a substance that combines with or displaces one gram of hydrogen. It is found by dividing the formula weight by its valence.

chemical feeder A device for dispersing a chemical at a predetermined rate to treat wastewater or solids. The change in feed rate may be effected manually or automatically by flowrate changes. Feeders are designed for solids, liquids, or gases.

chemical oxidation A process in which chemical compounds (e.g., ozone, chlorine, and potassium permanganate) are added to oxidize compounds in water or wastewater.

chemical oxygen demand (COD) A measure of the organic matter in water or wastewater that can be oxidized by a chemical agent (e.g., a solution of potassium dichromate).

chemical solution tank A tank in which chemicals are added in solution before they are used in a wastewater or solids treatment process.

chemical sludge Solids resulting from chemical treatment processes of organic wastes that are not biologically active. Formed by adding a chemical (e.g., an iron or aluminum salt) to wastewater to precipitate phosphorus.

chemical treatment Any treatment process (e.g., precipitation, coagulation, flocculation, sludge conditioning, disinfection, or odor control) involving the addition of chemicals to obtain a desired result.

chisel plow A static plow shank used to slice soil during the installation of subsurface drip tubing.

chloramines Disinfecting compounds of organic or inorganic nitrogen and chlorine.

chlorinated (1) Water or wastewater that has been treated with chlorine. (2) An organic compound to which chlorine atoms have been added.

chlorination The process of adding chlorine to a water or wastewater, typically to disinfect it.

chlorinator A metering device used to add chlorine to water or wastewater.

chlorine An oxidant typically used as a disinfectant in water and wastewater treatment. Chemical formula is Cl_2.

chlorine contact chamber A vessel in which chlorine is diffused through water or wastewater; enough contact time is provided for disinfection.

chlorine demand The difference in the amount of chlorine added to water or wastewater and the amount of residual chlorine remaining after a specific contact time (typically 15 minutes).

chlorine dioxide Chemical frequently used in disinfection. Chemical formula is ClO_2.

chlorine dose The amount of chlorine added to a liquid, typically expressed in milligrams per liter or pounds per million gallons.

chlorine residual The amount of chlorine remaining in water after application and contact time. See *free chlorine residual.*

chlorine tablets A common term for pellets of solidified chlorine compounds (e.g., calcium hypochlorite) used to disinfect water.

chlorine toxicity A measure of chlorine's detrimental effects on biota.

chlorophenols A group of toxic, colourless, weakly acidic organic compounds in which one or more of the hydrogen atoms have been replaced by chlorine atoms. Most applications of chlorophenols are based on their toxicity: they and compounds made from them are used to control bacteria, fungi, insects, and weeds.

cholera A highly infectious disease of the gastrointestinal tract caused by waterborne bacteria.

clarifier A quiescent tank in which suspended solids are removed from wastewater via gravity. It typically is equipped with a motor-driven chain-and-flight or rake mechanism to collect settled sludge and move it to a final removal point. Also called *sedimentation* or *settling basins.*

clarification Any process or combination of processes whose primary purpose is to reduce the concentration of suspended matter in a liquid.

Class A biosolids Biosolids that contain less than 1 000 most probable number (MPN)/g of fecal coliforms and less than 3 MPN/4g of *Salmonella* bacteria and meet one of six stabilization alternatives given in 40 CFR 503. The material also must meet the pollutant limits and vector-attraction reduction requirements set forth in Part 503.

Class B biosolids Biosolids that contain less than 2 million colony-forming units (CFU) [most probable number (MPN)] of fecal coliforms per gram of dry biosolids. The material also must meet the pollutant limits and vector-attraction reduction requirements set forth in 40 CFR 503.

Clean Air Act Initially passed in 1963, the Clean Air Act is the law that defines U.S. EPA's responsibilities for protecting and improving the nation's air quality and the stratospheric ozone layer. It has been amended several times since it was first enacted; the most recent amendment was passed in 1990.

clean filter headloss Initial headloss value at the start of the filtration cycle (after the backwash cycle), when the medium is clean.

Clean Water Act Enacted in 1972, the Clean Water Act is the primary federal law regulating water pollution. It was last updated in I987.

clearwell A tank or reservoir of filtered water that may be used to backwash a filter.

climate change Any long-term, significant change in the weather patterns of an area.

coagulant Chemical added to wastewater to destabilize, aggregate, and bind together colloids and emulsions to improve the settleability, filterability, or drainability of solids. A simple electrolyte, typically an inorganic salt containing a multivalent cation of aluminum, iron, or calcium [e.g., $FeCl_3$, $FeCl_2$, $Al_2(SO4)_3$, and CaO]. Also, an inorganic acid or base that induces coagulation of suspended solids.

coagulation (1) The destabilization and initial aggregation of finely divided suspended solids by the addition of a polyelectrolyte or a biological process. (2) The conversion of colloidal (<0.001 mm) or dispersed (<0.001 to 0.1 mm) particles into small visible coagulated particles (0.1 to 1 mm) by the addition of a coagulant, compressing the electrical double layer surrounding each suspended particle, decreasing the magnitude of repulsive electrostatic interactions between particle, thereby destabilizing the particle.

coalesce The merging of two droplets to form one large droplet.

coarse sand Sand with particles typically larger than 0.5 mm in diameter.

coarse screen A screening device that typically has openings between 6 and 36 mm (0.25 and 1.4 in.).

cocci Sphere-shaped bacteria.

co-digestion A process in which two or more types of substrates (feedstocks) are digested together in the same reactor. Frequently refers to anaerobic digestion of solids together with food wastes, food processing wastes, FOG, or other organic waste material.

coefficient A numerical measure of a physical or chemical property that is constant for a system under specified conditions (e.g., the coefficient of friction).

coefficient of viscosity A numerical factor that is a measure of the internal resistance of a fluid to flow; the greater the resistance to flow, the larger the coefficient. It is equal to the shearing force in dynes per square centimeter (dyne/cm^2) transmitted from one fluid plane to another parallel plane 1 cm distant, and is generated by a difference in fluid velocities in the two planes of 1 cm/s in the direction of the force. The coefficient varies with temperature. Also called *absolute viscosity*. The unit of measure is the poise, a force of 1 dyne/cm^2.

cogeneration See *combined heat and power*.

coliform bacteria Rod-shaped bacteria living in the intestines of humans and other warm-blooded animals.

collector A mechanism used in clarifiers to collect and remove settled solids from the tank bottom.

colloid A suspended solid with a diameter less than 1 μ, which cannot be removed by sedimentation alone.

colony-forming units The number of bacteria present in a sample, as determined in a laboratory plate-count test. In this test, the number of visible bacteria colony units present is counted.

colorimetric The use of color change as an indicator.

combined heat and power A process in which one fuel simultaneously or sequentially produces useful heat and power. The power is typically electrical power, but can be direct mechanical-drive power.

combined sewer overflow A mixture of stormwater and sanitary wastewater in a combined sewer whose volume exceeds the collection system's capacity during a storm event and is discharged (untreated) directly to a receiving water.

comminutor A circular screen with cutters that grind large wastewater solids into smaller particles.

compensation A trade-off action intended to make a project more acceptable to stakeholders.

completely-mixed, stirred-tank reactor An ideal reactor in which the concentrations are uniform throughout.

compliance standards The water-quality and biosolids-quality requirements specified in a treatment plant's NPDES permit that must be met before the effluent can be discharged and the biosolids beneficially used (or disposed).

composite variable A combination of state variables typically to form variables that can actually be measured in the plant (e.g. BOD_5, total COD, TKN, total phosphorus, TSS, VSS).

composting Stabilization process relying on the aerobic decomposition of organic matter in solids by bacteria and fungi.

compressible medium filter A filter that uses a synthetic-fiber compressible porous material as the filter medium instead of conventional granular material.

compression settling A sedimentation phenomenon in which particles in a concentrated suspension will only settle if the existing structure of settled particles is compressed.

computional fluid dynamics A series of algorithms and calculations used to predict or validate the behavior of liquid or gases flowing over or through constructed surfaces.

concentration (1) The amount of a substance dissolved or suspended in a unit volume of solution. (2) The process of increasing the amount of a substance per unit volume of solution.

conditioning A chemical, physical or biological process designed to improve the thickening or dewatering characteristics of a solids.

conduction A transfer of heat energy via direct contact of the materials. In biosolids drying, heat is conducted to biosolids via contact with heated surfaces. Sometimes called *indirect drying*.

connected load The total load of all of the electrical equipment in a facility.

constant-rate filtration Filter operation in which flow through the filter is maintained at a constant rate by an adjustable effluent-control valve or influent-control weir.

constant-velocity channel A channel configuration that has a constant or nearly constant velocity, regardless of flow depth. Used in grit removal.

constructed wetlands A wastewater treatment system that uses the aquatic root system of cattails, reeds, and similar plants to treat wastewater applied either above or below the soil surface.

contact stabilization process A modification of the activated sludge process in which raw wastewater is aerated with activated sludge for a short time before solids removal and continued aeration in a stabilization tank. Also called the *biosorption process*.

contaminant Any foreign component present in another substance.

contamination The degradation of natural water, air, or soil quality

resulting from human activity.

contingency A reserve in a cost estimate for events that experience has shown will likely occur. The greater the engineering detail provided, the lower the contingency.

continuous element screen A type of fine screening device.

contractor The company responsible for constructing the facilities in accordance with contract documents; typically selected through a public bidding process.

convection Transfer of heat energy via the motion of a mass of fluid (e.g., air). In biosolids drying, this refers to the transfer of heat from hot gas directly to biosolids. Sometimes called *direct drying*.

core blow A method of clearing the feed ports in a recessed plate-and-frame filter press.

corner sweep A scraper used to remove solids from the corner of' a square clarifier.

corrosion The process of breaking down a metal via a chemical or electrochemical reaction with the surrounding medium.

corrosive The characteristic of a chemical agent that reacts with the surface of a metal, causing it to deteriorate or wear away.

criteria pollutants Compounds for which a National Ambient Air Quality Standard has been established.

critical shear stress The pressure required to move or scour particles from a surface.

cross-collector A mechanical solids-collection mechanism that extends the width of one or more longitudinal sedimentation basins; used to consolidate and convey accumulated solids to a final removal point.

cross-connection A physical connection in a plumbing system through which a potable water supply could be contaminated by wastewater.

cross-flow filtration A filtration method in which the influent flows parallel to the surface of the filter medium.

crypto See *Cryptosporidium.*

cryptosporidiosis Gastrointestinal disease caused by ingesting waterborne *Cryptosporidium parvum,* often the result of drinking contaminated runoff from pastures or farmland.

Cryptosporidium parvum A species of *Cryptosporidium* known to be infectious to humans. A protozoan parasite that can live in the intestines of humans and animals.

culture A microbial growth developed by furnishing enough nutrients in a suitable environment.

cut-throat flume Essentially, a Parshall flume with the center throat section "cut out". The typical converging inlet section of the Parshall flume is directly connected to the typical outlet diversion section. The flume works well when submerged. Useful for distributing flow to basins or reactors.

cyclone degritter A conical device that uses centrifugal force to separate grit from organics in grit slurries.

cycle time The time between pump starts; it includes the pump running time and the pump "off" time.

cyclone A conical vessel used to initially separate grit from water in the cyclone–classifier grit slurry-dewatering device.

cyst A resting stage formed by some bacteria and protozoa in which the whole cell is surrounded by a protective layer.

decant The act of separating liquid from settled solids by pouring or drawing off the upper layer of liquid after the solids have settled.

dechlorination A chemical of physical process in which residual chlorine is partially or completely reduced.

declining-rate filtration Filter operation in which the rate of flow through the filter declines throughout the length of the filter run. The level of the liquid above the filter bed rises throughout the length of the filter run due to the increase in headloss that occurs as solids accumulate on the media.

deep-bed filter A granular media filter with a sand or anthracite bed up to 1.8 m (6 ft) deep.

definitive estimate A capital cost estimate prepared from very well-defined engineering data. Typically prepared at the end of the design delivery process when the construction documents have been completed.

deflagration Combustion. Typically referring to a fire or explosion in a thermal drying system.

degree day A measure of heating used when determining the volatile solids reduction in an aerobic digester based on published curves. It is computed as the product of the aerobic digester liquid temperature (°C) multiplied by the digester's solids retention time (days).

delta (Δ) P Differential pressure.

delta (Δ) T Differential pressure.

demand charge A cost per kVa based on the peak power flow into a facility. This charge is a result of the electric utility's need to size its facilities to serve the maximum power that a facility will require at any given point in time.

demand load The total power load of all equipment that would be expected to operate at the same time.

denitrification A biological process in which nitrates are converted to nitrogen.

denitritation The bacterial reduction of nitrite in the absence of free oxygen.

density The ratio of the mass of an object to its volume.

depth filtration A classification for filters in which significant solids removal occurs within the filter medium.

design criteria (1) Engineering guidelines specifying construction details and materials. (2) Objectives, results, or limits that must be met by a facility, structure, or process in the performance of its intended functions.

design standards Standards established for the design of equipment and structures. These standards may or may not be mandatory.

detection threshold The odor concentration at which, statistically, half of the odor panel could correctly select the odorous sample from carbon-filtered (odor-free) air.

detention time The time theoretically required to displace the contents of a tank or unit at a given rate of discharge.

detritus tank A square grit chamber with a revolving rake to scrape settled grit to a sump for removal.

dewater (1) To extract a portion of the water in a sludge or slurry. (2) To drain or remove water from an enclosure.

dewatering A process (e.g., filter press or centrifuge) that removes a portion of the water contained in solids. Dewatering is distinguished from thickening in that the resulting dewatered cake may be handled as a solid, not a liquid.

dewatering lagoon A lagoon constructed with a sand and underdrain bottom to draw water away from solids.

dew point The temperature to which air with a given concentration of water vapor must be cooled for the vapor to condense.

diffused aeration A system that injects air under pressure through submerged porous plates, perforated pipes, or other devices to form small air bubbles from which oxygen is transferred to the liquid as the bubbles rise to the water surface.

diffused air A system that forms small air bubbles below the surface of a liquid to transfer oxygen to the liquid. See *diffused aeration.*

diffused-air aeration A system that introduces compressed air to water via submerged diffusers or nozzles.

diffuser A porous plate, tube, or other device through which air is forced and divided into minute bubbles for diffusion in liquids. Also can be a perforated tube through which a solution of a chemical is introduced. In the activated sludge process, it is a device for dissolving air into mixed liquor. It also is used to mix chemicals (e.g., chlorine) through perforated holes.

digested solids Solids in which the concentration of volatile solids has been significantly reduced via oxidation by microbes in an aerobic or anaerobic reactor. The digested material is now relatively non-putrescible and inoffensive.

digester A tank or other vessel used to store and anaerobically or aerobically decompose the organic matter in solids. See also *aerobic digestion* and *anaerobic digestion.*

digestion The process of biologically oxidizing the organic matter in solids, thereby reducing the concentrations of volatile solids and pathogens.

dilution (1) Lowering the concentration of a solution by adding more solvent. (2) The engineered mixing of discharged water with receiving water to lessen its immediate aesthetic and/or biochemical effects.

dilution-to-threshold (D/T) ratio A non-dimensional, volumetric ratio measure of odor concentration. It is expressed as the sample volume plus the volume of dilution air divided by the sample volume.

direct costs The component of capital cost estimates that includes land and site development costs, costs of services to the site, and relocation costs. These costs also include contractor's overhead, profit, mobilization, bond and insurance, and construction contingencies.

direct osmosis A natural phenomenon in which water flows from the lower osmotic pressure solution to the higher osmotic pressure solution.

direct potable reuse Supplying highly treated wastewater effluent directly to the intake source of a potable water treatment system (eliminating any intermediate natural storage system).

Disc screen A type of fine screening device.

discharge The release of effluent, by any means, to the environment.

discrete particle settling Phenomenon referring to sedimentation of particles from a suspension that has a low solids concentration.

disk filter A filtration system in which a cloth or steel disk is the filter medium.

disinfectant A substance used to disinfect water, wastewater, or solids.

disinfection The selective destruction of disease-causing microbes via the application of chemicals or energy.

dispersal A method of spreading effluent over and into the final receiving environment.

dispersion model A mathematical tool used to characterize the fate of pollutants released from the emission point and predict concentration at selected downwind receptors.

dissolved air flotation A clarification process in which minute bubbles attach themselves to flocculated material, float to the surface, and are removed via an overflow weir. Heavier solids settle to the bottom and also are periodically removed.

dissolved organic carbon The fraction of total organic carbon that is dissolved in a water sample.

dissolved oxygen The oxygen dissolved in a liquid.

dissolved solids Solids in solution that cannot be removed via filtration. See *total dissolved solids.*

diurnal flow A daily fluctuation in flowrate or composition that is similar from one 24-hour period to another.

doctor blade A scraping device used to remove or regulate the amount of material on a belt, roller, or other moving or rotating surface.

dolomite A natural mineral consisting of calcium carbonate and magnesium carbonate. Chemical formula is $CaMg(CO_3)_2$.

dolomitic lime Lime containing 35 to 40% magnesium oxide.

domestic wastewater Wastewater originating in sanitation devices (e.g., sinks and toilets) in residential dwellings, office buildings, and institutions. Also called *sanitary wastewater*.

dose A specific quantity of a substance applied to a unit quantity of liquid to obtain a desired effect.

downwash The downward movement of a plume caused by the aerodynamic affects of wind moving around structures and on the lee side of tall stacks.

draft tube A centrally located vertical tube used to promote mixing in a solids digester or aeration basin.

drip irrigation Distribution of effluent water over an infiltrative surface via pressurized low flow discrete emitters on or within plastic tubing (lines) supplied by associated devices and parts (pump, filters, controls, and piping). Drip irrigation can be at the surface or subsurface.

Drum screen A type of fine screening device.

dry bulb temperature The air temperature measured by a conventional thermometer.

dry pit The chamber in a conventional wastewater pumping station where the pump, motor, piping, and valves are located.

dry weather flow A measure of total flows in the sanitary collection system under dry weather conditions. It is the sum of the volume of wastewater discharges and the volume of groundwater infiltrating the collection system.

drying Using thermal or radiant energy to evaporate moisture from biosolids or solids.

drying bed A partitioned area consisting of sand or other porous material on which solids are dewatered via drainage and evaporation.

dual-media filter A granular media filter that uses two types of filter media (typically silica sand and anthracite).

duckweed See *Lemnaceae*.

duplex pump A reciprocating pump that has two side-by-side cylinders connected to the same suction and discharge lines.

dynamic head See *total dynamic head.*

dynamic simulation A simulation in which the inputs to the model vary with time.

dysentery A disease of the gastrointestinal tract typically resulting from poor sanitary conditions; transmitted by consuming contaminated food or water.

E. coli See *Escherichia coli.*

energy conservation measure (ECM) A physical improvement, plant operation, or equipment maintenance practice that reduces the consumption of energy or improves the management of energy demand..

effective collector size A major parameter for characterizing filter media. In the case of granular filter media, the effective size is equal to the sieve size, in millimeters, that will pass 10% (by weight) of the sand.

effluent Partially or completely treated water or wastewater flowing out of a basin or treatment plant.

ejector A device that passes steam, air, or water through a Venturi to develop suction to move another fluid. Sometimes called an *eductor* or *jet pump.*

electrical distribution system The network of electrical wires and equipment in the wastewater treatment plant.

electrolyte A chemical substance or mixture, typically liquid, containing ions

that migrate in an electric field.

emission isolation flux chamber A sampling device that allows the collection of contaminants from liquid or solid surfaces by isolating a sampling area from the ambient air. It uses a carrier gas to deliver the contaminants to the sample collection or measurement device so that a contaminant concentration can be determined for a given surface area.

Emission point The location of the stack, vent, or area were a pollutant is released into ambient air.

empty bed contact time The time that a substance theoretically remains in a specified volume, vessel, or media.

environmental management system (EMS) A set of processes and practices that enable an organization to reduce its environmental impacts and increase its operating efficiency.

endogenous respiration Bacterial growth phase during which microbes metabolize their own protoplasm without creating more protolasmt because not enough food is available.

endothermic A process or reaction during which heat is absorbed.

energy audit A study to identify opportunities for improving plant efficiency and reducing operating costs, as well as quantify the cost and savings of these options.

energy star An international standard for energy-efficient consumer products.

enhanced biological phosphorus removal The biological removal of phosphorus through the cultivation and wasting of bacteria that retain excess phosphorus.

enteric bacteria Bacteria that inhabit the gastrointestinal tract of warm-blooded animals.

environmental justice The practice of ensuring that minority groups do not bear an inequitable environmental burden as a result of a project.

Electric Power Research Institute (EPRI) An organization that conducts research and development on technology, operations, and the environment for the global electric power sector.

EQ (1) Effluent quality index; a measure of the entire effluent pollution load to a receiving waterbody. (2) "Exceptional Quality" biosolids, which meet 40 CFR 503 Table 3 pollutant limits, Class A pathogen requirements, and vector-attraction reduction requirements.

equalization The process of dampening hydraulic or organic variations in a flow so nearly constant conditions can be achieved.

equalization basin A basin or tank used to equalize flow.

equitable distribution of regional resources The practice of ensuring that public facilities are equally dispersed among the communities they serve.

energy services contracting (ESC) A project delivery process in which savings brought about by an energy-conservation project, renewable-energy project, or other facility improvement are used to pay for the cost of the capital improvement.

energy services company (ESCO) A business that provides energy services contracting.

Escherichia coli A fecal coliform bacteria species used as an indicator of wastewater pollution.

estuary A semi-enclosed coastal waterbody at the mouth of a river in which rivers's current meets the sea's tide.

euthrophic lake A lake with an abundant supply of nutrients, excessive growth of floating algae, and an anerobic hypolimnion.

eutrophication Nutrient enrichment of water causing excessive growth of aquatic plants and eventual deoxygenation of the waterbody.

evaporation pond A natural or artificial pond that converts solar energy into heat to evaporate water.

evaporation rate The mass or quantity of water evaporated from a specified water surface per unit of time

evapotranspiration The sum of evaporation and plant transpiration.

exceptional quality (EQ) biosolids Biosolids that meet the Class A pathogen-reduction requirements in 40 CFR 503, meet the vector-attraction reduction requirements [503.33(a)(1) through (8)], and have low concentrations of regulated pollutants (503.13, Table 3).

exclusionary areas Existing site features that should be avoided during plant siting and layout.

exothermic A process or reaction that **produces** heat.

extended aeration A modification of the activated sludge process that uses long aeration periods to promote aerobic digestion of the biological mass via endogenous respiration. The process also stabilizes organic matter under aerobic conditions and emits gaseous products. The effluent contains finely divided suspended matter and soluble matter.

extended aeration process A variant of the activated sludge process with a longer detention time to allow endogenous respiration to occur.

facultative bacteria Microbes that can survive with or without oxygen.

facultative lagoon A lagoon or pond in which wastewater is stabilized by aerobic, anaerobic, and facultative bacteria.

facultative ponds See *facultative lagoons.*

fall A sudden change in the water surface elevation.

fatigue life The number of hours or revolutions at which 10% of the bearings are likely to fail because of fatigue.

fecal coliform Coliforms present in the feces of warm-blooded animals. Aerobic and facultative, Gram-negative, non-spore-forming, rod-shaped bacteria capable of growth at 44.5°C (112°F), and associated with then fecal matter of warm-blooded animals.

fecal indicators Fecal coliform, fecal Streptococci, and other bacterial groups originating in human or other warm-blooded animals, indicating contamination by fecal matter.

feces Excrement of humans and animals.

fermentation The conversion of organic matter into carbon dioxide, methane, and other low-molecular-weight compounds.

ferric chloride A soluble iron salt often used as a solids conditioner to enhance precipitation, bind up of sulfur compounds, or improve the thickening and/or dewatering properties of solids. Chemical formula is $FeCl_3$.

ferric sulfate A water-soluble iron salt formed by the reaction of ferric hydroxide and sulfuric acid or by the reaction of iron and hot concentrated sulfuric acid; also obtainable in solution by the reaction of chlorine and ferrous sulfate. A commonly used coagulant, it often is used as a solids conditioner to enhance solids precipitation to improve the thickening and/or dewatering properties of residuals. Chemical formula is $Fe_2(SO_4)_3$.

ferrous chloride A soluble iron salt often used as a solids conditioner to enhance precipitation, bind up sulfur compounds, or improve the thickening and/or dewatering properties of solids. Chemical formula is $FeCl_2$.

ferrous sulfate A water-soluble iron salt that is a commonly used coagulant. It is used with lime as a solids conditioner to enhance solids precipitation to improve the thickening and/or dewatering properties of residuals. Sometimes called *copperus* or *iron vitriol*. Chemical formula is $Fe(SO_4) \cdot 7H_2O$.

fertilizer Materials (typically those containing nitrogen and phosphorus) that are added to soil to provide essential nutrients for plant growth.

fiber-reinforced concrete Concrete reinforced with fiberglass materials.

filamentous growth The hair-like biological growth of some species of bacteria, algae, and fungi that results in poor solids settling.

fill fraction **(%)** The total bulk volume (L^3) of carriers installed in a reactor, expressed as a percentage of the wet reactor volume.

fill systems Above-grade soil treatment area designed and installed so the entire infiltrative surface is located above the original ground elevation; suitable imported soil material is used for fill.

filter A device using a granular material, woven cloth, or other medium to remove suspended solids from water, wastewater, or air.

filter aid A polymer, coagulant, or other material added to improve the effectiveness of filtration.

filter bottom See *underdrain*.

filter cycle The filter's operating time between backwashes. Also called *filter run time*.

filter fly See *Psychoda fly*.

filter gallery A passageway to provide access for installing and maintaining underground filter pipes and valves.

filter loading, hydraulic The volume of liquid applied per unit area of the filter bed per day.

filter loading, organic The quantity of biochemical oxygen demand applied per unit area of the filter bed per day.

filter press A dewatering device in which water is forced from solids under high pressure.

filter run See *filter cycle.*

filter-to-waste An operating procedure in which the filtrate produced immediately after backwash is wasted.

filtrate Liquid remaining after solids were removed via filtration.

filtration rate A measurement of the volume of water applied to a filter per unit of surface area in a stated period of time.

final clarifier See *secondary clarifier.*

final effluent The effluent from the final unit treatment process at a wastewater treatment plant.

fine-bubble aeration A method of diffused aeration that uses fine bubbles to take advantage of their high surface areas to increase oxygen transfer rates.

fines Particles at the lower end of a range of particle sizes.

fine sand Sand particles with diameters that typically range from 0.3 to 0.6 mm.

fine screen A screening device with openings greater than 0.5 mm to 6 mm.

first flush Surface runoff at the beginning of a storm, which often contains much of the solid matter washed from streets and other surfaces.

first-order reaction A reaction in which the rate of change is directly proportional to the concentration of the reactant raised to the first power.

fixed-film process A biological treatment process in which the microbes are attached to an inert medium (e.g., rock or plastic). Also called *attached-growth process.*

fixed suspended solids The inorganic content of suspended solids in a water or wastewater sample, determined after heating the sample to 600°C.

flange A projecting rim or edge used for attachment to another object.

flap valve A valve that is hinged on one edge and opens in the direction of normal flow and closes with flow reversal.

flash mixing Process using a motor-driven stirring device designed to disperse coagulants or other chemicals instantly, before flocculation.

flight (1) The horizontal scraper on a rectangular solids collector. (2) The helical blade on a screw pump.

floatables Materials (e.g., oil, scum, paper, and plastic) that do not settle; instead, they float at the surface of a discharge.

float switch An electrical or pneumatic switch operated by a float in response to changing liquid levels.

floc (1) Small, gelatinous masses formed in water by adding a coagulant or in wastewater via biological activity. (2) Collections of smaller particles agglomerated into larger, more easily settleable particles via chemical, physical, or biological treatment.

flocculant Water-soluble organic polyelectrolytes that are used alone or with an inorganic coagulant (e.g., aluminum or iron salts) to agglomerate solids into large, dense floc particles that settle rapidly and improve the thickening and dewatering properties of solids.

flocculating tank A tank used to form floc via gentle agitation of liquid suspensions, with or without the aid of chemicals.

flocculation (1) In wastewater treatment, the agglomeration of colloidal and final suspended matter after coagulation via gentle agitation by mechanical or hydraulic means. (2) Gentle stirring or agitation to accelerate the agglomeration of particles to enhance sedimentation or flotation.

flocculation agent A coagulation substance that, when added to water, forms a flocculant precipitate that will entrain suspended matter and expedite sedimentation, as well as solids thickening and dewatering.

flocculant aid An insoluble particulate used to enhance solids–liquid separation by providing nucleating sites or acting as a weighting agent or sorbent; also used colloquially to describe the action of flocculants in wastewater and biosolids treatment.

flocculant settling A phenomenon referring to the sedimentation of particles in a dilute suspension as they coalesce (flocculate).

flocculants Organic polyelectrolytes used alone or with metal salts to coagulate solids particles.

flocculator A device used to enhance the formation of floc via gentle stirring or mixing.

flotation A treatment process in which gas bubbles are introduced to water and attach to solid particles, creating bubble–solid agglomerates that float to the surface, where they are removed.

flotation thickening Solids thickening via dissolved air flotation.

flow-control valve A device that controls the rate of fluid flow.

flow equalization Transient storage of wastewater for later release to a collection system or treatment process at a controlled rate to provide a reasonably uniform now.

flowrate The volume or mass of a gas, liquid, or solid material that passes some point in a stated period of time.

flow splitter A chamber that divides incoming flow into two or more streams.

fluid Any material or substance that flows or moves, whether in a semisolid, liquid, solids, or gaseous form or state.

fluidization The upward flow of a gas or fluid through a granular bed at enough velocity to suspend the grains.

flume A channel used to carry water.

flux The volumetric filtration rate for a given area of membrane, expressed as flowrate per unit area per time. A typical unit of flux is liters per square meter per day (gallons per square foot per day) of membrane area per day.

flux chamber See *emission isolation flux chamber*.

fly ash One of the residues generated when coal is combusted; it typically is captured from the chimneys of coal-fired power plants. Depending on the source and makeup of the coal being burned, the components of the resulting fly ash vary considerably; however, all fly ash includes substantial amounts of silicon dioxide (SiO_2) and calcium oxide (CaO).

force main A pipeline through which flow is transported from a point of higher pressure to a point of lower pressure.

forward osmosis A natural phenomenon in which water flows from a lower osmotic pressure solution to a higher osmotic pressure solution (the same as direct osmosis).

fouling factor A design criterion used to allow for some variation in equipment performance resulting from fouling.

free available residual chlorine The portion of the total residual chlorine in water or wastewater after a specified contact period that will react chemically and biologically as hypochlorous acid or hypochlorite ion.

freeboard The vertical distance between the normal maximum liquid level in a basin and the top of the basin; it is provided so waves and other liquid movements will not overflow the basin.

free chlorine The amount of chlorine available as dissolved gas, hypochlorous acid, or hypochlorite ion.

free chlorine residual The portion of total residual chlorine remaining after a specific contact time that will react as hypochlorous acid or hypochlorite ion.

free oil Non-emulsified oil that separates from water, typically in 5 minutes or less.

free settling The settling of discrete, nonflocculant particles in a dilute suspension.

free water Suspended water covering the surface of solid particles or the walls of fractures as a film. The amount of water present in the film is in excess of pellicular water. The water is free to move in any direction under the pull of the force of gravity and unbalanced film pressure.

friction factor A measure of the resistance to liquid flow caused by the texture of a pipe or channel wall.

full-cost pricing Prices for products and services that fully support the cost of providing them.

fully submerged disk filter The disk filter and the backwash mechanism are submerged in the tank.

fuzzy filter See *compressible medium filter*.

fuzzy logic A process-control system intended to replace a skilled human operator by using multilevel logic to adjust process operations based on a set of approximate, rather that exact, rules.

garnet A dense mineral often used as a filtration media.

gas Of the three states of matter, the state with no fixed shape or volume and capable of expanding indefinitely.

galvanize An electrolytic or hot dipping process to coat steel products with zinc to increase corrosion resistance.

gastroenteritis An inflammation of the stomach and intestinal tract.

gastrointestinal Related to the stomach or intestines.

gate valve A valve with a disk that slides over the opening through which water flows.

gear pump A positive-displacement pump in which cavities created between the teeth of two meshing gears move fluid from the suction to the discharge side of the pump.

Giardia lamblia A protozoan parasite; responsible for giardiasis.

giardiasis A gastrointestinal disease caused by ingesting waterborne *Giardia lamblia*, often resulting from the activity of beavers, muskrats, or other warm-blooded animals in surface water used as a potable water source.

globe valve A valve that can be closed by lowering a horizontal plug onto a matching seat in the center of the valve.

grab sample A single water or wastewater sample taken at a time and place representative of total discharge.

grade (1) The finished surface of a civil structure. (2) The inclination or slope of a surface or structure. (3) To rate according to a standard or size.

gradient The rate of change of an elevation, velocity, pressure, temperature, or other parameter.

granular media filtration A tank or vessel filled with sand or other granular media to remove suspended solids and colloids from water or wastewater as it flows through the media.

granule Small particle of dried product; typically within 0.5 to 5.0 mm in diameter. See *pellet*.

gravel Rock fragments measuring 2 to 70 mm in diameter often used as support material in granular media filters.

gravitational acceleration The acceleration of a free-falling body caused by the force of the Earth's gravity; equal to 9.81 m (32.2 ft) per second per second.

gravity belt thickener A solids dewatering device that uses a porous filter belt to promote water drainage via gravity.

gravity filter A granular media filter that operates at atmospheric pressure.

gravity system A hydraulic system that relies on gravity flow and does not require pumping.

gravity thickening A process in which a sedimentation basin is designed to operate at high solids loading rates to thicken residuals. It typically includes vertical pickets mounted onto the revolving solids scrapers to help release entrained water.

gray water All non-toilet household water (e.g., water from sinks, baths, and showers). Also called *sullage*.

grease Common term for the fats, oils, waxes, and related constituents found in wastewater.

grease trap A receptacle used to collect grease and separate it from wastewater.

greenfield construction A piece of previously undeveloped land in a city or a rural area.

greenhouse gas (GHG) Any of the gases (e.g., carbon dioxide, methane, ozone, and fluorocarbons) whose absorption of solar radiation is responsible for the greenhouse effect.

Green Globes™ An online tool that offers an assessment protocol, rating system, and guidance for green building design, operation, and management.

grit Sand, gravel, cinders, and other heavy solid matter with settling velocities substantially higher than those of organic (putrescible) solids in wastewater.

grit chamber A settling chamber used to remove grit from organic solids via sedimentation or air-induced agitation.

grit classifier A mechanical device that uses an inclined screw or reciprocating rake to wash putrescible organics from grit.

grit removal A preliminary wastewater treatment process to remove grit from organic solids.

grit washer A device used to wash organic matter from grit.

groundwater Subsurface water found in porous rock strata and soil.

guide vane A device used to direct or guide the flow of a liquid or vapor.

Gujer matrix A table of state variables and their interactions in the model. Also called a *Peterson matrix*.

gypsum A mineral consisting primarily of fully hydrated calcium sulfate. Chemical formula is $CaSO_4 \cdot 2H_2O$.

half-life The time required for half of the atoms of a particular radioactive substance to transform, or decay, to another nuclear form.

half-saturation concentration The concentration at which the processing rate is half of its maximum rate. Used in the Monod equation and switching functions.

halide A compound containing a halogen.

halogen One of the chemical elements of the group consisting of fluorine, chlorine, bromine, iodine, and astatine.

hammermill A device with hammerlike arms used to shred or grind solids to facilitate further treatment or disposal.

hazardous air pollutants (HAPs) The 188 compounds or chemical groups that are listed in Section 112 of the Clean Air Act.

head A measure of the pressure exerted by a fluid, expressed as the height of an enclosed column of the fluid that could be balanced by the pressure in the system.

header A pipe manifold fitted with several smaller lateral outlet pipes used to collect or distribute flow.

headloss (1) The difference in water level between the upstream and downstream sides of a treatment process that is attributed to friction losses. (2) Energy loss for a fluid moving through a conduit caused by friction, turbulence, and other energy uses.

headworks The initial structure and devices located at the receiving end of a water or wastewater treatment plant.

heat drying The process by which sludge is thermally heated to remove moisture and produce a dry product that can meet the Class A biosolids criteria.

heat value The quantity of heat that can be released from residuals per unit mass of the solids.

heavy metals Metals that can be precipitated by hydrogen sulfide in an acid solution and that may be toxic to humans in excess of certain concentrations .

hedonic tone A relative measure of pleasantness or unpleasantness; typically pertains to odors.

Heliobacter pylori A bacterium that causes stomach ulcers and has been identified as an emerging waterborne health threat.

helminth A parasitic worm.

Hemicellulose Heteropolymers present in plant cell walls (along with cellulose).

Henry's law The principle that at a constant temperature, the concentration of a gas dissolved in a fluid (with which it does not combine chemically) is almost directly proportional to the partial pressure of the gas at the surface of the fluid.

hepatitis An acute viral disease that results in liver inflammation; it may be transmitted via water directly contaminated by wastewater.

heterotrophic bacteria A type of bacteria that derives its cell carbon from organic carbon; most pathogenic bacteria are heterotrophic bacteria.

high-calcium lime Lime containing 95 to 98% calcium oxide.

high-density polyethylene A synthetic organic material often used as a landfill liner because of its low permeability.

high-pressure membranes Nanofiltration and reverse osmosis membranes.

hindered settling Phenomenon referring to sedimentation of particles in a suspension of intermediate concentration. The interparticle forces of the particles hinder the settling of neighboring particles.

horizontal benchmark Fixed point for horizontal control of construction works.

hyacinth Floating aquatic plants whose roots provide a habitat for a diverse culture of aquatic organisms that metabolize organics in wastewater.

hydrated lime Limestone that has been "burned" and treated with water under controlled conditions to convert calcium oxide into calcium hydroxide.

hydraulic grade line The piezometric surface in a pressure conduit; the water surface in open-channel flow conditions.

hydraulic gradient The slope of the hydraulic grade line, which indicates the change in pressure head per unit of distance.

hydraulic jump A sudden rise in water surface level that occurs when water with high velocity (supercritical velocity) transitions to low velocity (subcritical flow).

hydraulic loading Total volume of liquid applied per unit of time to a tank or treatment process.

hydraulic radius The ratio of the area of flow (wetted area) to its wetted perimeter.

hydraulic residence time Vessel volume divided by the liquid throughput rate, expressed in minutes, hours, or days (depending on the situation).

hydraulic retention time The length of time that a given hydraulic loading of wastewater or solids will be retained in a pipe, reactor, unit process, or facility.

hydrocyclone A conical-shaped device that uses centrifugal force to separate grit and other solids from a liquid.

hydrogen peroxide An oxidizing agent used for odor control and disinfection. Chemical formula is H_2O_2.

hydrogen sulfide A toxic and corrosive gas formed by the decomposition of organic matter containing sulfur. Chemical formula is H_2S.

hydrolysis Enzyme-mediated reactions that convert complex organic compounds (i.e., particulate) into simple compounds or reduced-mass materials. Often refers to solids particle breakdown in the first step of anaerobic digestion.

hydrophilic Having a strong affinity for water.

hydrophobic Having an aversion to water.

hydrostatic pressure The pressure exerted by water as a result of depth alone.

hydrotest A method for testing the integrity of piping, tubing, or vessels by filling them with water and pressurizing them.

hypochlorite Chlorine anion typically used as an alternative to chlorine gas for disinfection. Chemical formula is OCl_3^-.

Imhoff tank A two-story wastewater treatment tank; sedimentation occurs in the upper compartment, and anaerobic digestion occurs in the lower compartment.

incineration The process of reducing the volume of a solid by burning of organic matter.

incinerator A furnace or device for incinerating solids.

inclined cylindrical screens A type of fine screening device.

indicator organism A microbe whose presence indicates the absence or presence of a specific pollutant.

indirect costs The component of capital cost estimates that includes engineering, permitting, and legal services during construction, as well as contingencies and any other associated costs.

indirect potable reuse The use of highly treated reclaimed water to augment a surface water or groundwater intended to serve as a potable water supply.

industrial waste Waste generated by manufacturing or industrial practices that is not a hazardous waste regulated under Subtitle C of the Resource Conservation and Recovery Act.

inert Having no inherent power of action, motion, or resistance.

inerts Constituents that are assumed not to react with anything in the model. Inerts may be soluble or insoluble, organic or inorganic.

infectious agent Any organism that can be communicated in body tissues and can cause disease or other adverse health effects in humans.

infiltrate To filter into or through; permeate (e.g., effluent into soil).

infiltration chamber A pre-formed, manufactured distribution medium with an open bottom that typically is used in soil-treatment areas.

influent Water or wastewater flowing into a basin or treatment plant.

influent characterization See *influent stoichiometry*.

influent fractionation See *influent stoichiometry*.

influent stoichiometry The breakdown of influent constituents into state variables. Also called *influent fractionation* and *influent characterization*.

injector A mechanical device for feeding gaseous chemical into a stream whereby pressurized water creates a vacuum that draws the chemical.

inorganic matter Substances of mineral origin (not containing carbon) that are not subject to decay.

innovative technology A process or technique that represents an advancement over the existing state of the art but has not been fully proven under the circumstances of its contemplated use.

insoluble Incapable of being dissolved.

instrumentation Technology used to control, monitor, or analyze physical, chemical, or biological parameters.

interceptor A pipe that receives flow from a number of other pipes or outlets for disposal or conveyance to a treatment plant.

interface model A model that describes how output variables are passed from one type of model to a different type of model, which uses different variables for its inputs.

intermediate pumping station A pumping station created at a location in the treatment process other than influent or effluent pumping.

invert The lowest point of the internal surface of a drain, sewer, or channel at any cross-section.

International Organization for Standardization (Organisation internationale de normalisation) (widely known as ISO) An international standards-setting body composed of representatives from various national standards organizations.

jar test A test procedure using laboratory glassware to evaluate coagulation, flocculation, and sedimentation in a series of parallel comparisons.

jet A stream of pressurized liquid or vapor from a nozzle or orifice.

jet aeration Wastewater aeration system using floor-mounted nozzle aerators that combine liquid pumping with air diffusion.

kinematic viscosity A fluid's absolute viscosity divided by its mass density.

L10 Life (fatigue life, rating life) The number of hours or revolutions at which 10% of the bearings are likely to fail because of fatigue.

lagoon An excavated basin or natural depression that contains water, wastewater, or solids.

land application The process of spreading of biosolids on land to improve and maintain productive soils and stimulate plant growth.

land disposal The process of spreading large volumes of solids on land in a dedicated disposal site.

landfill A land-based disposal site designed to collect and store solid wastes while minimizing environmental hazards and protecting the quality of surface water and groundwater.

lateral A secondary pipe that branches off from a main water pipe, or header.

launder A trough used to transport water.

Leadership in Energy and Environmental Design (LEED®) A Green Building Rating System™ developed and administered by the U.S. Green Building Council to encourage and accelerate the global adoption of sustainable green buildings, and to develop sustainable design, construction, management, and operation practices through the creation and implementation of universally understood and accepted tools and performance criteria.

leachate Fluid that percolates through solid materials or wastes; it typically contains suspended solids, dissolved materials, or products of the solids.

leach line (lateral) A pipe, tubing, or other conveyance method used to carry and distribute effluent.

lift station A chamber that contains the pumps, valves, and electrical equipment necessary to pump water or wastewater.

lime Any of a family of chemicals [e.g., calcium hydroxide, limestone (calcite), or a mixture of calcium and magnesium carbonate] used to increase the pH of wastewater or solids to promote precipitation, improve solids thickening and/or dewatering characteristics, or kill pathogens.

lime recalcining The process of recovering lime from water or wastewater solids; it typically involves a multiple-hearth furnace.

lime slaker A device used to hydrate quicklime.

lime stabilization A process in which lime is added to solids to raise the pH to 12 for at least 2 hours to chemically inactivate pathogens.

limestone A sedimentary rock composed primarily of calcium carbonate.

liner (1) A barrier of plastic, clay, or other impermeable material that prevents leachate from contacting surface water or groundwater. (2) A protective, corrosion-resistant layer attached or bonded to the inside of a tank, pipe, or other equipment.

liquid chlorine A chlorine compound that contains no water; it occurs when gaseous chlorine is put under high pressure. Stored in steel drums and cylinders.

loss of head A decrease in total energy that results from friction, bends, obstructions, etc. in a pipeline or channel. Also called *headloss*.

low-pressure membranes Microfiltration and ultrafiltration membranes.

luminaries Light fixtures.

lysis A cell rupture that results in loss of its contents.

malodor An odor that is offensive or creates a nuisance.

marine Of or pertaining to the sea; existing in or produced by the sea (ocean).

mass balance A method for analyzing physical systems based on the law of conservation of mass. By accounting for all material entering and leaving a system, mass flows can be identified that otherwise might have been unknown or difficult to measure.

mass transfer The movement of atoms or molecules via diffusion or convection from an area of high concentration to one of low concentration.

materials balance See *mass balance.*

maturation Sufficient decomposition of organic matter to achieve a stable material.

maximum achievable control technology The level of air pollution control technology required by the Clean Air Act.

maximum contaminant level The maximum permissible level of a contaminant in water delivered to the free-flowing outlet of the ultimate user of a public water system.

maximum contaminant level goal The maximum level of a contaminant (including an adequate safety margin) at which no known or anticipated adverse effect on human health would occur.

maximum daily peaking factor Ratio of the maximum daily flow or constituent mass to the annual average value.

maximum monthly peaking factor Ratio of the maximum monthly flow or constituent mass to the annual average value.

minimum daily peaking factor Ratio of the minimum daily flow or constituent mass to the annual average value.

mean cell residence time The average time that a microbial cell remains in an activated sludge system. It is equal to the mass of cells divided by the rate at which cells are wasted from the system.

mean velocity The average velocity of a fluid flowing in a channel, pipe, or duct, determined by dividing the discharge by the cross-sectional area of the flow.

mechanical aeration A system that mechanically agitates water to promote mixing with atmospheric air.

mechanically activated solar drying beds Drying beds that use a robot to break up the surface of drying solids to increase the evaporation rate.

mechanistic model A model that describes a process' behavior in a mechanistic manner (i.e., equations are used to directly describe outputs based on input variables in an attempt to match the observed behavior of the process). Also called *physics-based models* because they attempt to describe the physical behavior of the process. Most process models typically used for design are mechanistic models.

media displacement volume **(L^3/L^3)** The volume of a reactor displaced by the installed carrier media, calculated as the bulk *specific volume* multiplied by the *fill fraction*.

medium compression The medium properties (e.g., porosity, collector size, and depth) are all adjusted by the applied medium compression ratio. This action occurs in a compressed medium filter.

membrane bioreactor A wastewater treatment process that basically combines the activated sludge process with membrane filtration. The membranes are suspended in an activated sludge reactor to separate liquid from solids.

membrane brine A concentrated slurry containing the solids rejected by high-pressure membranes during filtration. Also called *membrane concentrate*.

membrane concentrate A concentrated slurry containing the solids rejected by membranes during filtration.

membrane diffuser A fine-bubble aeration diffuser with perforated flexible plastic membranes.

membrane filter (1) A filtration process in which membranes separate solids from liquid. (2) A paperlike filter, with small pore sizes, that can retain bacteria for use in laboratory examinations of water.

membrane reject A slurry containing the solids rejected by low-pressure membranes during filtration.

mesh (1) The number of openings per lineal inch, measured from the center of one wire or a bar to a point 25.1 mm (1 in.) distant. (2) A type of screening media.

mesophilic An operating temperature range (typically 30 to 40°C) for anaerobic digestion; it affects the microbial population in the digester and the reaction rates.

mesophilic digestion A process in which solids are digested by microorganisms that thrive in the mesophilic temperature range (about 30 to 40°C).

metabolic models Models developed to describe the metabolic processes of biological treatment by evaluating the rates of transformations occurring, including intermediate compounds.

metabolism The biological conversion of organic matter to cellular matter and gaseous byproducts.

metal In general, the chemical elements that easily lose electrons to form positive ions.

metal salt coagulants Alum and iron (III) salts, which typically are used to coagulate solids particles.

metering pump A pump that provides a specific volume of fluid; used to add treatment chemicals to water or wastewater.

methanogenesis The metabolic conversion of organic acids or hydrogen and carbon dioxide to methane. The primary methanogenic populations associated with anaerobic digestion are acetoclastic methanogens and hydrogenotrophic methanogens.

methanogens A group of anaerobic bacteria responsible for converting organic acids into methane gas and carbon dioxide.

methanol A solvent often used as a supplemental carbon source during denitrification. Chemical formula is CH_3OH.

method detection limit The lowest quantity of a substance that, when analyzed using a given method, can be unequivocally distinguished from a blank that has undergone the same process.

methylotrophic methanogenesis Methanogenic populations that can convert methylated compounds (e.g., methanethiol and triemethyl amine) into methane. These organisms are thought to be significant in controlling odors from biosolids.

microconstituents Natural and anthropogenic substances (e.g., elements and inorganic and organic chemicals) detected in water and the environment, for which a prudent course of action is suggested for the continued assessment of the potential effect on human health and the environment.

microfiltration A low-pressure membrane filtration process that removes suspended solids and colloids larger than 0.1 μ from wastewater.

microfloc Destabilized floc particle that permits in-depth penetration of a granular media filter bed to optimize the filter's solids retention capacity.

microorganism Organisms observable only through a microscope. Also called *microbes*.

microscreens A filtration device consisting of a rotating drum with a fine-mesh screen fixed to its periphery. As water flows through the interior of the drum, solids are retained by the mesh for later removal via a high-pressure wash.

mist eliminator A physical obstruction in an air stream designed to collect suspended liquid droplets.

misting scrubber An air treatment technology in which a scrubbing liquid is sprayed into the contaminated air stream. The contaminant is removed either by physical impingement or via chemical reaction with the scrubbing liquid.

mitigation Changes or additions to a treatment plant design to reduce or eliminate identified impacts.

mixed liquor The mixture of wastewater and activated sludge being treated in an aeration basin.

mixed-liquor suspended solids The suspended solids concentration in the mixture of wastewater and activated sludge being treated in the aeration basin.

mixed-liquor volatile suspended solids The volatile fraction of mixed-liquor suspended solids.

mixed-media filter A granular media filter using two or more types of filter media with different sizes and specific gravities (typically silica sand, anthracite and ilmenite or garnet).

mixing tank A tank equipped with a device for agitating or mixing wastewater or solids to increase the dispersion rate of applied chemicals.

mixing zone A limited area of a natural waterbody where highly treated wastewater is received and diluted. The intent of mixing zones is to prevent discharged effluent from harming the aquatic environment and its designated uses (e.g., drinking, fishing or swimming). In theory, this zone allows for efficient, natural pollutant assimilation. In practice, mixing zones can be used as long as the waterbody's integrity is not impaired.

model An equation or set of equations used to describe a process or several connected processes.

mole The molecular weight of a substance, typically expressed in grams.

molecular weight The sum of the atomic weights of all the atoms in a molecule.

Monod equation An equation typically used in wastewater treatment models to describe the kinetics of biological growth. Identical in form to the Michaelis–Menten equation often referred to in industrial applications.

monomedia filter A granular media filter using only one size and type of filter media.

most probable number A statistical analysis technique based on the number of positive and negative results acquired when testing multiple portions of equal volume; typically used to count pathogens in solids samples.

motor control center A structure housing the controls for electrical equipment (multiple motor starters, variable-frequency drives, circuit breakers, etc.).

mound An above-grade soil treatment area designed and installed with at least 305 mm (12 in.) of clean sand (ASTM C-33) between the bottom of the infiltrative surface and the original ground elevation. It uses pressure distribution to distribute effluent across the mound. A final cover of suitable soil material stabilizes the surface and supports vegetative growth.

moving-bed filter A granular media filter that continuously cleans and recycles filter media while the filter continues to operate.

mudballs Agglomerations of floc, solids, and filter media in a filter bed that may grow into a larger mass and reduce filtration efficiency.

mud valve A valve used to drain sediment from the bottom of a sedimentation basin or to drain the contents of an aeration tank.

mudwell Common name used for a backwash reject water storage basin.

multi-criteria decision analysis A tool that helps decision makers take multiple criteria into account when comparing design alternatives.

multimedia filter A granular media filter that uses two or more types of filter media with different sizes and specific gravities (e.g., silica sand, anthracite, and ilmenite or garnet).

multiple-hearth furnace An incinerator consisting of numerous hearths that is used to combust organic solids or recalcinate lime.

multiple rake screen A type of fine screening device.

municipal waste The combined solid and liquid waste from residential, commercial, and industrial sources.

municipal wastewater treatment plant Collectively, the buildings, processes, and equipment needed to treat municipal wastewater.

Nasal Ranger® A field olfactometer developed by St. Croix Sensory.

National Ambient Air Quality Standards (NAAQS) Outdoor standards for air pollutants (e.g., ozone, sulfur dioxide, particulates, nitrogen dioxide, carbon monoxide, and lead) shown to cause adverse health effects. An ambient standard has been established for each of these pollutants; they can be found in 40 CFR 50.

National Pollutant Discharge Elimination System As authorized by the Clean Water Act, this is a permit program that controls water pollution by regulating point sources that discharge pollutants into waters of the United States.

needle valve A valve that controls flow by means of a tapered needle that extends through a circular outlet.

negative head Filter operating condition that occurs when the pressure in the filter bed is less than atmospheric pressure during a filter cycle.

neighborhood advisory committee (NAC) A committee comprised of community stakeholders that have some level of contribution to the design of the facility.

neoprene A synthetic rubber with a high level of resistance to oils, ozone, oxidation, and flame; made by polymerizing chloroprene.

nephelometric turbidity unit A measure of turbidity via instrumentation.

net environmental benefit The sum of the positive benefits minus the sum of the negative impacts of a particular project or treatment system on the environment.

net positive suction head The difference between the total absolute pressure head and the vapor pressure of the liquid being pumped.

net specific surface area (L^2/L^3) The resulting *specific surface area* within a reactor based on the *bulk specific surface area* and the installed *fill fraction*.

net yield The net mass of solids produced in a biological process divided by the mass of substrate removed, typically BOD or COD. It is equal to the synthesis yield minus decay.

neutralization A chemical process that produces a solution that is neither acidic or alkaline.

New York State Energy Research and Development Authority (NYSERDA) A public-benefit corporation created in 1975 and focused solely on research and development with the goal of reducing the state's petroleum consumption.

nitrate A stable, oxidized form of nitrogen. Chemical formula is NO_3 .

nitrate formers See *nitrobacter*.

nitrate-nitrogen The nitrate concentration reported in terms of the nitrogen in the nitrate. See *nitrate*.

nitric acid Chemical is used in cleaning and preservation. Chemical formula is HNO_3.

nitrification A biological process in which ammonia is converted first to nitrite and then to nitrate.

nitrifying bacteria Bacteria that can oxidize nitrogenous material.

nitratation The biological conversion of nitrite to nitrate.

nitritation The biological oxidation of ammonia to nitrite.

nitrite An unstable, easily oxidized nitrogen compound. Chemical formula is NO_2^-.

nitrite formers See *Nitrosomonas*.

nitrite-nitrogen See *nitrite*.

nitrobacter Nitrifying bacteria that convert nitrites to nitrates. Also called *nitrate formers*.

nitrogen fixation The biological conversion of atmospheric nitrogen to nitrogen compounds.

nitrogen oxides (NO_X) Compounds formed during combustion.

nitrogen oxide compounds (NO_X) Pollutants formed during combustion.

nitrogen removal Physical, chemical, and biological processes that remove nitrogen from wastewater.

nitrogenous biochemical oxygen demand The portion of biochemical oxygen demand whereby oxygen is consumed as a result of the oxidation of nitrogenous material; measured after the carbonaceous oxygen demand has been satisfied. Also called *second-stage biochemical oxygen demand.*

nitrogenous oxygen demand The portion of oxygen demand associated with the oxidation of nitrogenous material, and the oxidation of free ammonia and ammonia released from nitrogenous material.

Nitrosomonas Ammonia-oxidizing bacteria that convert ammonia to nitrite under aerobic conditions and derive energy from the oxidation reaction.

Nocardia Bacteria than can accumulate to create a nuisance foam in aeration basins and secondary clarifiers.

non-ionic polymer A polyelectrolyte with no net electrical charge.

nonpoint source A source of air or water pollutants whose discharges are more diffuse (e.g., fertilizer runoff from farms).

nonsettleable solids Suspended solids that typically remain in suspension for more than 1 hour.

nuisance The condition created when an odor concentration injures human health, is offensive to the senses, or interferes with the reasonable or comfortable enjoyment of life and property.

numerical solver Software used to solve the multiple differential equations in a model using numerical methods.

nutrient (1) Any substance that is assimilated by organisms to promote or facilitate their growth. (2) Nitrogen and phosphorus, when considering their potential to result in excess biological growth in the environment.

nutrient trading A structured credit trading system for pollutant contributors to a specific receiving waterbody. The system allows entities to earn nutrient removal credits for treating beyond permit requirements and trade those credits to other entities to allow them to cost-effectively meet their nutrient removal requirements.

odor character A reference vocabulary used to describe the perceived characteristics of an odor.

odor concentration An odor measurement expressed as a dilution-to-threshold ratio (D/T), a nondimensional volumetric ratio.

odor dispersion The dilution of an odor via turbulent mixing of the odorant with ambient air from the point of release to downwind locations.

odor intensity A measurement of the relative strength of an odor, expressed in parts per million by volume (ppm) of 1-butanol in air.

odor intensity referencing scale A tool commonly used to evaluate odors that is standardized upon a given reference chemical.

odor persistency The rate at which an odor's intensity decreases with dilution (i.e., the slope factor in Steven's law).

odor panel A group of five or more individuals selected and trained to assess odors in accordance with established procedures and guidelines. An odor laboratory will draw an odor panel from a pool of trained assessors.

odor thresholds The odor concentration that, statistically, half of the odor panel would detect or recognize the odor.

odor unit (OU) Originally defined as a unit mass of odor per unit volume, it now typically is used to describe an odor concentration, expressed as a dilution ratio.

offgas The gaseous emissions from a process or equipment.

olfactometer A device used by a panel of testers to compare the odor from an ambient air sample to reference samples of varying dilutions to determine odor strength.

olfactometry The measurement of odors using the sense of smell.

onsite (wastewater treatment) system A system designed to collect and treat wastewater from one or more dwellings, buildings, or structures; the resulting effluent is dispersed on property owned by the individual or entity.

open channel A natural or artificial channel in which fluid flows with a free surface open to the atmosphere.

open drip proof A designation for electrical motor enclosures in which the ventilating openings are constructed so successful operations are not affected by drops of liquid or solid particles that strike or enter the enclosure at any angle from 0 to 15° downward from vertical.

open-path optical transect method A method for measuring odorants across an open surface or basin.

operation cost index (OCI) A parameter that integrates effluent criteria, energy costs, and solids treatment costs.

order-of-magnitude estimate A capital cost estimate made without detailed engineering data; typically prepared at the end of the schematic design phase of the design delivery process.

organic loading The amount of organic matter applied to a treatment process per day.

organic nitrogen Nitrogen that is bound to carbon-containing compounds. Measured as the difference between TKN and ammonia-nitrogen.

organic phosphorus Phosphorus that is bound to carbon-containing compounds.

orifice plate (1) A flow meter that measures flow as a function of differential pressure across a flow-restricting orifice. (2) A flow-limiting device.

outfall The location where stormwater, wastewater, or reclaimed water is discharged to a receiving waterbody. Also refers to the pipeline or conduit that conveys flow to a receiving water.

overflow rate A measure of the upward water velocity in a sedimentation tank, expressed as flow per day per unit of basin surface area ($L^3/L^2/T$ which is equal to L/T, a velocity). Also called *surface loading rate.*

ovum A mature egg ready to be fertilized.

owner The entity that possess property, or in a construction project, pays for the design, modifications, and/or construction. The owner may be public (e.g., a government agency) or private (e.g., a commercial land-development firm).

oxic A biological environment that contains molecular oxygen.

oxidation (1) A chemical reaction in which an element or ion loses electrons. (2) The biological or chemical conversion of organic matter to simpler, more stable forms.

oxidation ditch An extended aeration waste treatment process that occurs in an oval-shaped channel or ditch (also called a *racetrack*); aeration is provided by a mechanical brush aerator or by diffusers with mechanical mixers.

oxidation pond An earthen wastewater basin in which organic matter is biologically oxidized naturally or with the assistance of mechanical oxygen-transfer equipment.

Oxidation-reduction potential The potential required to transfer electrons from an oxidant to a reductant; it indicates the likelihood that an oxidation–reduction reaction will occur.

oxygen transfer (1) The exchange of oxygen between a gaseous and a liquid phase. (2) The amount of oxygen absorbed by a liquid compared to the amount fed into the liquid through an aeration or oxygenation device; typically expressed as percent.

oxygen uptake The amount of oxygen used during biochemical oxidation.

oxygen uptake rate The oxygen used during biochemical oxidation, typically expressed as mg $O_2/(L \cdot h)$ in the activated sludge process.

ozonation A treatment process that uses ozone for oxidation, disinfection, or odor control.

ozonator An ozone generator.

ozone A strong oxidizing agent with disinfection properties similar to chlorine. Also used in odor control and solids processing. Chemical formula is O_3.

ozone generator A device used to produce ozone by passing air or oxygen through an electric field.

packed bed scrubber A type of mist scrubber that uses internal packing to provide a contact surface for contaminants and the scrubbing liquid.

Palmer Bowlus flume A flow meter particularly suited for insertion into pipes and manholes.

partially submerged disk filter The disk filter is partially submerged in the tank with backwash mechanism remains above the water surface.

particle size distribution A method for characterizing the size spectrum of suspended solids particles in wastewater.

particulate Typically considered to be a solid particle larger than 1 μ; large enough to be removed from water or wastewater via filtration.

pasteurization A process in which heat is applied for a specific period of time to kill pathogens.

pathogen Highly infectious, disease-producing microbes typically found in sanitary wastewater.

peak flow Flow experienced during hours of high demand; typically determined to be the highest 2-hour flow expected under any operating conditions.

peaking factor The ratio of peak to mean value of a measured quantity; the mean value typically is the annual or yearly average.

pellet See *granule*. The term *pellet* often is used to refer to granules that are more uniform in shape and size.

perforated plate A type of screening media.

peristaltic pump A type of positive-displacement pump whereby the fluid is squeezed through a flow tube by external rollers.

permeate Filtered or treated effluent from a filtration or reverse osmosis membrane. Filtration membrane effluent also is called *filtrate*.

Peterson matrix See *Gujer matrix*.

pH The reciprocal of the logarithm of the hydrogen ion concentration in gram moles per liter. On the 0 to 14 pH scale, a value of 7 at 25°C (77°F) represents a neutral condition. Decreasing values indicate increasing hydrogen ion concentration (acidity), and increasing values indicate decreasing hydrogen ion concentration (alkalinity).

phosphate A salt or ester of phosphoric acid.

phosphorus A nutrient that is an essential element of all life forms.

physical treatment A water or wastewater treatment process that uses only physical methods (e.g., filtration or sedimentation).

physical–chemical treatment A water or wastewater treatment process that uses both physical and chemical methods.

Physics-based model See *mechanistic model.*

phytotoxic Poisonous to plants.

piezometer An instrument fitted to the wall of a pipe or container to measure pressure head; it consists of a small pipe and manometer.

piezometric head The elevation plus pressure head.

pig A water-propelled internal pipe cleaner.

pigment Finely ground, natural or synthetic, inorganic or organic, insoluble dispersed particles that, when dispersed in a liquid vehicle to make paint, may provide several beneficial characteristics (e.g., color, opacity, hardness, durability, and corrosion-resistance).

pilot plant A water or wastewater treatment plant that is smaller than full-scale; it is used to test and evaluate a treatment process.

pinch valve A valve with one or more flexible elements that can be pinched to stop flow.

piston pump A reciprocating pump whose piston typically incorporates a sliding seal with the cylinder wall.

plate-and-frame press A batch-process dewatering system in which solids are pumped through a series of parallel plates fitted with filter cloth.

plenum An air-filled space in a structure in which air is distributed evenly.

plug flow A flow condition characterized by the fact that fluid and fluid particles discharge a system in the same sequence they enter.

plume A plume is a concentration distribution caused by the release of pollutants as they are transported down gradient.

plunger pump A reciprocating pump whose plunger does not contact the cylinder wall; instead, it enters and withdraws from the cylinder through packing glands, that hold a deformable material used to control leakage around the plunger.

point source A source of pollutants characterized by the existence of a specific discharge point (e.g., stack, vent, or outfall pipe).

pollutant A substance, organism, or energy form present in amounts that impair or threaten an ecosystem to the extent that its current or future uses are precluded.

polyelectrolytes Complex polymeric compounds typically composed of synthetic macromolecules that form charged ions in solution. Water-soluble polyelectrolytes are used as flocculants; insoluble polyelectrolytes are used as ion exchange resins.

polyelectrolyte flocculant A polymeric organic compound used to induce or enhance the flocculation of suspended and colloidal solids, and thereby facilitate solids thickening or dewatering.

polymers Synthetic organic compounds with high molecular weights and repeating chemical units (monomers); these polyelectrolytes may be water-soluble flocculants or water-insoluble ion exchange resins.

polymer injection ring A device with four or more equally spaced ports designed to inject polymer into a solids pipeline. Installed as part of the piping, this device should promote thorough, even distribution for thorough mixing with solids.

polyphosphates Phosphate compounds used as sequestration agents to prevent the formation of iron, manganese, and calcium carbonate deposits.

pond depth The radial dimension of the water that has been separated from thickened solids in a centrifuge. Also called *pool depth*.

pool depth The radial dimension of the water that has been separated from thickened solids in a centrifuge. Also called *pond depth*.

porosity The ratio of the void space to the total volume of the porous medium.

porous disk diffuser A circular, fine-bubble aeration device made of porous plastic or ceramic.

positive-displacement pump A pump in which liquid is drawn into a cavity and the pressure is increased, which forces the liquid through an outlet port into the discharge line.

postaeration A process in which oxygen is added to effluent before it is discharged to a receiving water.

post-treatment Treatment of water or wastewater treatment plant effluent to further enhance its quality.

potassium permanganate Chemical frequently used in odor control. Chemical formula is $KMnO_4$.

powdered activated carbon A powdered form of activated carbon that is slurried and fed to water to absorb organics (e.g., taste- and odor-causing constituents).

power grid The network of electrical wires and equipment owned and operated by an electrical utility.

preaeration A preliminary treatment process in which wastewater is aerated to remove gases, add oxygen, promote the flotation of grease, and/or aid coagulation.

precoat Applying an inert material to a filter cloth to prevent solids from blinding the cloth and facilitate cake release.

precipitation (1) Any chemical reaction in which a dissolved substance becomes a solid. (2) Any form of water (e.g., rain, snow, sleet, or hail) that falls to the earth's surface.

preliminary treatment Treatment steps (e.g., comminution, screening, grit removal, preaeration, and/or flow equalization) that prepare wastewater influent for further treatment.

pressate The liquid wastestream from a filter press.

pressure filter A filter enclosed in a vessel that may be operated under pressure.

pretreatment (1) The initial water or wastewater treatment process that precedes primary treatment processes. (2) The treatment of industrial wastes to reduce or alter the characteristics of pollutants before the wastes are discharged to a wastewater treatment plant.

primary clarifier A sedimentation basin that precedes secondary wastewater treatment.

primary distribution The high-voltage section (at least 2 400 V) of the electrical distribution system.

primary sedimentation A gravity-based process for removing settleable suspended solids from water or wastewater; typically occurs in a quiescent basin or clarifier. The principal form of primary wastewater treatment, which is used to reduce the solids loading on subsequent treatment processes.

primary sludge Solids produced during primary wastewater treatment (e.g., sedimentation).

primary residuals Solids produced via sedimentation.

primary treatment Treatment processes (e.g., sedimentation and/or fine screening) designed to produce an effluent suitable for biological treatment.

privatization The involvement of nonpublic and entrepreneurial interests in project development, ownership, and/or operation of municipal facilities (e.g., water and wastewater treatment systems).

process/equipment vendors Companies that sell equipment or supplies for a project to the contractor or owner.

procurement The process of legally obtaining equipment or services; often requires a contract.

program manager The person hired by the owner to oversee and manage the design consultant's day-to-day project activities.

progressing cavity pump A pump used for viscous fluids (e.g., solids) that consists of a single-threaded shaft rotor rotating in a double-threaded rubber stator.

progressive cavity pump See *progressing cavity pump*.

proportional weir A weir whose discharge is directly proportional to the head. Sometimes called a *Sutro weir*.

psychrometric chart A graph of the thermodynamic properties of moist air.

publicly owned treatment works (POTW) Wastewater treatment works [both treatment plant(s) and collection system] owned by a state or municipality.

pug mill A device with rotating blades that simultaneously mixes and grinds two materials to reduce the size of the mixture to facilitate further treatment or disposal.

pump curves Graphs of pump characteristics (e.g., total discharge head, net positive suction head, power required, and efficiency relative to capacity) that indicate pump performance.

pumping station A facility that contains the pumps, valves, and electrical equipment necessary to move water or wastewater through a distribution or collection system, respectively.

pure-oxygen process A variant of the activated sludge process in which pure molecular oxygen is used rather than atmospheric oxygen.

putrescible Organic matter that is likely to become a rotten, foul-smelling product when it decays or decomposes.

pyrophoric Capable of spontaneous combustion.

quality assurance Planned, systematic production processes that provide confidence in a product's suitability for its intended purpose

quality control A system for verifying and maintaining a desired level of quality in a product or process by careful planning, use of proper equipment, continued inspection, and corrective action as required.

quicklime A calcined material, the major part of which is calcium oxide, or calcium oxide in natural association with a lesser amount of magnesium oxide. This material can be slaked (i.e., chemically react with water or moist air).

radial flow A flow pattern in which the flowing materials are directed either from the center to the periphery or from the periphery to center.

radiant Transfer of heat via electromagnetic radiation (e.g., solar energy used to dry biosolids).

rapid mix Any method of quickly, thoroughly blending water, wastewater, or solids with coagulants or conditioning chemicals to ensure a complete reaction.

rapid sand filter A granular media filter in which water flows downward through the sand bed at a rate typically ranging from 80 to 320 $L/min{\cdot}m^2$ (2 to 8 $gal/min/ft^2$) of surface area.

rat holing Compaction on the side of a silo that occurs when the material being handleed has sufficient cohesive strength to resist gravity flow through a silo. Typically, a channel will form where the material will fall through the silo; however, once the channel has emptied, all flow from the silo stops.

rate schedules The electric utility's list of charges for providing power to a wastewater treatment facility.

rating life The number of hours or revolutions at which 10% of the bearings are likely to fail because of fatigue.

raw sludge Untreated wastewater solids.

reaeration The absorption of oxygen into water under oxygen-deficient conditions.

reasonable potential analysis (RPA) An analysis of effluent constituents that have a reasonable potential to cause or contribute to a violation of a water quality standard.

recarbonation the reintroduction of carbon dioxide to water, typically during or after lime soda softening.

recalcining The process of recovering lime from solids; typically performed in a multiple-hearth furnace system.

receiving water A surface waterbody that receives effluent from a wastewater treatment plant.

receptor In odor modeling, a grid coordinate where pollutant concentrations are predicted.

recessed plate-and-frame press A separation device that uses pressure to force water through a series of filter cloths mounted between plates. Meanwhile, solids collect on the cloths.

reciprocating rake screen A type of coarse screening device.

reclaimed wastewater Wastewater treated to a level that allows its reuse for a beneficial purpose. Also called *recycled water*.

recognition threshold An odor concentration that, statistically, half of the odor panel would be able to characterize and distinguish from carbon-filtered (odor-free) air.

recreational waters Any waterbody used for recreational activities (e.g., swimming, boating, or fishing).

rectifier An electrical device that converts alternating current voltage to direct current voltage.

recycle ratio The recycled flowrate divided by the influent flowrate; applicable to an activated sludge system or other treatment system.

recycling The process of converting recovered materials into new products.

reed bed A treatment system in which wastewater or solids are used to grow reeds, which use the water, nitrogen, and other nutrients provided by the material. Reed beds have been used for both tertiary wastewater treatment and solids disposal.

refractory Temperature-resistant material used in high-temperature areas of furnaces, incinerators, and boilers to protect the metal housing or ducting. Refractory may be produced with insulating and abrasion-resistant qualities

refractory organics Organic substances that are difficult or impossible to metabolize in a biological system.

regenerative thermal oxidation (RTO) High-temperature oxidation of contaminants in an air stream; the system uses its waste heat to preheat the incoming airstream.

regenerative thermal oxidizer An emissions-control device that uses heat to oxidize volatile organic compounds.

residence time The period of time that a volume of liquid or solids remains in a tank or system.

residuals The non-liquid components of wastewater that are removed from the liquid during various treatment processes.

resin General term applied to a wide variety of transparent and fusible products that may be natural or synthetic.

resistivity (1) A material's resistance to current per unit length for a uniform cross-section. (2) The reciprocal of conductivity.

respiration Intake of oxygen and discharge of carbon dioxide as a result of biological oxidation.

retention time The length of time that water or wastewater will be retained in a unit treatment process or facility.

return activated sludge Settled activated sludge that is returned to the beginning of the activated sludge process to mix with raw or primary settled wastewater.

return sludge See *return activated sludge.*

reverse osmosis Technological process (and HPM membrane used to achieve this process) that uses forward osmosis phenomena to retain salts in the higher osmotic pressure solution when displacing water from this solution through the semipermeable HPM membranes (reverse osmosis membranes).

rotameter A variable-area, constant-head, rate-of-flow volume meter in which the fluid flows upward through a tapered tub, lifting a shaped weight to a position where upward fluid force just balances the weight.

rotarv collector A rotating mechanism used in circular clarifiers to collect and remove settled solids.

rotary drum thickener A rotating cylindrical screen used to thicken sludge.

rotary kiln An incinerator consisting of a long, horizontal, slowly rotating cylinder in which material is fed at one end and tumbled by the kiln to promote drying as it is conveyed to the other end.

rotary press A device that dewaters sludge by passing flow through a channel that is bound between two revolving screens. The filtrate passes through the screens while the dewatered sludge continues through the channel.

rotating biological contactor A fixed-film biological treatment device in which microorganisms are grown on circular disks mounted on a horizontal shaft that slowly rotates in wastewater.

sack screen A coarse screening device.

Salmonella An aerobic bacteria that is pathogenic in humans; chiefly associated with food poisoning.

salmoneliosis A common type of food poisoning caused by eating food contaminated with *Salmonella* bacteria; it is characterized by a sudden onset of gastroenteritis.

sand filtration See *granular media filtration.*

sanitary wastewater Domestic wastewater that originated in sanitation devices (e.g., sinks, toilets, and washing machines).

Scentometer A field olfactometer developed by Barnebey-Cheney Company in 1974.

Schistosoma A parasitic flatworm or blood fluke that is drawn to freshwater snails during one phase of its life and to humans during another.

schistosomiasis A waterborne disease common to tropical and subtropical regions; it is transmitted to humans who wade or bathe in water infested by *Schistosoma*. The life cycle of human schistosomes involves freshwater snails that act as an intermediate host to produce new parasites in the water phase.

screen retention value Percentage of total influent solids retained by a screening device.

screening (1) A physical separation process in which a screen is used to remove particles from a fluid. (2) The process of systematically examining multiple items to determine their suitability.

screenings The material removed by a screening device.

screenings conditioning A physical process in which collected screenings are dewatered, washed, and compressed.

screenings organics test A method for determining the organic content of collected screenings.

screenings washer/compactor Device used for screenings conditioning.

screw conveyor A device that uses a helical screw rotating within a trough to convey material from one location to another.

screw press A device in which a rotating helical screw presses solids against a cylindrical or conical screen to remove water from them.

screw pump A low-lift, high-capacity pump that uses a helical screw rotating slowly in a trough or pipe to raise water from one elevation to another.

scrubber A device used to remove particulate or pollutant gases from air or gas streams (e.g., exhaust gas).

scrubbing A process in which impurities are removed from an air or gas stream by entraining the pollutants in a water spray.

scum Buoyant materials (e.g., food wastes, grease, fats, paper, and foam) often found floating on the surface of primary and secondary settling tanks.

scum collector A mechanical device that removes scum from the surface of a settling tank.

scum trough A trough used to collect scum and convey it to another location for treatment or disposal; typically used in primary sedimentation processes.

seal Anything that tightly or completely closes or secures a thing (e.g., packing or a mechanical device used to prevent leakage around a rotating motor shaft).

seal water Water (typically treated effluent) pumped into a pump seal to reduce wear on the pump shaft; typically used in applications where grit or abrasive solids might be present.

secondary clarifier A vessel in which suspended matter is removed from wastewater via gravity; this vessel is located after a secondary treatment process.

secondary effluent Wastewater that has received preliminary, primary, and secondary treatment.

secondary distribution The low-voltage (480 V) section of the electrical-distribution system.

secondary sludge Solids generated during a secondary treatment process.

secondary treatment Any process designed to degrade the biological content of wastewater; typically follows primary treatment.

sedimentation A gravity-based process for removing settleable suspended solids from water or wastewater; typically occurs in a quiescent basin or clarifier.

sedimentation basin A quiescent tank in which suspended solids are removed from water or wastewater via gravity; they typically are equipped with a motor-driven rake to collect settled solids and move them to a central discharge point. Also called *clarifiers* or *settling tanks*.

selective catalytic reduction (SCR) A combustion process in which ammonia (or urea) and a catalyst react with NO_X to produce nitrogen, metal oxide, and water. The catalyst allows for lower-temperature operation than SNCR.

selective non-catalytic reduction (SNCR) A combustion process in which ammonia or urea is introduced at high temperatures and reacts with NO_X to form nitrogen, carbon dioxide, and water.

self-priming pump A pump designed to retain enough liquid in its casing to clear its passages of air, when necessary, so it can resume delivering liquid without outside attention. The design allows it to be placed above the level of the liquid being pumped.

sensitivity analysis A method for evaluating how changes in a model's input parameters affect its output.

septage The contents of a septic tank.

sequencing batch reactor A biological treatment process with five phases: fill, react, settle, draw, and idle. It typically includes a system of multiple tanks so one can be filled while another is treating wastewater or being drained.

service factor A measure of how much over the nameplate rating (i.e., maximum operating parameters) any given electric motor can be driven without overheating.

settleability The tendency of suspended solids to settle to the bottom of a tank or Imhoff cone.

settleable solids The portion of suspended solids that are the right size and weight to settle to the bottom of an Imhoff cone in 1 hour.

settling tank A quiescent tank in which suspended solids are removed from water or wastewater via gravity; they typically are equipped with a motor-driven rake to collect settled solids and move them to a central discharge point. Also called *clarifiers* or *sedimentation tanks*.

settling velocity The rate at which particles collect on the bottom of a tank or Imhoff cone.

sewage See *wastewater*.

sewer An underground pipe used to transport wastewater.

sewerage A system of underground piping used to transport wastewater to one or more treatment plants.

sewer gas A gaseous mixture produced when the organic matter in wastewater decomposes anaerobically in the collection system; typically contains high percentages of methane and hydrogen sulfide.

sewershed The area that drains into a sewerage system.

short-circuit current The amount of electrical current that will flow in an electrical system during a fault (short-circuit).

short-circuiting Uneven flow through a vessel; it occurs when density currents or inadequate mixing allows some of the flow to leave the vessel more quickly than the rest.

sidestream Liquid streams generated during solids processing or odor control that typically are returned to the head of a plant for re-processing in the wastewater treatment train.

side water depth The depth of water in a basin or tank, as measured at the vessel's interior wall.

sieve analysis A method for analyzing the particle size distribution of a granular material (e.g., filter sand); it typically involves pouring a sample through a standard series of sieves with increasingly smaller openings.

sieve size A measure of the diameter of the openings in a sieve.

silica A mineral composed of silicon and oxygen.

silicate Any compound containing silicon, oxygen, and one or more metallic compounds.

silo A tall, cylindrical vessel used to process or store solids.

siloxanes A family of anthropogenic organic compounds containing silicon, carbon, and oxygen that are becoming increasingly common in wastewater solids. When the solids are anaerobically digested, volatile siloxanes become part of the digester gas. When this gas is combusted, the siloxane compounds form tough, abrasive silicon dioxide deposits on the interior surfaces of boilers, engines, and other combustion-related equipment.

simulator Software used to run a model of a treatment process.

simulation A model run providing outputs based on model inputs.

single line diagram A drawing depicting the electrical distribution system in a facility.

siphon A closed conduit, a portion of which lies above the hydraulic grade line, resulting in a pressure less than atmospheric and requiring a vacuum within the conduit to start flow.

skimming The process of removing or diverting water and/or floating matter from the surface of a liquid.

slake To combine chemically with water or moist air.

slaked lime See *hydrated lime.*

slime (1) Viscous organic substances, typically formed via microbiological growth, that attach themselves to other objects, forming a coating. (2) The coating of biomass that accumulates in trickling filters or sand filters and periodically sloughs away to be collected in clarifiers.

slow sand filter A sand filter with low flowrates designed to promote the formation of a solids layer on top of the sand bed; this solids layer provides most of the filtration.

sludge Any residual produced during primary, secondary or advanced wastewater treatment that has not undergone any process to reduce pathogens or vector attraction. Also called *raw sludge*. The term *sludge* should be used with a specific process descriptor (e.g., primary sludge, waste activated sludge, or secondary sludge).

sludge age See *solids retention time.*

sludge blanket An accumulation of solids hydrodynamically suspended in a basin or tank.

sludge volume index The volume (in milliliters) occupied by 1 g of settled sludge after settling for 30 minutes in a 1-L graduated cylinder.

slug load A hydraulic or organic load suddenly added to a treatment process.

sluice gate A manually or power-operated gate used to isolate a channel from flow.

slurry A suspension of a relatively insoluble chemical in water; it typically has a suspended solids concentration of 5 000 mg/L or more.

sodium bisulfite A liquid dechlorinating agent. Chemical formula is $NaHSO_3$.

sodium chlorite A frequently used chemical. Chemical formula is $NaClO_2$.

sodium hydroxide Caustic soda. Chemical formula is NaOH.

sodium hvpochlorite A liquid chlorine solution frequently used as a water or wastewater disinfectant. Chemical formula is NaOCl.

sodium metabisulfite A crystalline form of sulfur dioxide used to remove chlorine. Chemical formula is Na_2SO_3.

solid-bowl centrifuge A centrifuge consisting of a conical bowl and an internal helical scroll that rotate at slightly different speeds to separate solids from water via centrifugal force; designed for continuous operations.

solids Any residual produced during wastewater treatment.

solids balance A mathematical representation of a treatment system that tracks the amount of solids entering and exiting each unit or process.

solids content The percentage of dry matter (by weight) in a mixture.

solids disposal The act of getting rid of solids via incineration, landfilling, surface disposal, etc.

solids retention time The average period of time that solids have remained in a process or system.

solids stabilization The act of reducing the number of pathogens in solids to meet the requirements of 40 CFR 503.

solubility The amount of a substance that can dissolve in a solution under a given set of conditions.

soluble Capable of being dissolved in a fluid.

solution A liquid that contains dissolved solute.

solvent A substance that dissolves another to form a solution.

specific gravity The ratio of the density of a substance to the density of water.

specific oxygen uptake rate A measure of the microbial activity in a biological system, expressed in mg O_2/g·h of VSS. Also called *respiration rate* or *oxygen consumption rate*.

specific resistance A measure of how strongly solids oppose the draining of their liquid component.

splitter box A chamber that divides incoming flow into two or more streams.

spore A reproductive cell or seed of a microbe, often dormant or environmentally resistant.

spray irrigation A method of spreading reclaimed water on agricultural land.

stability The degree of oxidation or decomposition of organic matter.

stabilization A treatment process designed to reduce the number of pathogens in solids to meet the requirements of 40 CFR 503; the resulting biosolids also are less odorous and less likely to attract vectors.

staged digestion A treatment process in which solids are digested in phases. It consists of two or more tanks arranged in series, typically divided into primary digestion (where solids are mixed) and secondary digestion (where quiescent conditions prevail and supernatant liquor is collected).

stair screen A type of fine screening device.

stakeholder A person or group that has an investment, share, or interest in something (e.g., a business or industry).

Standard Methods for the Examination of Water and Wastewater (Standard Methods) A publication published jointly by the American Public Health Association, the American Water Works Association, and the Water Environment Federation; it contains descriptions of analytical techniques commonly accepted for use in water and wastewater treatment.

stapling The entanglement of stringy or fibrous debris on a mesh or bar rack.

state variable An element of the set of variables (e.g., organics and nutrients) that describe the state of a dynamic system. Some state variables can be measured directly (e.g., ammonia), but many have to be determined via special sampling and test methods.

static head The vertical distance between a fluid's supply surface level and free discharge level.

static screen A type of fine screening device.

statistical model An alternative to mechanistic models that uses statistical rules based on historical data to determine the likely behavior of a process rather than using fixed equations to do so. Also called a *black box model*.

steady-state simulation A simulation in which the inputs are fixed and do not change with time.

step aeration A variant of the activated sludge process in which settled wastewater is introduced at several points in the aeration tank to equalize the food-to-microorganism ratio.

sterile Free from bacteria or other microorganisms.

Steven's law A proposed relationship between the magnitude of a physical stimulus (e.g., odor concentration) and its perceived intensity or strength.

stilling well A tube or chamber used to dampen waves or surges in a large body of water or a flume; typically used for the purpose of water-level measurement.

stoichiometric Pertaining to the quantitative relationship between reactants and products in a chemical reaction (e.g., the ratio of chemical substances reacting in water that corresponds to their combined weights in the theoretical chemical reaction).

stoichiometric coefficients The coefficients given before substances in a balanced chemical equation that are used to convert mass units for the different substances in the reaction.

storm sewer An underground pipe used to transport stormwater, not wastewater.

stormwater Water produced during precipitation events (e.g., snowmelt and storm-water runoff).

Streptococcus A genus of bacteria that includes some of the most common human pathogens.

struvite A crystalline solid or tenacious scale (typically whitish in color). Chemical formula is $MgNH_4PO_4 \cdot 6H_2O$. Also called *magnesium ammonium phosphate*.

submerged launder A pipe and control valve constructed below the normal water level used to remove liquid from sedimentation basins. Frequently used in primary sedimentation basins to avoid the odors associated with free fall launders and weirs.

submerged weir A weir where the water level on the downstream side is as high or higher than the weir crest. Also called a *drowned weir*.

submodel A model within a model; typically used to describe a particular facet of a larger or complex process.

subnatant Liquid underneath the surface of floating solids.

substrate (1) Wastewater or solids constituents used to promote biological growth. (2) Any surface to which a coating is applied.

sulfur dioxide Chemical frequently used in dechlorination. Chemical formula is SO_2.

sulfur-oxidizing bacteria Bacteria that can oxidize hydrogen sulfide to sulfuric acid.

sump A pit or reservoir that collects water or wastewater for subsequent removal from the system.

supermodel A mechanism for passing variables between different types of models. A supermodel includes all of the process variables in every process, even if some are not thought to be significant. This makes it easier to pass variables between models without requiring an interface model.

supernatant (1) The liquid remaining above a sediment or precipitate after sedimentation. (2) The most liquid stratum in a solids digester.

Support Center for Regulatory Air Models (SCRAM) A U.S. Environmental Protection Agency Web site where guidance on disperision models can be obtained.

support gravel Layers of graded gravel between the underdrain openings and filter media to prevent media from leaking into the underdrain.

surcharge (1) The height of wastewater in a sewer manhole above the crown of the sewer when the sewer is flowing completely full. (2) Loads on a system that are greater than typically anticipated. (3) An extra monetary charge imposed when set quantity or quality limits are exceeded, especially on flows discharged to a wastewater collection system.

surface aerator Mechanical aeration device consisting of a partially submerged impeller attached to a motor; it is mounted on floats or a fixed structure.

surface area loading rate ($M/L^2/T$) The subtrate mass loading rate to a reactor divided by *net specific surface area*.

surface area removal rate ($M/L^2/T$) The mass of substrate removed in a reactor divided by *net specific surface area.*

surface loading rate A criterion used when designing sedimentation tanks, expressed as flow per day per unit of basin surface area. Also called the *overflow rate*.

surface profile Contour of a blast-cleaned or substrate surface, viewed from the edge (cross-section of the surface).

surface tension The force acting on a liquid surface that tends to minimize the area of the liquid surface. Produced by the unbalanced inward pull exerted on the layer of surface molecules by molecules below the liquid surface.

surface wash An auxiliary high-pressure water spray system used to agitate and wash the surface of granular media filters.

suspended-growth process A biological treatment process in which the microbes and substrate are suspended in the wastewater.

suspended solids Solids captured via filtration through a glass wool mat or 0.45-µ membrane.

sustainability Development that meets today's needs without compromising the ability of future generations to meet their own needs.

switching function An equation used to describe how reactions change based on environmental conditions (e.g., a switching function for dissolved oxygen can be used to describe different rates for anoxic versus aerobic conditions).

syntrophic A condition in which two or more organisms combine metabolic activities to degrade a substrate that typically cannot be as well-degraded by one organism. Each organism's activities are critical to substrate degradation, but they are not performed for the organisms' mutual benefit.

system curve A plot of total dynamic head versus flow for a piping system or network.

system SRT Solids retention time within a processing system. In an activated sludge process, system SRT will include the length of time that solids are held in the anaerobic, anoxic, and aerobic portions of aeration tanks, as well as in the secondary clarifiers.

tapered aeration A variant of the activated sludge process in which the amount of air supplied in an aeration basin is tapered to match the demand exerted by the microbes.

Tedlar® bags These bags, which are made of a non-reactive material, are used to collect odor samples.

Ten States Standards Common name for *Recommended Standards for Wastewater Facilities,* a report of the Wastewater Committee of the Great Lakes-Upper Mississippi River Board of State Public Health and Environmental Engineers.

terminal headloss The headloss that occurs at the end of a filter run cycle, signifying that the filter bed is filled with solids.

terminal settling velocity The maximum sedimentation rate of an unhindered suspended solids particle.

tertiary filtration A filtration process used to improve the quality of secondary effluent.

tertiary treatment A physical, chemical, or biological process used to improve the quality of secondary effluent.

theoretical oxygen demand A calculation of the amount of oxygen required to oxidize a compound to its final oxidation products; used to estimate the amount of organic matter in water or wastewater.

thermal conditioning A process in which heat and pressure are applied to solids simultaneously to enhance their dewaterability without the addition of conditioning chemicals.

thermal oxidation A process in which organic solids are converted to oxidized products by heating them in the presence of oxygen or air.

thermal oxidizer An emissions-control device that uses heat to oxidize volatile organic compounds.

thermal plastic A material that can be repeatedly softened by heat and hardened by cooling.

thermal-set A material that becomes relatively infusible once it has undergone a chemical reaction via heat, catalysts, or ultraviolet light.

thermophiles Bacteria that grow best at temperatures between 45 and 60°C.

thermophilic An operating temperature range (typically 50 to 60°C) for an aerobic digester. It affects the microbial population in the digesters, as well as the reaction rates.

thermophilic aerobic digestion An aerobic digester that operates in the range of 40 to 80°C.

thermophilic digestion A process in which solids are digested by microorganisms that thrive in the thermophilic temperature range (about 50 to 60°C).

thickener A tank vessel, or apparatus where residuals or a slurry is thickened by reducing its water content.

solids thickening A process designed to increase the solids concentration in residuals by removing a portion of the liquid; such processes include a sedimentation tank, DAF, gravity thickener, centrifuge, gravity belt thickener, and membrane thickener.

thixotropy The time-dependent ability of some emulsions and solids to change viscosity when left at rest.

tide gate A swinging gate in a collection-system pipe that prevents seawater from entering the system during high tides.

ton container A 1-ton storage container; typically used to store treatment chemicals (e.g., chlorine or sulfur dioxide).

top dressing In landscaping, the application of a thin layer of soil or organic material to improve the turf surface by stimulating new growth, filling minor depressions, and improving drainage.

total capital cost The sum of direct and indirect capital costs for a project.

total dissolved solids The sum of all volatile and nonvolatile solids dissolved in a water or wastewater. Experimentally determined as solids remaining after a sample is filtered through a standard glass-fiber filter, placed in an evaporation dish, and heated to 180°C for 1 hour.

total dynamic head The total energy that a pump must impart to water to move it from one point to another, measured as the difference in height between the free-water surface level on the discharge and suction sides of a pump.

total inorganic nitrogen (TIN) Typically the sum of ammonia, nitrite, and nitrate-nitrogen concentrations in wastewater.

total Kjeldahl nitrogen The sum of organic nitrogen plus ammonia-nitrogen.

Total nitrogen Typically the sum of TIN and organic nitrogen.

total oxygen demand A measure of the organic matter in a water or wastewater that can be converted to stable products in a platinum-catalyzed combustion chamber.

total solids (1) The sum of dissolved and suspended solids in a water or wastewater. (2) The matter remaining on a weighed dish after the fluid has been evaporated at 103 to 105°C.

total maximum daily load (TMDL) A calculated pollutant load typically developed based on water quality models and field testing; it identifies the maximum amount (in pounds per day) of one or more pollutants that can be discharged to a receiving waterbody from sources within a defined area.

total suspended solids A measure of particulate suspended in a sample of water or wastewater. After filtering a sample of a known volume, the filter is dried and weighed to determine mass of the residue.

totally enclosed fan cooled A designation for a motor enclosure that is not airtight but does not allow free exchange of air between the inside and outside of the motor case. Exterior cooling is provided by an integral external fan.

totally enclosed explosion proof A designation for a motor enclosure that is not airtight but does not allow free exchange of air between the inside and outside of the motor case. Exterior cooling is provided by a nonsparking fan.

TOXCHEM A mass emissions model that predicts the fate of compounds in the liquid phase of a wastewater treatment plant and predicts mass emission rate into the air phase; it also includes some preliminary algorithms that estimate emissions from solids-handling processes.

toxicity The property of being poisonous or causing an adverse effect on a living organism.

transmembrane pressure The difference between feed pressure and permeate pressure (i.e., the driving force required to achieve a given flux through a membrane).

trash rack A coarse screening device that uses a set of parallel, stationary bars typically spaced 38 to 150 mm (1.5 to 6 in.) apart.

traveling-bridge clarifier A rectangular clarifier in which the solids-removal mechanism is supported by a mobile bridge.

traveling-bridge filter A granular media filter with multiple compartments that can be individually cleaned by a movable, bridge-mounted backwashing device without taking the entire filter out of service.

trickling filter An aerobic, fixed-film treatment process in which wastewater flows across a bed of highly permeable media. As the wastewater disperses, its organic matter is degraded by microorganisms in the slime on the media surface.

trihalomethanes Disinfectant byproducts formed when chlorine reacts with organic compounds in water. These halogenated organics are named as derivatives of methane and include suspected carcinogens.

triple bottom line A method of evaluating organizational performance by measuring economic success, environmental sustainability, and social responsibility.

Trommel screen A large rotating screen (i.e., a large metal mesh drum).

turbidimeter An instrument used to measure water turbidity by detecting the intensity of light scattered at angles from a beam of light projected through a water sample.

turbidity Suspended matter in water or wastewater that scatters or otherwise interferes with the passage of light through the water.

turbidity breakthrough The point when particles begin appearing in filtrate; an indication that filter effluent quality has deteriorated to an unacceptable level.

turbidity unit See *nephelometric turbidity unit.*

turbulence (1) A situation characterized by irregular variations in the speed and direction of movement of individual particles or elements of the flow. (2) A state of flow in which water is agitated by cross-currents and eddies, as opposed to laminar or streamlined flow.

two-tray clarifier Space-saving clarifier arrangement in which one longitudinal clarifier basin is located above another; they can be operated in parallel or in series.

ultimate biochemical oxygen demand The amount of oxygen required to completely satisfy carbonaceous and nitrogenous biochemical oxygen demand.

ultrasonic Sound waves at frequencies that are greater than or equal to 20 kHz. These frequencies are beyond the range of human hearing.

ultraviolet light The portion of the electromagnetic spectrum extending from the violet end of visible light to the X-ray region. Its wavelengths are between about 400 and 10 nm, corresponding to frequencies of 7.5×10^{14} to $3{:}10^{16}$ Hz.

unburned lime Another term for calcium carbonate.

underdrain A flow-collection and backwash water-distribution system used to support the filter bed in most granular media filters. Also called *filter bottom.*

underflow The concentrated solids removed from the bottom of a tank or basin.

uniformity coefficient A method of characterizing filter sand. It is equal to the sieve size, in millimeters, that will pass 60% of the sand divided by the size that will pass 10%.

uninterruptable power supplies Typically, a battery back-up system that will generate alternating current power during a power outage.

upflow filter A filtration system in which fluid flows upward through the filter bed.

U.S. Environmental Protection Agency The governmental agency in the United States with primary responsibility for enforcing federal environmental laws.

utility energy service contract (UESC) An agreement to partner with the electric utility to implement energy-conservation or renewable-energy projects.

vacuum chamber sampler A sample-collection device in which an empty sample bag is placed in a closed chamber, and then air is pumped out of the chamber to create a vacuum that expands the bag and draws gas into it.

vacuum filter A dewatering system in which a cloth-covered cylindrical drum slowly rotates in a tank of solids. An internal vacuum draws water through the filter cloth, while solids remain in the tank.

valve A device used to regulate the flow of fluids through a piping system.

vapor flux The rate at which mass is released through a unit surface area, commonly used to measure air emissions.

variable declining-rate filtration Filter operation whereby the rate of flow through the filter declines and the level of liquid above the filter rises throughout the filter run..

variable-frequency drive (1) A piece of electrical equipment that takes standard electrical power (typically 480 V, 50 to 60 Hz) and converts it into a variable-frequency power source to vary the speed of standard induction motors. (2) A method of controlling the rotational speed of an alternating current electric motor by controlling the frequency of the electrical power supplied to the motor.

vector An insect or other organism capable of transmitting pathogens to other species.

velocity gradient A measure of the degree of mixing imparted to water or wastewater during flocculation. Also called *G value*.

velocity head The kinetic energy in a hydraulic system.

venture scrubber See *Venturi scrubber*.

Venturi meter A meter used to measure flows in closed conduits by registering the difference in velocity heads at the entrance and outlet of a contracted throat.

Venturi effect An increase in a fluid's velocity as it passes through a constriction in a pipe or channel.

Venturi scrubber A misting scrubber for gases. Its constriction is designed to emit miniscule droplets of a fluid at high velocity, thereby maximizing the scrubbing liquid's physical contact with contaminants.

vertical benchmark The fixed point for vertical control of construction. The elevation datum should be indicated for the benchmark.

vibratory plow An oscillating plow shank used to install subsurface drip tubing and utility lines.

virus The smallest biological structure capable of reproduction; it can only grow and reproduce inside a host organism; infects its host, producing disease.

viscosity The internal friction of a fluid that resists the force tending to cause the fluid to flow.

vivianite Particulate crystals or scale that is blue, green, or gray-black in color. It is soluble in hydrochloric acid or nitric acid (HNO_3) and turns opaque or dark when exposed to light. Chemical formula is $Fe_3(PO_4)_2 \cdot 8(H_2O)$. Also called *hydrated iron phosphate*.

V-notch weir A weir with a V-shaped notch; used to measure flow.

volatile A substance that evaporates or vaporizes at a relatively low temperature.

volatile organic compounds (VOC) The term, volatile organic compounds, is often used generically to mean total organic carbon. In the context of air quality, the term means total nonmethane hydrocarbons.

volatile solids Organic matter subject to decomposition; it is ignitable at 550°C. Typically this is used to represent the organic fraction of the sludge or other solids material.

volatile suspended solids Organic, biodegradable matter suspended in water or wastewater. The percentage of this matter in a total suspended solids sample is determined by heating the sample to 600°C.

volume sources A three-dimensional source of pollutant emissions; typically refers to a situation (e.g., fugitive emissions) that is not easily characterized as a point or area source.

volumetric feeder A device that delivers a constant preset or proportional volume of dry chemical; it is not affected by changes in material density.

vortex A whirling mass of water or air with a cavity in the center toward which particles are drawn.

vortex flow regulator A funnel-shaped flow-control device used to withdraw fluid from a storage basin, tank, or cistern at a uniform flowrate.

vortex grit removal A treatment process that uses a mechanically or hydraulically induced vortex to capture grit in the center hopper of a circular tank.

waste activated sludge Excess activated sludge that is discharged from an activated sludge treatment process.

wasteload allocation (WLA) The maximum load of pollutants that each point source and nonpoint source contributor is allowed to release into a particular waterway.

wastewater Water containing the wastes from households, commercial facilities, and industrial operations; it may be mixed with surface water, stormwater, or groundwater that infiltrated the collection system.

WATER9 An emissions model available from U.S. EPA that predicts which compounds in wastewater will volatilize during collection, storage, treatment, and disposal, and estimates the rate at which they will be released to the atmosphere.

water hammer A rapid increase in pressure in a closed piping system that suddenly changes the velocity of its contents; this change may damage or rupture the piping system.

Water Quality-Based Effluent Limits (WQBELs) Regulatory limits for pollutant discharges that are established based on a waterbody's ability to assimilate pollutants while maintaining the water quality appropriate for its established beneficial uses; these limits are partially based on an antidegradation approach to pollution control.

Water Quality Standards (WQS) Regulatory limits for pollutant discharges that are established based on the receiving waterbody's designated uses, the criteria set to protect such uses, and other provisions established to avoid backsliding. These standards typically are addressed in a wastewater treatment plant's NPDES permit.

water reclamation plant (WRP) A wastewater treatment plant designed to produce water suitable for reuse.

water reclamation The process of removing solids and pollutants from wastewater and purifying the water so it is suitable for reuse.

watershed The area drained by a given waterbody.

wedge wire A type of screening media.

weir A baffle over which water flows.

weir loading The rate at which fluid flows out of a basin, expressed as the volume of liquid passing over a stated length of weir within a given timeframe.

weir overflow rate A measurement of the volume of water flowing over each unit length of weir per day.

wet air oxidation A process in which solids and compressed air are pumped into a pressurized reactor and heated to oxidize volatile solids without vaporizing the liquid.

wetlands Areas of land that are inundated or saturated by water (e.g., swamps, marshes, and bogs); they often support vegetation that thrives in saturated-soil conditions.

wetlands treatment A wastewater treatment system in which the aquatic root system of cattails, reeds, and similar plants are used to treat wastewater applied either above or below the soil surface. The vegetation, soil, and microbial environment filter and remove many contaminants from wastewater through natural processes.

wet pit See *wet well.*

wet scrubber An air pollution-control device that removes particulates and fumes from air by entraining them in a water spray.

wet weather flow The flow in a collection system that is the result of rainfall or snowmelt.

wet well A chamber in which water or wastewater is collected and to which the suction side of a pump is connected.

wetted perimeter The length of wetted contact area between a stream of flowing water and the channel that contains it.

whole plant model A model used to describe all unit processes in a treatment plant and all of their interconnections in the mass balance.

wind-tunnel sampler A sampling device that moves air across a surface to create the mass transfer of contaminants from liquid or solid surfaces and conveys the contaminants to a sample collection chamber or directly to a measurement device.

windrow A long, triangular-shaped pile of material.

windrow composting A composting method in which solids are mixed with a bulking agent and arranged in windrows that periodically are turned and remixed mechanically.

wound-rotor motor A motor whose rotor is designed to allow varying external resistances to be imposed. This allows the speed of the motor to be varied.

zero liquid discharge Description of a facility that discharges no liquid effluent to the environment.

Index

A

D

E

F

G

H

I

M

P

S

T

U

V

W

Y

Z